New products and finance for industry

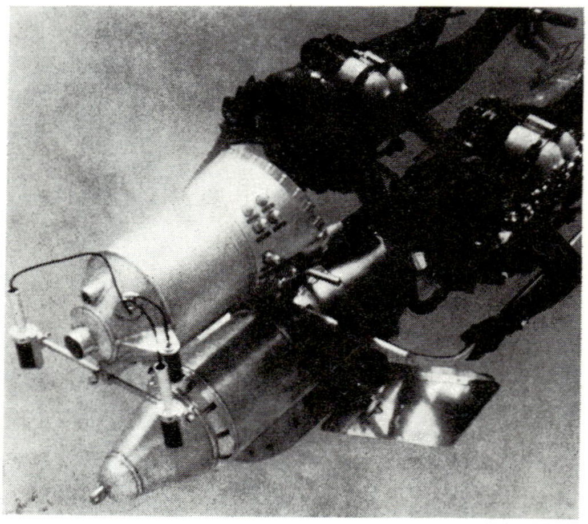

For information on how NRDC can support new ocean technology, write to

National Research Development Corporation
Kingsgate House
66-74 Victoria Street
London SW1E 6SL
Telephone 01-828 3400
Telex 23580

There's a lot in it for you

JANE'S

OCEAN TECHNOLOGY

Edited by **Robert L. Trillo**
CEng, FIMechE, FRAeS, AFAIAA, AFCASI

Order of Contents

World Sales Distribution

Jane's Yearbooks,
St. Giles House, 49/50 Poland Street,
London W1A 2LG, England

All the World
except

United States of America and Canada:
Franklin Watts Inc.,
730 Fifth Avenue,
New York, NY 10019.

Editorial communication to:

The Editor, Jane's Ocean Technology
Jane's Yearbooks, St. Giles House, 49/50 Poland Street
London W1A 2LG, England
Telephone 01-437 9844

Advertisement communication to:

Jane's Advertising Department
Jane's Yearbooks, St. Giles House, 49/50 Poland Street
London W1A 2LG, England
Telephone 01-437 9844

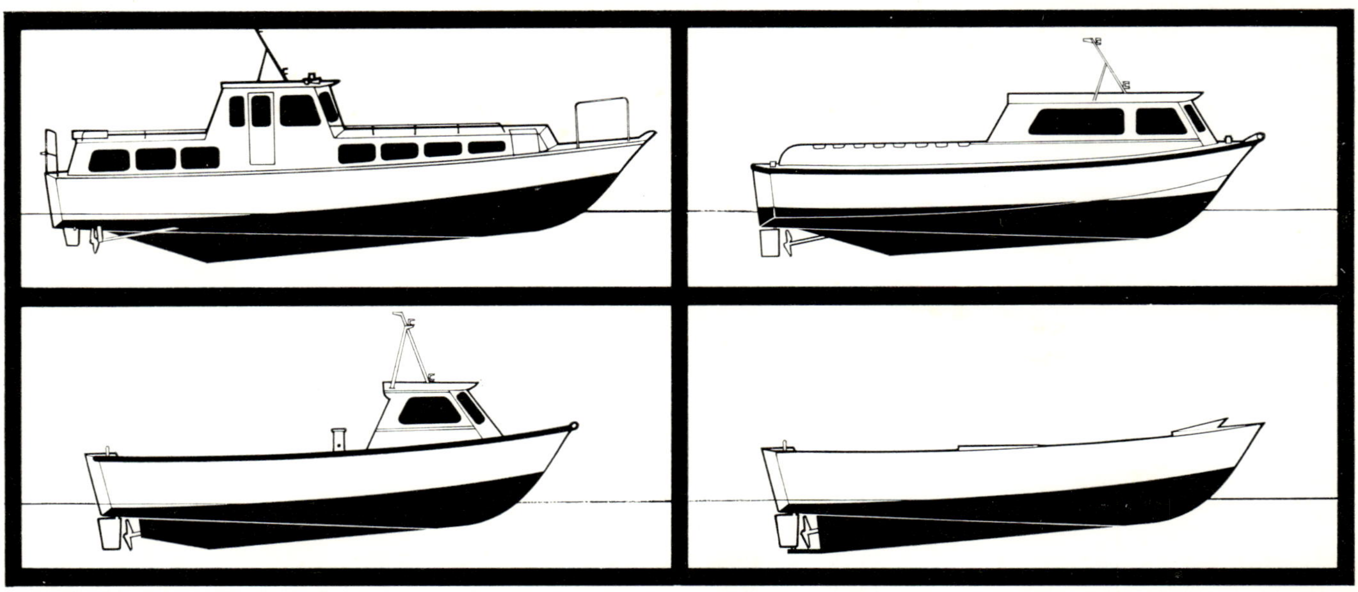

ALPHABETICAL LIST OF ADVERTISERS

1974/75

Halter, No. 1

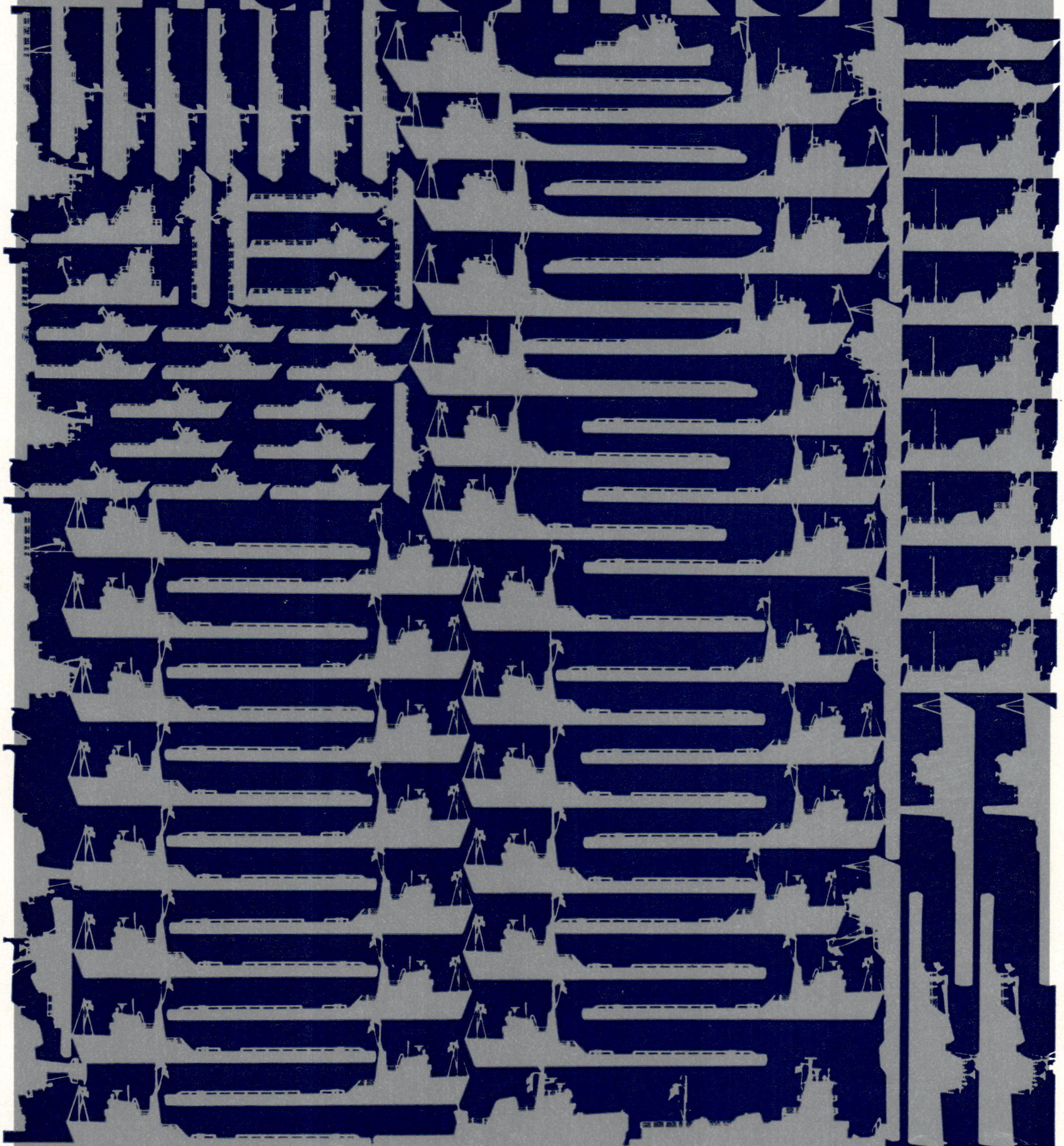

Above is a graphic representation of the number and type of vessels we have delivered or had under construction in 1973, making us the number one builder in the world of offshore support vessels. These include 22 crewboats, 53 supply and tug/supply boats, 14 pushboats, and 7 tugs.

HALTER MARINE SERVICES INC.

HALTER MARINE SERVICES, INC., P.O. Box 29266, New Orleans, Louisiana 70129, 504/254-1222 Telex: 48-4200 Cable: HALMAR

A 1. AGENTS
Stewart Offshore Services

A 2. AIR CUSHION VEHICLES
British Hovercraft Corporation

A 3. ALUMINIUM FABRICATIONS
Allday Aluminium Ltd.

A 4. AUTOMATIC CONTROL SYSTEMS
British Hovercraft Corporation
Decca Survey Ltd.

B 1. BUOYS
Halter Marine Services Inc.

B 2. BROKERS
Stewart Offshore Services

B 3. BURY BARGES
Brown & Root

C 1. CABLE LAYING VESSELS
Brooke Marine Ltd.

C 2. CARGO HANDLING FACILITIES
Milford Dock Co.

C 3. CASTINGS, HIGH DUTY IRON
Seaforth Maritime Limited

C 4. CASTINGS, STEEL
Seaforth Maritime Limited

C 5. COMPRESSED AIR STARTERS FOR GAS TURBINES AND DIESEL ENGINES
Hatch & Kirk

C 6. COMPRESSORS
Swift Aqua Supplies

C 7. COMPUTER SERVICES
Decca Survey Ltd.

C 8. CRANES—MARINE
Marathon

C 9. CREWBOATS AND WATER TAXIS
Halter Marine Services Inc.

C 10. CURRENT METERS
NBA (Controls) Ltd.

D 1. DATA ACQUISITION SYSTEMS
NBA (Controls) Ltd.

D 2. DERRICK BARGES
Brown & Root

D 3. DEVELOPMENT OF INVENTIONS
National Research Development Corporation

D 4. DIESEL ENGINES, AUXILIARY
Hatch & Kirk

D 5. DIESEL ENGINES, MAL PROPULSION
Hatch & Kirk

D 6. DIESEL FUEL INJECTION EQUIPMENT
Hatch & Kirk

D 7. DIESEL, SPARE PARTS
Hatch & Kirk
A.G. Schoonmaker

D 8. DIVING SERVICES
Marex
Swift Aqua Supplies

D 9. DIVING TENDERS
Halter Marine Services Inc.

D 10. DIVER TRANSPORT VEHICLES
Marex

D 11. DOCK OWNER AND OPERATORS
Milford Dock Co.

D 12. DRILLING RIG PREFABRICATION WORK
Brooke Marine Ltd.

D 13. DYNAMIC POSITIONING
British Hovercraft
Decca Corporation

E 1. ELECTRONIC EQUIPMENT
Decca

E 2. ENGINE PARTS, DIESEL
Hatch & Kirk
A.G. Schoonmaker
Seaforth Maritime Limited

E 3. ENVIRONMENTAL DATA ANALYSIS
Marex

F 1. FUEL OIL INJECTORS
Hatch & Kirk

F 2. FUNDING RESEARCH AND DEVELOPMENT IN MARINE TECHNOLOGY
National Research Development Corporation

G 1. GEARS, REVERSE, REDUCTION
Hatch & Kirk

G 2. GEOPHYSICAL RESEARCH VESSELS
Brooke Marine Ltd.

G 3. GOVERNORS ENGINE SPEED
Hatch & Kirk
A.G. Schoonmaker

H 1. HEAT EXCHANGES
Hatch & Kirk
A.G. Schoonmaker

H 2. HYDROGRAPHIC SURVEY EQUIPMENT
Brooke Marine Ltd.
Decca Survey Ltd.

I 1. INJECTORS
Hatch & Kirk Corporation

I 2. INSTRUMENT PRECISION
British Hovercraft Corporation

I 3. INSTRUMENT TEST EQUIPMENT
British Hovercraft

L 1. LANDING CRAFT AND LIGHTERS (DESIGN AND CONSTRUCTION)

L 2. LAUNCHES AND TENDERS
Halter Marine Services Inc.

M 1. MARINE BIOLOGICAL RESEARCH VESSEL
Brooke Marine Ltd.

M 2. MARINE INSTRUMENTATION
Brooke Marine Ltd.

M 3. MINERAL RECOVERY CRAFT
Brooke Marine Ltd.

M 4. MODEL MAKERS AND DESIGNERS
British Hovercraft Corporation

M 5. MOORING LAUNCHES
Brooke Marine Ltd.
International Offshore Services
N.B.A. (Controls) Ltd.

O 1. OCEANOGRAPHIC INSTRUMENTATION
Brown & Root
Marex

O 2. OCEANOGRAPHIC SURVEY SHIPS
Decca Survey Ltd.
Halter Marine Services Inc.

O 3. OFFSHORE PRODUCTION PLATFORM
Brown & Root

O 4. OFFSHORE SUPPLY BASE OWNERS AND OPERATORS
Milford Dock Co.

O 5. OIL RIG SUPPLY/ANCHOR HANDLING VESSELS AND WORKBOATS
Brooke Marine Ltd.
Halter Marine Services Inc.
Marathon
Seaforth Maritime Limited
Stewart Offshore Services

O 6. OIL RIG SUPPLY/PERSONNEL CARRIERS
International Offshore

O 7. OIL RIG SUPPLY/TOWING VESSELS
Brooke Marine Ltd.
International Offshore

P 1. PATROL BOATS, LAUNCHES, TENDERS
Allday Aluminium Ltd.
Halter Marine Services Inc.

P 2. PILOT BOATS
Halter Marine Services Inc.

P 3. PIPELINE BARGES
Brooke Marine Ltd.
Brown & Root

P 4. PROPELLERS, SHIPS RESEARCH

R 1. RESEARCH STUDIES
Marex

R 2. REVERSE REDUCTION GEARS, OIL OPERATED
Hatch & Kirk

S 1. SCIENTIFIC INSTRUMENTS
N.B.A. (Controls) Ltd.

S 2. SHIP AND BOAT BUILDERS (STEEL AND ALUMINIUM)
Halter Marine Services Inc.
Marathon

S 3. SHIP REPAIRERS AND MARINE ENGINEERS
Milford Dock Co.

S 4. SLIP RING ASSEMBLIES

S 5. SOCKETS AND PLUGS, ELECTRIC WATERTIGHT
Swift Aqua Supplies

S 6. SPECIALIZED ELECTRIC MOTORS, WINCHES, AND JACKING SYSTEMS
Marathon

S 7. STD SYSTEMS
N.B.A. (Controls) Ltd.

S 8. SUBMARINES-SUBMERSIBLES
International Offshore Services

S 9. SUBMERSIBLE SUPPORT VESSELS
Halter Marine Services Inc.

S 10. SUPERSTRUCTURES
Allday Aluminium Ltd.

S 11. SUPPLY WORKBOATS
Brooke Marine Ltd.

S 12. SUPPORT VESSELS FOR UNDERWATER SURVEY
Brooke Marine Ltd.

S 13. SURVEY BOATS (IN SHORE)
Allday Aluminium Ltd.

T 1. TIDE RECORDERS
N.B.A. (Controls) Ltd.

T 2. TUGS
Halter Marine Services Inc.
International Offshore Services
Seaforth Maritime Limited
Stewart Offshore Services

U 1. UNDERWATER CABLE CONNECTORS
Glenair
Swift Aqua Supplies

W 1. WELDING ARGON ARC
Allday Aluminium Ltd.

BUGGE SUPPLY SHIPS

OPERATORS OF OFFSHORE TUGS, SUPPLY AND ANCHORHANDLING VESSELS

TØNSBERG – NORWAY TEL. (033) 16 270 CABLE: EGGUB TELEX: 11 556 P.O. BOX: 160

OFFSHORE SUPPLY VESSELS SERVING THE OIL INDUSTRY WORLDWIDE – NORWEGIAN/BRITISH FLAG AND CREW TRAINED AND EXPERIENCED FOR NORTH SEA SERVICE

MULTIPURPOSE TUG / SUPPLY / ANCHORHANDLING / PERSONNEL TRANSPORTATION

SEA PEARL		185 x 38 x 16 feet	4500 HP	15 knots
SEA SUPPLIER SEA GARNET		183 x 38 x 16 feet	4000 HP	14 knots
SEA FORTUNE		190 x 36 x 18 feet	5300 HP	15 knots
SEA SCOUT SEA VIKING		183 x 38 x 16 feet	4000 HP	14 knots
SEA STORK		173 x 36 x 11 feet	2300 HP	13 knots
SEA BRUSE SEA BURE	Autumn 1974	190 x 38 x 16 feet	6300 HP	15,5 knots
SEA BRISK SEA BØR	1975	218 x 42' 6" x 19 feet	8000 HP	15,5 knots
SEA PIPER	1975	203 x 45 x 21 feet	5000 HP	14,5 knots
SEA TOPAZ	1976	203 x 45 x 21 feet	5000 HP	14,5 knots

ANCHORHANDLING TUGS:

SEA CRYSTAL SEA WORKER SEA DIAMOND	Mid 1974	130 x 32' 6" x 17' 5"	6000 HP	14 knots
SEA HUSKY SEA SETTER	End 1974	130 x 32' 6" x 17' 5"	6000 HP	14 knots

PLEASE CALL US ANY TIME

After office:

Finn A. Bugge (033) 13 118
Sven Gjeruldsen (033) 21 175
Terje Reiersen (033) 26 414

OPERATIONS OFFICE

18 A Regent Quay Phone: 575133/4
Aberdeen Telex: 73 497
Marine Superindendent J. A. Hesselberg
Ass. Superintendent S. Kvala
Office Manager A. Smith

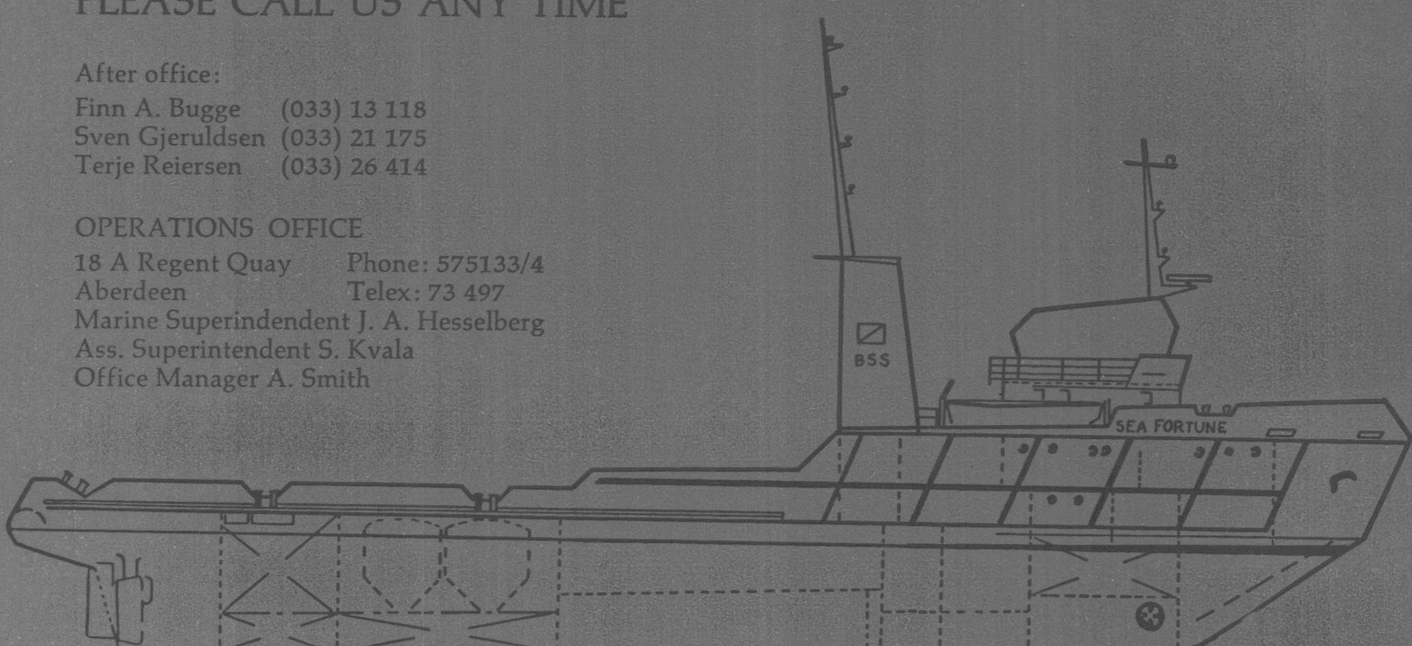

Retrieval experiments with a dolphin at the Makapuu Ocean Center in Hawaii (*Courtesy Sea Life Park*)

JANE'S
OCEAN TECHNOLOGY

FIRST YEAR OF ISSUE

EDITED BY
ROBERT L. TRILLO CEng, FIMechE, FRAeS, AFAIAA, AFCASI

1974-75

I.S.B.N. 0 354 00500 6

JANE'S YEARBOOKS
LONDON

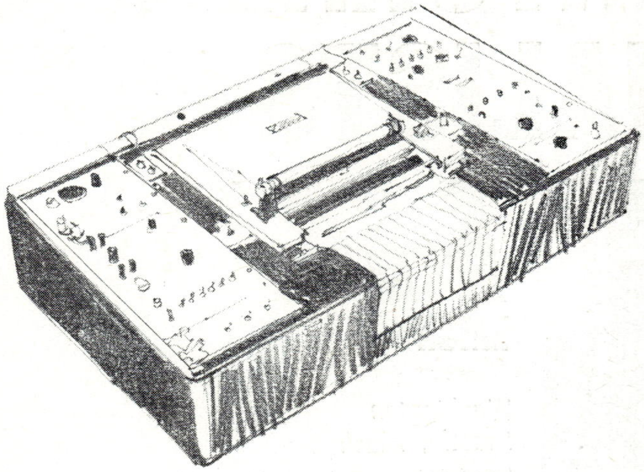

FOREWORD

This yearbook from Jane's is concerned with some of the major hardware activities associated with the exploration and exploitation of the oceans. This field of activity is already vast and expanding rapidly on a world-wide scale. Heavy investment in practical commercial exploitation and fundamental oceanographic research, has resulted in immense progress in instrument technology and equipment, and in the development of totally new techniques for use on and under the sea. JANE'S OCEAN TECHNOLOGY is designed to provide a comprehensive, authoritative, single source of reference for the engineer or scientist involved in designing or working with underwater systems or structures; it supplies information which would otherwise require extensive and time-consuming research.

It is just 27 years since the first offshore exploration for oil began. The Kerr McGee company of the United States drilled in 5·5m (18 ft) of water, 10·5 miles off the Louisiana coast on 14th November 1947. By 1972 this small rig (Rig No. 16) had produced over 900,000 barrels of oil and 7·62 million cubic metres (270 million cubic ft) of gas. Shallow-water exploration continued in the 1950s leading to the development of jack-up rigs. Drill ships, dynamically positioned, enabled operations to be carried into deeper waters, with Shell carrying out the first deep-water exploration off North West Borneo in 420m (1,380 ft) of water.

Perhaps the most exciting ocean exploration development in progress in the world today is that of the North Sea. Britain's preparedness to exert itself to engage in the massive ocean engineering business opportunities provided by the North Sea has yet to be shown. The exploration and exploitation of the North Sea could not have accelerated as rapidly as it has without the concentration of international company experience which has been brought to bear. This obvious fact should not, however, be allowed to divert attention from the missed opportunities. The Norwegian Aker H-3 rigs, a standard sized rig design (26 on order at the time of writing, including sub-contracting), are the result of Norwegian technology and production capability. The H-3 rig has more repeat orders than any other. Aker now has an H-4 project, an exploration, self-propelled, semi-submersible rig 113m (370 ft) long, 73m (240 ft) wide, and with a displacement at drilling draught of 31,530 metric tons. This rig will be capable of drilling in water depths of 500m (1,640 ft). In Finland too rigs are being built in numbers, under licence, by Rauma Repola. Steel shortages and limited production capacities have been quoted as reasons for British companies not showing interest in the building of such equipment as exploration rigs, pipe-laying barges, and undersea pipe lines. Typically, rigs are in the £10m to £15m bracket, pipe-laying barges £10m to £20m, and undersea pipe lines £0·6m per mile. Perhaps the inability to get to grips with the enormous opportunity presented is due fundamentally to a lack of combined technology and management expertise, which in itself may arise from the low status of the engineer in British society as compared with his counterpart in America or Russia. Many ideas are produced in Britain and a considerable amount of fundamental research of value to offshore oil extraction is undertaken, but is not put into saleable hardware. The engineer's status cannot be changed in the short term, so if Britain is to be in the business she must continue to work with other countries to get the oil ashore. On a happier note, the first crude oil from the North Sea should be flowing in from the Argyll Field by the time this book is published.

The ocean salvage tug *Lloydsman* built by Robb Caledon Shipbuilders Ltd for United Towing Ltd of Hull

The Mk-L, 4.25m (14 ft) diameter, 2-stage oil skimmer, developed by Reynolds International Inc. for oil pollution control

The first AKER H-3 oil drilling rig under construction at the Aker group's Verdal yard. This rig was launched in July 1973

The North Sea oil extraction scene changes almost daily with announcements of new strikes, new rigs, new submersibles, new oil production platforms, new companies, new supply vessels, new weather recording and research facilities, and so on. The opportunities are there but the rate of activity is such that there will be no opportunity for those who put extended argument, dispute and discussion before action. Looking further ahead, the exploration and production of oil from beneath deeper waters will call for continuing developments and innovations and again only those companies with drive, determination and high productivity will be involved. The civil engineering industry is coming into the oil production platform business in a big way and concrete structures designed to sit on the sea bed are being developed or projected in a number of forms. Companies involved in design and/or fabrication of production platforms include Laing Pipelines, Brown and Root and Wimpey, Hoyer Ellefsen, Selmer, Aker, Mowlem, Sir Robert McAlpine, Sea Tank Co., and Redpath, Dorman, Long (RDL). Further designs for production platforms are proposed by Christiani & Nielsen, Costain and Halcrow and Posford, Pavey and Partners.

Early in 1974 Norwegian shipbuilders were building ever-increasing numbers of supply boats, and Norwegian owners have now ordered over 100. Britain has approximately 10 per cent of the orders for supply vessels now on order, and this promises to be a good steady business. These vessels, capable of operating in the most severe environments, are being built in many countries but competition will rule out any shipbuilders who suffer delays. By October 1973 forty-five supply vessels were reported to be operating in the North Sea. The demand for supply vessels and tugs is increasing rapidly. The Robb Caledon Shipbuilding company which built the 10,000 b.h.p. *Lloydsman* (bollard pull 130 tons), the biggest British ocean-going salvage tug, is now engaged on the design of a giant 19,200 b.h.p. (150-ton bollard pull) tug for the South African Marine Corporation (Safmarine). This latter vessel will be the biggest and most powerful tug in the world.

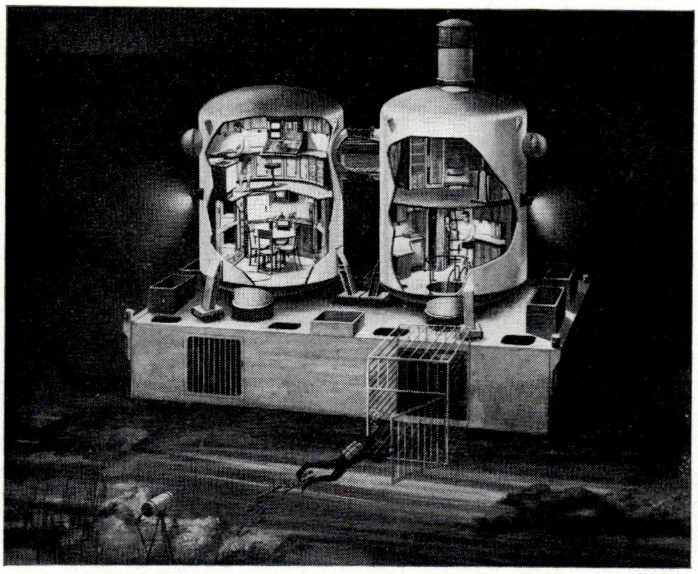

An artist's impression of the U.S. underwater habitat *Textite II*

The research vessel HMS *Endurance* in the Lemaire Channel Antarctica (*Courtesy Royal Navy*)

A notable development in international co-operation was announced in March 1974 with the formation of Ben-Odeco Ltd., a British offshore drilling company with access to the extensive experience of the American ODECO (Ocean Drilling and Exploration Co.) organisation. The new company is to have a new drill ship built by Scott Lithgow in association with IHC Gusto N.V, at a cost of £16m, for working in water depths of 916m (3,000 ft). This vessel will incorporate dynamic positioning and will come into use in 1976, and will be an example of yet another international collaborative effort directed towards providing effective equipment in the minimum possible time.

The exploitation of the resources of the sea and sea-bed can obviously lead to spoliation and sometimes disaster. The pollution of the sea and coast lines by oil is a problem likely to increase in the rush to obtain new supplies of fuel. The intensity of effort in the offshore oil exploration and production business is racing ahead of anti-pollution and safety measures. Many effective oil spill containment and collection devices now exist but they are limited to relatively calm water operations and are not yet in widespread use. There is an immediate need now to develop the systems compatible with the scale of operations being undertaken in offshore areas around the world. The ultimate purpose of extracting energy from the sea bed is presumably to enable man to enjoy life more fully, but man does not live by fuel alone. The quality of life depends so much on the preservation of the natural environment.

The need for facilities for working underwater safely and efficiently is going to increase enormously in the next decade. At the present time there is a discernable trend towards the increasing use of submersibles, following a decade of relative inactivity, and new vessels are being ordered by new groups in the business, such as P & O and Cunard. The unfortunate incident which occurred to the Vickers-owned *Pisces III* submersible in the Atlantic off the southern Eire coast in 1973, and earlier the tragedy which befell two crew members of the *Sea Link* submersible, trapped 60 fathoms down in Caribbean waters, had one major aspect in common—in neither case were rescue facilities instantly available. Should not submersibles be designed as part of a system which includes the means for rescue? The cost of such an approach will obviously be questioned but this should be looked at against the cost of mounting an unplanned rescue involving additional hazards.

For some time now the prospects of living underwater have been discussed in various media, and sea-bed habitats have been developed. Progress in this direction has not been particularly rapid, the costs and complications being considerable to support just several people. The requirement to work in comfort and for long periods underwater is, however, becoming more and more essential and widely differing approaches are being studied. D.H.B. Construction Limited of Great Britain have developed an articulated diving suit with a depth capability in excess of 300m (1,000 ft), while maintaining a normal atmosphere for the operator inside. Russian designs of habitats are reported on from time to time and there have been a number of American developments. (Of these the American *Tektite* and the Russian *Chernomor* habitats are perhaps two of the most notable projects.) An aspect of underwater working which has been studied both in America and Russia is the possibility of employing trained dolphins for such duties as data collection, sampling and the carrying of tools. One hopes that they will not be exploited in the appalling way that

The *Star III* submersible carrying a Benthos Model 199 Triple Plankton Net (*U.S. Navy Underwater Sound Laboratory Photograph*)

NAVY
54
630
NAVAIR
USN
POLAR STAR
ELLSWORTH TRANS-ANTARCTIC FLIGHT
NR-12269

whales have been. Experimental work on using dolphins to work with men in habitats has been undertaken by the Makapuu Oceanic Center, Hawaii, and by the U.S. Naval Missile Center, California.

On the oceans' surface, the world's fleet of research ships continues to expand. These vessels tend to be of conventional form but it seems likely that greater diversity in design will occur as more specific and intense tasks are undertaken. At the present time, oceanography and fisheries research perhaps dominate the scene, but research of the sea-bed and materials beneath is gathering momentum. Hydrographic survey work continues and the need exists for this not only in uncharted areas but also for the constant revision of existing data.

The spread of resources in and beneath the oceans is obviously not uniformly distributed. The unintelligent pushing of national interests not only hinders progress in the exploitation of the oceans but is also totally incompatible with the nature of the oceans where boundaries of resources cannot be defined. The recent conference in Caracas on the Law of the Sea, at which 149 nations attended, is hopefully indicative of the "getting together" of nations which interests in the resources of the oceans are bringing about. International co-operation between companies engaged in the ocean technology business is now commonplace and it would seem likely that the pressure for governments to co-operate in ocean exploitation will grow.

The Editor is extremely grateful to those organisations which have given such painstaking attention to providing useful information, and also offers his apologies to those organisations which have contributed material which could not for various reasons be included in the first edition. For instance, the subsea working systems of Lockheed Petroleum Services and of SEAL (Subsea Equipment Associates Ltd.), the immense oil production platforms now building for the North Sea, pipe-laying barges, and developments directed towards mineral extraction from the sea bed.

Preliminary work has already begun towards compilation of a second edition and the Editor welcomes comments and suggestions for inclusion. All editorial matter should be sent to:

The Editor,
Jane's Ocean Technology,
St. Giles House,
49–50 Poland Street,
London, W1A 2LG, England.

MARINE LINE SEATING LIMITED

4/6 SOUTH STREET,
DORKING.
SURREY.

TEL: DORKING 3710/81344

MANUFACTURE OF: HOVERCRAFT SEATING TO 6 G. C.A.A. SPECIFICATION.
HYDROFOIL SEATING TO D.T.I. SPECIFICATION
MANUFACTURE & FITTING OF HOVERCRAFT & HYDROFOIL INTERIORS.

"OUR SEATS ARE IN WORLDWIDE USE."

Here's the Fast, Economical Way to Make Slings with Beaver Brand Riggers' Forming Vises

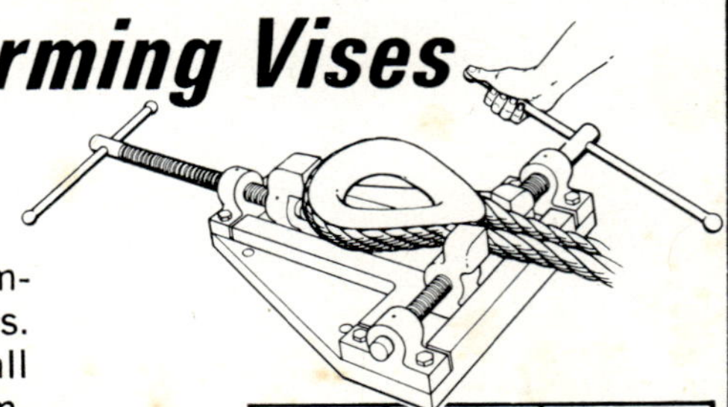

Handle 1/16" to 3½" Cable Without Adapters

Also indispensable for forming cable or wire rope around thimbles or fixtures or making "U" bolt clamp or clip installations. Easy, efficient operation. Rigid, durable construction. Meets all standard federal specifications. Proven in maritime, railway, lumber, manufacturing and other industrial uses.

Sold through your export agent.

BEAVER Tool & Machine Co.

Phone 405, 634-4148
P.O. Box 94717 • 540 S.E. 29th Street, Oklahoma City, Okla. 73109, U.S.A.

	Front Jaw Spread	Base Dimensions	Wire Line Capacity
SUPER SUPREME BEAVER	to 23"	35"x37"	1½"x3½"
SUPER BEAVER	to 23"	35"x37"	1½" to 2½"
BIG BEAVER	to 9½"	19"x19½"	⅜" to 1½"
LITTLE BEAVER	to 4¼"	10"x11½"	1/16" to ¾"

Acme threads used in all models.

1 SUPER SUPREME BEAVER

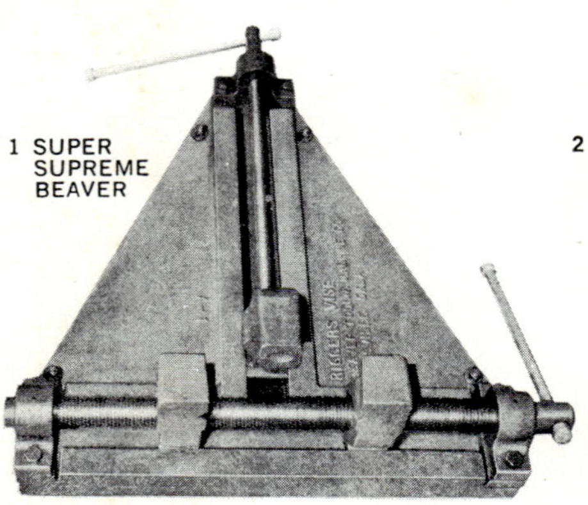

2 SUPER BEAVER

3 BIG BEAVER

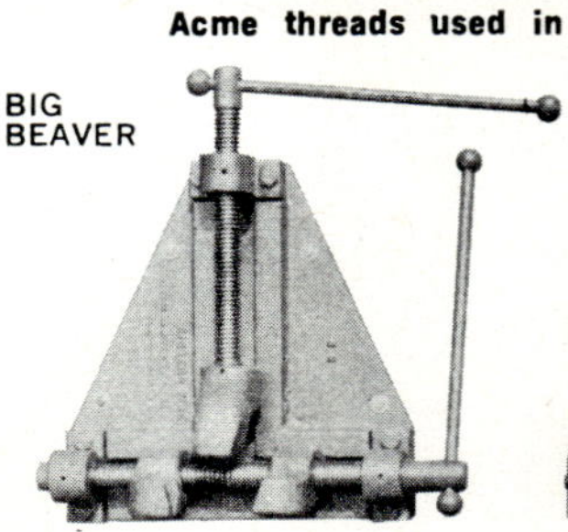

4 LITTLE BEAVER

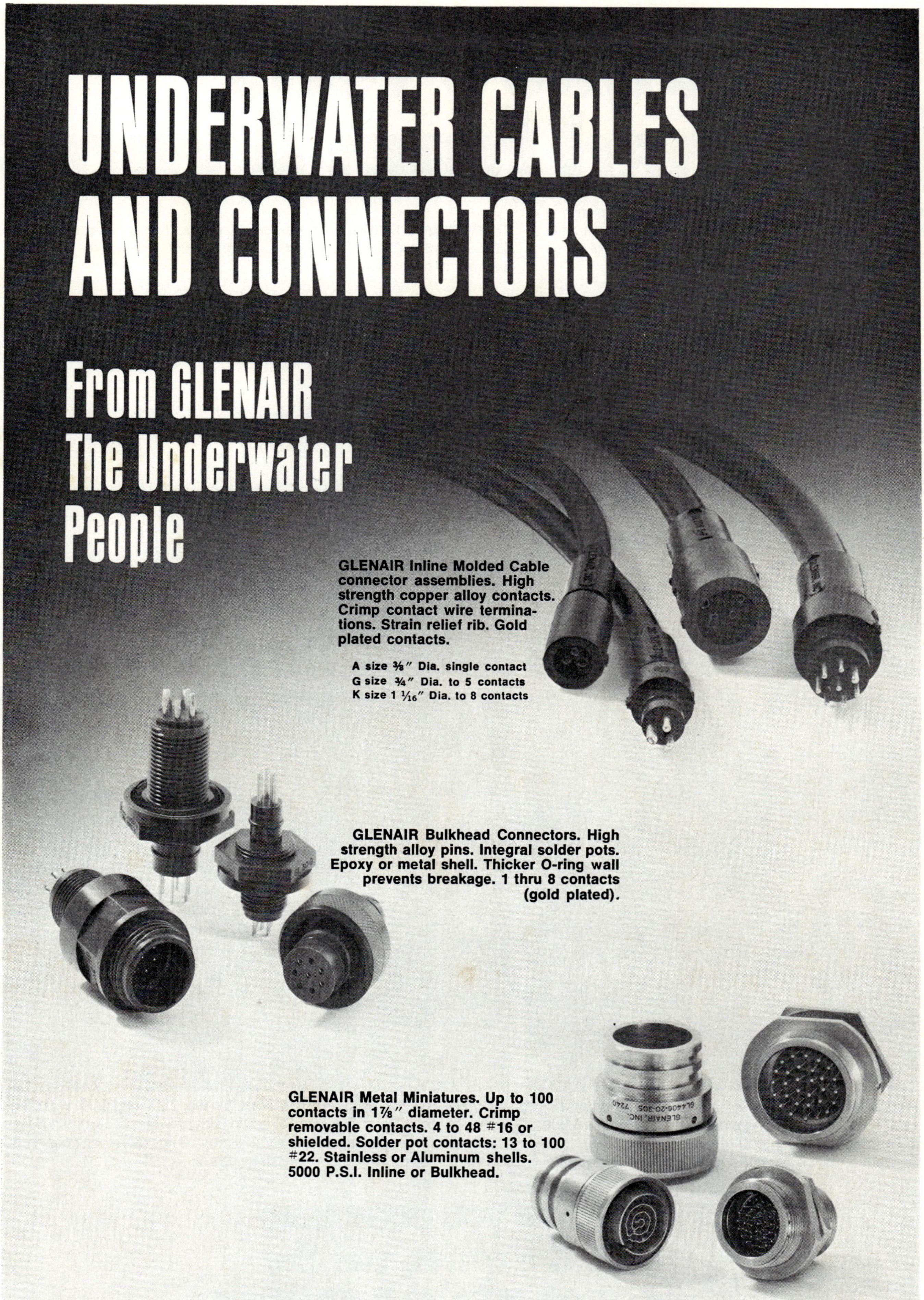
UNDERWATER CABLES AND CONNECTORS
From GLENAIR The Underwater People
GLENAIR Inline Molded Cable connector assemblies. High strength copper alloy contacts. Crimp contact wire terminations. Strain relief rib. Gold plated contacts.
A size ⅜″ Dia. single contact
G size ¾″ Dia. to 5 contacts
K size 1 1/16″ Dia. to 8 contacts
GLENAIR Bulkhead Connectors. High strength alloy pins. Integral solder pots. Epoxy or metal shell. Thicker O-ring wall prevents breakage. 1 thru 8 contacts (gold plated).
GLENAIR Metal Miniatures. Up to 100 contacts in 1⅞″ diameter. Crimp removable contacts. 4 to 48 #16 or shielded. Solder pot contacts: 13 to 100 #22. Stainless or Aluminum shells. 5000 P.S.I. Inline or Bulkhead.
Glenair®
Glenair Inc.
1211 Air Way, Glendale, Ca. 91201
(213) 247-6000
TWX 910-497-2066 TELEX 67-3485
Glenair (U.K.) Ltd.
Priory Works, Mansfield Woodhouse
Mansfield, Notts, England
(0623) 27157 Telex 377224

31,200 tons launched from 4 yards

BROWNSVILLE, TEXAS *Pentagone 82,* five-column semisubmersible, 325′ long, 338′ wide, overall height, 317′, 10,200 tons. Each column is 31′ in diameter. Crew of 74. Drilling in North Sea.

CLYDEBANK, SCOTLAND *Penrod 64,* jackup, hull dimensions of 230′ x 200′ x 26′; 6,000 tons. Designed for TD of 30,000′. Crew of 78. Scheduled to operate in North Sea.

REPUBLIC OF SINGAPORE *Margie,* semisubmersible, twin hull, measures 202′ long x 182′ wide x 110′ high; 9,000 tons. Designed to drill in 600′ of water. Crew of 90. Scheduled to drill off the coast of Northern Australia.

VICKSBURG, MISSISSIPPI *Key West,* jackup, 230′ long, 200′ wide, legs 467′ high; 6,000 tons. Designed to drill in 300′ of water. Crew of 97. Notice the three 45-ton marine cranes, usually on almost all rigs Marathon constructs. Scheduled to drill in waters off Belem, Brazil.

marathon

When you need help in the water, call the guys who've been there.

MARATHON LeTOURNEAU OFFSHORE (713) 224-8265/1700 Marathon Building, P.O. Box 61865/Houston, Texas 77061/Cable: LeToff; TWX: 910-881-3710
A Subsidiary of Marathon Manufacturing Company

Brooke Marine Limited

SHIPBUILDERS · ENGINEERS AND CONSULTING NAVAL ARCHITECTS

100
YEARS OF
ENGINEERING &
SHIPBUILDING

Oceanographic, Hydrographic, Research and Survey Ships.

H.M.S. Bulldog
Length overall 57·76 m (189'-6"), Beam 11·43 m (37'-6"),
Draft mean 3·73 m (12'-3").

R.V. Corella
Length overall 41·5 m (136'-0"),
Beam 9·7 m (32'-0"), Draft 3·7 m (12'-0").

Oceanographic Research Ship
Length overall 78·00 m (255'-11"),
Beam 13·25 m (43'-6"), Draft mean 4·60 m (15'-1").

Now building Survey Ship for the Nigerian Navy

A DOWSETT COMPANY

Shipyards: Lowestoft · Suffolk · England

TELEGRAMS: 'BROOKCRAFT' LOWESTOFT TELEPHONE: LOWESTOFT 5221 TELEX 97145

JANE'S OCEAN TECHNOLOGY
1974-75

CONTENTS

BEVIS MARKS HOUSE
BEVIS MARKS
LONDON
EC3A 7BS

For 24 Hour Service
Telephone: 01-283 2651
Telex: 888435 or 886001
Cable: STEWARLOND London

MEET EXECUTIVES OF THE GROUP AT THE 1973 OFFSHORE CONFERENCE STAND 1335-37

Watch this space
...with N.B.A....
For all your Environmental Data Acquisition requirements:
Met. Ocean Systems - Instrumentation - Installation - Data Processing
NBA
N.B.A. (CONTROLS) LTD.
Progress House · Albert Road · Aldershot
Hants GU11 ITD · England
Telephone: Aldershot 28028/9. Telex No: 85510
Cable Address: ENBEEAY Aldershot

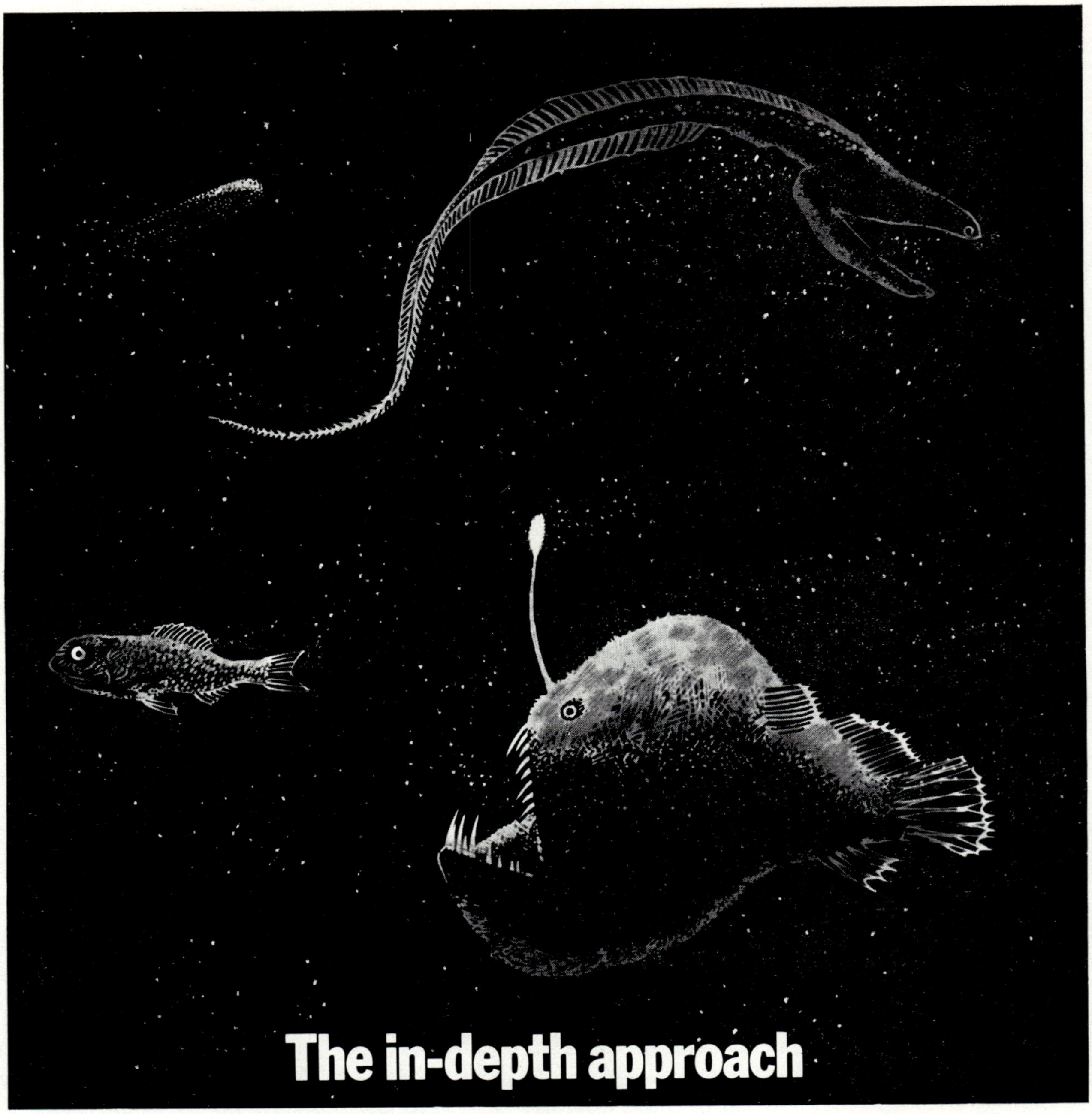

The exploration and the development of undersea resources are stretching man's technical resources to the limit – a limit imposed by the severe pressures met with increasing depths.

The depth of STC's underwater cable technology knows no such limits, the company having designed and manufactured a cable to operate in the world's deepest ocean trench.

The breadth, too, of STC's experiences in this technology is illustrated by the many other types of cable that have been designed and manufactured. Cables for underwater vehicles, submersibles and underwater habitats. Cables for a saturation diving system operating at depths of 1,000 feet. For exploration and detection, cables for all types of sonar devices, for sound range arrays and other long haul telemetry systems. And new developments for offshore oil include a cable used in the remote control and logging of a seabed well-head.

Each application has its own special requirements. STC has always recommended that the cable should be carefully considered in the early stages of design for any undersea project to ensure complete compatability between the requirements of the cable and the system.

To obtain further information on the full depth and breadth of STC's cable technology write, phone or telex.

STC Telephone Cable Division,
Wednesbury Street, Newport, Monmouthshire
NPT OWS. Telephone: 0633 52131.
Telex: 49368.

Standard Telephones and Cables Limited

A British Company of ITT

SUBMERSIBLES

CANADA

INTERNATIONAL HYDRODYNAMICS COMPANY LIMITED (HYCO)

General background
HYCO has designed three groups of submersibles:
(1) Basic submersible;
(2) Diver lockout submersible;
(3) Dry transfer submersible.

Group (1), comprising *Pisces I, II, III, IV, V* and *Aquarius I*, is not designed for diver lockout operations, but can be tooled for particular operations.

Group (2), comprising *SDL-1* and *Aries I* has a diver lockout capability whereby divers can be transported to a work location and dive from a pressurised compartment which is separate from the crew compartment. The divers can return to the submersible and commence decompression on the ascent to the surface.

Group (3), comprising the proposed *Aries II* type which is capable of a dry transfer system, i.e. the ability to lock onto another submersible or submerged pressure chamber, and transfer crew from one vessel to another independent of sea pressure.

All HYCO submersibles can be tooled for survey, salvage, rock sampling, ordnance retrieval, cable burying, etc., and the following details are generally applicable.

Special purpose tools
(1) Articulated manipulator:
Has six degrees of motion, 63·8 kg (150 lb) lift capability, 'Joy-stick' control. Useful for picking up small samples or holding tools.
(2) Seawater sample system (*Pisces IV* only):
The system is always installed but external lines are normally capped. Large (10 gal) samples may be taken at 1 atmosphere pressure or multiple 300 cc samples maintained at ambient sea pressure. The system can also be used for dye-trace studies.
(3) Sensor frame:
A general purpose sensor mounting frame installed on the bow complete with two 1·2 m (4 ft) arms pivoted 2·12 m (7 ft) apart and hydraulically swung through approximately 120° horizontally. Up to 13·6 kg (30 lb) may be hung on each arm.
(4) Pan and tilt:
Hydraulically operated 340° pan and 180° tilt. TV and still cameras and lights may be mounted. Unit is mounted on Sensor Frame.
(5) Sample tray:
Size 61 × 38 × 19·1 cm (24 × 15 × 7½ in). Hydraulically extended, covered in retracted position.
(6) Rock core drill:
Hydraulically operated and mounted in the heavy duty manipulator. Hard rock cores possible 15·9 mm (0·625 in) diam. by 10 in long.

Pisces electrical equipment normally aboard
The equipment that is normally attached to or on board a *Pisces* submersible will depend on the operation, but certain items can be considered essential to any missions and will always be included.

Echo sounder: Two TCDR's, one up and one down. Range 100 m. Used while approaching the bottom or surface.

Underwater telephone: Wireless acoustic communication with surface. Range 2 miles.

VHF radio: For use on surface, 10 watt 171·96 MHz presently supplied. 10 channels possible.

Tape recorder: Stereo cassette player/recorder for music or note taking, as required.

External lights: Two quartz-iodine lights 1,000 watt each, may be operated at 250 watt each, or individually at 1,000 watt.

Gyro: Directional gyro, non magnetic or true North reference. Drift rate 5°/15 min.

Sonar: Western Marine Electronics model SS120. As mounted can sweep 200° forward and from 4° above horizon to 90° down. Range 4,875 m (1,600 ft) max. (cliffs, etc.) resolution 10°. Used for search and obstacle avoidance, also as long range echo sounder.

Video recorder/camera: Sony 'Rover' AV3400 portable VTR and TV camera. Camera hand held. Recorder can be used with other cameras with minor modification.

Inverter: Solid state inverter, 115 volts AC ± 10% 60 Hz ± 3 square wave. Primarily intended for mercury vapor lights and video equipment.

Zenon flasher

R & F transmitter

Safety systems
Life support: CO_2 scrubbers and emergency rebreathers.

Ballast jettison: Drop weight, motors, manipulators jettisonable.

Internal environmental conditions
Crew space: Prone position normal.
Temperature control: None.
Pressure control: None.
Air purification: By lithium hydroxide or soda-sorb scrubbers.

Basic structure
Crew compartment: Pressure hull type, steel spheres.
Machinery compartment exterior pods: Pressure compensated.
Buoyancy compartment: Hard tanks—steel spheres: soft tanks—GRP.
Outer skin: FRP.

Handling
Lifting points, *Pisces I*: (Trifular Sling) to lugs on hull, all others by permanent frame structure one point.
Normal launching: As above with heaving lines.

Support requirements
Mother ship facilities include:
Air compressor
Battery charging unit
Crane derrick arrangement for launching and recovering submersible
Crew accommodation

Mobility
Air freight possibilities—all submersibles
Deck cargo
Propulsors and sail normally removed

Availability
All craft may be purchased
All craft may be leased; charges depending on application, duration and location
Charter contracts available

Range of submersibles
In August 1973 the production programme of submersibles by HYCO stood as follows:

Completed construction:	Construction engaged upon:
Pisces I	Aquarius I
Pisces II	Aries I
Pisces III	Future construction:
SDL-I	Aquarius II
Pisces IV	Aries II
Pisces V	Pisces VI
	Pisces VII

PISCES I
This is a manned, self-propelled submersible for use in depths down to 456 m (1,500 ft).

Design history
July 1964, construction of first prototype.
December 1965, construction completed of prototype.
January, March 1966, trials completed.
First prototype not certified.

Basic dimensions

Length, overall	4·87 m (16 ft)
Width, overall	3·50 m (11·5 ft)
Height, overall	3·20 m (10·5 ft)
Draught	2·27 m (7·5 ft)

Weight

Dry weight	6,800 kg (15,000 lb)
Payload	360 kg (800 lb)
Crew	1 to 3

Performance

Max. operating depth	456 m (1,500 ft)
Life support	72 hours

Main machinery

Power source	Lead acid batteries
Motors	Two 5 h.p. 120 volt DC 13,600 rpm
Propellers	Two 38 cm (15 in) dia. × 38 cm (15 in) pitch, 3 blade
Pumps	One ballast pump

Controls
Manoeuvring, forward and reverse

Depth	Ballast
Trim	Ballast

Operators
Vickers Oceanics Ltd.

PISCES II
A manned, self-propelled submersible for use in depths down to 1,080 m (3,600 ft).

Design history
Construction commenced in 1968
Construction of *Pisces II*, completed 1969.
Sea trials completed 1969
November 1969 American Bureau of Shipping certified

Basic dimensions (*applicable to Pisces' III, IV and V as well*)

Length, overall	5·82 m (19 ft 2 in)
Width, overall	3·00 m (10 ft)
Height, overall	3·64 m (12 ft)
Draught	2·65 m (8 ft 9 in)

Weights

Dry weight	10,950 kg (24,100 lb)
Submerged weight	9,700 kg (21,400 lb)
Payload	680 kg (1,500 lb)
Crew	1 to 3

Performance

Max. operating depth	1,080 kg (3,600 ft)
Speed:	
surface, cruise	1·2 knots
submerged, max.	2 knots
Duration at cruise	4 hours at 1·2 knots
Duration at speed	3 hours at 1·2 knots
Life support	200 hours

Main machinery

Power source	Lead acid batteries
Motors	Two 5 h.p. 120 volt DC 13,600 rpm
Propellers	Two 38 cm (15 in) dia. × 38 cm (15 in) pitch, 3 blades
Pumps	One ballast pump

Controls

Manoeuvring forward, reverse and tilt

Depth	Ballast
Trim	Ballast

Operators

Vickers Oceanics Ltd (owners).

This submersible, known as *Pisces I*, is manufactured by International Hydrodynamics and operated by Vickers Oceanics

PISCES III

A manned, self-propelled submersible for use in depths down to 1,080 m (3,600 ft).

Design history

As for *Pisces II.*

Basic dimensions, weights and performance

As for *Pisces II.*

Controls

As for *Pisces II.*

Operators

Vickers Oceanics Ltd.

Technical Paper on an operation

Core Sampling in the Hudson Bay by James R. McFarlane and Allan R. Trice. Offshore Technology Conference Paper No. OTC 1631, 1972.

The submersibles *Pisces I, II* and *III*

PISCES IV

A manned, self-propelled submersible for use in depths down to 2,000 m (6,600 ft).

Design history

April 1971 construction commenced
June 1973 construction completed
September 1972 sea trials completed
June 1973 American Bureau of Shipping certified

Basic dimensions, weights and performance

As for *Pisces II.*

Hulls

HY 100 steel

Controls

As for *Pisces II.*

Operators

Canadian Department of the Environment, 1973.

PISCES V

A manned, self-propelled submersible for use in depths down to 2,000 m (6,600 ft).

Design history

November 1972 construction commenced
February 1973 construction completed
April 1973 sea trials completed
August 1973 American Bureau of Shipping certified

Basic dimensions, weights and performance

As for *Pisces II*.

Controls

As for *Pisces II*.

Operator

International Hydrodynamics Ltd, 1973.

Pisces V, produced by Hyco

Launch of Hyco's SDL-1

SDL-I SUBMERSIBLE DIVER LOCKOUT-I

A manned, self-propelled, diver lockout submersible for use in depths down to 610 m (2,000 ft) and for diver lock out to 300 m (1,000 ft).

Design history

May 1970 construction commenced
September 1970 construction completed
October 1970 sea trials completed
December 1970 American Bureau of Shipping certified

Basic dimensions

Length, overall	7·62 m (25 ft)
Width, overall	3·64 m (12 ft)
Height, overall	3·64 m (12 ft)
Draught	2·88 m (9 ft 6 in)

Weights

Dry weight	11,350 kg (25,000 lb)
Submerged weight	10,000 kg (22,000 lb)
Payload	910 kg (2,000 lb)
Crew	3 and 3

Performance

Max. operating depth	610 m (2 000 ft)
Speed:	
surface, cruise	1·2 knots
submerged, max.	1·5 knots
Duration at cruise	3 hours at 1·2 knots
Duration at speed	3 hours at 1·2 knots
Life support	200 hours

Operators

Canadian Dept. of National Defence 1973.

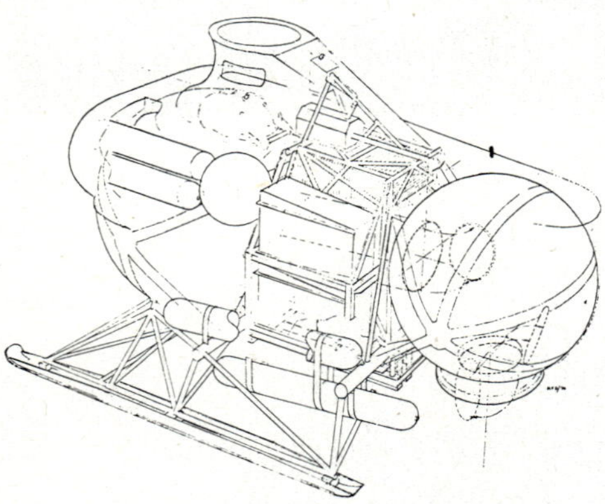

Internal arrangement of SDL-1

Aquarius 1, manufactured by International Hydrodynamics Ltd

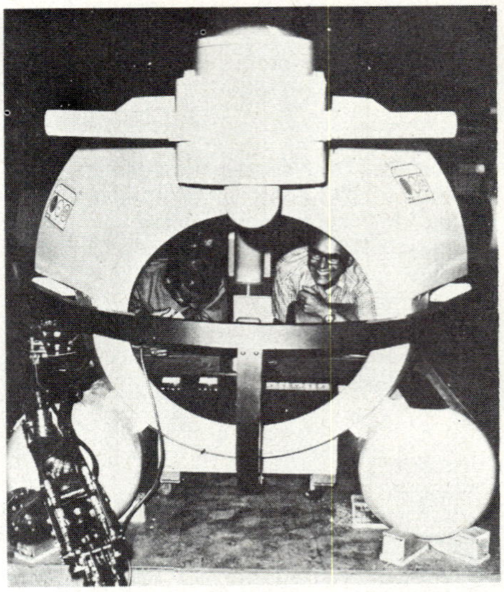

Rear view of *Aquarius 1*, showing viewing port for manipulator operations

AQUARIUS I

This is a manned, self-propelled submersible for use in depths down to 365 m (1,200 ft).

Design history

July 1973 construction of first prototype commenced
8 September 1973 launched

Basic dimensions

Length, overall	4·1 m (13 ft 6 in)
Width, overall	1·8 m (6 ft)
Height, overall	2·02 m (6 ft 8 in)
Draught	1·50 m (5 ft)

Weight

Dry weight	3,600 kg (8,000 lb)
Submerged weight	3,150 kg (7,000 lb)
Payload	4,500 kg (1,000 lb)
Crew	3

Performance

Max. operating depth	365 m (1,200 ft)
Speed:	
surface, cruise	3·75 knots
submerged, max.	4·5 knots
Duration at cruise	2 hours at 3·5 knots
Duration at speed	2 hours at 3·0 knots

Pressure hull

Diameter	0·91 m (3 ft)
Window diameter	0·91 m (3 ft)

Main machinery

One 5 h.p. 120 volt DC motor, 13,600 rpm
One propeller, 38 cm (15 in) dia. × 38 cm (15 in) pitch, 4 blades

Battery	225 amp hours

Controls

Manoeuvring, forward, reverse	
Depth	Ballast, dive planes
Trim	Ballast, dive planes

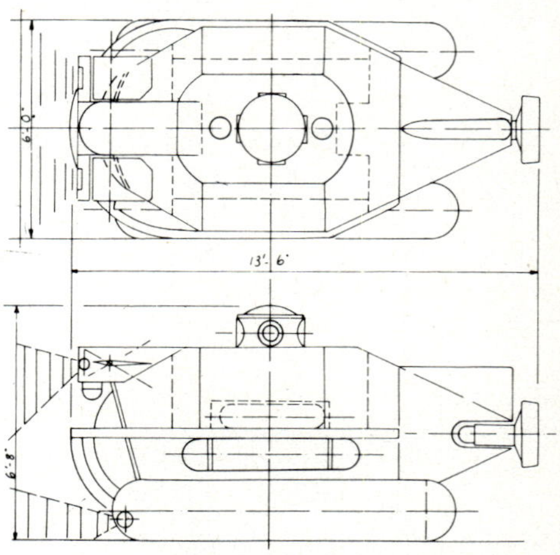

Plan and side elevation layout of *Aquarius 1*

PERRY SUBMERSIBLES LIMITED

PS-I NARWAL

The 'PS-I' is basically a submersible of similar form to the Perry Oceanographics 'PC-8' and has been built for ACCESS (Arctic Canadian Continental-Shelf Exploration Services Company Limited) of Toronto. ACCESS will be using the vehicle for under-ice operations in the Arctic.

Length, overall	5·95 m (19 ft 6 in)
Height, overall	1·95 m (6 ft 6 in)
Beam	1·22 m (4 ft)
Max. operating depth	313 m (1,025 ft)
Speed:	
max.	4 knots
cruising	2 knots
Range	15 nautical miles
Weight, dry	4,273 kg (9,400 lb)
Payload, including crew	398 kg (875 lb)
Power source	Nickel cadmium batteries
Propulsion motors	Two 2 h.p.

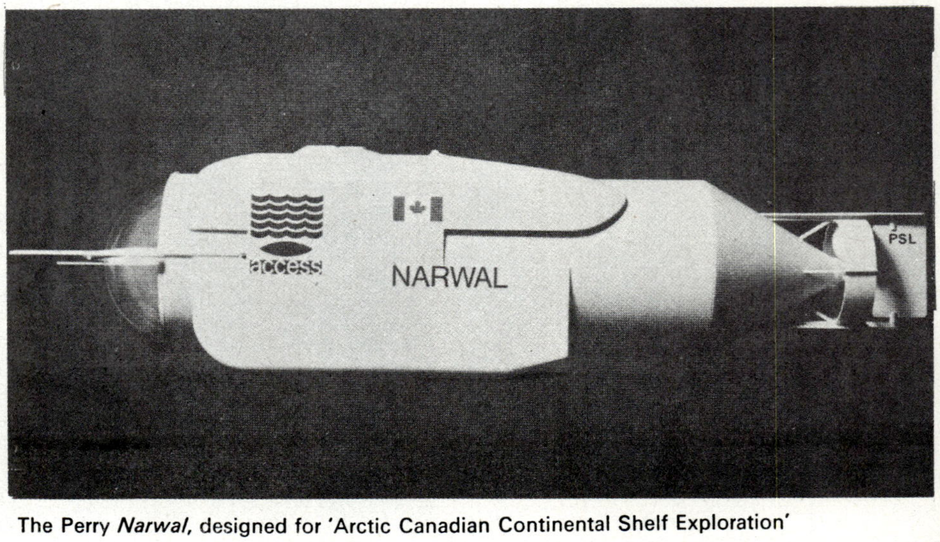

The Perry *Narwal*, designed for 'Arctic Canadian Continental Shelf Exploration'

Capabilities

Though it can operate in open water, *Narwal* was specifically designed to operate under arctic ice. Its modular construction permits changes in payloads and instrumentation packages which will provide the following capabilities:

Package A—sub-bottom profiling; side scan sonar; stereophotography.

Package B—seismic refraction.

Package C—seismic reflection. In addition hard and soft coring and sample recovery capabilities are available.

PS-2

The 'PS-2' is a Perry Oceanographics 'PC-8' submersible, but constructed by their subsidiary in Canada, Perry Submersibles Limited, for ACCESS.

The leading particulars are the same as for the 'PC-8' except that the maximum operating depth has been quoted by ACCESS as 313 m (1,025 ft).

Range	8 nautical miles
Power source	Lead acid or nickel cadmium batteries
Propulsion motor	7·5 h.p. DC electric motor

The Perry Oceanographics PC-P, or PS-2 as built by Perry Submersibles Ltd

FRANCE

MINISTERE DES ARMEES

DIRECTION TECHNIQUE DES CONSTRUCTIONS NAVALES
Centre d'Etudes et de Recherches Techniques Sous Marines

GRIFFON

A manned, self-propelled submersible designed for underwater survey and intervention and also for research and retrieving of sunken submersibles. Only a prototype has been built so far.

Design history

Date of construction of prototype started: 1970

Date of prototype completed: February 1973

Date trials completed: November 1973

The submersible *Griffon*, built and operated by Direction des Constructions et Armes Navales Centre d'Etude et de Recherches Techniques Sous Marines

Basic dimensions

Length, overall	7·8 m (25 ft 7 in)
Width, overall	2·3 m (7 ft 6·5 in)
Height, overall	3·1 m (10 ft 2 in)
Draught	2 1 m (6 ft 9·5 in)

Weights

Dry weight	12·5 metric tons
Submerged weight	14·1 metric tons
Payload	350 kg (770 lb)
Crew	2

Performance

Maximum operational depth	600 m (1,968 ft)
Speed:	
surface	2 knots
submerged cruise	2 knots
max.	4 knots
Duration at cruise speed	10 hours
Duration at max. speed	hours
Range	24 miles
Life support endurance (air supply)	80 hours

Main machinery description

Power source	192 Ni Cd cells—220 volts
Motors	220 volts DC 4 kw
Propellers	One 1,008 mm diameter (3 ft 6·7 in)

Controls:

manoeuvring	1 horizontal rudder, 2 lift propellers, 1 orientation propeller
depth	2 auxilliary tanks
trim	2 auxilliary tanks

Electrical on-board equipment

Navigational	A gyro compass
Vehicle function instrumentation	Propulsion and power control
Communication	1 underwater telephone —VHF radio
Search and recovery	1 side scanning sonar—1 mechanical arm
Television and visual	TV, tilt and pan

Safety systems

Life support	CO_2 absorbers—oxygen bottles
Ballast jettison	Compressed air, droppable batteries, droppable manipulator, droppable mass

Handling

The submersible has two lifting points and is normally launched by crane.

Support requirements

Mother ship: *Biesm Triton*

Operator

Marine Nationale GERS.

CENTRE D'ETUDES MARINES ADVANCEES (CEMA)

ARGYRONETE

A diver lock out submersible referred to as a mobile bottom habitat. Construction started 1969, halted October 1971. Design maximum operating depth 600 m (1,980 ft).

Length	28 m (92 ft)
Weight	255 tons
Crew	4 divers plus 6 passengers

Estimated cost $6·5 m (1971).

Built for Institute Français du Petrole (IFP) and CNEXO. Hull built by Neyrpic-Alsthom, Grenoble.

Purpose of vessel: saturation diving, oceanographic research, industrial work (oil) and rescue work.

The hull is in two sections, one to be maintained at atmospheric pressure, 16·3 m (54 ft) long and 3·7 m (12 ft) in diameter, and the other forming the habitat, 5·0 m (16·5 ft) long and 2·5 m (8 ft) in diameter.

CENTRE NATIONAL POUR L'EXPLOITATION DES OCEANS (CNEXO)

ARCHIMEDE

Archimede is a manned, self-propelled submersible, owned and operated (logistical) by the French Navy. CNEXO is responsible for all instrumentation (except life-support and propulsion system).

Prototype: built by La Direction Technique des Constructions Navales (D.T.C.N.) de la Marine Nationale

Prime purpose of craft: scientific work, exploration

Additional uses: search and recovery

Design history

Date of construction of prototype started: 1958

Date of prototype completed: 1961

Date trials completed: 1962

Basic dimensions

Length, overall	22·1 m (72·49 ft)
Beam, overall	5 m (16·40 ft)
Height, overall	9·1 m (29·85 ft)
Draught	5·2 m (surface) (17·06 ft)

Weights

Dry weight	203·4 tons
Submerged weight	208·9 tons
Payload	3 tonnes
Crew	3

Performance

Maximum operational depth	11,000 m (36,080 ft)
Speed:	
surface towed	8 knots
submerged cruise	0·7 knots
maximum	2 knots
Duration at cruise speed	10 hours
Duration at max. speed	5 hours
Range	10 kms
Life support endurance (air supply)	36 hours

Main machinery description

Power source	Cd-Ni battery 110 volts 860 Ah
Motors	DC motor 1 × 25 hp 2 × 5 h.p.
Propellers	1 vertical, 1 horizontal, 1 transversal
Pumps	1 hydraulic for mechanical arm
Controls	
manoeuvring	motors
depth	shot ballast jettisoning and gasoline ballast

The bathyscape *Archimede* (*Courtesy Marine Nationale*)

Electrical on-board equipment

Navigational	CTFM sonar 'gyro compass' loch
Communication	telephone
Television and visual	3 portholes, 3 TV cameras

Safety systems

Life support	O_2 bottles, CO_2 absorber, H_2O absorber (dessicant)
Quick ballast jettisoning	
Buoy markers	
Warning systems	acoustic

Internal environmental conditions

Crew space	sphere volume 4·8 m^3 (169 ft^3)
Temperature control	
Pressure control	
Air purification	

Basic structure

Crew compartment	spherical—steel
Machinery compartment	pressure compensated, oil filled
Buoyancy compartment	gasoline
Outer skin	steel
Collapse depth	30,000 m (98,400 ft)

Handling

Lifting points	for use only when gasoline is pumped out
Normal launching method	crane when empty

Support requirements

Minimum ship displacement: 600 tons

Operators

Captain de Froberville, Groupe des Bathyscaphes, 83800 Toulon Naval, France.

S.P. 3000 DIVING SAUCER (SOUCOUPE PLONGEANTE)

The S.P. 3000 is a manned, self-propelled submersible prototype, designed for scientific and exploration work and is also suitable for search and recovery. The vessel was built by Centre d'Etudes Marines Avancées, (C.E.M.A.).

Design history
Date of construction of prototype started: 1969
Date of prototype completed: 1970
Date trials completed: 1973

Basic dimensions

Length, overall	5·70 m (18·7 ft)
Beam, overall	3·04 m (9·97 ft)
Height, overall	2·10 m (6·89 ft)
Draught	2·00 m (surface) (6·56 ft)

Weights

Dry weight	8·3 tonnes
Submerged weight	8·3 tonnes
Payload	200 kg (440 lb)
Crew	3

Performance

Maximum operational depth	3,000 m (984 ft)
Speed:	
submerged	1 knot
cruise, max.	2·5 knots
Duration at cruise speed	10 hours
Duration at max. speed	5 hours
Range	12·5 km
Life support endurance (air supply)	48 hours

Main machinery description

Power source	lead-acid battery 110 volts, 360 Ah
Motors	DC motors 2× 2 h.p.
Propellers	horizontal
Pumps	hydraulic
Controls:	
manoeuvring	motors
depth	shot ballast and water ballast jettisoning
trim	mercury trim

Electrical on-board equipment

Navigational	CTFM sonar, gyrocompass, loch
Communication	telephone
Television and visual	2 portholes, 1 TV camera

Safety systems

Life support	O_2 bottles, CO_2 absorber, H_2O absorber (dessicant)
Quick ballast jettisoning	
Buoy markers	
Warning systems	acoustic

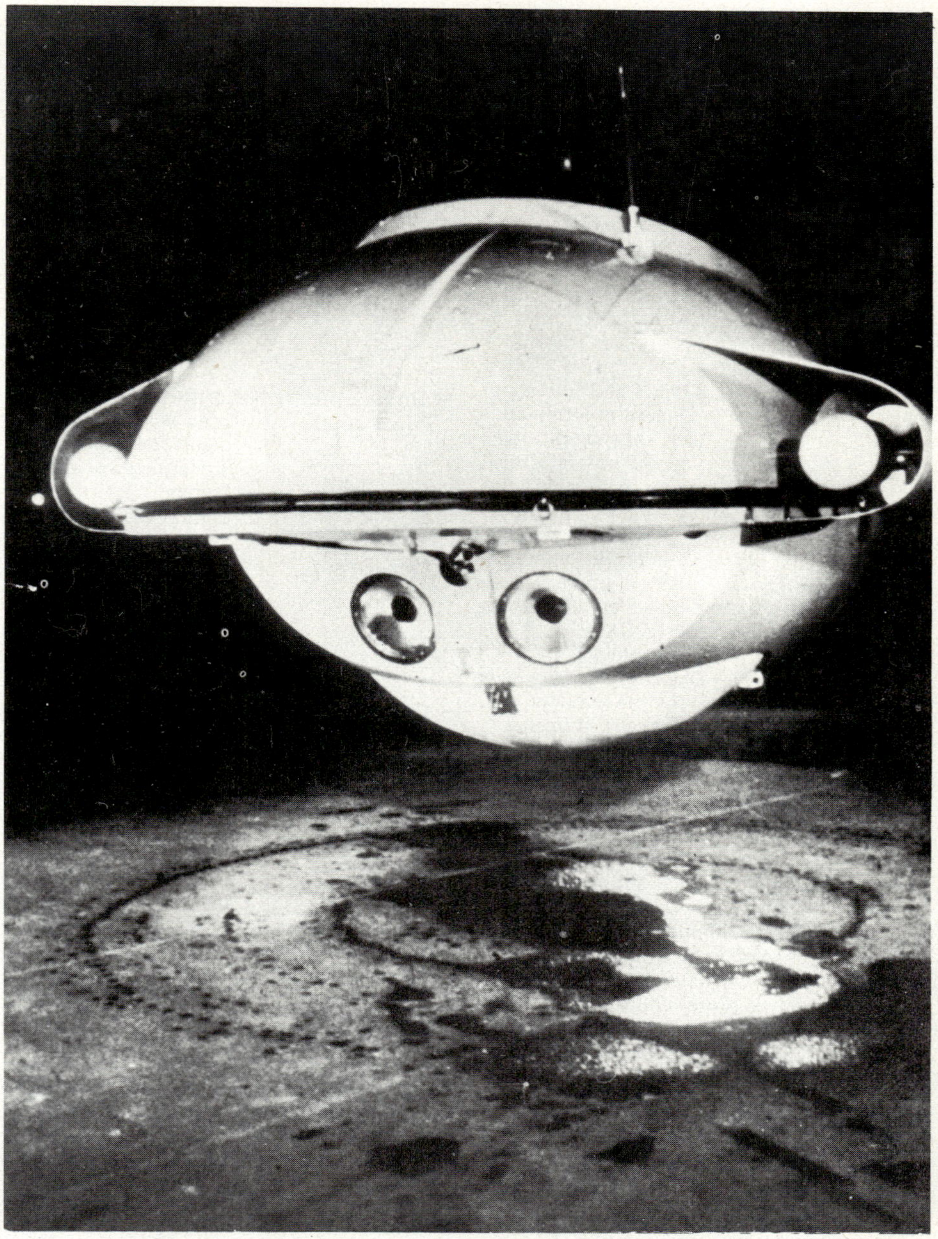

The diving saucer *S.P. 3000,* built by Le Centre d'Etudes Marines Avancées (C.E.M.A.) (*Courtesy C.N.E.X.O.*)

Internal environmental conditions

Crew space	sphere volume 4·1 m^3 (144·5 ft^3)
Temperature control	
Pressure control	
Air purification	

Basic structure

Crew compartment	spherical steel
Machinery compartment	pressure compensated oil filled
Buoyancy compartment	synctactic foam
Outer skin	plastic
Collapse depth	5,000 m (16,400 ft)

Handling

Lifting points	1 lifting point
Normal launching method	crane or gantry

Support requirements

Mother ship facilities required	crane and manoeuvrability, barge or ship

Mobility
May be air freighted

Availability
Leasing or chartering possible

Operators
Mr R. Kientzy, CNEXO, Section des Engins, Z.I. de Toulon Est, 83130 La Garde, France

GERMANY

BRUKER-PHYSIK AG

MANTA

This project is a one-man wet submersible for professional divers and pleasure purposes. A normal diving suit is worn by the occupant who sits in the closed but flooded cockpit.

Description

Weight in air	250 kg (555 lb)
Length, overall	3·2 m (10 ft 6 in)
Diameter	1·1 m (3 ft 7 in)
Operating depth, maximum for pressure-tight ports	80 m (262 ft)
Endurance	8 hours at 2·5 knots or 1 hour at 5·0 knots
Hull	Two shells of reinforced polyester resin with removable acrylic dome covering the cockpit, all streamlined for minimum battery power

The vehicle is almost non-magnetic which may be of interest for special tasks.

Batteries	Acid tight
Controls	One single control level for the rudders, and a single propeller switch

MERMAID I

A manned self-propelled submersible, one having been built.

Model or design number: 001

Prime purpose of craft: Underwater inspection, observation, pipeline survey

Additional uses: Coral-cracking, tourist

Variations: hydraulic manipulator, hydraulic tools, additional thrusters to increase manoeuvrability, underwater television.

Design history

Date of construction of prototype started: 1970

Date of prototype completed: 1971

Date trials completed: Fresh-water trials in 1971; sea-trials 1972 (3 months in the Mediterranean)

Certification authority clearance: classification by Germanischer Lloyd Hamburg

Basic dimensions

Length, overall	5·3 m (17 ft 5 in)
Width, overall	1·8 m (6·0 ft)
Height, overall	2·7 m (8 ft 11 in)
Draught	1·8 m (6·0 ft)

Weights

Dry weight	6·3 tons
Payload	500 kg (1,100 lb) instead of hard ballast; 100 kg (220 lb) lift-capacity
Crew	Two, 160 kg (360 lb)

Performance

Maximum operational depth	300 m (1,000 ft)
Speed:	
cruise, surface	3 knots
submerged	1·5 knots
max.	3 knots
Duration at cruise speed	about 8 hours
Duration at max. speed	about 3 hours
Range	12 to 15 miles
Life support endurance (air supply)	48 hours
Manoeuvring capability	1 stern-propeller, 2 horizontal thrusters, 2 vertical thrusters, trim-capability

Main machinery description

Power source	Lead-acid batteries 24 volt, 660 amp hour, charging voltage, 52 volts
Motors	Central DC-Motor, hydraulic pump
Propellers	All propellers driven by hydraulic motors
	One main driving motor, 3 h.p.; two steering motors, 1·60 h.p.
Controls: manoeuvring	remote control of all propeller-motors (speed and direction)
depth	vertical thrusters and pressurized ballast tanks
trim	batteries moved by hydraulic cylinders

Electrical on-board equipment

Navigational	Directional gyro, gyrocompass, speedometer, ultrasonic location equipment on demand
Vehicle function instrumentation	Echosounder, obstacle warning sonar and special instrumentation on demand
Communication	Ultrasonic communication system; range 3 miles
Search and recovery	instrumentation on demand
Television and visual	instrumentation on demand

Safety systems

Life support	CO_2 absorber; oxygen-feeding
Ballast jettison, pumping, etc.	300 kg (670 lb) ballast jettison; 80 kg (177 lb) pressurized ballast tank, 600 kg (1,340 lb) main ballast tanks
Buoy markers	
Rescue facilities	from surface or free escape
Warning systems	Leak-indicators in the battery-pods
Function failure	main circuits twice

Internal environmental conditions

Crew space	1·25 m (2·75 ft) × 2·5 m (5·5 ft)
Pressure control	manometers, barometers
Air purification	CO_2-scrubber
Compressed air supply	96 litres (58·6 in³) at 200 atm
Oxygen supply	24 litres (14·6 in³) at 200 atm

Basic structure

Crew compartment	pressure hull type, dia. 1·25 m (4 ft), length 4·3 m (14·1 ft): cylindrical pressure-hull, modular design (3 compartments) conning tower for pilot
Machinery compartment	machinery in the aft section
Buoyancy compartment	ballast tanks in the centre of buoyancy, outside ballast tanks, 300 kg (670 lb) in fresh water, 450 kg (1,000 lb) in salt water
Outer skin	fibreglass
Collapse depth	600 m (1,970 ft)

Handling

Lifting points	6 lifting points (normal gear for 4 points)
Normal launching method	crane

Support requirements

Ship with 7-ton crane capability, range 4 miles, catamaran-type vessel with central lifting point

Mobility

The submersible may be shipped as air cargo or as deck cargo

Availability

Purchase, capital cost: with standard equipment DM 460.000 (1973). The submersible may be leased or chartered.

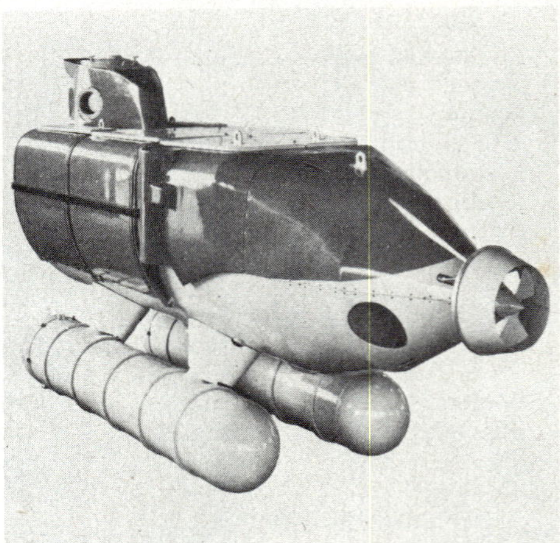

The Bruker *Mermaid 1 1B*, rear view

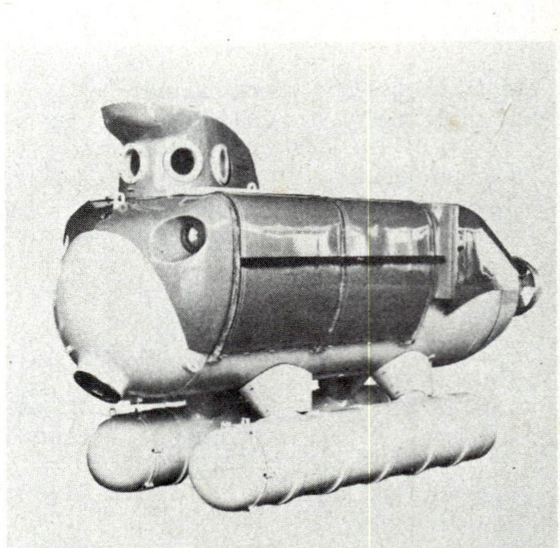

Forward view of the Bruker *Mermaid 1 1B*

BRUKER—SUBMARINE TYPE B-AT 2/300 U

The following is a technical description of the prototype tested in July 1971.

Construction details

The pressure-tight part of the boat consists of three cylindrical sections capped by two hemispheres. The main body serves to accommodate the crew, propulsion system, steering and control systems, compressed air and breathing air supplies, and also any desired equipment necessary for navigation, communication, measuring or working purposes. The two pressure-tight battery containers located underneath serve as stable 'legs' which can support the vessel both on land and on the sea bed. The battery containers also house the trimming system. The low centre of gravity guarantees excellent lateral stability both when floating and when submerged. Access to the battery containers is via cover plates on the front ends.

The main pressurised body consists of three sections, which can easily be taken apart for major servicing or re-equipping jobs. The structure is stiffened by substantial ribs.

The crew space is entered through a conical tower, which also contains observation windows for the pilot. The operator can control his instrumentation from his seat in the middle of the craft or alternatively, while lying prone, can carry out observations through the bow window and, if necessary, operate a manipulator.

The ballast tanks are arranged outside the main pressurised body and, like the bow, stern, and tower-cladding, consist of solid corrosion resistant parts made from glass-fibre-reinforced plastic. The top deck is made of the same material and is divided up into collapsible plates. Solidly constructed fenders protect the ballast tanks against larger shock .

Propulsion and steering

The submarine is propelled by means of a pusher screw in the stern and is steered by means of two transversely mounted thrusters, one of which is mounted in the stern cladding and the other within the tower cladding. When above water, steering is effected by means of the thrusters in the stern. By giving equal power to both thrusters it is, for example, possible when submerged to propel the craft sideways.

The propulsion and steering propellers are controlled independently and continuously in speed and direction by means of a single control lever. The fact that the transverse thrusters are situated far away from each other, together with the continuous adjustability of the propeller speed, guarantees extraordinarily good manoeuvrability and permits any given position to be reached and maintained against currents from any direction.

Power supply

The power supply of the craft consists of two lead-acid accumulators of high energy capacity. The batteries are enclosed in boxes of acid-tight glass fibre reinforced plastic, which are pushed into the cylindrical battery containers from the front end and are thus easily accessible for servicing. The batteries can be charged when the craft is surfaced. During the entire charging procedure, and especially during the final gassing period, the battery containers are forcibly ventilated.

Diving equipment

The ballast tanks attached to the outside of the pressurised section permit sufficient freeboard. When the craft is surfaced, the distance from the top edge of the tower to the water surface is about 75 cm (29·5 in). When diving, the tanks are flooded. The flooding slits underneath the tanks remain open and equilibrium forms between the pressures inside and outside the tanks. The ballast tanks are connected to the compressed air system and can be blown out for surfacing.

The slight buoyancy still remaining after flooding the ballast tanks is counteracted by controlled flooding of the pressure-tight regulating tanks. When submerged the buoyancy is controlled via the regulating tanks, thus determining the rate of ascent or descent.

The water level in the regulating tanks, which is an instantaneous measure of the bouyancy, is shown electronically on a meter.

Trimming arrangement

The craft is equipped with a trimming system to compensate for changes in weight distribution in the longitudinal direction, which could occur either when equipment is moved or when the operator moves forward from his seat to a lying position.

When the central hydraulic power supply is working, trimming is effected via the hydraulic network and is hand controlled. When the motor is stationary or if a fault occurs in the hydraulic system, the craft can be trimmed by means of a hand operated pump.

The system permits the longitudinal inclination to be adjusted very easily by about $\pm$ 20°. This way it is thus possible to dive and to emerge dynamically without operating the depth-regulating tanks.

Compressed air equipment

The compressed air required for blowing the ballast tanks and the depth regulation tanks is stored in 8 pressure tanks at 200 atmospheres. Four of these tanks are placed under the deck-hatch where they are easily exchangeable, under the deck-hatches. The others are attached in the aft part of the pressure hull. The compressed air network is divided into a high pressure circuit for blowing the depth regulation tanks and a low pressure circuit for blowing the ballast tanks. Some emergency connexions are possible as explained under 'safety and emergency devices'.

Life support

The air-renewal device contains a CO_2-washer with air ventilator and oxygen feeding valve. The latter allows a continuous adaptation to consumption. The oxygen tanks are also placed under the deck hatches and are easily exchangeable. They can be independently connected with the oxygen feeding equipment.

The ratio of carbon dioxide is controlled by means of a gas detector using reactive tubes. By request an electrical CO_2-indicator can be added. Depending on the operating conditions, an air dehumidifier may be necessary. It can be installed on request.

Safety and emergency devices

Pressure hull

The pressure hull is calculated with a safety factor of 2,0. The pressure tight compartments are made of a high tensile steel (St. 53,7).

All welding seams of the pressure hull are fully X-rayed.

All sections of the pressure hull have been subjected to a special heat treatment after welding in order to eliminate welding stresses.

The crew compartment and the battery containers are connected only with pressure-tight electrical and hydraulic connexions. In case of a leakage, no water can enter from one compartment into the other.

A leakage in one of the battery containers is signalled by an electronic sensor.

All penetrators of the hydraulic and compressed-air equipment can be shut by means of emergency valves in order to prevent entrance of water in case of damage of an external tube.

Except for the mechanism for the hatch of the conning tower and the mechanical device for dropping ballast, no rotating parts penetrate the pressure hull. This way all rotating seals which may be subjected to damage are eliminated.

The transfer of energy to the propulsion system and transverse thrusters is performed by means of static seals.

Valves

All important valves for flooding, ventilating and blowing depth regulation tanks and ballast tanks are duplicated, i.e. an electromagnetic valve and a hand valve.

All magnetic valves can be operated by the pilot by means of electrical switches or push-buttons.

An emergency flooding valve is also installed.

Compressed air and life-support device

For normal operating conditions, the compressed air network is divided into a high and a low-pressure system.

Fifty per cent of the air storage volume is in reserve. In case of damage of an air tube, different possibilities for reserve connexions are present, guaranteeing the availability of the necessary air for surfacing.

INSPECTION AND WORK SUBMARINE WITH DIVER LOCKOUT TYPE B-AT 2/2/200 U

General description

An independent work submersible designed for 200 m (655 ft) working depth with self-contained energy and gas supply. The submersible contains a diver lock-out compartment with access from the pilot room via a transfer chamber.

Basic dimensions

Length, overall	6·20 m (20 ft 4 in)
Height, overall	2·60 m (8 ft 6·3 in)
Beam, overall	1·80 m (5 ft 7·2 in)
Weight	9·1 tons

Divers can leave and re-enter the lock-out compartment underwater via an exit tunnel after pressure equilibrium has been accomplished. The pilot room of the standard version seats the pilot and an operator, who depending on the task will supervise the divers, operate a manipulator or other tools or replace the pilot.

The diver lock-out chamber accommodates two divers including equipment.

Visibility

The pilot has an all-round view through six port holes in the tower and one in the hatch looking upwards. In the front hemisphere there is the option of several small view-ports, for instance in a three + two distribution in the lower hemisphere, or one larger spherical acrylic window. The lock-out chamber has one view-port at the bottom for viewing the anchoring system, two that view oblique aft and one in the exit hatch. Another window gives a view into the transfer chamber.

Propulsion and steering system

The submersible is powered by one stern-mounted propeller with a Kort-nozzle and is steered by two horizontal thrusters, one mounted within the stern, the other within the tower lining.

When on the surface, all steering is done by the aft thruster. Underwater with both thrusters pushing in the same direction, the submersible will traverse and with each running in the opposite direction it will rotate.

In cases where obstacles must be quickly avoided, the regulating tanks might be insufficient. Therefore the installation of vertical thrusters is advisable. They are optional for the Bruker-diver-lock-out submersible.

Because of the modular design, subsequent installation is also possible.

The main and the two horizontal propellers are controlled by a single one-hand-lever for individual stepless speed and direction control. The two horizontal thrusters mounted far apart give excellent manoeuvrability and enable the pilot to keep position against currents from any direction.

Example of diver lockout operation

The surfaced submersible is towed to her operational area. Pilot, operator and divers board the sub. While descending there is atmospheric pressure in the pilot room as well as in the divers' compartment. The regeneration of the atmosphere is accomplished by standard CO_2-absorption with the addition of oxygen. On reaching the destination, the pilot will choose a spot to anchor the sub.

In the meantime the divers' compartment is pressurized by compressed air to the ambient water pressure, while the pilot room stays at normal pressure. In this phase the divers breath a gas mixture out of installed face-masks, depending on pressure and working conditions. The anchoring system consists of a 60 kg (132 lb) weight, that is lowered from below the centre of gravity. The sub's height above the seafloor is set by the length of anchoring line. When the sub is safely anchored, the divers will open the exit hatch hydraulically. Then fully equipped, the divers will exit via the tunnel and go to the work area. For long dives there is a possibility of supporting the divers via tubes out of sub mounted mixed gas supply cylinders and thus extending their working time considerably. Having accomplished their mission the divers will enter the lock-out chamber and close the exit hatch. They will take off their diving gear and use the sub installed face masks. The pilot will add lift to the sub's buoyancy to overcome the anchor weight and ascend. While surfacing the decompression of the divers starts, and is monitored according to standard procedures.

During decompression the divers will stay in their chamber. The surfaced sub can be supplied by additional gases from her support vessel. There will be a heating system for the diver's compartment and a helium-oxygen mixture is used.

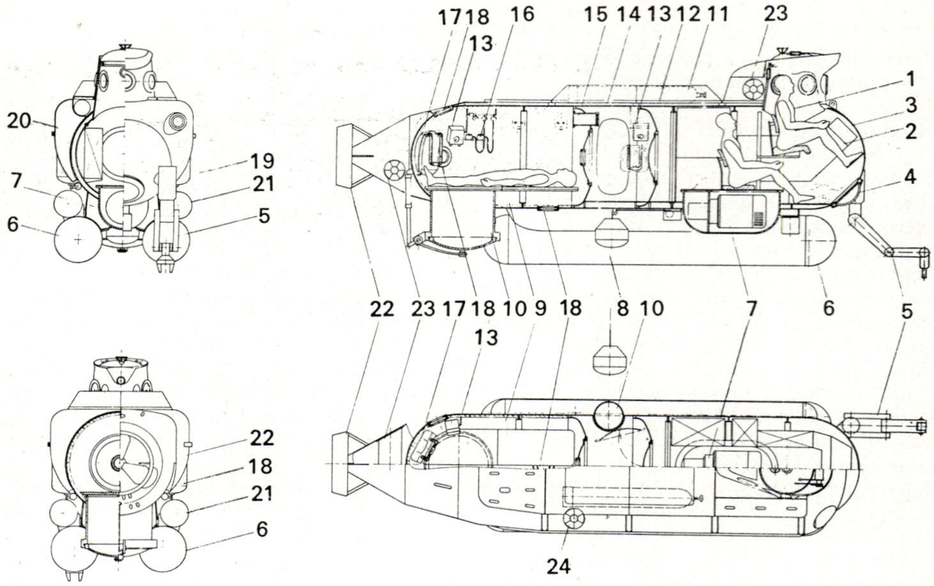

The Bruker Worksub, B-AT 2/2/200. Key: 1. Coxswain; 2. Assistant coxswain; 3. Control desk; 4. Bow port; 5. Manipulator; 6. Battery box; 7. Electric motor; 8. Anchor; 9. Pressure chamber (diving chamber); 10. Exit trunk; 11. Helium-oxygen bottles; 12. Breathing apparatus; 13. Mixed gas supply; 14. Lock; 15. Hatches; 16. Reserve breathing position; 17. Diving gear; 18. Porthole; 19. Compensating tanks; 20. Main tanks; 21. Compressed air bottles; 22. Main propeller; 23. Side thrust propellers; 24. Vertical thrust propellers

A flange on the diver's compartment is optional to allow for a pressure-tight connection with a transfer chamber that allows the diver to use a more spacious decompression chamber. The dimensions of this flange will correspond to the transfer system that will be employed.

Delivery volume

The delivery volume contains the ready-to-dive submersible as described above, including all equipment necessary for its operation:

—propulsion and steering systems consisting of a 10 kw main motor and two 1·5 kw horizontal thrusters, all remotely controlled for continuous rpm ratings from full ahead to full astern.
—one set of batteries with a 36 kw/h capacity mounted in two pressure proof battery pods. There is forced air circulation in the battery pods while charging.
—complete dive, trim and depth control system for electric push-button control.
—breathing gas system including CO_2 absorption, oxygen metering valves and gas control devices for the pilot room, as well as for the diver compartment including air and oxygen cylinders.
—compressed air system to pressurize diving and regulating cells including control and monitoring systems with high pressure air cylinders.
—compressed air system to pressurize diver compartment and transfer chamber including all control systems and air cylinders.
—mixed gas system for the diver's, hose-fed, breathing gas supply, including control systems and gas cylinders.

The capacity of compressed air, oxygen and mixed gas supply can be adjusted according to needs.

Additional accessories

—communication line from pilot room to diver compartment with optical or acoustical indicator.
—viewport in transfer chamber.
—ultra sonic underwater telephone for communication between sub and support vessel and divers.
—two front search lights.
—hydraulic manipulator for outside work or sampling.
—radio transmitter for communication between sub and support vessel.
—navigation and locating systems, includes echo sounder and gyro compass.
—electric heating system for diver compartment or wet suits.
—depth manometer for pilot room, or an encased one for diver compartment.
—safety device when exceeding set diving depth by alarm and automatic surfacing system.
—complete electric equipment including all safety devices and instrumentation necessary for operation and charging.
—anchoring system necessary for safe use of the submerged sub by divers, consisting of a hydraulically lowered weight.

Additional buoyancy is available independently of the compressed-air system by means of a hydraulic working-depth regulating tank which can be added on request.

Electrical equipment

Both main batteries can be connected independently with the electrical network. The installation is fully operational with only one of the batteries, but with half of the endurance.

An emergency battery is included. All life-support equipment (CO_2-washer, internal lights and all magnetic valves for surfacing and the leakage sensors) can be connected in the event of an emergency by one single switch.

Because of the separate battery containers, the crew is not at risk of intoxication from the battery gases. A safety-device assures that while the batteries are being charged, the ventilator is on, thus preventing any explosion from the battery gases.

Diving and surfacing equipment

A part of the external ballast can be dropped by the crew.

It is possible, to blow ballast tanks operating at maximum diving depth. In this case the submarine surfaces at relatively high speed.

The maximum diving depth can be adjusted at will by means of an electrical pressure switch. The submarine rises automatically, if maximum depth is exceeded.

MERMAID II

A manned self-propelled submersible with diver lock-out, primarily designed for repair work of submarine structures and also for rescue and research.

Model or design number: 002/003

Prime purpose of craft: repair of sub-marine structures with divers, diver lock-out, mobile diving chamber

Additional uses: rescue, submarine research

Variations: hydraulic manipulator, tools; different mixed gas equipment depending on operational depth

Design history

Date of construction of prototype started: 1972

Date of prototype completed: 1973

Trials: 1974

Certification authority clearance: classification by Germanischer Lloyd Hamburg

Production

Two under construction

Basic dimensions

Length, overall	6·5 m (21·4 ft)
Width, overall	1·8 m (5·95 ft)
Height, overall	2·7 m (8·90 ft)
Draught	1·8 m (5·95 ft)

Weights

Dry weight	about 11 tons
Payload	about 500 kg (1,120 lb)
Crew	about 350 kg (775 lb)

Performance

Maximum operational depth	250 m (825 ft)
Speed:	
cruise, surface	about 3 knots
submerged	2 knots
max.	4·5 knots
Duration at cruise speed	about 12 hours
Duration at max. speed	about 4 hours
Range	24 miles
Life support endurance (air supply)	48 hours
Manoeuvring capability	1 stern propeller, 2 horizontal thrusters, 2 vertical thrusters, trim-capability

Main machinery description

Power source	Lead-acid batteries 110 volt
Motors	Central electric DC-Motor
Propellers	All propellers driven by hydraulic motors
Controls:	
manoeuvring	remote control of all propeller-motors (speed and direction)
depth	vertical thrusters and pressurized ballast tanks
trim	batteries moved by hydraulic cylinders

Electrical on-board equipment

Navigational	Directional gyro, gyrocompass, speedometer, ultrasonic location equipment on demand
Vehicle function instrumentation	Echosounder, obstacle warning sonar and special instrumentation on demand
Communication	Ultrasonic communication system; range 3 miles
Search and recovery	Instrumentation on demand
Television and visual	Instrumentation on demand

Safety systems

Life support	CO_2-absorber, oxygen-feeding
Ballast jettison, pumping, etc.	500 kg (1,120 lb) jettison, 160 kg (350 lb) pressurized ballast tanks, 800 kg (1,755 lb) main ballast tanks
Buoy markers	
Rescue facilities	From surface or free escape
Warning systems	Leak indicators for battery-pods, oxygen-control
Function failure	Main circuits twice

Internal environmental conditions

Crew space	crew: 1·25 m (2·75 ft), length 2·5 m (5·5 ft), sluice: 1·25 m (2·75 ft), length 0·6 m (1.32 ft), decompression chamber: 1·25 m (2·75 ft), length 2·0 m (4·45 ft)
Temperature control	Thermometer
Pressure control	Manometer, barometer
Air purification	CO_2-scrubber for pilot-compartment and decompression chamber

Basic structure

Crew compartment, pressure hull type	cylindrical with stiffeners and hemispheres bow/stern
Machinery compartment	main pressure hull
Buoyancy compartment	two pressurized ballast tanks in the centre of buoyancy. Main ballast tanks outside the pressure hull
Outer skin	fibre glass
Collapse depth	500 m (1,640 ft)

Handling

Lifting points	7 lifting points; normal gear for 4 points or single point
Normal launching method	crane

Support requirements

Mother ship with 12-ton crane.

Mobility

The submersible may be shipped as air cargo or as deck cargo.

Availability

Purchase, capital cost	with standard equipment, mixed gas installations DM680.000 (1973)

The submersible may be chartered.

TOURISTSUB 14/150—005

The Bruker Tourist Sub is a self-propelled submersible specially developed for passenger trips at fairly great depths. It has seating capacity for 14 passengers and a diving depth of 150 m (492 ft). When dived, the vessel is completely independent of surface facilities and has its own power, compressed air and oxygen supply. It is thus ideal for sightseeing trips to view coral reefs, sunken wrecks, submarine parks and other interesting features on the ocean bed.

The passengers occupy comfortable armchair seats arranged on either side of the midship gangway. Each passenger has his own view-port through which to observe the submarine environment.

Access to the passenger area is reached by a ladder through the after access tower. The forward tower is the coxswain's conning tower and also provides a reserve exit. The coxswain has all-round vision through six ports.

The four propellers, two forward and two aft, arranged on each side of the pressure hull in slewable nascelles fitted with thrust nozzles, serve for propelling the vessel, for manoeuvring in confined spaces and as an aid to diving and surfacing. The drive is electrohydraulic.

The batteries are housed beneath the midship gangway in a pressure-tight well that also holds solid ballast. The battery capacity is sufficient for one day's diving trips.

The diving installation comprises the main tanks outside the pressure hull, which give the vessel adequate freeboard when surfaced, and the ballast tanks are used for weight compensation in different conditions of loading.

The pressure hull is constructed of high-tensile fine-grained structural steel and calculated with a two-fold safety factor. All welds are tested by radiographic and ultrasonic inspection. The sectional method of construction allows the building of larger models, with seating for 20 or 26 passengers, and a smaller model for only 8 passengers can also be supplied.

Design history

In design stage 1973

Certification authority clearance: classification by Germanischer Lloyd Hamburg

Basic dimensions

Length, overall	8·6 m (28·2 ft)
Width, overall	3·6 m (11·8 ft)
Height, overall	3·6 m (11·8 ft)
Draught	2·5 m (8·2 ft)

Weights

Dry weight	about 24 tons
Crew	1·3 tons

Performance

Maximum operational depth	150 m (500 ft)
Speed:	
cruise, surface	about 3 knots
submerged	about 2 knots
max.	about 4 knots
Duration at cruise speed	about 20 hours
Duration at max. speed	5 hours
Range	about 64·5 km (40 miles)
Life support endurance (air supply)	48 hours
Manoeuvring capability	four thrusters, pitch control

Main machinery description

Power source	Lead acid batteries 110 volts DC
Motors	One central electric motor with hydraulic pump
Propellers	Four hydraulic propeller-motors, slewable propellers
Pumps	Ballast pump for variable ballast
Controls:	
manoeuvring	remote control for all propeller-motors: speed, direction, pitch
depth	variable ballast tanks, vertical component of propeller motors
trim	control by ballast pumping

Electrical on-board equipment

Dependent on special operational tasks

Safety systems

Life support	CO_2-scrubbers
Ballast jettison	
Buoy markers	on demand
Rescue facilities	from surface
Warning systems	obstacle warning sonar
Function failure	main circuits twice

Internal environmental conditions

Crew space	2·0 m × 8·0 m (6 ft 6½ in × 26 ft 3 in)
Pressure control	manometers, barometers
Air purification	CO_2-scrubbers, oxygen feeding equipment

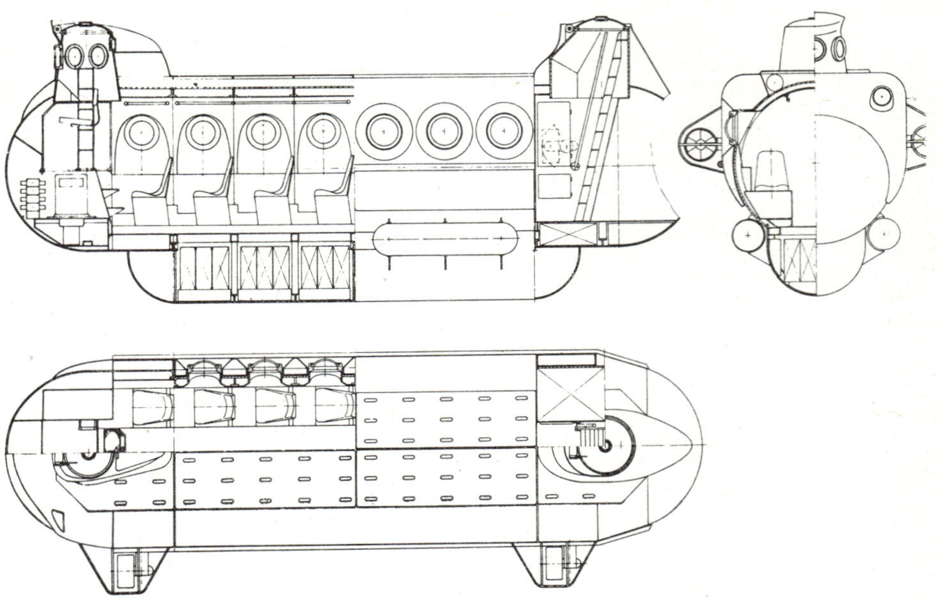

General arrangement diagrams of the Bruker Tourist-sub B-TS 150

Basic structure

Passenger compartment, pressure hull type	cylindrical pressure hull, 2·0 m (6 ft 6·5 in) with hemispheres, gas tight battery compartment underneath
Headroom in passenger space (in midship gangway)	1·75 m (5 ft 9 in)
Diameter of forward acess hatch	0·65 m (2 ft 1·5 in)
Width of midship gangway	0·6 m (2 ft 0 in)
Diameter of dome windows	0·3 m (1 ft 0 in)

Machinery compartment, main pressure hull
Buoyancy compartment, variable ballast tanks, outside ballast tanks
Outer skin: fibre glass

Collapse depth	about 300 m (1,000 ft)

Handling

Lifting points	4 lifting points
Normal launching method	crane

Support requirements

Mother ship facilities required	tugboat, catamaran-type mothership

Mobility

Can be air-freighted in single sections or shipped as deck cargo.

DEUTSCHE BABCOCK AND WILCOX AG

UNDERWATER WORK-BOAT

Description

The underwater work-boat is normally manned by two pilots and 4 research workers. The pilots sit in the front cabin which is always maintained at atmospheric pressure, so that no decompression times are required. The astern work-cabin, for the research workers, is provided with an underwater exit and may be maintained at atmospheric pressure. The research workers observe through the windows and may, through the manipulator attached to the boat, carry out underwater work or deliver bottom samples and wreckage parts to the surface.

By means of compressed air or a helium/oxygen mixture the work-cabin may also be placed under the pressure of the surrounding water. Then the lower exit hatch may be opened and the research workers may dive from the cabin into the water as well as return into the cabin. In the working cabin the divers are decompressed according to generally accepted rules.

The energy stored suffices for six hours. The boat reaches an underwater speed of approximately 3 km per hour (1·7 knots). The six amply sized over-pressure ballast tanks allow a large service load weight.

The swivelling seat allows the pilot orientation through the four windows of the entry port. The turret sheathing consists of plexiglass and allows free sight. In the submerged boat the observer lies by the side of

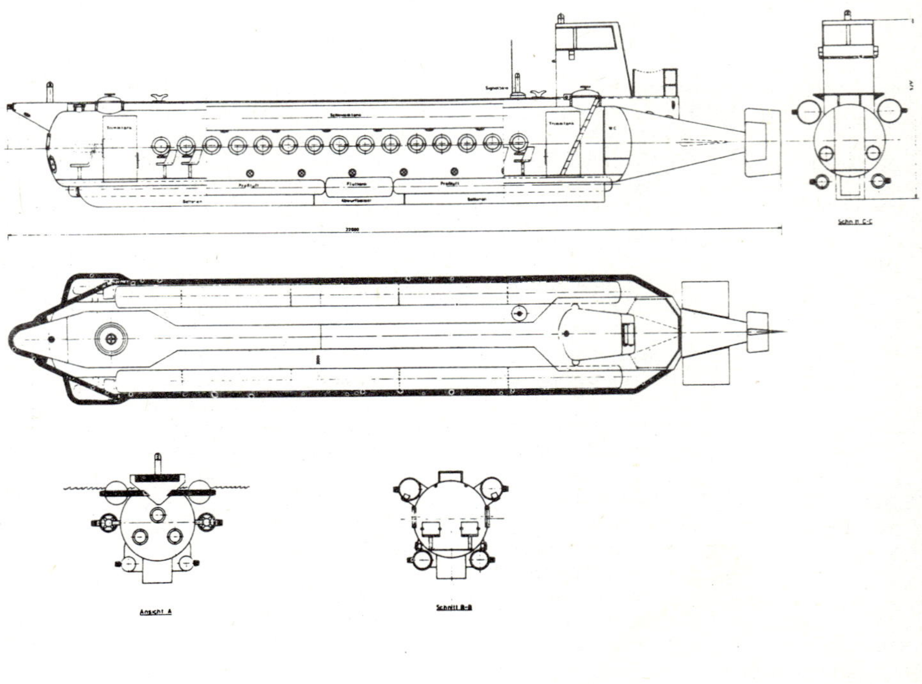

Layout diagrams of the underwater observation boat designed by Deutsche Babcock and Wilcox

the pilot for observation fore and aft through the bow windows. The trimmed boat is steered by two horizontal and two vertical motors.

UNDERWATER OBSERVATION BOAT

Description

The underwater observation boat accommodates a maximum of 30 passengers for underwater excursions and is operated by a three-man crew. The vessel is designed to permit ten underwater excursions per day over a distance of 3 km (1·7 nautical miles) and for a duration of 45 minutes each. It is also possible to make an excursion of 40 km (22 nautical miles) in approximately seven hours. For observation 32 projectors cast light into the sea. The boat will travel at a speed of 4 km per hour (2·2 knots) and the normal depth for service use will be 40 m (130 ft), but the boat may also dive to 60 m (200 ft).

The design is simple; only elements free from maintenance have been used. Greatest attention has been paid to the safety of the passengers and of the crew. Three different variable-ballast weights have been incorporated.

The seats are comfortable, and the visual angle through the windows is wide and the windows are arranged at eye level.

The turret, mounted astern and having an 800 mm (31·5 in) hatch, permits entry and exit by way of a ten-step stair. Projectors and television cameras at the bow and astern permit a good view for the helmsman. A sonic depth finder and depth gauge indicate obstacles and the diving depth. The batteries and compressed air tanks are charged during the night. After each dive the boat can be ventilated by means of a fan.

MASCHINENBAU GABLER GmbH

ARGUS I

TOURS 64 DGK/300 (Tourist Observation and Underwater Research Submersible)

A manned and self-propelled submersible, designed and developed by Prof. Dipl. Ing., Ulrich Gabler, Ingenieurkontor Lubeck.

Uses are as follows

Coral cracking; oceanographic research; plotting the route for the laying of cables and pipelines; underwater photography and television; aid in salvaging, diving and rescue missions; gathering of valuable oceanic products; inspection, surveillance and execution of work under water on submerged structures, moorings, cables, pipelines, drilling systems and underwater production plants; and control of oceanic pollution and sediments.

Design history

Date of construction of prototype started: 1970-1971

Date of prototype completed: 1971

Date trials completed: 1971

Certification authority clearance: Hull: GL + 100A2 Unterwasserfahrzeug (Submersible) Machinery: MC U

Production

Two Tours types built (64 and 66).

Basic dimensions

Length, overall	6·9 m (22·6 ft)
Width, overall	3·64 m (11·9 ft)
Height, overall	3·2 m (10·5 ft)
Draught	2·0 m (6·56 ft)

Weights

Dry weight: 14·0 metric tons ready for diving

Submerged weight: 16·0 metric tons

Payload: 200 kg (440 lb) by freeing compensating tanks; 100 kg (220 lb) by slipping emergency ballast; 500 kg (1,100 lb) by blowing out main ballast tank at max. diving depth; 300 kg (660 lb) vertical propeller thrust

Performance

Maximum operational depth: 300 m (984 ft)

Speed, surface 4·5 knots, submerged 3 knots cruise, 5 knots max.

Endurance: 7 hours submerged

Range: 21 nautical miles submerged at cruise speed with diesel electric propulsion; 400 nautical miles surfaced

Life support endurance (air supply): 30 hours.

Manoeuvring capability: 2 reversible side propellers, slewable by 210°; 1 hydraulically operated steering rudder

Main machinery description

Power source: Lead acid batteries, 1·9 tons (metric), 120 cells 210 Ah at 5-hours discharge, reduced H_2-formation

2 × 5·5 kw (7·5 h.p.) DC motors, 1,500 rpm, 220 volts, revolution control via thyristor pulse control, emergency control via dropping resistors

Propellers: 2 side propellers

Pumps: 2, one low-pressure ballast pump, one high pressure ballast pump

Controls: 2 depth gauges, one with automatic max. diving depth limitation; 1 hydraulically operated trim weight; 1 trim tank (water)

Electrical on-board equipment

Navigational: 1 gyro-compass with repeater, horizontal—echo-sounder, vertical—log

Vehicle function instrumentation: Diesel control panel. Battery control panel, main switchboard. Hydraulic control panel

Communications: Transmitter and receiver, underwater telephone, radio-telephone on surface

Search and recovery: UT-direction-finder, searchlight, horizontal echo-sounder

Television and visual: 9 view-ports (4 at 150 mm (5·91 in) dia. and 5 at 100 mm (3·8 in) dia.)

Safety systems

Life support: Oxygen supply equipment. Air regeneration device with CO_2-absorber. Gas detector device

Ballast jettison, pumping: 100 kg (220 lb) emergency ballast. High pressure ballast pump for freeing compensating tank and HP air for blowing main ballast tank

Buoy/rescue facilities: Two safety buoys connected with a safety line. Submerged escape apparatus

Warning systems: Optical and acoustical warning system by exceeding the diving depth and diving time by means of 'Dead Man Switching'

Function failure: Reset circuit breaker for the two propeller motors leakage indicator for (+) and (−), charging current exceeded, diesel function control, thyristor control switch-over into emergency control, check-set for electronic equipment

Internal environmental conditions

Crew space	8 m³ (282 ft³) air volume
Temperature control	thermometer
Pressure control	barometer
Air purification	CO_2-absorber

Basic structure

Crew compartment, pressure hull: Frame stiffened cylindrical pressure hull with hemispherical end bulkheads, material G 36 steel, 13 mm (0·51 in) thick. Forehead with four pressure-proof viewports, cylindrical tower with access hatch and five pressure-proof viewports

Buoyancy compartments: 2 compensating tanks inside the pressure-hull; 2 main ballast tanks outside the pressure-hull

Outer skin: Steel St 52-3, 3 to 4 mm (0·1 to 0·15 in) thick

Collapse depth: 600 m (1,968 ft)

Handling

Two lifting points. Ship-cradle or crane for normal launching.

Support requirements

Fishing vessel with additional equipment: UT/UT-direction-finder HP-air compressor, air bottles, Secondary diesel generator-set.

Mobility

May be air-freighted or shipped as deck cargo.

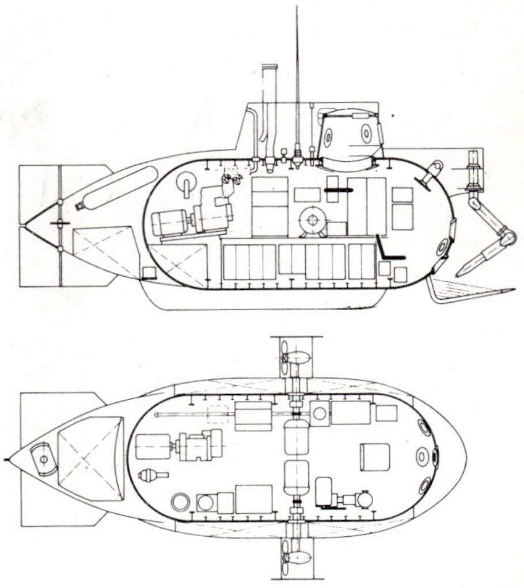

Layout of the (Maschinenbau Gabler GmbH) *Tours* submersible, type DGK/300

Typical equipment
Hydraulic grab arm, six motions, hydraulically controlled. Swivelling underwater search light, 1,000 watt. Collecting bucket.

Availability
Standard: 1,600,000 DM (1973).

Operators
KFODC Kno Feng Oceanic Development Co Ltd, Taipei/Taiwan, delivered in Spring 1971.

Relevant technical paper
To Harvest Coral with the Submarine, by Dipl. Ing. Peter Kruger, *Marine Technology* No. 1/1973.

Tours 64C DGK 300C (*Courtesy Ingenieurkontor Lubeck*)

ANTONINO MAGLIULO (Type: 002/Tours 66 DGK/300)

A manned and self-propelled submersible designed and developed by Prof. Dipl. Ing. Ulrich Gabler, Ingenieurkontor, Lubeck.

Uses
As for Tours 64 ("Argus I').

Design history
Date of construction of prototype started: 1971-1972
Date of prototype completed: 1972
Date trials completed: 1972
Certification authority clearance: Hull: GL+100 A2 Unterwasserfahrzeug (Submersible) Machinery: MC U

Production
Two Tours types built (64 and 66)

Basic dimensions

Length, overall	7·3 m (23·9 ft)
Width, overall	3·8 m (12·4 ft)
Height, overall	3·2 m (10·5 ft)
Draught	2·0 m (6·56 ft)

Weights

Dry weight	14·5 metric tons ready for diving
Submerged weight	16·75 metric tons
Payload	300 kg (660 lb) by freeing compensating tanks; 100 kg (220 lb) by slipping emergency ballast; 550 kg (1,210 lb) by blowing out main ballast tank in max. diving depth; 300 kg (660 lb) vertical propeller thrust; max. 1,250 kg (2,750 lb)

Performance and other details as for *Argus I.*

Antonino Magliulo (*Tours 66*, DGK/300 C) manned submersible designed by Maschinenbau *Gabler*

Operator
SELMAR Sarda Estrazione Lavaratione Prodotti Marini Sp. A. Cagliary/Italy, delivered July 1972.

TOURS TYPE PROJECTS

Designation of Type		Main Dimensions				Cruising Range nm			Standard Equipment							
Type		Diving Depth m	Crew	Weight tons (metric)	Length m	5 kn submerged	3 kn submerged	4·5 kn surfaced	Diesel Generator	Divers Lockout	Manipulator	Gyro Compass	Echo Sounder	UT	Radio Set	Searchlight
48 TDL/200	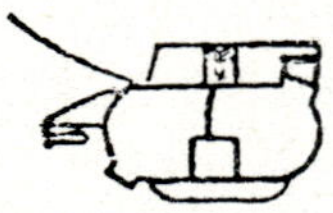	200 (656 ft)	1+2	8·2	4·8 (15·7 ft)	dependent on power supply of tender				●		●	●	●	●	●
60 DC/300	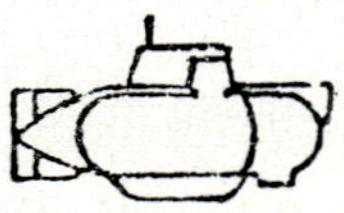	300 (984 ft)	2+2	9·5	6·4 (21 ft)	8	15	—		●		●	●	●	●	●
64 DGK/300	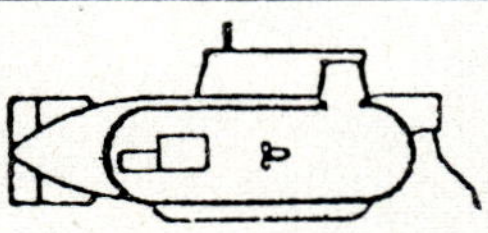	300 (984 ft)	2	14	6·9 (22·6 ft)	10	21	400	●		●	●	●	●	●	●
66 DGK/300	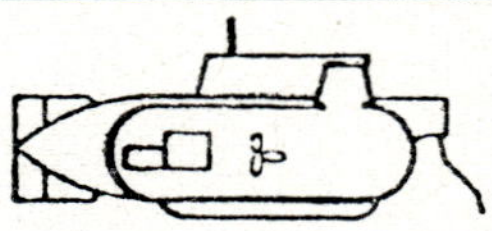	300 (984 ft)	2	14·5	7·3 (23·9 ft)	10	21	400	●		●	●	●	●	●	●
66 BK/100	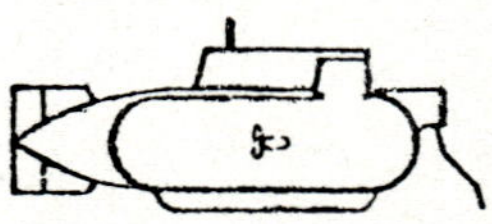	100 (328 ft)	2/3	14·5	7·3 (23·9 ft)	17	33	—			●	●	●	●	●	●
66 DC/300	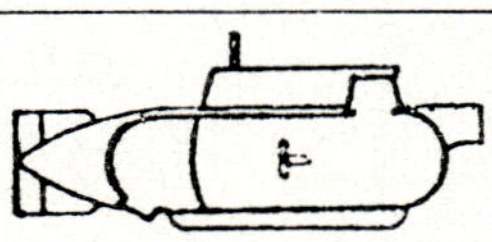	300 (984 ft)	2+2	14·5	7·3 (23·9 ft)	10	21	—		●		●	●	●	●	●
68 DCD/100	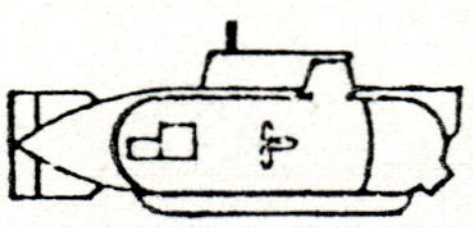	100 (328 ft)	2+2	16	7·5 (24·6 ft)	10	21	400	●	●		●	●	●	●	●
70 DCD/300	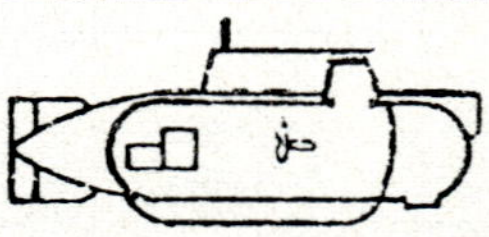	100 (328 ft)	2+2	17	7·7 (25·2 ft)	10	21	400	●	●		●	●	●	●	●
73/300	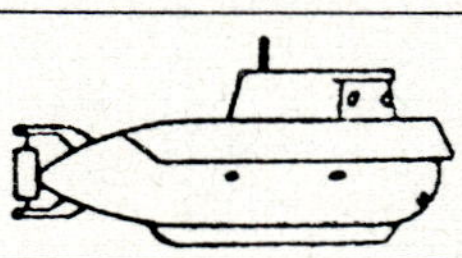	300 (984 ft)	2+4	15·5	8 (26·2 ft)	—	18	—				●	●	●	●	●
170 DGK/100	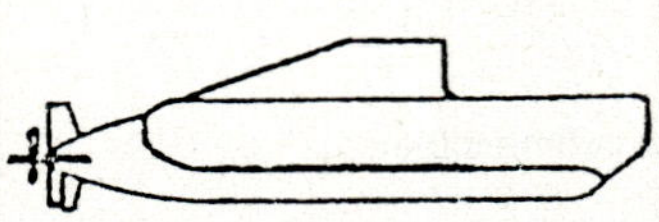	100 (328 ft)	5+8	77	18 (5·9 ft)	51 max. 11 kn	61	800	●	●		●	●	●	●	●
110/50	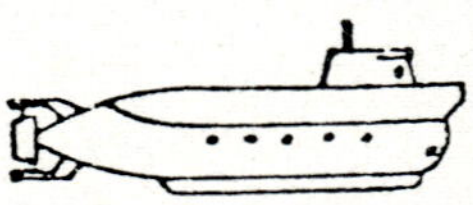	50 (164 ft)	2+10	32	11·4 (4·26 ft)	7	12	—				●	●	●	●	●
180/50	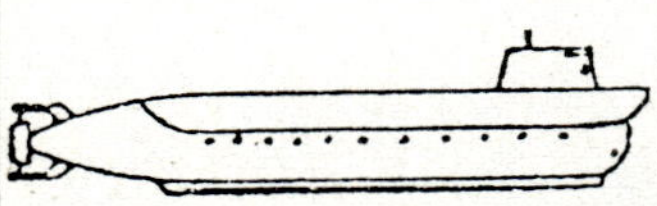	50 (164 ft)	2+28	57	18·8 (6·16 ft)	11	25	—				●	●	●	●	●

JAPAN

KAWASAKI HEAVY INDUSTRIES LTD

HAKUYO

A manned, self-propelled submersible, delivered by Kawasaki Heavy Industries Ltd on March 31, 1971 to Ocean Systems Japan, Ltd. The vessel was built with a subsidy from the Japan Ship Machinery Development Association. *Hakuyo* is fully certified by 'the special rule for manned submersibles' of the Japanese Government.

Hakuyo, the Kawasaki Submersible

Principal particulars

Length, overall	6·4 m (21 ft)
Breadth	1·6 m (5 ft 3 in)
Depth	2·7 m (8 ft 10 in) (from the bottom of battery pod to the top of hatch cover)
Displacement	6·6 t
Pressure hull diameter	1·4 m (4 ft 7 in)
Maximum operating depth	300 m (984 ft)
Crew:	
normal	2 men
maximum	3 men
Payload	150 kg (330 lb)
Speed (maximum)	3·5 knots
Endurance	5 hours
Life support	48 hours for 3 men

Main equipment and apparatus

Main motor	One
Horizontal thruster	One
Vertical thruster	Two
Hydraulic swivel device for main motor	One set
Diving plane	One set
View ports	Fourteen 150 mm (5·9 in) (inside diameter)
One point and four point lifting equipment	One set each
External light	Two
Magnetic flux gate compass	One set
Obstacle avoidance sonar	One set
Depth sonar	One set
Underwater telephone	One set
Wireless hand talkie (with antenna)	One set
Surface beacon light	One
Manipulator (hydraulic)	One set
Battery pod detaching device	One set
Emergency life support equipment	One set

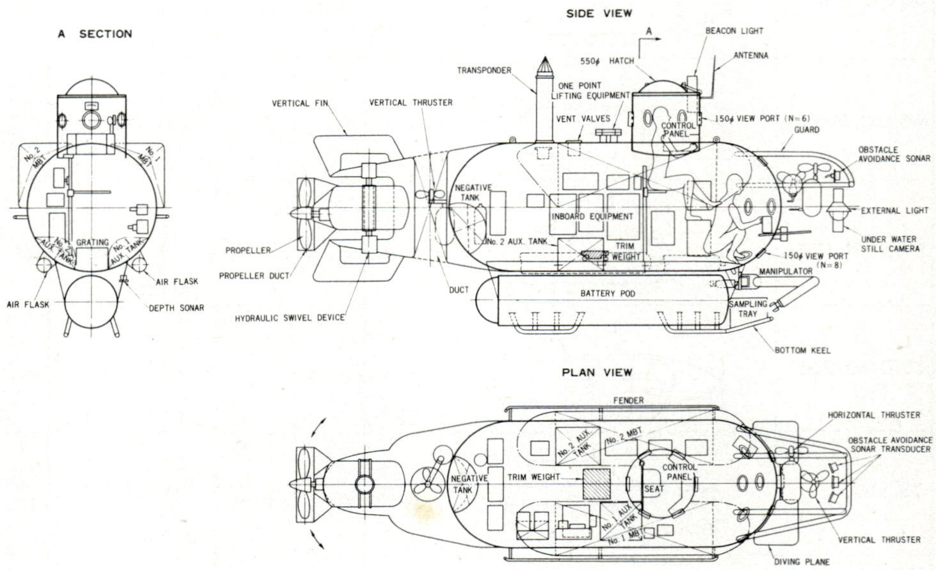

General arrangement of the Kawasaki *Hakuyo* submersible

MITSUBISHI HEAVY INDUSTRIES LTD

Kobi Shipyard
YOMIURI-GO

Yomiuri at sea

Early in 1963, Mitsubishi Heavy Industries Ltd. started the construction of a small submersible (*Yomiuri*) for research on the continental shelves of Japan. At that time

Kuroshio, a tethered submersible capable of reaching depths of 200 m (656 ft), was the only craft available for this mission. Therefore the need for a free running work submersible was emphasised and the following requirements were established:

(1) Operating depth to 300 m (984 ft); to conduct research on the continental shelves (it was desired to reach the upper part of the slope off the continental shelf).

(2) High efficiency in operation and ease of maintenance. With regard to main items of equipment, only some improvements of the existing techniques were made, and the construction of a vessel of high reliability and economy at an early date was sought.

(3) In view of the characteristic features of the coastal areas of Japan and in consideration of the purpose of investigation of fisheries, sufficient protection for operation in rocky places was provided.

(4) Emphasis was laid on the collection of samples by means of a manipulator and on observation and photography through the view ports.

(5) For operation at the bottom of the sea, a manipulator, view ports, etc. were provided, and the principal purposes of the craft were envisaged as coral collection, investigation of fishing grounds, and the investigation and observation of the sea bottom by means of photography and television.

Yomiuri was built on the basis of these requirements and has dived 435 times during the five and a half years from its completion in July 1964, with a total of 1,325 hours of diving time.

Yomiuri, built by Mitsubishi Heavy Industries

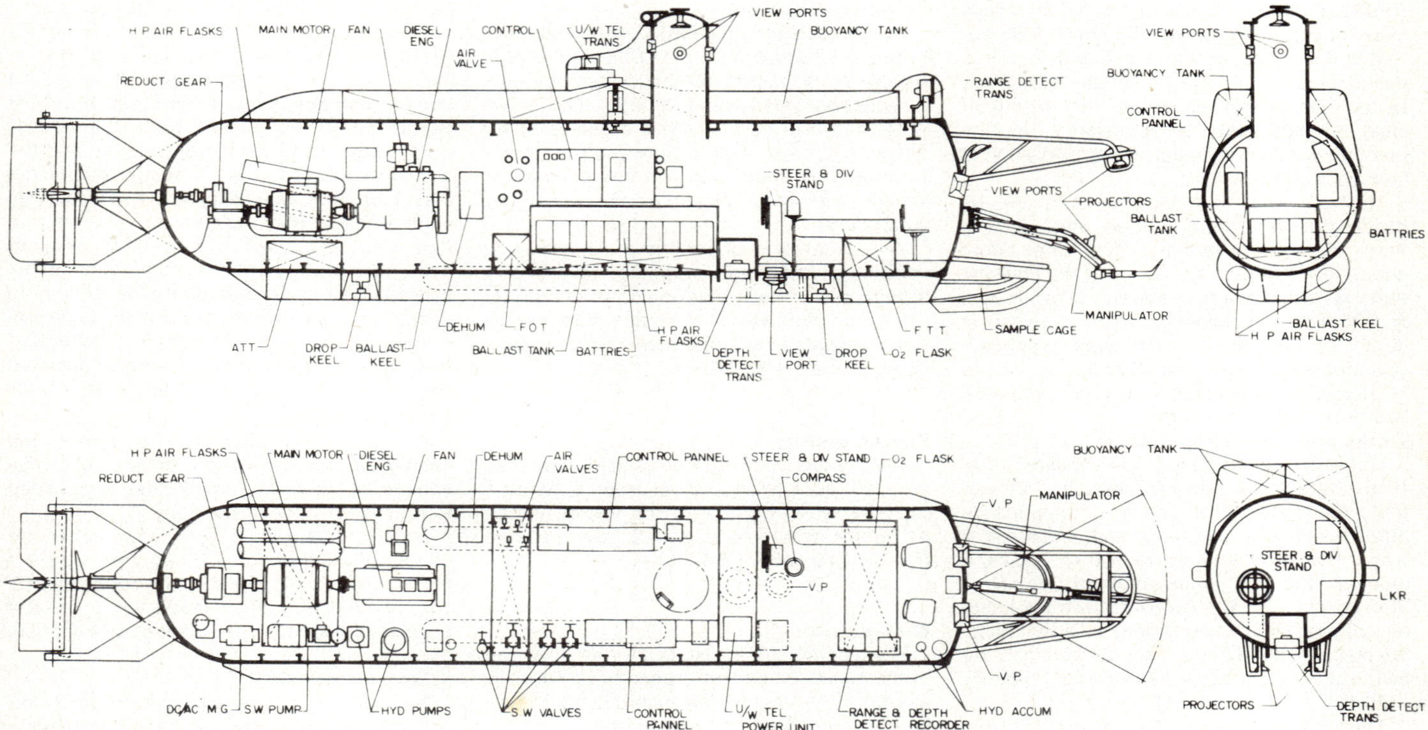

General arrangement diagrams of *Yomiuri*

Design history

Date of construction of prototype started: October 18, 1963
Date of prototype completed: May 15, 1964
Date trials completed: June 17, 1964
Certification authority clearance: July 4, 1964

General description

This submarine vessel is of a single-hull type, consisting of a cylindrical pressure hull with a spherical end plate on the front and rear ends and a buoyancy tank in the upper part. Inside the pressure hull there is a single compartment without a bulkhead. For facilitating operation on a rocky sea floor a steel fender is provided at the bow for protecting the body of the vessel and the manipulator, a wooden fender on both sides and a steel fender at the stern to protect the rudder and propeller. The front end of the ballast keel at the bottom of the vessel is extended like a sleigh, so that it will facilitate the operation of the vessel over a rocky sea floor.

There is a single crew compartment in the vessel, but the arrangement of equipment is functionally divided into the pilot and observation space, control space and machinery space.

Principal particulars

Length, overall	14·50 m (47 ft 6·7 in)
Breadth (including side fender)	2·45 m (8 ft 0·4 in)
Depth (to superstructure top above bottom of keel)	2·80 m (9 ft 2·2 in)
Draught on surface	2·20 m (7 ft 2·6 in)
Displacement on surface	36·3 tons
Submersible depth	300 m (984 ft)
Pressure hull diameter (outside of hull)	2·05 m (6 ft 8·7 in)
Speed (max.)	3 knots
Endurance	2 knots 6 hours
Life support	48 hours
Power:	
Main batteries	450 AH–50 cells
Diesel engine	25 PS/900 rpm × 1
Propulsion main motor	12 kW/1,000 rpm × 1
Complement (crew and observers)	6

Particulars of main motor (generator)

	Motor	*Generator*
Output (kw)	3–12	12
Voltage (volts)	100 (85–130)	110 (110–140)
Current (A)	45–155	109
Revolution (rpm)	500–1,000	900
Rating	Continuous	Continuous
Winding	Stabilised shunt	Stabilised shunt
Housing	Drip proof	Drip proof

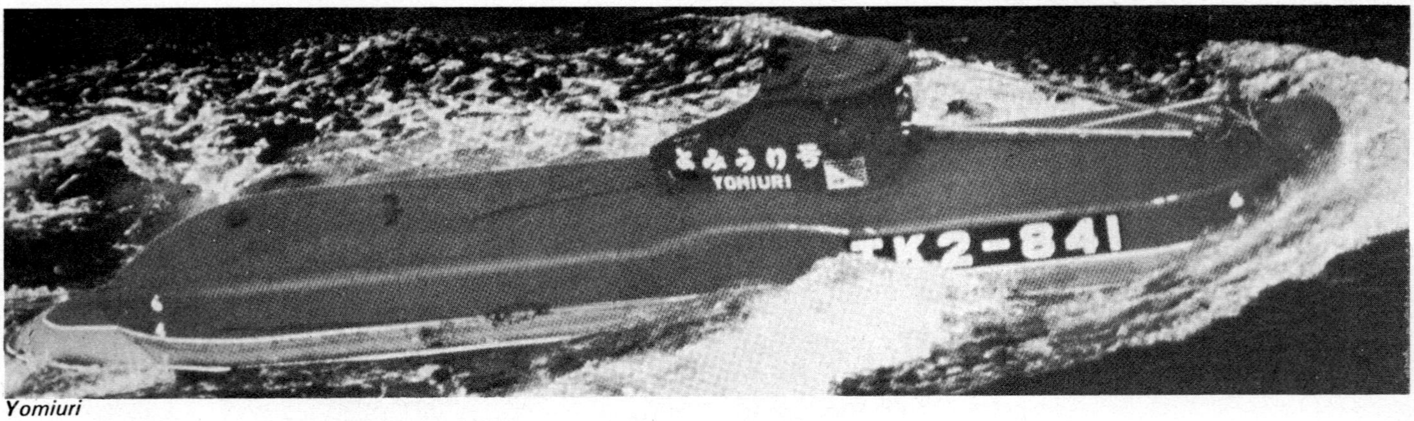

Yomiuri

Pressure hull

The pressure hull is in the form of a cylinder 2·05 m (6 ft 8·7 in) in outer diameter. Its front end is provided with a spherical end plate of 2·00 m (6 ft 6·7 in) in radius and its rear end with a hemispherical construction of 1·025 m (3 ft 4·3 in) in radius. Its total length is 10·683 m (35 ft 0·5 in). The pressure hull is made of a high tensile steel of 46 kg mm^2 (66,000 lb/in^2) in yield point. The plate thickness of the cylindrical part is 16 mm (0·63 in) and the plate is reinforced by means of an inner frame, spacing 500 mm (19·69 in).

The end plate at the front end is not a hemisphere but a part of a sphere to obtain a better efficiency of space, so that a ring of steel forging material is inserted in the junction with the cylinder in order to prevent concentration of stress.

The pressure hull was constructed by dividing it into three blocks, and welding them together after mounting the larger equipment on them. The safety of the pressure hull was ascertained in a water pressure test in which an equivalent pressure to that at a depth of 300 m (984 ft) was produced. (Conducted on land after all parts that had to be welded such as penetrators and foundation had been welded).

Tanks and ballasting system

The submersible vessel has a ballast tank (capacity 2·0 tons) and forward and aft trim tanks (capacity 0·4 tons each) in the pressure hull. For filling, draining and transferring water into, from and between these tanks, a plunger type sea water pump of 2 $m^3/h \times$ 30·5 kg/cm^2 (70·4 $ft^3/h \times$ 435 lb/in^2) is used. In ordinary diving, descending speed of about 10 m/min (33 ft/min) can be obtained by giving a negative buoyancy of about 100 kg (220 lb)

The superstructure in the upper part of the pressure hull is used as the buoyancy tank, which is partitioned into four quarters and has a total capacity of about 8 tons. One vent valve in the middle of the four tanks is used in common for all of them. The draining of the buoyancy tanks is conducted by means of high pressure air, and in the pressure hull are provided 5 air flasks of 125 kg/$cm^2 \times$ 46·7l (1,780 lb/$in^2 \times$ 2,850 in^3), and in the outboard 2 air flasks of 125 kg/cm^2 (1,780 lb/in^2).

Steering and tilting system

The vessel has a rudder and a stern plane at the stern and has no bow plane. The steering and tilting system is a manual chain-drive type operated by one person. Although equipping the vessel with a side propeller was considered, it was not so equipped in view of operation in rocky areas.

Navigation system

The vessel has two ultrasonic transducers, one mounted on the upper part of the bow, the other on the bottom of the vessel. The former is capable of being turned upwards, to right angles to the horizontal direction, from within the vessel, and is used as an early warning system for discovering obstacles forward and above, while the latter is used as an echo depth sounder with a distance of detection of 400 m (1,312 ft) in the forward and upward directions and 1,000 m (3,280 ft) downwards.

The vessel is equipped with an underwater telephone system of 1,000 m (3,280 ft) range for communicating with the mother ship while diving, and carries portable transceivers for surface communication.

Propulsion system

The vessel is propelled by means of its main motor of 12 kW/1,000 rpm. This main motor drives a 3-bladed propeller at the stern through a reduction gear (ratio 1:2·5). The main motor can also be utilized as a generator by disengaging the clutch between the main motor and the reduction gear and coupling it directly with a diesel engine for recharging the battery.

The adjustment of rotation, adjustment of voltage, starting and stopping, reversing or returning to normal rotation and switching for use as the generator of the main motor can be easily operated by a one man control system using a delay relay system.

Power system

The vessel has 50 lead-acid batteries installed as a main power source, with a capacity of 45 kWh. And the vessel has an AC motor generator as an AC power source (60 Hz, 110 volts, 1·5 kVA).

Manipulator

The manipulator is installed at the front end of the pressure hull and is operated hydraulically from the vessel. The manipulator can be bent, extended, turned and twisted by means of 6 hydraulically operated cylinders, and its range of operation is 33° to the right and left and 60° upward and downward, and its end when fully extended can reach to a distance of 2·5 m (8·2 ft) off the front end of the pressure hull.

The hydraulic unit for the manipulator is formed of a plunger type hydraulic pump (with a motor of 1·5 kW/1,800 rpm) of 140 kg/cm^2 $\times$ 1·5 l/min (2,020 lb/$in^2 \times$ 91·5 in^3) and an accumulator of 10 l (610 in^3).

View port

There are two view ports in the front end piece of the pressure-hull, one view port at the bottom of the pressure-hull and four view ports on the sides of the access trunk.

The port provided in the pressure-proof shell is 120 mm (4·72 in) in diameter, and the view port provided in the access trunk, 60 mm (2·36 in). The port pane is made of optically homogeneous glass designed particularly for photographic work. The thickness of the two types of view ports are respectively 62 mm and 40 mm (2·44 in and 1·57 in).

External lighting

Two underwater light projectors of 1 kW are fixed on the bow, and two others on the bottom of the vessel. One of the two at the bow is mounted on the arm of the manipulator and the direction of light projection can be changed. Furthermore, a provisional light projector can be mounted on the fender of the bow for use in the case of occasional colour photography that requires a high level of brightness.

Environmental control and life support

A fan of 7 m^3/min (243 ft^3/min) $\times$ 15 mmAq (0·59 inWg) is provided for air circulation in the pressure hull. During the anchorage and battery recharging, a provisional hose for ventilation is conducted from the access trunk to the outside for supplying air, and the air inside the vessel is exhausted out of the vessel by first absorbing it with the diesel engine. The battery used in the vessel is of a type that generates a very low level of hydrogen gas, and the slight amount of the hydrogen gas generated during the recharging operation is also exhausted out of the vessel through absorption by the engine. The level of hydrogen concentration is constantly detected by means of a hydrogen sensor to ensure complete safety.

A rise in humidity in the vessel will deteriorate the insulation of the electrical equipment, so that a dehumidifier is provided to prevent any rise in humidity.

For life support, LiOH for CO_2 scrubber and one Oxygen flask of 125 kg/$cm^2 \times$ 46·7 l (1,780 lb/$in^2 \times$ 2,850 in^3) for supplying Oz are provided. Breathing masks for 6 persons are also provided in case of air contamination.

Safety Devices

In addition to the above-mentioned life support systems, two jettisonable ballast blocks of 400 kg (880 lb) are provided for an emergency. By releasing them from the vessel, buoyancy can be gained. Moreover, in the case of a relatively shallow level, a skirt can be mounted under the access trunk for use as an escape trunk to enable the crew to escape from the vessel, and the scuba gear for this is also provided.

Operation

Upon its completion in July 1964, the training of the crew was started, and the vessel was used for investigation, in August of the same year, of the epicentre of the Niigata Earthquake that took place in June. Since then it has been in use for investigations at various spots along the coastal area of Japan. Its range of operation was greatly expanded in 1966 when it went to Amami Islands and then to Okinawa with a larger mother ship (235 tons). Furthermore, in the period from November 1968, to June 1969, the vessel participated in the joint research programme of Japan and Australia in the survey of Australia's Great Barrier Reef and brought about valuable results.

Major activities

Date	*Location*	*Purpose*
(1964)		
Mid. Aug.–early Sept.	Niigata	Scientific observation of the bottom around the epicentre of the Niigata Earthquake.
Oct.–Nov.	10 sites off the coast around Kyushu	Observation of fishery resources, artificial fish 'nests' and geological features of the bottom. Archaeological study.
Dec.	Uwanoumi (off the west coast of Shikoku), candidate area for the proposed under water park	Photography (Stills & Cine).
(1965)		
Mid. Jan.–Feb.	Suruga Bay	Searching for deep-sea fishes.
May	Off the south coast of Shikoku	Collecting coral
July–Aug.	Off Kurihama	Survey of shipwreck
Nov.	Akashi Straits	Data collection for the 'longest bridge' construction project.
(1966)		
Feb.–April	Amami Islands	Research and collection of coral
Mid. May–July	Okinawa Islands	Research and collection of coral
Nov.	Off Matsuyama	Rescue and salvage of aircraft
(1967)		
June–July	Amami Islands	Research and collection of coral
Sept.–Nov.	Okinawa Islands	Observation of fishery resources and geological features of the bottom. Measurement of water, temperature and salinity.
(1968)		
April–May	Amami Islands	Research and collection of coral
June	Ogasawara Islands & Iwojima	Observation of fishery resources
July–Aug.	Off Muroran (Hokkaido)	Observation of fishery resources, artificial fish 'nests' and geological features of the bottom.
Dec. 23	Arrived at Townsville (East Coast of Australia)	
(1969)		
Jan.–April	Great Barrier Reef	Observation of fishery resources, reef, geological features of the bottom. Collection of mud and minerals from the bottom. Measurement of temperature, salinity, current velocity and visibility.
April 26	Arrived Darwin	
May	Arafura Sea	Geological researches by sparker and observation of the bottom.
Oct.–Nov.	Osumi and Amami Islands	Taking photographs for EXPO '70
Dec.	Off the southern coast of Shikoku	Taking photographs for EXPO '70

Mitsubishi have reported that *Yomiuri* was wrecked in a storm while moored in 1970.

UNITED KINGDOM

WIGHT PLASTICS LIMITED

DOLPHIN MK. II

The *Dolphin Mk. II* is a self-propelled, two-man wet diver transport vehicle. It is designed for use by divers carrying out various underwater activities such as survey work, underwater photography and general marine study. By October 1973 three of these vehicles had been produced and another five were under construction.

Basic dimensions

Length, overall	4·27 m (14 ft)
Width, overall	0·86 m (2 ft 9·8 in)
Height, overall	1·37 m (4 ft 6 in)
Draught	0·69 m (2 ft 3 in)
Dry weight	approx. 223 kg (490 lb)
Payload	2 crew, approx. 205 kg (450 lb)

Performance

Maximum operational depth	24 to 30 m (80 to 100 ft)
Speed:	
normal submerged	2 knots
max.	3 knots
Duration at cruise speed	approx. 3 hours
Duration at maximum speed	approx. 2 hours

Main machinery description

Power source	seven 12 volt batteries of 52 amp/h each
Motor	one 2 h.p. electric motor
Propeller	one three-blade propeller in acetal resin
Controls	rudder, and forward and rear elevators for manoeuvring and depth control
Crew compartment	open cockpit, 2 seat in tandem
Machinery compartment	sealed battery compartment
Buoyancy compartment	2 ballast tanks
Outer skin	glass reinforced plastics

Handling

Lifting points	fore and aft
Normal launching method	trolley or lifting

Dolphin MK II wet-submersible manufactured by Wight Plastics Ltd

UNITED STATES OF AMERICA

GENERAL DYNAMICS, ELECTRIC BOAT DIVISION

STAR II

This craft (formerly *Asherah*) was used in 1972 by Makai Range, Inc (now dissolved) to harvest coral.

Design maximum operating depth	366 m (1,200 ft)
Speed	3 knots
Duration	8 hours
Life support duration	48 man hours
Crew	2 (pilot and observer)
Payload	114 kg (250 lb)

Star II is equipped with a mechanical arm, catch basket, navigational system and high capacity camera and strobe.

Star II, formerly *Asherah*

STAR III

Design maximum operating depth	600 m (2,000 ft)
Launched	May 3, 1966
Length	5·18 m (17 ft)
Weight, dry	4,454 kg (9,800 lb)
Life support	12 hours
Power source capacity	20·9 kw/h

The Institute of Oceanology of the USSR was interested in acquiring this submersible at one time.

GENERAL OCEANOGRAPHICS, INC

NEKTON ALPHA, BETA AND GAMMA

Manned, self-propelled submersibles, designed, built and operated by General Oceanographics, Inc, a company having experience of over a dozen submersibles. Designed primarily for scientific surveys and marine inspection, these vessels may also be used for search and recovery work amongst other duties. Four Nekton-type submersibles have been built.

ALPHA design history

Date of construction of prototype started: July 1967

Date of prototype completed: May 1968

Date trials completed: 23 May 1968

Certification authority clearance: None

BETA design history

Date of construction of prototype started: December 1969

Date of prototype completed: August 1971

Date trials completed: August 18, 1971

Certification authority clearance: Classification by the American Bureau of Shipping; 17 April 1973; Class A 1 Manned Submersible.

General Oceanographics' *Nekton Alpha* in mid-water off Oahu

GAMMA design history

Date of construction of prototype started: December 1969

Date of prototype completed: October 1971

Date trials completed: October 20, 1971

Certification authority clearance: Classification by the American Bureau of Shipping; Class A 1 Manned Submersible; 17 April 1973

The following details are common to all three vessels, *Alpha, Beta* and *Gamma,* except where indicated otherwise.

Basic dimensions

Length, overall	4·72 m (15 ft 6 in)
Width, overall	1·05 m (3 ft 6 in)
Height, overall	1·81 m (6 ft)
Draught	0·915 m (3 ft)

Weights

Dry weight	2,100 kg (4,700 lb)
Submerged weight	0
Payload	*Alpha*: 136 kg (300 lb); *Beta* and *Gamma*: 200 kg (450 lb)
Crew	Two: one pilot; one observer

Nekton Beta, manufactured by General Oceanographics

Performance

Maximum operational depth	300 m (1,000 ft)
Speed:	
cruise, surface	3 knots
submerged, forward and reverse, normal	2 knots
max.	3·5 knots
Duration at cruise speed	3·5 hours
Range	11·3 km (7 miles)
Life support endurance (air supply)	24 hours
Manoeuvring capability	4·55-m (15-ft) turning radius

Main machinery description

Power source	Lead acid batteries (in sealed compartments
Motors	Westinghouse 3·5-h.p. motor
Propellers	*Alpha* and *Beta* one two-blade; 25·4 cm (10 in) diameter; *Gamma* one three-blade, 25·4 cm (10 in) diameter
Controls:	
manoeuvring	Rudder, bow plane
depth	Water ballast, bow plane, drop weights
trim	Internal auxiliary trim tank

Electrical on-board equipment

Navigational	Magnetic compass, directional gyro, depth gauge, plus retractable marker buoy, pinger on *Beta* and *Gamma*
Vehicle function instrumentation	
Communication	Citizens' band radio; underwater telephone
Search and recovery	Sonar, mechanical arm, external lights
Television and visual	Videotape system

Safety systems

Life support	Oxygen, baralyme
Ballast jettison, pumping, etc.	Drop weights
Buoy markers	Retractable buoy marker
Warning systems	Oxygen meter; altimeter

Internal environmental conditions

Crew space	Pilot on seat; observer can sit or lie flat
Pressure control	Altimeter, oxygen bleed valve
Air purification	Baralyme in scrubber

Basic structure

Crew compartment, pressure hull type	cylindrical, *Alpha,* 20 viewports, *Beta* and *Gamma,* 17 viewports
Machinery compartment	Motor housing
Buoyancy compartment	Two main exterior ballast tanks
Pressure hull length	*Alpha*: 2·26 m (7 ft), *Beta* and *Gamma*: 2·42 m (8 ft)
Pressure tested depth	455 m (1,500 ft)
Collapse depth	762 m (2,500 ft)

Handling

Lifting points	Stainless shackles fore and aft; lifting yoke fits over conning tower
Normal launching method	Crane lifts submersible off deck and into water; two minute operation

Support requirements

Mother ship facilities required: Have ranged from a 4·9 m (16-ft) runabout towing the submersible to a 48 m (158-ft) Gulf Coast supply ship with 40-ton crane. Special vessel not required.

Mobility

Have been air freighted on standard commercial airlines and carried as deck cargo, also car trailer transported by road.

Availability

May be leased or chartered

Licensed builders: Nekton, Inc.

Operations

Nekton submersibles have made over 1,250 dives since 1968.

Nekton Alpha has mase 621 dives on surveys off southern California, Alaska, Hawaii, northern California, and the Great Lakes.

Nekton Beta has made 263 dives including 134 in 1972 and has been used on surveys off southern California, British Honduras, Florida, Spain, Morocco and Algeria.

Nekton Gamma has made 363 dives including 252 in 1972. 'Nekton Gamma' has been used on surveys off southern California, British Columbia, Jamaica, Virginia, Massachusetts, the Great Lakes and the Gulf of Mexico.

Costs (1972)

Typical day rate charge with crew $US1,000 per day.

LITTON INDUSTRIES

ALVIN

A manned, self-propelled deep submergence research vehicle. *Alvin* was commissioned June 5, 1964, and is now operated by Woods Hole Oceanographic Institution and has a depth capability of 3,506 m (11,500 ft) It is owned by the US Navy.

Basic dimensions

Length, overall	7·01 m (23 ft)
Width, overall	2·40 m (8 ft)
Height, overall	3·04 m (10 ft)
Draught, on surface	2·13 m (7 ft)

Displacement

16 tons (dry)

Performance

Depth capability	3,506 m (11,500 ft)
Full speed	$2\frac{1}{2}$ knots
Cruising speed	$1\frac{1}{4}$ knots
Range at cruise	submerged 10 to 15 miles
Endurance at low low speed	50 hours for three people

Main machinery description

Power source	3 banks of lead acid batteries
Propellers	Two for lift and a large reversible stern propeller that turns from side to side for steering
Pumps	Hydraulic motors powered by pumps
Controls—manoeuvring	Mercury trim system

Pressure hull

In 1973 a new pressure hull in titanium was fitted extending the depth capability from 1,829 m (6,000 ft) to 3,506 m (11,500 ft).

Equipment

Navigational	gyro compass, repeater, magnetic compass, horizontal scanning sonar
Survey and search	radar, sonar, TV, seismic, closed circuit TV, precision depth indicator
Other	sonar telephone, lights, cameras, mechanical arm, tape recorder, water temperature monitor, echo sounder, current speed meter

***Alvin*, produced by Litton Industries for Woods Hole Oceanographic Institution**

Crew

Pilot, co-pilot, scientific observer

Safety systems

Batteries can be dropped, trim system may be dumped, mechanical arm can be dropped and personnel sphere may be disconnected from hull.

LOCKHEED MISSILES AND SPACE COMPANY

DEEP QUEST

This research submersible, designed to operate at more than a mile and a half beneath the ocean surface, was launched in San Diego in June 1967. With a displacement of 50 tons *Deep Quest* has a crew of two and space for two to three scientist-observer passengers. Payloads of up to 3,180 kg (7,000 lb) can also be carried.

The free-flooding outer hull is 12 m (40 ft) long and is constructed in welded 5083 aluminium alloy. The inner pressure hull consists of two interconnecting maraging steel spheres, each 2·13 m (7 ft) in diameter.

Principal characteristics

Length, overall	12·189 m (39 ft $11\frac{3}{4}$ in)
Beam, maximum (stern planes)	5·79 m (19 ft)
Depth, maximum	4·04 m (13 ft 3 in)
Draught surfaced	2·62 m (8 ft 7 in)
Displacement	50 long tons
Operating depth, maximum	2,439 m (8,000 ft)
Speed, submerged maximum	4 knots
Personnel	Crew, 2 Observers, 2 or 3
Payload	
Pressure hull after sphere	0·568 m^3 (20 ft^3)
Modulator, outer hull	909–3,182 kg (2,000–7,000 lb) (depending on location)
Trim angle variation	±30 degrees
List angle variation	±10 degrees
Endurance:	
Power	12 hours at 3 knots or greater 18 hours at 2 knots
Life support	Normal 192 man-hours Emergency 12 man-hours
Power plant:	
Main	Two 115 Vdc lead-acid batteries totalling 230 kWh
Emergency	Two independent 28 Vdc silver-zinc batteries totalling 3·6 kWh
Propulsion:	
Axial	Two $7\frac{1}{2}$ h.p. AC motors and nozzle-propellers
Vertical	Two $7\frac{1}{2}$ h.p. AC motors and ducted propellers
Lateral	Two $7\frac{1}{2}$ h.p. AC motors and water jet nozzles

Observation features

Two viewports, one with a remote optical viewing system

Two pan-and-tilt mounted TV cameras—port and starboard

One sail-mounted 360-degree vision periscope and scanning TV camera

One bow-mounted, forward looking TV camera

One fixed-mount TV camera, optional location

Two TV monitor screens with TV camera selector switches

One 70-mm camera with synchronised strobe light

One motion picture camera, 24 frames/min, and viewport mount (internal)

CTFM sonar—sail mounted

Fathometer—top and bottom scan, 6,000 fathoms

Fathometer—bottom only, 200 fathoms

Floodlights:

4 fixed	500 watts
w fixed	750 watts
2 fixed	1,000 watts
1 fixed	2,500 watts
2 pan-and-tilt mounted	500 watts

Weight jettison systems:

Shot ballast (fail safe)	750 kg (1,650 lb) wet weight
Mercury, trim system	655 kg (1,440 lb) wet weight
Mercury, list system	376 kg (828 lb) wet weight
Batteries (two cell packs)	1,386 kg (3,050 lb) wet weight

Entanglement jettison systems:

Manipulator assemblies

Bow instrument rack

Communications equipment:

Marine band transceiver

Underwater telephone (top and bottom transducers)

Acoustic tracking equipment:

Omnidirectional pinger

Parabolic reflector pinger

Work systems:

Dual manipulator system

Piston coring device (six barrles)

Vane shear device
Object recovery system (prototype)

Capabilities

Deep Quest is designed to support a wide range of research and development projects applicable to both commercial and military ocean programmes. It can be equipped with a variety of work systems and interchangeable payload modules, individually tailored to specific mission requirements. Ample payload space is available, both internal and external to the pressure hull.

A 48·3 cm (19 in) wide, 150 cm (59 in) high, standard EIA electronics rack is available in the aft sphere of the pressure hull for payload-peculiar equipment to be operated by the scientists/observers. Additional racks can be readily installed if required. Total normal reserve payload capacity internal to the pressure hull is approximately 0·568 (20 ft^3).

Interchangeable payload modules in excess of 910 kg (2,000 lb) wet weight can be installed outside the pressure hull, but in protected areas within the outer hull. Significantly larger payloads can be carried internally or externally to the outer hull, dependent on total mission equipment requirements. Typical payload modules are:

An array of oceanographic-data-gathering equipment mounted on an extendable/retractable rack;
A recovery module;
An additional pressure hull for special electronic devices;
A diver support module for total saturation lockout diving;
A replenishment module for resupplying underwater dry habitats.

In addition, a variety of smaller devices can be installed, either internally or externally to the outer hull. The *Deep Quest* system at present includes the following work devices:

Two manipulator assemblies mounted on common transverse rails in a manipulator bay in the bow section. Each manipulator assembly can be individually controlled in all of its movements, including telescoping and lateral translation, and can be jettisoned in an emergency. The manipulator hands can be interchanged with alternative designs to perform specialised functions. The unique dual installation of the manipulator assemblies permits great versatility and increased lift capacity by operation in unison. The lifting capacity of each manipulator is 227 kg (500 lb) wet weight in water for a vertical lift, or 45·4 kg (100 lb) horizontally at a radius of 1·8 m (6 ft). The gripping force exerted by each hand is 227 kg (500 lb) with a maximum jaw opening of 20·3 cm (8 in). When not is use, both manipulators are fully housed in their athwartships bay.

A six-barrel coring device mounted on the bow. The corer can obtain six 6·35 cm ($2\frac{1}{2}$ in) diameter by 1·22 m (4 ft) long cores per dive. It is operated hydraulically, functions to depths of 2,440 m (8,000 ft) and can be jettisoned in an emergency.

A vane shear device capable of measuring the shear strength of sea floor sediment to a sediment depth of 1·22 m (4 ft). The device is contained in a cylindrical envelope 2·13 m (7 ft) long and 25·4 cm (10 in) in diameter. With *Deep Quest* resting firmly on the sea floor, the vane can be extended at a rate of 38·1 cm (15 in) per minute and can be stopped at any point to a maximum of 122 cm (48 in). Penetration depth and axial load on the vane shaft are measured during extension and retraction. A permanent strip chart record is made of axial loads and shear strength.

A prototype, self-contained large object recovery and salvage unit capable of recovering from the ocean bottom objects as large as 1·52 m (5 ft) in diameter and 6·1 m (20 ft) long. The device is attached to the underside of the submarine and is powered and controlled from inside the pressure hull via an electrical umbilical. Once the target has been captured, it can be lifted to the surface in either of two modes: (1) by using the available buoyancy of *Deep Quest*—approximately 909 kg (2,000 lb) lift capability, or (2) by releasing the recovery device and its target and then hoisting it by surface vessel—approximately 5,793 kg (19,000 lb) lift capability.

Navigation and control

Deep Quest's advanced integrated control and navigation system allows the submarine to be controlled in either a manual or a fully automatic mode with five degrees of freedom in both hover and cruise speed regimes. The automatic control mode allows the submarine to follow a predetermined course or pattern during search missions or while obtaining samples or measurements at specific locations relative to a known reference point. In addition, the automatic control mode allows the submarine to hover over a fixed location in head currents of up to 4·0 knots or broadside currents of up to 0·5 knot.

The navigation system consists of a strapdown gyro system and an analogue computer which processes the vehicle speed, direction, and orientation inputs, and plots the submarine's course on an X-Y plotter. Direction with respect to the deck is obtained from a gyrocompass and converted to true heading by means of attitude information from a vertical reference gyro. A doppler sonar is used to measure velocity relative to the ground when the submarine is at distances of less than 45·7 m (150 ft) from the bottom. Water speed and direction sensors are used to measure the vehicle's velocity with respect to the water. These two velocities are resolved into longitudinal and lateral components and subtracted to compute the water current.

The Lockheed submersible *Deep Quest*

In addition to velocity and heading, the navigation system provides electrical signals to indicate pitch and yaw angles, pitch and yaw rates, and depth and depth rate. The X-Y recorder will create a permanent record of the submarine's course including reference marks for map making or will enable the pilot to follow a planned course.

Propulsion

Deep Quest is equipped with two $7\frac{1}{2}$ h.p. axial thrusters, two $7\frac{1}{2}$ h.p. vertical thrusters and two $7\frac{1}{2}$ h.p. lateral water jets; AC power and speed control for each of these units are provided by an individual solid state inverter-controller. These propulsion units, in conjunction with stern planes and a rudder, provide a five-degree-of-freedom dynamic manoeuvering capability throughout the entire speed range from 0 to 4·0 knots.

Depending upon forward speed, dynamic yaw rotational moments are applied by differential action of the variable thrust lateral water jets, by differential action of the variable speed axial thrusters, or by rudder action. Dynamic pitch rotational moments are applied by differential action of the variable speed vertical thrusters, or by stern plane action. Static roll and pitch rotational moments are applied by weight transfer with the mercury list and trim systems, respectively.

Electrical systems

Main power of *Deep Quest* is supplied by two 120 volt DC (nominal) pressure-compensated, lead-acid batteries connected in parallel, so that full line voltage is maintained in the event of loss of either battery. The total available main power of 230 kWh enables the submersible to operate at a speed of approximately 3 knots for 12 hours or at a speed of 2 knots for 18 hours. Two independent 28 volt DC silver-zinc batteries located inside the pressure hull provide 3·6 kWh of emergency power. All safety-critical circuits may be operated with either normal or emergency power.

For payload operation, the following electrical power is available either inside or outside the pressure hull:

120 volt DC (nominal)
29 volt DC rms $\pm$2 per cent
115 volt AC rms $\pm$2·5 volt, 60 Hz, single phase
115 volt AC rms $\pm$2 per cent, 400 Hz, single phase

The only type of penetrations through the pressure hull (other than hatches and viewports) are electrical, and these presently provide 26 two-wire No. 18 AWG circuits, four two-wire No. 16 AWG circuits, and one RG58U coaxial lead for payload usage. Additional conductors can be made available to satisfy customer requirements.

Passage of all normal service and payload circuits by terminal boards immediately inside the forward sphere of the pressure hull lends great versatility for circuit interchange and facilitates diversion of normally used circuits to other purposes when not essential to a specific mission.

The pressure hull of *Deep Quest* being inserted into the outer hull

Provisions for observation

The bispherical pressure hull contains two viewports, a forward/downward looking viewport in the forward pressure hull sphere and a downward looking viewport in the bottom hatch of the after pressure hull sphere. Observation through the forward viewport is accomplished from a prone position between the Pilot and Navigator stations; a body pad and chin rest are provided for this purpose. The integrated control system is designed to permit the control of the submersible by the pilot in this position. Photography by other than externally mounted cameras is also accomplished through this viewport. The downward looking viewport in the after pressure hull sphere is equipped with an optical remote viewing system which has a 180-degree field objective in the vertical plane, and a 360-degree field objective in the horizontal plane. Photography is also possible through this viewport. An external fixed-mount, variable-focus 70-mm camera with synchronized strobe light is used for obtaining photographic records to supplement those obtained through the viewports.

Television monitoring of the environment is accomplished by port and starboard pan-and-tilt-mounted cameras, a sail-mounted 360-degree-vision periscope scanning camera, and a bow-mounted forward-looking fixed camera. A fifth television camera can be mounted at optional locations to meet special mission requirements. Two television monitor screens, one each for the Pilot and Navigator, are used to monitor the output of any of these cameras on selection. Signals selected by the Pilot can be recorded on video tape. Nine fixed lights, ranging in power from 500 to 2,500 watts, and two pan-and-tilt-mounted 500-watt floodlights are used for illumination. These lights can be individually selected for optimum lighting combinations.

Rapid-scanning and high-resolution sonar detection and viewing of underwater objects at all bearings and ranges of 10 to 1,500 yards are performed by a Model 500 CTFM sonar system. The system generates both audio and visual outputs and provides the capability to maintain contact with multiple targets and markers or transponders simultaneously. In addition to calibrated range and bearing marks on a 17·78 cm (7 in) cathode ray tube sweep, the operator is provided with a digital readout of the target range in yards by means of a range bug. When operating in combination with transponders, the CTFM sonar system can also be used for navigation purposes. A sonar sounding system with upward and downward looking transducers measures the water depth above and below the submarine to a maximum range of 6,000 fathoms, presents the information on a cathode ray tube indicator and two depth indicating meters, and records it on calibrated paper. Additional depth information is obtained by means of two pressure transducers and by means of a direct-reading depth gauge mounted near the after sphere lower hatch viewport.

Communications

The communication system provides for an efficient flow of audio information between *Deep Quest* crew members and between the submarine and support vessels. It also provides a means by which various sensors with audio outputs can be monitored. The system has worked effectively in tests to the maximum operating depth of 2,440 m (8,000 ft). A radiophone, which can operate on four channels, is used for surface communication. An 8-kHz, high-powered underwater telephone is used for transmitting and receiving single-sideband, suppressed-carrier, voice and CW communication. Over 100 acoustic watts are transmitted into the water by either the top or the bottom underwater transducer.

To facilitate surface tracking of the submerged submarine, *Deep Quest* is equipped with two acoustic pingers—a 27-kHz omnidirectional and a vertically oriented 27-kHz parabolic reflector. Both pingers operate on self-contained power supplies and are activated immediately on submergence. In the event of surfacing during inclement weather, *Deep Quest* can be readily located by means of a radio direction finder homing on the 2,182-kHz voice broadcast or by means of a Coast Guard aircraft homing in on a 121·5-MHz signal transmitted by a self-powered omnidirectional emergency beacon. A strobe beacon is also installed topside to assist in locating *Deep Quest* during night operations.

Buoyancy and depth control

Neutral buoyancy of *Deep Quest* has been achieved by the installation of 15,450 kg (34,000 lb) of syntactic foam with an average density of 580 kg/m^3 (36 lb/ft^3). The freeboard required for safe operation of the submarine on the surface is provided by a main ballast system, which produces a reserve buoyancy of 12 percent. This is achieved by compressed air in tanks located in each of the main ballast tanks.

Precise depth control in both water column and the near-bottom hovering positions is readily accomplished with the aid of forward and after vertical thrusters, lateral water jets, a shot ballast system, and the integrated control system. Attitude changes of up to ± 30 degrees trim and ± 10 degrees list, which may be required for certain operations, can be obtained by transferring mercury between appropriately located tanks.

Environmental control system

Deep Quest's life support and environmental control system can sustain four men for a total of 48 hours. A gas supply system monitors and automatically regulates the oxygen level within the pressure hull. A closed loop, oxygen-demand system with four full-face masks provides for a 3-hour survival period in a contaminated atmosphere. Both normal and emergency gas supply systems are equipped with manual by-pass valves as backup for the automatic oxygen supply. Temperature, humidity, and contaminant control are provided by a system of blowers, heat exchangers, and lithium hydroxide/charcoal cannisters. A comfortable, shirt-sleeve environment results.

Emergency jettison features

Safe return of the submarine to the surface is ensured by several weight and equipment jettisoning features:

A shot ballast system provides 750 kg (1,650 lb) (wet weight) of fail-safe ballast. This ballast can be released in increments, as desired, during operation of the submarine. In case of a total loss of both main and emergency power, the entire amount of shot ballast is automatically jettisoned.

The trim and list systems contain provisions for jettisoning 655 kg (1,440 lb) and 376 kg (828 lb) (wet weight) of mercury, respectively.

Two cell packs of the forward main battery can be jettisoned for a total wet weight reduction of 1,386 kg (3,050 lb). In this event, all systems remain operative using the after main battery.

The two manipulator assemblies and the bow instrument rack can be individually jettisoned in the event they become entangled.

Jettisoning of all the items listed above will result in a total weight reduction of approximately 3,180 kg (7,000 lb).

NORTH AMERICAN ROCKWELL
(Space Division)

BEAVER MK IV (ROUGHNECK)

A five-man self-propelled lock-in/lock-out submersible designed and built by Ocean Systems Operations of the North American Rockwell Space Division.

Launched	26 September 1968
Owned by	International Underwater Contractors
Design maximum operating depth	610 m (2,000 ft)
Lock-out capability to	305 m (1,000 ft)
Length	7·3 m (24 ft)
Weight, dry	12,629 kg (27,784 lb)
Life support	36 hours
Power source capacity	44 kw/h, lead-acid batteries
Hull construction	HY-100 steel for pressure hull, glass-reinforced plastic for outer hull
Max. speed	5 knots. Normal submerged cruising speed 2½ knots
Propellers	Three 45·7 cm (18 in) diameter, pod mounted
Propulsion motors	Three oil-filled 5 h.p. AC
Manipulators	Two, having 1·83 m (72 in) reach and 22·7 kg (50 lb) vertical lift

OCEAN SYSTEMS, INC

OPSUB

A manned submersible, tethered with a strength-power cable. Designed for sea bottom inspection and survey, and may be used for recovery of equipment.

Prototype construction started 22 June 1971, completed 19 December 1971

Trials completed 15 December 1971

American Bureau of Shipping clearance 1 August 1972

Basic dimensions

Length, overall	5·45 m (17 ft 11 in)
Width, overall	2·55 m (8 ft 5 in)
Height, overall	2·30 m (7 ft 6 in)
Draught	2·10 m (6 ft 10 in) in ready-to-dive condition

Weights

Dry weight	4,700 kg (10 400 lb) in dive condition
Submerged weight	Neutrally buoyant
Payload	340 kg (750 lb) including 2 passengers
Crew	Two

Performance

Maximum operational depth	610 m (2,000 ft)
Speed:	
cruise, surface	5 knots
submerged	2 knots
max.	3 knots
Duration at cruise speed	Unlimited
Duration at max. speed	Unlimited
Range	Tethered—1,000 or 2,000 ft tether
Life support endurance (air supply)	24 hours normal + 2 hours emergency
Manoeuvring capability	Athwartship thrusters 1 knot; rise and dive ¼ knot

Strength Power cable

The 440 volts AC power supply is carried from the surface through a special strength cable and terminated at *Opsub.* This umbilical cable is provided with a special winch, and the power is taken off through slip rings. The strength of the cable is rated at over 4,500 kg (10,000 lb) working 13 conductors; 6 power, 6 communication and 1 coaxial (TV signal to surface and hardware communication).

Main machinery

Power source: 440 volts, 3 phase AC

Motors: 5 to 10 h.p., 440 volts, 3 amps AC, 2-speed reversible, 2-side thrusters, 1-stern thruster rotatable 90° of centerline. 1 vertical and 1 horizontal thruster

Propellers: Five 0·381 m × 0·203 m (15 in × 8 in)

Controls: Manoeuvring—Individual controls to each motor on panel. Stern motor rotatable 90° on each side of centerline. Depth—Variable buoyancy system and vertical thruster 91 kg (200 lb) variable buoyancy 15·0 m (50 ft) tank. Trim—Vertical thruster

Pumps: Variable ballast pump, steering hydraulic pump

Electrical on-board equipment

Navigational: Precision scanning-fathometer sonar. Directional gyro compass, 3 view ports, depth gauge, altimeter

Vehicle function instrumentation: Motor voltage, ground detector

Communication: Hardware phone, emergency underwater telephone

Search and recovery: Tether and 270 kg (600 lb) droppable keel, lights, TV, scanning sonar

Television and visual: Underwater TV with hardwire to sortace-mercury vapour lights, EG&G camera on pan and tilt units

Safety systems

Life support: 24 hours, 2 hours emergency, O_2 supply, CO_2 scrubber, CO_2 monitor, CO monitor

Ballast jettison, pumping, etc.: 270 kg (600 lb) droppable keel, variable ballast pump, droppable keel weight exceeds weight of water in hardcan in case of flooding

Buoy markers: Tether only

Rescue facilities: Cable cutter hydraulically actuated padeye on hull for surface line

Warning systems: Ground detector on power supply, ground detector with kill switch

Function failure: (1) TV system; (2) EG&G camera; (3) Power available

Internal environmental conditions

Crew space	None
Temperature control	None
Pressure control	None
Air purification	CO_2 scrubber

Basic structure

Crew compartment, pressure hull type	Sphere
Machinery compartment	None—hardcan containing motor and power controls
Buoyancy compartment	None—syntactic foam
Outer skin	Aluminium
Collapse depth	915 m (3,000 ft)

Handling

Lifting points	1 lift point at c.g. by using tether
Normal launching method	Crane over stern or side with quick release
Turn round time	Less than 2 hours

Support requirements

Mother ship facilities	Support ship *Duplus* has been used with special lifting feature incorporated into crane

TV monitor on surface for real time observation and recorder for playback.

Mobility

May be air freighted or carried as deck cargo.

Ocean Systems' *Opsub*

Availability

Purchase capital cost	$300,000

May be chartered or leased

Licensed builders, if applicable: ABS certified, built by Perry Submarine

Operators

Dave Penson—Stavanger, Norway

Training and operation tests only

PERRY OCEANOGRAPHICS, INC

SUBMARINE REEF HUNTER

A battery-powered wet submersible which can carry two scuba divers down to 45 m (150 ft). Basically recreational vehicle but suitable for commercial, industrial and scientific tasks.

Length	3·65 m (12 ft)
Width	1·97 m (6 ft 6 in)
Height	0·71 m (28 in)
Weight	360 kg (800 lb)
Speed	2 knots
Duration	2 hours at cruise
Purchase cost (1973)	$US2,950 2-motor version
	$US3,936 4-motor version

May be rented or leased

Joystick-type control for movement in four directions

PC-8C

A manned, self-propelled submersible.

Hull diameter	1·06 m (42 in)
Length	5·63 m (18 ft 6 in)
Height	2·05 m (6 ft 9 in)
Beam	1·72 m (5 ft 9 in)
Weight	5·5 tons
Maximum operating depth	365 m (1,200 ft)
Payload (in addition to 2 crew)	336 kg (750 lb)
Speed	4 knots for 1 hour; 1 knot for 6 to 8 hours
Power	two bolt-on battery pods with quick-charge batteries
Purchase cost (1973)	$US275,000 with twin mechanical arms and sonar installed; also includes underwater telepnones, CB radios, fully operational including training

Ocean Systems have been using a modified PC-8 submersible for use with the Ekofisk project in the North Sea.

Sub Sea Oil Services purchased a PC-8 in 1973 for close-up inspection of water-jet burying of 120-mile pipeline off the Scottish coast at depths of 130 m (420 ft) and for other North Sea operations.

In the autumn of 1973 Perry Oceanographics was awarded a $149,256 contract for the construction of a submersible for Texas A and M University.

Maximum operating depth: 366 m (1,200 ft)

Maximum speed: 2 knots

Crew: Two

PC-14C

A manned, self-propelled submersible.

Hull diameter	1·06 m (42 in)
Length	3·97 m (13 ft)
Height	2·27 m (7 ft 6 in)
Beam	1·50 m (5 ft)
Weight	3·5 tons
Maximum operating depth	365 m (1,200 ft)
Payload (in addition to 2 crew)	102 kg (225 lb)
Speed	2 knots for 3 hours; 1

	knot for 4 to 6 hours
Power	bolt-on battery pod with quick-charge batteries batteries, approximately ten kw/h 36 volts DC
Purchase cost (1973)	$US114,000, includes underwater telephone, CB radios, fully operational including pilot training

Pressure hull

Constructed of 11·1 mm ($\frac{7}{16}$ in) SA-516 Grade 70 normalized steel. Hull is modular in that it includes a bolt flange.

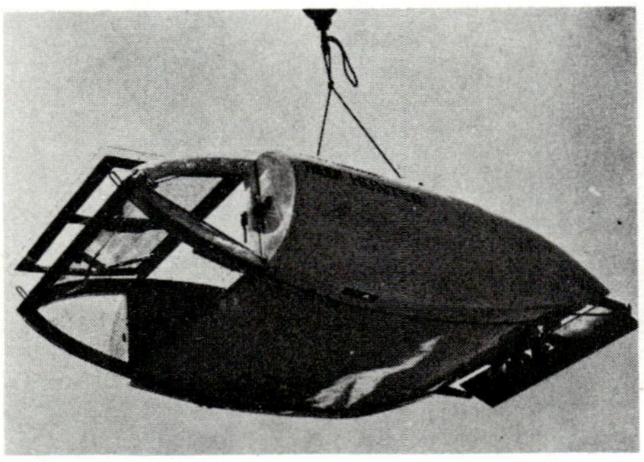

Reef Hunter

Life support

CO_2 scrubbing with O_2 make-up in accordance with American Bureau of Shipping requirements.

Spare hull penetrations

One 27·9 cm (11-in) diameter removable spare penetration disk.

Safety features

The 'PC-14' will be provided with an independent ballast blow system capable of blowing ballast tanks dry at maximum operating depth.

The 'PC-14' is fitted with a droppable battery pod designed to give the vessel positive buoyancy should all power and ballast blows fail under normal trim conditions.

In a normal surface condition, the 'PC-14' will have approximately 5·08 cm (2·0 in) of freeboard measured from the waterline to the top of the conning tower hatch seat.

'PC-14' will be provided with an emergency breathing regulator, life jacket, and face mask for each person.

Included equipment

The 'PC-14' will be delivered with the following listed equipment or equivalent installed:

Underwater lights—one 250 watt underwater light
Directional gyro installed
Simple manipulator with four functions:
Extend—retract
Left—right
Fore—aft
Jaw open—close
Citizens band radio

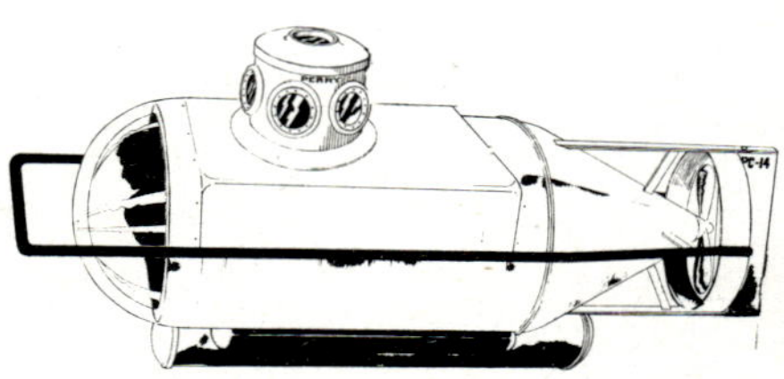

An artist's impression of the Perry PC-14 submersible

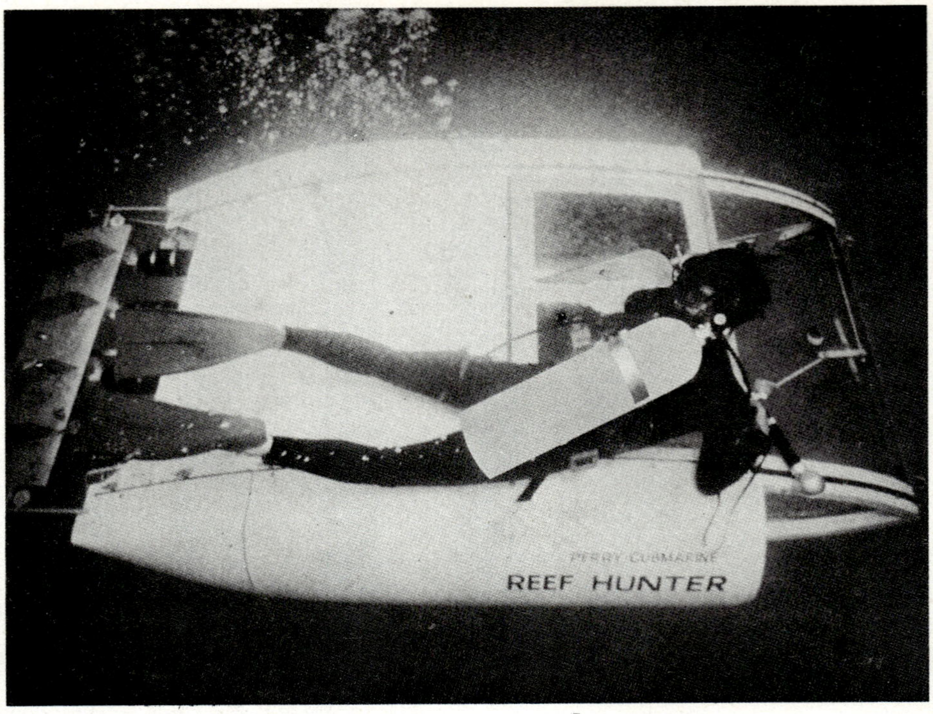

The Perry Oceanographics' Cubmarine *Reef Hunter*

One complete set of batteries
External bumper rail forward
Ground detector
Main propulsion motor speed and direction control
Propulsion power volt and ampere meters
Altimeter
Depth gauge

VOL L1

Originally called 'PC-15', 'VOL L1' was built by Perry Oceanographics, Inc for Vickers Oceanics Ltd (VOL) for use in the North Sea and designed for use in depths down to 365 m (1,200 ft).

Main sections of the submersible are of modular construction and bolted together. For special missions, the entire lockout section and gas sphere may be removed and the tail section bolted to the front section, making a smaller 3-place conventional submersible. Modular construction also assures against obsolescence as new power and propulsion systems are developed in the future.

The Perry PC-14C submersible

Specification

Length	9·75 m (32 ft)
Weight	14 tons
Power	Two 10 h.p. 120 volt DC Reliance motors
Speed	1st motor only, 4 knots at 232 rpm; 1st and 2nd motor, 5 knots at 310 rpm
Payload	910 kg (2,000 lb)
Endurance	Cruise at 1 knot for 15 hours

The Perry Cubmarine *Reef Hunter*

Battery system, lead-acid	40 batteries for 120 volts
Gould 'Power Breed'	8 batteries for 24 volts 12 volts rated at 93 amp hours
Total boat power	54·8 kw/h
Battery weight	1,155 kg (2,540 lb)
Clutches	Four, electric, by Carlyle Johnson
Crew	Pilot, observer, 2 divers
Propeller	0·91 m (36 in) dia., 0·86 m (34 in) pitch; thrust: 515 kg (1,150 lb)
Hull	516 grade 70 steel 1·27 cm (½ in) thick
Observation window	Acrylic, 1·07 m (42 in) dia., 5·08 cm (2 in) thick
Thrusters	Three, electro-hydraulic, 1,200 rpm, 22·9 cm (9 in) propellers, 34 kg (75 lb) thrust each
Breathing supply:	
Compressed air	13·5 m^3 (480 std ft^3) at 168 kg/cm^2 (2,400 lb/in^2) 2 cylinders
Oxygen	7·04 m^3 (248 std ft^3) at 152 kg/cm^2 (2,200 lb/in^2) 4 cylinders
Helium	185·0 m^3 (6,500 std ft^3) at 172 kg/cm^2 (2,500 lb/in^2) 1 sphere
Cost	$US750,000

Batteries are in twin droppable pods below the hull, giving good stability and providing natural skids for landing on deck or on the bottom. Pipelines can be straddled, bringing the pipe within inches of the hemispherical 1·07 m (42 in) diameter, 5·08 cm (2 in) thick, acrylic window.

Pilots' controls are mounted in a box which can be moved with the pilot, giving him freedom of movement.

Quick-change trays enable battery exchange and full turn-around in two hours. A removable conning tower has six viewports for horizontal viewing plus an upward-looking viewport in the conning tower hatch. The aft, or diver compartment, has a downward-looking viewport in the lockout hatch, and four viewports for horizontal viewing to each side.

The diver lockout feature enables two divers to be delivered to the bottom at one atmosphere, pressurized to depths of 365 m (1,200 ft) with the submersible's gas storage sphere, and exit from the submersible to execute work at pre-determined locations. On completion of their seabottom tasks, the divers return to the surface aboard the submersible.

The sub is lifted aboard the surface support vessel and placed atop the decompression system. The divers transfer into the system under pressure. The sub then takes aboard new divers and re-submerges.

The submersible is fitted with an extendable landing gear to give bottom stand-off for divers during lockout missions.

The submersible was delivered classified by the American Bureau of Shipping. Equipment includes gyro-compass, auto-pilot, underwater telephones and sonar. Manipulators will be fitted in due course.

The PC-8 showing the transparent acrylic nose, which is a typical fitting on Perry submersibles

The Perry VOL L1 diver-lockout submersible

Commercial diver-lockout submersible operations, dated from 1967, have been conducted with Perry submarines and these have included the deepest lockout at 222 m (730 ft).

Perry's commercial submarines, chambers, habitats and other underwater hardware are in use in many of the oceans of the world.

Shortly before being chartered to Vickers, the Perry PC-8 was operated in the Bering Sea, where she was used by US scientists of the National Oceanic and Atmospheric Administration (NOAA) to study walrus habitats beneath the Arctic ice. She also has been on missions from Eastern Canada to the Gulf of Mexico.

REYNOLDS INTERNATIONAL, INC

ALUMINAUT

An 81-ton manned, self-propelled submersible designed for exploration, research, engineering development and salvage. Constructed in aluminium, the building of the vessel called for some of the largest aluminium ingots and forgings ever to be made.

Design history

Date of construction of prototype started: 1963

Date of prototype completed: December 1964

Date trials completed: December 1965

Certification authority clearance: US Navy

Production

One built

Basic dimensions

Length, overall	15·5 m (51 ft)
Width, overall	4·87 m (16 ft)
Height, overall	5·02 m (16·5 ft)
Draught	2·88 m (9·5 ft)

Weights

Dry weight	81 tons
Payload	2,680 kg (6,000 lb) (includes 4 persons)

Crew	2 to 3
Observers	3 to 4

Performance

Maximum design operational depth	4,550 m (15,000 ft), safety factor 1·5
Operating depth	3,340 m (11,000 ft), safety factor 2·0
Speed:	
cruise, surface	3·0 knots
submerged	3·0 knots
max.	3·8 knots
Duration at cruise speed	30 hours
Duration at max. speed	6 hours
Range	73 km (40 nautical miles)
Life support endurance (air supply)	72 hours

Main machinery description

Power source	Silver zinc alkaline batteries
Motors	Three DC 115/230 volt
Propellers	Two 3-bladed, 5 h.p. motor drive each; one 2-bladed
Controls:	
depth	Shot ballast
trim	Internal trim tanks

Electrical on-board equipment

Navigational	DRT, CTFM sonar
Communication	8·08 kHz
Television and visual	Trainable camera + TV tape

Safety systems

Life support	$Li(OH)_2 + O_2$
Ballast jettison	2 ton weight
Warning systems	Battery scanner

Internal environmental conditions

Crew space	4·25 m³ (1,500 ft³)
Temperature control	Humidity control + heaters
Air purification	CO_2 absorbent

Basic structure

Crew compartment, pressure hull type; machinery compartment; buoyancy compartment	16·5 cm (6·5 in) thick 2·42 m (8 ft) dia. cylinders with 3·00 m (1 ft) internal flanges -bolted together. 7079-T6. Aluminium alloy
Outer skin	Alclad 5083 alloy
Collapse depth	6,850 m (22,500 ft)

Handling

Lifting points	2-Doeron Straps
Normal launching method	Railway

Support requirements

Mother ship	M/V *Privateer* 41·46 m (136 ft) ex mine-sweeper

Mobility

May be carried as deck cargo.

Availability

Purchase, capital cost	$US4 million
Lease, charges	Variable, on request
Charter availability	Available
Licensed builders	General Dynamics Corporation

Aluminaut in operation on the sea-bed

Operator

Arthur L. Markel. V.P. Reynolds Submarine Services Corporation.

General description

Payload

A scientific instrument payload capability of 2,727 kg (6,000 lb) is provided. Reductions in passenger complement and/or life support equipment loads can provide proportionate increases in the scientific payload.

(a) **Internal payload stowage**

(1) A total of approximately 1·68 m³ (60 ft³) is available in the spaces shown by the *Aluminaut* space diagram without encroachment upon any of the normal submarine equipment or personnel spaces.

(2) A significant addition to the defined internal space is available in the central

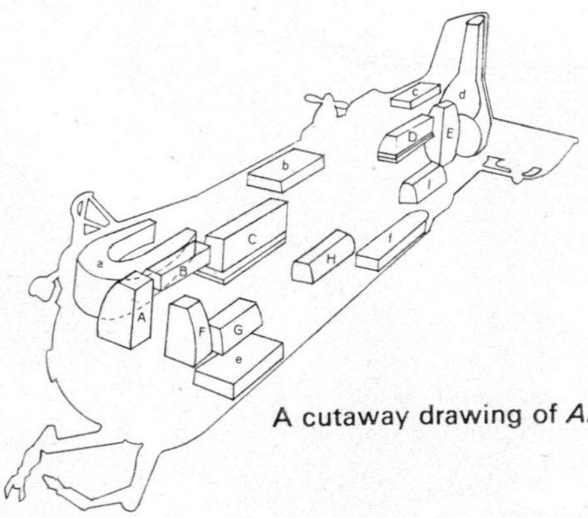

A cutaway drawing of *Aluminaut*

Space available inside pressure hull:
A. Volume 0·126 m³ (4·5 ft³).
B. Volume 0·097 m³ (3·5 ft³).
C. Lab Bench 0·56 m × 1·78 m, Volume over 0·425 m³ (15·0 ft³).
D. Lab Bench 0·56 m × 1·78 m, Volume over 0·425 m³ (15·0 ft³).
E. Volume 0·170 m³ (6·0 ft³), Starboard sector of stern Hemi Head.
F. Volume 0·051 m³ (1·8 ft³).
G. Volume 0·079 m³ (2·5 ft³).
H. Volume 0·225 m³ (8·0 ft³).
I. Volume 0·225 m³ (8·0 ft³).
Space available in protected, free-flooding areas:
a. Volume of approx. 0·170 m³ (6·0 ft³) in bow access trunk.
b. Volume of approx. 0·111 m³ (4·0 ft³) in forward deck areas.
c. Volume of approx. 0·140 m³ (5·0 ft³) in after deck areas.
d. Volume of approx. 0·340 m³ (12·0 ft³) in stern section.
e. Volume of approx. 0·225 m³ (8·0 ft³) in forward section of keel.
f. Volume of approx. 0·340 m³ (12·0 ft³) in stern section of keel.

Key:
1. Stern access trunk
2. Vertical propulsion motor
3. T.V. Monitor
4. Forward sonar display
5. Gyrocompass
6. Deck superstructure
7. Bow access trunk
8. Bow access hatch
9. Scanning C.T.F.M. sonar
10. Extensible illuminator
11. T.V. Camera and illuminator
12. Manipulator device
13. Observer
14. Oxygen flask
15. Port illuminator
16. Battery
17. Side-looking sonar (under ballast tank)
18. Keel superstructure
19. Shot ballast solenoid
20. Skipper
21. Ballast tank
22. Port propulsion motor
23. Stern access hatch

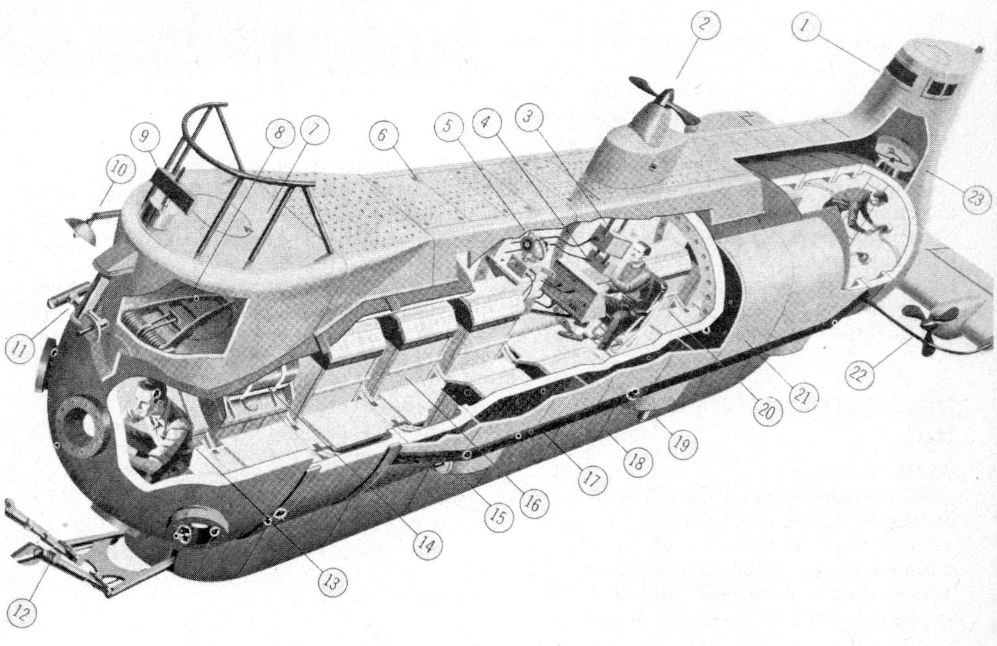

passageway and in the forward hemihead areas. The extent of use of these areas will be determined by the nature of the mission and the submarine operating requirements.

(b) **External wet payload stowage**

(1) A total of approximately 1·4 m^3 (50 ft^3) of installation and stowage space is afforded by the free-flooding exo-structures. Universal foundation members have been provided in the bow and stern structures.

(2) Deck and keel space for additional equipment and instrumentation installations can be provided to the extent permitted by the submarine operating and safety requirements.

(3) A suitably configured, jettisonable exterior instrument or equipment load of 1,950 kg (4,400 lb) could be substituted for the present keel-mounted ballast bar, in excess of the stated 2,700 kg (6,000 lb) payload.

Speed

Submerged speeds ranging from 0 to 3·8 knots are possible with the stated crew and passenger complements and full scientific payload. The addition of external equipment and instrumentation without special hydrodynamic housing treatment, will naturally reduce the maximum speed of the submarine. Present external transducer and equipment installations, categorized as scientific payload and as listed later as auxiliary equipment, permit a maximum speed of about 3·5 knots.

Manoeuvrability

Aluminaut is capable of hovering at any depth of water for periods up to the maximum mission time. The combination of an essentially incompressible pressure hull (which is the buoyancy element) and the shot-ballast depth control, provide a means for non-power consuming, highly precise depth controlled hovering in the absence of currents.

(a) Precision control of the widely spaced horizontal propulsion motors and the depth control planes (diving planes) provides an attitude control capability in the face of ocean currents.

(b) Precision control of the vertical propulsion motor provides additional dynamic maintenance of hovering depth or accurately controlled variations in the hovering depth without movement in the horizontal plane.

Duration of dive

The normal maximum submerged mission time of thirty-two hours is adequately provided for as defined herein, with extended life support provisions for seventy-two hours. A decrease in the amount of electrical power consumed and a decrease in the payload will allow an increase of life support supplies, thereby increasing the duration.

Safety

Aluminaut designed features provide safe operating characteristics.

(a) The buoyancy element is the 16·5 cm (6·5 in) thick walled pressure hull, designed for a maximum operating depth of 4,550 m (15,000 ft). The hull is essentially incompressible, and accordingly becomes more buoyant as depth and sea water density increase.

Aluminaut submerging

(b) The diving negative buoyancy characteristic is provided by steel shot ballast contained in two saddle tanks amidships. Shot ballast is retained by a magnetic field established in the 5·08 cm (2 in) diameter shot discharge valve orifice. The shot discharge orifice to shot diameter ratio is approximately 40 with the resultant discharge of shot at the rate of 8·15 kg (18 lb) per second in the absence of the retaining influence of the magnetic field. When the 28·0 volt DC shot valve solenoid circuit is opened manually or in the event of power failure for example, the maximum shot load required for attainment of a 4,575 m (15,000 ft) depth, will be discharged in approximately 3·3 minutes. In the event of shot solenoid circuit power failures that may not justify aborting a dive and other conditions which may contrive to make it desirable to delay ballast release, a manually applied, short-term ballast solenoid power supply is provided in the form of a separate battery. Finally, a reverse polarity demagnetizing current can be applied to the shot release valve in the event of reduced discharge rate due to residual magnetism.

(c) A 1,950 kg (4,400 lb) ballast bar housed in the keel is retained by an electro-mechanical system for instant release in the event of emergency. This ballast weight exceeds the weight of the maximum shot ballast tank capacity and when released, makes the submarine positively buoyant regardless of the status of the controlled ballast system.

(d) *Aluminaut*'s pneumatically controlled water ballast tank system, normally used for providing excess positive buoyancy at the surface, may also be considered as a safety feature at depths to the order of 1,200 m (4,000 ft). The 0·34 m^3 (12·0 ft^3) supply of 317 kg/cm^2 (4,500 lb/in^2) of water in the ballast tanks at a depth of 1,200 m (4,000 ft), providing approximately 910 kg (2,000 lb) of positive buoyancy. This system may also be employed to conserve shot ballast.

Pressure hull

Access hatch

Both bow and stern hatches have clear diameters of 50 cm (19·75 in).

Optical view ports

Four high density optical quality methacrylate view ports, provide observers in the bow hemispherical head compartment, with a total horizontal field of view athwart the centre line of 160° and a vertical field extending from 25° above the hull centre line to 80° below it.

Internal arrangement

All primary controls and instruments fundamental to the control and navigation of the submarine are located on Ship's Control Panel Number 1, the Ballast Control Panel and the Dead Reckoning Tracer. The Ship's Control Panel Number 1 is the principal component of the pilot's console amidships on the starboard side of the centre passageway. Immediately across the passageway, at pilot's shoulder height on the port side, the Ballast Control Panel is within easy reach. The Dead Reckoning Tracer is located immediately below the Ballast Control Panel on a swing-out mounting.

A portable remote propulsion control panel is located in the forward hemihead viewing port area for use during scientific observations.

Electrical

1. *Aluminaut*'s electrical hull penetrations provide 249 conductors for exterior scientific instrument and equipment use in excess of the operating electrical service conductors. The following listed conductor configurations are presently available for use:

2 Pair AWG 12, Cable Type SO12/2, Connector Type RM2SFS
4 Pair AWG 16, Cable Type SO16/2, Connector Type RM2SFS
4 Shielded Pair AWG 20, Cable Type 2N5S, Connector Type RM3SFS
4 Pair AWG 20, Cable Type 2N4, Connector Type RM2FS
6 Single AWG 20, Cable Type 1N4, Connector Type RM1FS

2. A special 'scientific penetrator' provides 176 conductors that are completely isolated from the submarine's electrical power, control and other instrumentation conductors.

48 Shielded 16 AWG Pairs, MIL Grade Cable, Connectors as Required
16 Shielded 16 AWG Singles, MIL Grade Cable, Connectors as Required

These conductors are available in eight separate groups of 6 TPSJ and 2 SJ conductors, externally cabled to reach any part of the submarine's structure.

3. **Conductor termination**

(a) **Outboard**

Conductors terminate at the Scientific Outboard Panel located in the Stern Access Trunk. Underwater connectors are installed in this panel and special waterproof covers are fitted to all connectors not in use. Extension cables installed in deck and keel wireways make these conductors available at any desired point throughout the length of the submarine. The maximally isolated conductors are available in eight (8) groups of six (6) shielded pairs and two (2) shielded single conductors, externally cabled to reach any exterior part of the submarine. Each of the eight cables is equipped with a Marsh and Marine MS20-CCP connector on the outboard end. A mating FCR connector terminates each of these cables in a moulded eight circuit breakout assembly ready for instalation of connectors of any type required.

(b) **Inboard termination**

All scientific conductors are extended from the inboard ends of the hull penetrations to the forward scientific spaces in the submarine by appropriate colour coded conductor configurations.

(1) Unassigned scientific conductors will be terminated with the requisite connector at the time of their assignment.

Power

(a) Four 115 volt, 400 ampere hour silver zinc alkaline batteries, provide power for all operating, hotel and scientific electrical loads. Sustained submerged cruising missions of 30 hours are adequately implemented by this most efficient form of electrical energy storage. Nearly 200,000 watt hours are available from the *Aluminaut*'s battery.

(b) The following forms of electric power are available from existing *Aluminaut* sources and available additional sources.

(1) 115 volts DC direct from *Aluminaut*'s batteries, long-term potential range 118 to 140 volts.

(2) 230 volts DC direct from *Aluminaut*'s batteries, long-term potential range 236 to 280 volts.

(3) 28 volts DC regulated (±) 10%, 5 amps continuous from solid state converter. Additional regulated DC power can be provided on the basis of approximately 45 kg (100 lb) per kw from solid state sources and approximately 68 kg (150 lb) per kw from rotating converters.

(4) 28 volts, 400 Hz regulated ((±) 10% voltage, (±) 5% frequency) alternating current, 200 watts continuous, available from solid state inverter. Additional 400 to

1,000 Hz power can be provided from solid state or rotating inverters within the above noted power and weight estimates.

(5) 115 volts, 60 cycle regulated ((±) 10% voltage, (±) 5% frequency) alternating current, 2·5 kw continuous, available from two rotating inverters.

(6) 115 volts, 60 cycle super regulated ((±) 1·0% voltage, (±) 0·03% frequency) alternating current, 1 kw continuous, available from special solid state inverter.

Attachments

External scientific equipment installation provisions have been made and additions to these provisions that are in consonance with safety and the submarine operating characteristics may easily be made.

Auxiliary equipment

Acoustic and ultra-acoustic equipment

Underwater telephone, Straza Model ATM502. This equipment provides continuous telephonic or CW communications with the UQC-1B underwater telephone equipment installed aboard the tending ship, *Privateer*. The transmission and reception characteristics are as follows:

Range: In excess of 13,500 m (45,000 ft)
Power to transducer: In excess of 100 watts
Carrier frequency: 8·0875 kHz
Circuitry: Completely silicon transistorized
Transmission modes: Voice and 800 Hz/s CW (Single Sideband Suppressed Carrier)
Acoustic sound pressure: 95 dB above 1 micro bar at 1 m
Transducer sensitivity: 77 dB relative to 1 volt/micro bar

Communications equipment

A 75 watt radio telephone transmitter and receiver, 6 channel, 2 to 6·5 mHz is installed.

Electronic fathometer equipment

Two identical fathometer/recorder units are installed at the pilot's station. Three transducers are installed to provide distance to the surface, distance to the bottom, and distance to objects ahead of the submarine. Normally, distance to surface and distance to the bottom transducers are connected to the fathometer/recorders. In event of failure of the CTFM forward area scanning sonar, the forward transducer could be switched to either fathometer/recorder unit. The characteristics of these equipments are as follows:

Range scales: 0 to 91·5 m (0 to 300 ft), 91·5 to 183 m (300 to 600 ft), 183 to 366 m (600 to 1,200 ft)
Frequency: 5 kHz
Circuitry: Completely transistorized
Chart duration: 14 hours
Chart paper: Dry, electrosensitive

Continuous transmission frequency modulated (CTFM), Scanning Sonar,

Straza Model SM502A

This equipment is installed on the bow exostructure providing continuous scanned 22·5° or 90° sectors ahead of the submarine. A full 360° scan is available for tracking associated transponder units that are employed as precision bottom navigation benchmarks or for position determination relative to the tending ship. The principal characteristics of this equipment are as follows:

Range scales: 45·5 m (150 ft), 91·5 m (300 ft), 183 m (600 ft), 366 m (1,200 ft), 732 m (2,400 ft)
Range resolution: 0·92 m, 1·84 m, 3·68 m, 7·36 m, 14·72 m (3 ft, 6 ft, 12 ft, 24 ft, 48 ft)
Displays:
Visual: Plan position indicator cathode ray tube 17·78 cm (7 in) CRT
Aural: Loudspeaker and headphones
Operating frequency: 72·5 to 87·5 kHz
Circuitry: Completely transistorized

Bottom scanning sonar

Westinghouse *Aluminaut* side looking sonar equipment. This equipment provides a high resolution chart record of the topographical features of the ocean bottom. It is a bimodal sonar providing 762 m (2,500 ft) wide and 120 m (400 ft) wide mapping and classification scan patterns. Equipment characteristics are as follows:

Frequency: 225 kHz
Pulse repetition rates: Mapping mode: 2 pulses/s; classification mode: 12 pulses/s
Transducer altitude: Mapping mode: 68·6 m (225 ft); classification mode: 6·1 m (20 ft)
Traverse velocity: Mapping mode: 2·5 knots; classification mode: 1·5 knots
Circuitry: Completely transistorized
Display: Alden special 48·3 cm (19 in) wide continuous helixblade recorder
Chart paper: Current sensitive 48·3 cm (19 in) wide continuous

Underwater illumination equipment

Six Edgerton Germeshausen and Grier, Inc tungsten filament, iodine atmosphere quartz lamp units are installed at the bow to provide uniform illumination of the water volume and the ocean floor immediately ahead of the submarine. The average illumination level measured on a 9 m² (100 ft²) horizontal plane area coinciding with the bottom of the keel, the near edge of the area coinciding with a perpendicular line dropped from the bow, is 377 candles/m² (35 candles/ft²). Smaller areas may be illuminated to provide levels on the order of 807 to 1,075 candles/m² (75 to 100 candles/ft²). The tungsten filament source provides a colour temperature in the 3,200 to 3,400° Kelvin range. Underwater illumination provisions may be varied, within limits, to conform to any reasonable requirements.

The remote controlled pan and tilt television mount provides controlled directivity of an illuminator or strobe light to permit optimum utilization of the wide angle viewing port fields and the underwater television camera.

Underwater television equipment

A high resolution, remote controlled pan and tilt television camera housed in an Edgerton Germeshausen and Grier, Inc water pressure proof housing is installed outboard on the forward end of the bow exo-structure. The camera is equipped with a remote controlled 1·27 cm to 12·7 cm (0·5 in to 5·0 in), variable focal length lens which provides wide angle and close-up viewing at display units in the forward scientific area and at the pilot's station amidships. Television images may be recorded by a completely transistorized video tape recorder or the 16 mm Arriflex motion picture camera.

Aluminaut Manipulator Device (AMD)

A special hydraulically powered, two-arm manipulator, developed by the General Electric Company for installation in the forward section of the keel, has the following characteristics:

Configuration: Two fully independent arms mounted on the lower forward end of the *Aluminaut*'s keel
Size: 2·75 m (9 ft 1 in) reach from the shoulder pivot
Load capacity: Each hand 91 kg (200 lb), total 175 kg (400 lb) lift
Motions:
Shoulder azimuth: 212° total travel from stow astern to straight ahead (180°) and 50° inboard
Shoulder elevation: 120° total travel, 60° up and 60° down
Elbow azimuth: 270° total travel from stow above the upper arm to straight out in line with the upper arm (180° swinging inboard to unfold) to 90° outboard from the upper arm
Wrist bend: 180° total travel, 90° up and 90° down
Hand grasp: 15·24 cm (6 in) open and close

Construction

Welded aluminium and stainless steel—piston type rotary actuators for shoulder and elevation motions.

Power supply

Enclosed and pressure compensated hydraulic system—pump driven by an AC induction motor—motion control valves actuated by DC solenoids.

Vehicle control station

Single portable control box with two joy sticks for each arm—fine control stick for elbow azimuth and wrist bend with hand close trigger on top. Course control stick for shoulder elevate and azimuth. Separate rotary switches for wrist rotate.

Each arm can be fully controlled with one hand.

Magnetic recorder equipment

A Uher Model 4000S transistorized tape recorder is installed to provide a record of operations and scientific data. The speech recording mode provides a continuous eight hour period of operation. Other operating modes permit high fidelity recording of frequencies from 20 Hz to 25 kHz for use with hydrophones, etc. Due to the compactness of this equipment, additional units can be installed for any future requirement.

Camera equipment

A 200/400/1,200 ft magazine Arriflex 16 mm motion picture camera is available for wide angle porthole photography. A 250 frame capacity automatic transport 35 mm still camera is also furnished. An Edgerton 35 mm external camera is mounted on the *Aluminaut* pan and tilt mechanism complete with strobe light system.

US NAVAL CIVIL ENGINEERING LABORATORY

CONSTRUCTION ASSISTANCE VEHICLE (CAV)

In conjunction with the US Naval Facilities Engineering Command's Ocean Engineering Research and Development programme, the Naval Civil Engineering Laboratory has developed equipment to increase the working capability of Navy divers. NCEL's new CAV (Construction Assistance Vehicle) is an experimental craft designed to provide free-swimming scuba divers with an 'underwater pickup truck' capable of delivering up to 910 kg (2,000 lb) of wet-weight cargo, equipment and divers to and from underwater construction sites.

All controls in the electro-hydraulically powered craft are actuated mechanically by the diver-operator so that there is no electrical circuitry in the cockpit vicinity. CAV's constant speed electric motors, which power the hydraulic system, are controlled by the operator in the cockpit, using mechanical linkage.

The craft's hydraulic and pneumatic power is available to operate diver tools. The salt water ballast system and compressed air system, serving both life support and trim systems, are operated by valves in easy view and reach of the operator.

Instrumentation has been kept simple. There are gauges for monitoring pressure in the compressed air flasks and measuring water in the port and starboard auxiliary

The US Naval Civil Engineering Laboratory's Construction Assistant Vehicle (CAV)

ballast tanks. The only other displays are for pitch, roll and depth.

The CAV is a catamaran-hulled vehicle. The main tubes (or hulls), are fabricated from 66 cm (26 in) int. dia. rolled-steel plate. An important feature of the design and construction of the craft is the fact that it is built primarily with off-the-shelf components.

The CAV's cargo bed holds 910 kg (2,000 lb) of cargo. Awkward loads may be slung beneath the vehicle for delivery to the seafloor.

Concept

Naval Civil Engineering Laboratory.

Design

Under contract with M. Rosenblatt & Son, Naval Architects & Marine Engineers, San Francisco, California.

Fabrication

Marine Resources Inc, Fern Park, Florida.

General

Weight (dry)	8,150 kg (18,000 lb)
Length	8·23 m (27 ft)
Beam	3·00 m (10 ft)
Draft	1·20 m (4 ft)
Height (skeg to propulsion motors)	2·13 m (7 ft)
Submerged speed	2·5 knots maximum 4-hour duration)
Cargo bed	3·35 m × 1·37 m × 0·46 m (11 ft × 4·5 ft × 1·5 ft)
Maximum operating depth	36·5 m (120 ft)
Cargo capacity	910 kg (2,000 lb)

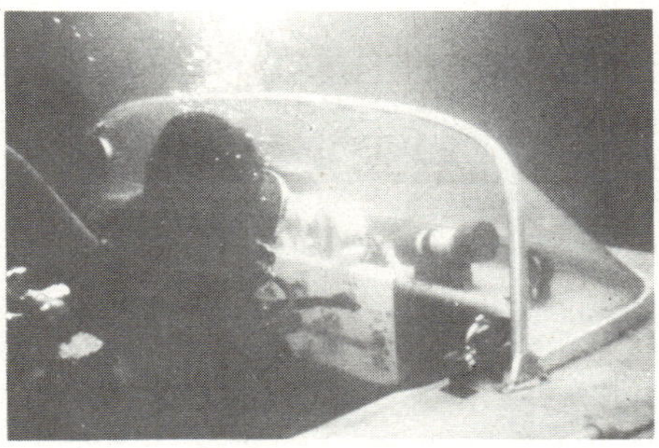

Divers at the CAV control panel during an operation

Power train

Batteries

Two independent 30-cell sets, each 60 volts, 18·5 kWh 95 amps at 77°F. Lead-acid batteries, oil immersed, pressure compensated.

Electric motors

Two 60 volt, 6 h.p., 1,850 rpm, 100 amp (compound wound). Motors and hydraulic pumps are sealed together in dry, one-atmosphere containers.

Hydraulic pumps

Variable volume, lever-controlled 6 gpm at 84·2 kg/cm^2 (1,200 lb/in^2) to operate propulsion motors.

Propulsion motors

Two main and two auxiliary hydraulic propulsion motors. Each motor: 6 gpm at 84·2 kg/cm^2 (1,200 lb/in^2) 429 rpm, approximately 3 h.p. (delivered through the propeller).

Life support

Five compressed air bottles located below the cargo deck. Each bottle 220 ft^3 diver-pure air. Air supply sufficient for 3 round trips to 36·5 m (120 ft) depth (approximately 1½ hours supply). En route, divers breathe from CAV's air supply system, although each diver carries his own tanks. These tanks are also used in the trim and ballast systems.

Trim-ballast

Auxiliary ballast control and port and starboard trim are achieved by a two-tank inter-connected system. The tanks, which are an integral part of the tubular catamaran hull structure, are located on each side of the cargo deck. Their size gives CAV a 910 kg (2,000 lb) variable ballast capability. Salt water may be pumped from one tank to the other (to provide port and starboard trim) as well as to and from the sea (8 gpm at 5·25 kg/cm^2 (75 lb/in^2) over ambient).

Fore and aft trim is achieved by two 136·5 kg (300 lb) weight assemblies which are moved mechanically inside the vehicle skegs from 1·5 m (5 ft) forward to 1·5 m (5 ft) aft of the craft's centre of gravity.

Port and starboard trim tanks can, in an emergency, be blown free of water by means of the compressed air system.

Main ballast tanks, port and starboard, are open to sea pressure. They are filled with salt water when CAV is submerged. On the surface they are blown dry to provide 2,230 kg (5,000 lb) of buoyancy to increase freeboard.

Propellers

Main (submerged)

61 cm (24 in) diameter, 38·1 cm (15 in) pitch, clipped to 50·8 cm (20 in) diameter. Each is mounted in a Kort nozzle.

Auxiliary (surface)

50·8 cm (20 in) diameter, 38·1 cm (15 in) pitch.

The propellers are counter-rotating.

The main propulsion units swivel through an arc of 190°.

Steering

Steering is accomplished by independent control of the speed and thrust direction of the main propulsion propellers. There are no movable rudders or control planes.

THE NAVAL UNDERSEA CENTER (NUC)

MAKAKAI

Makakai was developed and built by the Hawaii Laboratory of the Naval Undersea Research & Development Center, now known as the Naval Undersea Center. *Makakai* employs the demonstrated advantages and practicability of using a clear, acrylic plastic sphere as a control module. The total, panoramic visibility thus made possible is an enormous advantage over viewports in a traditional submersible. In *Makakai*, an operator and one observer-passenger can descend to depths of 180 m (600 ft) in a completely comfortable 'shirtsleeve' environment. The submersible enjoys unlimited, efficient translational mobility in addition to its superior visibility capability.

Makakai is propelled by two sets of oppositely arranged cycloidal thrust units. Each set mounts four short, variable-pitch propeller blades. The system enables the craft to move in any direction merely by altering the pitch of the propeller blades.

Cycloidal propulsion has several advantages over traditional propellers. The system generates relatively large control forces while maintaining high overall operating efficiency, even at slow speeds. This is important in all submersible propulsion and control where manoeuvrability and energy/fuel economy are such critical factors. Cycloidal propulsion has been built into surface ships, chiefly in Europe, for a number of years. However, *Makakai* is thought to be the first submersible to embody this principle.

This craft's second outstanding innovation is the location of most of its control system 'outside' the control module. One reason for consideration of such a revolutionary measure was to reduce heat within the plastic sphere. Under changing thermal conditions within the hull during operations, the surface of the sphere fogs over. Reducing the amount of heat tends to reduce fogging. Placement of controls outside the sphere also provided more room within, though operator comfort was a secondary consideration. Modulated light beams are aimed from within the sphere at contact points in the console 'outside' (another advantage of the transparent hull). The system's chief advantage is that it greatly reduces the number of support system penetrations in the bottom of the sphere.

Makakai was certified on 15 September 1972 by the US Naval Ship Systems Command for operations to 182 m (600 ft).

***Makakai* (*US Naval Undersea Center*)**

Basic dimensions

Length, overall	5·63 m (18 ft 6 in)
Width, overall	2·42 m (8 ft)
Height, overall	2·26 m (7 ft 6 in)

Weights

Weight in air	4,600 kg (10,200 lb)
Displacement	5,000 kg (11,070 lb)
Payload (including crew)	390 kg (870 lb)
Crew	1 operator, 1 observer

Performance

Maximum operational depth	182 m (600 ft)
Speed:	
cruise	0·5 to 0·75 knots
maximum	3 knots
Propulsion endurance	2 hours at 3 knots; 6 hours at 5 knots
Life support endurance	normal 8 hours; emergency 40 hours

Main machinery

Power source

External (lead acid cells)

(a) Two 120 volt strings (twenty 6 volts, 190 amp-hours)

(b) Two 30 volt strings (two 12 volts, plus 1 each 6 volt in series, 53 amp-hours)

(c) 6 volt batteries of the 30 volt strings (190 amp-hours)

Internal

(a) Twenty-two 6 volt globe GEL/CEL GC 680 (7·5 amp-hours) batteries providing 30 volts, 12 volts and 6 volts supplies

Hydraulic systems

(a) Main hydraulic system

15 gpm hydraulic pump driven by 10 h.p. electric motor

Motors

(a) Two 4 h.p. hydraulic (axial piston, variable displacement) for propulsion

(b) 10 h.p. electric to drive hydraulic pump (15 gpm)

Ballast

Buoyancy and Pitch Trim System (combined) ± 175 kg (400 lb), air/water displacement

Emergency

(a) Ballast blow, single valve allows all ballast tanks to be blown

(b) Battery pod drop, explosive cutter allows either or both battery pods to drop (545 kg (1,200 lb) buoyancy each pod)

(c) Pressure hull release, mechanical release through hull or explosive cable

Communications

(a) Surface: CB radio

(b) Subsurface: Underwater telephone (8·1 kHz); diver unit available

Pressure hull

Material	Acrylic
Shape	Spherical
Outside diameter	1·65 m (66 in)
Wall thickness	6·35 cm (2·5 in)
Weight (incl. hatch)	680 kg (1,500 lb)
Safety factors:	
(a) Hull design depth	300 m (1,000 ft)
(b) Proof depth	340 m (1,120 ft)
(c) Implosion depth	1,250 m (4,150 ft)

Environmental

External:	
Ocean temperature envelope	1°C to 21°C (34 to 70°F)
Depth	300 m (1,000 ft)
Maximum sea state	2
Internal:	
Partial pressure of oxygen	18 to 23% of one atmos
Partial pressure of carbon dioxide	less than 1% of one atmos
Internal temperature	85° with proper precautions
Internal humidity	low enough to prevent hull fogging

Design factors, stability and safety

Dynamic load (rough lifting, launching operations, etc.)	fore-aft 3 g's lateral 1·5 g's up-down 3 g's
Stability	CB-CG separation: 0·3 m (1 ft)

Subsystem

Life support (2 men for 12 hours, 36 hours additional)

(a) O_2 supply

(b) CO_2 absorbent fan forced

(c) Humidity (silica gel and ice)

(d) O_2 meters (Teledyne 330 and Johnson-Williams K2500)

(e) CO_2 meter (Fyrite 10-5053)

(f) Two closed circuit rebreathers (emergency)

(g) Fyrite 10-5046 (O_2 back-up)

(h) Mine Safety Appliances 83500 (CO_2 back-up)

Applications

Inspections

Harbours

Ships

Moorings to 180 m (600 ft)

Underwater work platforms, arrays, manned and unmanned stations

Support

Photographic, stereo

Directing diver work (U W telephone available)

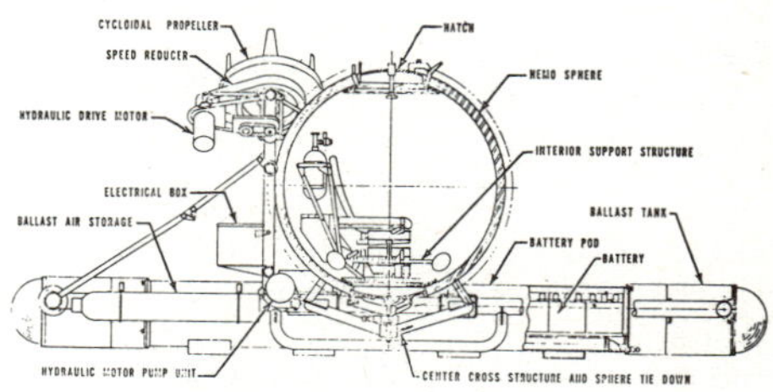

A side cross sectional view of *Makakai*

Co-ordination of operations (diver, ship, other submersibles)

DSRV operations, matings, SDS

Experimental stations (Seacon, etc.)

Evaluation

Handling systems studies

U/W power tools

U/W cable performance

U/W power sources

Subsystems (manipulators, lift packages, lighting, cameras, etc.)

Mooring performance

Hull-type model studies

Buoy studies

Diving systems

Cavitation studies

Pollution control

Oil leaks

Sewage outfall

Radioactivity measurements

Artificial reef studies

Fish die-offs

Disposal areas

Ecology studies

Search capabilities

Ocean research

Deep scattering layer

Fish studies (recordings, anti-shark devices, black cod studies, etc.)

Borings

Corrosion and fouling

Continental shelf sea floor

Currents and turbidity-currents
Sand movement, beach erosion, sediments
Light transmission
Laser studies
Magnetic and gravitational fields
U/W radio wave propagation
Hydrodynamics
U/W sound studies
Polar applications
Sea animals (walrus, sea lion, etc., as in Manned Undersea Science and Technology Program with Alaska Dept. of Fish and Game, NUC, NSF, NOAA)

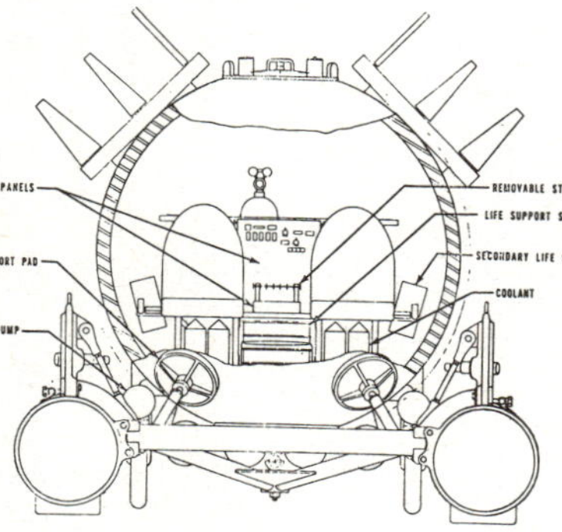

A front cross sectional view of *Makakai*

DEEP VIEW

Deep View, a two-man self-propelled submersible with a transparent bow, designed and built by the US Naval Undersea Center, is the first submersible to incorporate glass as a significant portion of the pressure hull. The trend towards transparent hulls was based on the needs and desires of oceanographic investigators and vehicle operators for the ultimate in maximum visibility. *Deep View* differs from other submersibles in that the forward end consists of a large glass hemisphere having a thickness of 3·8 cm (1·5 in) and a diameter of 1·13 m (3 ft 8·5 in). The hemisphere of borosilicate glass and weighing 137 kg (300 lb) was manufactured by Corning Glass Works.

First sea trials took place in the Spring of 1972, the project having begun in 1966.

Deep View was certified on 20 November 1972 by the US Naval Ship Systems Command for operations to 182 m (600 ft).

Basic dimensions

Length, overall	5·02 m (16 ft 6 in)
Width, overall	1·88 m (6 ft 2 in)
Height, overall	2·29 m (7 ft 6 in)

Weights

Weight in air	5,450 kg (12,000 lb)
Payload	224 kg (500 lb) including crew

Crew

One operator, one observer

Performance

Maximum operational depth	182 m (600 ft)
Speed	1 to 3 knots submerged 5 knots surfaced
Propulsion endurance	6 hours under average operating conditions
Life support endurance	6 hours normal, 18 hours emergency

Power source

(a) Main battery

Sixteen 6 volt lead-acid batteries connected in series-parallel and tapped at 12, 24 and 48 volts DC, 100 amp-hour

(b) Emergency battery

Eight 6 volt globe gel cell (lead-acid) connected in series-parallel to provide 5·2 connected in series-parallel to provide 5·2 amp-hours at 24 volts DC

Propulsion and manoeuvring

Five 5 h.p. Mark 28 motors (DC) (2 vertical, 1 lateral thruster, 2 horizontal) speed controlled by voltage stepping in forward and reverse

Ballast

(a) Main: 2·27 kg (5 lb) steel plates 91 kg (200 lb); sixteen 1·36 kg (3 lb) buoyancy blocks (releasable)

Deep View **(*US Naval Undersea Center*)**

(b) Emergency: Battery pods are releasable (500 kg (1,100 lb) in H_2O)

Communications

(a) Surface: Main—FM radio; aux—sound powered phone on surface

(b) Submerged:

(a) underwater telephone with slant range of 3,650 m (12,000 ft) compatible with AN/UQC-1

(b) diver-operated unit available

Pressure hull

Material	HY-100 steel with pyrex glass hemisphere (1 end)
Shape	rib stiffened cylinder, hemisphere ends
Outside diameter	1·13 m (44·5 in) (glass), 1·10 m (43·2 in) (cyl)
Wall thickness	1·65 cm (0·65 in) (cylinder), 3·18 cm (1 25 in) nominal (glass)

Design parameters
Environmental

External:	
Ocean temperature envelope	1°C to 27°C (30°F to 80°F)
Maximum sea state	2
Internal:	
Partial pressure of oxygen	18 to 23% of 1 atmos
Partial pressure of carbon dioxide	less than 1% of atmos
Internal temperature	approx. 22°C (72°F)
Internal humidity	45 to 55%
Type of material	non-toxic, non-flammable and non-outgassing materials to be used for internal subsystems

Design and test

Safety factors:	
(a) Hull design	140 kg/cm² (2,000 lb/in²) hydrostat (cyl); 70 kg/cm² (1,000 lb/in²) hydrostat (glass)
(b) Support system design	pressure compensated except emergency access trunk blow bottle
Dynamic load (rough lifting, launching operations, and wave slap)	5:1 based on 2 g loading for yield of metal
Static stress	calculations include weight of entrapped water
Stability, submerged	GB = 6 cm (2·37 in)
Safety	US Navy certified

Subsystems

Life support (2 men for 6 hours, 18 hours emergency additional)

(a) O_2 supply
(b) CO_2 absorbent fan forced
(c) Humidity—silica gel
(d) Teledyne pO_2 meter
(e) Cabin pressure—barometer
(f) 2 scuba 1·4 m³ (50 ft³) outfits
(g) Drager kit (back up for CO_2 pressure)
(h) Pocket teledyne for O_2 back-up

Applications

Inspections

Harbours
Ships
Moorings to 183 m (600 ft)
Underwater work platforms, arrays, manned and unmanned stations

Support

Photographic, stereo
Directing diver work (U/W telephone available)
Co-ordination of operations (diver, ship, other submersibles)
DSRV operations, matings, SDS
Experimental stations (Seacon, etc.)

Evaluation

Handling systems studies
U/W power tools
U/W cable performance
U/W power sources
Subsystems (manipulators, lift packages, lighting, cameras, etc.)
Mooring performance
Hull-type model studies
Buoy studies
Diving systems
Cavitation studies

Pollution control

Oil leaks
Sewage outfall
Radioactivity measurements
Artificial reef studies
Fish die-offs
Disposal areas
Ecology studies

Search capabilities

Ocean research

Deep scattering layer
Fish studies (recordings, anti-shark devices, black cod studies, etc.)
Borings
Corrosion and fouling
Continental shelf sea floor
Currents and turbidity-currents
Sand movement, beach erosion, sediments
Light transmission
Laser studies
Magnetic and gravitational fields
U/W radio wave propagation
Hydrodynamics
U/W sound studies
Polar applications
Sea animals (walrus, sea lion, etc., as in Manned Undersea Science and Technology Program with Alaska Dept. of Fish and Game, NUC, NSF, NOAA)

WESTINGHOUSE ELECTRICAL CORPORATION

'DEEPSTAR 2000
(DS-2000)

A manned, self-propelled submersible, for scientific oceanographic work, also ocean bottom recovery duties and sediment study, etc.

The *Deepstar/Searchstar* submersible research system is a versatile, fully transportable system for use in the study of the ocean, inland waterways, lakes and estuaries throughout the world.

A unique feature of this system is the closely integrated design of the support vessel *Searchstar* with its submersible, *Deepstar 2000*. The *Searchstar* is so constructed that it is hydrodynamically matched to the surface motion of 'DS-2000' and is also designed to take full advantage of the mobility attributes of *Deepstar*. The *Searchstar* can be easily dismantled with standard dockside equipment for loading, together with *Deepstar*, on standard road trailers or aboard an air transport. Result: widely separated short-term diving operations become economically feasible, operations which are generally impracticable for currently existing types of submersible systems.

Searchstar is self-propelled and provides surface support and monitoring for the 'DS-2000' for up to 10 hours of submerged operations, as well as performing manoeuvres needed for launch and retrieval.

Deepstar 2000 is designed to carry three men to depths of 610 m (2,000 ft) for conventional submersible operations up to 12 hours. The submersible has good manoeuvrability and elevation control, along with demonstrated high stability in pitch and roll and is an ideal tool for geological research, biological research, physical and chemical research, and as a test bed for oceanographic equipment and for environmental and life testing of materials.

The 'DS-2000' is fitted with the highly sensitive Westinghouse A-38 scanning sonar for observing the ocean floor.

Minimum disturbance of sediment or water-borne particles is caused in the forward viewing area due to propeller slipstream effects.

The *Deepstar/Searchstar* research submersible system

Basic dimensions

Length, overall	6·1 m (20 ft)
Width, overall	2·27 m (7 ft 6 in)
Height, overall	1·67 m + 0·915 m 'sail' (5 ft 6 in + 3 ft 'sail')

Weights

Dry weight	7,950 kg (17,500 lb)
Submerged weight (gross displacement submerged)—includes entrained water	15,500 kg (34,000 lb)

Deepstar 2000, manufactured by the Westinghouse Electric Corporation

Payload	222 kg (500 lb) of scientific equipment net weight *in situ* distributed as 227 kg (500 lb) in the brow and pressure hull (available buoyancy) 680 kg (1,500 lb) air weight attachable
Crew	one
Observers	two

Performance

Maximum operational depth	610 m (2,000 ft) upgradable to 762 m (2,500 ft)
Speed (continuously controllable up to):	
cruise	1 knot
maximum	3 knots
Propulsion endurance	8 hours for normal diving for 3 people
Life support endurance	for 3 people, normal 12 hours, additional reserve 36 hours, emergency (rebreathers), capacity 1 hour/person and reserve of about 1 hour

Manoeuvring capability

Variable ballast system 136 kg (300 lb) capacity

Stop distance from speed of 1 knot: 6·1 m (20 ft)

Time to pivot 180° about point in front of view ports: 30 seconds

Time to make a vertical excursion of 1·8 m (6 ft) and come to a stop: 15 seconds

Maintain position in a transverse current of 0·5 knot

Trim stability: 136 kg (300 lb) moment causes less than 6° of roll or pitch

Controllable steering by variable rpm on main propulsors and horizontal thrusters

Pitch trim ± 89·5 kg/m (650 ft/lb) fore and aft giving pitch change at least ± 15°

Hold constant depth without thrusters to within 0·3 m (1 ft) per 10 minutes

Descent rate using variable ballast only: 0·6 knot

Main machinery

Power source	Lead-acid batteries: one 120 volt/150 amp/h; two 28 volt/150 amp/h (total power = 26·5 kw/h)
Propulsion	2 main thrusters (50·8 cm (20 in) shrouded propellers); 2 vertical thrusters (22·86 cm (9 in) ducted propellers; 2 horizontal thrusters (22·86 cm (9 in) ducted propellers

Pressure hull type

A 1·5 m (5 ft) diameter cylinder capped by hemispherical ends of HY80 steel 1·9 cm (¾ in) thick.

External structure

A removable brow of sufficient strength and volume to hold up to 222 kg (500 lb) (in water) of instrumentation

Brow interchangeable with special equipment modules

Brow jettisonable in case of emergency

Design features and attachments consistent with various methods of handling, including the surface support vessel, *Searchstar*.

Emergency buoyancy

Air blowing surface buoyancy tanks at depth: 222 kg (500 lb)

Dropping high pressure air bottles: 222 kg (500 lb)

Dropping batteries: 568 kg (1,250 lb)

Dropping instrument brow: 222 kg (500 lb)

Total available buoyancy: 1,250 kg (2,750 lb)

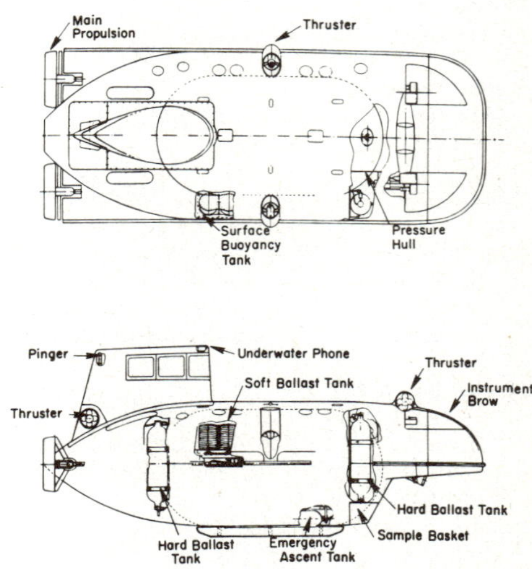

Plan and side elevation drawings of *Deepstar*

Basic power system
Dual DC electric motor driven hydraulic pumps at 10 h.p. total

Basic control system
Proportional-controlled servo valves and solenoid valves operating in a constant pressure hydraulic system.

Availability
The *Deepstar 2000* had, up to June 1973, been operated by Westinghouse Electric Corporation, Oceanic Division, but a training programme is available should the user choose to operate it himself.

***Deepstar 2000* operating and design characteristics**

Maximum operating depth	610 m (2,000 ft)

Typical dive schedule
A. Single dive day
1 hour predive checkout
$\frac{1}{2}$ hour dive time (to 610 m (2,000 ft))
6 hour bottom time
$\frac{1}{2}$ hour ascent (from 610 m (2,000 ft))
1 hour recovery and post dive checkout

B. Multiple dives to intermediate depths
0 to $\frac{1}{2}$ hour dive
1 hour bottom
0 to $\frac{1}{2}$ hour ascent and turn-around
5 dives per day (maximum)

Personnel transfer at sea (operational in sea state 3)
Personnel acces through 'sail'
Sail height sufficient to avoid flooding in design sea state and short enough to be within allowable limits for road clearance and handling at sea
Weather hatch and windows in sail to permit surface manoeuvrability using remote control system

Buoyancy change and adjustment
Fast-acting variable displacement system
Differential buoyancy control of ± 68 kg (150 lb)

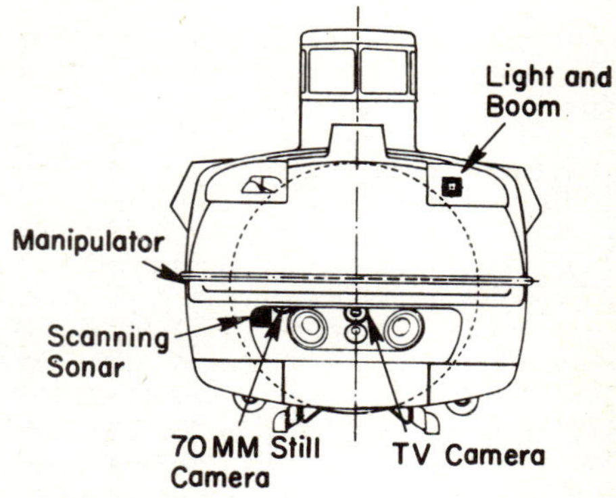

General configuration diagram of *Deepstar 2000*

Capability of changing buoyancy through full range in 12 minutes
Capability of maintaining vertical position (hovering) to within ± 0·03 m (0·1 ft) per minute

DEEPSTAR 4000
A manned, self-propelled submersible, capable of operating at depths to 1,200 m (4,000 ft). The design is based on the highly successful 300 m (1,000 ft) 'Diving Saucer' concept with techniques of the saucer extended to the deeper diving *Deepstar*. The increased diving depth was accomplished by changing the hull shape from ellipsoidal to spherical and using modern high strength steel for the pressure resistant capsule. The crew's compartment is a spherical pressure chamber, 2 m (6 ft 7 in) in outside diameter, fabricated from two hemispheres of high-strength steel 5,600 kg/cm² (80,000 lb/in²). Each hemisphere was formed from a 6·35 cm ($2\frac{1}{2}$ in) thick steel plate. After forming, the hemispheres were welded together to form a sphere which was then machined. Batteries, motors, weights, pumps, hydraulic system, and other auxiliary equipment are located outside of the pressure sphere in a free-flooded area protected and supported by metal structural members. The structural members are covered by a streamlined fibreglass outer hull or fairing which minimizes vehicle-flow resistance, optimizes vehicle handling capability, and protects the auxiliary equipment from wave action when operating on the surface.

***Deepstar 4000* (*Westinghouse*)**

Basic dimensions

Length, overall	5·5 m (18 ft)
Width, overall	3·5 m (11 ft 6 in)
Height, overall	2·13 m (7 ft)

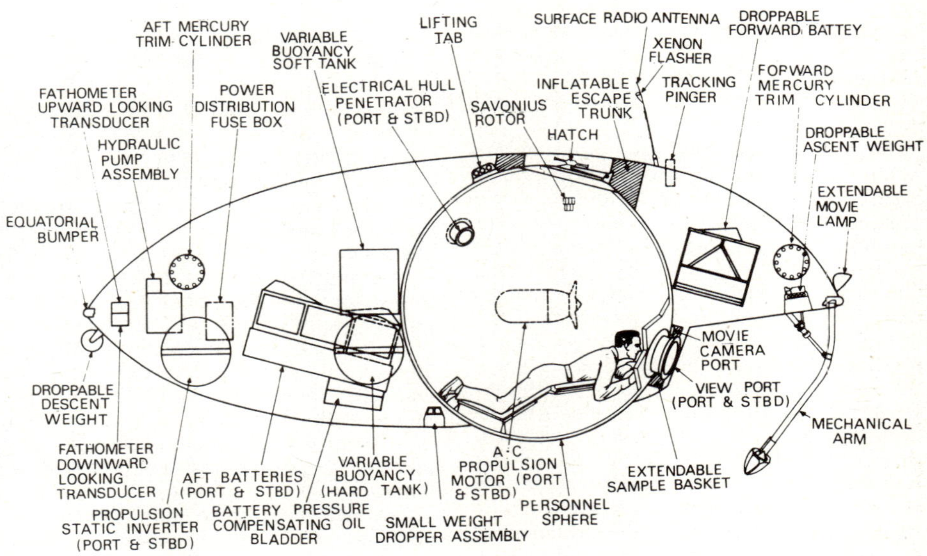

Deepstar 4000, equipment arrangement

Weights

Total weight (in air)	9 tons
Payload	136 kg (300 lb) inside, 136 kg (300 lb) outside
Crew	one
Observers	two

Performance

Maximum operational depth	1,220 m (4,000 ft)
Speed, submerged	3 knots
Propulsion endurance	8 hours, depending on use of electrical power
Life support endurance	48 hours for 3 men (12 hours + 36 hours emergency)

Main machinery
Power source
Power for the vehicle is provided by three lead-acid batteries, mounted outside the pressure sphere and encased in containers which permit covering the cells and electrical connections with oil at ambient sea pressure. This design obviates the need for a pressurized container. Utilizing the full battery capacity of 400 ampere hours 50 kw/h at 125 volts, propulsion may be maintained for 6 to 12 hours, depending upon vehicle speed and the power used for lamp illumination.

Each battery weighs 450 kg (1,000 lb) and is constructed and mounted to permit maximum obtainable tilt of the vehicle without spilling electrolyte. Two batteries are located in the aft section of the vehicle and one battery is mounted in the forward section.

Internal power is 120 volts DC and 28 volts DC.

Propulsion and manoeuvring
Two 5 h.p., free floodable, squirrel cage AC electric motors which are direct-coupled to 35·5 cm (14 in) diameter shrouded propellers propel the vehicle. Attached in a fixed horizontal position, one on each side of the vehicle, the propulsion units provide manoeuvring ease in gradual turning, and forward or reverse movement by varying the speed of the individual propellers. By reversing the rotation of one unit with respect to the other, the vehicle can be turned about in place.

Housed in pressure resistant containers located in the free-flooded structural framework, static inverters supply variable voltage/variable frequency to the induction motors.

Continuous propeller speeds ranging between 30 rpm and 900 rpm in either forward or reverse rotation are possible.

Hydraulic system

All components of the hydraulic system are located outside the pressure sphere. The hydraulic pump assembly, driven by a 0·6 h.p. electric motor, is mounted outside the oil reservoir. A diaphragm maintains the oil reservoir at ambient sea pressure. The hydraulic pump maintains system pressure 70 kg/cm² (1,000 lb/in²) above ambient pressure. Hydraulic piston actuators controlled by electric solenoid valves are used to operate weight release mechanisms during normal trimming, as well as for manipulating the mechanical arm and photographic lamps. Mercury trim ballast is pumped from one reservoir to another by means of hydraulic pressure.

Depth control

Vertical movement of the *Deepstar* vehicle is controlled by the pilot, using a system of droppable weights and a variable ballast system, thus conserving propulsion power. The vehicle descends and ascends in a helical path, stern downward, at a maximum tilt of 60°. The helical path permits a nearly vertical placement of the surface tender ship nearer to the beginning and end points of the bottom exploration than would be possible using a straight line path at an equivalent angle of descent.

Two major weights are provided for control of descent and ascent: one descent weight of 85 kg (187 lb) and one ascent weight of 103 kg (226 lb). Depth of dive is controlled by releasing the descent weight upon reaching the approximate depth desired.

Fine buoyancy is controlled by dropping small lead trim weights and by control of the variable ballast system. The lead trim weight, totalling 74·5 kg (162 lb), consists of a set of 54 weights, 1·47 kg (3 lb) each, which may be dropped one at a time. The variable ballast system permits changes in vehicle displacement of $\pm$ 24·8 kg (55 lb). The trim weights and the variable ballast system may be used in combination to make fine adjustments in operating level to compensate for variations in seawater density, or aid in ranging up or down.

Pitch of the vehicle is controlled by pumping 82 kg (180 lb) of mercury back and forth between two storage tanks, one located forward and the other aft in the vehicle. Dropping of the 103 kg (226 lb) ascent weight will cause the vehicle to rise in a manner similar to that of descent (forward and upward at a maximum tilt of 30° and in a helical path). The helical path in both cases results from the distribution of weight in the vehicle imparting a slight constant roll. Speed under these conditions is 2 knots. If quicker descent or ascent (at the expense of total power endurance) is desired, the propellers may be used to double the speed during this mode of operation.

Life support system

Air purification of the pressure sphere is patterned after the method proven successful in the 'Diving Saucer'. Carbon dioxide gas is absorbed by granular lithium hydroxide. A blower is used to pass air over lithium hydroxide packets to maintain the carbon dioxide level at 1 or 2 per cent for the duration of the mission. Should the concentration exceed this level, the carbon dioxide monitoring instrument will so indicate and the crew may then replenish the absorbent from the reserve supply. Internal pressure in the sphere is indicated by an aneroid barometer. Make-up oxygen is released at a rate sufficient to maintain normal internal pressure. The operational supply of absorbent and oxygen is sufficient for three men for 48 hours. Storage space is provided for food and water, as well as a sanitary container for personal relief.

Auxiliary equipment

Communication devices include an underwater telephone with a range of 6,850 m (22,500 ft) for contact with the surface vessel or with other submersibles and a two-way radio with a range of 16·1 km (10 miles) for surface communications. A precision echo sounder, consisting of two transducers with an indicator and recorder, permits examination of bottom terrain beyond the crew's normal viewing distance when using the viewing ports. Lights are grouped at the forward end of the vehicle for running, observation, and still and movie photography. These lights include a 500 watt and a 1,000 watt spot for running, a 1,000 watt and a 500 watt flood for close observation, and a 2,500 watt light for photography. The lamps use special pressure-resistant, filament type bulbs.

A camera port, located midway between the viewing ports, accommodates a 16-mm movie camera with a 10-mm or 25-mm focal length lens, or a 70-mm still camera with a 50-mm focal length lens. To facilitate gathering of information, a tape recorder is installed for use as a log book.

Fitted externally on the sphere, a hydraulically operated mechanical arm and claw assembly permits gathering of samples or performance of other work.

Safety systems

Make-up oxygen and carbon dioxide absorbent is carried in sufficient supply to provide life support for an additional 36 hours beyond the normal 12 hour period.

In case of emergency, the descent and ascent weights may be jettisoned by hydraulic actuators. If the hydraulic system fails, the weights may be jettisoned by a manually operated mechanical linkage system. If additional jettisoning is required, the forward battery may be detached from the vehicle. As an additional safety provision, mercury in the trim system may be jettisoned. The access hatch may be opened from either the inside or outside of the sphere. To minimize compartment flooding during moderately rough sea conditions, a collapsible conning tower surrounding the access hatch may be inflated. A two-way radio provides emergency communications and aid in vehicle location.

Surface handling requirements

A surface support vessel of approximately 120 tons displacement, with a 9·15 m $\times$ 6·1 m (30 by 20 ft) clear deck space and a shallow freeboard is used in handling the *Deepstar* vehicle. In addition, a crane with a boom reach of 3·0 m (10 ft) from the stern of the surface ship and capable of lifting 20 tons is used to lift the vehicle between the deck and the surface of the water. Auxiliary equipment aboard the surface vessel provides for battery recharging, replenishment of air and oxygen tanks, and testing of electronic equipment. In addition, small portable shops for field maintenance in mechanics, electronics, and photography are used.

Applications

Some of the tasks which can be performed by *Deepstar 4000* are:

1. Provide direct and systematic observation in the deep ocean by marine biologists and geophysical scientists.
2. Conduct underwater photography missions at operating depth.
3. Conduct deep sea environmental studies.
4. Place, observe, and recover instruments and equipment on the ocean floor.
5. Advance the development of sensors, underwater lighting, vehicle propulsion, sonar equipment and deep sea instrumentation.
6. Conduct ocean bottom exploration, sampling and coring.
7. Advance the techniques of mapping of the oceans.
8. Conduct systematic studies of operations in the deep sea environment.
9. Assist in military and commercial salvage operations.

UNION OF SOVIET SOCIALIST REPUBLIC

AIRCRAFT ENGINEERING INSTITUTE, MOSCOW

UNDERWATER RESEARCH BOAT

The craft is designed to carry two people and is equipped to remain at a depth of 50 m (162·5 ft) for 2 hours. Speed under water is 7 knots.

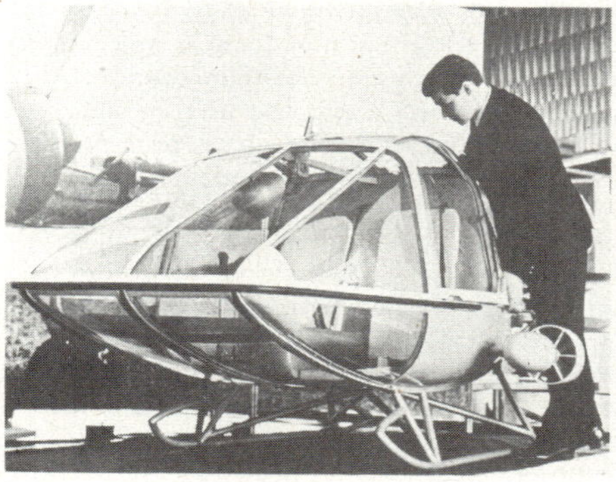

A view of the underwater research craft designed and built at the Moscow Engineering Institute in 1968 (*Courtesy Novosti Press*)

SUBMARINE GLIDER

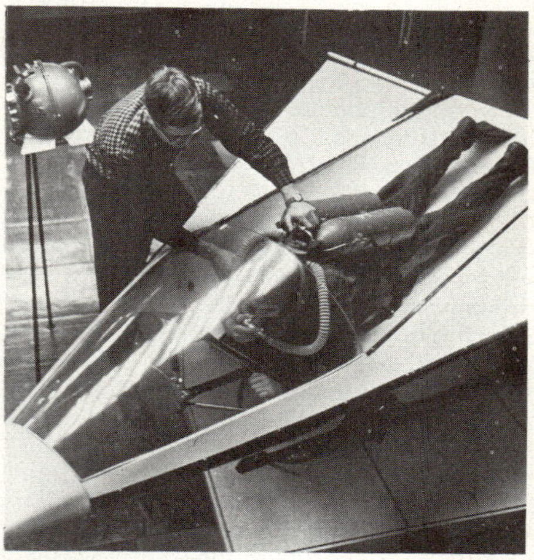

The *Submarine Glider*, built by students of the Moscow Aircraft Engineering Institute in 1963 (*Courtesy Novosti Press*)

ALL-UNION SCIENTIFIC RESEARCH INSTITUTE OF SEA FISHING AND OCEANOGRAPHY (VNIRO)

GVIDON

Gvidon is a manned apparatus designed mainly for studying fish shoals, fishing tackle and for conducting soil research in inland reservoirs and at a definite depth in any ocean fishing area.

The apparatus is capable of submerging to a depth of 250 m (820 ft) and is highly manoeuvrable. It has two propulsion devices, ballast and stabilising cisterns. In the lower part is a section for emergency ballast and for the batteries. Its electricity reserve is sufficient for autonomous operation during five hours. The air regeneration system ensures the vital activities of a crew of 2 to 3 scuba divers for three days. The displacement of *Gvidon* is just over 3 tons.

The *Gvidon* unit has a cylindrical pressure hull with exterior circular ribs. The inlet port coaming, with a diameter of 640 mm (25·2 in), is welded onto the upper elliptic partition, and the body of the propeller drive mechanism and the emergency ballast delivery device is welded onto the lower partition. The vertical orientation of the hull simplifies control and gives the vessel the required stability in both underwater and above-water positions. A circular ballast tank with a capacity of 0·36 m^3 (12·7 ft^3) and consisting of three sections joined by a manifold is installed in the upper part of the pressure hull. Under the ballast tank there are 28 air cylinders with a capacity of 7 litres each at an air pressure of 150 kg/cm^2 (2,100 lb/in^2) Detachable adapters are placed under the cylinders with immersion accumulator batteries. Between the adapters under the bottom of the vessel there is a bunker for emergency ballast. Lead or steel shot is used as ballast— diameter 6–8 mm (0·24–0·31 in).

The vessel has three supports fixed to the bottom and inclined to the vertical axis. The supports, which are hollow, are sections of a compensating tank with a capacity of 0·12 m^3 (4·3 ft^3). Two rotating propellers are installed in the centre part of the hull. Movement of the unit is controlled along the course and reverse is obtained by tumbler switches at the control panel. One general mechanism serves to rotate the engines and shed emergency ballast.

Gvidon, The Soviet self-propelled, manned, underwater apparatus in preparation for use in the Black Sea (*Courtesy Novosti Press*)

The immersion depth of the unit is controlled by two depth gauges. The depth gauges have a signalling device which operates in the event that the given immersion depth becomes too great. There is accommodation for two men in the vessel: the pilot—below, and a lookout—at the top. The crew members enter and leave the unit on the deck of ship; in exceptional (emergency) cases it is possible to leave while afloat. In the hull there are ten illuminators made of organic glass (the lookout has eight, the pilot—two) giving all-round vision. The internal diameter of six of the illuminators is 154 mm (6·06 in) and of the remaining four, 66 mm (2·6 in). The vessel is fitted with eight lamps: four of them are installed in the centre part of the hull, one on the inlet port coaming, two near the propellers and one in the lower part of the hull on the extensible telescopic rod. Twin-thread large lamps are used as sources of light with an output of up to 600 watts. An emergency signal beacon and guide rope with electric winches with two-way action have devices for delivery and shedding in emergencies.

Inside the pressure hull are the air regeneration system, microclimate control devices, emergency breathing apparatus, a radio transmitter, a directional gyro, a log, two sounding devices, an emergency source of electrical energy, a fire extinguisher, a set of instruments for registering parameters of the outside medium, etc. The pilot's illuminator, the lower illuminator and the observer's swivel chair have cine and photographic equipment attached to them. The vertical orientation of the unit and the low moving speed make it possible to use interchangeable sets of instruments on it and hanging devices without impairing the hydrodynamic characteristics.

The first tests on the *Gvidon* unit were carried out in autumn 1970 in the Crimea. Working as a self-contained unit, moving in the depth of the water and on the bottom, the unit showed excellent manoeuvrability in vertical and horizontal directions. With synchronous operation of two propellers it had steady, straight movement. When one of its propellers was shut off (without reverse) a turn was made with a gyration diameter of no more than 6 m (20 ft). It is of special significance that the vessel has great manoeuvrability and minimal inertia. During the tests the possibility of taking the unit off the seabed was tried. With this aim an air bubble was created under the hull with a lifting capacity of 140–150 kg (308–330 lb).

Principle particulars

Height	4·5 m (14·76 ft)
Breadth	2·45 m (8·04 ft)
Internal diameter of pressure hull	1·22 m (4 ft)
Height of pressure hull (including inlet port coaming)	3·42 m (11·22 ft)
Immersion	3·2 m (10·5 ft)
Displacement:	
above-water	3·9 tons
underwater	4·3 tons
Reserve buoyancy	17%
Working immersion depth	250 m (820 ft)
Emergency ballast mass (fraction)	300–320 kg (660–704 lb)
Capacity of immersion accumulator battery	400 amp hours
Capacity of electric-drive propellers	
nominal	2·8 h.p.
maximum	5·2 h.p.
Running speed	0·7–1·0 knots
Self-sufficiency:	
in reserves of electrical energy	5 hours
in reserves of regeneration agents	48–72 hours
Crew, no. of men	2–3

STATE DESIGN INSTITUTE OF THE SOVIET FISHING FLEET

ACADEMICIAN N M KNIPOVICH POLAR INSTITUTE OF FISHERIES AND OCEANOGRAPHY

SEVER-2 (NORTH-2)

Within a relatively short time span, Soviet specialists have designed, built, and successfully tested the first Soviet deep-sea submersible *Sever*-2 intended for national-economic research applications. The vehicle's particulars and the set of scientific equipment and instrumentation installed allow performance of a wide variety of research missions associated with fish finding and the quest for other commercial marine life, bottom charting, and the selection of regions suitable for working by bottom trawls. Principal particulars are as follows:

Length	12 m (3 ft 8 in)
Breadth	2·5 m (7 ft 8½ in)
Depth	4 m (1 ft 2½ in)
Displacement	40 tonnes
Economical speed	2·5 knots
Submerged endurance	10 hours
Reported diving depth	2,000 m (6,560 ft)
Crew	3–4

The pressure hull comprises a framed cylinder with hemispherical extremities.

The vehicle commander (operator) and engineer are located at consoles in the midbody on port and starboard sides. Between the main control consoles is a vertical vision tube with built-in cine camera. Current converters, standby storage battery, galley and sanitary services, ventilation facilities, and switchboards are arranged aft. In the false hull forward are a rotating light, hydroacoustic sensors, and hydrological instrumentation. Flanking these at the sides are ballast tanks, while the stern contains a wheelhouse providing vehicle control in the submerged condition. To the stern extremity, with its horizontal stabilisers, is attached the propulsion arrangement, while below, storage batteries, hydraulic devices, an electric drive, and other equipment are positioned. If required, storage battery containers may be isolated from the hull.

Main propulsion is by a hydraulically-driven steerable ducted propeller. Vertical transfer of the submersible is accomplished by two vertical-thrust ducted propellers. Special arrangements for remote-controlled and automatic release of emergency ballast and heavy equipment provide an emergency surfacing capability.

As tests have shown, the submersible's equipment and systems ensure a high level of operational safety. Submergence is managed from *Odissei*, a specially designed 3,800-tonne displacement oceanographic research vessel equipped with a vehicle deployment/recovery arrangement, hangar, workshops and laboratories necessary for submersible handling and scientific data evaluation.

Sever-2, the first Soviet deep-diving submersible for 'national-economic research duties' has entered experimental service with the Academician N. M. Knipovich Polar Institute of Fisheries and Oceanography.

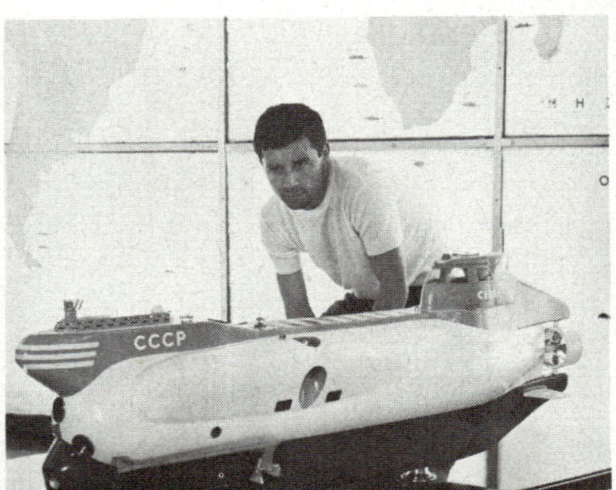

A scale model of *Sever 2.* This deep-water craft was designed by the Soviet State Design Institute (*Courtesy Novosti Press*)

Sever-2 **operating from the support ship** ***Odissei***

The Soviet submersible *Sever-2*

THE USSR RESEARCH INSTITUTE OF THE FISHING INDUSTRY AND OCEANOGRAPHY

SEVERYANKA

The Soviet research submarine *Severyanka* (*Courtesy Novosti Press*)

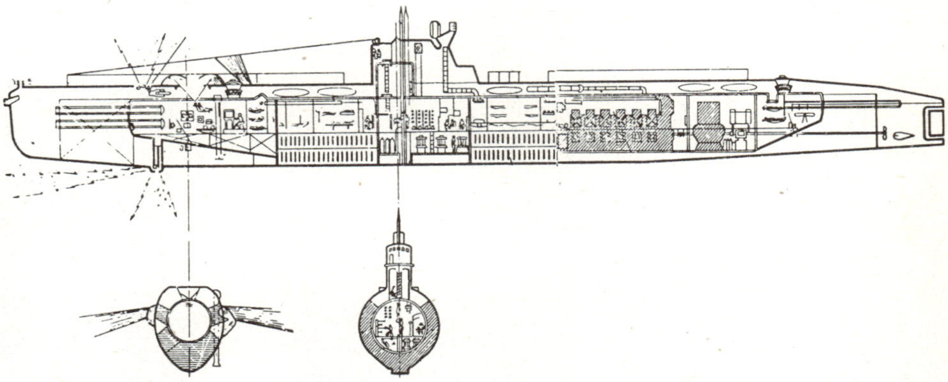

Diagram of the Soviet research submarine *Severyanka* designed for oceanographical and biological studies

USSR PACIFIC FISH LABORATORY

TINRO I

1965

Designed for fish-shoal search and fish-behaviour studies.

Weight	35 to 40 tons
Design maximum operating depth	300 m (1,000 ft)
Length	11·0 m (36·08 ft)
Width	2·7 m (8·85 ft)
Power	diesel engine 64 h.p.
Speed:	
surface	8 knots
submerged	6 knots

TINRO II

Similar to 'Tinro I' but smaller.

This submersible has accommodation for two. The hull has nine viewports. Trials started in December 1973 in the Black Sea prior to further trials in the Atlantic and Indian Oceans. The vessel was built in Leningrad and is intended for investigations of the shelf zone of the world's oceans.

An artists' impression of *Tinro II*. This Soviet submersible was projected in 1965

SUBMERSIBLE SUPPORT VESSELS

CANADA

INTERNATIONAL HYDRODYNAMICS CO LTD (HYCO)

HUDSON HANDLER

The M/V *Hudson Handler* is a portable, self-contained, self-powered, modular barge, specially designed by HYCO, to be dismantled and assembled quickly for shipment by either truck, rail or ship anywhere in the world. It provides a stable working platform to maintain, launch and recover (through the use of a patented capturing device) the *Pisces* class submersibles. The barge has been C.S.I. approved for open ocean work (deep sea trade) and has fuel and support facilities to sustain 10 men at sea for a 30-day period.

Length	27·43 m (90 ft)
Beam	12·19 m (40 ft)
Draught	1·98 m (6 ft 6 in)
Displacement	343 tons
Propulsion	Two GMC diesels

Hudson Handler operated by International Hydrodynamics (HYCO) of Canada.

Hudson Handler

UNITED KINGDOM

VICKERS OCEANICS LTD

VICKERS VENTURER

As an ex-stern trawler, *Vickers Venturer* possesses a sheltered well-deck aft, providing an ideal platform close to the water-line from which submersibles or instruments weighing up to 12 tons can be handled by the hydraulically operated lifting gear. A special winch incorporated in the handling system automatically compensates for ship/sea motions.

On-board back-up facilities for operations are provided by workshops which are designed to cope with almost all possible breakdowns on-site, thereby avoiding costly downtime caused by returning to harbour for repairs. These workshops also allow a flexibility for the fitting and adapting on-site of customers' equipment which may be required over and above that provided by the vessel's extensive equipment fit.

Vickers Venturer is fitted with all the requisite navigation equipment, in addition to which are fitted an echo-sounder, side-scan sonar and the Vickers-developed SPATE system (Submersible Position And Tracking Equipment).

Communication systems include marine band radio and radio telephone, plus hand-portable sets on private and emergency frequencies for submersible operations, etc, carried out within the vicinity of the ship. In addition, an underwater telephone using acoustic transmission with a range of up to $1\frac{1}{2}$ miles is provided for communication with the submersible.

Vickers Venturer operated by Vickers Oceanics

Basic details

Length, b.p.	35·9 m (117 ft 8 in)
Length, overall	37·2 m (122 ft)
Breadth, moulded	7·6 m (25 ft 1 in)
Draught	4·58 m (15 ft)
Displacement	640,000 kg (630 tons)
Freeboard:	
stern	0·9 m (3 ft)
midships	2·4 m (8 ft)
Propulsion	Two 394 h.p. diesel via single screw variable pitch propeller
Accommodation	30 including crew
Normal operating crew	13
Speed	9 knots

Special features

Deck decompression chamber

Fully air conditioned

Dark room

Mechanical and electronic workshop

Sheltered well-deck for apparatus

Surplus AC and DC electrical power. Full view aft from bridge

Excellent manoeuvrability

Over-stern lift capability of 11·8 tonnes (12 tons)

Submersible launching up to sea state 4

Submersible recovering up to sea state 5

SPATE II equipment designed to permit the surface support ship to locate position and track the submersible on sea bed

VICKERS VOYAGER

Vickers Voyager was commissioned to supplement the underwater work service and improve the deep-sea operating and submersible support capability.

Voyager, a large and stable ex-fish-factory ship, has been fully fitted-out to support and handle two submersibles by day or night in adverse weather and sea conditions, such as those encountered in the North Sea.

The submersibles are handled over the stern of the ship from the sheltered well-deck aft by means of the Vickers-developed handling system. Ship/sea motions are automatically compensated for by a specially developed winch.

Voyager is fitted with comprehensive workshop facilities which provide on-site submersible maintenance plus a capability to overcome most breakdowns, thereby avoiding the need to return to harbour for repairs. In addition, the workshops are equipped to develop and adapt underwater tools to meet customers' particular requirements which cannot be catered for by existing equipment.

Navigation equipment carried by *Voyager* includes a satellite navigation system and the Vickers-developed SPATE System (Submersible Position and Tracking Equipment). Communication systems include marine band radio and radio telephone, plus underwater telephones using acoustic transmission for communication with both submersibles and divers at ranges up to 1½ miles.

Vickers Voyager handling a Pisces submersible

Basic details

Length:	
b.p.	71·6 m (235 ft)
overall	83·4 m (274 ft 8 in)
Breadth, moulded	14·6 m (48 ft)
Draught	6·7 m (22 ft)
Displacement	Approx. 4,500 tons
Propulsion	Diesel electric via single screw
Diesels	Three Ruston Hornsby
Generators	GEC
Output from main engines	1,410 b.h.p. at 500 rev/min
Speed	12½ knots
Normal operating crew	29

Special features

Fully air-conditioned
Dark room
Electronic and mechanical workshops
Over-stern lift of 18 tonnes (17·7 tons)
Over-side derrick lift of 18 tonnes (17·7 tons)
SPATE II submersible tracking equipment
Satellite navigation: Deck decompression chamber
Telephone to ensure rapid contact between all major areas of vessel
Single-berth accommodation for up to 20 supernumeraries
Submersible launching up to sea state 5
Submersible recovery up to sea state 6

THIRD SUPPORT VESSEL

Conversion of a third ship for submersible support work is now in hand and the ship is expected to be in operation later this year.

Submersible Handling System

The Vickers system represents completely new thinking on the problem of recovering large floating bodies from rough seas. Previous known handling philosophies are based upon snatching the submersible from the sea at just the right instant, limiting safe handling to little more than 0·915 m (3 ft) waves, putting a tremendous responsibility upon the operator, and resulting in frequent enforced idleness for many submersibles.

The handling gear on the *Vickers Venturer*

In contrast, Vickers *use* sea induced motion whenever it is favourable to the operation, rather than trying to fight it, gradually transferring from a sea-dominated motion regime to ship-dominated one.

A single braided nylon rope passes to the submersible lift point—a very important feature minimising the time required to make a connection, eliminating the need for a heavy block and using only a light shackle easily handled by one diver. It also eliminates any danger of injury to the diver or of damage to the submersible from a swinging block.

Using an 'A' frame extending over the stern, control of pendulation is ensured before the submersible is brought near to any hard structure when the 'A' frame is being pivoted inboard. Provision of a pendant frame, freely pivoting fore and aft from the top of the 'A' frame, but providing restraint athwartships, permits large outreach with transverse control. Ship's stability problems associated with a high suspension point are also minimised and it is possible to use a much smaller support ship than would otherwise be the case.

Finally, a wave-compensating winch is used, in which stored energy from hydraulic accumulators is continuously available to a small, low torque, potentially high-speed motor, which can overrun the main hoist motors and maintain a taut rope when wave-induced relative upward motion of the submersible exceeds normal hoisting speed.

This handling system has proved effective in 4·25 m (14 ft) waves.

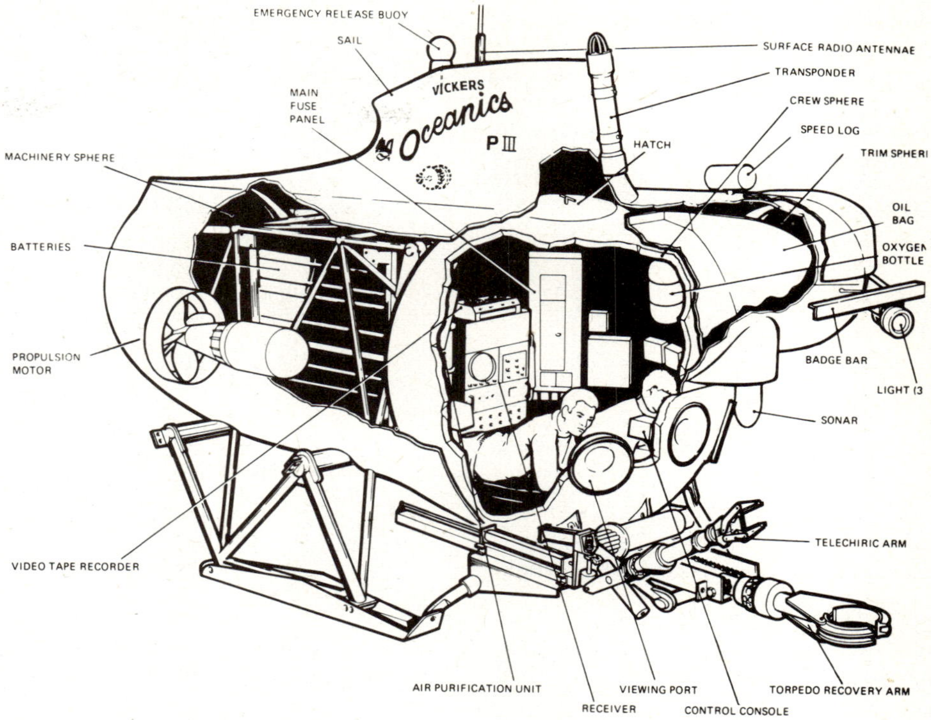

The HYCO *Pisces III* as operated by Vickers Oceanics

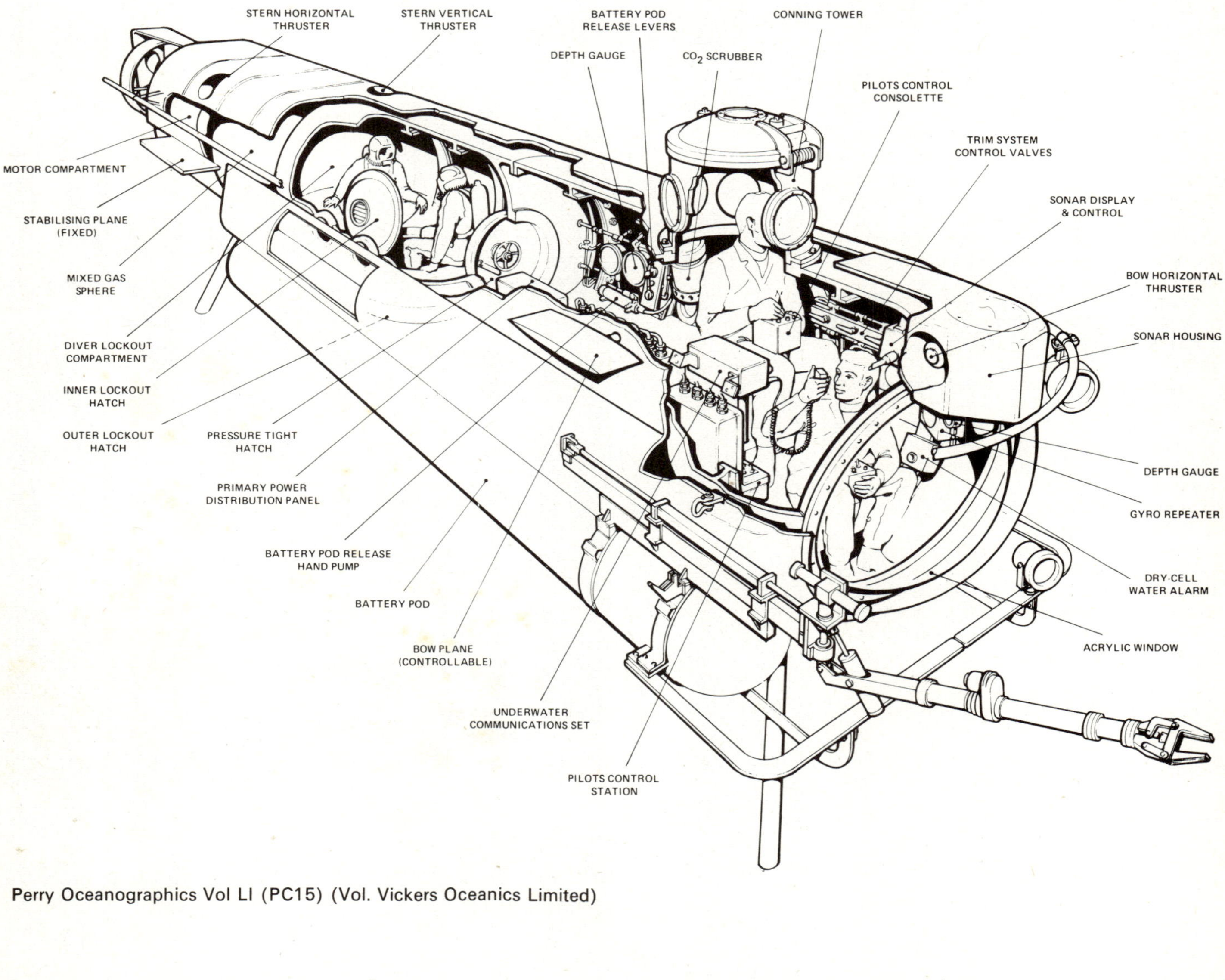

Perry Oceanographics Vol LI (PC15) (Vol. Vickers Oceanics Limited)

SUBMERSIBLES OPERATED BY VICKERS OCEANICS LTD

Vickers Oceanics Ltd (V.O.C.) own four submersibles, the International Hydrodynamics *Pisces I, II* and *III* and The Perry Oceanographics *PC-15*, re-named *VOL L1*. Major modifications made by Vickers Oceanics to the Pisces submersibles are as follows:

PISCES I

(a) New 5 h.p. propulsion motors;

(b) New forward air ballast tank;

(c) New single point lift frame;

(d) New 12 volt and 24 volt battery system;

(e) Emergency release buoy fitted;

(f) New Vickers manufactured buoyancy material;

(g) Transponder fitted for Vickers submersible tracking system;

(h) Ergonomically re-designed command sphere;

(j) Tail sphere pressure tested to 49 kg/cm² (700 lb/in²);

(k) Vessel re-wired and re-piped;

(l) Life support increased to 100 crew hours;

(m) Sonar modifications to include two navigation transponders.

PISCES II and III

(a) New 3 h.p. propulsion motors;

(b) New Vickers manufactured buoyancy material;

(c) Emergency release buoy fitted;

(d) Sonar modification to include two navigation transponder channels;

(e) Transponder fitted for Vickers submersible tracking system;

(f) Lift hook modified for Vickers submersible handling system;

(g) Main command sphere ergonomically re-designed;

(h) CCTV and still photographic system fitted with externally mounted cameras on pan and tilt units, Thallium lighting and strobe light. Video recorders fitted;

(j) Hydraulic underwater tool pack manufactured and fitted as required;

(k) Jetting equipment fitted for cable trenching;

(l) Life support increased to 100 crew hours;

(m) Vessels rewired and re-piped;

(n) Vessels ultrasonically tested and radiographed annually.

UNITED STATES OF AMERICA

GENERAL OCEANOGRAPHICS

LRT (LAUNCH, RECOVERY, AND TRANSPORT VEHICLE)

Designed and invented by Mr. Henry M. Horn, President of Makai Ocean Engineering Co., the LRT was developed in 1969 for the purpose of launching and recovering submersibles in very rough sea state conditions. The prototype LRT was successfully tested in operations with the Nekton Alpha submersible and more recently has been in operation with the Star II submersible.

Normal operating depth: 39·6 m (130 ft)
Crew: 3–4 men (divers)
Payload: 6,364 kg (14,000 lb)

LRT-8

This version was built on a very tight budget and has been in service for nearly six years. She operates normally from Makai Pier in Waimanalo, Oahu, Hawaii with the submersible STAR II, averaging three dives per week to 366 m (1,200 ft) and in sea states up to sea state 3.

LRT-88A

This design was drawn up as a support vehicle for the US Navy Deep Submergence Rescue Vehicle (DSRV). Of aluminium construction, the design was such that the vehicle could be dismantled and transported in a C-141 aircraft.

General Oceanographics LRT (Launch Recovery and Transport Vehicle) shown handling *Nekton Alpha*

General Oceanographics LRT (Launch, Recovery and Transport Vehicle) shown handling *Nekton Alpha*.

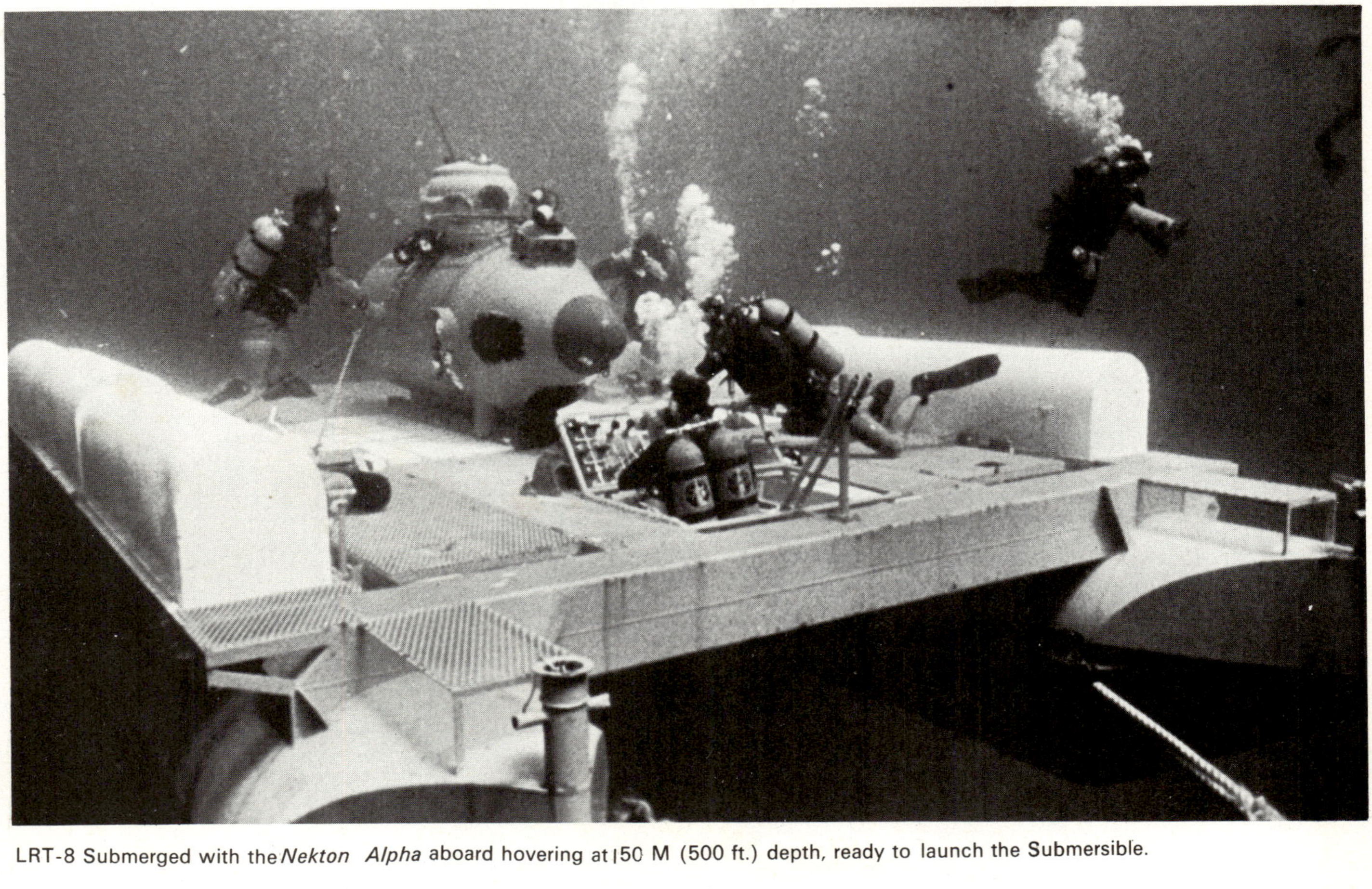

LRT-8 Submerged with the *Nekton Alpha* aboard hovering at 150 M (500 ft.) depth, ready to launch the Submersible.

NAVAL UNDERSEA CENTER (NUC)

SUPPORT SHIP YFNX-30

The support ship used at the Center's San Clemente Island and Long Beach ranges is the converted lighter YFNX-30. It has a flat hull, a length of 33·3 m (110 ft), a beam of 10 m (34 ft), a draught of 1·52 m (5 ft) and a displacement of 420 tons. Its main feature is its two cycloidal propellers with variable-pitch blades. These propellers allow manoeuvering in any direction and are compatible with automated station-keeping. The ship's maximum speeds are $5\frac{1}{2}$ knots ahead or astern, 2 knots athwartships, and it can change its heading at $1\frac{1}{2}$ rpm.

The YFNX-30 is capable of berthing a crew of 16 military and 8 civilian personnel and can support their activities for a month without refurbishing. Its operation is limited to Southern California waters because of its low cruising speed. For assignments in other waters the vehicle is deployed from an appropriate ship of opportunity.

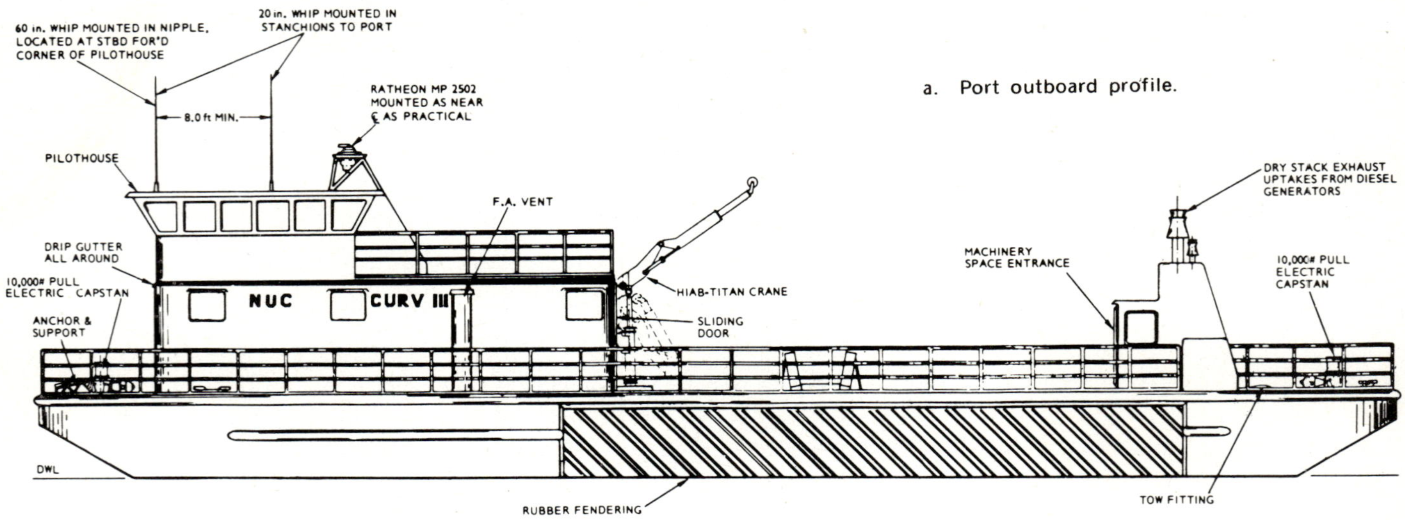

a. Port outboard profile.

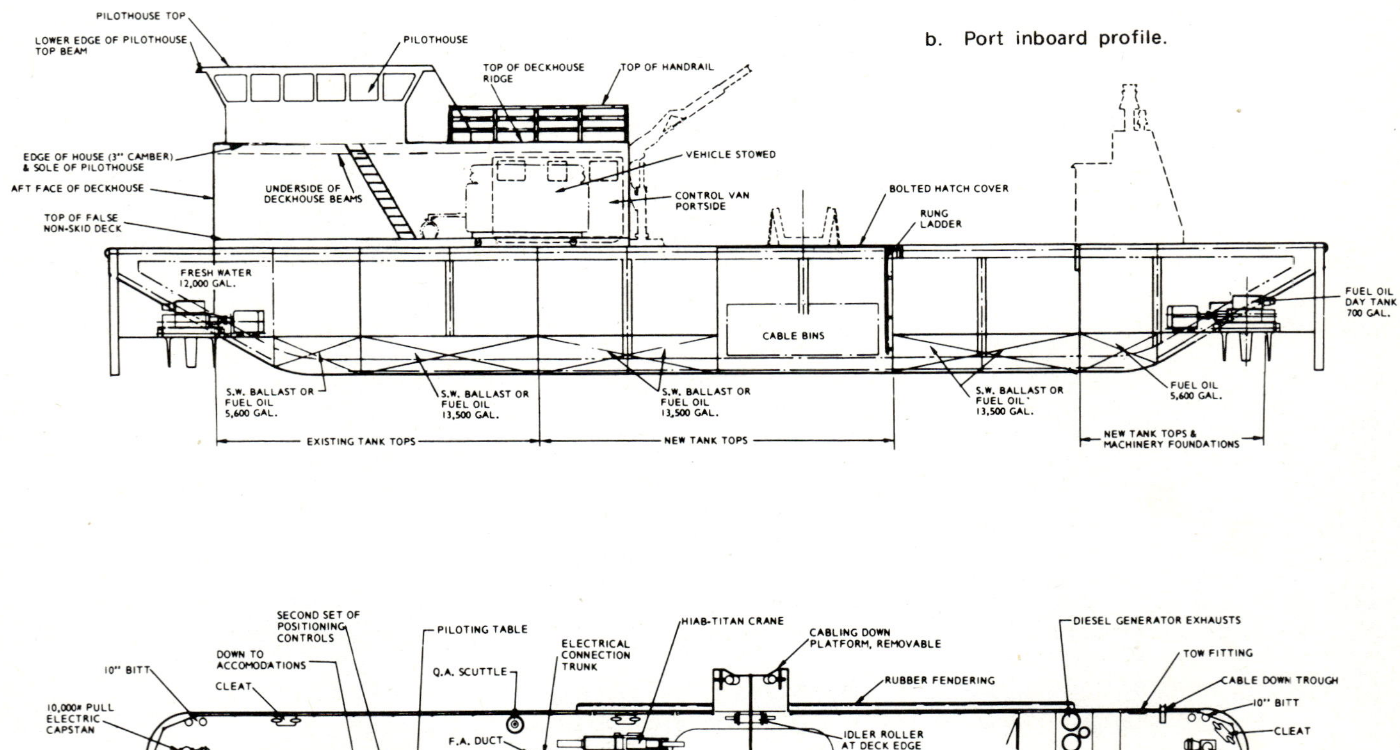

b. Port inboard profile.

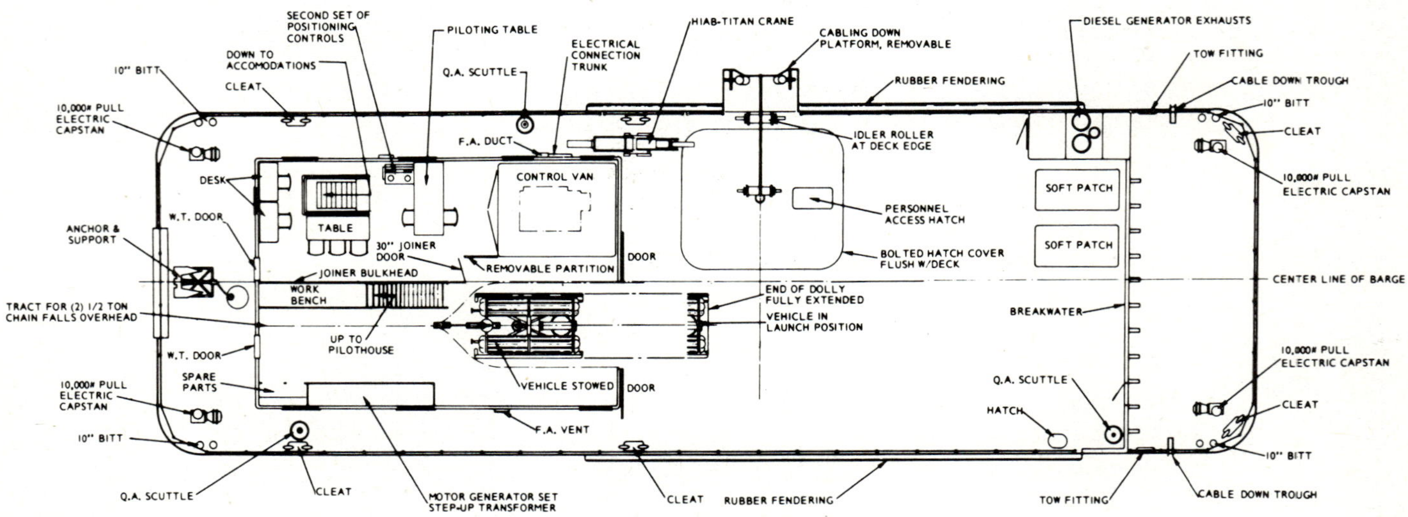

a. Main-deck plan.

The Naval Undersea Centres Submersible Support Ship for CURV III

b. Second-deck plan.

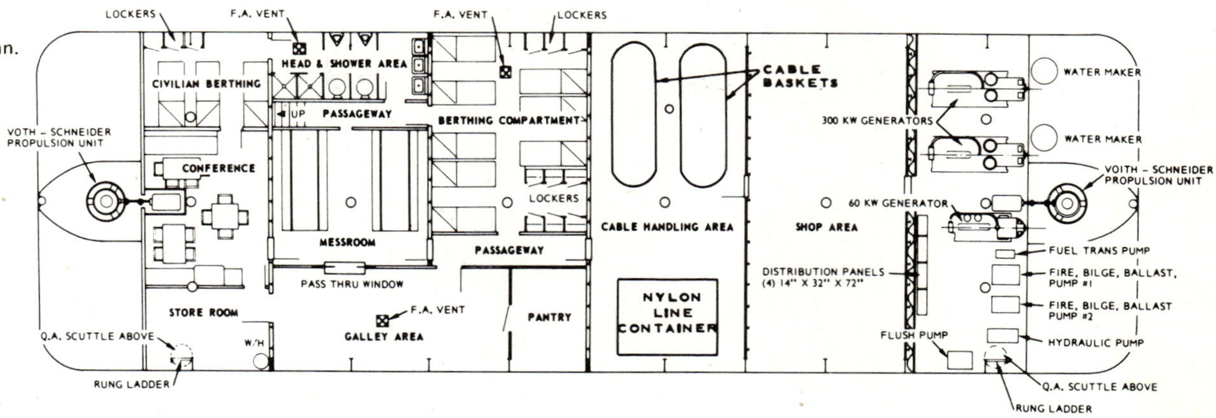

The Naval Undersea Centres Submersible Support Ship for CURV III

WOODS HOLE OCEANOGRAPHIC INSTITUTE

LULU

A catamaran oceanographic research vessel built in 1965 and designed to carry the 25 ton deep submersible *Alvin*.

Length, overall	30·17 m (99 ft)
Beam, overall	13·41 m (44 ft)
Draught	30·5 m (10 ft)
Gross tonnage	290
Speed: maximum	7 knots
cruise	6 knots

Range at cruise	5,000 miles
Max. anchoring depth	30·5 m (100 ft)
Main engine	600 s.h.p diesel, 1,800 rpm
Auxiliary engine	200 s.h.p. diesel, 1,800 rpm
Propellers	3

Equipment

Crane: 1½ ton maximum, 3·05 m (10 ft) above main deck

Winches: BT portable

Trainable sonar

One S-band radar, Loran A, RDF, position fixing associated with deep submersible, one fathometer, one gyrocompass, one 12 kc echo sounder

UNION OF SOVIET SOCIALIST REPUBLICS

ACADEMICIAN N.M. KNIPOVICH POLAR INSTITUTE OF FISHERIES AND OCEANOGRAPHY

ODISSEI

A specially designed 3,800-tonne displacement oceanographic research vessel equipped with a vehicle deployment/recovery arrangement, Langor, workships and laboratories necessary for submersible handling and scientific data evaluation. The submersible *Sever-2* has been operated from this vessel.

Woods Hole Oceanographic Institutions Support Ship *Lulu* shown carrying *Alvin*

UNDERWATER HABITATS

GERMANY

DEUTSCHE BABCOCK AND WILCOX AG

BAH I

An underwater habitat planned and designed in close colaboration with The Helgoland Biological Institute. The cylindrical hull is unpressurised and is designed to be used down to depths of approximately 200 m (660 ft).

Four straddling legs together with removable lead weights support the habitat on the sea bottom.

The inside consists of two compartments, each 2 m (6·6 ft) long; these serve the aquanauts as accommodation and work rooms. The so-called 'wet compartment' has an entry-well at the bottom and a surface entry cover at the top. The lower cover may invariably be left open under water, because the pressure between habitat and water is always balanced. The so-called 'dry compartment' is used as living room, bedroom, and work room.

The ballast has been sized so that there will be only a slight buoyancy. The rest of the total length is comprised of the flooding tanks. When these are flooded, the ground pressure amounts to 6 tons.

Four views of the Babcock and Wilcox BAH 1 habitat

The air supply is autonomous. The gas mixture for breathing is purified in an automatic installation which removes all foreign matter and maintains the humidity at a certain value. An additional deodorizing plant is installed on the WC. The odours are removed biologically by the Babcock biofilter process. Sufficient space is available for technical and scientific equipment. Shower bath, sleeping facilities, table, shelves, heating, WC etc. are provided for the convenience of the crew.

First submergence
20th September, 1968.

Accommodation

Wet rooms	one
Dry rooms	one

Dimensions

Length, overall	6 m (19·75 ft)
Width, overall	5·1 m (16·70 ft)
Height, overall	3·9 m (12·80 ft)

Weights

Weight in air	20 tons
Maximum weight submerged	6 tons
Crew	two

Operational conditions

Normal maximum depth permitted	100 m (328 ft)
Anchorage	one ton
Endurance submerged	12 months
Buoyancy and sinking systems	windlass and flooding tanks

Life support systems
Air supply: regenerative and replenished from external source

Commercial (1973)

Cost	DM 200 000
Delivery	6 months

Notable operations to date

1968 Ostsee	Dr Jurgen Dorschel
1969 Bodensee	Dr Jurgen Dorschel

BAH II

The BAH II habitat is in the design stage and provides considerably more working space for the crew than the BAH I.

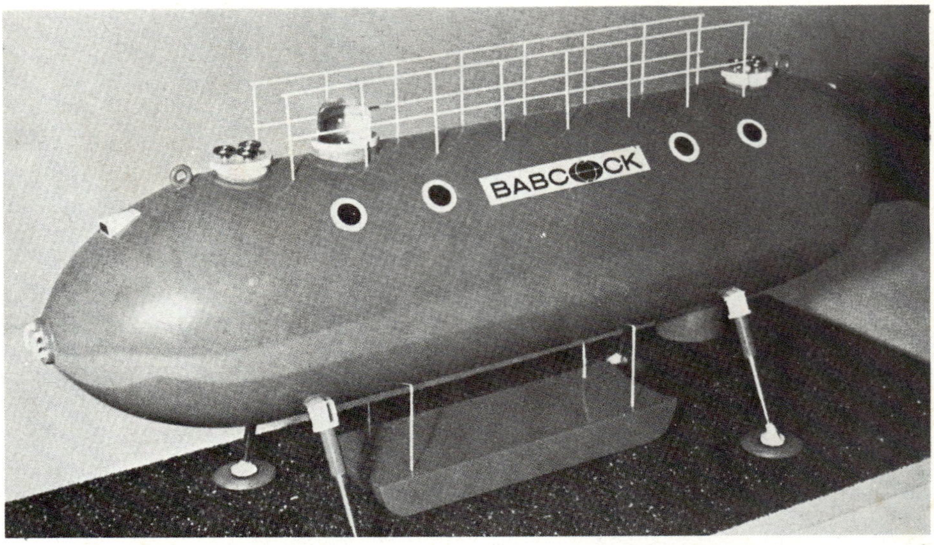

A model of the Babcock BAH II habitat showing the streamlined hull configuration

A cutaway drawing of BAH II

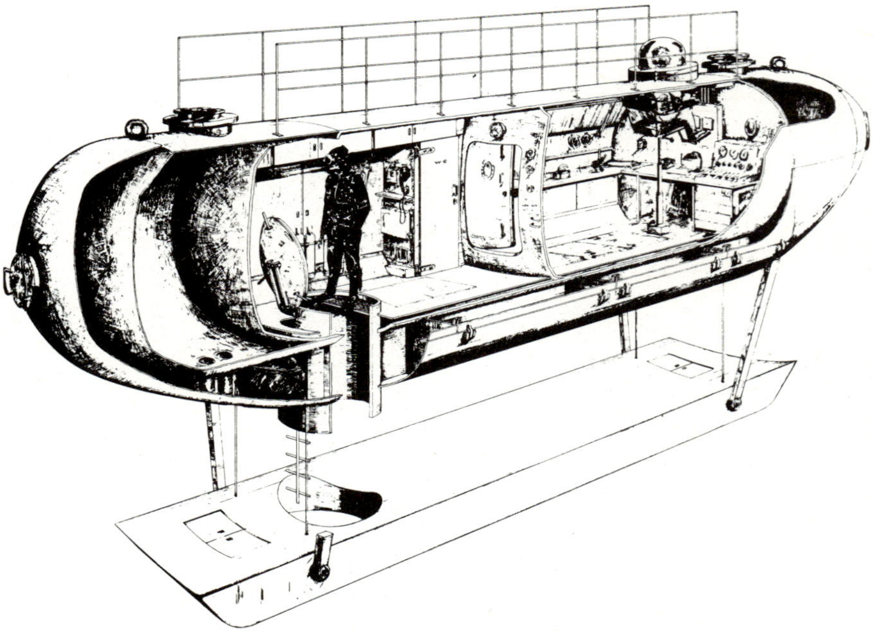

The BAH II habitat. This drawing differs in some details from the illustration above

DRÄGERWERK AG

HELGOLAND

The Underwater Laboratory (UWL) *Helgoland* has been designed for marine-biological research and for medical-psychological tests on men working under water and was positioned on 28 July, 1969 in the North Sea.

Description

The UWL *Helgoland* has a pressure-proof body of cylindrical shape, the length being approximately 9 m (29 ft 6 in) with a diameter of approximately 2·5 m (8 ft 2½ in). The hull stands on four adjustable legs which are fitted with two ballast weights for stability. The pressurisation capacity of the underwater complex is designed for an equivalent water depth of 100 m (328 ft). The total volume amounts to approximately 43 m³ (1,520 ft³). The weight of the UWL including ballast amounts to approximately 60 tons.

The complex consists of two separate compartments, one wet-room and one work-room. The frontal walls of the UWL are fitted with two beds each. The wet-room is fitted with an oval hatch through which the Aquanauts enter the water. This hatch has a length of approximately 2·4 m (7 ft 10 in) and is designed to compensate the water level (as caused by tides).

Both sides of the underwater complex are fitted with ballast tanks which may be flooded and released, thus facilitating swimming, transportation and a comparatively simple change of stationing. This ballast also provides the necessary weight for stability under water. The straddled legs have been fitted with boxes to take up additional ballast for constant level positioning on the seabed.

The power source of the UWL and the breathing gas supply are provided by a supply buoy on the water surface, which is moored adjacent to the Underwater Laboratory. The buoy enables constant communication with the outside by radio and television. An additional cable provides for teletype and telephone communication with the shore control station. The Underwater Laboratory is provided with an independent means of life support and power for an emergency period of two weeks.

Stock replenishment is performed by way of a pressure container (soup bowl). Supplies may be stored in an underwater depot immediately adjacent to the laboratory in order to economise on space in the UWL.

The atmosphere within the UWL is continuously controlled by metering units. Exhaled carbon dioxide is absorbed by canisters, the exhausted oxygen is continuously replenished and the temperature as well as the humidity are controlled automatically. The atmospheric conditions are transmitted to the shore control station by way of television.

It is a special advantage of this project that decompression may be performed within the Underwater Laboratory, and that automatic supply of power and breathing gas is provided from an unmanned buoy.

Numerous safety and rescue facilities have been provided: such as a 'one man' pressure chamber which can be released from the UWL and subsequently recovered on the surface by a helicopter or boat, which are constantly stationed near ***Helgoland.***

Dragerwerk has made valuable contributions as to the design and construction of the 60 ton Underwater Complex. The wide established experience of this firm in environmental conditioning, life support and diving systems has been the base for the construction of auxiliary equipment such as diving chambers, decompression chamber, rescue chamber and underwater depot (the so-called soup bowl).

Port and starboard views of *Helgoland*, the Dragerwerk underwater laboratory

Dragerwerk's habitat *Helgoland*

In co-operation with other prominent firms Messrs Dragerwerk have solved the different problems which are encountered in the technical equipment of the various fittings, such as a special WC, a hot water shower and a complete electric kitchen (galley) including an emergency breathing system. Prior to final delivery each single item is checked by simulated diving tests.

Together with the UWL Dragerwerk supplies a water 'Igloo' which has been initiated and financed only by this firm, and which is intended as a 'satellite' to the bigger brother' UWL.

Helgoland—general view (*Photograph by Schilling, Lubeck*)

General arrangement drawings of *Helgoland*.

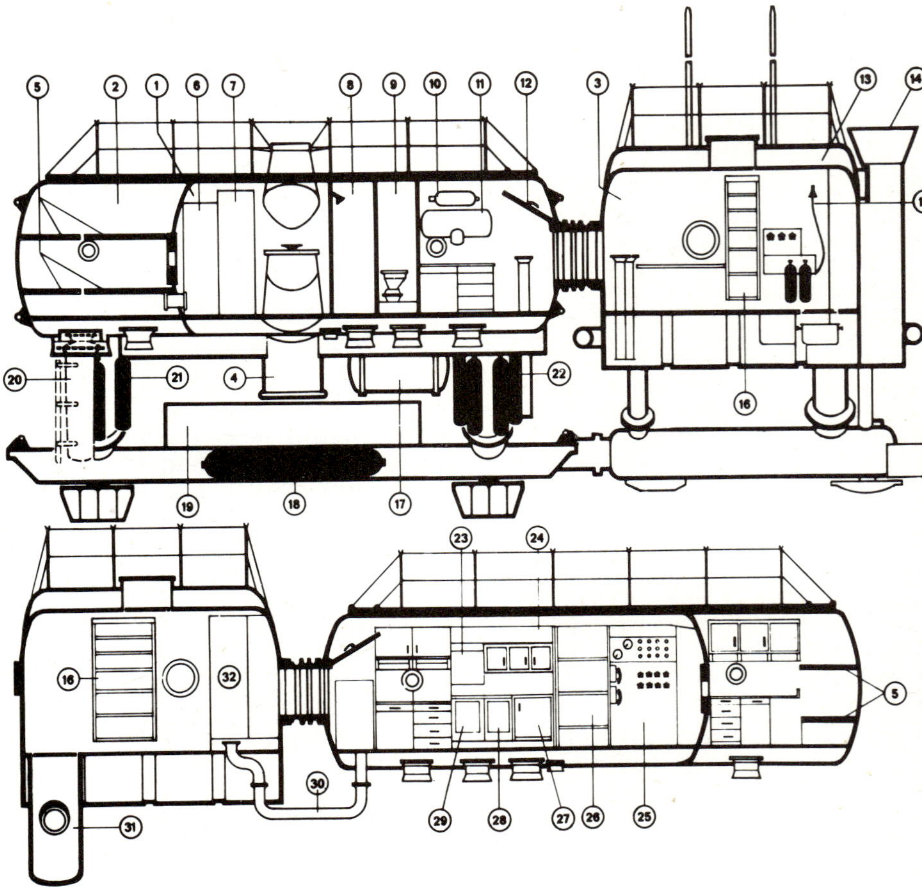

Key:

(1) Wet-room;
(2) decompression chamber;
(3) extension;
(4) exit/entrance shaft;
(5) bunks;
(6) E-switch cupboard;
(7) CO_2 absorption;
(8) shower room;
(9) WC;
(10) gas cylinder storage point;
(11) hot water storage tank;
(12) hatch;
(13) tank;
(14) vent funnel;
(15) shower room;
(16) storage space;
(17) wastage tank;
(18) compressed air bottle;
(19) emergency electricity supply;
(20) one-man survival chamber;
(21) nitrogen bottles;
(22) oxygen bottles;
(23) 'defreezing' room;
(24) drinking water
(25) gas control desk;
(26) measuring equipment;
(27) flushing equipment;
(28) freezer;
(29) refrigerator;
(30) air circulation connection;
(31) observatory chamber;
(32) drying cupboard

The Dragerwerk 'Igloo', use in conjunction with the habitat *Helgoland*

JAPAN

MITSUBISHI HEAVY INDUSTRIES LTD. KOBE SHIPYARD AND ENGINE WORKS

SEATOPIA

The *Seatopia* habitat was ordered by the Japanese Science and Technology Agency through the Association of Underwater Exploration and was delivered on 31 March 1971.

The habitat itself is one of four parts constituting the 'Man in the Sea' project.

(1) Undersea habitat.
(2) Surface support vessel.
(3) Deck decompression chamber (DDC).
(4) Personnel transfer capsule (PTC).

Using this system, four aquanauts may spend a month at a depth of 100 m (328 ft).

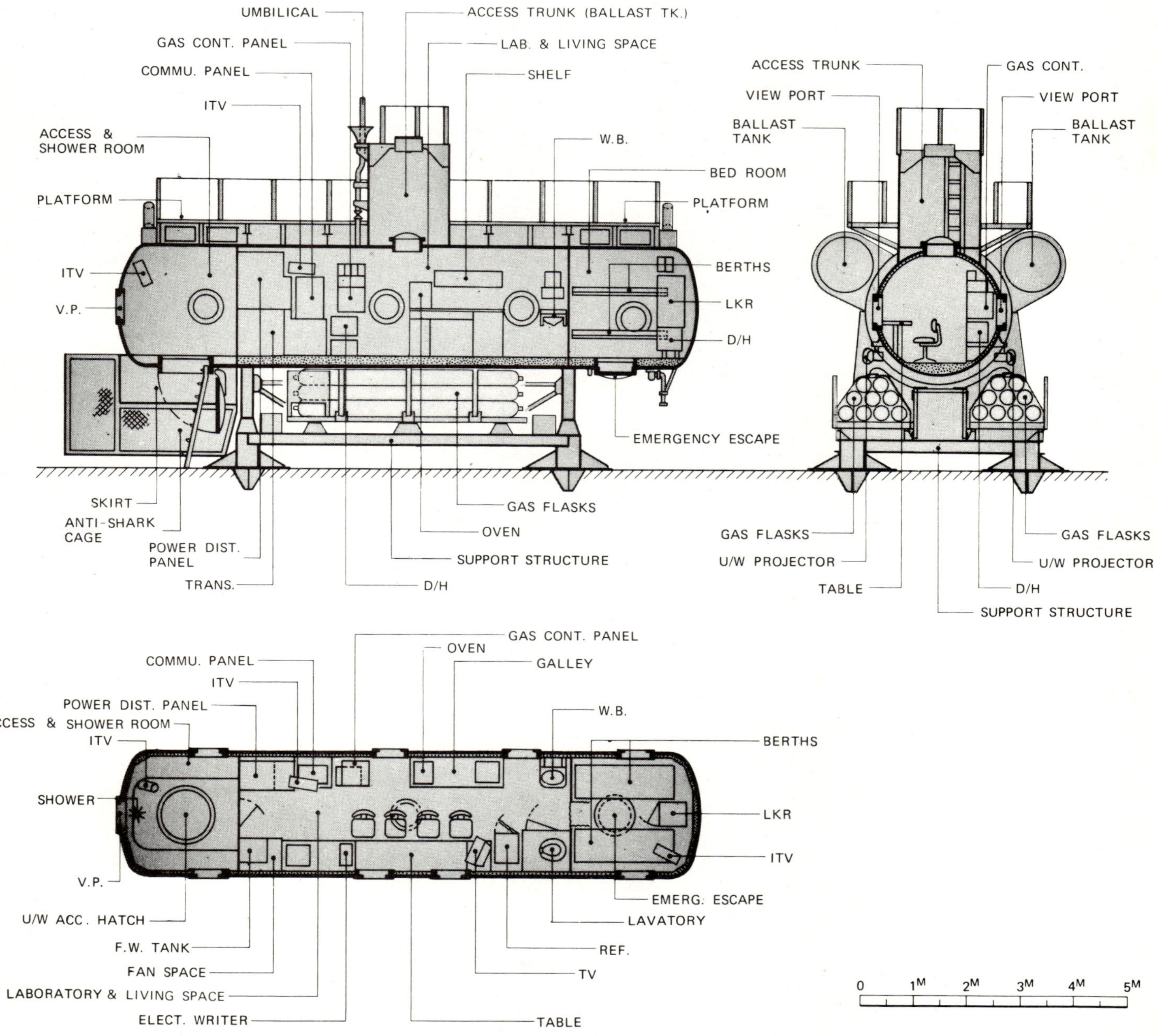

General arrangement diagrams of the Mitsubishi habitat *Seatopia*

Description of the habitat
Surface support

(1) Electric power, water, food and other expendables are supplied from the surface support vessel.

(2) The life in habitat is monitored around the clock from the surface support vessel although the PTC is always at anchor near the habitat for the safety of the aquanauts.

(3) The submersion and recovery of the habitat is carried out after aquanauts' departure from the habitat.

Principal particulars

Length, overall	11·85 m (38 ft 10 in)
Breadth, without fender	4·70 m (15 ft 5 in)
Height	6·53 m (21 ft 5 in)
Pressure hull:	
Resisting to internal pressure	11·1 kg/cm² g (160 lb/in²)
Diameter outside	2·30 m (7 ft 6½ in)
Length	10.85m (35 ft 7 ins)
Weight in air	about 65 tons
Occupants	4 aquanauts
Maximum depth to be placed	100 m (328 ft)
Maximum duration	30 days

Equipment
All instruments to be used in the habitat are made and checked with special care so that they are able to function under the ambient operational environment (hyperbaric helium rich environment).

The Mitsubishi underwater habitat *Seatopia*

Seatopia undergoing a submergence test

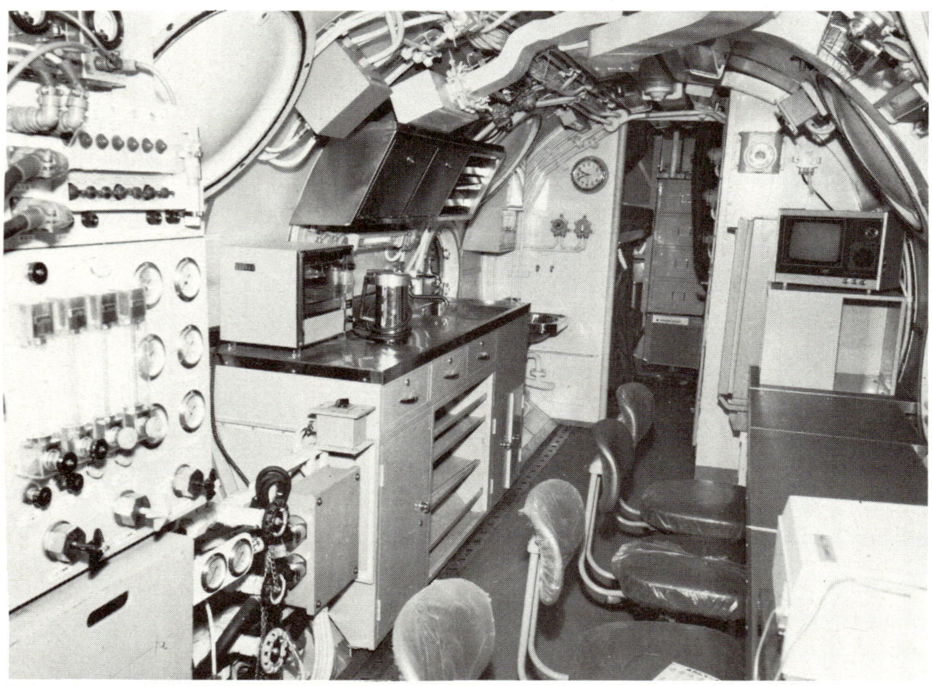

Laboratory and living space aboard *Seatopia*

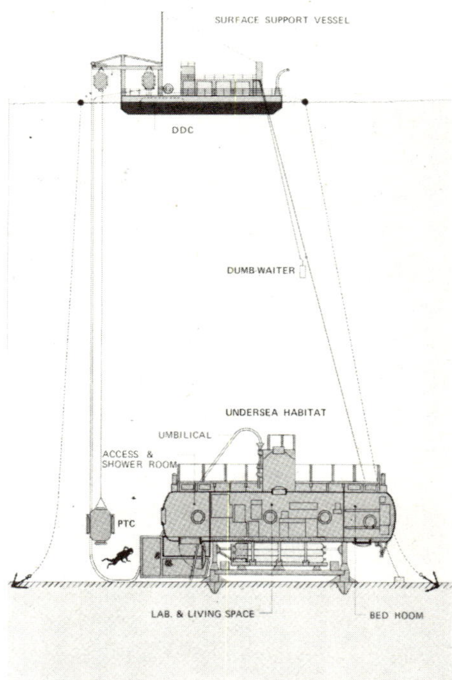

The total system of the 'Man in the Sea' project, based on the *Seatopia* habitat

(1) Atmosphere control and treatment.
(a) Ventilation and air purification.
There is a circulation fan (4 m^3/min × 140 mmAq) (141 ft^3/min × 5·5 in Wg) to uniformly distribute air. For the purification of the air, a CO_2 scrubber (LiOH cannister) and charcoal filter are mounted on the suction side of the circulation fan.

(b) Insulation and heating.
The interior of the pressure hull is lined with cork board of 50 mm (2 in) in thickness and electric floor heaters (0·9 kW × 3) and radiant heaters (0·8 kW × 6) are provided to maintain the temperature inside the habitat at 30°C when the temperature of the seawater is 10°C.

(c) Dehumidification.
There are two dehumidifiers of equal capacity (350 cc/hour) to maintain the relative humidity inside the habitat below 70%.

(d) Gas control system.
There is gas control system to maintain the partial pressure of oxygen and the total pressure inside the habitat constant. This procedure is carried out by releasing gas into the habitat from gas flasks. The gases are discharged into the habitat through the automatic action of the electromagnetic valves coupled with both PO_2 sensor inside the habitat and the water level indicator in the skirt.

(2) Electric power.
All the electric power for use in the habitat is supplied from the surface support vessel. The electrical rating are three-phase, 440 volts and 60 Hz. The maximum power available is 40 kVA.

(3) Illumination.
There are fluorescent lamps (20W × 12, 10W × 5) for interior lighting and pressure resisting and water tight metal halide lamps (400W × 13) for exterior lighting.

(4) Communication.
Cable telephone with helium speech unscrambler, electrowriter and supersonic underwater telephone for communication between the habitat and the support vessel are provided.

(5) Monitoring.
The aquanauts inside the habitat are always monitored by four ITV cameras.
Atmospheric gases are monitored by telemetering the values on the PO_2 sensor and by continuous gas-chromatographic analyses on samples taken through the gas sampling hose.

Access and shower room of *Seatopia*, showing TV camera, door over central view-port and shower above

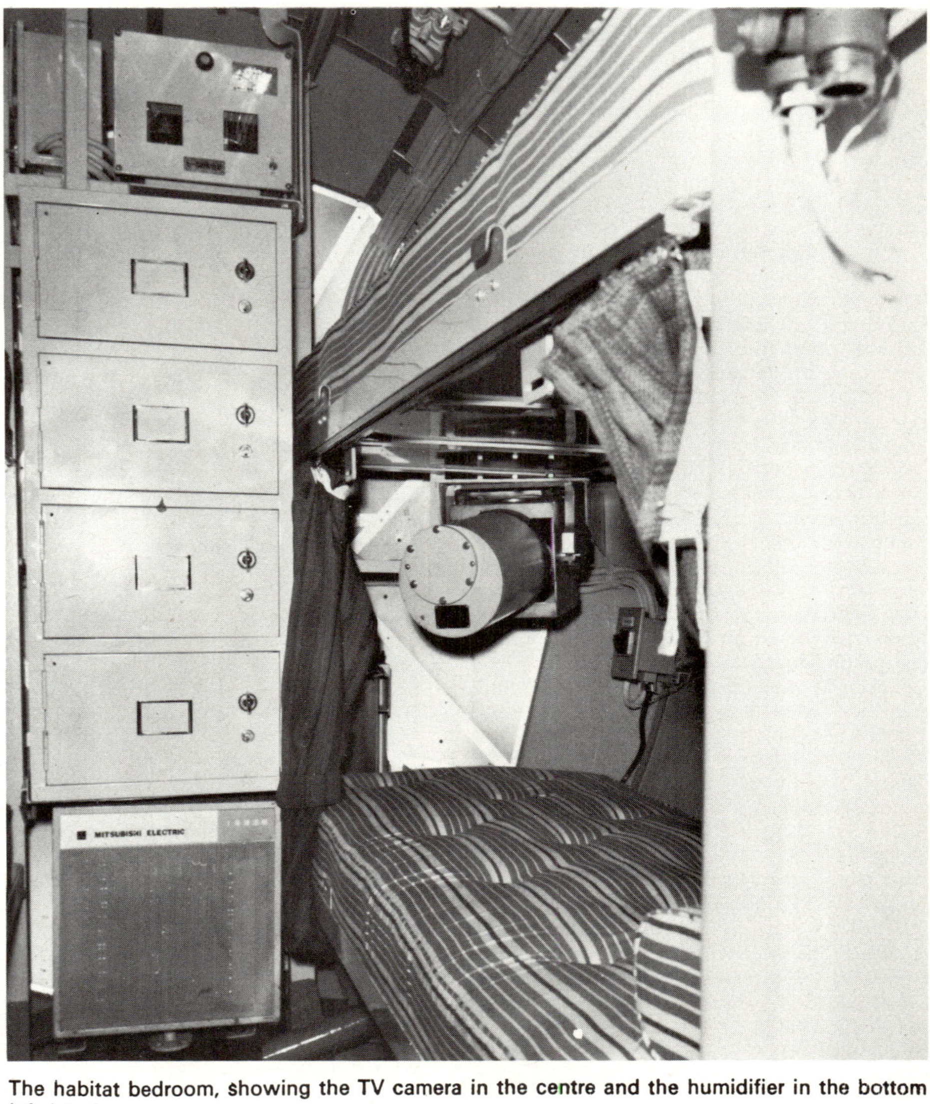

The habitat bedroom, showing the TV camera in the centre and the humidifier in the bottom left-hand corner

Laboratory and living space aboard *Seatopia*

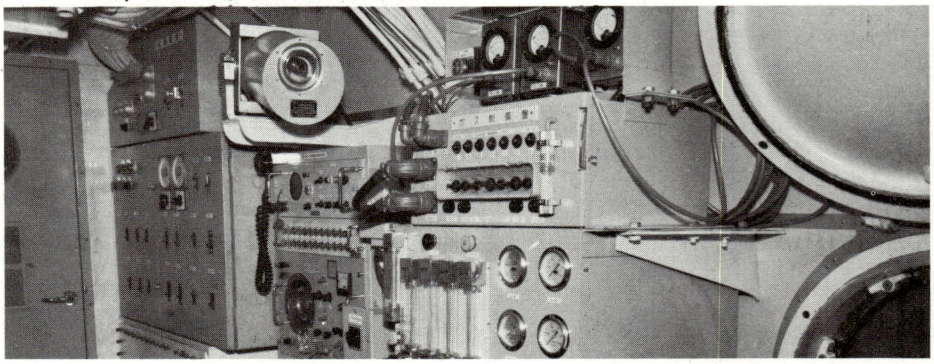

A TV monitoring camera can be seen perched above some of *Seatopia*'s laboratory equipment

Operations

The habitat is towed afloat to the area where it is to be submerged. The ballast tanks of the habitat are filled with sea water, and the habitat, still unmanned, is lowered (by a crane installed on the surface support vessel) to the bottom of the sea.

The aquanauts enter the DDC in the surface support vessel, where pressure is gradually increased to a pressure equal to the hydrostatic pressure at the specified depth.

The aquanauts, after pressurisation in the DDC, transfer to the PTC which is kept at the same pressure level, and descend to the bottom of the sea.

The aquanauts then leave the PTC and enter the habitat, now set in position on the bottom of the sea.

The aquanauts live in the habitat and can freely go out of the habitat to carry out specific tasks. Electric power, fresh water and other goods are supplied from the surface support vessel, from where the situation in the habitat can be monitored. To return to the surface the procedures mentioned above are reversed.

UNITED STATES OF AMERICA

GENERAL ELECTRIC COMPANY

TEKTITE

A habitat designed to support underwater scientific studies requiring long-term habitation.

The habitat, designed and built by the General Electric Company, consists of two vertical cylinders (3·81 m (12 ft 6 in) in diameter and 5·64 m (18 ft 6 in) high mounted on a rectangular base. Each cylinder contains an upper and a lower compartment, which provide a total of four rooms within the habitat. The lower compartment of one cylinder contains bunks and a small galley. The upper compartment serves primarily as the bridge or communications centre but also houses atmospheric monitoring equipment, alarm devices, and a complex array of equipment required for biomedical measurements. It also contains a modicum of space for compiling data and for microscopy. A tunnel 1·37 m (4 ft 6 in) in diameter connects the bridge with the upper compartment of the second cylinder, which is largely occupied by the air-conditioning and air-purification system. The lower compartment of the second cylinder provides storage space for scuba and research equipment and provides counter space for preparing and analysing biological and geological specimens. This compartment contains the entry trunk through which the team enters the water on daily sorties into the sea. Each compartment contains hemispheric ports that allow broad, undistorted visibility of the undersea area adjacent to the habitat. A cupola on top of one of the cylinders provides 360° visibility.

During the project, umbilical lines from a nearby barge supplied air, water, and electricity to the habitat. Air was added as required to maintain a constant level of oxygen and air pressures. The breathing mixture was 92 percent nitrogen and 8 percent oxygen; this mixture, at the ambient pressure of the experiment, closely approximates the partial pressure of oxygen at atmospheric pressure. Carbon dioxide was removed within the habitat by chemical absorption with barium hydroxide.

The habitat is owned by the General Electric Co and the scientific programmes have been financed by the United States Department of the Interior, NASA, the National Science Foundation and the US Navy. Two major programmes have been undertaken, Tektite I and Tektite II, as summarised below.

Event	*Tektite I*	*Tektite II*
Location	St John, US V.I.	St John, US V.I.
Depth	15 m (50 feet)	15 m (50 ft)
Crew Size	4	5
Number of Missions	1	10
Programme Duration	58 days	14-20-30 days
Operational Time Frame	Feb. 1-April 1, 1969, 58 days	April 4-Nov. 6, 1970, 215 days
Frequency of Crew Change	0	14-20-30 days
Diving Equipment	Open Circuit Scuba	Open Circuit Scuba Closed Circuit Scuba
Participating Countries	USA	USA, West Germany, France, Australia, Canada
Scope of Science	Ocean Floor Biology Geology Physical Oceanography	Same as Tektite I *Plus* Extensive Oceanographic Surveys from Research Vessels *Plus* NASA Overflight Programme

Description

Accommodation

Wet rooms	1
Dry rooms	3

Dimensions

Length, overall	11·2 m (37 ft)
Width, overall	4·55 m (15 ft)
Height, overall	7·62 m (25 ft)

Internal dimensions of each compartment: 3·81 m dia X 2·42 m high (12 ft 6 in X 8 ft)

Weights

Weight in air: 70 tons plus 70 tons ballast

Maximum weight submerged: 9,100 kg (20,000 lb) negative buoyancy

Crew 4-5

Construction

Principles: ASME Code, Vertical Cylinders with Dished Heads

Materials: 1·77 cm (0·5 in) Steel SA 285C

Operational conditions

Maximum depth permitted: 300 m (100 ft)

Anchorage method and forces: 9,100 kg (20,000 lb) negative buoyancy

Endurance submerged: Limited only by surface support and crew endurance. 1968 mission endurance was 4 men—60 days.

Buoyancy and sinking systems: Winch down to bottom anchors while 2,220 kg (5,000 lb) buoyant then flood ballast tanks to achieve 9,100 kg (20,000 lb) negative buoyancy

Life support systems

Air: surface supply, habitat air treatment. Built-in emergency air supply and bottle charging storage

Fresh water: surface supply, internal water heater

Self-contained surface-powered electrical humidity and temperature control.

Lighting: self-contained surface-powered incorporating lamps

Monitoring: surface and self-contained read-outs and alarms

Machinery

Generators: 120/240/480 V power distribution

High pressure air storage.

External lighting: optional

Laboratory facilities

Aural and visual communications between compartments and with surface

Wet work compartment equipped with counters, sink, shower, and storage

Dry work compartment equipped with counters and storage

Living compartment equipped with bunks, stove, refrigerator, storage, radio, tape player and TV

0·61 m (2 ft) dia hemispherical windows for external observation

Shark cage

Internal and external TV

Surface support and equipment

Normal operation is manned surface support from shore or ship. Umbilical connections carry power, air, hi-pressure air, water communications

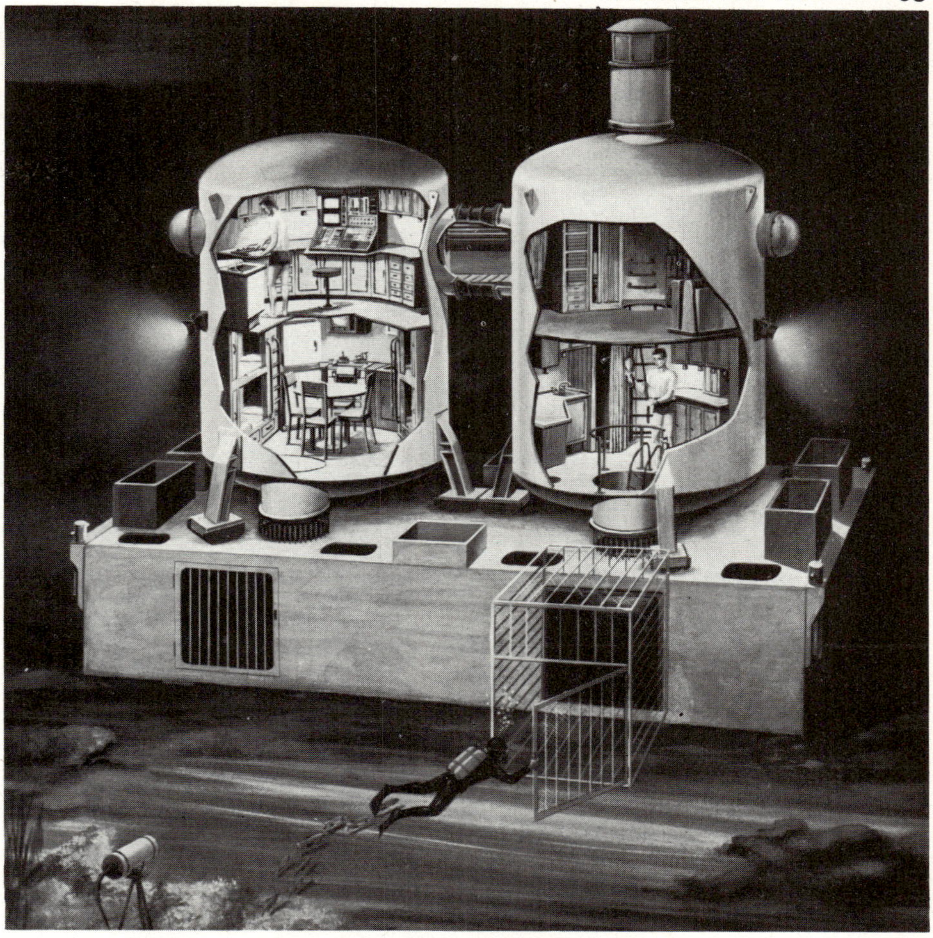

An artist's impression of *Tektite*, the underwater habitat produced by General Electric for scientific research under saturated diving conditions

Safety

Emergency life support: Built-in-breathing system, self-contained 72 hour air supply

Escape facilities: Normally open hatch plus normally closed emergency hatch in opposite cylinder

Function failure warning: Visual and audible alarms on all primary systems

Decompression requirements: 20 hours from 12·5 m (42 ft) saturation

Operations

In 1968 a 4-man, 60-day mission was funded by the U.S. Department of the Interior, NASA, NSF and the U.S. Navy, to study behaviour, physiology and scientific usefulness at a 15·0 m (50 ft) N^2-O^2 saturation. In 1969 a total of 53 aquanauts were saturated in Tektite at a 15·0 m (50 ft) depth over a 9 month program of intensive marine scientific study funded by the same agencies

MAKAI OCEAN ENGINEERING COMPANY (operators)

AEGIR

A mobile saturation diving complex, co-inventors being Mr Gosta Fahlman and Mr Henry M. Horn.

Details

Displacement	200 l tons
Length	21·3 m (70 ft)
Draught (when on surface)	2·44 m (8 ft)
Accommodation	6 aquanauts
Life support	14 days plus 7 days reserve
Design maximum operating depth	183 m (600 ft)

In June 1970, *Aegir* performed a record 159 m (520 ft) dive for 5 days off Hawaii

Aegir is to be used for a dry saturation dive by the University of Hawaii in 1974

***Aegir* at the pier in Hawaii**

***Aegir* beginning a dive to a depth of 152 m (500 ft) (*Courtesy Makai Ocean Engineering Co*)**

OCEAN SYSTEMS INC.

UNDERWATER WELDING CHAMBER SYSTEM

Description

The equipment consists of a welded steel trusswork and hydraulic clamping mechanisms which apply axial restraint to the pipeline, often stressed during installation and storm activity, and holds the ends in close alignment while the damaged section is removed and replaced with a 'pup'. An integral welder's habitat provides a 'captured bubble' of inert but respirable gas mixture which prevents weld contamination and insures the welder/divers of the safe and comfortable dry working environment necessary to produce API code quality welds. To date the deepest field operation has been at 67 m (220 ft), however, the TIG welding process and equipment have been satisfactorily tested to 91·5 m (300 ft) and unmanned tests run at 183 m (600 ft). In all cases weld quality was compatible with API standards.

The system is surface supported through an umbilical bundle which supplies, monitors and controls the welders' life support and required services throughout the entire operation.

Once clamped to the pipeline the welding fixture is independent of surface sea state effects. In the event of heavy weather the welders will abandon the welding chamber, leaving it firmly in place on the pipeline, disconnect and buoy off the service umbilical. When favourable weather returns the umbilical is reconnected and operations are resumed.

Specifications

Pipe sizes accommodated: 15 cm to 61 cm (6 in to 24 in)

Clamping system capability = $5{\cdot}88 \times 10^6$ cm kg ($5{\cdot}1 \times 10^6$ in lb) Moment—any plane.

Pipe alignment (bending) capabilities = $5{\cdot}88 \times 10^6$ cm kg ($5{\cdot}1 \times 10^6$ in lb) Moment—Horizontal; $4{\cdot}85 \times 10^6$ cm kg ($4{\cdot}2 \times 10^6$ in lb) Moment—Vertical.

Maximum length of repair pup joint = 2·42 m (8 ft)

Depth: limited only by welding process

Dimensions

Overall length	7·62 m (25 ft)
Overall beam	3·65 m (12 ft)
Overall height	2·42 m (8 ft)
Working Chamber	
Length	2·74 m (9 ft)
Width	2·42 m (8 ft)
Height	2·12 m (7 ft)

Weights

Dry	17,181 kg (37,798 lb)
Excursion	4,757 kg (10,466 lb) (negative)
Working on bottom	597 kg (1,313 lb) (negative)

Crew Two

Stability

BG = 9·53 cm (3·75 in) (worst case)

Clamping and alignment mechanism

The pipe is gripped by a pair of opposing clamping shoes at either end of the chamber. Each shoe is actuated by two 209 kg/cm², 15·2 cm (3,000 lb/in², 6 in) bore hydraulic cylinders. A hydraulic control circuit superimposed over the clamping circuit provides pipe fit-up alignment in the horizontal plane. Vertical alignment is provided by a 209 kg/cm², 12·7 cm (3,000 lb/in², 5 in) bore hydraulic cylinder at either end of the chamber, each connected to the pipe by a separate wrap-around shoe. Positioning is completely controlled by manual hydraulic control valves situated inside the working chamber.

Life support

Two-way electrical communications with surface

Four open-circuit surface-supplied breathing masks. Chamber atmosphere and diving masks serve as backup

Five 100-watt, 30 volt lights

Oxygen levels of breathing and chamber gas continuously monitored and controlled topside

When necessary the chamber atmosphere is maintained at oxygen content which will support life but not combustion

Weight and buoyancy control

Two 'soft' buoyancy tanks are situated on either side of the working chamber and can be blown down with air or flooded by means of manual control valves inside the chamber. Separate piping systems are provided so that the chamber may be tilted about the pipe if necessary. Each tank has a liquid level gauge, situated inside the working chamber.

Control module

Operations are controlled and monitored from a topside control module which contains communications, instrumentation, and gas and electrical controls.

Services

The hose bundle and control module provide the following services:

- Breathing Gas
- Welding Power
- Electrical Power (lights)
- Low Pressure Air
- Welding Shield Gas
- Hydraulic Power for chamber operation and tools
- Communications
- Closed Circuit Television (optional)
- Analysis Gas
- Inert Chamber Dilutent Gas

Safety

Emergency life support: Emergency gas cylinders and CO_2 scrubber system

Emergency surfacing: Surface via submersible decompression chamber

Escape facilities: Escape trunk with mud removal system provided

Function failure warning: Monitored from Surface

Decompression requirements: Function of depth and time under pressure. Must be used with separate decompression facilities

Miscellaneous working chamber features

Entry trunk (for access through mud)

Pressure—resisting dry compartment (tool storage)

Tool racks at either end of chamber

Folding deck grating

Air and hydraulic supply connections for tools

Auxiliary equipment

2 Hydraulic supply units
2 Hydraulic grinders
2 Hydraulic chippers
2 Hydraulic pipe saws
2 Hydraulic impact wrenches
2 Air chipping hammers
2 Needle scalers
1 Air lift template
3 Volume tanks
10 No. 10 air whips
4 Flast lights
5 Radios
8 Short high pressure whips
8 Long high pressure whips
2 Jet pumps
91·5 m (300 ft) Jet hose
2 Jet nozzles
23 kg (50 lb) Topside welding rods
4 02 Regulators
1 500 amp knife switch
1 Topside oxygen and acetylene cutting outfit
10 Bottles argon
2 Beckman 0-2 analyzers
2 N-2 manifolds
1 Welding power supply

Sufficient N_2 or HEO_2 supply to support the operation

Commercial use

Cost (1973)	$400,000
Delivery	1 month (Currently manufactured); 4 months (New construction)

Available for charter.

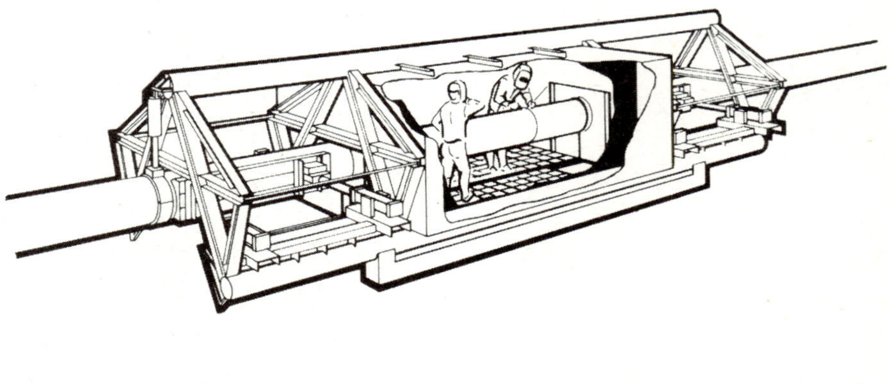

Drawing shows the Ocean Systems underwater welding chamber in operation

PERRY OCEANOGRAPHICS INC.

HYDRO-LAB

Hydro-Lab was initially put into service off Palm Beach in 1966, after being built for Florida Atlantic University as a contribution from the inventor, John H. Perry, Jr.

Except for brief periods ashore for refitting and modification, including installation of an acrylic window 1·2 m (48 in) in diameter, the lab has been submerged continuously, amassing, it is believed, more sea-bottom time than any other underwater laboratory in the world. Hydro-Lab is located in 15 m (50 ft) of water approximately 1·9 km (1·2 miles) off Bell Channel Inlet, Lucaya, Grand Bahama Island.

The habitat has been used for a number of experiments, for instance, dry transfers of personnel from lockout submersibles, at both ambient and one atmosphere pressures, and operation of an Apollo-type fuel cell to supply all electrical needs during a 48-hour 'live-in' by four aquanaut-scientists.

The Hydro-Lab activity is a long-range scientific and educational programme centred around the lab, which operates at either ambient or surface pressure. Hydro-Lab is available to individuals and organizations for underwater research and educational dives. Most of the saturation dives in the last two years have been done by marine scientists funded by the National Oceanic and Atmospheric Administration (NOAA). The Russell Foundation also has given considerable support.

The habitat is 4·87 m (16 ft) long and 2·42 m (8 ft) in diameter and can house 3 to 4 persons for up to one week. The lab is supported on a non-profit basis by the Perry Foundation, Inc. The programme also has the co-operation of the International Underwater Explorers Society, headquartered in Lucaya.

Most scientific teams using Hydro-Lab stay saturated for 5 to 7 days.

The Hydro-Lab site is sand and coral bottom, immediately adjacent to the outer reef off the southern coast of the island and about 242 m (800 ft) inshore of a deep drop-off.

The site was selected after an extensive baseline survey, which included establishing a grid pattern on the bottom for mapping purposes and an analysis of the benthic flora and fauna. The survey was conducted by scientists of the Rosentiel School of Marine and Atmospheric Sciences, University of Miami, and the Perry Foundation.

The life-support system maintaining the habitat from the surface is built into a 7 m (23-ft) fibreglass boat hull. This boat contains a 16·5 h.p. Lister diesel generator (7·5 kw AC), air compressor 13·8 kg cm^2 (200 lb/in^2), and 250 (US) gallon fuel and water tanks. The unit is completely self-sufficient, eliminating the need for a manned support vessel overhead. This unit maintains internal and external breathing systems, supplies fresh water and generates electrical power to the lab. Life support is designed to operate more than seven days without maintenance, and an emergency system, within the habitat, will support three persons for several days.

A trunk with intermediate hatches allows diver lock-outs should the habitat be at atmospheric pressure. Decompression takes place and is maintained from within the habitat itself. Interior accessories include bunks, showerhead, dehumidifier, air conditioning, tables, AC and D3 lighting and CB radio for shore communications. The main structure contains eight small portholes plus the 1·22 m (4 ft) viewing window.

A captured air stand-up booth large enough for three men, and a plexiglass hemisphere are used outside the habitat as 'talking stations'. There also is a small stand-up ferrocement booth. This structure can be used as a way station for deep excursions from Hydro-Lab.

An aquanaut/scientist photographs reef formations near the Perry *Hydro-Lab* on the ocean bottom off Freeport, Grand Bahama Island

A view of the 1·22 m (4 ft) diameter acrylic window of *Hydro-Lab*

A four-man decompression chamber is part of the programme and is shore-based at the International Underwater Explorers Society headquarters in Lucaya. Lab space is also available to researchers at the Society's headquarters.

Other associated facilities besides a 7 m (23 ft) support boat include two-man wet submersibles, called *Shark Hunter* and *Reef Hunter*; underwater scooters, and pinger acoustic homing equipment. Full face masks with 'hookah' air hoses and communications are available.

Hydro-Lab is available to any recognised researchers. When the habitat is not being used for scientific studies it is available for educational purposes to familiarise individuals, especially students, with the problems of living and working in an underwater habitat. The lab can also be made available to recreational divers.

The Hydro-Lab Underwater Research Program has an advisory committee of prominent scientists and engineers to continuously review and advise on the habitat programme. Dr. Joseph MacInnis and Dr. David Youngblood, widely known diving physicians, are among those available for medical consultation.

Between February 1971, when Hydro-Lab was emplaced off Freeport, and June 1972, 368 persons entered the habitat. Their man hours totalled an estimated 5,544.

Commitments from July 1972 to July 1973 total some 12,961 man hours with 313 persons involved. Most projects will be funded by the Office of Manned Undersea Science and Technology, National Oceanic and Atmospheric Administration (NOAA).

Projects have covered a broad range: biological oceanography, physical oceanography, sediments, currents, acoustics, pollutants, instrumentation. One team, from the University of California, made both EEG brainwave readings and EKG recordings on divers and telemetered the information to recording instruments in the lab.

Missions off Freeport have included: Eight dives by the University of Miami to study ecology of coral reefs; Four dives by the University of Miami to study chemistry of coral reefs; Two dives by Texas A & M to study plankton crops over coral reefs; One dive by Texas A & M to study sound production of reef fishes; One dive by the Marine Fisheries Service to study behaviour of fishes related to fish traps; One dive by Central Michigan University to study degassing of water samples at different pressures; One dive by the US Environmental Protection Agency to study dispersal rates of digested sewage over the ocean bottom; One dive by Marine Resource Consultants to study the behaviour of oil on the ocean bottom; One dive by the American Institute of Research to study the behaviour of management people under stress conditions (non-saturation); One dive by the University of Alabama for training; One dive by the American Museum of Natural History to study changes in behaviour of reef fishes related to time; and two dives by Fairhaven College, State of Washington, to study cosmic rays coming from space (non-saturation).

PERRY OCEANOGRAPHICS INC.

LA CHALUPA

La Chalupa is a new type of undersea laboratory that can readily be towed from site to site and which includes an underwater 'garage' for accommodating a research submarine. The laboratory is in operation off the coast of Puerto Rico as an international undersea laboratory and was built for the Marine Resources Development Foundation of Puerto Rico by Perry Oceanographics, Inc., Riviera Beach, Florida, which designed

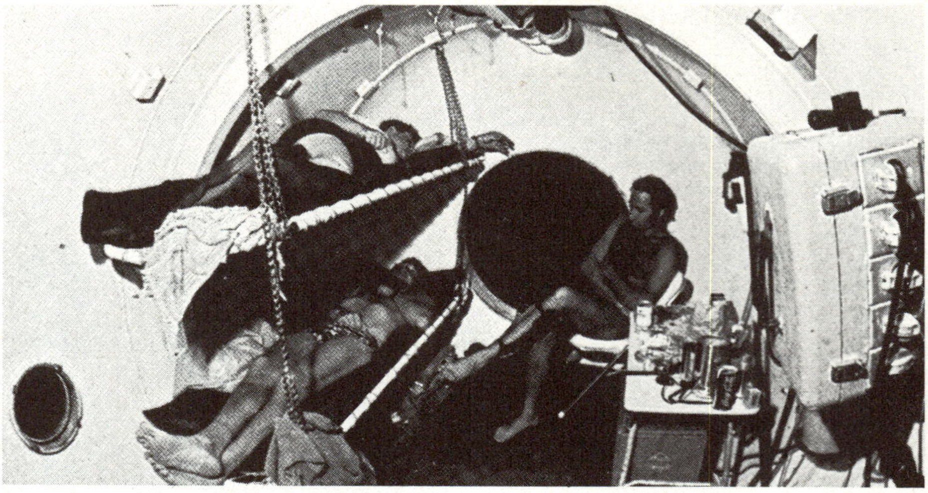

The interior accommodation of *Hydro-Lab*

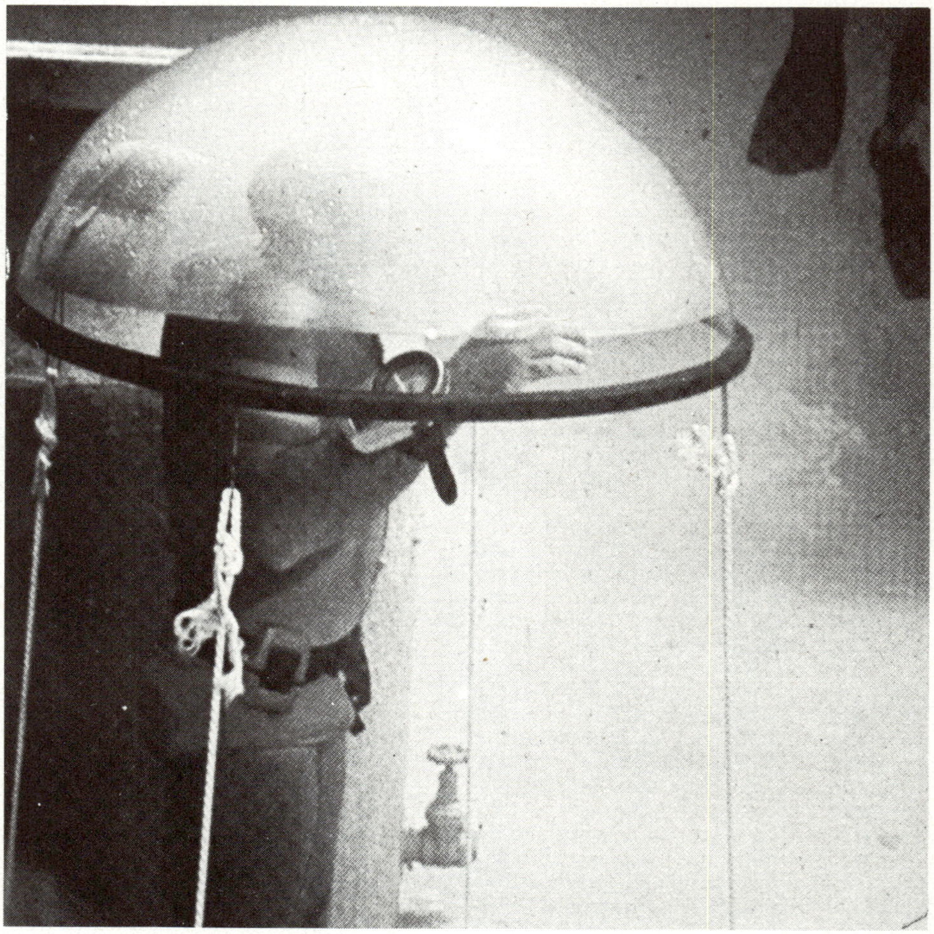

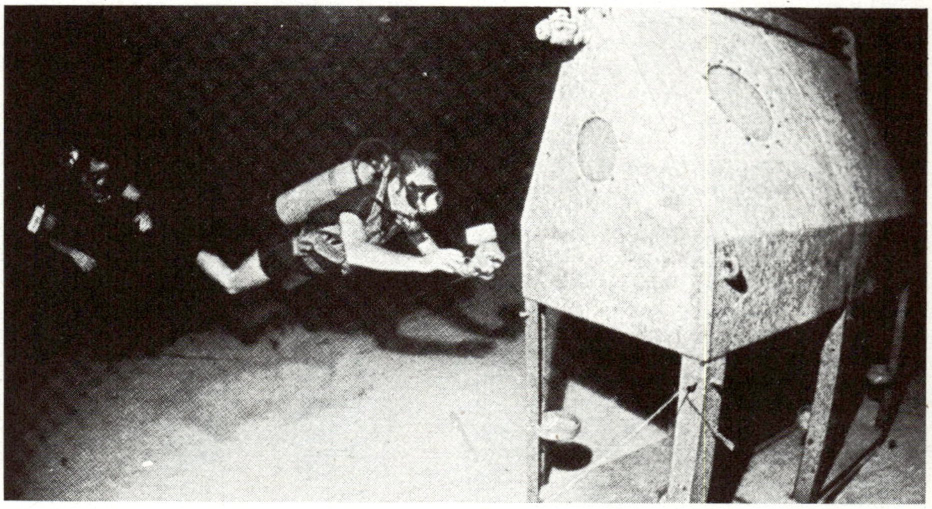

Captured air-booths used outside *Hydro-Lab* as 'talking stations'

and engineered the habitat to the specifications of the Foundation.

The Foundation was formed under leadership of Ian G. Koblick, former special assistant for undersea programmes to the governor of the US Virgin Islands. He participated in Tektite I and Tektite II projects at St John, Virgin Islands.

The new lab's design, Koblick says, resulted from his Tektite experiences and 'represents a complete systems approach to an economical and versatile undersea laboratory.'

The programme is supported by the Commonwealth of Puerto Rico's Undersea Research and Development Corp., a government-funded public corporation. The project's home base is at Punta Ostiones, 26 km (16 miles) from the city of Mayaguez on a 400-acre government tract specifically set aside for marine resources development. The overall project has the name 'PRINUL,' an acronym for Puerto Rico International Undersea Laboratory. A number of 'live-ins' already have been conducted by marine scientists and researchers, some of them funded by the National Oceanic and Atmospheric Administration (NOAA).

Through use of an all-weather 10·8 m (36 ft) life-support boat tethered above the lab., high costs associated with customary landbased support systems have been sharply reduced. A similar life-support boat is currently used with the Perry Hydro-Lab off Freeport, Grand Bahama Island, where the boat supplies air, electrical power and fresh water to the lab and its occupants.

La Chalupa is basically a submersible steel barge, 14·5 m (48 ft) long, 3·25 m (10 ft 8 in) high and with a 6·1 m (20 ft) beam. It has four adjustable steel legs, each 0·61 m (2 ft) in diameter, to enable level positioning on the seabottom. The hydraulically operated legs also provide jack-up capability for cleaning and repairs. Each leg is 3·33 m (11 ft) long and attaches to a swivel 'foot' on the bottom.

Normal operational capacity enables four occupants to live in the habitat for up to two weeks. The barge contains two steel living chambers, each 2·43 m (8 ft) in diameter and 5·8 m (19 ft) long. One chamber is fitted out as a living compartment with two 1·042 m (41 in) windows, four bunks, a marine toilet, wet lab., desk, etc. The living compartment is capable of withstanding 3·5 kg/cm^2 (50 lb/in^2) internal pressure and is used as a decompression chamber when the laboratory is surfaced at the completion of a mission.

The other chamber is capable of withstanding 3·5 kg/cm^2 (50 lb/in^2) external pressure and is a control compartment, with a 1·042 m (41 in) window, a decompression control station, galley, electrical controls, TV monitors, etc. In an emergency this compartment can be used as a submerged decompression chamber.

Both compartments have a transfer lock attached to the upper hatch, permitting transfer of personnel and equipment when the chamber is at pressure other than ambient.

The barge structure also contains negative tanks, high pressure air cylinders, and a trim tank for fine control during surfacing and submerging operations.

A captured air space located between the living and control chambers and open at the bottom is used for dry entrance into the habitat, dry access between living and control compartments, storage space for SCUBA gear, a shower, and docking facilities for diver delivery vehicles such as the Perry Submarine *Reef Hunter*. In addition, it provides more than 9·3 m^2 (100 ft^2) of working space for specimen sorting or equipment handling.

In case of loss or removal of the life support boat, the habitat is equipped with emergency batteries which will operate CO_2 scrubbers and emergency lights for 48 hours.

Perry Oceanographics' *La Chalupa* undersea laboratory

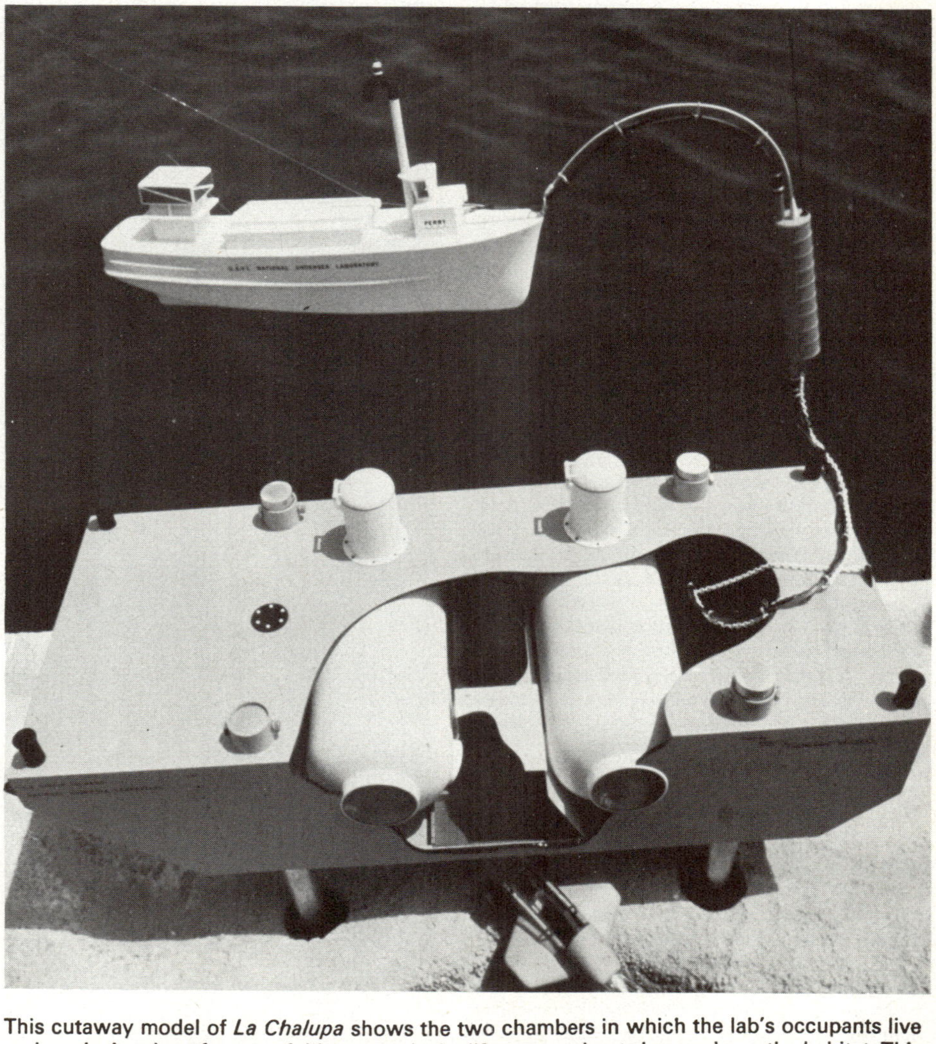

This cutaway model of *La Chalupa* shows the two chambers in which the lab's occupants live and work. A unique feature of this system is the life support boat shown above the habitat. This permanently tethered boat supplies all the electrical, air and water needs of the lab during missions. Shown below the lab is a wet submersible about to enter the lab's 'garage'

During a typical operation, the lab is towed to the operational site in a dry or de-ballasted condition. At this time the barge structure has approximately 1·37 m (4·5 ft) of freeboard. After arriving on site, an anchor is dropped to hold the habitat in position during the implanting operation.

The forward or internal pressure compartment is pressurised a little over bottom pressure, the legs extended, and the control compartment equalised. The barge structure is flooded by opening flood and vent valves.

The actual submergence and descent are controlled by one diver operating the valves. In order to surface, the reverse of this procedure is followed.

'The mobility feature,' Koblick says, 'adds a new dimension to man-in-the-sea efforts by providing the ability to study sea biological and ecological problems when they occur.'

The undersea laboratory *La Chalupa* surfaces during trials off the coast of Florida prior to being shipped to Puerto Rico, where the lab is now operational in 30 m (100 ft) of water near Mayaguez

UNIVERSITY OF NEW HAMPSHIRE ENGINEERING & DESIGN ANALYSIS

EDALHAB

This underwater habitat is a scaled-down version of a previously conceived habitat design called Oscillab which EDAL undertook in 1966.

The project became what is believed to be the first student experiment in underwater living and included the conceptual design of the underwater habitat, followed by detailed engineering of sub-systems, construction and testing of the habitat. A minimal cash budget required a carefully conducted campaign for donated components and materials and careful attention to logistics.

EDALHAB living quarters are housed in a horizontal cylindrical tank 2·44 m (8 ft) in diameter and about 3·66 m (12 ft) long. The habitat is entered through a trunk open to the water. Equipment has included two bunks, a cabinet, a small collapsible table, and folding chairs. Communication to shore is provided by a field telephone and a remote television monitor which keeps the crew under surveillance at all times. A household type electrical supply line, protected by an electronic ground fault interruptor, supplies light and heat. Air is supplied by a commercial compressor and is filtered to remove oil vapour and other contaminants, and is fed through 152 m (500 ft) of hose to which communication and power lines are lashed to make an umbilical cable running from the shore control station to the habitat. Food cooked ashore can be delivered hot in a pressure cooker by divers.

EDALHAB was in the water at Alton Bay, New Hampshire, from April 26 to 29, 1968. During this period the dive crew occupied the habitat for forty-eight hours, including decompression time (eight hours at half the test depth). During the dive period the group had many scuba-visitors, among them Dr Joseph B. MacInnis of Ocean Systems and Frank Scalli of the Boston Sea Rovers.

Specific purpose of design: Education and Research, it was designed and built by students as a student educational project. It is a self-lowering, semi-mobile habitat. Financial support—Office of Naval Research and the University of New Hampshire funded the initial construction. The habitat is owned by the University of New Hampshire.

Accommodation

Wet rooms: Entrance-Exit trunk serves as a wetroom

Dry rooms: One multi-purpose room

EDALHAB's first dive in Florida waters

EDALHAB and the Woods Hole R.V. *Lulu* saturation diving system in Florida waters

A diver/scientist photographing specimens outside EDALHAB

EDALHAB being launched at Alton Bay, Lake Winnipesukee, New Hampshire on 26 April, 1968

Dimensions

Length, overall	7·39 m (24 ft 3 in)
Width, overall	4·92 m (16 ft 1·5 in) (over lifting frame)
Height, overall	3·96 m (13 ft)
Main compartment	2·44 m (8 ft) diameter, 3·66 m (12 ft) long

Weights

Weight in air: 18,640 kg (41,000 lb)

Maximum weight submerged: This is varied by addition of ballast to suit the operating mode

Crew: 273 kg (600 lb)

Operational conditions

Maximum depth permitted: 12·96 m (42·5 ft) on air during saturation dives.

Anchorage method and forces: EDALHAB is positioned over two main railroad wheel-anchor clusters of 2,091 kg (4,600 lb) each. Buoyancy is adjusted to 909 kg (2,000 lb) force on each anchor. Four Danforth anchors, size depending on current conditions, are deployed laterally

Endurance submerged: When supported from shore, duration is practically unlimited. When ship supported duration is dependent on surface conditions

Buoyancy and sinking systems: as for anchorage

Life support systems

Air supply: replenished from external source

Fresh water: from surface supply source

Heat: from surface electric generator

Lighting: from surface

Monitoring: closed circuit TV. Sound powered telephone

External lighting: 12 volt flood lights

Instrumentation

Brought aboard as needed

Laboratory facilities

Small laboratory work bench plus a ventilated hood for chemical studies

Surface support and equipment

Mode of operation: EDALHAB is towed or carried to the site, then set afloat. The main anchors are lowered to the bottom, the buoyancy is adjusted by partially flooding the floatation tanks. EDALHAB is winched to working depth using her own winches

Surface requirement

Requires continuous support from a surface ship or shore station to supply breathing air and electric power, and to carry decompression equipment and the monitoring station

Safety

Emergency life support: Two 300 S.F.C. breathing air tanks. Emergency low voltage lighting. These provisions were deemed ample in view of the shallow operating depth

Emergency surfacing: From present design depth of 13 m (42·5 ft), divers proceed directly to the surface and enter the decompression chamber which is always available

Escape facilities: Entrance trunk which is always open, or the top hatch which can be opened either from the inside or from the outside

Function failure warning: System relies mainly on continuous monitoring of life support parameters by the surface support. The habitat has a trunk water level alarm and an electrical system ground fault interrupter

Decompression requirements: Aquanauts are usually prepared for ascent from the maximum depth of 13 m (42·5 ft) by administration of oxygen. Following this, divers ascend to surface and immediately enter the decompression chamber

EDALHAB is available for charter or sale

Shipping/handling requirements: Edalhab's carrying frame fits the Woods Hole's R/V *Lulu*; without the frame it has been moved over the highway

Operations

Notable operations to date:

(1) April, 1968—Two-day 9 m (30 ft) saturation dive at Lake Winnipesaukee, N.H., four students participating. Operator—Eng. Des. & Analysis Lab. (EDAL)

(2) April, 1971—Four day 12 m (40 ft) saturation dive off the Isles of Shoals, N.H., three students participating. Operators—EDAL, University of New Hampshire and Woods Hole Oceanographic Institute (WHOI)

(3) January to April, 1972—Florida Aquanaut Research Expedition (FLARE); to demonstrate site-to-site mobility of a mobile habitat system for scientific diving. A total of nine missions were conducted off the Florida east coast. EDALHAB was supported by RV *Lulu*. FLARE was sponsored by the National Oceanic and Atmospheric Administration with WHOI personnel in charge of operations and EDAL in technical supervision.

THE UNION OF SOVIET SOCIALIST REPUBLICS

GIPRORYBFLOT INSTITUTE

BENTOS 300

This submarine laboratory was designed by the Giprorybflot Institute of Leningrad. Accommodation is provided for ten research scientists and includes cabins, kitchen, mess room, shower bath and radio. The main working place is the observation compartment which is equipped with a TV unit, still and cine cameras. A lock-out chamber is provided. The laboratory is capable of moving freely in water at depths down to 300 m (984 ft) and the occupants may live in it for up to two weeks before surfacing.

Externally, *Bentos 300* when working in the sea bed, is held by two anchors, and will carry adjustable searchlights and electronic-optical transformers to study objects in the infra-red spectrum.

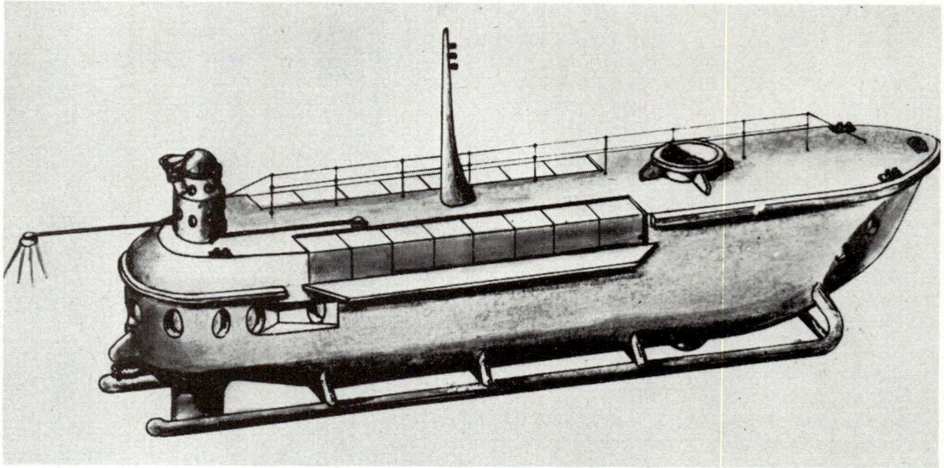

An artist's impression of *Bentos 300* (*Courtesy Novosti Press*)

INSTITUTE OF OCEANOLOGY ACADEMY OF SCIENCES SOUTHERN BRANCH

CHERNOMOR

Chernomor (operational since 1968) takes the form of a horizontal cylinder mounted on keel blocks; ballast in the form of cast iron slugs is stored in special bins in the keel blocks. Tanks for the water ballast are mounted in the upper part of the 2·95 m (9·68 ft) diameter hard hull to make a flat deck area.

Chernomor is about 4·5 m (14·76 ft) high, 3 m (9·84 ft) wide and 8 m (26·24 ft) long and has a submerged displacement of 62 tons. It has a hull designed to act as a decompression chamber for the crew after it has surfaced.

The laboratory's independent sinking and surfacing system places no requirements on the lifting gear of the laboratory's support ships and makes it fairly easy to handle. This is of considerable importance as the laboratory is intended for underwater research over a period of several years in different areas of the sea.

The laboratory is equipped with a surface-linked ventilation system and an independent air recirculating system. It also has electrical draining and sewage disposal systems, showers, hot and cold running fresh water and telephone intercom systems.

The on-board resources, breathing compound (oxygen-75 and nitrogen-150 m^3 (2,640 to 5,280 ft^3) and compressed air for technical purposes (240 m^3 (8,500 ft^3), *Chernomor*, submerged to a depth of 25–30 m (82–98 ft) to be independent of supply ships and coastal services for 15 days, and the emergency electric current supply can last for seven days. Under ordinary conditions, electricity is supplied by cable from the shore or from a ship.

During early experiments, the laboratory's support ship was a 60-ton catamaran pontoon which could supply it with air, DC and AC power and two-way telephone. The pontoon was linked to the shore base by an intercom.

The laboratory is completely independent, with power supply and a store of breathing gas mixtures sufficient for one submerged working cycle. This capability simplifies the procedure of submersion to the seabed and increases the reliability and safety of the experiment in conditions of heavy seas.

Chernomor being lowered into Black Sea waters in Golubaya Bay, near Gelenjik (*Courtesy Novosti Press*)

Soviet divers checking and preparing *Chernomor*'s underwater site (*Courtesy Novosti Press*)

The living accommodation on the laboratory is 7·5 × 2·8 m (18·45 × 9·18 ft) in area and 2 m (6·56 ft) high. The bow section of the compartment contains living quarters with four bunks (two on the port and two on the starboard sides), a folding table, a wall-mounted cupboard and four emergency aqualungs mounted at the head of each bunk. The central control post is situated on the port side, adjacent to the sleeping area and contains three control panels for checking and controlling the laboratory's systems. Next to the central control post there is a work table under which there is a cabinet for drying divers' equipment, a shelf for eight sets of working, self-contained, underwater breathing apparatus, and, finally, a water-closet separated from the living quarters by a water tight bulkhead.

The carbon dioxide scrubbing apparatus is housed under another work table on the starboard side. The diving area, with three entrance hatches, is in the stern on the starboard side. The upper hatch is used only when the laboratory is on the surface. The bottom hatch, which seals off the divers' well, is used by the crew for underwater exit.

The above details are taken from a paper by Pavel Borovikov of the USSR Institute of Oceanography.

A view of *Chernomor* prior to submergence (*Courtesy Novosti Press*)

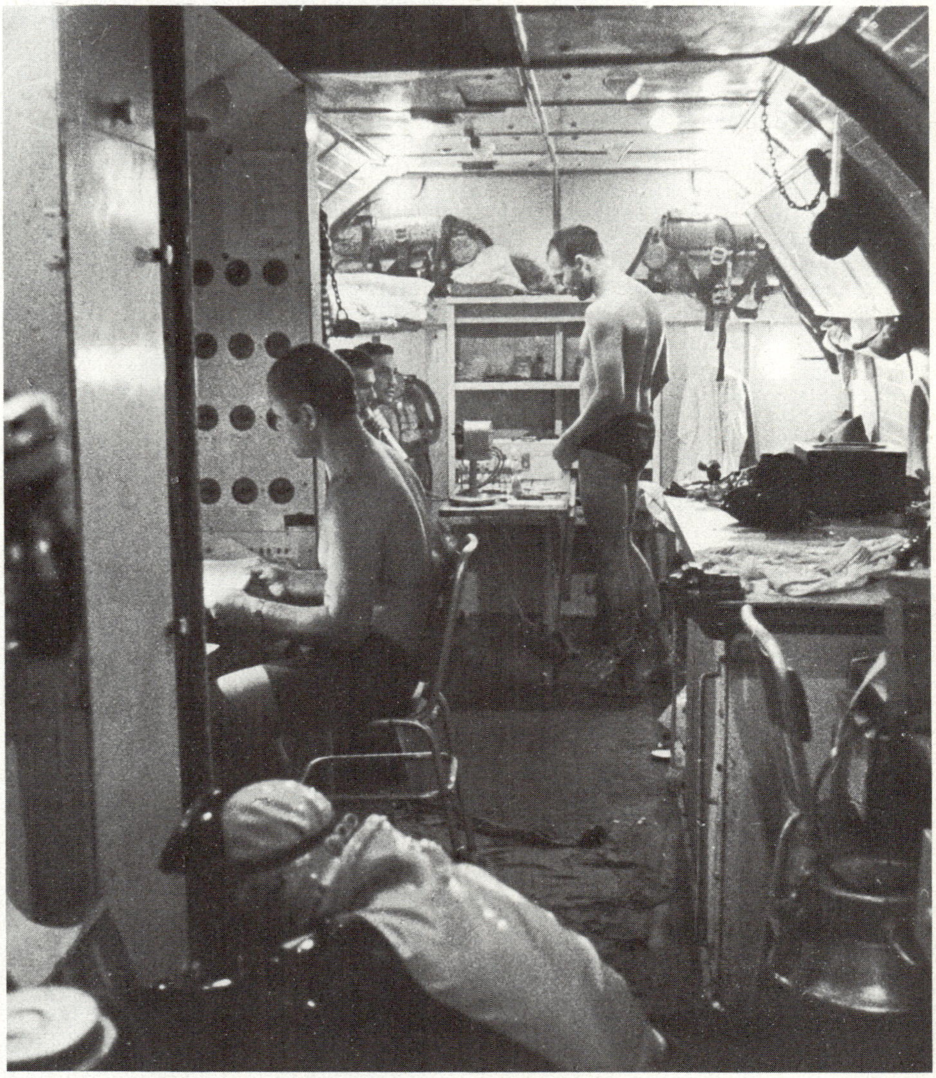

The interior of *Chernomor* (*Courtesy Novosti Press*)

Chernomor before a 14-metre descent
(*Courtesy Novosti Press*)

INSTITUTE OF OCEANOGRAPHY

SPRUT

This laboratory facility has operated in the Black Sea at Sukhumi Harbour, and is used mainly as a divers' underwater resting and communication station.

Sprut was designed at the USSR Institute of Oceanography and is believed to have been in operation since 1970.

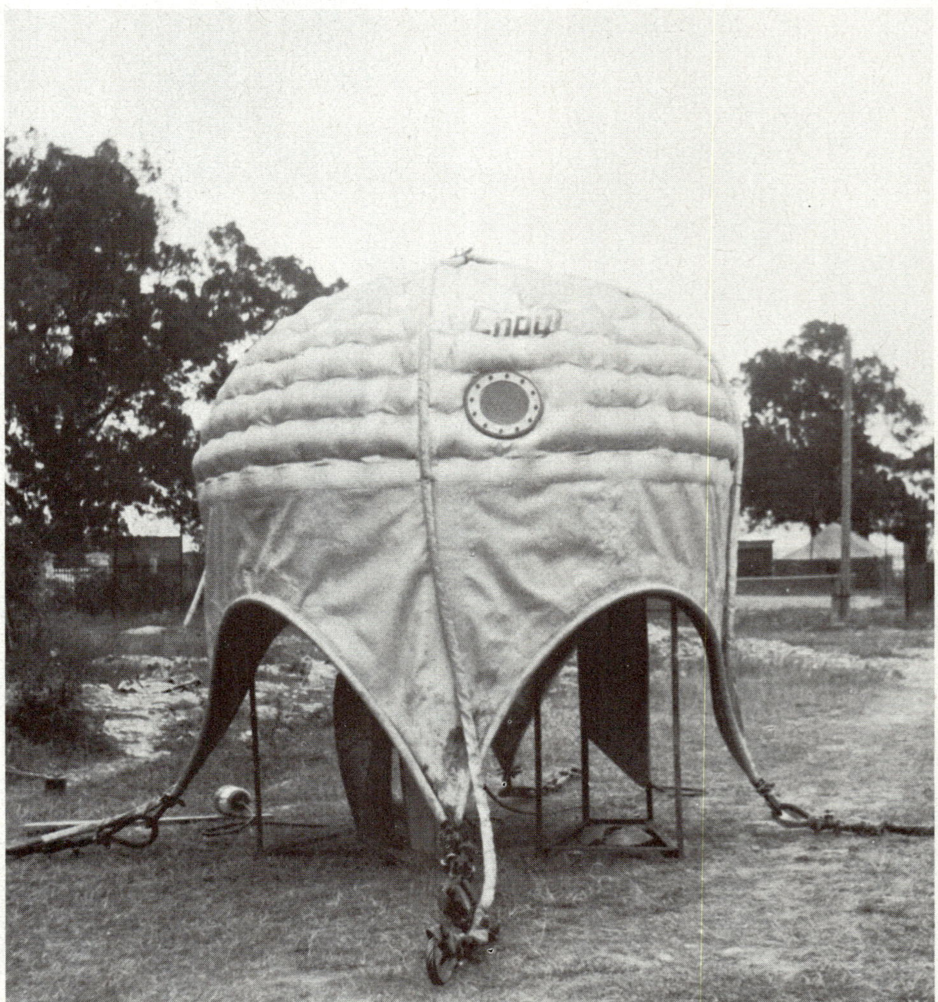

The Soviet underwater laboratory *Sprut* (*Octopus*) at the Black Sea Harbour of Sukhumi (*Courtesy Novosti Press*)

LENINGRAD HYDRO-METEOROLOGICAL INSTITUTE UNDERWATER RESEARCH LABORATORY

An underwater studies laboratory constructed of two rigidly-connected spheres, *Sadko 2* has been in use since 1967. Power and life support breathing gases are provided by means of cable and pipe lines to the shore.

Entry into the laboratory is through a hatch in the lower sphere which contains a 'cloakroom', a shower and a toilet. Aquanauts' work room, upper sleeping room and dining room are in the upper sphere. The laboratory is equipped with an electric heater to maintain the necessary temperature. The laboratory is also equipped with a lab-to-surface telephone and numerous scientific instruments.

Sadko-2 was installed at a depth of 25 m (82 ft) off Sukhumi in the Black Sea. Two aquanauts manned the laboratory to study hydrophysical processes in the coastal zone.

The Leningrad Hydro-Meteorological Institute's underwater laboratory, SADKO 2 (*Courtesy Novesti Press*)

SADKO 2 before submersion (*Courtesy Novesti Press*)

SADKO 3

A more advanced model of Sadko 2, apparently enclosed in a cylindrical shell. This apparatus has been in operation since 1970.

SADKO 3 in preparation for submergence (*Courtesy Novesti Press*)

UNMANNED UNDERWATER VEHICLES

FRANCE

DIRECTION DES CONSTRUCTIONS ET ARMES CENTRE D'ETUDES ET DE RECHERCHES TECHNIQUE SOUS MARINES

ERIC

An unmanned, tethered vehicle for survey work, research and the retrieving of underwater wrecks.

Eric, an unmanned underwater research vehicle

Design history

Date of construction of prototype started: 1970

Date of prototype completed: October, 1971

Date trials completed: June, 1972

Basic dimensions

Length, overall	4 m (12·2 ft)
Width, overall	2 m (6·1 ft)
Height, overall	2 m (6·1 ft)

Weights

Dry weight: 2,000 daN

Submerged weight: positive buoyancy 50 daN

Payload: 50 daN, parts on the bottom (max. dim. approx. 0·3 m × 0·3 m × 0·3 m (0·9 ft × 0·9 ft × 0·9 ft))

Operating crew

One pilot aboard the ship, one sonar and navigation operator

Performance

Maximum operational depth: 500 m (1,640 ft)

Speed: surface cruise 1 m/s (3 ft/s), submerged cruise 1 m/s (3 ft/s), max. 3 m/s (9 ft/s)

Range tether: 2 lengths, 500 m (1,640 ft) each

Main machinery description

Power source: 60 hz, triphase, diesel-electric generator 90 KVA, on the ship

Motors: Three 10 kw, continuous variable speed

Propellers: diam. 0·680 m (26·76 in)

Controls

Manoeuvring: differential control of the two horizontal propellers

Depth: vertical propeller

Trim: dead weights

Electrical on-board equipment

Navigational: Electrical magnetic compass

Vehicle function instrumentation: main power volts, propellers intensity

Communication: bi-directional digital data link

Search and recovery: pressure gauge, depth sounder attitude sensor, 5 degrees freedom telemanipulator

Television and visual: industrial type TV, tilt and pan

Buoyancy

Aluminium cylinders

Collapse depth: 800 m (2,624 ft)

Handling

Lifting points: one, spherical

Normal launching method: hydraulic crane, special automatic groping device

Support requirements

Mother ship: *Gabarre Luciole,* crane equipped; 440 V, triphase 60 hz, 90 kVA electrical power

Operators

Délégation Ministérielle de l'Armement;

Direction des Recherches et Moyens d'Essais;

Service d'Equipement des Champs de Tirs.

TROIKA

An unmanned, towed sled for photographic survey work. Three had been built up to September 1973 and show only slight differences with Commander J. Y. Cousteau's craft—he being the original designer of the *Troika* concept. The sled slides on the bottom and takes photographs at predetermined intervals of from 4 to 40 seconds; each passage of the Troika can survey a bottom strip 8 m (26 ft) wide.

Troika, an underwater towed sled for photographic survey work. This craft is based on a design by Jacques Cousteau

Basic dimensions

Length, overall	4·35 m (14 ft 3 in)
Width, overall	2·10 m (6 ft $10\frac{1}{2}$ in)
Height, overall	2·20 m (7 ft $2\frac{1}{2}$ in)

Weights

Dry weight	850 kg (1,870 lb)
Submerged weight	700 kg (1,540 lb)
Payload	200 kg (440 lb)

Performance

Maximum operational depth: 2,500 m (8,200 ft)

Speed, submerged: 1·3 knot/2,500 m (8,200 ft)

Operator

D.C.A.N.-C.E.R.T.S.M. Troika was successfully used in 1969 to find the wreck, in 2,200 m (7,216 ft) of water, of the Caravelle lost between Ajaccio and Nice.

INSTITUT FRANCAIS DU PETROLE

TELENAUTE

Télénaute is a remote controlled submarine robot, designed for underwater observation by means of a television camera and a 16 mm motion picture camera, and for performing simple handling operations, such as picking up samples from the ocean floor.

Télénaute is composed of three main parts:

(1) The submarine vehicle, made of aluminium alloy, and equipped with floats adjusted so that it has no apparent weight in sea water.

(2) The surface cabin contains the testing and control equipment.

(3) The multi conductor cable connects the underwater instruments and motors to the surface cabin. In particular, the energy necessary for the vehicle is transmitted from a surface generator through the cable which carries a 380 V triphase current. An underwater 12 kVA electrical motor produces hydraulic energy.

The vehicle is equipped with three independent propellers driven by three 3·5 h.p. hydraulic motors, two for horizontal motion and the third one for vertical motion.

The following navigational instruments are installed on *Télénaute*: A magnetic compass transmitting the course indication to surface; A pressure gauge transmitting the depth indication; An ultrasonic device measuring the distance either to the ocean floor or to surface. The vertical thrust propeller can be controlled, if desired, by this device so that the machine remains at any given constant distance from the ocean floor; Two flood lights.

A remote-controlled arm can move small objects lying on the sea floor or pick up samples and take them up to surface. This articulated arm is operated by several hydraulic valves.

Other ultrasonic devices are being added to increase the radius of investigation beyond the light range and to permit an accurate positioning of the submarine vehicle with respect to the support ship or to the sea bottom.

Télénaute can thus perform continuous underwater observations during very long periods of time and instantaneously transmit valuable data to the surface. These two features make *Télénaute* quite different from existing submarine devices and open a large field of applications.

Télénaute, a remote controlled observation submersible

Technical characteristics

	Length	*Width*	*Height*	*Weight*
Submarine vehicle	4·12 m (13 ft 6 in)	1·52 m (5 ft)	1·52 m (5 ft)	1,000 kg (2,200 lb)
Cabin	2·59 m (8 ft 6 in)	2·01 m (6 ft 7 in)	2·01 m (6 ft 7 in)	1,000 kg (2,200 lb)
Cable storage	3·96 m (13 ft)	0·99 m (3 ft 3 in)	0·71 m (2 ft 4 in)	400 kg (880 lb)

Technical characteristics

Electrical power supply: 30 kw 380 V triphase 50 cycles

Hydraulic power plant: 12 kw

Power of each propeller engine: 3·5 h.p.

Maximum thrust of the propellers: 81·8 kg (180 lb)

Power of each flood light: 1 kw

Cable length: 305 m (1,000 ft) (a 915 m (3,000 ft) cable is under study)

Cable diameter: 28·58 mm (1·125 in)

Maximum operating pressure: 98·0 kg/cm^2 (1,400 lb/in^2) corresponding to a depth of 915m (3,000 ft)

Maximum horizontal speed: 3 knots

Maximum vertical speed: 1·5 knots

Telemanipulator

Operating range	1·22 m (4 ft)
Lifting capacity at 1·22 m (4 ft)	50 kg (110 lb)
Grip	5 to 50 kg (11 to 110 lb)

SOCIETE ECA

PAP

A self-propelled vehicle which may be used for exploration of the sea bed and for the recovery of underwater objects.

Propulsion	Two variable pitch propellers, also providing manouvering control
Length	2·70 m (8·8 ft)
Weight	800 kg (1,760 lb)
Maximum speed	6 knots
Maximum depth of immersion	150 mm (492 ft)

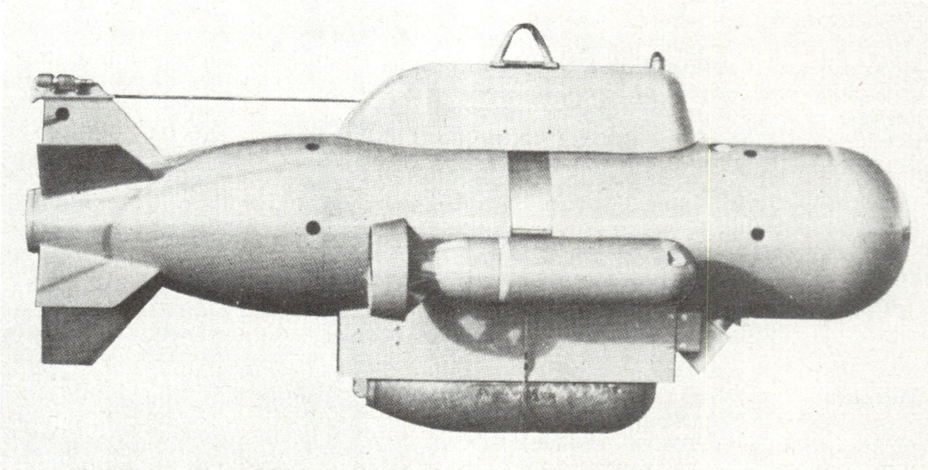

Societe ECA's (PAP) self-propelled underwater vehicle

GERMANY

DEUTSCHE BABCOCK AND WILCOX

UNDERWATER OBSERVATION BUOY

The Babcock observation buoy is designed to accommodate three people. In operation the craft can be flooded to two diving depths; 1 m or 2 m (3·28 ft or 6·56 ft) depending on visibility and the depth of the sea bed from the surface. When submerged the buoy is towed over the area to be observed by a motor boat. In this way the buoy can be used for observation of shallow areas of the Continental Shelf e.g. fish, plants, coral reefs and other benthic phenomena.

The buoy has good stability and cannot sink when the hatch cover on the turret is closed. The hatch cover can be opened from inside and outside, allowing convenient and safe entry and exit from the buoy. Thirty large observation ports allow all-round, upwards and downwards, visibility.

When the buoy is operated in calm conditions, it is ventilated by means of a battery driven fan and air intake. Special equipment includes compressed air storage and oxygen supply.

In normal operations the crew of the towing vessel is in charge of the procedures necessary for submerging and surfacing the buoy, but surfacing may also be accomplished by the occupants of the buoy.

The Babcock underwater observation buoy

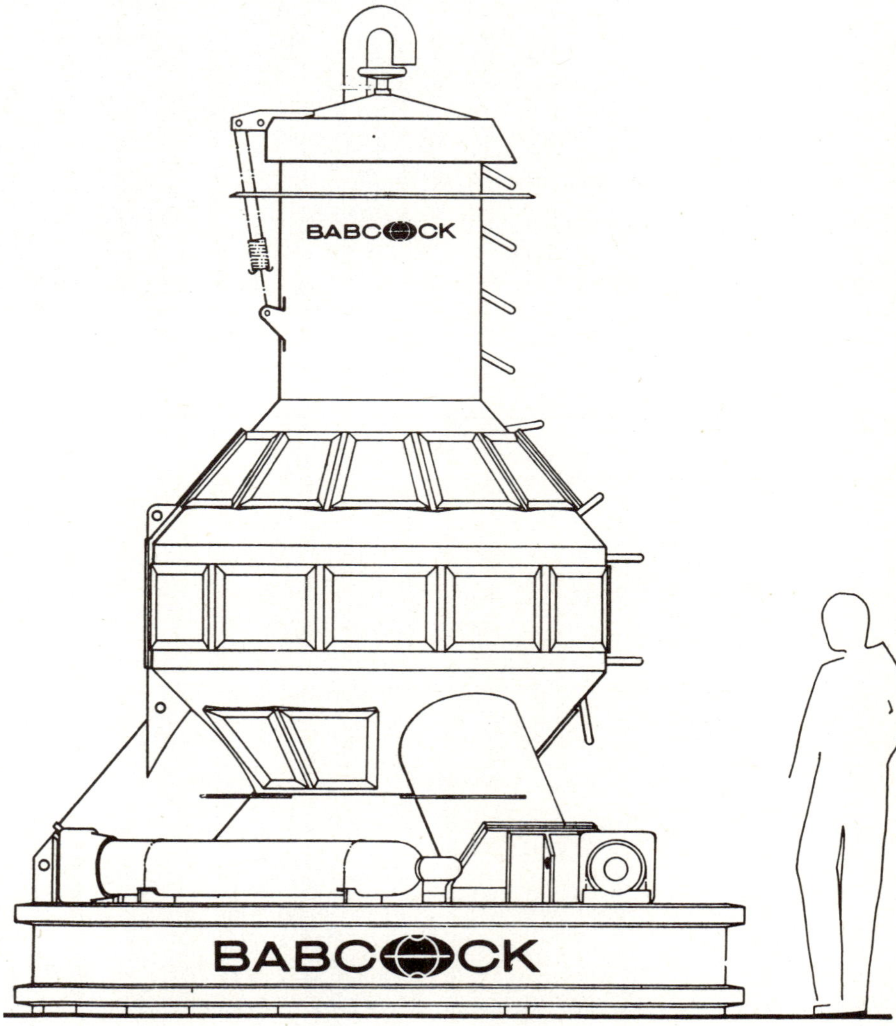

A schematic diagram of the Babcock underwater observation buoy

IBAK HELMUT HUNGER

DEEP SEA PHOTO AND TV TOWING SYSTEM

This system has been designed to allow systematic mapping of manganese module deposits on a large scale in depths down to 6,000 m (19,680 ft) and at a constant prefixed distance from the sea bed. Towing speeds of up to 6 knots are envisaged.

All underwater units are housed in nickel-chromium steel and arranged in a sledge shaped towing frame. When towing is in progress, two fins in the rear of the frame determine and stabilise the viewing direction. The low light level TV camera as well as the two photographic cameras, intermittent TV illumination, photo-strobe light, echo sounder, parallel light scale, current meter, buffer batteries, and remote control unit are all fed through one single armoured coaxial camera towing cable of 18 mm (0·7 in) diameter. A pressure-tight electromechanical swivel with coaxial connector between towing cable and underwater unit compensates the cable twist. Operating at 6,000 m (19,680 ft) depth the research vessel is towing a cast iron weight of about 2 tons at the end of the 8,000 m (26,240 ft) long cable over the ocean floor with the 'sledge' suspended at the same cable approximately 5 m (16·5 ft) above the ground. The cameras view forward and down. An echo sounder suspended from the towing cable warns of dangerous obstacles further ahead so that the unit can be lifted in time.

From the live TV transmission on the first screen individual pictures may be singled out at random and observed for several seconds as stand-stills on the second monitor. At the same time they are photographed automatically for documentary purposes. This method of photographing the screen with its data field at intervals already provides a good sampling of the total area investigated and reduces the amount of storage material. In addition interesting TV sequences are recorded on video tape at normal towing speeds for later interpretation on the third monitor screen using slow motion or stand-still.

The two underwater photography cameras can be released by remote control as soon as an interesting picture appears on the TV screen, thus providing selected direct photographs in either black and white, colour, or stereo. Each camera takes 1,500 pictures 24 × 36 mm per dive, all of them marked with date, time, photo number and additional written information.

The fourth monitor is equipped with a variable mask and wedge range selector, identifying the manganese nodules being counted automatically. A periphical computer and printer transfer data about nodule size, number and coverage to the main board computer, where the information merges with the exact geographical data and is stored for future analysis.

The IBAK Deep Sea Photo and TV Towing System has undergone first trials on the West German research vessel *Valdivia* off Hawaii. The system will be used for systematic exploration of manganese nodule deposits in the Pacific Ocean during 1974. The development of the so-called TFF-System was sponsored by the Federal Ministry for Research and Technology. IBAK sub contracted parts of the system to the Røbert Bosch Fernsehanlagen GmbH, the Felten & Guilleaume Kabelwerke AG, Ernst Leitz GmbH and the Electroacustic GmbH.

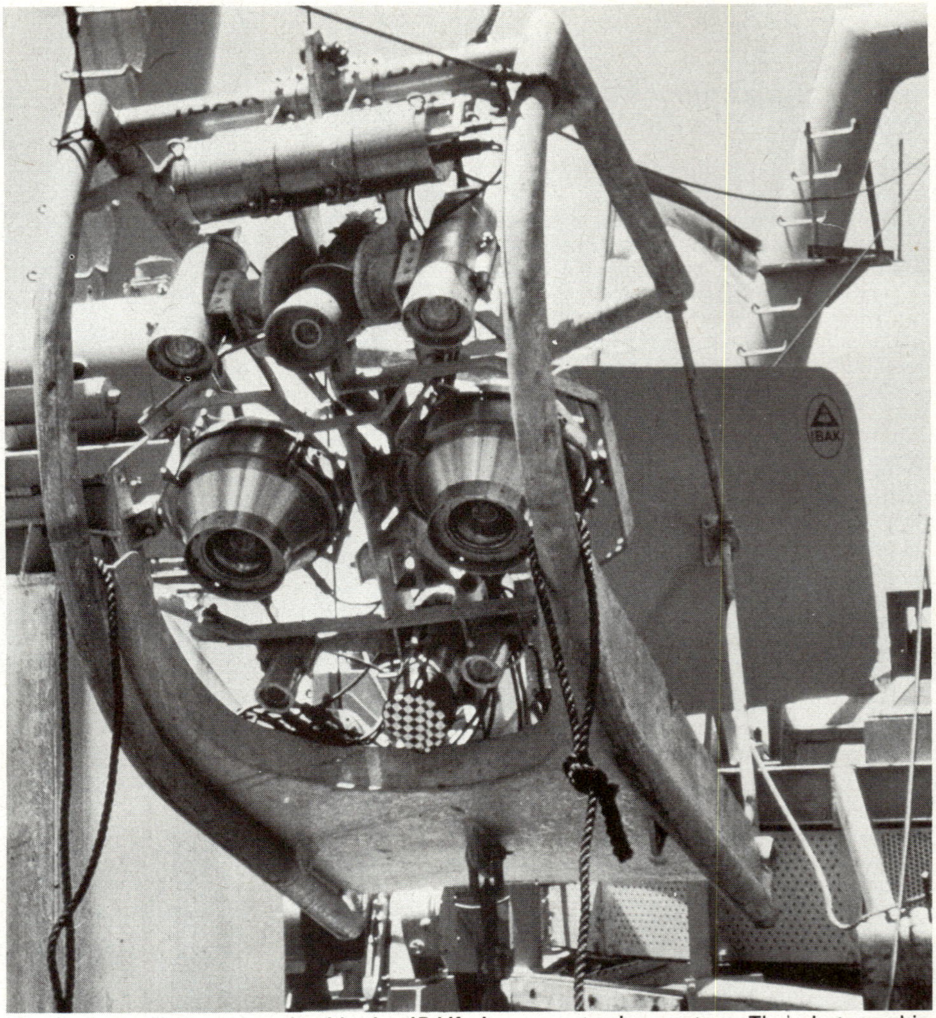

The underwater frame used with the IBAK deep sea towing system. The photographic equipment can be clearly seen on the front of the frame

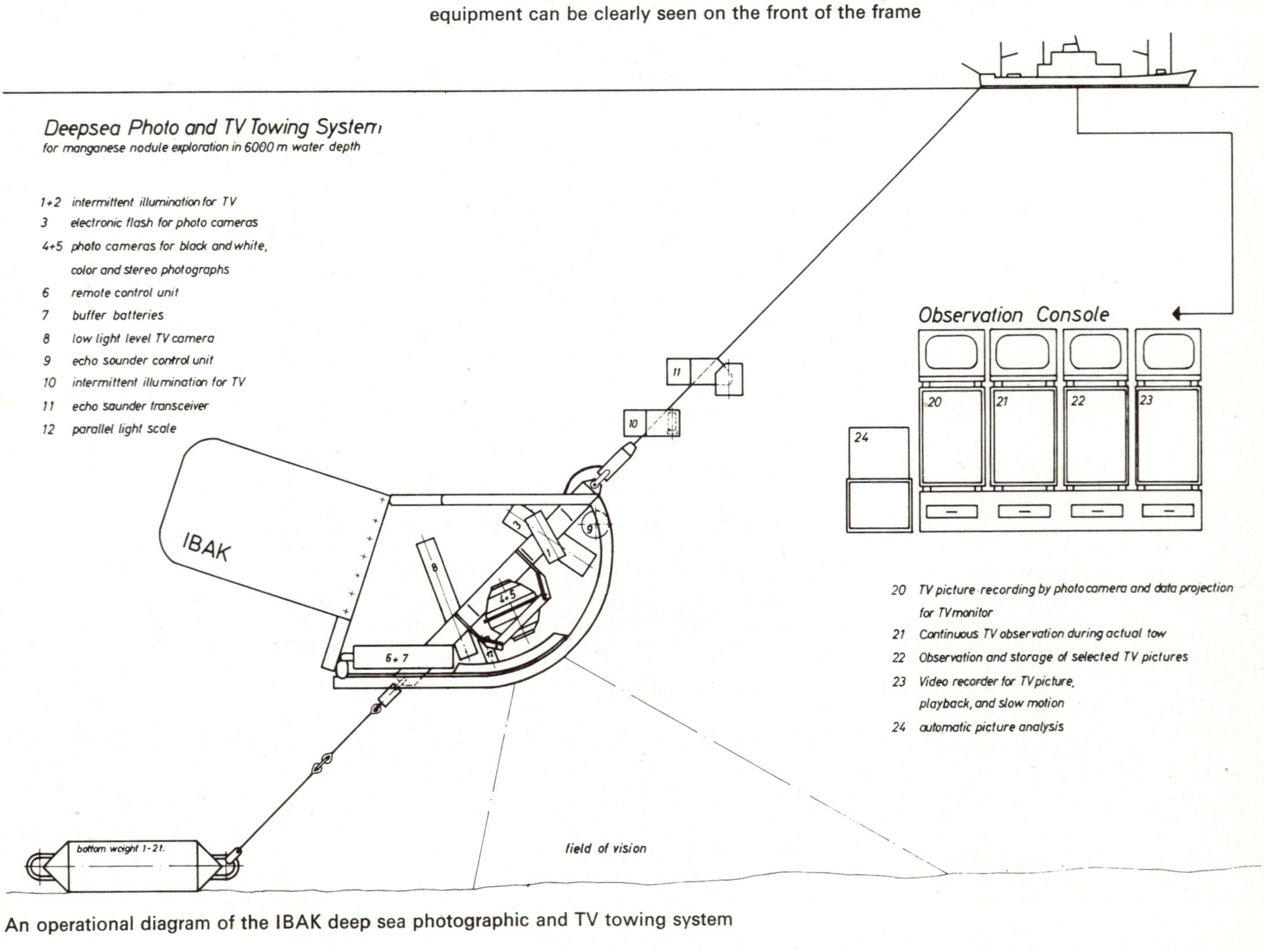

An operational diagram of the IBAK deep sea photographic and TV towing system

The observation console for use with the IBAK deep sea towing system

UNITED KINGDOM

BRITISH AIRCRAFT CORPORATION LTD

BAC 1

An unmanned, self-propelled, cable-controlled submersible designed and built by the British Aircraft Corporation for the Institute of Geological Sciences (UK) Continental Shelf Unit II, West Granton Road, Edinburgh. The specification called for a very low cost, simple vehicle which would operate in water currents of 2 knots and down to depths of 610 m (2,000 ft). The prime purpose of the vehicle is for use in a survey of the United Kingdom's continental shelf and for obtaining rock core samples. Other types of sampling may be carried out, various work payloads can be fitted and a mobile remote-controlled power pack can be installed. The design was based on a mathematical model and hence size, performance, function, etc. can be changed readily within the scope of a cable-controlled vehicle.

Design history

Date of construction of prototype started: April 1973

Date of prototype completed: January 1974

Date trials completed: July 1974

Basic dimensions

Length, overall	2·41 m (8·00 ft)
Width, overall	1·66 m (5·50 ft)
Height, overall	1·25 m (4·10 ft)
Draught	1·20 m (4·00 ft)

Weights

Dry weight	800 kg (1,760 lb)
Payload	132 kg (290 lb) in air for Pan and Tilt Head fitted with stereo film cameras and flash, TV camera and flood light, and rock coring drill

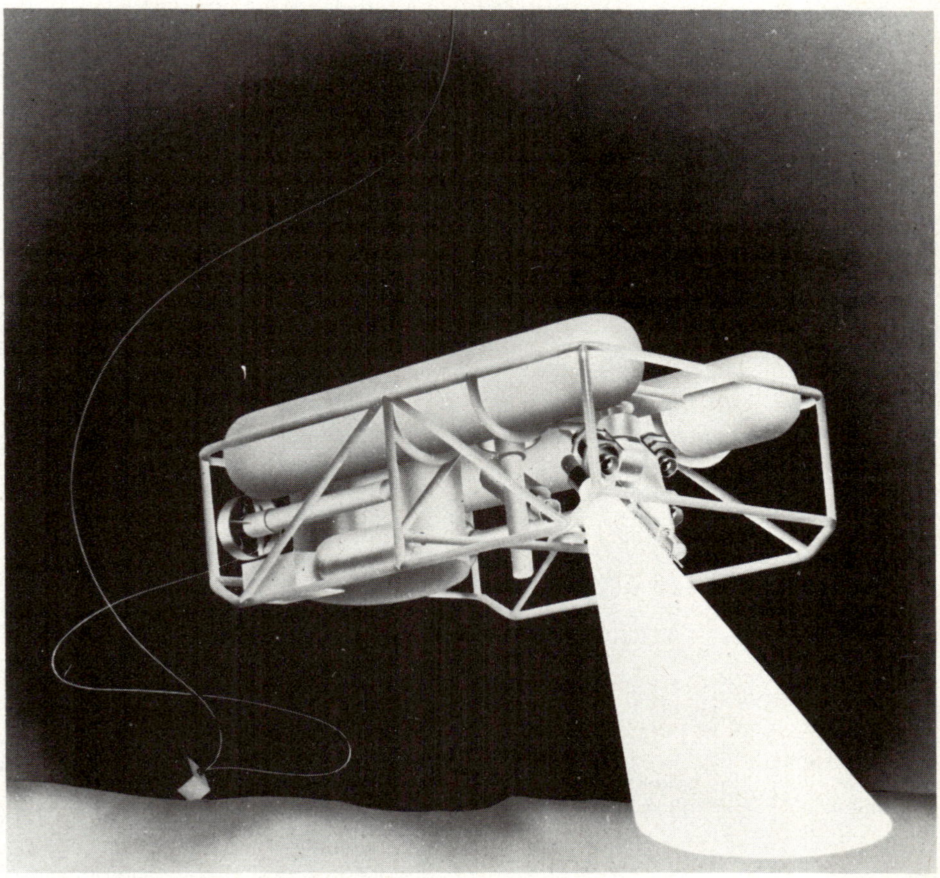

An artist's impression of BAC 1 in operation above the sea bed

Performance

Maximum operational depth	610 m (2,000 ft)
Speed, submerged	2·3 knots
Duration at cruise speed	Indefinite
Range	60 m (200 ft) about lowering point
Manoeuvring capability	By thrusters to give forward or reverse thrust, steer and vertical motion

Main machinery description

Power source	Umbilical cable to ship supply
Motors	Carron Hydraulic
Propellers	3-bladed, fixed pitch, guarded
Controls:	
manoeuvring	Joystick
depth	Joystick

Electrical on-board equipment

Video link for direct observation

Safety systems

Ballast jettison. Umbilical cable as buoy rope. Positive buoyancy to give fail safe

Basic structure

Machinery compartment: Space framework with individual pressure resistant housings for components

Buoyancy compartment: Separate monocoque tanks

Collapse depth: Below 900 m (2,950 ft)

Handling

Lifting points: Single central lifting point

Normal launching method: Vehicle is fitted into handling cradle and combination lifted by ship's gantry and lowered to sea bed on umbilical cable

Support requirements

20 kVa, 440 volt, 3 phase supply;

Deck space for stowage of vehicle and handling cradle;

Lifting gantry;

Special winch for umbilical cable;

Deck house, cabin, or Portakabin for vehicle operator and observer

Availability

Prototype only being built

Operators

Institute of Geological Sciences
Continental Shelf Unit II
West Granton Road,
Edinburgh EH5 1JA
United Kingdom

Costs

Reported value of UK Department of Trade and Industry Research Division 8-month contract placed with B.A.C. was £36,000

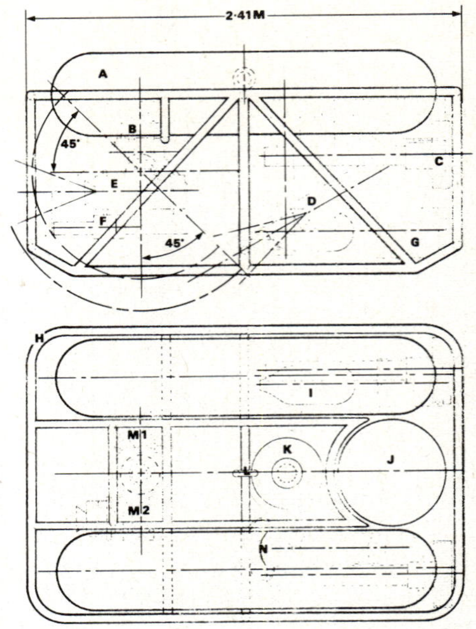

Side elevation and plan views of BAC 1

A Buoyancy Tank (2 off)	**H** Vehicle Frame
B Pan & Tilt Head	**I** Compressed Air Bottle
C Horizontal Thruster (2 off)	**J** Pop-up Buoy
D Fixed TV Camera	**K** Vertical Thruster
E TV Camera	**L** Cable Attachment Point
F Rock Drill Assembly	**M1 & M2** Stereo Camera
G Vehicle Electrical Junction Box	**N** Compass/Attitude Indicator & Depth Gauge

MOBELL MARINE DEVELOPMENT CO LTD

MOBELL DIVING BELL MK II

The Mobell Diving Bell is a wet submersible and may be towed or self-propelled. The vehicle is primarily intended for diver support work, though it can be adopted for various purposes such as underwater surveys, TV and visual inspection, transport of tools, etc. This vehicle took part in the rescue of survivors from the B.P. North Sea rig disaster.

Design history

Date of construction of prototype started: 1964

Date of prototype completed: 1966

Date trials completed: 1966

Production

Two: Prototype scrapped, working model now in use

Basic dimensions

Length, overall	3·66 m (12 ft)
Width, overall	1·45 m (4 ft 9 in)
Height, overall	1·68 m (5 ft 6 in)

Weight

Dry weight	309 kg (680 lb)
Crew	2 to 3
Ballast	68 kg (150 lb)

Bell volume

0·255 m^3 (9 ft^3)

Performance

Maximum operational depth	183 m (600 ft)
Speed:	
surface	towed
submerged	3 knots
maximum	3 knots
Life support endurance (air supply)	Two hours

Main machinery description

Power source	Battery/surface cable; 2,000 watt hours, 2½ h.p. compound motor
Motors	Electric (torpedo)
Propellers	One
Controls	Manual

Electrical on-board equipment

Communication	Diver phone
Search and recovery	As required
Television and visual	Power source when required, either battery or surface supply

Safety systems

Life support	Reserve air
Ballast jettison	Weight jettison—slip pin
Buoy markers	Carried
Function failure	installed

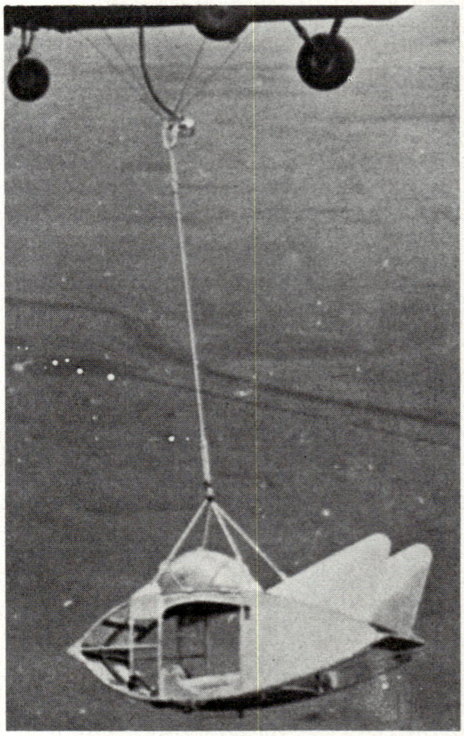

The air-portable Mobell Mk 11 diving bell

The Mobell Marine diving bell Mk 11

RECORDING DESIGNS (EMI) LTD

UNMANNED SUBMERSIBLE VEHICLE (U.S.V.)
(Based on Military Deep Mobile Target (D.M.T.))

Recording Designs (EMI) Limited was commissioned by the UK Ministry of Defence to develop a free-ranging, unmanned submersible vehicle with the primary role of a recoverable test target. In its conception, the vehicle replaces the submarine which until then had performed this function on underwater weapons trials and exercises. The high operating cost of a submarine and its inability to operate in this capacity in confined areas during inclement weather conditions, was now considered economically unsuitable for this secondary task. Added to this are the attendant hazards for the submarine as a weapon target.

A development programme was completed in three years in collaboration with the UK Underwater Weapons Establishment. This was followed by highly successful proving trials and the vehicle became known as the Deep Mobile Target (DMT).

The DMT simulates the dynamic and acoustic characteristic of present day submarines and provides sonar operators and command teams in both aircraft and ships with a simulated full-scale target during exercises.

The Company is currently studying the development, from the DMT, of a basic training target providing greater endurances at lower speeds. RDL were commissioned by the UK Ministry of Technology (now the Department of Trade and Industry) to execute a feasibility study for the development of a self-propelled Unmanned Submersible Vehicle (USV) for civil use. This study has been completed and the Company is now in a position to build, for sale or lease, civil USV's with instrument packages to customer requirement.

Instrument packages could include: Side Scan sonar; Photographic site inspection; Magnetometry; Physico-chemical analysis; Bottom profiling and seismic sonar.

Design history
Date of construction of prototype started: 1966—DMT prototype
Date of prototype completed: 1968
Date trials completed: 1969
Certification authority clearance: First purchase after trials by the UK Admiralty Underwater Weapons Establishment

Production
27 DMT's built
2 vehicles under construction. 16 further vehicles presently planned

Basic dimensions

Length, overall	3·28 m (10 ft 9 in) in DMT role, longer in USV roles
Width, overall	Diameter 32·4 cm (12·75 in)

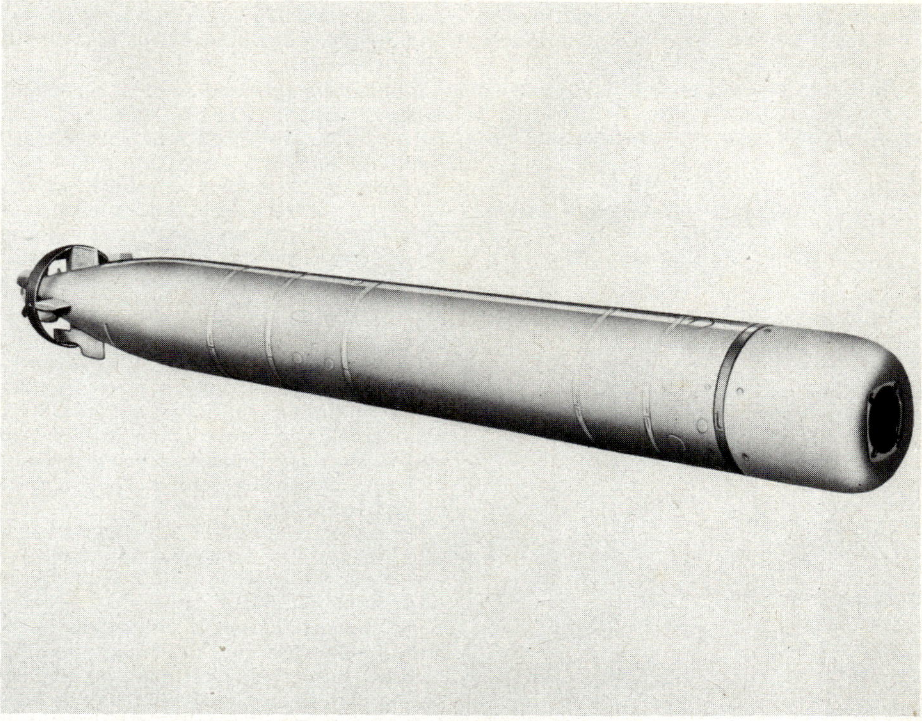

The unmanned submersible vehicle (U.S.V.) built by R.D.L.

Weights

Dry weight	236 kg (520 lb) approx.
Submerged weight	Neutral
Payload	Approx. 45 kg (100 lb) Instrumentation package
Crew	None

Performance
Maximum operational depth: 366 m (1,200 ft)
Speed, submerged, cruise DMT 8 knots, max. DMT 22 knots. A study is in hand to develop the USV for speeds from tide stemming up to a maximum of 8 knots, coupled with greater endurance.
Duration at cruise speed at 8 knots—100 minutes
Duration at maximum speed at 22 knots: 10 minutes
Range: 13½ miles
Manoeuvring capability: Typical turn radius: 76 metres (250 ft). Climb/dive angles: up to 30 degrees

Main machinery description

Power source	Electrical battery
Motors	Contra rotating DC motor
Propellers	Contra rotating
Controls:	
manoeuvring	Azimuth—integrally linked rudders
depth	Independantly variable elevators, controlled by pressure sensitive transducer, or potional bottom following height sensor
trim	Fixed

Electrical on board equipment

Navigational	Orthogonally mounted fluxgate compass. Compensation is provided to minimise the effects of pitch and roll on vehicle heading
Vehicle function instrumentation	Magnetic tape recorder
Communication	Acoustic link
Search and recovery	Acoustic beacon
Television and visual	Could be fitted in instrument package

Safety systems

Ballast jettison	Dump of lead ballast
Buoy markers	Optional
Function failure	Safety feature—run about and surface on detection of certain fault conditions

Basic structure

Pressure hull	Cylindrical
Outer skin	Anodised aluminium
Collapse depth	Proof pressure tested to 457 m (1,500 ft). Collapse depth about 760 m (2,500 ft)

Handling

Lifting points	Single
Normal launching method	Simple chute from ship or boat deck level

Support requirements
Mother ship for launch and recovery and acoustic link if fitted. A wide variety of tenders, boats and rubber dinghies may be used.

REYNOLDS INTERNATIONAL INC

SHRIMP

A towed, un-manned underwater search and survey vehicle which operates to a depth of 455 m (1,500 ft). The vehicle is installed in the United States M/V *Privateer*, a 38·5 m (126 ft) research vessel, and is basically an aluminium sled bearing pressure-tight-equipment cylinders horizontally fore and aft.

Design history

Date of construction of prototype started: 1970

Date of prototype completed: 1971

Production

One built

Basic dimensions

Length, overall	2·75 m (9 ft)
Width, overall	1·36 m (4·5 ft)
Height, overall	1·20 m (4 ft)

Weight

Total weight in air	540 kg (1,200 lb)

Performance

Maximum operational depth	610 m (2,000 ft)

Handling

Associated handling equipment includes a stern cable guide, a cable puller and a cable storage reel. It is readily lifted with a small deck crane or an 'A' frame

Support requirements

Mother ship facilities required; Vessels of 10·5 m (35 ft) length or larger

The total system is composed of the following components

1. Towed vehicle with: a. Television system; b. Bottom mapping sonar; c. Magnetometer; d. Altimeter; e. Steering and depth control surfaces

2. Cable reel with slip-ring assembly. The reel is hydraulically operated

3. Cable Puller—hydraulic operation

4. Instrumentation consoles for: a. Television monitoring; b. Sonar recording; c. Magnetometer recording; d. Altimeter recording

5. Controls for: a. Altitude control; b. Steering control

The towed vehicle *Shrimp* may be fitted with other equipment such as sub-bottom profiler, still camera—strobe light system, fish traps, bottom samplers, etc

Television

The television system consists of a camera by Concord Electronics Corporation carried in *Shrimp*, a Concord VTR monitor for the camera and a Concord VTR 600-1 video tape recorder located in the ship control centre.

The system, which produces direct broadcast quality 525 line presentation, employs the latest low-light-level, solid state vidicon type camera. A special long transmission line video signal processing amplifier is installed in the underwater housing. By processing the video signal before transmission, as opposed to the more convenient post transmission amplification and equalization, the highest image resolution and lowest noise levels are obtained at the surface for high quality video tape recording and viewing. The camera housing is equipped with an optical quality, plexiglass 120° truncated cone window that permits the use of the widest field angle lenses available. A 9·0 mm focal length lens is employed to give full field angle coverage of 72° in the water.

Night operations, and daylight operations at depths beyond 110 m to 180 m (60 to 100 fathoms), are provided for by special metal halide, gaseous arc lamps that produce an optimum visible energy spectrum for sea water transmission. The spectrum is sufficiently broad to permit colour motion picture or televisual photography when suitable compact colour television camera equipment is available.

Sonar

The bottom mapping sonar is by Westinghouse. An Alden recorder is located in the ship control room. 'Flying' at 6·1 m (20 ft) 'altitude' this sonar covers a width of 135 m (450 ft) and the recorder provides a permanent 48·3 cm (19 in) wide-paper shadowgraph of the path surveyed.

The system frequency is 225 kHz and beam width is 1 degree. The normal high resolution will reproduce objects 15·3 cm (6 in) or larger within the search path.

The continuous chart record is a rectilinear presentation of the sea floor topography. The normal chart production rate permits an operator to make manual notations of televisual, magnetometric and navigational data in support of fully correlated search and identification operations or marine biological and geophysical studies.

The permanent chart record, when produced by submerged vehicle operation at a constant altitude, may be photogrametrically analysed for dimensional data for all sea floor irregularities depicted. The accuracy of these data will be essentially in proportion to the navigational accuracy of the vehicle.

Magnetometer

The Magnetometer is the HeleFlux Magnetic Null Sensor by the Schonstedt Instrument Company. It is a single component, saturable-core (flux-gate) type of magnetometer. The recorder for the magnetometer is the Model G-11A Strip Chart Recorder by Varian Aerograph. It is a true rectilinear recorder.

The system is primarily designed for high sensitivity detection of magnetic anomalies of the earth's vertical field structure. Minute variations as small as one tenth of one gamma will be clearly indicated by the continuous strip recorder at the surface *Shrimp* control station. The anomaly search sensitivity is equal to that of the most complex proton precession magnetometers commonly used for ocean search and research. The system involves only a small fraction of the electronic components common to conventional high sensitivity magnetometers. The heart of the system is a special, earth's field compensated flux gate magnetometric sensor developed for NASA space exploration applications. The control station recorder has provisions for navigational data marking of the chart to permit correlation of magnetic anomalies with the ship's geographic search pattern.

Altimeter

The *Shrimp* altimeter system is primarily a navigational aid to permit the surface operator to control the *Shrimp* vehicle distance above the bottom and thereby maintain planned search patterns and response sensitivities of the *Shrimp* data collection and recording systems. The instrument operates on 50 KHz and records to 365 m (1,200 ft).

The altimeter continuous calibrated display and record is an essential component of any post mission evaluative analysis. This record also contributes to a thorough knowledge of the sea floor area traversed.

Controls

The 'flying' controls are electro-hydraulic to control the hydraulic power to rudder and planes

Profiler

The sub-bottom profiler is a 3·5 KHz unit. The recorder for this unit is a Hydro Products Gifft Depth Recorder which produces a permanent paper record

Camera

The still camera and strobe light are by EG&G. The 35 mm camera has a 500 frame capacity

Cable guide

The cable guide prevents the power and control cable from chafing and bending on a short radius

Cable puller

The cable puller is built with two pairs of aircraft wheels and tires under tension, and hydraulically powered to grasp the cable. Line pull is 1,770 kg (4,000 lb)

Storage reel

The cable storage reel contains 1,372 m (4,500 ft) of pressure tested power and control cable. It is equipped with a watertight slip-ring connector system which connects the underwater cable directly to the presentation and control instruments. This ensures uninterrupted power for control and signals during periods when the reel is rotating. The cable is a product of Victor Cable, tested by SWRI

All this equipment is designed and balanced to operate at the end of a 1,500 m (5,000 ft) power and control cable

The *Shrimp*'s control centre is in an air conditioned room adjacent to the pilot house. Control and recording equipment is arranged for efficient operation by two persons

Flexibility

The *Shrimp* data transmission system has provisions for addition of bi-directional control and data transmission channel for remote control of, and signal transmission from, additional sensor systems and equipments

Additional sensor units, provided by Reynolds or the purchaser, may be included in the *Shrimp* suit or exchanged with units normally installed

Speed

Towing speed of the *Shrimp* will be determined by the type survey and the sensors to be emphasised. Television and bottom mapping should be towed at about 1·6 knots whereas the magnetometer and some other type sensors may be towed at 3 knots or more.

THE UNITED STATES OF AMERICA

US NAVAL CIVIL ENGINEERING LABORATORY

BUOYANCY TRANSPORT VEHICLE

The Buoyancy Transport Vehicle (BTV) provides free-swimming divers with an 'underwater fork-lift' capable of lifting and moving up to 450 kg (1,000 lb) of wet-weight cargo at depths to 260 m (850 ft) in the ocean.

The Naval Civil Engineering Laboratory (NCEL) is evaluating the BTV concept. The BTV under test was designed and built by the Naval Undersea Research and Development Center (NURDC), Hawaii, for the Naval Facilities Engineering Command's Ocean Engineering Research and Development Programme.

The BTV is built around its central lifting sphere, which is a variable ballast compartment. Two upper pods provide additional buoyancy and a nitrogen filled battery storage area. Two lower pods house the electro-hydraulic power unit and the space for the hydrazine fuel tank. Two hydraulic thrusters propel BTV horizontally and two move it vertically, once neutral buoyancy is achieved.

There is an outlet on BTV which permits divers to plug in underwater tools, using BTV's hydraulic power supply.

The BTV descends and rises by controlling the ballast in the sphere. Flooding and de-watering is accomplished through the flood and vent valves at the top and bottom of the sphere.

Hydrazine gas or compressed air is used for de-watering the sphere. When liquid hydrazine is used, it is converted to a gaseous state by a small generator on BTV. In shallow water 3 m (10 ft) the BTV is capable of increasing its lift at the rate of 38 kg (84 lb) buoyancy per second. High pressure air replaces hydrazine for operations at depths less than 45·5 m (150 ft).

Specifications

Weight (dry)	815 kg (1,800 lb)
Length	2·4 m (8 ft)
Width	1·8 m (6 ft)
Height	1·75 m (5 ft 9 in)

Submerged Speed (unloaded) 2·8 knots maximum (about 1 hour duration).

The US Naval Civil Engineering Laboratory's Buoyancy Transport Vehicle (BTV)

Power train

Batteries: Eight independent 52 cell, 100 volt, 12·5 amp-hour sets of silver zinc alkaline batteries.

Electric motor: One modified Mk 43 torpedo motor. Motor and hydraulic pump are sealed in oil filled ambient pressure containers.

Hydraulic pump: Variable displacement piston pump, 10 (US) gpm (1,800 lb/in^2)

Propulsion motors: Each motor 3 (US) gpm, 400 rpm

Propellers: 45·7 cm (18 in) diameter, 35·6 cm (14 in) pitch, counter-rotating

Steering

Steering and pitch control is accomplished by independent control of the four propulsion motors. There are no movable rudders or control planes.

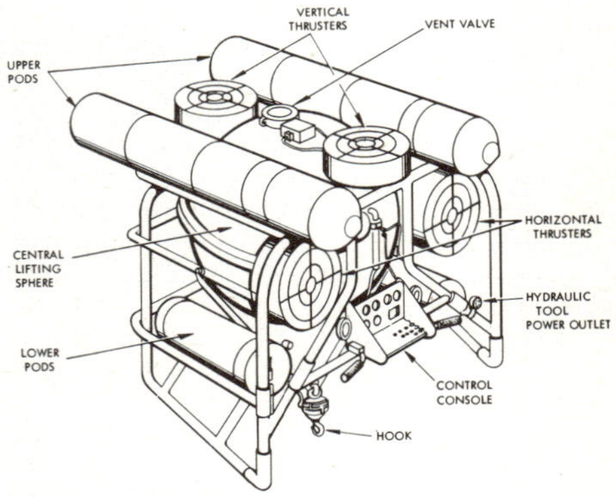

Layout diagram of the NCEL's Buoyancy Transport Vehicle (BTV)

CURV III

(Cable-Controlled Underwater Recovery Vehicle III)

CURV III is a cable-controlled underwater recovery vehicle for deep operation and is an engineering prototype model of an unmanned submersible system. It has an operational depth of 2,100 m (7,000 ft), 1,350 m (4,500 ft) greater than that of its predecessors CURV I and CURV II. The basic system consists of underwater vehicle, surface support ship, control console, and cable. The vehicle is equipped with sensors and a tool arm that enable it to perform a variety of underwater work and inspection tasks. The sensors include two television cameras, a still camera,

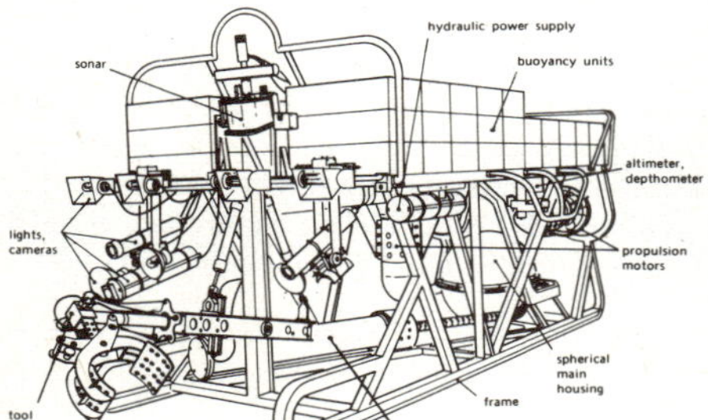

Layout diagram of CURV 111

A view of CURV 111 on the deck of its support ship

and active and passive sonar. The six-function tool arm can be fitted with grasping claws, cable cutters, grapnels, and other tools. Propulsion is provided by three reversible electric motors with infinite speed control. The support ship is a converted lighter, YFNX-30, designed for use at the US Navy's underwater test ranges off the coast of Southern California. It has two cycloidal propellers, allowing accurate manoeuvring and station-keeping, and can support a crew of 24 for 30 days without refurbishing. The vehicle, control console, and cable are portable and can be transferred to a suitable ship of opportunity for deployment in other parts of the world. Development, begun in 1966, is complete, and the system has now proven its capability in a number of operations. A major rescue operation was undertaken by CURV III on September 1st, 1973 when it was used to fix a lifting line onto the Pisces III submersible which had sunk in 477 m (1,565 ft) of water in the Atlantic 100 miles south west of Eire.

The Cable-Controlled Underwater Recovery Vehicle (CURV) programme at the US Naval Undersea Center was organised to develop unmanned systems to recover ordnance at the Center's underwater test ranges at Long Beach and San Clemente Island. The first fully operational system, now referred to as CURV I, was designed for depths of up to 610 m (2,000 ft). During the years when it was in service, 1965 to 1967, it made many successful recoveries.

When the hydrogen bomb was lost in the Mediterranean off Palomares, Spain, in April 1966, CURV I was modified for deeper operation and participated in the salvage work, making the final recovery from 868 m (2,850 ft). This successful recovery demonstrated the versatility and potential of unmanned systems, and a project to develop a CURV with a deeper capability was undertaken.

In the course of planning for the deeper CURV, it became apparent that because of wear and reliability problems CURV I could not continue to meet the Center's range needs until the new development was completed. It was thus decided to undertake two new CURV projects, fabrication of a CURV II vehicle for 760 m (2,500 ft) depths and development of a CURV III system for greater depths.

CURV II was to retain most of the features of CURV I, with modifications as necessary to replace inadequate components and improve reliability. The CURV I vehicle was to be 'cannibalised' wherever possible, and the same cable-support gear and support ship were to be used. No increase in performance was planned other than the additional 150 m (500 ft) depth capability.

CURV III was to use the successful concepts of CURV I but was otherwise to be an entirely new system. The goal was initially an operational depth of 1,800 m (6,000 ft) but was changed after early design studies to 2,120 m (7,000 ft).

The need for a system capable of making recoveries at depths greater than those attainable by CURV I was established by Specific Operational Requirement 46-16. This Requirement appeared in 1964, when CURV I was already in development. It called for a system capable of recovering small objects at depths of up to 6,100 m (20,000 ft).

By 1966 need also existed for a self-propelled, manoeuvreable, and economical recovery system at the Atlantic Underwater Test and Evaluation Centre, West Palm Beach, Florida; the Atlantic Fleet Weapon Range, St Croix, Puerto Rico; and the Naval Torpedo Station, Keyport, Washington, (Existing methods of recovery included divers (shallow depths only) and snares and grapnels manoeuvred by means of surface ships or complicated devices deployed on the bottom).

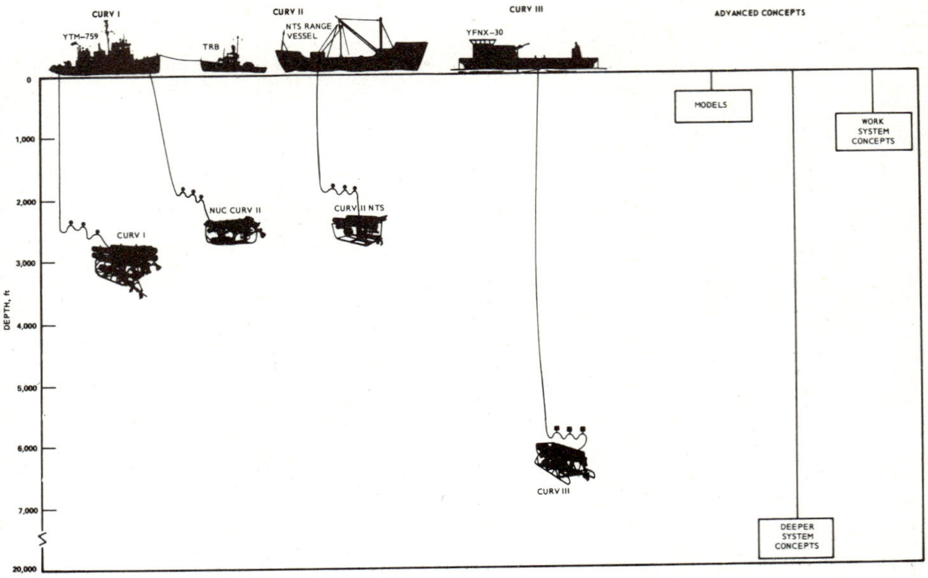

A summary, in diagramatic presentation, of the development and planned progress of the CURV Programme

The depth at St Croix, the deepest of these ranges, exceeded 1,800 m (6,000 ft). The trend on all ranges was to deeper testing.

There was, finally, a foreseeable need for deeper recoveries at the Centre's San Clemente Island range, where the maximum depth is 1,800 m (6,000 ft).

An unmanned recovery system was considered suitable to meet these needs for the following reasons:

(1) CURV I had demonstrated that sensor information and control sensitivity were comparable to those of a manned system.

(2) Development would be relatively quick and economical.

(3) The vehicle would have unlimited underwater endurance and would be able to perform hazardous tasks without endangering personnel.

(4) The system would be modified to meet the needs of particular tasks without affecting its safety or certification.

Other CURV projects

Concurrent with CURV III development were the fabrication and assembly of CURV II, the development of CURV II NTS for the Naval Torpedo Station, Keyport, Washington, and various exploratory development projects. The most significant of the latter were investigations of a 20,000 ft recovery system and of shallow-water work systems for underwater tasks more complicated than recovery.

System description

Vehicle

The vehicle is strictly functional in design; components and subsystems are mounted in an open frame. It is 2 m × 2 m × 4·55 m ($6\frac{1}{2} \times 6\frac{1}{2} \times 15$ ft), it weighs approximately (4,500 lb) in air and is slightly buoyant in water. The principal subsystems are the lights and cameras, the search and navigation equipment, the propulsion motors, the hydraulic system, the tool assembly, and the structural units.

Lights and cameras

Underwater viewing and documentation are provided by two Vidicon television cameras, four mercury-vapour headlights, two mercury-vapour spotlights, and a 500-exposure, 35 mm colour camera and strobe light. These components are mounted on two independent pan-and-tilt units, each with 360 degrees of lateral and 180 degrees of vertical movement. The vehicle thus has full viewing and self-inspection capabilities.

Search and navigation equipment

The search and navigation equipment includes a compass; a continuous-transmission, frequency-modulated active sonar with a range of 730 m (800 yards) and scanning sector of 120 degrees; a transponder mode with an 730 m (800-yards) range; a passive sonar capable of tracking a 45-kilohertz pulse; a 37-kilohertz, vehicle-mounted pinger locator which operates in conjunction with a boat-mounted acoustic locating device (BALD) on the support ship; a pulse-measurement altimeter for determining the distance to the bottom; and a pressure-transducer depth-ometer.

Propulsion

Three reversible electric motors, two mounted horizontally and one vertically, provide propulsion. They are pressure-compensated, oil-filled units rated for continuous service even if flooded with salt water. They operate on a 3-phase alternating-current power supply of 440 volts. Their speed is infinitely variable. Each has a capacity of 10 horsepower and provides 180 kg (400lb) of forward and 90 kg (200 lb) of reverse thrust. The combination of port, starboard, and vertical thrust, together with infinite speed control, allows very fine manoeuvering at close quarters. The vertical motor is run constantly to overcome the vehicle's 4·5 kg to 22·0 kg (10 to 50 lb) of positive buoyancy.

Power equipment

The CURV III system draws its power from the support ship's power supply. The system requirement is 40 kilowatts. Power is converted by a portable 5-kilowatt motor generator and three variable transformers with a capacity of up to 1,000 volts.

Storage equipment

Storage equipment includes an electrically powered dolly for the vehicle; two cable baskets welded together, each 1·2 × 0·915 × 4·25 m (4 × 3 × 14 ft); a storage bin for the nylon support line; and a spare-parts van.

Handling equipment

Handling equipment includes a 5·8 m (19 ft) articulating crane for deploying and retrieving the vehicle, a hydraulic traction winch for deploying and retrieving the support line, and a davit and block for handling the line and cable.

Tracking equipment

Tracking equipment includes the boat-mounted acoustic locating device (BALD) used to track the 37-kilohertz pinger on the vehicle or a similar pinger on a target. This device is a directional hydrophone that scans a horizontal sector of 360 degrees and a vertical sector of 90 degrees. It is lowered over the side and below the hull of the support ship on a retractable mount with a 6·1 m (20 ft) shaft.

Hydraulic system: The hydraulic system operates the pan-and-tilt units and the tool assembly. It is pressure compensated and provides a supply pressure of 70 kg/cm² (1,000 lb/in²) at all depths. Components include a ½-horsepower electric motor, a 0·64-gal/min pump, control valves, and actuators.

Tool assembly: The tool assembly is a hydraulically powered, five-function unit with vertical and rotational but no lateral movement. The interchangeable tools include the snap-hook manipulator, the 33 cm (13 in) diameter grasping claw, snares, hooks, and grapnels. The tools can be ejected from the arm by a remotely controlled release.

Structural units: The structural units include the frame, made of 6061-T6 aluminium I-beams and channel-beams; fifty buoyant, syntactic-foam blocks, each 14 × 28 × 56 cm (5½ × 11 × 22 in), held to the frame by tie-rods and brackets; and the one-atmosphere electronic housings. The main housings are a 0·71 m (28 in) cast-aluminium sphere for electronic and hydraulic components and a 1·5 m (60 in) aluminium cylinder for sonar components. There are also small aluminium housings for the television cameras, the still camera, the strobe light, and the compass. To eliminate maintenance problems only pressure-balanced connectors are used.

Cable

The cable assembly is composed of the main cable and whip, connectors, nylon support line, and floats. The main cable is 2,100 m (7,000 ft) long, 3·8 cm (1½ in) in diameter, with 58 conductors and an epoxied fibreglass strain member; the whip is 180 m (600 ft) long, 3·18 cm (1¼ in) in diameter, with 55 conductors and a braided stainless-steel support member. Both are non hosing and weigh approximately 0·73 kg/m (½ lb/ft) in water. The nylon support line is 2·54 cm (1 in) in diameter, tied at 9·15 m (30-ft) intervals to the main cable and whip, it prolongs their life by taking the load off the conductors. Syntactic-foam floats 30 × 30 × 18 cm (12 × 12 × 6 in) are spaced at 9·15 m (30 ft) intervals along the length of the whip, giving it a slightly positive buoyancy and supporting it in a horizontal position.

Surface equipment

The surface equipment consists of the control console and the power, storage, handling, and tracking equipment.

Control console: The control console contains six bays. The operator's bay includes the controls for the vehicle's motors, lights, and tool assembly; it also includes the two television monitors. The other bays contain the controls for the active and passive sonar and pinger locator. The console is housed in a 4·55 m (15 ft) air-conditioned van.

NEMO

Nemo is the first underwater vehicle to utilise acrylic plastic for its hull structure. The vehicle was designed and developed at the Naval Civil Engineering Laboratory, Point Mugu, California and has been transferred to the US NUC to become a part of the NUC family of submersibles. The vehicle has the capability of either operating in a tethered or untethered mode.

The *Nemo* system provides a panoramic view of hydrospace and functions as a comfortable in-situ centre for observation, documentation, communication and control. *Nemo* is an untethered submersible, completely self-contained, that provides a one-atmosphere environment for the operator and passenger.

The spherical acrylic plastic pressure hull is surrounded by a structural cage that carries the entire weight of the *Nemo* system during lifting and protects the acrylic plastic hull from impact loads. Directly beneath the hull is the unit which contains the main ballast tank, service module, winch, and the main battery pack. The lower unit and structural cage are constructed of mild steel covered with an epoxy-based paint.

Nemo was certified on 10 December 1970 by the US Naval Ship Systems Command for operations to 182 m (600 ft).

Basic dimensions

Length, overall	2·3 m (7 ft 6 in)
Width, overall	2·3 m (7 ft 6 in)
Height, overall	2·8 m (9 ft 2·4 in)

Weights

Weight in air	3,750 kg (8,300 lb)
Payload, including crew	383 kg (850 lb)
Crew	1 operator, 1 observer

Performance

Maximum operational depth	182 m (600 ft)
Speed, maximum	0·75 knots
Propulsion endurance	8 hours
Life support endurance	normal 8 hours, additional 24 hours
Winch-down speed	18 to 30 m/min (60 to 100 ft/min)

Nemo before a dive

Power
Main battery (series connected): Twenty-one, 6 volt, 150 amp-hour, lead acid (120 vdc)
Secondary battery (internal): 25·5 vdc, 20 amp-hour, silver zinc to operate all emergency equipment.

Hydraulic systems
Main hydraulic system: Lear Ziegler 5 h.p., 120 vdc motor driven Vickers variable displacement hydraulic pump; 2·8 gpm at 3,000 lb/in²; Manual hydraulic system (internal) used to: cut anchor cable and drop battery in emergency; and to control mechanical winch brake.

Rotation motors
Two Vickers 1½ h.p. 3,000 rpm hydraulic motor thrusters with 20·4 cm (8 in) diameter outboard motor screws; limited ¾ knot translation possible.

Pressure hull

Material	Acrylic
Shape	Spherical
Outside diameter	1·65 m (66 in)
Wall thickness	6·35 cm (2·5 in)
Implosion depth	1,250 m (4,200 ft)

Safety factors
Hull design: 7 on collapse (instantaneous loading); 4·5 on collapse (50-hour load duration); 2·1 on yield (50-hour load duration)

Environmental
External:

(a) Ocean temperature envelope: 1°C to 21°C (34°F to 70°F)

(b) Pressure: 18·5 kg/cm² (267 lb/in²)

(c) Maximum sea state: Beaufort 3: wind 7 to 10 knots, waves 0·61 m (2 ft)

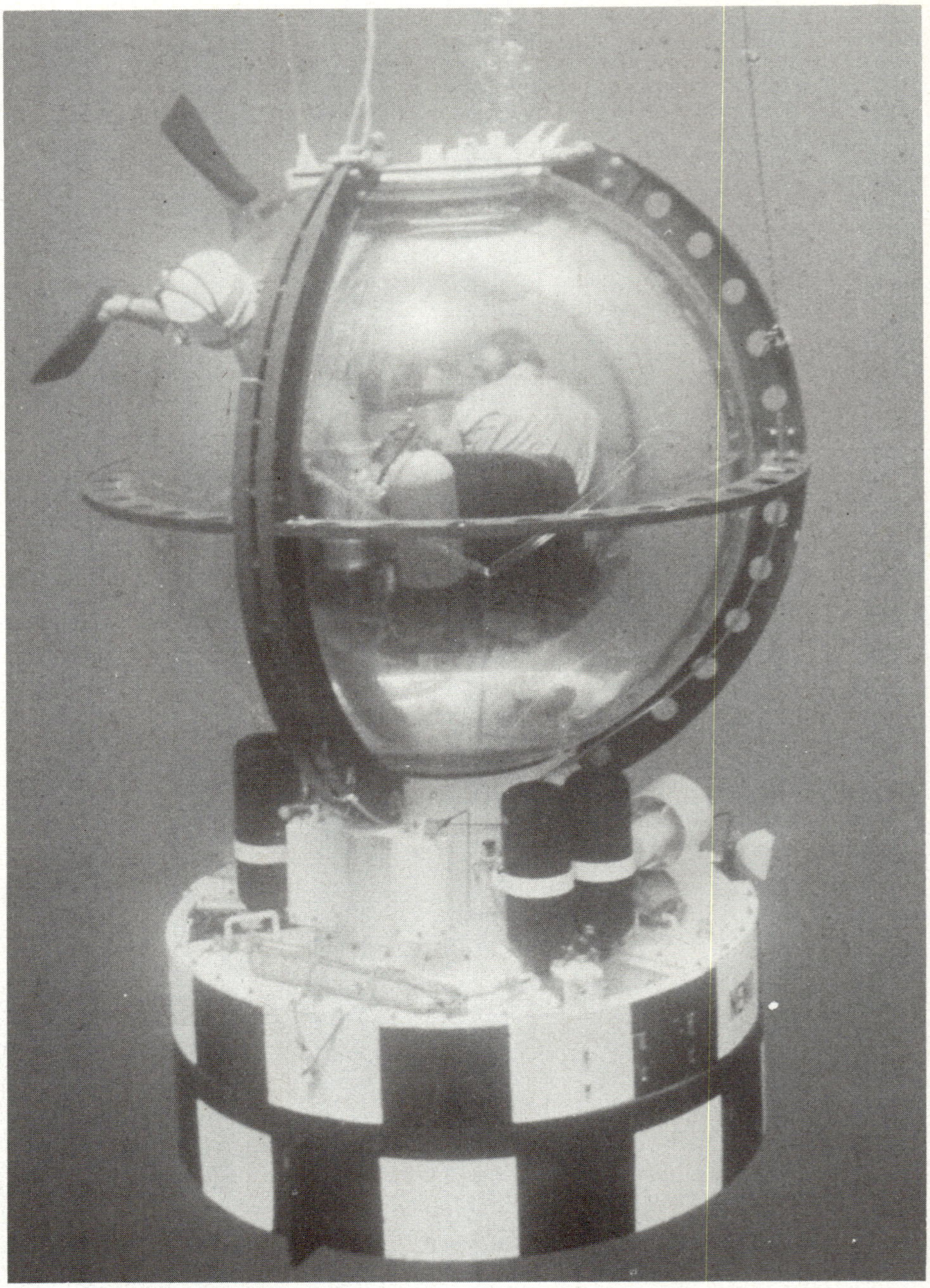

Scuba divers in attendance on the submerged *Nemo*

Internal:

(a) Partial pressure of oxygen: between 18 and 23 per cent of one atmosphere.

(b) Partial pressure of carbon dioxide: less than 1 per cent of one atmosphere

(c) Internal temperature: less than 24°C (75°F)

(d) Internal humidity: sufficient amount of moisture to be removed to prevent fogging and condensation on the inside hull walls; internal equipment and materials to withstand and be operable under 90 to 100 per cent humidity conditions.

(e) Type of material: non-toxic, non-flammable and non-outgassing materials used for internal subsystems

Design factors, stability and safety
Dynamic load (rough lifting, launching operations, and wave slap): 3g
Static stress: calculations include weight of entrapped water during lifting

Stability

Condition	*Distance between centre of buoyancy and centre of gravity*
(a) Submerged, Main Ballast Tank (MBT) flooded:	79·8 cm (31·4 in)
(b) Submerged, MBT dry:	61·5 cm (24·2 in)
(c) Surface, MBT dry:	54·8 cm (21·6 in)
(d) Surface, MBT dry, anchor on the bottom:	49·0 cm (19·3 in)
(e) Surface, MBT dry, anchor and battery dropped:	38·3 cm (15·1 in)

Safety factors
Structural cage design: 5
Support system design: pressure compensated except for lights and air ballast supply tanks
System tests: 1·5 times working pressure

Life support (2 men for 8 hours, 24 hours additional)
(a) O_2 supply.
(b) CO_2 absorbent fan forced.
(c) Humidity (silica gel and ice) 70 to 80 per cent rel. humidity.
(d) Beckman pO_2 meter.
(e) Kolisman cabin pressure gauge.
(f) Two closed circuit rebreathers (emergency).
(g) Drager Kit (backup for CO_2 and pO_2 pressure).
(h) Relief and check valve to cabin overpressure relief.

Anchor winch
(a) Winch drive: Vickers fixed displacement, full reversible 2 h.p. hydraulic motor.
(b) Drum capacity: 365 m (1,200 ft), 6·4 mm (¼ in) wire rope

(c) Winch down speed: 18 to 30 m/min (60 to 100 ft/min); 175 kg (400 lb) anchor

Ballast

(a) Main: Cylindrical ballast tank, 0·25 m^3 (8 ft^3)

(b) Auxiliary: Aft in service module, 0·057 m^3 (2 ft^3)

(c) Six 1·4 m^2 (50 ft^3) one 0·2 m^3 (71 ft^3), 155 kg/cm^2 (2,250 lb/in^2) air tanks

(d) Lead weights in battery pack.

Communications

(a) Surface: Johnson HF transceiver

(b) Subsurface: Aquasonic underwater telephone with slant range of 3,650 m (4,000 yds), compatible with AN/UQC-1

(c) Diver-operated unit available.

Applications

Inspections: Harbours; Ships; Moorings to 183 m (600 ft); Underwater work platforms, arrays, manned and unmanned stations

Support: Photographic, stereo; Directing diver work (U/W telephone available); Coordination of operations (diver, ship, other submersibles); DSRV operations, matings, SDS; Experimental stations (Seacon, etc.)

Evaluation: Handling systems studies; U/W power tools; U/W cable performance; U/W power sources; Subsystems (manipulators, lift packages, lighting, cameras, etc.); Mooring performance; Hull-type model studies; Buoy studies; Diving systems; Cavitation studies

Pollution control: Oil leaks; Sewage outfall; Radioactivity measurements; Artificial reef studies; Fish die-offs; Disposal areas; Ecology studies

Search capabilities

Ocean research: Deep scattering layer; Fish studies (recordings, anti-shark devices, black cod studies, etc.); Borings; Corrosion and fouling; Continental shelf sea floor; Currents and turbidity-currents; Sand movement, beach erosion, sediments; Light transmission; Laser studies; Magnetic and gravitational fields; U/W radio wave propagation; Hydrodynamics; U/W sound studies; Polar applications; Sea animals (walrus, sea lion, etc., as in "Manned Undersea Science and Technology Program" with Alaska Dept. of Fish and Game, NUC, NSF, NOAA).

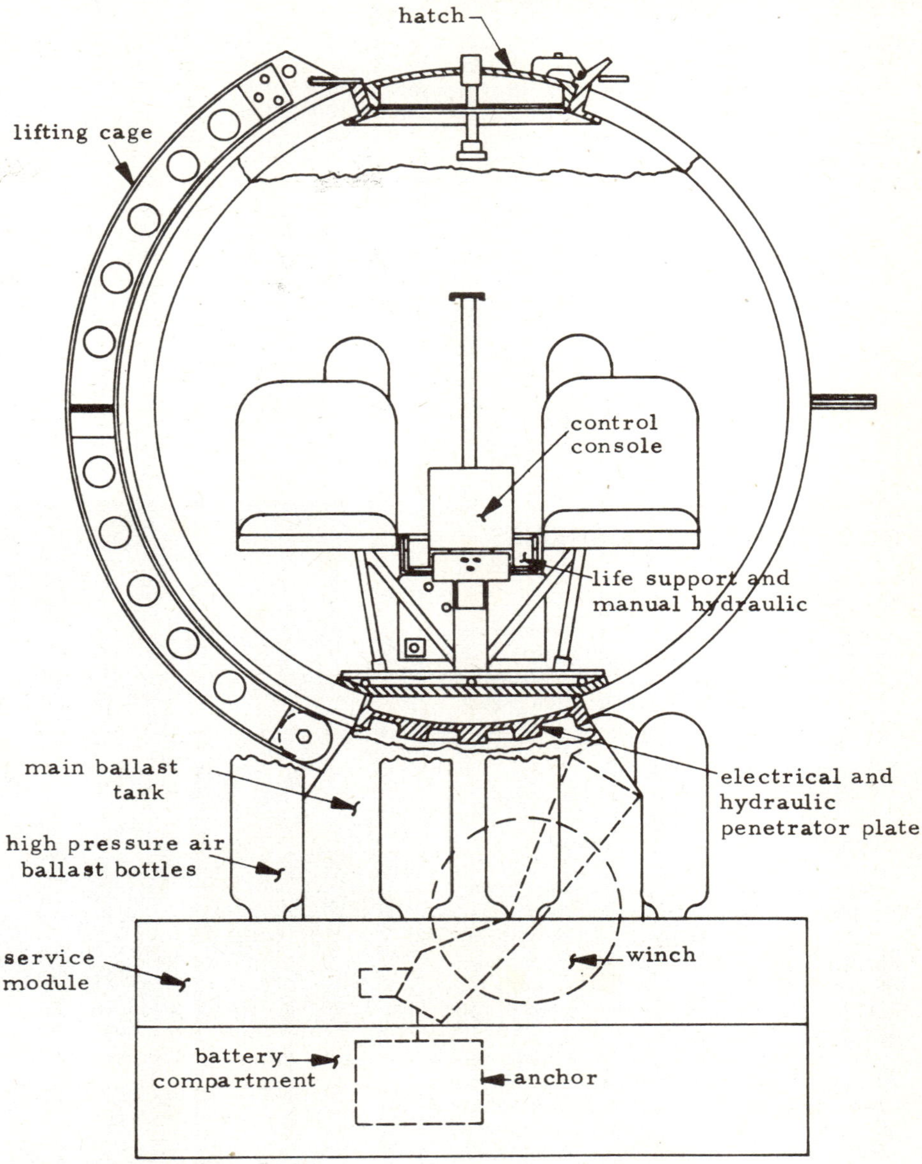

An elevation diagram of *Nemo*

THE NAVAL UNDERSEA CENTER

SEA-SEE

The *Sea-See* is a catamaran hull-type craft outfitted with an underwater observation compartment. The craft is designed for operations in Southern California water. It is capable of fully self-contained support for periods of one week or longer. With the observation compartment extended, the vessel is capable of maintaining 3 knots and is an excellent platform for the investigation and research of marine mammals, sharks, near surface ecology. In addition, the fibreglass hull provides an ideal platform for acoustic studies.

Characteristics

Hull	Catamaran type
Length	15 m (50 ft)
Beam	6·1 m (20 ft)
Power	Two GMC 353-N diesel engines
Draught	1·05 m (3 ft 6 in) with observation capsule up; 3 m (10 ft) with observation capsule down
Speed	7·5 knots with observation capsule up; 3 knots with observation capsule down
Range	480 km (300 miles)

Electronic equipment: Sonar Corp. Model 115 marine radiotelephone; Raytheon ADF;

A Scuba diver in attendance on *Sea-See* while the craft is submerged

Raytheon Model D-120 m depth indicator, 76 m (250 ft) maximum; Raytheon Model 1900 radar; Motorola 35-watt UHF FM transmitter-receiver; Motorola 15-watt UHF FM portable transmitter-receiver; Binaural hydrophone array and system frequency range, flat ±3 dB, 20 Hz to 100 kHz; Lockheed Model 417 instrumentation tape recorder-reproducer, seven-channel direct record-reproduce, frequency response ±3 dB, 200 Hz to 100 kHz at 30 in/sec

Electrical power: 12-volt and 24-volt DC, 130 amps; Kohler 110 to 115-volt AC, 60-cycle, 2·5-kW motor generator; CML Model MRS 500 24-volt DC to 110 to 115 AC inverter (500 watts)

Accessories: Electrical anchor winch; Four-transducer underwater pass-through in observation capsule; 4·12 m (13 ft 6 in) Boston Whaler, 20-h.p. Johnson outboard motor; Two 250-watt thallium iodide underwater lights; Instrumentation console

Personnel: Four scientists, two crew, observation capsule seats two; In addition, sleeping accommodations are provided for the crew and two scientists with a galley, head, for hotel services

Photographic capability: Still and motion picture

Applications

Marine mammal investigation; Shark studies; Kelp bed investigation; Shallow water artificial reef investigation; Acoustic studies; Support for diving operations

UNION OF SOVIET SOCIALIST REPUBLICS

ATLANTIC INSTITUTE OF ORGANISATION OF FISHERIES

ATLANT 1

Referred to as a bathyplane, this underwater sailplane or glider is used by the staff of the Atlantic Institute of Organisation of Fisheries for examining the behaviour of fishing gear.

Characteristics

Length	4·5 m (14 ft 9½ in)
Weight	2 tons
Design maximum operating depth	200 m (656 ft)
Illumination for filming	5 searchlights of 500 watts each

The craft was designed in the city of Klaipeda, Lithuania. Designated in the main for fishing exploration, it will help control the work of trawlers, study the behaviour of fish and carry out filming operations.

The apparatus is equipped with wings and operates on the principle of a glider. Lowered on the water by its tugboat, the 2-ton bathyplane freely maintains itself on the surface. When moving, the increasing downwards pressure on the wings counters its buoyancy and it 'dives' into the water. The depth to which it sinks is 200 m (656 ft).

The pilot-observer, in a hermetically sealed chamber, can manoeuvre the apparatus in the vertical and horizontal planes and can watch through four portholes how the trawlers are filled, and takes still pictures and films. Five searchlights of 500 watts each provide sufficient illumination for observation. In taking pictures a photo-flash connected to the camera operates synchronically.

The bathyplane's electric supply and the pilot's telephone communication with the tugboat are provided by means of cables running through the towline.

The first deep-sea apparatus passed its tests in the Mediterranean successfully and has since been employed for submarine research in the Atlantic. Plans existed in 1968 for the construction of several more 'submarine gliders' but it is not known if these were implemented.

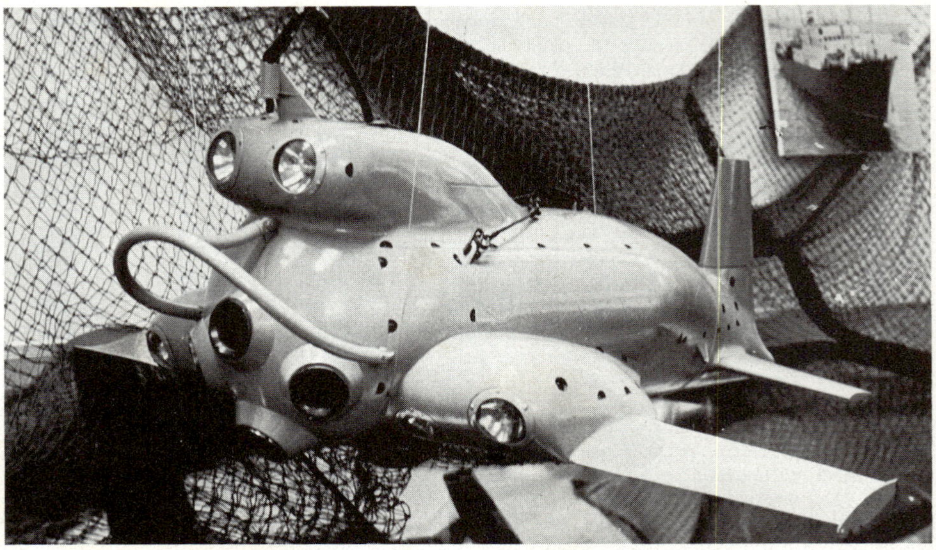

A model of *Atlant 1*, the Soviet underwater research submersible

Atlant 1 on display at a Soviet fisheries and oceanographic equipment exhibition in 1968

The recovery of *Atlant 1* after a diving operation

Atlant 1 before a dive in 1966. This Soviet observation submersible is designed for fisheries research and underwater photographic work

The Soviet 'bathyplane' *Atlant 1*, is designed for fisheries research, but can also take geological samples from the ocean bed. The submersible has been used to study underwater fauna and plant-life down to a depth of 200 m (656 ft), and is used in archaeological investigations of the sea bottom

LENINGRAD HYDRO-METEOROLOGICAL INSTITUTE UNDERSEA RESEARCH LABORATORY

SCORPENA

This vehicle was designed by the All-Union Sea Fisheries and Oceanography Research Institute (VNIRO) and was built in 1971 as a joint venture with a group of engineers from the Novosibirsk Scientific Production Association 'Faked' under the leadership of G. M. Chornykh at a Pilot Plant of the USSR Academy of Sciences, Siberian Division.

The submersible described belongs to the category of unmanned self-contained deep-submergence vehicles for mid-water and sea/ocean floor research, which may be deployed from oceanographic research and fish-finding vessels. The unit comprises a self-contained unmanned submersible intended for general-purpose scientific apparatus transportation in a deep-sea environment. Carrying movie- still- and stereo photogrammetric cameras, video recorders, facilities for sensing and recording underwater illumination, bio-luminescence, and other optical and hydrological parameters, etc., the vehicle is purpose-built for conducting a variety of research programmes, is never in contact with the surface by either cable or rope, and features emergency surfacing and surface-detection systems, etc. The submersible's operational path may be controlled without application of hydrodynamic stabilisers or rudders, and may vary from a steep sawtooth-like configuration to virtually level passage by altering the direction of propulsion-motor thrust and the time of its operation. Requisite operational silence in transferring to working locations is obtained in the following manner: descent, e.g. to bottom or, in sawtooth operating mode, to reascent layer, occurs with the propulsion motor shutdown under the effect of gravity, the predetermined submergence rate (down to minimum value) being sustained by controlling negative buoyancy.

Scorpena's watertight hull is of sectional design, each section being an independent functional unit. Some serve as floats, while others house the automatic control facilities and support systems. Sections are arranged one above the other from bottom to top in order of increasing buoyancy to provide a large metacentric height, small cross-sectional area (which diminishes perturbation of the environment), and convenience of static trim. The craft is easily dismantled into individual sections to aid transportation. The sectional feature is also meritorious in facilitating variation in carrying capacity when restocking with scientific equipment. This is accomplished by varying the number of float sections. On the other hand, replacement of the sections containing automatic control modules enables the craft's operational programme to be varied. *Scorpena* is fitted with a rotating electric motor, whose variable-angle installation relative to the craft's vertical axis enables, in conjunction with control of the time of its operation and propeller speed, a variable path to be pursued. Scientific equipment is carried in individual containers.

The pressure-hull sections comprise hollow cylinders horizontally arranged one above the other. Float sections are arranged above and sections housing automatic control modules and support systems below. The number of sections may be varied. Increase or decrease in the number of float sections changes the vehicle's carrying capacity. Sections containing automatic control modules and support systems are interconnected by cables and watertight joints. High up in the structure is a liquid-ballast tank forming part of the false hull. Surmounting the tank are a radio beacon, a flashing signal light, and a frame to which is attached the rotating electric motor. This motor may be turned in the frame at a specific angle relative to the craft's vertical axis. At any installation angle, the thrust line will be oriented in such a way as to prevent inclination of the submersible during motor propulsion. A protective jacket and fairing are held on catches and, being part of the false hull, double as solid ballast, which can be jettisoned in emergency situations. These fairings bear attachments for additional loads —controllable solid ballast for partial adjustment of vehicle buoyancy and static trim. The emergency ballast dumping system is duplicated. There is also a liquid-ballast blowing system (compressed-air bottles in the version shown). Below are struts for supporting the submersible on deck and brackets for attaching replacement scientific equipment. Standard items of equipment are pressure and hydro-acoustic working-depth sensors.

Scorpena operates in the following manner: Prior to deployment, the requisite number of float sections are determined in accordance with the weight of the scientific apparatus to be carried. Depending on the operational path envisaged, the rotating motor is inclined, the per-cycle time of its operation preset by appropriate tuning of a time relay, and the motor's rotational speed preselected to control the craft's speed of ascent along the indentations of the course it will steer. A variety of operating modes are possible: both midwater passage with pressure-sensor control and at a fixed distance from the bottom with hydro-acoustic sensor control. The appropriate mode is set by means of a switch. Programmed surfacing occurs in one of the following ways: (a) the vehicle surfaces after shooting the required number of frames; (b) the vehicle surfaces when the camera film or recorder tape has finished; (c) the vehicle surfaces after a given time has elapsed. A special signal light indicates the operational readiness of all vehicle systems. Then, following installation and arrangement of the controllable solid-ballast load, the submersible is trimmed in the afloat condition, and negative buoyancy imparted to it in such a way as to provide the necessary submergence rate.

After recovery of the deployment hook and flooding of the ballast tank, *Scorpena* initiates its descent, approaching the objects to be photographed in virtual silence with all equipment switched off. If the object of investigation is a shoal of fish detected by echo sounder and the depth-response programme is operational, the photographic apparatus will be triggered and the propulsion motor engaged either in response to the hydro-acoustic sensor when the image of the object sought is close enough to be defined by the camera lens, or in response to the pressure sensor when a given depth is reached. On engagement of its motor, the submersible transfers into oblique ascent for some distance and then at its new location, on functioning of the time relay shutting down its motor, is turned again into descent, inaugurating a new operating cycle. A similar operational pattern is enacted during passage near the sea bed only the motor is started up at attainment of a given distance from the ground on signal acquisition from the relevant sensor. Operating cycles of the submersible recur until the surfacing programme is actuated. Liquid ballast is blown and *Scorpena* rises to the surface. When 1·5–2·0 m (5–6½ ft) depth is reached, a submergence sensor causes the flashing signal light and radio beacon to be energised. Surface detection of the submersible is either visual or by radio detection finder.

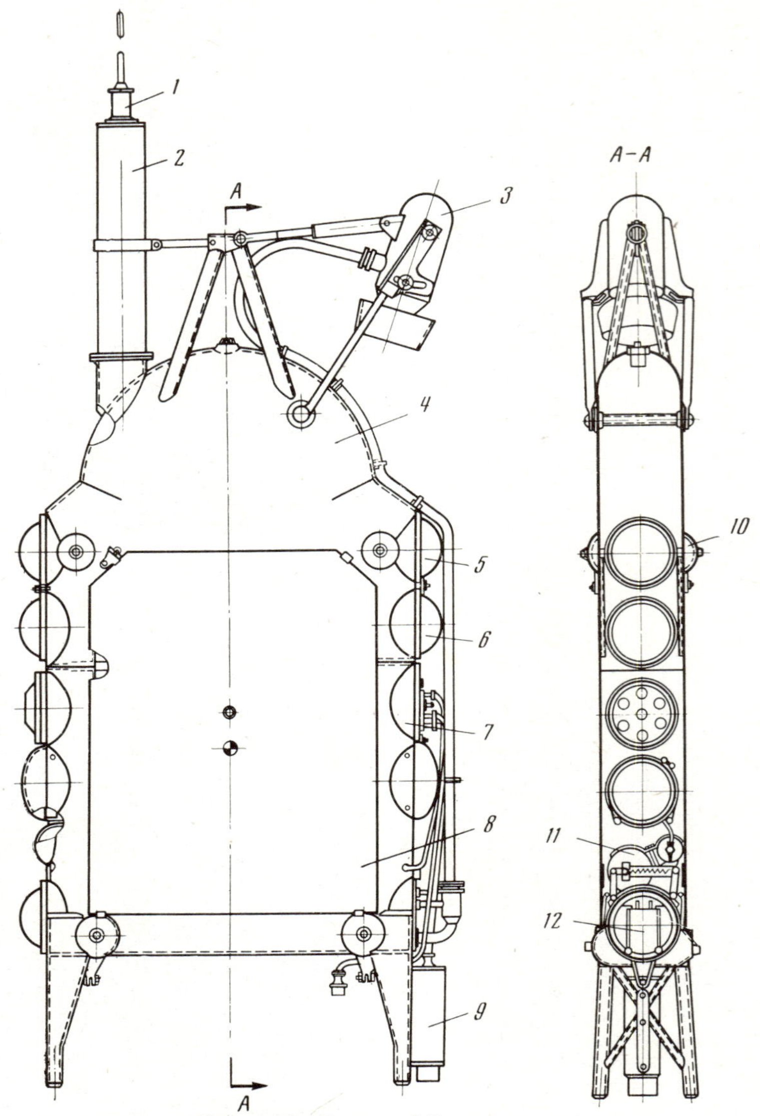

Schematic diagram of *Scorpena*

Key: (1) flashing signal light; (2) radio beacon; (3) rotating electric motor; (4) ballast tank; (5 and 6) float sections; (7) sections containing automatic control modules; (8) protective jacket doubling as solid ballast; (9) hydro acoustic ground-distance sensor; (10) supports for horizontal stowage of submersible; (11) compressed-air bottles and solid-ballast dumping arrangement; (12) sections containing support system modules

The design concept incorporates an emergency surfacing system, which functions in the following cases: (1) where the craft's watertight integrity is in any way impaired; (2) where a hazardous depth is reached; (3) where the vehicle exceeds a critical angle of heel; (4) where the vehicle outstays its scheduled submergence time; (5) where feed voltage drops below a critical level; (6) where there is a pressure drop in air bottles intended for ballast-tank evacuation and emergency-ballast release.

In an emergency situation, both the ballast-tank evacuation system and emergency-ballast release system are actuated.

Principle particulars

Out-of-water weight	410 kg (902 lb)
Submerged displacement with ballast tank flooded	515 kg (1,133 lb)
Length	1·5 m (4 ft 11 in)
Breadth	0·38 m (1 ft 3 in)
Depth without antenna	3·25 m (10 ft 8 in)
Height of flashing light afloat with liquid ballast blown	1·2 m (3 ft 11 in)
Overall (out-of-water) weight of emergency ballast	40 kg (88 lb)
Design submergence depth	1,000 m (3,280 ft)
Capacity of power-circuit battery during continuous operation of motor at medium speed	1 hour
Radio beacon detection range using associated radio direction finder	20 km (12·4 miles)
Flashing light detection range from SRT medium trawler	15 km (9·3 miles)

Preliminary analysis of the scope for submersible application indicates its suitability for investigation of the structure of deep-sea fish shoals and accumulations, interpretation of echo soundings, acquisition of data on the distribution of demersal and pelagic marine life, detection of fish accumulations unregistered by echo sounders, and for research in hydrology, marine physics, and marine geology.

POLAR INSTITUTE OF FISHERIES AND OCEANOGRAPHY

SEVER 1

This vessel is designed for deep water observation purposes and is believed to be also referred to as the GG 57 hydrostat. The depth capability is 610 m (2,000 ft) and endurance, 6 hours.

Sever 1 aboard the research ship *Tunets* in the Barents Sea.

SEA BED VEHICLES

GERMANY

DEUTSCHE BABCOCK & WILCOX AG

DEEP SEA BOTTOM SAMPLER

In order to obtain more information about the sediment deposits and the nature of the sea floor, Deutsche Babcock & Wilcox AG, has built a bottom sampler for the German Federal Ministry for Science and Research.

The apparatus is towed by ship over the sea floor and can take bottom samples from a trench 300 mm (11·8 in) wide, the penetration depth being variable between 50 and 150 mm (1·96 and 5·9 in). The samples are stored in 18 buckets arranged around the periphery of the excavating wheel. Prior to lowering the apparatus to the floor of the sea an excavating plan is set on the machine. On an excavating route of approximately 100 m (328 ft) it is possible to fill one bucket after another, or if needed a variable interruption of storing can be planned between bucket filling operations and the excavating route can be extended to approximately 6,000 m (19,700 ft). Every 300 m (1,000 ft) one bucket of sediment will then be collected. Each bucket has a capacity of 120 litres (4·23 ft^3). After one excavating run it is possible to take 2,100 litres (74·2 ft^3) of sediment on board. The buckets may be taken individually out of the bucket rim suspension so that it will be possible to evaluate the nature of the sea floor by sampling routes.

The apparatus is made from stainless steel and aluminium resistant to sea water. The sampler has been submerged to the deep points of the oceans. The strength of all materials used on the apparatus withstands a pressure over 500 kp/cm^2 gauge. The sampler is 5 m (16·4 ft) long, 3·7 m (12·2 ft) wide and 2·2 m (7·3 ft) high and its weight is 4 tons when empty and approximately 8 tons when filled with sediment and water. The excavating wheel and the driving wheels have a diameter of 1·6 m (5·3 ft). The bucket rim and the control mechanism of the sampler are driven by the rolling force of these wheels.

It will be possible, at a later date, to provide this unmanned underwater apparatus with its own power source and self propulsion unit. It is also possible to use the apparatus, in its present design or in a larger version, for collecting manganese nodules from the floor of the sea.

A view of the Babcock Deep-Sea Bottom Sampler

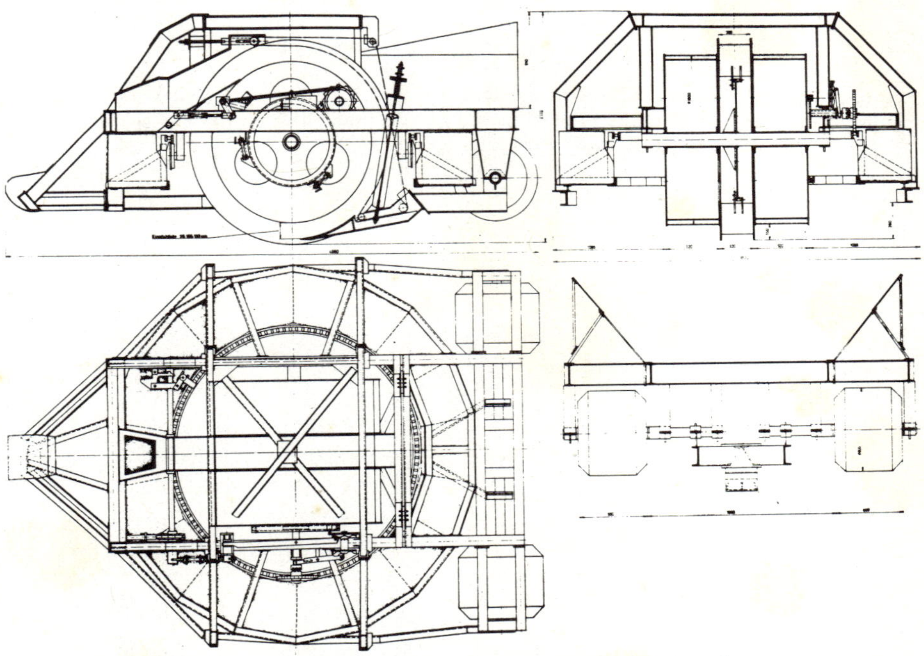

Construction diagrams of the Babcock Deep-Sea Bottom Sampler

HAGENUK VORM NEUFELDT AND KUHNKE

SEA BED VEHICLE

The Hagenuk sea bed vehicle is remote-controlled via a floating cable and can be used as an equipment carrier for various tasks. An electrical sprocket drive enables all manoeuvres to be performed. A remote-controlled cable winch on the vehicle releases the required length of cable. The vehicle can be equipped with underwater television, underwater cameras, side-scan-sonar, manipulators and oceanographic sensors. The vehicle transmits its own data as well as the corresponding operating data to the shore or to the experimental vessel.

The Hagenuk seabed vehicle

JAPAN

HITACHI CONSTRUCTION MACHINERY CO LTD

JH 160 SHALLOW TYPE UNDERWATER BULLDOZER

Two types of underwater bulldozer systems have been developed by Hitachi. The first system completed in October, 1969 is called the shallow-type underwater bulldozer system JH 160, and can operate from 0 to about 5 m (16 ft) depth.

The second system completed in September, 1971 is called the deep-type underwater bulldozer system JH 360. This system is utilized for underwater bulldozing at depths of 5 to 60 m (16 to 200 ft).

The shallow-type bulldozer, JH 160 system, is composed of a power barge, an unmanned underwater bulldozer, and connecting bar.

The drive system of the underwater bulldozer has a hydrostatic transmission. The travelling speed of the bulldozer can be varied continuously from zero to the maximum forward and backward speeds by adjusting the oil flow rate.

Two air tanks are mounted on the bulldozer. Air is supplied to the tanks from an air compressor, and the bulldozer can be floated or lowered by adjusting the amount of air in the tanks. The bulldozer can be floated and towed to a work-site. The air tanks also serve to adjust the operating weight when the bulldozer also is underwater.

A special 'sign' pole is attached to the unmanned bulldozer which enables the operator to control the vehicle from the control cabin on the support craft (power barge). The marked pole protrudes above the surface of the water, indicating the inclination and depth of the bulldozer, as well as other operational conditions such as the position of the bulldozer's blade.

The power barge is a self-propelled catamaran equipped with a diesel-driven hydraulic power unit, an air compressor, and a control cabin. Operating devices are fitted inside the cabin.

The connection bar does not interfere with being connected to the top of the bulldozer by a universal coupling, and the other end to the bottom centre of the power barge by a chain.

The connection bar does not interfere with the motion of the bulldozer and the appropriate distance is always kept between the power barge and the bulldozer. The air and oil hoses are fixed to the sides of the bar so that they do not interfere with work in progress.

Leading particulars

Underwater bulldozer

Operating weight	
Ground weight	16·0 tonnes
In water:	13·0 tonnes
Overall dimensions	
Length, width, height	6·2 m × 4·0 m × 2·8 m (20·3 × 13·1 × 9·2 ft)
Operating speed	Forward and reverse, 0–3 km/h (0–1·9 mph)
Hydraulic motor	Radial piston motor, Max. 210 kg/cm² (3,000 lb/in²)
Working depth	0–5 m (Standard) (0–16 ft)
Drawbar pull (in water)	10·0 tonnes

Catamaran power barge

Weight	14·0 tonnes
Dimensions	
Length, width, height	11·6 m × 6·2 m × 1·3 m (38·0 × 20·3 × 4·3 ft)
Prime mover	90 h.p. 2,000 rpm diesel engine
Speed	2 knots

Power barge equipment

(a) Power unit	
Engine model	160 h.p./1,600 rpm Hitachi B-60 diesel engine
Hydraulic pump	Variable axial plunger pump, Max. 210 kg/cm² (3,000 lb/in²)
(b) Air compressor	Hitachi PS-3 Portable compressor 7 kg/cm² (100 lb/in²)

Applications

The system can be used for leveling and grading of the sea bottom, sea-bed 'weeding', sea-bed ditching and levelling and compacting of the rubble bed of sea walls.

THE JH360 UNDERWATER BULLDOZER

This system is equipped with various up-to-date devices such as acoustic positioning plotter, acoustic forward surveyor, depth finder, pitch and roll indicator, etc., and can perform bulldozing in dark, muddy seabottom areas. The bulldozer can float or be lowered in the sea by its own power. It can be towed floating to a new working site thereby eliminating the need for a largescale hoisting system such as a floating crane. The system is electrically operated under remote control through an electronic and acoustic indicator, and with many types of attachments may be applied to a wide variety of tasks.

The power barge and bulldozer are connected with hoisting wire rope, power and control cable and air hose. The bulldozer tracking sonar is connected to the power barge. The construction of the bulldozer is completely watertight and withstands water pressure at maximum design depth. The operator in his cab on the power barge controls the bulldozer electrically by observation of its movement on the cathode-ray tube. Air tanks are installed on the bulldozer to facilitate its floating or lowering through the water by a winch on the power barge. The power barge is towed or moved by a tug boat, pusher or anchor wire take-up system.

The Hitachi JH 360 underwater bulldozer

Specification

Underwater bulldozer (with air tanks)

Operating weight: 32 tons on land, 24 tons in water

Length, width, height: (7·2 × 5·1 × 4·2 m) (23·6 × 16·7 × 13·8 ft)

Operating speed: (0 to 3 km/hr) (0 to 1·9 mph) forward and reverse

Travelling device: Hydraulic unit; Variable displacement pumps and low-speed motors, 200 kg/cm² (3,000 lb/in²) 100 HP × 2

Electric motor: 150 KW, 3,300 V

Max. working depth: 60 m (197 ft)

Acoustic indicator and monitor devices: Acoustic positioning plotter for bulldozer in water; Acoustic forward surveyor for bulldozer in water; Monitor for bulldozer heading, roll and pitch in water; and other instruments

Power barge (bulldozer commander with 360 h.p. diesel electric generator plant) Non-self propelled barge (made by Watanabe Steel Works, Ltd.)

Fully equipped weight: approx. 100 tons
Length, width, height: 16·5 × 10·2 × 1·8 m (54·1 × 33·5 × 5·9 ft)
Equipment on board: Electric power generating plant
Main diesel engine: 360 h.p.
Main electric generator: 300 kw, 3,300 V
Bulldozer hoisting winch: electro-hydraulic type, 1 set
Air compressor: 7 kg/cm^2, 10·3 m^3/min (99·4 lb/in^2, 363·8 ft^3/min)
Power and control cables: 110 m (361 ft) long, special compound cab-tire cables made by Hitachi Cable Co

Control system

The following devices are installed on the control desk on the power barge:

Acoustic positioning plotter

The interogator connected to the power barge transmits a supersonic signal into the water and receives the response from the transponder installed on the underwater bulldozer, thereby indicating the plan position of the bulldozer in the water on the cathode-ray tube on board the barge

Acoustic forward surveyor

Supersonic scanning is performed by the transducer installed on the front arm of the bulldozer, and the profile of the sea bottom ahead of the bulldozer blade is displayed on the cathode-ray tube on board the barge.

By means of the above-mentioned cathode-ray tubes as well as the bulldozer monitoring gyroscope, depth finder, blade lifting and lowering meter, etc., arranged around the cathode-ray tubes, the operator can manipulate or control the bulldozer accurately for its underwater operations, observing the machine in the water on the instrument panel

JH 360 working on the sea bed

Applications

The optimum working range of the underwater bulldozer is from 5 to 60 m (16·4 ft to 197 ft) in water depth, and a circular range of approximately 110 m (328 ft) in diameter, centring on the power barge

The following underwater operations are possible: Subsea foundation construction work (levelling ripraps) along the quay walls in harbours or breakwaters; Foundation work in land reclamation or bank protection work; Foundation work for construction of underwater pier, submerged tunnelling and underwater pipe-laying work. (Excavation, levelling and filling); Ripping or breaking of rocks, hard ground of sea bottom in harbours, or navigational line, etc

By relocating the power supply and control devices on a rugged structure or work stand on water, the bulldozer can be operated undersea in stormy or adverse weather conditions

KOMATSU LTD

UNDERWATER BULLDOZER

Komatsu's underwater bulldozer has been developed to meet the requirements of underwater exploitation, and it can be successfully operated to a depth of 60 m (197 ft). The Company developed the world's first amphibious bulldozer, operable to a depth of 3 m (10 ft) in March 1969, on orders from the Ministry of Construction of Japan. The amphibian fleet has since been engaged in river construction works throughout the country.

Characteristics

The underwater bulldozer has a dual control system; one is by radio remote control and the other by direct control by a diver, and either method can be selected according to operational conditions. The control panel may be installed on land or in ships and has a complete set of instruments to keep the operator constantly informed of the bulldozer's location, performance and inclination, and of the performance of the attachments. Electric power is utilized for the operation of the bulldozer. A water leak alarm and a circuit breaker are provided and are automatically actuated whenever leakage occurs.

The Komatsu underwater bulldozer

Specification
Bulldozer
Crawler type with blade and hydraulic ripper
Operating weight: 34,000 kg (74.960 lb) on land; 27,000 kg (59,520 lb) in water
Length, width, height: 8·4 × 3·7 × 3·2 m (27·56 × 12·3 × 10·7 ft)
Distance between track centres: approx. 2·14 m (7 ft)
Length of track on ground: approx. 2·93 m (9·62 ft)
Width of track shoe: approx. 7·67 m (2·34 ft)
Ground pressure: 0·6 kg/cm² (8·53 lb/in²) in the water
Travel speeds: Forward: 1st speed 2·0 km/h (1·2 mph); 2nd speed 3·5 km/h (2·2 mph)
Effective operating depth: 2 to 60 m (6·6 to 197 ft)
Generator output: 125 kw/1,800 rpm
Blade capacity: 7 m³ (248·4 ft³)

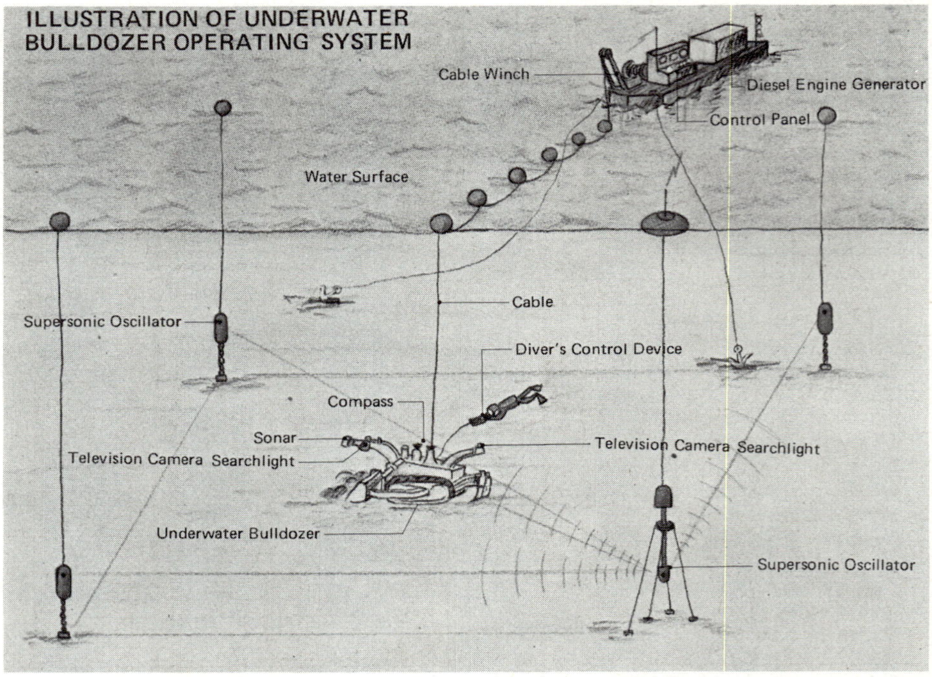

An artist's impression of the operating system associated with the Komatsu underwater bulldozer

Generator and engine
Portable diesel engine-driven type.
Engine: Komatsu-Cummins NRTO-6
Output 300 h.p./1,800 rpm
Generator: 170 KVA/1,800 rpm

Control panel
Cabinet type with control lever, instruments and monitoring device

Cable winch
150 m (492 ft) of generator cable can be retained on the motor-driven drum. Slip rings are attached to the ends of drum shaft, with sheaves and guide rollers

Other equipment
Equipped with transponder, compass, sonar, TV camera, etc

Applications
Suitable for the ripping of bedrock, digging and excavating, dozing and grading, at underwater civil engineering and construction sites. As with conventional land-use bulldozers, the underwater bulldozer is designed to accept various attachments to extend its utility. Operation at depths of up to 60 m (197 ft) should play an active part in underwater civil engineering and construction works

Future projects
Komatsu is planning to add two more amphibious bulldozers capable of underwater operation at 7 and 15 m (23 and 49 ft) to their underwater vehicle series, to give four differently featured bulldozers in their range.

UNITED STATES OF AMERICA

SCRIPPS INSTITUTION OF OCEANOGRAPHY

RUM

Rum is a remotely controlled, tracked sea-floor work vehicle equipped with television cameras and a claw for bringing back samples. It has operated down to depths of 1,250 m (4,100 ft) and is normally connected to an oceanographic research buoy (ORB) by a co-axial cable.

RESEARCH VESSELS

FRANCE

CENTRE NATIONAL POUR L'EXPLOITATION DES OCEANS (C.N.E.X.O.)

CAPRICORNE

Class of vessel
BV ✠ 1-3-3 F. (haute mer) I.I.A. and CP.

Specific purpose of vessel
Oceanographic vessel and fishing research (tropical seas)

Design history
Start of construction: 1969
Date of launching: 12 November 1969
Date trials were completed: 10 March 1970

Basic dimensions

Length, overall	46·55 m (152·7 ft)
Beam, overall	9·30 m (30·5 ft)
Height, overall	4·25 m (13·94 ft)
Draught	4·47 m (14·6 ft)

Weights

Displacement	710 tons
Crew	27 (plus 12 oceanographers)

Performance

Full speed	13 knots
Cruising speed	11 knots
Endurance at low speed	7,000 nautical miles at 10 knots (30 days)
Position holding capability	4 Beaufort
Towing capacity	15 tons

Main machinery description

Main engines	Baudouin DVX 12. Diesels 600 h.p. each (1,250 rpm)
Auxiliary engines,	GE. One 135, one 100 kvA
Propeller type	One controllable pitch
Controls— manoeuvring position holding	Wheel house

Passive stabilization system

On-board equipment
Sampling equipment: Trawl, purse seine, plankton nets, water sampling bottles
Craneage facilities: 1 stern gallow for trawling
Winching facilities, capacities: Fishing winch (2 × 2,700 m (8,856 ft) dia 22 mm (0·87 in))—Hov—Efeco.
(Lower Block) Line hawler Hydrology 1 × 7,000 m (22,960 ft) Dia. 4 mm (0·16 in) Electric sound STD (1 × 2,000 m (6,560 ft) Dia. 4·775 mm (0·19 in)

N/O *Capricorne* (*Courtesy delacroix – Poidvin*)

Plankton winch 8,000m (26,240 ft) Dia. 6.35mm (0.25 in)
Navigational equipment: Radar, Decca, RM 326, gyro anochütz and autopilot, gonio GFB 6
Survey and search equipment: Sonar Simrad SB. 3
Echo sounders Simrad EK 120, Atlas Tiefseelot Atlas Fish-Finder—Netzsond—Fish Echo Integrator

Laboratories

1	Physics	30 m² (325 ft²)
1	Chemistry—Biology	30 m² (325 ft²)
1	Electronic room with computer	12 m² (130 ft²)
1	Fish hold (0° Celsius)	
2	Fish wells	

Crewing requirements
Sea-going complement: officers 7, men 20
Scientific complement: 12
Accommodation: 16 cabins
Crew safety facilities and equipment: 1 lifeboat, life rafts, 1 Zodiac

Purchase, capital cost: 6·5 million francs (1970)

Operators
Office de la Recherche Scientifique et Technique d'Outre Mer (ORSTOM), 24 rue Bayard, 75008 Paris (Telephone BAL.31.52)

JEAN CHARCOT

Class of vessel
I—3/3 E Glace I
✠ Haute Mer

Specific purpose of vessel
Oceanographic research

Design history
Start of construction: 20 April 1964
Date of launching: 19 January 1965
Date trials were completed: November 1965

Basic dimensions

Length, overall	74·50 m (244·4 ft)
Beam, overall	14·10 m (46·25 ft)
Height, overall	7·45 m (upper decks) (24·4 ft)
Draught	5·05 m (16·56 ft)

Weights

Displacement	2,100 tons
Crew	48 (+22 scientists)

Performance

Full speed	15 knots
Cruising speed	12 knots
Endurance at low speed	10,000 nautical miles at 10 knots (40 days)
Position holding capability	5/6 Beaufort
Towing capacity	25 tons

Main machinery description

Main engines	three MAN diesel 1,120 h.p. each, 750 rpm
Auxiliary engines	GE. three 375, one 250 1 × 100 kvA
Propeller(s) type	2 fixed blade
Controls: manoeuvring position holding	wheelhouse 1 bow thruster, 1 stern thruster

Passive stabilisation system

On-board equipment

Facilities for carrying major equipment: 2 main cranes

Facilities for diving: decompression chamber

Sampling equipment: water sampling bottles, trawls, plankton nets, coring equipment

Craneage facilities: 1 crane, 12 tons; 1 crane, 3 tons; 1 crane, 1·5 ton

Winching facilities: trawl winch 2 × 3,000 m (9,840 ft) Dia. 13·4 mm (0·53 in), Dredge winch 2 × 12,000 m (39,360 ft) Dia. 13·4 mm (0·53 in), hydrological winch 1 × 6,000 m (19,680 ft) Dia. 5·3 mm (0·21 in) 1 × 8,000 m (26,240 ft) Dia. 5·3 mm (0·21 in)

Navigational equipment: Radars Decca, TMS 2,400 and RM. 426 SNS 'Transit' I.T.T./FL 4,007 AB, Decca navigator, gyro and auto-pilot A.O.I.L. gonio plath loran A/C Loch Sal, Electro-Loch Ben. Taiyo Fax TF 783

Survey and search equipment: Sonar Simrad SK 2A Echo sounder ELAC/ENIF EDO sonar, Gravity meter Askania

N/O *Jean-Charcot* (*Courtesy CNEXO*)

Laboratory and technical facilities

1 Geophysical laboratory

1 Scientific headquarters with data aquisition package (PDP8.E)

Electronics room, hydrological store and laboratory, chemistry and physical laboratories Wet laboratory, common laboratory, diving room

Geological laboratory with X-ray gravitymeter room

Portable containers 6·1 m × 2·4 m × 2·4 m (20 ft × 8 ft × 8 ft) laboratories:

—1 CTD profiling system

—1 Biological = Autoanalyser Technicon

—1 Seismic Data Recorder

Reel for streamer

N/O *Jean-Charcot* (*Courtesy MAZO*)

Crewing requirements

Sea-going complement: 15 officers, 33 men

Scientific complement: 22

Accommodation: 41 cabins

Crew safety facilities and equipment: 1 lifeboat, 1 hydrographic boat, life-rafts, 1 zodiac

Purchase, capital cost

20 million francs (1965)

Operators

Direction des Cables sous Marins, 20 rue Las Cases, 75007 Paris. (Telephone: 551 46 46).

CORIOLIS

Class of vessel

B.V. ✠ I 3/3 F.II.A and CP

Specific purpose of vessel

Oceanographic vessel and fishing research vessel (tropical seas)

Design history

Date of launching: 30 October 1963

Date trials were completed: February 1964

Basic dimensions

Length, overall	37·60 m (123·3 ft)
Beam, overall	8·00 m (26·24 ft)
Height, overall	3·80 m (12·46 ft)
Draught	3·75 m (12·30 ft)

Weights

Displacement	450 tons

Crew

18 (+12 oceanographers)

Performance

Full speed	11·5 knots
Cruising speed	10·0 knots
Endurance at low speed	10,000 nautical miles at 9·5 knots
Position holding capability	4 Beaufort
Towing capacity	8 tons

N/O *Coriolis* (*Courtesy Ville/CNEXO*)

Main machinery description

Main engines	Two Baudouin DV8 diesel 350 h.p. each, 1,100 rpm
Auxiliary engines	two GE, 55 kvA
Propeller type	one controllable pitch
Controls:	
manoeuvring	wheel house
position holding	

On-board equipment

Equipment for streamer (seismic)

Equipment for nodule research (manganese nodules)

Sampling equipment: trawl, long line, water sampling bottles, plankton nets

Craneage facilities: stern gallow 3 tons

Winching facilities, capacities: trawl winch 2 × 6,000 m (19,680 ft), dia. 9·5 mm (0·374 in), Plankton and hydrology 8,000 m (26,240 ft) dia. 6·35 mm (0·25 in)), 1 line hauler 1 G.E.K.

Navigational equipment, radar, position fixing: Radar Raytheon, gyro and auto-pilot Anschutz, Gonio. CRM, 1 loch log, 1 loch Ben SNS 'Transit' 6000 ITT.

Survey and search equipment, radar, sonar, TV, seismic: Echo sounder EDO 12 KHz 1 ECO sounder Atlas 10,000 m.

Laboratory and technical facilities

Wet laboratory, electronic room, fish ponds, fish hold (0°C to 2°C), refrigerated fish hold (−18°C to −20°C)

Crewing requirements

Sea-going complement: 5 officers, 13 men

Scientific complement: 12

Accommodation: 13 cabins

Crew safety facilities and equipment: 1 lifeboat, life rafts

Operators

Compagnie des Moyens de Surface adaptés a l'Exploitation des Oceans, (SURF) 148 rue Sainte, 13 Marseille, France.

CRYOS

Class of vessel

BV. ✠ 1.3/3E (Pêche Haute Mer) 1.1.A.CP Glace II

Specific purpose of vessel

Fishing vessel (Research and fishing in cold seas)

Design history

Start of construction: 1969

Date of launching: 7 February 1970

Date trials were completed: 12 April 1970

Basic dimensions

Length, overall	48·70 m (160 ft)
Beam, overall	10·00 m (32·8 ft)
Height, overall	6·55 m (upper deck) (21·5 ft)
Draught	4·55 m (14·9 ft)

Weights

Displacement	800 tons

Crew

22 (+9 oceanographers)

Performance

Full speed	13·5 knots
Cruising speed	11·0 knots
Endurance at low speed	7,500 nautical miles at 11 knots (30 days)
Position holding capability	5 Beaufort
Towing capacity	16 tons

Main machinery description

Main engines	2 diesel Duvant type 6, 1,380 h.p. at 430 rpm
Auxiliary engines	GE 1 × 220 kvA, 1 × 85 kvA, 1 × 330 kvA (coupled)
Propeller type	one controllable pitch
Controls:	
Manoeuvring	wheel house
position holding	active rudder 75 h.p.

On-board equipment

Sampling equipment: trawl, water sampling bottles, plankton nets

Craneage facilities: stern gallow for trawl, 2 booms (3 tons)

Winching facilities: hydrology = 1 (5,000 m, dia. 4)—1 (5,000 m, dia. 6), dredge (2,300 m, dia. 12·6)—fishing (2 × 3,000, dia. 22) — GEK, 1 Netsond

Navigational equipment: Radars = Decca AC 626 (true motion), Decca RM 314, Decca Navigator Mark XII, Gonio GFB 6 gyro and auto-pilot Arma Brown, loch sal 64, loran A/C

Survey and search equipment: sonar Simrad SB3, echo sounder Simrad EH2E and 'Scientific' 10,200 metres, 1 netsond

Laboratories

1 hydrological store

1 plankton room

1 fish processing plant

N/O *Cryos* (*Courtesy Lecture*)

N/O *Cryos* (*Courtesy Lecture*)

1 fish loading and sorting area

1 sorting room

1 physical and 1 biological laboratory

1 fish well

Room facilities

1 fish room (0°C)

1 Air-blast freezer (−40°C)

1 frozen fish room (−28°C)

Crewing requirements

Sea-going complement: 8 officers, 14 men

Scientific complement: 9

Accommodation: 18 in cabins

Crew safety facilities and equipment: 1 life boat, life rafts

Purchase, capital cost

6·5 million francs (1970)

Operators

Institut Scientifique et Technique des Pêches Maritimes, la Noe, Route de la Jonelière 44 Nantes (France) tel = 15(40) 74-99-81

LE NORDIT

N/O *Le-Noroit* (*Courtesy Photo Technic*)

Class of vessel
BV ✠ 1-3/3-E. Pêche Haute Mer 'Glace III'

Specific purpose of vessel
Oceanographic vessel

Design history
Start of construction: 8 July 1969
Date of launching: 13 November 1970
Date trials were completed: 17 June 1971

Production
Le Suroit, second vessel of 'Norois' type building, 1973

Basic dimensions

Length, overall	50·55 m (165·8 ft)
Beam, overall	10·66 m (34·96 ft)
Height, overall	6·12 m (20·07 ft) (upper decks)
Draught	4·32 m (14·17 ft)

Weights
Displacement 870 tons

Crew
20 (+10 scientists)

Performance

Full speed	14 knots
Cruising speed	12 knots
Endurance at low speed	7,500 nautical miles at 12 knots
Position holding capability	4 to 5 Beaufort
Towing capacity	15 tons

Main machinery description
Main engines: Two MGO type 12 ASHR diesels 825 h.p. each: 1,320 rpm
Auxiliary engines: GE one 150 kvA, one 625 kvA, coupled
Propeller type: one controllable pitch
Controls:
manoeuvring position holding: wheel house, 1 bow thruster, 200 h.p. 1 active rudder, 150 h.p.
Passive stabilization system

On-board equipment
Facilities for carrying major underwater equipment: 1 stern gallow (10 tons)

Facilities for carrying major surface equipment: 1 stern gallow, 1 crane 3 metric tons at 10 metres

Sampling equipment: water sampling bottles, trawls, plankton nets, coring equipment
Craneage facilities: 1 crane
Winching facilities, capacities: 1 winch dredge (7,500 m, (24,600 ft) dia. 12·7 mm (0·5 in), 2 winches hydrologic (2 × 8,000 m (26,240 ft) dia. 5·5 mm (0·217 in), 1 deck winch (3 tons)
Navigational equipment: Radar Decca RM 426 and ACS 630, Decca Navigator Mk 21. 'SNS' transit system CERW/CSEE/SAT—Cronio LNG Plath (Radio) Gryo and autopilot Arma Brown. Electromagnetic log Ben Mk. 5. Tavyofax TF 783
Survey and search equipment: Simrad echo sounder EP2BN EDO sounder 4025

Laboratory and technical facilities
1 Wet laboratory 30 m² (323 ft²)
Scientific headquarters 35 m² (377 ft²)
Powder magazine (for seismic work) gravimetry room

Crewing requirements
Sea-going complement: 6 officers, 1 electronic engineer, 13 men
Scientific complement: 10
Accommodation: 18 cabins
Crew safety facilities and equipment: 1 life boat, 1 zodiac, life rafts

Purchase, capital cost
10 million francs (1971)

Operators
Direction des Cables sous Marins, 20 rue Las Cases, 75007 Paris. (Telephone: 551 46 46).

LA PELAGIA

N/O *La Pelagia* (*Courtesy ISTPM*)

Class of vessel
Pêche Haute Mer

Specific purpose of vessel
Oceanographic and fishing vessel

Design history
Start of construction: October 1964
Date of launching: 14 September 1965
Date trials were completed: 7 December 1965

Basic dimensions

Length, overall	32·43 m (106·4 ft)
Beam, overall	7·35 m (24·1 ft)
Height, overall	4·06 m (13·3 ft)
Draught	4·00 m (13·1 ft)

Weights
Displacement 242 tons

Crew
12 (+7 scientists)

Performance

Full speed	11·5 knots
Cruising speed	10·0 knots
Endurance at low speed	7,000 nautical miles at 10 knots
Position holding capability	3/4 Beaufort
Towing capacity	8 tons

Main machinery description

Main engines	Two diesel Baudouin D.V.8, 400 h.p. each, 1,200 rpm
Auxiliary engines	Two GE 80/90 KVA, One 17 KVA
Propeller type	One variable pitch
Controls:	
manoeuvring and position holding	Wheel house, active rudder

On-board equipment

Sampling equipment: water sampling bottles, plankton nets, trawls

Craneage facilities: Portuguese gallow, 2 cranes (2 tons each)

Winching facilities, capacities: Fishing winch (6 tons) (2 × 2,000 m (6,560 ft) Dia. 18 mm (0·71 in), Purse winch seine (1 × 300 m 984 ft) Dia. 16 mm (0·63 in), net winch, hydrologic winch (3,000 m (9,840 ft) Dia. 5 mm (0·20 in)

Navigational equipment: Rodan CRM-1226—Decca Navigator Mk. 21, Omega Radio, Receiver Sercel, Gyro and Auto-Pilot Arma Brown, Cronio CRM, Loch Ben, Tavyo Fax TF 783

Survey and search equipment: sonar SIMRAD SK2A, echo-sounder 3, SCAM 610, 1 EK 38 Simrad, 1 Atlas echo sounder

Laboratory and technical facilities

1 Fish hold 0° to −5°C 25 m³ (888 ft³)
2 Fish wells, capacity 20 m³ (710 ft³) (1 refrigerated)
1 common laboratory (10 m²) (355 ft²)

Crewing requirements

Sea-going complement: officers 4, men 8
Scientific complement: 7
Accommodation: 10 cabins
Crew safety facilities and equipment: life rafts, 1 Zodiac

Operators

Institut Scientifique et Technique de Peche Maritimes, La Noe, Route de la Joneliere, 44 Nantes, France.

FRENCH NUCLEAR RESEARCH CENTRE BIOLOGICAL DEPARTMENT

MARARA

A research vessel for biological experiments and controls, built by Ateliers et Chantiers de la Manche

Construction started: 6 November 1972
Launched: 3 May 1973

Specifications

Length, overall	43·00 m (141 ft 1 in)
Length, between pp	37·50 m (123 ft)
Moulded breadth	8·90 m (29 ft 2 in)
Depth to main deck	4·10 m (13 ft 5 in)
Draught aft	4·30 m (14 ft 1 in)
Bunkers	130 m³ (4,650 ft³)
Fresh water	35 m³ (1,240 ft³)
Propelling machinery	Two Baudouin diesels Mk. DVX 12 MJ HR coupled on a reduction gear driving a Baudouin variable pitch propeller
Horsepower	1,200 h.p.
Trial speed	12·2 knots

Accommodation to be provided for 28 including five officers and ten scientists.

The new vessel to be fitted with two well-equipped laboratories for physical and chemical measurements and biological research; there will be an hydraulically operated four-ton oscillating portico aft and five electrically operated winches for the fishing and recovery equipment.

UNITED KINGDOM

DECCA SURVEY LIMITED

DECCA SURVEYOR

This vessel was built by Appledore Shipyard in 1963, and was originally known as *Britania*. Towards the end of 1971 she entered the service of Decca Survey Limited for use in the North Sea. She was, at that time, claimed to be one of the most comprehensively equipped and versatile commercial survey vessels operating in European waters.

Specifications

Classification	Lloyds 100 A1
Signal letters	EBBC
Port of registry	London
Reg. nett tonnage	75·18
Reg. gross tonnage	567·97
Main propulsion	Type MN18 8 cylinder Polar diesel 8 cyls.
I.H.P.	2,000
B.H.P.	1,680
Length B.P.	45·2 m (148 ft 6 in)
Moulded beam	9·6 m (31 ft 6 in)
Service speed	12 knots
Operational range	22,000 nautical miles
Operational endurance	75 days

Decca equipment fitted

6 Hi-Fix receivers 9217/111
2 Sea search receivers 375
2 Decca Navigator survey receivers 1844
2 Remote Indicators 9370
2 Lane Identification Units 9361
1 Pattern Progress Recorder 9995
1 Data Logger 9661
2 Track Plotter 8991C
1 Track Plotter 9232
1 Track Plotter 350 C 384
1 Decca Pilot Sea Track Unit
1 Decca radar D202
1 Decca Pilot Type 550G

Decca Surveyor

Other equipment

1 Atlas DESO 10 AN 1021 hydrographic echo sounder
1 Atlas digitiser EDIE AZ 6017
1 Creed teleprinter 75 PK

NATURAL ENVIRONMENT RESEARCH COUNCIL (N.E.R.C.)

FLEET

A paper describing the vessels of N.E.R.C. was published in 1971 by the Society of Underwater Technology, *Research Vessels of the Natural Environment Research Council* by Rear Admiral G. S. Ritchie, at a meeting held on 10 November, 1971. Apart from the current fleet as described below, other vessels are chartered from time to time and over the period 1967 to 1971 the following charter vessels were used:

Researcher
Vickers Venturer
Pisces
Moray Firth
Arran Firth
Fathomer
Surveyor

The N.E.R.C. fleet now consists of four ocean-going vessels, RRS *Discovery*, RRS *Shackleton*, RRS *John Murray* and RRS *Challenger*, and two smaller coastal vessels, the RV *Edward Forbes* and the RV *Jane*.

CHALLENGER

Designed by Y-ARD, Glasgow, in collaboration with NERC and SMBA and constructed by James Lamont of Port Glasgow, the vessel was launched 7 September 1971 and became operational in April 1973. She is primarily for biological work but also has a capability for general oceanographical research.

RRS *Challenger* (*Courtesy James Hall*)

General

Length, overall	54·8 m (180 ft)
Beam	11·3 m (37 ft)
Draught	4·1 m (13·5 ft)
Displacement	1,463 metric tons 1,440 tons
Gross tonnage	approx. 750
Speed	10½ knots cruising
Cruise endurance	10½ days
Operational capability	UK continental shelf and approaches

Propulsion

Peter Brotherhood steam turbine geared to single three-bladed manganese bronze propeller in Kort rudder; White Gill bow thruster

Complement

25 crew, 9 scientists

Electrical supplies

415 volts AC 3 phase 50 Hz 300 kw
240 volts AC 1 phase 50 Hz 60 kvA
240 volts AC stabilised 50 Hz 11 kvA
115 volts AC stabilised 60 Hz 5·5 kvA
220 volts DC 8 kw
24 volts DC 0·6 kw

Navigation equipment

Decca radar TM629 with relative plotter
Decca radar RM916
Decca Mk 21 navigator
Atlas echo sounder 800 (33 KHz)
Precision echo sounder MS38 (10 KHz)
Kelvin Hughes echo sounder MS32M
Grenway Xtal clock and timing system
Decca Arkas auto-pilot and Brown gyro
Plessey Sippican bathythermograph
SAL log and distance run unit
2 component ship's log } (These are two
LORAN C/A } separate equipments)

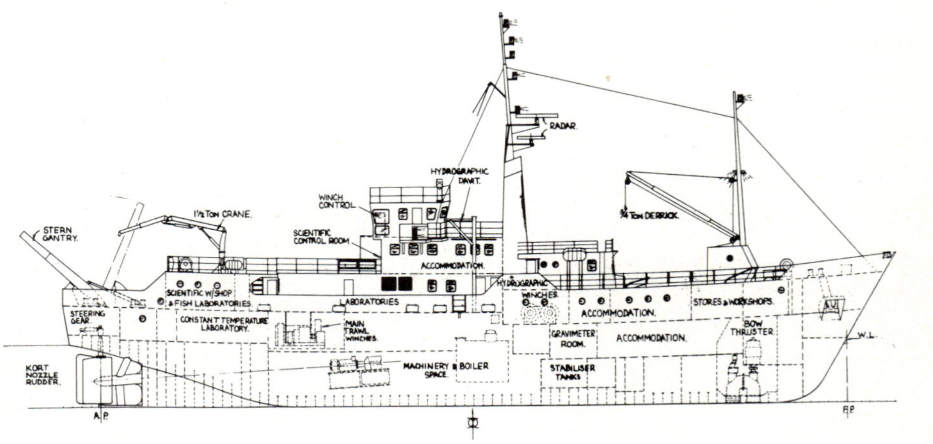

RRS *Challenger*, elevation diagram

Scientific spaces

Control and plotting room 18 m² (194 ft²), chemical laboratory 7·18² (77 ft²), hydrographic laboratory 12·3 m² (134 ft²), general laboratory 30 m² (323 ft²), biological laboratory 8·75 m² (94 ft²), gravimeter laboratory 6·7 m² (72 ft²), scientific workshop 218 m² (77 ft²), dark room 3·15 m² (34 ft²), controlled temperature laboratory 10·98 m² (188 ft²).

Handling facilities

Stern gantry 20 ton SWL

Hydraulic Atlas crane (starb. aft) 1½ ton Derrick (fwd) and winch 15 cwt

Hydrographic Davit (starb) 1 ton and winch (9,840 m of 6 mm wire and 5,500 m of 6 mm conducting cable)

Main trawl winch (7,561 m of 13 mm, 4,229 m of 16 mm and 1,995 m of 19 mm, total 13,785 m of wire)

Port and starboard trawl winches (3,400 m of 22 m wire)

Auxiliary trawl winches (2 on upper-deck) (each 1,000 m of 10 mm wire)

Helical capstan and storage reel.

Note: Main and trawl Winch wires pass over tension metre shares onto upper deck and these restrict shackles, etc.

DISCOVERY

Designed as an oceanographic research ship for the National Institute of Oceanography by T. R. Little & Co., consulting naval architects, and built by Hall, Russell & Co. Ltd., of Aberdeen in 1962.

RRS *Discovery* (*Courtesy Craelys Studios Ltd.*)

Gross tonnage	2,665
Speed	14 knots max: 10 knots cruising
Cruise endurance	30 days or 7,500 miles
Operational capability	World wide

General

Length, overall	79·7 m (261·25 ft)
Beam	14 m (46 ft)
Draught	4·7 m (15·5 ft)
Displacement	2,845 metric tons (2,800 tons)

Propulsion
3 × Diesel electric, single propeller and bow thruster (4 tons)

Complement
45 crew and 21 scientists in air conditioned laboratories and living spaces

Electrical supplies
415 volts AC 3 phase 62·5 kvA 50 Hz
230 volts AC 1 phase 52·5 kvA 50 Hz

Special equipment
SONAR trunk with removable bottom plate
IBM 1,600 data acquisition and control system
Special davit for handling GLORIA system

Navigation equipment
Sperry & Arma Brown gyro compasses
Decca radars Type 729, 729A and 1226 AC
Decca Navigator and Plotter
Magnovox satellite navigator
NIO 2 component ships log
Sperry auto pilot
Weather facsimile receiver
Echo sounder KH M526
Crystal clock and timing system
Loran type C receiver

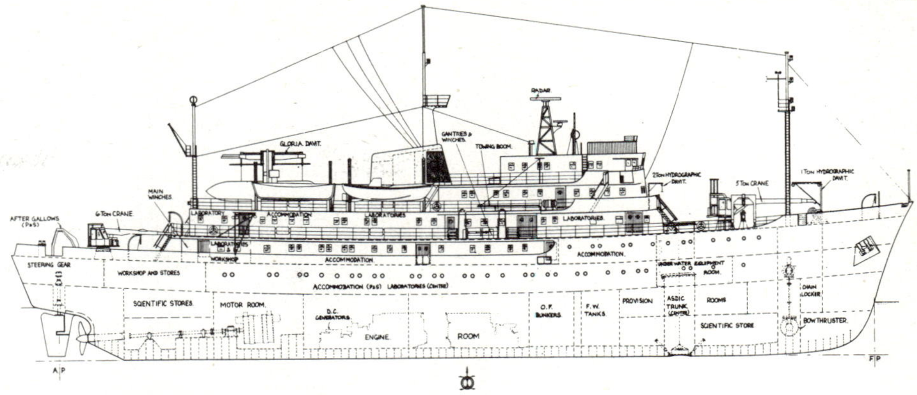

RRS *Discovery*, elevation diagram

Scientific spaces
Plotting office
Hydrographic, biological, chemical, photographic and electronic laboratories
Constant temperature laboratory and deep freeze room
Gravity room, scientific clean room
Scientific workshop
Total Area 418 m² (4,500 ft²)

Handling facilities
2 winches (stb) for sampling and instrumentation; trawl winch (aft) capacity 10 km and 2 × 2 km; 2 cranes 1·6 ton aft, 1·5 ton fwd.; 2 gallows (aft, std. and port) 8–12 tons SWL.

The vessel has an open well situated on the centre-line. A removable bottom plate is provided for this well which enables scientific instruments to be attached to the bottom of the vessel without the need to dry-dock the ship.

By October 1972 *Discovery* had completed 48 major scientific cruises, taking her to the Indian Ocean, the Mediterranean and all parts of the North Atlantic.

EDWARD FORBES

Built as a stern trawler in 1964 by Cochran & Sons, Ltd., of Selby to the design of her owner Ross Trawlers, Ltd., Grimsby, she was converted for chartering for North Sea offshore surveys in 1967 and purchased by NERC in November, 1971. After operating for a few months in early 1972 with limited capacity she was converted to her full research role by Vosper Thornycroft of Portsmouth, completing at the end of September, 1972 and operating from Barry in October, 1972.

RV *Edward Forbes* (*Courtesy Terence Soames Ltd.*)

General

Length, overall	30·6 m (100·5 ft)
Breadth	7 m (23 ft)
Draught	3·4 m (12·6 ft)
Displacement	295 metric tons (290 tons)
Speed	9·5 knots max.; 8·5 knots cruising
Cruise endurance	5 days or 1,100 miles
Operational capability	UK Coastal Waters

Propulsion
Diesel geared to single variable pitch propeller

Complement
9 crew and 4 scientists

Electrical supplies
440 volts 3 phase 15 kvA 50 Hz
220 volts DC 80 kw

Special equipment
Air compressor 0·46 m³/min (16 ft³/min) 246 kg/cm² (3,500 lb/in²)

Navigation equipment
Kelvin Hughes Tx magnetic compass and auto-pilot; Decca radar Type D202; NIO 2 component log and bridge repeater; Decca Mk. 12 navigator; Decca Mk. 21 navigator and plotter; Simrad echo sounder (38·5 KHz); Transit sonar (transducer only) Type MS47 (48 KHz); Echo sounder (transducer only) Type MS36 (32 KHz)

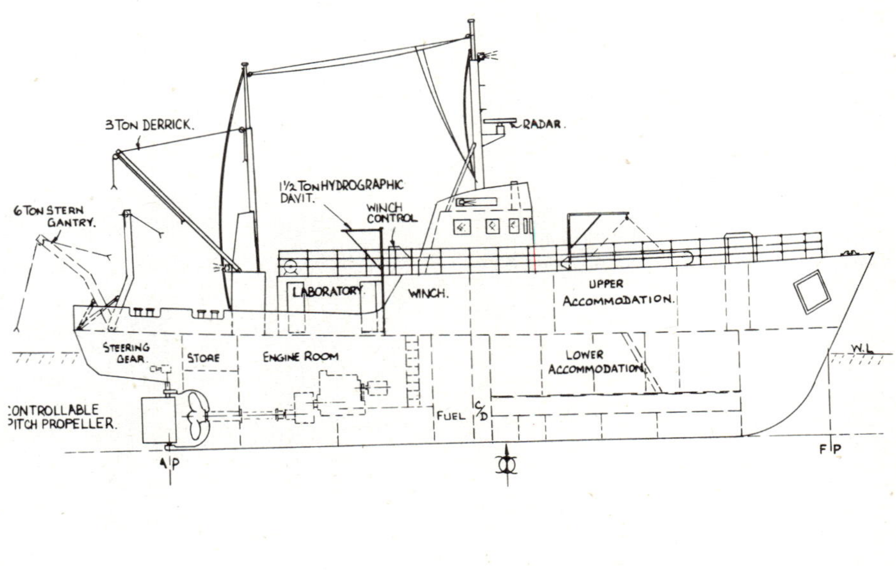

RV *Edward Forbes*, elevation diagram

Scientific spaces

Main laboratory	$12{\cdot}36\ m^2$ ($133\ ft^2$)
Sparker room	$3{\cdot}25\ m^2$ ($35\ ft^2$)

Handling facilities

6 ton SWL hydraulic gantry (aft); 1·5 ton SWL hydrographic davit (stbd); 2–3 section drum winches each loaded with (984 ft) 300 metres of 13 mm (0·5 in), 6 mm (0·23 in) and 4 mm (0·15 in) wire (Derrick temporarily removed)

JANE

Constructed in 1970 by Tees Marine Services, Ltd., Middlesborough, to the design of Keel Marine, Ltd.

General

Length, overall	16·1 m (53 ft)
Breadth	5·3 m (17·25 ft)
Draught	1·7 m (5·75 ft)
Displacement	61·5 metric tons (60·45 tons)
Speed	9 knots max.; 6·5 knots cruising
Cruise endurance	Day running
Operational capability	Inshore UK Coastal Waters

Propulsion

Two 110 h.p. diesel with twin propellers

Complement

4 crew and 4 scientists

Electrical supplies

230 volts AC 1 phase 12 kvA 50 Hz
24 volts DC (battery supply)

Navigation equipment

Decca Radar RM.914; Decca Mk. 12 navigator and track plotter; Furno 22D echo sounder/side scan sonar (200 KHz); Sestrel magnetic compass; Decca Arkas auto-pilot type 350; Harrier log

Working spaces

General laboratory fitted with sink unit and electrical supplies $2{\cdot}79 m^2$ ($30\ ft^2$)

Handling facilities

½ ton SWL manual stern gantry; small hydraulic winch (fwd) 1¼ ton; Derrick, ½ ton SWL (fwd, port or stbd).

RV *Jane*

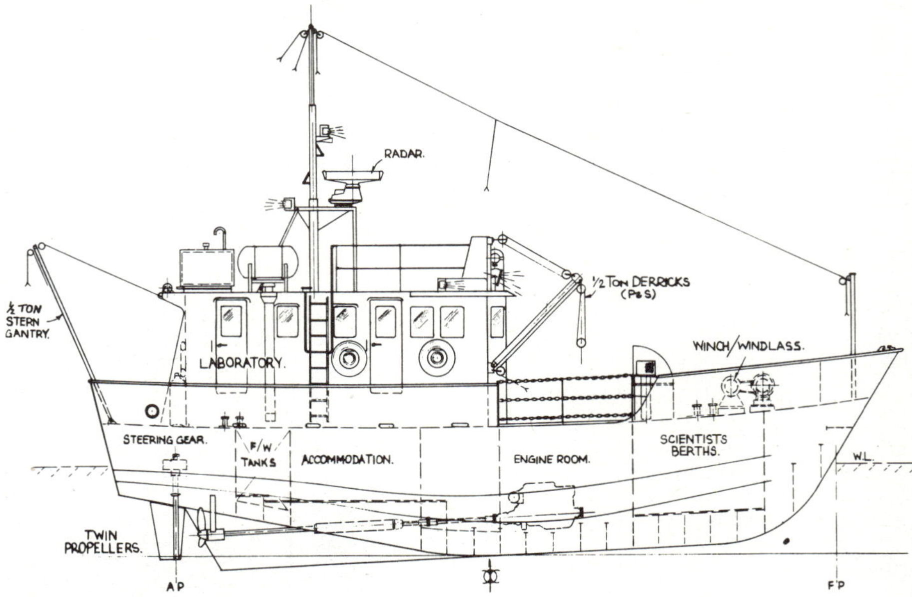

RV *Jane*, elevation diagram

JOHN MURRAY

Originally constructed as a stern trawler (*Noblesse*) in 1962 the vessel was purchased by NERC and converted in 1967 to a general purpose oceanographic research vessel by Willoughby's of Plymouth becoming operational in her new role in March 1967.

General

Length, overall	40·7 m (133·5 ft)
Beam	7·5 m (25 ft)
Draught	4·4 m (14·25 ft)
Displacement	605 m tons (595 tons)
Gross tonnage	441
Speed	11·5 knots max: 9 knots cruising
Cruise endurance	15 days or 3,500 miles
Operational capability	UK Continental Shelf

Propulsion

Two diesel geared to single variable pitch propeller

Complement

19 crew and 8 scientists in air conditioned laboratories and living spaces

RRS *John Murray* (Courtesy Robert Chapman)

Electrical supplies
230 volts AC 1 phase. 25 kvA 50 Hz
220 volts DC 110 kw

Special equipment
Air compressor 210 kg/cm² (3,000 lb/in²) at 0·34 m³/min (12 ft³/min)

Navigation equipment
Sperry gyro compass and auto-pilot
Decca radar type RM314 & JM629 with relative plotter
Decca Navigator Mk. 12 and plotter
Bergen Nautic ships log
Loran type A receiver
Muirhead Weatherfax receiver
Precision echo sounder MS38 (10 Kz)

Scientific spaces

Plotting laboratory	15·8 m² (170 ft²)
General laboratory	25·09 m² (270 ft²)
Wet laboratory	7·43 m² (80 ft²)
Biological laboratory	5·85 m² (63 ft²)
Chemical laboratory	4·18 m² (45 ft²)
Dark room	2·32 m² (25 ft²)
Gravity room	4·18 m² (45 ft²)
Scientific workshop	9·29 m² (100 ft²)

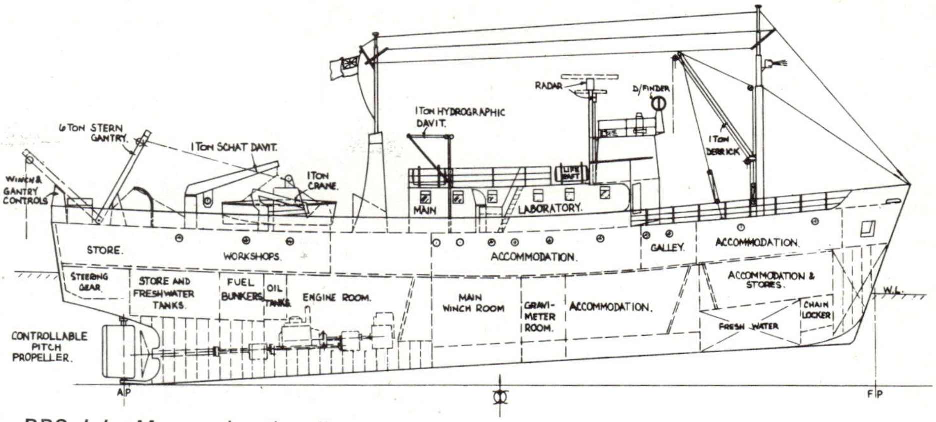

RRS *John Murray*, elevation diagram

Handling facilities
Main winch, single barrel capacity 9 km tapered wire
Buoy mooring winch, 2·1 km of 2 mm (0·08 in) wire
Bathythermograph winch, capacity 512 m
Hydrographic winch 1 km of conductor cable and 8 km of 4 mm (0·16 in) wire
Hydraulic stern gantry 6 tons SWL
Hydraulic Atlas crane (aft) 1 ton × 6·4 m (21 ft)
Hydrographic Davit 1 ton SWL
Derrick (fwd) 1 ton × 2·4 m (8 ft)

The *John Murray* had undertaken over 50 cruises for the NERC component institutions and university departments.

SHACKLETON

Constructed in 1954 by Solverborgs Varv and Rederi of Sweden, as a general cargo vessel and subsequently converted and operated by the British Antarctic Survey as a supply vessel for the South Antarctic bases. Further conversion by James Lamont of Greenock in 1970/1 brought the vessel into operation as an oceanographic research ship in October 1971.

General

Length, overall	61 m (200·5 ft)
Beam	11 m (36 ft)
Draught	4·4 m (14·25 ft)
Displacement	1,685 metric tons (1,658 tons)
Gross tonnage	1,103
Speed	11·5 knots max.: 10 knots cruising
Cruise endurance	25 days or 6,000 miles
Operational capability	World Wide

Propulsion
Diesel direct drive to single variable pitch propeller. White Gill bow thruster

Complement
32 crew and 13 scientists in air conditioned accommodation

Electrical supplies
440 volts AC 3 phase. 420 KVA 50 Hz
240 volts AC 1 phase. 25 KVA 60 Hz±1%
240 volts AC 1 phase. 5 KVA 50 Hz ±1%
220 volts DC 30 kw

Special equipment
Air compressors 210 kg/cm² at 0·59 m³ min and 245 kg/cm² at 0.45 m³ min, (3,000 lb/in² at 21 ft³/min and 3,500 lb/in² at 16 ft³/min)
Explosive store (30 tons of Geophex)

Navigation equipment
Micro-Technica Sirius gyro compass
Decca radar types RM916 and 729
Decca navigator Mk. 21
Decca Arkas auto-pilot
Loran type C receiver
Echo sounder type MS26J (10KHz)
Precision echo sounder MS38 (10KHz)
Magnavox satellite navigator
Walker and NIO 2 component ships logs
Weather facsimile receiver
Grenway Xtal clock and timing system

RRS *Shackleton* (*Courtesy Shepherd Photos*)

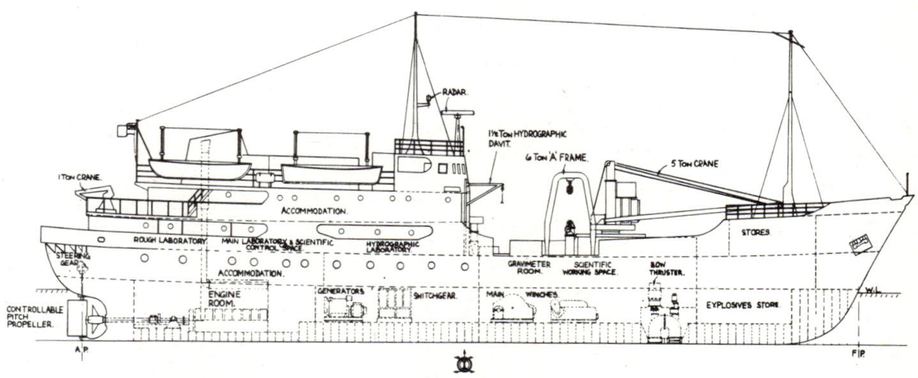

RRS *Shackleton*, elevation diagram (Starboard)

Scientific spaces

Hydrographic laboratory	14·68 m² (158 ft²)
Instrument laboratory	49·81 m² (536 ft²)
Rough laboratory	19·33 m² (208 ft²)
Gravity room	15·24 m² (164 ft²)
Electronic workshop	9·66 m² (104 ft²)
Mechanical workshop	11·89 m² (128 ft²)

Handling facilities
Crane (fwd) 3–5 tons
A Frame (stbd, side, fwd) 6-ton SWL
2—Hydrographic davits (Port and Stbd)
Main winch, 9 Km of tapered wire
Hydrographic winch 8 km of 4 and 6 mm (0·15 in and 0·23 in) wire also 1 km of conductor cable
Hydraulic crane (aft boat deck)

BRITISH ANTARCTIC SURVEY

JOHN BISCOE

Originally built as a survey vessel for the Falkland Islands Dependencies and now used as a research and supply vessel.

Designers: Graham and Woolnough

Builders: Fleming and Ferguson Ltd., Paisley

Built: 1956

Specifications

Length, overall	67·1 m (220 ft)
Length, b.p.	61·0 m (200 ft)
Breadth, moulded	12·2 m (40 ft)
Depth, moulded	6·1 m (20 ft)
Gross tonnage	1,584
Net tonnage	615
Service speed	12 knots
Engine	diesel electric; 1,350 b.h.p.
Class	Lloyds Class 3 for ice
Crew	33
Scientific accommodation	34

Cost: just over £0·5 m (1956)

RRS *John Biscoe* (*Courtesy British Antarctic Survey*)

BRANSFIELD

A research and supply vessel for Antarctic operation.

Designers: Graham and Woolnough

Builders: Robb Caledon Shipbuilders, Leith Division.

Launched: 4 September 1970

Specifications

Length, overall	99·25 m (325 ft)
Length, b.p.	90·00 m (295 ft)
Beam, moulded	18·30 m (60 ft)
Depth, moulded	9·50 m (31 ft) at 9·15 m (30 ft) half breadth
Draught, moulded	6·23 m (20 ft 5 in)
Gross tonnage:	4,816
Displacement, moulded	6,900 tons (m)
Service speed	13·5 knots
Endurance	50 days at service speed

RRS *Bransfield* (*Courtesy A. G. Ingram Ltd.*)

Engines: Two Mirrlees National KM8 Major diesel engines, each developing 3,200 b.h.p. at 500 rpm

Propeller: Stone Manganese Marine controllable-pitch, stainless steel propeller, dia. 3,100 m (10 ft 2 in), pitch 2·034 m (6 ft 8 in) at 0·7 radius. Propeller wt.=11 tons 16 cwt.

Cost: £1·75 m (1970)

Classed: Lloyds Ice Class 1*

The Bransfield sailed from Southampton on her first Antarctic voyage on 4 January 1971.

INDIA

OIL AND NATURAL GAS COMMISION OF THE GOVERNMENT OF INDIA

OCEANOGRAPHIC SEISMIC SURVEY VESSEL

Towards the end of 1973 a contract was placed with the Dallas based Whitehall Corporation for the building of a 53·3 m (175 ft) survey vessel. The vessel will have a crew of 24 and will have accomodation for 16 scientists. Capital cost is over $US 4,000,000 (1973).

NATIONAL INSTITUTE OF OCEANOGRAPHY

GAVESHANI

Formerly a twin-screw Calcutta Port Commission hopper vessel, the RV *Gaveshani* was converted by Calcutta's Garden Beach Workshop. The vessel will be operated by Shipping Corporation of India for the collection and analysis of chemical, geological, physical and ecological data, and have accomodation for 70.

The vessel will carry an 8 m (26 ft) diesel-engined launch for inshore and esturial work.

UNITED STATES OF AMERICA

NATIONAL OCEANIC AND ATMOSPHERIC ADMINISTRATION (N.O.A.A.) RESEARCH SHIP FLEET

Introduction

The National Oceanic and Atmospheric Administration (NOAA) was created within the US Department of Commerce on 3 October 1970, by executive order. Its formation brought together the functions of the Environmental Science Services Administration (ESSA), a Commerce agency; the Bureau of Commercial Fisheries, Marine Game Fish Research Program, and Marine Minerals Technology Center from the Interior Department; the National Oceanographic Data Center and National Oceanographic Instrumentation Center, from the US Navy; the National Data Buoy Development Project, from the Coast Guard, US Transportation Department; the National Sea Grant Program, from the National Science Foundation; and elements of the US Lake Survey, from the Army Corps of Engineers.

Within NOAA three of the Primary Organizational Elements, the National Marine Fisheries Service (NMFS), the National Ocean Survey (NOS), and the Environmental Research Laboratories (ERL), are the principal users of the ships needed to fulfill its broad national missions to improve man's understanding and uses of the physical environment and oceanic life.

Centralised management of the NOAA Fleet is the responsibility of the Office of Fleet Operations (OFO), an integral part of NOS. Operational support for the fleet is provided through the Atlantic Marine Center, Norfolk, Virginia, and the Pacific Marine Center, Seattle, Washington.

Uniform identification of the fleet will be accomplished when routine repainting is scheduled. NOAA emblems will be displayed on the port and starboard sides of the ships' stacks. The port and starboard sides of the bows of the ships will bear a letter-number identification on one line and NOAA beneath.

The letter designations used are:

OSS Oceanographic Survey Ship
MSS Medium Survey Ship
CSS Coastal Survey Ship
FRS Fisheries Research Ship
FPS Fisheries Provision Ship
FRV Fisheries Research Vessel
ARV Auxiliary Research Vessel
ASV Auxiliary Survey Vessel

OCEANOGRAPHER (OSS-01)
(sister ship is DISCOVERER)
Call letters WTEP

NOAA Class of vessel: Class I

Specific purpose of vessel: oceanographic research

The 92·4 m (303 ft), 3,800-ton vessel was built at Jacksonville (Florida) Shipyards under the supervision of the Maritime Administration. Begun in 1963, she was completed in April 1966. Including scientific equipment, the ship cost $10,000,000.

The vessel can be provisioned for 150 days at sea and has a cruising range of 16,000 miles. She carries a normal complement of 16 officers, 39 crew and 45 technical and scientific personnel, with additional accommodations for visiting scientists.

The ship has over 38 m² (410 ft²) of laboratory space. All living quarters and scientific areas are air-conditioned. Closed circuit television is provided throughout the machinery spaces, where Westinghouse's centralised engine room control (CERC) system provides remote starting and stopping of machinery, programming of fuel, and the automatic recording of operating data at a master control station.

A single computer serves both ship operation and the collection and processing of environmental data. Using this computer, the propulsion and other machinery is automated through CERC. The CERC system also permits remote control of main propulsion units and principal auxiliary machinery from a master control station in the engine room and from the bridge. In addition to automatic logging of ship operating data, CERC includes an alarm system which detects and locates malfunctions, gives a warning signal, and types out a description of the problem. The *Oceanographer* is the first American oceanographic survey ship to employ so advanced a concept of centralised engine room control. When fully developed in future years, this concept will permit a single operator to monitor and control the ship's engineering plant from a central control station.

The heart of the *Oceanographer's* automated controls is a Westinghouse Prodac 510 computer using a UNIVAC 1218 computer main frame. Because controlling and monitoring ship operations requires only about 25 per cent of the computer's total capacity—100,000 arithmetic calculations per second, 16,384-function memory core—the computer is used principally by the Data Acquisition System (DAS). When the ship is underway, DAS samples (via shipboard and towed sensors), records, and processes geophysical, oceanographic, hydrographic, and meteorological data on a routine basis; ship position is logged continuously; and the computer can be used for concurrent processing of non-routine data. When the ship is stationary, DAS samples and processes data sensed by shipboard instruments and by an underwater multisensor package, along with its handling of oceanographic data. The automatic data processing system will free many highly trained specialists from the routine task of sorting and analysing a great amount of data, a task which usually consumed months of painstaking effort.

The *Oceanographer's* equipment measures and records ship's course and speed, magnetic field intensity, gravity, surface current and

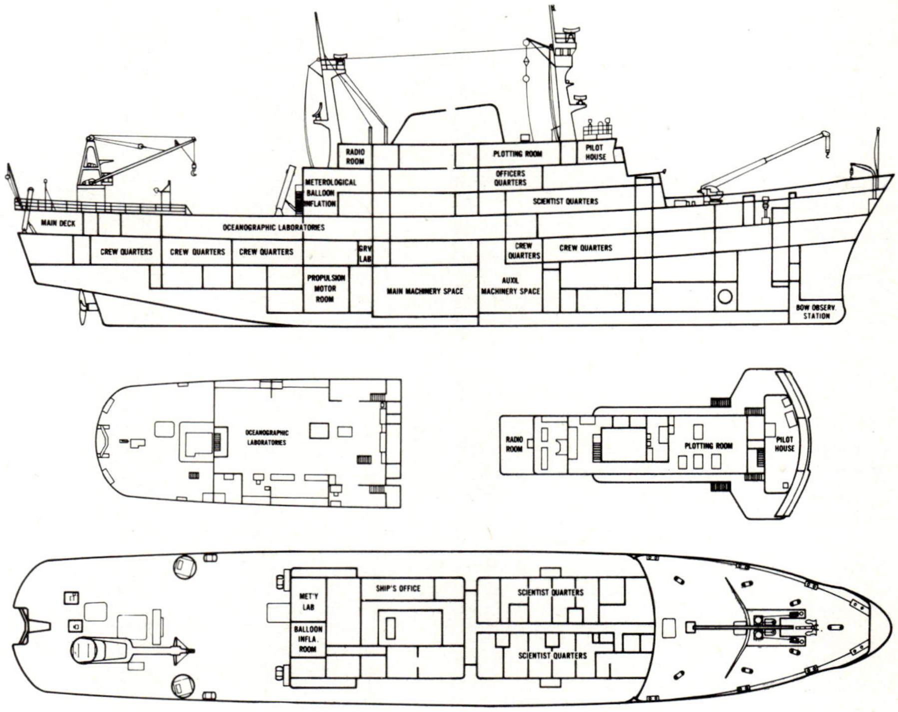

Layout diagrams of *Oceanographer*

Oceanographer (OSSOI) (*Courtesy Environmental Science Services Administration*)

temperature, temperature at depth, and ocean depth. Sub-bottom profiles are taken while the ship is underway and show the structure of the ocean floor beneath its bottom sediment. The ship takes water samples at various depths and is equipped to obtain 100-ft core samples from the deepest ocean floor. Meteorological data are gathered at regular intervals by ship-launched sounding balloons. Data and samples are studied and analysed in the ship's laboratories.

The ship's equipment also includes a newly-developed marine type automatic picture transmission (APT) which enables the ship to receive signals direct from ESSA satellites as they pass overhead. APT instantaneously reproduces the information in the form of a weather picture of the local area as seen by the satellite.

The *Oceanographer* can operate equally well in any area of the global sea, including polar waters. A special control feature—a bow thruster of 400 horsepower—enables the vessel to maintain a nearly constant heading when the ship is on station despite wind and wave conditions. The bow thruster is an underwater duct fitted with a reversible propeller that thrusts a water jet to either side as desired.

The ship can take samples anywhere in the world. Over 22 miles of wire rope are carried aboard the ship for oceanographic work. One continuous length of wire is over seven nautical miles long. The wires are reeled in by powerful hydraulic winches.

Another unusual feature of the ship is a 1·8 m × 2·4 m (6 ft × 8 ft) well near the centre of the ship which enables special experimental equipment to be lowered and scuba divers to enter and leave the vessel. An elevator carries the equipment and men directly from the oceanographic laboratory into the water 10·7 m (35 ft) below.

Six special glass-covered ports near the bow and stern, about 4·57 m (15 ft) below the water's surface, permit scientists to view underwater life and formations from within the ship.

The ship has an ample supply of fresh water. Normal consumption for all purposes is approximately 5,000 US gallons per day, with a storage capacity of about 27,000 US gallons and a sea water distillery capacity of 8,000 US gallons a day.

The *Oceanographer* has extensive communication facilities. These include radiotelephones, emergency receivers and transmitters, mobile transceivers, portable radiotelephones, facsimile equipment, standard frequency broadcast service, portable lifeboat transmitter receiver, and radio teletype.

An anti-rolling device (a passive rolling tank) enables the ship to conduct continuous operations, except in unusually heavy weather.

A notable outside feature is the conning tower, an enclosed crow's nest, on top of the bridge approximately 18 (60 ft) above the water. It is reached by an inside ladder. In addition to affording a commanding view of the sea, the ship can also be fully controlled from the conning tower. From this control station, the ship can be kept on a steady course whenever delicate instruments are being trailed over the sides or from the stern. Similar, but less elaborate, controls on the deck on each side of the bridge enable similar control to be maintained from these stations.

Ample storage facilities, including cold storage, enable scientists to bring home samples of their findings in their original organic state for further studies in laboratories ashore.

Design history

Start of construction: 22 July 1963 (Jacksonville (Florida) Shipyards, now Aerojet-General Shipyards)

Date of launching: 18 April 1964

Date trials were completed: 20 April 1966

Certification authority clearance: 13 July 1966.

Basic dimensions

Length, overall	92·38 m (303 ft)
Length, waterline	85·4 m (280 ft)
Beam, moulded	15·85 m (52 ft)
Height, overall	32·01 m (105 ft)
Draught, full load	5·49 m (18 ft)

Displacements

Displacement, full load	3,959 tons
Displacement, light	2,580 tons

Performance

Full speed	18 knots
Cruising speed	16 knots
Range at cruise	15,200 nm
Endurance at low speed	35 days
Position holding capability	by bow thruster
Maximum anchoring depth	135 fathoms
Towing capacity	moderate

Main machinery description

Main engines	Twin screw diesel electric, 2,500 shp each 850 rpm
Auxiliary engines	Bow thruster
Propeller(s) type	fixed-pitch
Pumps	salt, fresh water, fuel, bilge, etc.
Controls: manoeuvring	one 400 h.p. fixed pitch, through hull bow thruster 2/4,500 kg (10,000 lb) thrust

On-board equipment

Facilities for carrying major underwater equipment: Lifting, 5,040 kg (11,200 lb) at 10·67 m (35 ft) radius

Facilities for carrying major surface equipment: one 10·06 m (33 ft) launch

Facilities for diving: Scuba gear, through hull well, decompression chamber

Sampling equipment: biological tows and trawls, light and heavy coring, rock dredging Nansen bottles, STD

Craneage facilities: one oceanographic w/capacity of 5,040 kg (11,200 lb) at 10·67 m (35 ft) radius; one forward w/capacity of 3,285 kg (7,300 lb) at 12·20 m (40 ft) radius

Winching facilities: One deep-sea, 150 h.p. electrohydraulic w/13,719 m (45,000 ft) of 1·91 cm ($\frac{3}{4}$ in) to 0·97 cm ($\frac{3}{8}$ in) stepped wire, 2 double drum oceanographic, 30 h.p. electrohydraulic w/9,146 m (30,000 ft) of 0·48 cm ($\frac{3}{16}$ in) wire and 3,658 m (12,000 ft) of 6 conductor electrical logging cable, two BT winches

Navigational equipment: 2 Loran-C, 1 Satellite navigation, 1 Loran-A, 1 radio direction finder, 1 Omega receiver, 1 gyrocompass, 2 navigation radars, 1 EM underwater log and gyropilot

Survey and search equipment:

Two deep water sounders, three shoal water depth sounders

One 21 KH_3 Lodar w/recorder

Two precision depth recorders

One buoy tracking radar

Computer controlled acquisition/display system, non-real time processing

Current meters, magnetometer, gravity meter, sub-bottom profiling equipment, radiosonde receiver and recorder, surface meteorological equipment, heat probes, tide gages

One meteorological radar

Pumps, compressors and ancillary power equipment and services: Three 400-kw, 450-VAC, 60 Hz, 3 phase generators 220-VAC, 60 Hz, 3 phase, 115-VAC, 60 Hz, single phase. One 75-KE, 450-VAC, 60 Hz, 3 phase emergency generator DC power

Laboratory and technical facilities

Oceanographic:	Wet and dry work 316·20 m² (3,400 ft²)
Geological	(gravity) 7·44 m² (80 ft²)
Photographic	15·35 m² (165 ft²)
Data processing	49·29 m² (530 ft²)

Crew

Sea-going complement, officers men	87
Scientific complement	18
Accommodation	Comfortable quarters with air conditioning; 112 berths
Crew safety facilities and equipment	Life preservers, inflatable life rafts

Capital cost

$9,300,000 (1963)

Operators

Director, Pacific Marine Center
1801 Fairview Ave., East,
Seattle, Washington, 98102

DISCOVERER (OSS-02)
(sister ship is *Oceanographer*)

NOAA Class of vessel
Class I

Specific purpose of vessel
Oceanographic research

Design history
Start of construction: 10 September 1963 (Jacksonville, Florida Shipyards, now Aerojet-General Shipyards)
Date of launching: 29 October 1964
Date trials were completed: 15 December 1966
Certification authority clearance: 29 April 1967

Production

Basic dimensions

Length, overall	92·38 m (303 ft)
Beam, overall	15·85 m (52 ft)
Height, overall	32·01 m (105 ft)
Draught	5·49 m (18 ft)

Displacement
3,959 tons

Performance

Full speed	18 knots
Cruising speed	16 knots
Range at cruise	15,200 nm
Endurance at low speed	35 days
Position holding capability	bow thruster
Maximum anchoring depth	135 fathoms, 246 m (810 ft)
Towing capacity	moderate

Main machinery description

Main engines	Twin-screw diesel electric, 2,500 SHP each
Propeller type	fixed-pitch
Pumps	Salt and fresh water, fuel, bilge, fire, etc.
Controls: manoeuvring	One 400-h.p. fixed pitch, through hull bow thruster 4,500 kg (10,000 lb) thrust

On-board equipment
Facilities for carrying major underwater equipment: Lifting capability 5,040 kg (11,200 lb) at 10·67 m (35 ft)
Facilities for carrying major surface equipment: One 4·88 m (16 ft) Boston whaler, one 3·66 m (12 ft) Boston whaler
Facilities for diving: Through hull well. Decompression chamber, scuba gear aboard
Sampling equipment: Biological tows and trawls, light and heavy coring, rock dredging, Nansen bottles, STD
Craneage facilities: One forward w/capacity of 3,285 kg (7,300 lb) at 12·20 m (40 ft) radius. One aft w/capacity of 5,040 kg (11,200 lb) at 10·67 m (35 ft) radius
Winching facilities: One deep-sea, 150 h.p., w/drum capacity 13,719 m (45,000 ft) of 1·91 cm ($\frac{3}{4}$ in) to 0·97 cm ($\frac{3}{8}$ in) wire. Two BT winches. One 40 h.p. dredge and trawl w/ drum capacity 1,829 m (6,000 ft) of 0·97 cm ($\frac{3}{8}$ in) wire. One magnetometer winch. One with 9,161 m (30,050 ft) of 0·48 cm ($\frac{3}{16}$ in) wire
Navigational equipment: Two Loran-C, one satellite navigation, one Loran-A. One radio direction finder, one Omega receiver. One gyrocompass and gyropilot. One electromagnetic underwater log. Two navigation radars
Survey and search equipment: Two deep-water depth sounders, three shoal water depth sounders. One 21-KH_3 Lodar w/ recorder. One 12-KH_3 narrow beam stabilized. Depth sounder w/6,000 fathom range. One buoy tracking radar. Computer controlled acquisition/display system, non-real time processing. Current meters, magnetometer, gravity meter, sub-bottom profiling equipment, radiosonde receiver and recorder, surface meteorological equipment, heat probes, tide gages. Two precision depth recorders. One meteorological radar
Pumps, compressors and ancillary power equipment and services: Three 400 kw generators, 450-VAC, 60 Hz, 3 phase. 220-VAC, 60 Hz, 3 phase. 115-VAC, 60 Hz, single phase. One 75-kw, 450-VAC, 60 Hz, 3 phase emergency generator 24 VDC rectified power

Laboratory and technical facilities
Analytical

Oceanographic	316 m² (3,400 ft²), wet and dry work
Geological	(gravity) 14·88 m² (160 ft²)
Photographic	15·35 m² (165 ft²)
Data processing	44·29 m² (530 ft²)

Six ports in the bow observation chamber enable viewing forward and to the sides

Crew

Sea-going complement, officers, men	87
Scientific complement	18
Accommodation	Comfortable quarters with air conditioning; 115 berths

Crew safety facilities and equipment: Life preservers, inflatable life rafts

Capital cost
$9,700,000 (1963)
Vessel de-activated 7 March 1973

Operators
Director, Atlantic Marine Center,
439 W. York Street,
Norfolk, VA 23510

RESEARCHER (OSS-03)
Call letters WTER

Class of vessel
Class I

Specific purpose of vessel
Deep sea oceanographic research

Design history
Start of construction: 8 June 1966 (American Shipbuilding Co., Toledo, Ohio)
Date of launching: 5 October 1968
Date trials were completed: 18 June 1970
Certification authority clearance: 8 October 1970

Basic dimensions

Length, overall	84·76 m (278 ft)
Beam, overall	15·55 m (51 ft)
Height, overall	27·74 m (91 ft) (water to top aircraft light)
Draught	5·49 m (18 ft)

Displacement
2,772 tons

Performance

Full speed	15 knots
Cruising speed	14 knots
Range at cruise	13,000 nm
Endurance at low speed	35 days
Position holding capability	360° retractable bow thruster propels ship up to 7 knots
Maximum anchoring depth	120 fathoms, 183 m (600 ft)
Towing capacity	moderate

Main machinery description
Main engines: Two 1,600 shp geared diesel engines 1,000 rpm
Propeller(s) type: Two three-blade controllable pitch
Pumps: Salt water, fresh water, fuel, fire, bilge, etc.
Controls: Manoeuvring—One 450 h.p., 630 rpm retractable bow thruster

On-board equipment
Facilities for carrying major underwater equipment: 20 ton crane
Facilities for carrying major surface equipment: One 5·18 m (17 ft) Boston whaler. Two 7·32 m (24 ft) life boats
Facilities for diving: Scuba gear aboard
Sampling equipment: Biological tows and trawls. Deep-ocean cores and dredging. Nansen bottles, STD.

Craneage facilities: One 20-ton telescoping, electrohydraulic 18,000 kg (40,000 lb) at 4·27 m (14 ft) radius A-frame, fabricated, 8·54 m (28 ft) throat, w/9,000 kg (20,000 lb) lift. One telescoping boom, 4,500 kg (10,000 lb) at 3·05 m (10 ft) radius. Four service electrohydraulic telescoping boom, 2,475 kg (5,500 lb) at 8·54 m (28 ft)

Winching facilities: One deep sea, 150 h.p., w/drum capacity 12,195 m (40,000 ft) of 1·42 cm ($\frac{9}{16}$ in) wire. Two oceanographic, 40 h.p., 1 drum w/capacity 3,658 m (12,000 ft) of American 4-H-1. One drum w/capacity of 1,829 m (6,000 ft) of 0·97 cm ($\frac{3}{8}$ in) wire. Two BT, electric. One magnetometer, electric. One GEK, electric

Navigational equipment: One Loran A/C. Two surface search radar. Two Loran C w/passive ranging. One radio direction finder. One gyrocompass w/19 repeaters. Satellite navigation equipment. One EM underwater log

Survey and search equipment: Two depth recorder system, deep water. Three depth recorder system, shoal water. One echo sounder, narrow beam stabilized. One EM log. One buoy tracking radar. Complete real time data acquisition and processing system using two computers. Current meters, tide gages, magnetometer, gravity meter, sub-bottom profiling equipment. Seismic reflection profiler

Pumps, compressors and ancillary power equipment and services: Two 500 kw service generators, 450-VAC, 60 Hz, 3-phase generators; 220-VAC, 60 Hz, 115-VAC constant voltage system; 115-VAC single phase lighting 24-VDC rectified power. One 500 kw, diesel generator for bow thruster. 3-phase emergency generator

Laboratory and technical facilities

Analytical:

Oceanographic	218·55 m² (2,350 ft²)
Geological	16·74 m² (180 ft²) (gravity)
Photographic	16·28 m² (175 ft²)
Data processing	50·22 m² (540 ft²)

Crew

Sea-going complement, officers, men	67
Scientific complement	13

Accommodation: Air conditioned, accommodation for women scientists, 84 berths

Crew safety facilities and equipment: Life preservers, 4 inflatable life rafts, 2 life boats

Purchase, capital cost:
$9,200,000

Operators
Director, Atlantic Marine Center,
439 West York Street,
Norfolk, VA 23510

SURVEYOR (OSS-32)
Call letters WTES
Vessel deactivated 7 March 1973

NOAA Class of vessel
Class I

Specific purpose of vessel
Marine charting and geophysical surveys

Design history

Start of construction: 1958

Date of launching: 25 April 1959 (National Steel and Shipbuilding Co., San Diego, California)

Date trials were completed: 29 April 1960

Certification authority clearance: 30 April 1960

Basic dimensions

Length, overall	89·02 m (292 ft)
Beam, overall	14·02 m (46 ft)
Height, overall	34·76 m (114 ft)
Draught	5·49 m (18 ft)

Displacement
3,150 tons

Performance

Full speed	16 knots
Cruising speed	14 knots
Range at cruise	12,800 nm
Endurance at low speed	25 days
Maximum anchoring depth	150 fathoms, 274 m (900 ft)
Towing capacity	moderate

Main machinery description

Main engines: 1 steam turbine, 3,200 shp at 135 rpm

Auxiliary engines: 1 Harbormaster outboard. Electric motor 1,800 rpm max.

Propeller type: Single screw, fixed-pitch propeller

Pumps: fresh and salt water, bilge, etc.

Controls: Manoeuvring: 1 Harbormaster

On-board equipment

Facilities for carrying major surface equipment: Two boat davits for 7·93 m (26 ft) motor whaleboat. Two boat davits for 4·88 m (16 ft) launches. Four chocks to stow two 10·98 m (36 ft) launches and two LCVP

Facilities for diving: Scuba gear

Sampling equipment: Biological tows, trawls, and dredges. Light and heavy coring, rock dredging, Nansen bottles, STD

Craneage facilities: Two telescoping electrohydraulic cargo booms with capacity 2,250 kg (5,000 lb) at 5·49 m (18 ft). One heavy life boat crane with capacity 11,250 kg (25,000 lb) at 10·06 m (33 ft). One A-frame, capacity varies with block used

Winching facilities, capacities: One BT winch. One electrohydraulic oceanographic with drum capacity 6,097 m (20,000 ft) of 0·48 cm (3/16 in) wire. One electrohydraulic deep-sea double drum; 1 drum w/7,317 m (24,000 ft) of 0·48 cm (3/16 in) wire rope, 1 drum w/6,920 m (22,700 ft) of 1·91 cm ($\frac{3}{4}$ in) to 1·27 cm ($\frac{1}{2}$ in) wire rope. One electrohydraulic w/two interchangeable drums; 1 grooved drum w/capacity 9,146 m (30,000 ft) of 0·48 cm (3/16 cm) cable, 1 drum w/capacity 6,097 m (20,000 ft) of 0·292, 6 conductor cable

Navigational equipment: Two Loran A/C receiver indicators. One radio positioning system. One satellite navigation system. One gyropilot w/18 repeaters output. One electromagnetic underwater log. One main ship radar

Survey and search equipment: Two precision depth recorders. One 12-KHZ narrow-beam stabilized depth sounder. One shoal-water depth recorder aboard launch. One raw data logger for digital and analog inputs. One magnetic tape unit. Four teletypes. Current meters, magnetomer, gravity meter. Sub-bottom profiling equipment, seismic reflection profiler, radiosondes, heat probes, tide gages, nansen bottles, marine photographic equipment

One helicopter platform

Pumps, compressors and ancillary power equipment and services: Two 400-kw, 450-VAC, 60 Hz, 3 phase generator. 115-VAC, 60 Hz, single phase. One 100-kw, 450-VAC, 60 Hz, 3 phase emergency generator

Laboratory and technical facilities

Analytical:

oceanographic	8·65 m² (93 ft²)
Chart room	42·69 m² (459 ft²)
Photographic	7·53 m² (81 ft²)
Data processing	37·94 m² (408 ft²)
Sea-going complement, officers, men	90
Scientific complement	9

Accommodation

Comfortable quarters with air conditioning 129 berths

Crew safety facilities and equipment: Life preservers, 4 inflatable life rafts

Capital cost
$7,600,000 (1959)

Operators
Director, Pacific Marine Center,
1801 Fairview Avenue, East,
Seattle, Washington 98102

FAIRWEATHER (MSS 20) (sister ships *Rainier* and *Mt. Mitchell*)
Call letters WTEB

NOAA Class of vessel
Class II

Specific purpose of vessel
Hydrographic surveys

Design history

Start of construction: 30 November 1965 (Aerojet-General Shipyards, Jacksonville, Florida)

Date of launching: 15 March 1967

Date trials were completed: 31 January 1968

Certification authority clearance: 2 October 1968

Basic dimensions

Length, overall	70·43 m (231 ft)
Beam, overall	12·80 m (42 ft)
Height, overall	26·22 m (86 ft) above D.W.L.
Draught	4·27 m (14 ft)

Displacement
1,800 tons

Performance

Full speed	14 knots
Cruising speed	13 knots
Range at cruise	8,000 nm
Endurance at low speed	30 days
Maximum anchoring depth	100 fathoms
Towing capacity	moderate

Main machinery description

Main engines: Twin-screw diesel, 1,200 SHP each 850 rpm

Auxiliary engines: bow thruster

Propeller(s) type: 2 controllable-pitch

Pumps: Salt, fresh water, fuel, bilge, etc.

Controls: Manoeuvring—One 200 b.h.p. diesel engine bow thruster 2,250 kg (5,000 lb) thrust

On-board equipment

Facilities for carrying major surface equipment: Six boat davits, rated for wt. of boats 3,600 kg (8,000 lb) to 5,850 kg (13,000 lb). Four 7·62 m (25 ft) hydrographic survey launches. Two 7·93 m (26 ft) motor whale boats. Two 4·88 m (16 ft) Boston whalers. Two 3·96 m (13 ft) A1 skiffs

Facilities for diving: Scuba gear aboard

Sampling equipment: Bottom snapper, corers and dredging, nansen bottles

Craneage facilities: One non-telescoping boom, radius 3·66 m (12 ft) to 12·20 m (40 ft) w/capacity 2,250 kg (5,000 lb). Two telescoping booms, radius 4·88 m (13 ft) to 7·62 m (25 ft) w/capacity 1,350 kg (3,000 lb) min. radius 1,125 kg (2,500 lb) max. radius Two A-frames, radius varies w/block used

Winching facilities: One electric BT w/ drum capacity 1,067 m (3,500 ft) of 0·23 cm (3/32 in) steel wire rope. One electrohydraulic double-drum oceanographic, 1 drum capacity of 3,658 m (12,000 ft) of 0·76 cm (0·298 in) electric cable, 1 drum w/capacity of 9,146 m (30,000 ft) of 0·48 cm (3/16 in) wire rope. One electrohydraulic dredge and trawl w/drum capacity of 1,829 m (6,000 ft) of 0·97 cm (⅜ in) wire rope

Navigational equipment: Two Loran A/C. One radio direction finder. Two surface search radars. One radio positioning system (3 units). One gyrocompass system w/14 repeater. One EM underwater log

Survey and search equipment: Two deep-water depth recorders. Two shoal water depth recorders. One horizontal echo sounding system. One Wang 700A programmable calculator

Data Acquisition System: Two NOS hydrolog system. One NOS hydroplot system w/COMPLOT DP-3 plotter

Current meters, magnetometer, BT 0–274 m (0–900 ft)

Bottom profiler, tide gages

Pumps, compressors and ancillary power equipment and services: Two 300-kw, 450-VAC, 60 Hz, 3 phase generators. 115-VAC, 60 Hz, single phase. One 75-kw, 450-VAC, 60 Hz, 3 phase emergency generator

Laboratory and technical facilities

Analytical:

Oceanographic	22·32 m² (240 ft²)
Data processing	33·48 m² (360 ft²)

Crew

Sea-going complement, officers, men	74
Scientific complement	2

Accommodation: Comfortable quarters, air conditioned 78 berths

Crew safety facilities and equipment: 1 life boat, life preservers, 4 inflatable life rafts

Capital cost

$5,480,000 (1965)

Operators

Director, Pacific Marine Center,
1801 Fairview Ave., East,
Seattle, WA 98102

RAINER (MSS 21) (sister ships *Fairweather* and *Mt. Mitchell*)
Call letters WTEF

NOAA Class of vessel

Class II

Specific purpose of vessel

Hydrographic surveys

Design history

Start of construction: 20 January 1966 (Aerojet-General Shipyards, Jacksonville, Florida)

Date of launching: 15 March 1967

Date trials were completed: 24 April 1968

Certification authority clearance: 2 October 1968

Basic dimensions

Length, overall	70·43 m (231 ft)
Beam, overall	12·80 m (42 ft)
Height, overall	26·22 m (86 ft)
Draught	4·27 m (14 ft)

Displacement

1,800 tons

Performance

Full speed	14 knots
Cruising speed	13 knots
Range at cruise	8,000 nm
Endurance at low speed	30 days
Maximum anchoring depth	100 fathoms, 183 m (600 ft)
Towing capacity	moderate

Main machinery description

Main engines: Twin-screw diesel, 1,200 SHP, 850 rpm

Auxiliary engines: bow thruster

Propeller(s) type: Two controllable pitch

Pumps: Salt and fresh water, fuel, bilge, etc.

Controls: Manoeuvring—1,200 bhp diesel engine bow thruster w/2,250 kg (5,000 lb) thrust

On-board equipment

Facilities for carrying major surface equipment: Six boat davits, rate for weight of boats 3,600 kg (8,000) 5,850 kg 13,000 lb). One 9·45 m (31 ft) hydrographic survey launch. Three 7·62 m (25 ft) hydrographic survey launches. One 7·93 m (26 ft) motor whaleboat. Two 4·88 m (16 ft) Boston whaler. Two 3·96 m (13 ft) Boston whaler

Facilities for diving: Scuba gear aboard

Sampling equipment: Bottom snapper, corers and dredging Nansen bottles

Craneage facilities: One non-telescoping, radius 3·66 m (12 ft) to 12·20 m (40 ft) w/capacity 2,250 kg (5,000 lb). Two telescoping booms, radius 3·96 m (13 ft) to 7·62 m (25 ft) w/capacity 1,350 kg (3,000 lb) min. 1,125 kg (2,500 lb) max. radius

Winching facilities: Two electro hydraulic BT w/drum capacity 1,219 m (4,000 ft) of 0·23 cm (3/32 in) wire. One electrohydraulic double-drum, 1 drum w/capacity of 3,658 m (12,000 ft) of 0·76 cm (0·298 in) electric cable. One drum w/capacity of 9,146 m (30,000 ft) of 0·48 cm (3/16 in) wire rope. One electrohydraulic dredge and trawl w/ drum capacity of 1,829 m (6,000 ft) of 0·97 cm (⅜ in) wire rope

Navigational equipment: One Loran A receiver. Two Loran A/C receiver. Two surface search radar. One radio direction finder. One radio positioning system. One gyrocompass w/18 repeater outputs. One EM underwater log

Survey and search equipment: Two deep water depth recorders. Two shoal water depth recorders. One horizontal echo sounding system. One Wang 700A programmable calculator

Data Acquisition System: One NOS hydroplot system w/COMPLOT plotter. Two NOS hydrolog system

Bottom profiler, BT 0–274 m (0–900 ft)

Current meters, magnetometers, tide gages

Pumps, compressors and ancillary power equipment and services: Two 220-kw, 450-VAC, 60 Hz, 3 phase generator. 115-VAC, 60 Hz, single phase. One 60-kw, 450-VAC, 60 Hz, 3-phase emergency generator

Laboratory and technical facilities

Analytical:

Oceanographic	22·32 m² (240 ft²)
Data processing	33·48 m² (360 ft²)

Crew

Sea-going complement, officers, men	74
Scientific complement	2

Accommodation: Comfortable quarters, air conditioned 78 berths

Crew safety facilities and equipment: One life boat, life preservers, 4 inflatable life rafts

Capital cost

$5,610,000 (1966)

Operators

Director, Pacific Marine Center,
1801 Fairview Ave., East,
Seattle, Washington 98102

MILLER FREEMAN (FRS 21)
Call letters WTDM

NOAA Class of vessel

II, Stern trawler

Specific purpose of vessel

Fisheries research

Design history

Start of construction: 1965 (American Ship Building Co., Lorain, Ohio)

Date trials were completed: 1967

Certification authority clearance: 1967

Basic dimensions

Length, overall	65·24 m (214 ft)
Beam, overall	12·80 m (42 ft)
Draught	5·79 m (19 ft)

Displacement

1,782 tons

Performance

Full speed	16 knots
Cruising speed	14·5 knots
Range at cruise	16,000 nm
Endurance at low speed	46 days
Maximum anchoring depth	30 fathoms
Towing capacity	moderate

Main machinery description

Main engines: One V-16 cyl. diesel, 2,150 SHP

Auxiliary engines: Two diesel 250 h.p. 350 kw, 1 diesel 240 h.p. 125 kw

Propeller(s) type: Controllable pitch, 3 blades

Pumps: Water, fuel, bilge, fire, general service

Controls: Manoeuvring—pilot house pitch and engine controls

On-board equipment

Facilities for carrying major underwater equipment: Stern rampe and U-frame for small submersible support

Sampling equipment: Trawling gear, plankton and biological nets

Craneage facilities: Three 1,800 kg (4,000 lb) capacity each

Winching facilities: Split 2 drum trawl winches, one STD winch, one plankton trawl, one BT winch

Navigational equipment: One gyrocompass. One automatic pilot. Two radars. One Loran A/C. One radio direction finder. One electromagnetic underwater log.

Survey and search equipment: One Edo fathometer, 6,000 fm, one sonar, BTs, STD

Laboratory and technical facilities

Analytical:

Organic	
Chemical	17·30 m² (186 ft²)
Physical	16·46 m² (177 ft²)
Inorganic	
Wet	29·76 m² (320 ft²)
Biological	20·46 m² (220 ft²)
Electronis	11·72 m² (126 ft²)

Crew

Sea-going complement, officers, men	23
Scientific complement	9
Accommodation	35

Crew safety facilities and equipment: life preservers, 2 life boats

Capital cost
$3,450,000 (1965)

Operators
Director, Pacific Marine Center,
1801 Fairview Ave., East,
Seattle, Washington 98102

MT. MITCHELL (MSS 22)
Call letters WTEG
(sister ships *Fairweather* and *Rainier*)

NOAA Class of vessel
Class II

Specific purpose of vessel
Hydrographic surveys

Design history

Start of construction: 2 April 1965 (Aerojet-General Shipyards, Jacksonville, Florida)

Date of launching 29 November 1966

Date trials were completed: 30 October 1967

Certification authority clearance: 23 March 1968

Basic dimensions

Length, overall	70·43 m (231 ft)
Beam, overall	12·80 m (42 ft)
Height, overall	26·22 m (86 ft)
Draught	4·27 m (14 ft)

Displacement
1,800 tons

Performance

Full speed	14 knots
Cruising speed	13 knots
Range at cruise	8,000 nm
Endurance at low speed	30 days
Maximum anchoring depth	100 fathoms, 183 m (600 ft)
Towing capacity	moderate

Main machinery description

Main engines: Twin-screw diesel, 1,200 SHP, 850 rpm

Auxiliary engines: Bow thruster

Propeller(s) type: Two controllable pitch

Pumps: Salt and fresh water, fuel, bilge, etc.

Controls: Manoeuvring—One 200 BHP diesel engine bow thruster 2,250 kg (5,000 lb) thrust

On-board equipment

Facilities for carrying major surface equipment: Six boat davits, rated for weight of boats 3,600 kg (8,000 lb) to 5,850 kg (13,000 lb). One 9·45 m (31 ft) hydrographic survey launch. Two 7·62 m (25 ft) hydrographic survey launches. One 7·93 m (26 ft) motor whale boat. Two 4·88 m (16 ft) Boston whalers. Two 3·96 m (13 ft) Boston whalers

Facilities for diving: Scuba gear aboard

Sampling equipment: Bottom samples (coring), Nansen bottles, plankton casts

Craneage facilities: One non-telescoping boom, radius 3·66 m (12 ft) to 12·20 m (40 ft) w/wt. capacity 2,250 kg (5,000 lb). Two telescoping booms, radius 3·96 m (13 ft) to 7·62 m (25 ft) w/wt. capacity 1,350 kg (3,000 lb). Two A-frames, radius varies w/block used

Winching facilities: One electric BT w/ drum capacity 762 m (2,500 ft) of 0·23 cm (3/32 in) wire rope. Two electrohydraulic w/three interchangeable drums. One drum w/capacity 3,658 m (12,000 ft) of 0·76 cm (0·298 in). One drum w/capacity 9,146 m (30,000 ft) of 0·48 cm (3/16 in) wire rope. One drum w/capacity 1,829 m (6,000 ft) of 0·97 cm ($\frac{3}{8}$ in) wire rope

Navigational equipment: One Loran A, one Loran A/C, one Omega system. Two surface search radar, one radio direction finder. One radio positioning system. One gyrocompass w/18 repeater outputs. One horizontal echo sounding system. One electromagnetic underwater log

Survey and search equipment: Two deep water depth recorders. Two shoal-water depth recorders. One precision depth recorder. One NOS hydroplot system w/COMPLOT plotter. Two NOS hydrolog systems without plotter. Bottom profiler BT 0–274 m (0–900 ft). Current meters, magnetometer, tide gage

Pumps, compressors and ancillary power equipment and services: Two 300-kw, 450-VAC, 60 Hz, 3 phase generator. 115-VAC, 60 Hz, single phase. One 75-kw, 450-VAC, 60 Hz, 3 phase emergency generator

Laboratory and technical facilities

Analytical:

Oceanographic	22·32 m² (240 ft²)
Data processing	33·48 m² (360 ft²)

Crew

Sea-going complement, officers, men	74
Scientific complement	1

Accommodation: Comfortable quarters, air conditioned 78 berths

Crew safety facilities and equipment: One life boat, life preserver, 4 inflatable life rafts

Capital cost
$6,033,000 (1965)

Operators
Director, Atlantic Marine Center,
439 West York Street,
Norfolk, VA 23510

PEIRCE (CSS 28) (sister ship is *Whiting*)
Call letters WTEQ

NOAA Class of vessel
Class III

Specific purpose of vessel
Hydrographic surveys

Design history

Start of construction: 6 November 1961

Date of launching: 15 October 1962

Date trials were completed: 6 May 1963

Certification authority clearance: 6 May 1963

Basic dimensions

Length, overall	49·70 m (163 ft)
Beam, overall	10·00 m (33 ft)
Height, overall	21·80 m (71 ft 6 in) from B.L.
Draught	3·05 m (10 ft)

Displacement
760 tons

Performance

Full speed	12 knots
Cruising speed	12 knots
Range at cruise	4,500 nm
Endurance at low speed	10 days
Maximum anchoring depth	105 fathoms, 192 m (630 ft)
Towing capacity	slight

Main machinery description

Main engines: Twin screw diesel, 800 SHP, 800 rpm

Propeller(s) type: Controllable-pitch propellers, dia. 1·83 m (6 ft)

Pumps: Salt and fresh water, fuel, bilge, etc.

On-board equipment

Facilities for carrying major surface equipment: Two 7·93 m (26 ft) hydrographic survey launches. Two 4·88 m (16 ft) Boston whalers.One 3·96 m (13 ft) Boston whaler

Facilities for diving: Scuba gear aboard

Sampling equipment: Bottom sampling, plankton nets, Nansen bottles

Craneage facilities: One telescoping boom, radius 3·96 m (13 ft) to 7·62 m (25 ft) weight capacity 1,350 kg (3,000 lb) to 1,136 kg (2,500 lb). One A-frame, radius varies w/blocks used

Winching facilities, capacities: One electric BT w/drum capacity 1,219 m (4,000 ft) of 0·23 cm (3/32 in). One electrohydraulic double drum, 1 drum w/capacity 1,829 m (6,000 ft) of 0·76 cm (0·298 in), 1 drum w/capacity 4,573 m (15,000 ft) of 0·48 cm (3/16 in) wire rope

Navigational equipment: Two surface search radars. Two electronic positioners One EM underwater log. One gyrocompass system w/7 repeaters

Survey and search equipment: One deep water depth recorder. Two shoal water depth recorders. One precision depth recorder. Tide gage, magnetometer, BT 0–274 m (0–900 ft). Four teletype manual logging system. Current meters

Pumps, compressors and ancillary power equipment and services: Two 220-kw, 450-VAC, 60 Hz, 3 phase generators. One 60-kw, 450-VAC, 60 Hz, 3 phase emergency generators

Laboratory and technical facilities

Analytical

Organic	Chart and plotting room 23·81 m² (256 ft²)
Data processing	14·32 m² (154 ft²)

Crew

Sea-going complement, officers, men: 36

Scientific complement: 2

Accommodation: Messing and berthing areas, air conditioned 40 berths

Crew safety facilities and equipment: Life preservers, inflatable life rafts

Capital cost
$2,570,000 (1961)

Operators
Director, Atlantic Marine Center,
439 West York Street,
Norfolk, VA 23510

WHITING (CSS 29) (sister ship is *Peirce*)
Call letters WTEW

NOAA Class of vessel
Class III

Specific purpose of vessel
Hydrographic surveys

Design history
Start of construction: 29 November 1961 (Marietta Manufacturing Co., Pt. Pleasant, W. Virginia)
Date of launching: 20 November 1962
Date trials were completed: 8 July 1963
Certification authority clearance: 8 July 1963

Basic dimensions

Length, overall	49·7 m (163 ft)
Beam, overall	10·1 m (33 ft)
Height, overall	21·8 m (71 ft 6 in) from B.L.
Draught	3·05 m (10 ft)

Displacement
760 tons

Performance

Full speed	12 knots
Cruising speed	12 knots
Range at cruise	4,500 nm
Endurance at low speed	10 days
Maximum anchoring depth	105 fathoms, 192 m (630 ft)
Towing capacity	slight

Main machinery description
Main engines, type, shp and rpm: Twin screw diesel, 800 SHP, 800 rpm
Propeller(s) type: Controllable pitch, 1·83 m (6 ft) diameter
Pumps: Water, fuel, bilge, etc.

On-board equipment
Facilities for carrying major surface equipment: Two boat davits, rated for weight of boats 5,850 kg (13,000 lb). One 4·88 m (16 ft) Boston whaler. Two 7·93 m (26 ft) hydro launches. One 3·96 m(13 ft) Boston whaler
Facilities for diving: Scuba gear aboard
Sampling equipment: Bottom sampling, plankton nets, Nansen bottles
Craneage facilities: One telescoping boom, radius 3·96 m (13 ft) to 7·62 m (25 ft) w/ capacity 1,350 kg (3,000 lb) and 1,136 kg (2,500 lb). One A-frame, radius varies w/ block used
Winching facilities: One electric BTw/drum capacity 1,219 m (4,000 ft) of 0·23 cm (3/32 in) wire rope. One electrohydraulic double-drum, 1 drum w/capacity 1,829 m (6,000 ft) of 0·76 cm (0·298 in) cable, 1 drum w/ capacity 4,573 m (15,000 ft) of 0·48 cm (3/16 in) wire rope
Navigational equipment: One Loran-A, one gyrocompass w/7 repeaters. One electronic positioner, two surface search radars, one EM underwater log
Survey and search equipment: One deep water depth sounder. One precision depth sounder. Two shoal water depth sounders. BT 0–274 m (0–900 ft), current meters. Tide gage. One NOS hydroplot system w/ CALCOMP 502 plotter
Pumps, compressors and ancillary power equipment and services: Two 220-kw, 450-VAC, 60 Hz, 3 phase generator. One 60-kw, 450-VAC, 60 Hz, 3 phase emergency generator

Laboratory and technical facilities
Chart and plotting room: 23·81 m² (256 ft²)
Sea-going complement, officers, men: 36
Scientific complement: 2
Accommodation: Messing and berthing areas air conditioned 40 berths
Crew safety facilities and equipment: Life preservers, inflatable rubber rafts

Capital cost
$2,640,000 (1961)

Operators
Director, Atlantic Marine Center,
National Ocean Survey, NOAA,
439 West York Street,
Norfolk, Virginia 23510

McARTHUR (CSS 30) (sister ship of *Davidson*)
Call letters WTEJ

NOAA Class of vessel
Class III

Specific purpose of vessel
Hydrographic surveys

Design history
Start of construction: 15 July 1965 (Norfolk Shipbuilding & Dry Dock Corp., Norfolk, Virginia)
Date of launching: 15 November 1965
Date trials were completed: 9 November 1966
Certification authority clearance: 15 December 1966

Basic dimensions

Length, overall	53·3 m (175 ft)
Beam, overall	11·6 m (38 ft)
Height, overall	20·7 m (68 ft) above DWL, to aircraft warning light
Draught	3·35 m (11 ft)

Displacement
995 tons

Performance

Full speed	13 knots
Cruising speed	13 knots
Range at cruise	6,400 nm
Endurance at low speed	21 days
Maximum anchoring depth	105 fathoms, 192 m (630 ft)
Towing capacity	slight

Main machinery description
Main engines, type, shp and rpm: Twin screw diesel, 800 SHP, 800 rpm
Propeller(s) type: Controllable-pitch, 3 blades 2·07 m (81½ in) diameter
Pumps: Salt and fresh water, fuel, bilge

On-board equipment
Facilities for carrying major surface equipment: Two 7·62 m (25 ft) hydrographic launches, two 4·88 m (16 ft) Boston whalers, one 3·65 m (12 ft) Boston whaler
Facilities for diving: Scuba gear aboard
Sampling equipment: Bottom samples, plankton casts, Nansen bottles
Craneage facilities: One rotating self-contained hydraulic, w/capacity 1,136 kg (2,500 lb) at 8·20 m (27 ft) radius
Winching facilities: One BT electrohydraulic w/drum capacity 1,219 m (4,000 ft) of 0·23 cm (3/32 in) wire rope. One oceanographic, three interchangeable drums on fantail. One w/capacity 6,400 m (21,000 ft) of 0·76 cm (0·298 in). One w/9,146 m (30,000 ft) capacity of 0·48 cm (3/16 in) wire rope, one w/1,829 m (6,000 ft) capacity of 1,829 cm (⅜ in) wire rope
Navigational equipment: One Loran A. One Loran C. Two surface search radar. One radio direction finder. One gyrocompass w/8 repeaters. One electromagnetic underwater log
Survey and search equipment: One deep-water depth system. One precision depth system. Two shoal water depth system. One precision depth recorder. One Wang 700A programmable calculator. Two Flexowriter manual logging system. Tide gage, current meter
Pumps, compressors and ancillary power equipment and services: Two 220-kw, 450-VAC, 60 Hz, 3 phase generator. 115-VAC, 60 Hz, single phase. One 60-kw, 450-VAC, 60 Hz, 3 phase emergency generator

Laboratory and technical facilities

Oceanographic	16·0 m² (176 ft²)
Data processing	23·7 m² (258 ft²)

Crew
Sea-going complement, officers, men: 38
Scientific complement: 1
Accommodation: Air conditioned quarters —40 berths
Crew safety facilities and equipment: 80 life preservers, 2 inflatable life rafts, one life boat

Capital cost
$3,107,000 (1965)

Operators
Director, Pacific Marine Center,
National Ocean Survey, NOAA,
1801 Fairview Avenue, East,
Seattle, Washington 98102

DAVIDSON (CSS 31) (sister ship is *McArthur*)
Call letters WTEK

NOAA Class of vessel
Class III

Specific purpose of vessel
Hydrographic surveys

Design history
Start of construction: 22 November 1965 (Norfolk Shipbuilding and Drydock Corp., Norfolk, Virginia)
Date of launching: 7 May 1966
Date trials were completed: 7 February 1967
Certification authority clearance: 10 March 1967

Basic dimensions

Length, overall	53·4 m (175 ft)
Beam, overall	11·6 m (38 ft)
Height, overall	20·7 m (68 ft) above DWL to aircraft warning light
Draught	3·35 m (11 ft)

Weights

Displacement	995 tons
Crew	2·9 tons

Performance

Full speed	13 knots
Cruising speed	13 knots
Range at cruise	6,400 nm
Endurance at low speed	21 days
Maximum anchoring depth	105 fathoms (192 m (630 ft)
Towing capacity	slight

Main machinery description

Main engines, type, shp and rpm: Twin screw diesel, 800 SHP, 800 rpm

Propeller(s) type: Controllable pitch, 3 blades 2·07 m (81½ in) diameter

Pumps: Salt and fresh water, fuel, bilge

On-board equipment

Facilities for carrying major surface equipment: Two 4·88 m (16 ft) Boston whalers. One 3·96 m (13 ft) Boston whaler. Two 7·62 m (25 ft) hydrographic launches

Facilities for diving: Scuba gear aboard

Sampling equipment: Bottom samples, plankton casts, Nansen bottles

Craneage facilities: One rotating, self-contained hydraulic w/capacity of 1,136 kg (2,500 lb) at 8·23 m (27 ft) radius

Winching facilities: One BT, w/drum capacity 1,220 m (4,000 ft) of 0·23 cm (3/32 in) wire rope. One oceanographic w/three drums. One drum w/1,220 m (12,000 ft) of 0·76 cm (0·298 in) cable. One drum w/9,146 m (30,000 ft) capacity of 0·48 cm (3/16 in) wire rope. One drum w/1,829 m (6,000 ft) of 0·70 cm (⅜ in) wire rope

Navigational equipment: One Loran A/C. Two digital micro-circuit Loran-C receivers. Two surface search radars. One radio direction finder. One electronic navigation system. One gyrocompass w/8 repeaters

Survey and search equipment: One deep water depth system. One precision depth system. Two shoal water depth system. One precision depth recorder. One Wang 700A programmable calculator. Two Flexowriter manual logging systems. Tide gage, current meters

Laboratory and technical facilities

Analytical:

Oceanographic	16·0 m² (176 ft²)
Data processing	23·7 m² (258 ft²)

Crew

Sea-going complement, officers, men: 38

Scientific complement: 1

Accommodation: Air conditioned quarters 40 berths

Crew safety facilities and equipment: One life boat, 80 life preservers, 2 inflatable life rafts

Capital cost

$3,210,000 (1965)

Operators

Director, Pacific Marine Center,
1801 Fairview Avenue, East,
Seattle, Washington 98102

OREGON II (FRS 32)
Call letters WTDO

NOAA Class of vessel

Class III, large side trawler

Specific purpose of vessel

Fishery research

Design history

Start of construction: 1965 (Ingalls Shipbuilding Corp.)

Certification authority clearance: 1967

Basic dimensions

Length, overall	52·13 m (171 ft)
Beam, overall	10·3 m (34 ft)
Draught	4·3 m (14 ft)

Displacement

906 tons

Performance

Full speed	14·5 knots
Cruising speed	13 knots
Range at cruise	10,700 nm
Endurance at low speed	60 days
Maximum anchoring depth	40 fathoms 73 m (240 ft)
Towing capacity	moderate

Main machinery description

Main engines: Two, opposed piston, 720 h.p.

Auxiliary engines: Two, 320 BHP, 440 VAC 220 kw

Propeller(s) type: Single, variable pitch 4-blade

Pumps: Salt and fresh water, fuel, bilge, general serv. fire

Controls: Manoeuvring—Engine

On-board equipment

Sampling equipment: Trawl and seine nets, Nansen bottles, other fishing gear

Craneage facilities: Two swinging hydraulic booms

Winching facilities: One BT, one hydrographic, one trawl, two seine

Navigational equipment: One gyrocompass, one Loran A/C. One Pit log. One radar-Decca RM-326, radio direction finder

Survey and search equipment: Two sonars, two echo sounders—range to 600 fathoms

Pumps, compressors and ancillary power equipment and services: 440-220-110 AC power, 120-VAC lighting

Laboratory and technical facilities

Analytical:

Wet	25·2 m² (275 ft²)
Specimen	16·1 m² (175 ft²)
Data processing	18·8 m² (210 ft²)

Crew

Sea-going complement, officers, men: 15

Scientific complement: 14

Accommodation: 32

Crew safety facilities and equipment: One life boat, 4 inflatable rafts, life preservers

Capital cost

$2,673,000 (1965)

Operators

NOAA—National Ocean Survey,
National Marine Fisheries Service,
Southeast Fisheries Center,
Miami, Florida

ALBATROSS IV (FRS 42)
Call letters WZ 2158

NOAA Class of vessel

Class IV, Stern trawler

Specific purpose of vessel

Biological investigations

Design history

Start of construction: 1960 (Southern Shipbuilding Corp., Slidell, Louisiana)

Date of launching: 1961

Date trials were completed: 10 October 1962

Certification authority clearance: 1962

Basic dimensions

Length, overall	57·01 m (187 ft)
Beam, overall	10·06 m (33 ft)
Draught	4·88 m (16 ft)

Displacement

1,089 tons

Performance

Full speed	11 knots
Cruising speed	11 knots
Range at cruise	9,000 nm
Endurance at low speed	14 days
Maximum anchoring depth	30 tons
Towing capacity	moderate

Main machinery description

Main engines, type, shp and rpm: Twin V-8 cyl, 500 BHP

Auxiliary engines, type, shp and rpm: Three 6 cyl 150 h.p., diesel generators 220 volts DC

Propeller type: One, controllable pitch, 3 blades, kort nuzzle

Pumps: Salt, fresh water, fuel, bilge

Controls: Manoeuvring—Pilot house, pitch and engine. Position holding—Pilot house, bow thruster

On-board equipment

Facilities for carrying major underwater equipment: Ramp and U-frame for small submersible support

Sampling equipment: Fishing gear, dredging equipment, biological samples

Craneage facilities: One 5 ton and two 1 ton booms

Winching facilities: One double drum trawl, two hydrographic, one dredge, two BT

Navigational equipment: One gyrocompass, one electromagnetic log, two radars (3 CM and 10 CM), two Loran A, one radio direction finder

Survey and search equipment: Two echo sounders, one sonar

Pumps, compressors and ancillary power equipment and services: 220 volt 30 AC and 110 volt AC service. One 26 kw, emergency generator

Laboratory and technical facilities

Two biological. One electronic. One chemical. One hydrographic

Crew

Sea-going complement, officers, men: 18

Scientific complement: 13

Accommodation: 34

Crew safety facilities and equipment: 6·1 m (20 ft) standard life boat, 3 inflatable rafts, life preservers

Capital cost

$2,225,000 (1960)

Operators

National Marine Fisheries Service, Northeast Fisheries Center,

National Oceanic and Atmospheric Administration

Woods Hole, Massachusetts

Additional information

Radiotelephone

TOWNSEND CROMWELL (FRS 43)
Call letters WTDF

NOAA Class of vessel
Class IV, general purpose—side trawler

Specific purpose of vessel
Fishery research

Design history
Date of launching: 1963 (J. Ray McDermott Co., Amelia, Louisiana)
Certification authority clearance: 1963

Basic dimensions

Length, overall	49·7 m (163 ft)
Beam, overall	10·06 m (33 ft)
Draught	2·74 m (9 ft)

Displacement
652 tons

Performance

Full speed	13 knots
Cruising speed	12 knots
Range at cruise	10,000 nm
Endurance at low speed	60 days
Maximum anchoring depth	45·73 m (150 ft) 25 fathoms

Main machinery description
Main engines, type, shp and rpm: Two 6 cyl diesel, 400 BHP each, 714 rpm
Auxiliary engines, type, shp and rpm: Two 6 cyl diesel 510 h.p.—350 kw. One 6 cyl diesel 240 h.p.—125 kw
Propeller(s) type: Two controllable pitch, 3 blades, fixed pitch
Pumps: Fire, fuel, bilge
Controls: Manoeuvring—Pilot house, engine and pitch controls

On-board equipment
Facilities for diving: Diver tank stowage
Sampling equipment: Fishing gear for trawling, trolling and longlining
Craneage facilities: 3 ton cargo boom
Winching facilities: One double drum trawl, one STD, one hydrographic, one long-line
Navigational equipment: Gyrocompass, one radar, one Loran A, one radio direction finder
Survey and search equipment: One sonar, three echo sounders, STD, BT's
Pumps, compressors and ancillary power equipment and services: Hydraulic deck machinery. INVI bulbous bow with viewing ports. Passive flume stabilization

Laboratory and technical facilities
Analytical:
Organic:

Physical	16·74 m² (180 ft²)
Inorganic:	
Chemical	11·81 m² (127 ft²)
Biological	8·84 m² (95 ft²)

Crew
Sea-going complement, officers, men: 14
Scientific complement: 7
Crew safety facilities and equipment: Life preservers, 2 inflatable life rafts

Capital cost
$1,487,000 (1963)

Operators
Director, Pacific Marine Center,
National Ocean Survey, NOAA
1801 Fairview Avenue, East,
Seattle, Washington 98102

Additional information
Vessel deactivated: 21 May 1973

DAVID STARR JORDAN (FRS 44)
Call letters WTDK

NOAA Class of vessel
Class IV, special design

Specific purpose of vessel
Fishery research—oceanographic research

Design history
Start of construction: 1963
Date of launching: 19 December 1964 (Christy Corp. Sturgeon Bay, Wisconsin)
Date trials were completed: November 1965
Certification authority clearance: November 1965

Basic dimensions

Length, overall	52·13 m (171 ft)
Beam, overall	10·98 m (36 ft)
Draught	3·66 m (12 ft)

Displacement
890 tons

Performance

Full speed	12 knots
Cruising speed	11 knots
Range at cruise	9,000 nm
Endurance at low speed	33 days
Maximum anchoring depth	30 fathoms, 55 m (180 ft)

Main machinery description
Main engines, type, shp and rpm: Twin 6 cyl diesel, 459 BHP each, 900 rpm
Auxiliary engines, type, shp and rpm: Twin 8 cyl diesel, 200 h.p., 1,200 rpm, 200 kw 440 volts
Propeller(s) type: Controllable pitch, 3 blades
Pumps: Salt and fresh water, fuel, bilge, fire
Controls: Manoeuvring—Pilot house, pitch and engine controls. Position holding—Pilot house, bow propeller controls

On-board equipment
Facilities for carrying major underwater equipment: Ramp and V-frame for submersible support
Facilities for carrying major surface equipment: 4·88 m (16 ft) and 3·96 m (13 ft) work/rescue boats
Facilities for diving: Diving gear locker onboard
Sampling equipment: Bottom sampling, coring, biological and fishing, trawling
Craneage facilities: One articulated w/12 ton capacity. One small foredeck, 2 ton capacity
Winching facilities: Two hydraulic-hydrographic w/capacity of 7,622 m (25,000 ft) of 0·48 cm (3/16 in) cable, fishing net reel, one triple drum w/capacity of 1,000 fm of 1·27 cm ($\frac{1}{2}$ in) cable
Navigational equipment: One gyrocompass, one radio direction finder, one Omega receiver, one Loran A/C, two Decca radars
Survey and search equipment: Three echo sounders, two sonar
Pumps, compressors and ancillary power equipment and services: 220 volt and 110 volt AC service. One 30-kw, emergency generator

Laboratory and technical facilities
Analytical:

Organic	49·38 m² (531 ft²)
Inorganic	53·94 m² (580 ft²)
Photographic	3·72 m² (40 ft²)
Radiological	9·3 m² (100 ft²)

Crew
Sea-going complement, officers, men: 16
Scientific complement: 13
Accommodation: 35
Crew safety facilities and equipment: 4 inflatable life rafts, life preservers

Capital cost
$2,455,000 (1964)

Operators
National Marine Fisheries Service,
Southwest Fisheries Center,
National Oceanic and Atmospheric Administration,
La Jolla, California

The vessel has underwater viewing ports forward and port midship.

DELAWARE II
NOAA Class of vessel
Class IV, Stern Trawler

Specific purpose of vessel
Fisheries research

Design history
Start of construction: June 1966 (South Portland Engineering Co., South Portland, Maine)
Date of launching: 31 December 1967
Certification authority clearance: October 1968

Basic dimensions

Length, overall	47·26 m (155 ft)
Beam, overall	9·15 m (30 ft)
Draught	3·35 m (11 ft)

Displacement
680 tons

Performance

Full speed	12 knots
Cruising speed	12 knots
Range at cruise	8,000 nm
Endurance at low speed	30 days
Maximum anchoring depth	35 fathoms, 64·02 m (210 ft)
Towing capacity	moderate

Main machinery description
Main engines, type, shp and rpm: V-16 cyl diesel, 1,000 BHP
Auxiliary engines, type, shp and rpm: Two 8 cyl diesel, 150 h.p., 440 volt AC, 150 kw
Propeller(s) type: Fixed pitch, 4 blades
Pumps: Water, fuel, bilge, pumps, fire, general service
Controls—manoeuvring: Engine speed control in pilot house

On-board equipment
Facilities for carrying major underwater equipment: Stern ramp for launching/recovering submersibles
Sampling equipment: Fishing gear, 'Bongo' plankton net, Neutson nets, bottom samplers
Craneage facilities, capacities and clearances above decks: One 1-ton cargo boom forward. One 5-ton boom aft
Winching facilities: Trawl winch, hydraulic driven, capacity 3,660 m (2,000 fm) 2·86 cm (1⅛ in) wire. Two BT. One portable hydrographic winch
Navigational equipment: One gyrocompass w/3 repeaters. One magnetic compass w/autopilot. One EM log. Two radars. Two Loran A receivers. One radio direction finder
Survey and search equipment: Three echo sounders, one sonar

Laboratory and technical facilities
Analytical:
Wet 23·25 m (ft^2) (250 ft²)

Crew
Sea-going complement, officers, men: 15
Scientific complement: 8
Accommodation: 26
Crew safety facilities and equipment: 3·65 m (12 ft) rescue boat, 4 inflatable life rafts, life preservers

Availability
Purchase, capital cost: $1,645,000 (1966)

Operators
Middle Atlantic Coastal Fisheries Center,
National Marine Fisheries Service,
National Ocean Survey, NOAA
Sandy Hook, New Jersey
The vessel has an underwater viewing port.
The ship was deactivated on 1 May 1973

FERREL (ASV 92)
Call letters WTEZ

NOAA Class of vessel
Class IV

Specific purpose of vessel
Circulatory surveys

Design history
Start of construction: 23 October 1967 (Zigler Shipyards, Inc.)
Date of launching: 4 April 1968
Date trials were completed: 14 May 1968
Certification authority clearance: 4 June 1968

Basic dimensions

Length, overall	40·55 m (133 ft)
Beam, overall	9·76 m (32 ft)
Height, overall	18·29 m (60 ft)
Draught	2·13 m (7 ft)

Displacement
360 tons

Performance

Full speed	10 knots
Cruising speed	10 knots
Range at cruise	2,700 nm
Endurance at low speed	15 days
Towing capacity	slight

Main machinery description
Main engines, type, shp and rpm: Twin screw diesel, 375 SHP each
Propeller(s) type: Fixed pitch
Pumps: Water, fuel, bilge pumps
Controls: Manoeuvring—100 h.p. electric driven bow thruster

On-board equipment
Facilities for carrying major surface equipment: Buoys stowed on open after deck 41·85 m² (450 ft²) of enclosed deck area for buoy workshop. Carries one 17·99 m (59 ft) high-speed launch and one 8·54 m (28 ft) aluminium launch
Facilities for diving: Scuba gear aboard
Sampling equipment: STD system
Craneage facilities: Two heavy duty deck cranes
Winching facilities: One deep water for STD casts. Two buoys, motor driven, w/removable drums, one anchor windlass
Navigational equipment: One gyrocompass, one magnetic compass, one radar, one Loran A/C, one automatic pilot
Survey and search equipment: One shoal-water depth recorder, tidal current system—TICUS, general purpose digital computer
Pumps, compressors and ancillary power equipment and services: Two 150-kw, 450-VAC, 60 Hz, 3 phase generators

Laboratory and technical facilities
Electronic workshop 41·85 m² (450 ft²)

Crew
Sea-going complement, officers, men: 16
Scientific complement: 2
Accommodation: Air conditioned quarters —18 berths
Crew safety facilities and equipment: Life preservers, inflatable life rafts

Capital cost
$1,153,000 (1967)

Operators
Director, Atlantic Marine Center,
National Ocean Survey, NOAA,
439 West York Street,
Norfolk, Va. 23510

RUDE (ASV 90) (sister ship is *Heck*)
Call letters WTET

NOAA Class of vessel
Class V

Specific purpose of vessel
Specially built for wire drag surveys

Design history
Start of construction: 10 September 1965 (Jakobson Shipyard Inc., Oyster Bay, Long Island, N.Y.)
Date of launching: April 1966
Date trials were completed: December 1966
Certification authority clearance: 29 March 1967

Basic dimensions

Length, overall	27·44 m (90 ft)
Beam, overall	6·71 m (22 ft)
Height, overall	10·37 m (34 ft)
Draught	2·13 m (7 ft)

Displacement
170 tons

Performance

Full speed	11 knots
Cruising speed	11 knots
Range at cruise	800 nm
Endurance at low speed	3 days

Main machinery description
Main engines, type, shp and rpm: Twin screw diesel, 400 SHP
Propeller(s) type: Two-fixed pitch propellers
Pumps: Water, fuel, bilge
Controls: Manoeuvring—Main propulsion Position holding

On-board equipment
Facilities for carrying major surface equipment: One 6·10 m (20 ft) launch, small buoys, one 4·88 m (16 ft) launch
Facilities for diving: Scuba gear aboard
Craneage facilities: One hydraulic crane to handle boats, one hydraulic anchor windlass
Winching facilities: One wire drag w/drum capacity of 3,050 m (10,000 ft), 0·64 cm (¼ in) wire rope, one buoy boat
Navigational equipment: One gyrocompass, one surface search radar, one electronic positioning system
Survey and search equipment: One sonar system, one shoal water depth recorder, wire drag equipment
Pumps, compressors and ancillary power equipment and services: Two 60-kw, 230-VAC, 60 Hz, 3 phase generator

Laboratory and technical facilities
Data processing: 9·30 m² (100 ft²) in pilot house

Crew
Sea-going complement, officers, men: 10
Accommodation: Air conditioned, 10 berths
Crew safety facilities and equipment: Life preservers

Capital cost
$1,160,000 (1965)

Operators
Director, Atlantic Marine Center,
National Ocean Survey, NOAA,
439 West York Street,
Norfolk, Va. 23510

HECK (ASV 91) (sister ship is *Rude*)
Call letters WTEY

NOAA Class of vessel
Class V

Specific purpose of vessel
Specially built for wire drag surveys

Design history
Start of construction: 10 September 1965 (Jakobsen Shipyard, Inc., Oyster Bay, Long Island, New York)
Date of launching: 1 November 1966
Date trials were completed: 11 March 1967
Certification authority clearance: 29 March 1967

Basic dimensions

Length, overall	27·44 m (90 ft)
Beam, overall	6·71 m (22 ft)
Height, overall	10·37 m (34 ft)
Draught	2·13 m (7 ft)

Displacement
190 tons

Performance

Full speed	11 knots
Cruising speed	11 knots
Range at cruise	800 nm
Endurance at low speed	3 days

Main machinery description
Main engines, type, shp and rpm: Twin screw diesel, 400 SHP each
Propeller(s) type: Fixed pitch propellers
Pumps: Water, fuel, bilge

On-board equipment
Facilities for carrying major surface equipment: One 6·10 m (20 ft) launch, small buoys, one 4·88 m (16 ft) skiff
Facilities for diving: Scuba gear aboard
Craneage facilities: One hydraulic crane for boats. One hydraulic anchor windlass
Winching facilities: One wire drag w/drum capacity 3,050 m (10,000 ft) of 0·64 cm (¼ in) wire rope. One buoy hoist, one buoy retrieval capstan
Navigational equipment: One gyrocompass, one electronic positioning system, one surface search radar
Survey and search equipment: One sonar system, one shoal water depth recorder, wire drag equipment
Pumps, compressors and ancillary power equipment and services: Two 60 kw, 230-VAC, 60 Hz, 3 phase generators

Laboratory and technical facilities
Data processing: 9·30 m^2 (100 ft^2) in pilot house

Crew
Sea-going complement, officers, men: 10
Accommodation: Air conditioned, 10 berths
Crew safety facilities and equipment: Life preservers

Capital cost
$1,160,000 (1965)

Operators
Director, Atlantic Marine Center,
National Ocean Survey, NOAA
439 West York Street,
Norfolk, Virginia 23510

OREGON (FRV 51)
Call letters WZ 2022

NOAA Class of vessel
Class V, purse-seiner

Specific purpose of vessel
Fisheries research and biological investigations

Design history
Start of construction: 1946 (Astoria Marine Construction Co.)
Certification authority clearance: 1946

Production
Number of craft of this particular type built: Three originally—two converted to research ships

Basic dimensions

Length, overall	30·49 m (100 ft)
Beam, overall	7·93 m (26 ft)
Draught	4·27 m (14 ft)

Displacement
410 tons

Performance

Full speed	10 knots
Cruising speed	9 knots
Range at cruise	4,800 nm
Endurance at low speed	31 days
Maximum anchoring depth	54·88 m (30 fathoms)

Main machinery description
Main engines, type, shp and rpm: Single screw diesel, 600 BHP, 400 rpm
Propeller(s) type: Fixed pitch propeller, 3 blades
Pumps: Water, fuel, bilge, fire

On-board equipment
Facilities for carrying major surface equipment: 4·88 m (16 ft) Boston whaler, 3·96 m (13 ft) Boston whaler
Sampling equipment: Plankton nets, fishing gear for trawling, seining, dredging and longlining
Craneage facilities: One main boom
Winching facilities: Two BT w/capacity 1,666 fm of 0·23 cm (3/32 in) cable, one trawl, double drum w/capacity 1,500 fm 1·27 cm (½ in) cable each
Navigational equipment: 1 gyrocompass, 1 automatic pilot, 2 radars, 1 Loran A/C, 1 radio direction finder
Survey and search equipment: Two echo sounders, one fishfinder, BT's
Pumps, compressors and ancillary power equipment and services: Two 85-kw, 220-VAC, 60 Hz generator

Laboratory and technical facilities
Data processing: 9·3 m^2 (100 ft^2)

Crew
Sea-going complement, officers, men: 10
Scientific complement: 4
Accommodation: 15
Crew safety facilities and equipment: Life preservers, inflatable life rafts

Capital cost
$378,000 (1946)

Operators
National Marine Fishery Service—Alaska Region
National Oceanic and Atmospheric Administration,
Kodiak Marine Fishery Center,
Kodiak, Alaska

JOHN N. COBB (FRV 52)
Call letters WZ 2023

NOAA Class of vessel
Class V, Combination seiner/trawler

Specific purpose of vessel
Fishery research and biological investigations

Design history
Start of construction: 1949 (Western Boat Building Co., Tacoma, Washington)
Certification authority clearance: 1950

Basic dimensions

Length, overall	28·35 m (93 ft)
Beam, overall	7·62 m (25 ft)
Height, overall	3·66 m (12 ft)

Displacement
250 tons

Performance

Full speed	10·5 knots
Cruising speed	9 knots
Range at cruise	4,800 nm
Endurance at low speed	21 days
Maximum anchoring depth	36·59 m (20 fathoms)

Main machinery description
Main engines: Single screw diesel, 8 cylinder, 345 BHP, 376 rpm
Propeller type: Fixed pitch, 3 blades
Pumps: Water, fuel, bilge
Controls—manoeuvring: Pilot house engine controls

On-board equipment
Facilities for carrying major surface equipment: 4·27 m (14 ft) utility outboard. 4·88 m (16 ft) boat, inboard
Sampling equipment: Plankton and biological nets, fishing gear for trawling, seining, trolling and other fishing
Craneage facilities: One 1¼ ton electric boom. One 3 ton hydraulic boom
Winching facilities: One trawl, hydraulic, double drum w/capacity 1,000 fathoms of 1·27 cm (½ in) cable. One oceanographic, hydraulic w/capacity 1,830 m (6,000 ft) of 0·48 cm (3/16 in) wire. One BT w/capacity 183 m (600 ft) of 0·23 cm (3/32 in) wire
Navigational equipment: One magnetic compass, one automatic pilot, two radars, one Loran A/C, one radio direction finder
Survey and search equipment: One Sonar, two echo sounders, fish counting echo sounder, BT's
Pumps, compressors and ancillary power equipment and services: Two 30-kw, 110-220 VAC, 60 Hz generators

Laboratory and technical facilities
General: 13·95 m^2 (150 ft^2)

Crew
Sea-going complement, officers, men: 7
Scientific complement: 4
Accommodation: 13
Crew safety facilities and equipment: Life preservers and inflatable life rafts

Capital cost
$247,000 (1949)

Operators
Director, Pacific Marine Center,
National Ocean Survey, NOAA,
1801 Fairview Avenue, East,
Seattle, Washington 98102

This vessel discovered the Cobb Seamount, located off the coast of Washington

MURRE II (FRV 63)
Call letters: WZ 2003

NOAA Class of vessel
Class VI Wooden power barge

Specific purpose of vessel
Biological and fishing research

Conversion
1946

Design history
Built: Maritime Shipyards, Seattle, Washington
Certification authority clearance: 1943

Basic dimensions

Length, overall	26·22 m (86 ft)
Beam, overall	7·93 m (26 ft)
Draught	2·13 m (7 ft)

Displacement
295 tons

Performance

Full speed	8 knots
Cruising speed	8·2 knots
Range at cruise	1,800 nm
Endurance at low speed	8 days
Maximum anchoring depth	15 fathoms, 27·44 m (90 ft)

Main machinery description
Main engines: Twin screw diesel, 104 BHP each
Propeller type: Two fixed pitch, 3 blades
Pumps: Water, fuel, bilge
Controls—manoeuvring: Pilot house engine controls

On-board equipment
Facilities for carrying major surface equipment: 4·26 m (14 ft) Boston whaler
Facilities for diving: Diver gear and bottle stowage
Sampling equipment: Fishing gear
Craneage facilities: One cargo and one trawling boom
Winching facilities, capacities: One hydraulic trawl and cargo 0·97 cm ($\frac{3}{8}$ in) wire. One electric hydrographic w/capacity 1,000 metres 0·41 cm (5/32 in)
Navigational equipment: One magnetic compass. One radar. One radio direction finder. One autopilot
Survey and search equipment: Two echo sounders, BT's
Pumps, compressors and ancillary power equipment and services: One 20 kw, 110-VDC, 45 h.p. diesel generator. One 30 kw, 110-VDC, 63 h.p. diesel generator

Laboratory and technical facilities
Analytical:

Organic	18·60 m² (200 ft²)
Inorganic	12·09 m² (130 ft²)

Crew
Sea-going complement, officers, men: 3
Scientific complement: 2-6
Accommodation: 9
Crew safety facilities and equipment: Life preservers and inflatable life rafts

Capital cost
$135,000 (1943)

Operators
National Marine Fisheries Service,
National Oceanic and Atmospheric Administration,
Alaska Region,
Auke Bay Biological Lab.,
Auke Bay, Alaska

GEORGE M. BOWERS (FRV 65)
Call letters: WZ 2019

NOAA Class of vessel
Class VI, Shrimp Trawler

Specific purpose of vessel
Fishing and biological research

Conversion
1955 (converted from shrimp boat hull)

Design history
Start of construction: 1953 (Steamway Corp.)
Certification authority clearance: 1955

Basic dimensions

Length, overall	22·56 m (74 ft)
Beam, overall	6·10 m (20 ft)
Draught	2·44 m (8 ft)

Displacement
125 tons

Performance

Full speed	9 knots
Cruising speed	8 knots
Range at cruise	2,000 nm
Endurance at low speed	10 days
Maximum anchoring depth	15 fathoms 27·44 m (90 ft)

Main machinery description
Main engines: Single screw diesel, 200 BHP
Auxiliary engines: Two 6 cylinder 100 h.p., 30 kw, 220 volts AC
Propeller type: Controllable pitch, 3 blades
Pumps: Water, fuel, bilge
Controls—manoeuvring: Pilot house engine and pitch controls

On-board equipment
Facilities for carrying major underwater equipment: Tows remote underwater fisheries assessor (RUFAS)
Facilities for carrying major surface equipment: One life boat
Sampling equipment: Fishing gear
Craneage facilities: One boom and hoist
Winching facilities: One combination trawl-seine, 400 fathoms (731·71 m), mechanical power
Navigational equipment: One magnetic compass. One Loran A, one radar
Survey and search equipment: One sonar one echo sounder

Laboratory and technical facilities
General: 5·58 m² (60 ft²)

Crew
Sea-going complement, officers, men: 4
Scientific complement: 6
Accommodation: 10
Crew safety facilities and equipment: Life preservers, inflatable life raft

Capital cost
$148,000 (1953)

Operators
National Marine Fisheries Service,
Southeast Fisheries Center,
National Oceanic and Atmospheric Administration,
Miami, Florida

UNIVERSITY NATIONAL OCEANOGRAPHIC LABORATORY SYSTEM (UNOLS)

Organisation

The University National Oceanographic Laboratory System (UNOLS) is an adaptation of recommendations by the US President's Commission on Marine Science, Engineering and Resources (1969) and the National Council on Marine Resources and Engineering Development (1970). It was established in September 1971 following a year of development jointly by Federal Agencies and the Academic Community.

As stated in its charter UNOLS has as its objectives:

To create a mechanism for co-ordinated utilization of an planning for oceanographic facilities through an association of academic institutions in a national system whereby institutions can work together and with funding agencies to assist in the effective use, assessment and planning for oceanographic facilities.

To improve the level and stability of Federal support for academic oceanography, thereby continuing and enhancing the excellence of the US oceanographic programme.

UNOLS is organized according to the following structure:

The membership is defined as those academic institutions which operate Federally funded and significant oceanographic facilities. At the present time seventeen institutions comprise the membership. The Chairman of UNOLS is Dr. Arthur E. Maxwell, Provost of Woods Hole Oceanographic Institution. The UNOLS Office which provides staff and secretarial services is operated on behalf of the members by the Woods Hole Oceanographic Institution and is jointly funded by the National Science Foundation, Office of Naval Research, National Oceanic and Atmospheric Administration and the Environmental Protection Agency.

The Advisory Council is a group of seven scientists elected by UNOLS members. Its purpose is to monitor the activities of the System giving attention to the effective use of existing oceanographic facilities and to the performance of the member institutions in providing access to Federally supported University facilities for scientists from other institutions, and to evaluate the need for replacement and additional facilities and assess whether some facilities are outmoded or in excess of current needs.

The Research Vessel Operators Council (RVOC) comprises the marine superintendents and naval engineering personnel of university operating institutions. Its role is to exchange ship operating and technical information between members and to serve as a forum for working with the US Coast Guard on ship inspection and marine safety matters.

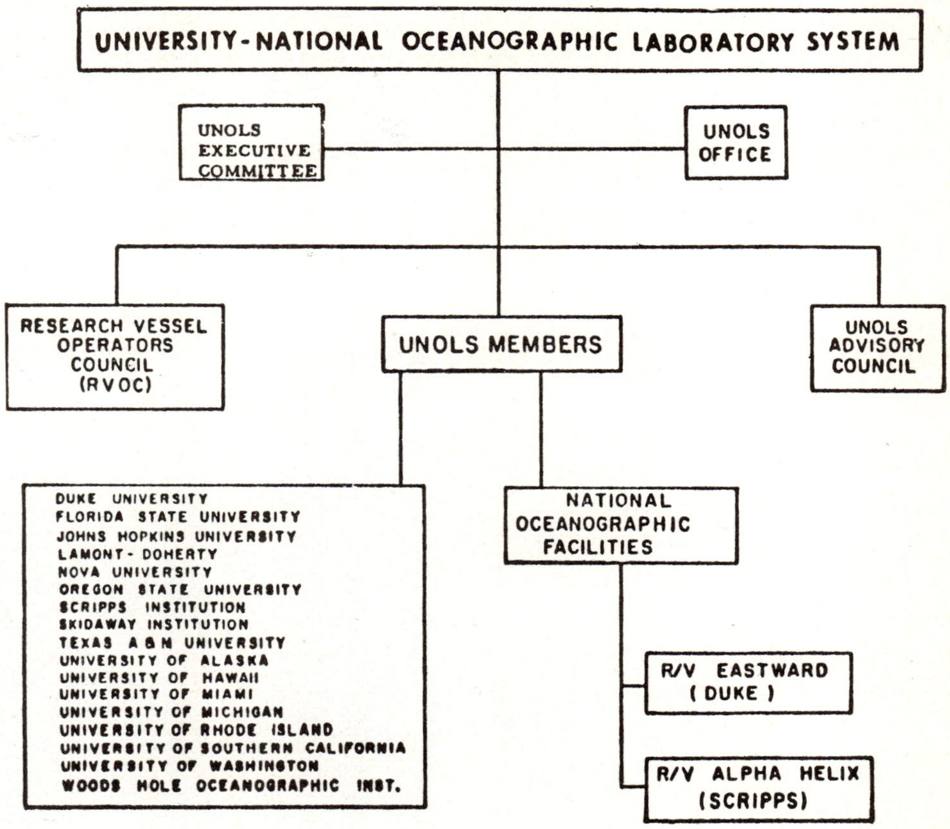

Research fleet

The Academic research ships funded directly by the Federal Government comprises a fleet of about thirty-four ships. These are operated by eighteen institutions (UNOLS Members) under 'block funding' grants by the Office of Naval Research and the National Science Foundation.

An inventory of these ships is given in the following table:

Of the nine ships over 61·0 m (200 ft) in length, four are relatively new (less than ten years old). Seven are owned by the Federal Government.

Two of seven ships in the 46–61 m (150–200 ft) class are new and the remaining are converted World War II cargo ships approaching thirty years of age. All are privately owned. Two new replacements funded in 1971 by the US Navy will enter service in 1974.

Of the seven ships 30–40 m (100–150 ft) long, three are new and the others are World War II conversions. One of these older ones is Navy owned.

The twelve ships from 15–30 m (50–100 ft) include five over 20 m (65 ft) of which three are of recent construction. Six 20 m (65 ft) vessels include four old conversions. Two under 20 m (65 ft) are both relatively new. Two of the 15–30 m (50–100 ft) class are Navy owned.

Geographically there is about an even distribution between the East and West Coasts for vessels over 30 m (100 ft). Below this size there is a preponderance of Atlantic and Gulf Coast based vessels.

In addition to the Federally funded university ships, other academic research vessels are operated either from project, state or other forms of funding. These are generally smaller ships but the great majority of academic research vessels over 23 m (75 ft) are Federally block-funded. At 20 m (65 ft), however, the comparison shifts. In this size range there are a total of about 35 boats engaged in university research and education of which eight are 'block-funded'. This relatively large number of 20 m (65 ft) boats results largely from the Federal motor boat laws and the availability of surplus Army 'T-Boat'.

There are an estimated seventy-five vessels between 8–15 m (26–50 ft) mostly used for estuarine research and training.

Replacement and new construction

As in any fleet of ships the effectiveness of future operations depends on replacements and new construction. This is especially true in oceanography where the growth and improvement of science at sea depend largely on the capability of and access to research vessels.

The number of university research vessels has grown from 12 ships in 1960 to 33 ships in 1973. This includes only the Federally funded vessels over 19·8 m (65 ft) in length. It does comprise almost all existing academic research vessels over 30·5 m (100 ft). A breakdown by age is given by the following table:

PROFILE OF FEDERAL FUNDED ACADEMIC RESEARCH FLEET

Age	Size (LOA)				
	Over 61 m (200 ft)	46–61 m (150–200 ft)	30–46 m (100–150 ft)	20–30 m (65–100 ft)	
Under 10	4	4	3	3	14
10–20	3	0	0	4	7
20–30	1	3	2	3	9
Over 30	1	0	1	2	4
Total	9	7	6	12	34

The above includes the new Navy owned AGOR ships (*Gyre* and *Moana Wave*) which entered service at the end of 1973.

The present ship construction rate averages about one ship per year. However, in terms of recent funding the fleet replacement cycle is closer to 60 years. By any reckoning there now appears to be inadequate provision to replace the existing fleet. Although the NSF long term ship construction plan does provide for increases in the current rate, the Navy, which has been the major provider of ships to the academic fleet in the past, has no plans for future ship construction in this category.

In planning for the future, the needs for future research ships to support academic research at sea is inextricably linked to the science to be conducted. Present trends appear to be in the direction of large inter-institutional, regional (and international) co-operative programmes, and towards the use of more sophisticated technology. This would imply larger and more capable ships as well as a departure from conventional hulls.

UNIVERSITY OF ALASKA INSTITUTE OF MARINE SCIENCE

ACONA

The research vessel *Acona* was built in 1962 as a North Sea trawler. In 1970 she was converted to operate as a research vessel and has been in use for oceanographic research in Arctic and Sub-arctic waters of the Pacific Ocean. The vessel is ice strengthened to break 0·3 m (12 in) of sea ice.

Basic dimensions

Length, overall	25·9 m (85 ft)
Beam, overall	6·61 m (21 ft 8 in)
Height	5·26 m (17 ft 3 in) keel to main deck (1·64 m (5 ft) freeboard) 9·76 m (32 ft) to top of bridge
Draught	3·73 m (12 ft 3 in) at maximum displacement

Gross tonnage
197 tons

Displacement
179 tons (full load)

Crew
Six (3 officers, 3 men)
Scientists: Accommodation for nine

Performance

Full speed	9·9 knots
Cruising speed	8·4 knots
Minimum speed	3·0 knots
Range	5,000 nm at cruise, 3,140 nm at maximum speed
Endurance at low speed	19 days
Position holding capability	Exact, Sea State 5
Maximum anchoring depth	1,098 m (600 fathoms)

Machinery
Main engines: Two 300 s.h.p. diesels
Auxiliary engines: One 125 s.h.p. diesel
Propellers: One stern propeller, fixed pitch. One bow thruster, adjustable pitch
Controls: Wheel and auto-pilot

Craneage facilities
Main crane: 2,727 kg (6,000 lb) 3·05 m (10 ft) height and 4·72 m (15 ft 6 in) depth capacity, 6·10 m (20 ft) clearance

Winching facilities
One 1,364 kg (3,000 lb). One 136 kg (300 lb)

On-board equipment
Navigational equipment: Two radars, Loran A, Loran A-C, Magnavox Satellite Navigator MX-706
Survey and search equipment: Two radars, seismic receivers
Ancillary power equipment: 120 volt, 1 or 3 phase AC, 25 or 60 kw. 240 volt, 3 phase, 60 kw, AC. 34 volt DC 1 kw (One Caterpillar and 2 Westinghouse ship's service generators)

RV *Acona* (*Courtesy University of Alaska*)

The engine room of RV *Acona* (*Courtesy University of Alaska*)

Instrumentation

Two depth sounders and recorders. PDR. Gyro compass. STD. B/T. Digital indicator. Two transducers, one wide and one narrow beam. Salinometer. Blast and regular freezers. Two Boston whaler small boats, with motors

Laboratories

Complete laboratory facilities are provided for organic, inorganic and geological work

Cost and availability

Purchase, capital cost: (1973) Original cost $US 300,000; overhauls and conversion $US 240,000.

Lease, charges: Currently $US 1,316 per sea day

Charter availability: Possible for scientific work, preferably in conjunction with scheduled programme

Remote controlled boat

A 6·1 m (20 ft) boat 'Firefish' operates remotely controlled from *Acona* for near-glacier exploration

Operators

University of Alaska,
Institute of Marine Science,
Box 95102, Fairbanks,
Alaska 99701 USA
Dr. Donald W. Hood, Director
Mr. John C. Garland, Facilities Manager

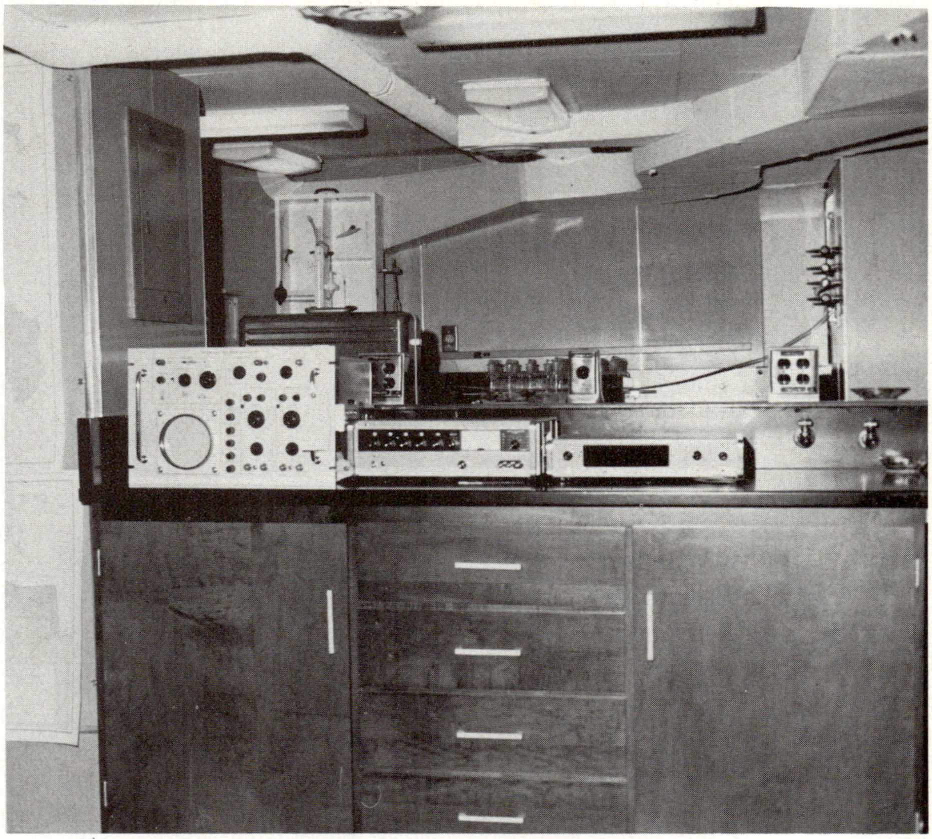

Part of a laboratory on the RV *Acona* (*Courtesy University of Alaska*)

COLOMBIA UNIVERSITY LAMONT—DOHERTY GEOLOGICAL OBSERVATORY (USER)

ROBERT D. CONRAD (AGOR 3)

Robert D. Conrad, built in 1962 by Gibbs Shipyard, Jacksonville, Florida, is a single screw diesel electric ship of welded steel construction designed for oceanographic research at sea.

Ship's main propulsion plant is diesel electric and has two 600 h.p. Caterpillar diesel generator sets driving a double armature DC (each armature rated at 500 h.p.) Westinghouse propulsion motors in a series loop circuit. The excitation for the diesel generator sets is supplied by a motor generator set and for the main motor by a rectifier. One motor generator set and one rectifier is used as a stand-by source of excitation. The plant is equipped with a main propulsion control board located in the EOS.

In addition to the main propulsion, a bow propulsion system is included as a vertical mounted 175 h.p. DC motor driving a bow propeller. A gas turbine generator set driving a 150 kw DC generator and a 150 kw AC generator provides power for the bow propulsion motor and stand-by AC ship service power. Means are provided on the main propulsion control board for energizing the main propulsion motor from the gas turbine generator set when the main propulsion diesel engines and bow propulsion motor are not in use.

Three 200 kw 440 volt AC Caterpillar diesel generator sets with attached exciters supply the vessel's requirements for electrical power and lighting and in addition 150 kw 400 volts is supplied from the gas turbine set as a stand-by source.

Two auto-fired 40 h.p. vapour water tube boilers for the steam heating system, the hot water heaters, and the galley requirements supply approximately 545 kg–682 kg (1,200–1,500 lb) per hour at 2/1 kg/cm: (30 lb/in:).

Two distilling plants, Maxim 4.5.50 operate under low pressure, producing 2,000 US gallons per day each. To reduce waterborne noise caused by machinery sound, isolation mountings have been installed under main propulsion and ship service diesel generators and all other noise producing machinery.

USS *Conrad* (AGOR-3)

Leading particulars

Length, overall	63·41 m (208 ft 4 in)
Navigational draught (full load)	5·11 m (16 ft 9 in)
Freeboard to weather deck (in way of work area at full load)	1·98 m (6 ft 6 in)
Beam	11·51 m (37 ft 5 in)
Displacement (full load)	1,370 long tons
Gross tonnage	1,425
Operating crew:	
officers	4
crew	3
Scientific accommodation	15
Enclosed laboratory area	126·2 m (1,357 ft^2)
Cruise speed	10 knots (bridge control)
Maximum speed	13·5 knots (bridge control)
Minimum speed	1·0 knots (bridge control)

Range at cruise speed	15,840 miles
Range at maximum speed	10,000 miles
Maximum endurance	60 days
Maximum sea state capability at cruising speed	Sea State 4
Propellers	One, fixed pitch, three-blade, manganese bronze. Coolidge propeller, dia. 2·74 m (9 ft), pitch 2·34 m (7·67 ft)
Shaft horse power	1,000
Fuel capacity	90,344 US gallons
Potable water capacity	2,860 US gallons
Main engines	Two 600 h.p. Caterpillar diesel D-397
Bow thruster	175 h.p. DC motor driven, Western Gear Corporation, Model 6147

COLOMBIA UNIVERSITY LAMONT—DOHERTY GEOLOGICAL OBSERVATORY

VEMA

Leading particulars

Length, overall	61·59 m (202 ft)
Navigational draft (full load)	4·87 m (15 ft $11\frac{3}{4}$ in)
Beam	10·06 m (33 ft)
Displacement (full load)	1,000 long tons
Operating crew:	
Officers	9
Crew	14
Scientific accommodation	14
Maximum speed	12 knots
Cruise speed	9·5 knots
Range at cruise speed	5,000 miles
Maximum endurance	31 days
Maximum sea state at cruise speed	Sea State 5
Minimum sustained speed	4 knots
Enclosed laboratory area	94·86 m^2 (1,020 ft^2)
Propellers	One, fixed pitch, four-blade manganese bronze, 2·1 m (7 ft) dia. 1·8 m (6 ft) pitch
Engine	One GM Model GOGIA-6-71 900 s.h.p.
Fuel capacity	27,000 US gallons
Potable water capacity	6,500 US gallons
Cranes	One, telescopic boom, hydraulic type
Capacity	4,500 kg (10,000 lb) at minimum radius of 1·52 m (5 ft) 900 kg (2,000 lb) at maximum radius 5·49 m (18 ft)

SCRIPPS INSTITUTION OF OCEANOGRAPHY

(See also under Universities)

ALEXANDER AGASSIZ

Call letters: KEYI

An Army FS type vessel, built in 1944 by Higgins, New Orleans and converted in 1962 for biological and fisheries research, hydrographic and limited oceanography work.

Leading particulars

Length, overall	54·88 m (180 ft)
Length, water line	52·44 m (172 ft)
Navigational draught (full load)	3·20 m (10·5 ft)
Depth to main deck	4·57 m (15 ft)
Freeboard to weather deck (in way of work area at full load)	1·52 m (5 ft)
Beam	9·76 m (32 ft)
Displacement (full load)	869·04 long tons
Operating crew:	
Officers	8
Crew	10
Scientific accommodation	13
Maximum speed	11·0 knots
Cruise speed	10·5 knots
Minimum speed	0–1·0 knots
Range at cruise speed	5,544 miles
Maximum endurance	22 days
Maximum sea state at cruising speed	Sea State 4
Enclosed laboratory area	35·34 m^2 (380 ft^2)

Alexander Agassiz (*Courtesy UCSD*)

Propellers	Two, fixed pitch
Shaft horse power	1,000, diesel
Fuel capacity	22,315 US gallons
Potable water capacity	21,750 US gallons
Cranes	A-frame (hydraulic) over stern: capacity 6,750 kg (15,000 lb) Two articulated cranes, portable, 3 tons at 2·13 m (7 ft) 1 ton at 6·10 m (20 ft)
Hydrographic davit, capacity	1,800 kg (4,000 lb)

ALPHA HELIX

The *Alpha Helix* was built during 1964/65 for research in medical and biological fields. She contains an analytical laboratory, a 'wet' laboratory that can be maintained at any temperature down to 0° Centigrade, a physiological laboratory, a photo laboratory, a walk-in freeze laboratory held at −20° Centigrade, and a special machine and instrument shop. Together with a supporting shore laboratory and experimental pools, the National Science Foundation donated the *Alpha Helix* to Scripps in 1966.

Leading particulars

Length, overall	40·47 m (132·75 ft)
Length, water line	3·20 m (10·5 ft)
Depth to main deck	1·37 m (4·5 ft)
Beam	9·45 m (31 ft)
Displacement (full load)	512 long tons
Operating crew:	
Officers	5
Crew	7
Scientific accommodation	12
Maximum speed	12 knots
Cruise speed	10 knots
Sustained minimum speed	1 knot
Range at cruise speed	7,200 miles
Maximum endurance	30 days
Maximum sea state at cruising speed	Sea State 3
Enclosed laboratory area	108·62 m^2 (1,168 ft^2)
Propellers	One, controllable pitch
Shaft horse power	820, diesel
Fuel capacity	30,600 US gallons
Potable water	5,000 US gallons

Crane: Marco articulated boom, Capacity 2,250 kg (5,000 lb) at maximum radius of 5·79 m (19 ft)

Capital cost	$US 1,300,000 (1965)

RV *Alpha Helix* (*Courtesy UCSD*)

RV *Alpha Helix* (*Courtesy UCSD*)

MELVILLE (AGOR 14)

Leading particulars

Length, overall	74·70 m (245 ft)
Navigational draught (full load)	4·98 m (16 ft 4 in)
Beam	14·02 m (46 ft)
Displacement (full load)	1,915 long tons
Operating crew:	
Officers	10
Crew	15
Scientific accommodation	25
Maximum speed	13·95 knots
Cruise speed	11·20 knots
Range at cruise speed	9,540 miles
Maximum endurance	41 days

Melville (AGOR-14) (*Courtesy UCSD*)

Maximum sea state at cruising speed	Sea State 4
Enclosed laboratory area	181·35 m² (1,950 ft²)
Propellers	Two, cycloidal
Shaft horse power	2,500 diesel
Fuel capacity	105,000 US gallons
Potable water capacity	8,000 US gallons
Cranes:	
Pettibone (Telescoping boom)	Capacity 16,200 kg (36,000 lb) at min. radius of 1·52 m (5 ft) 1,593 kg (3,540 lb) at max. radius of 18·29 m (60 ft)
Daybrook (Articulated Boom)	Capacity limited to 900 kg (2,000 lb) single weight load
Capital cost	$US 6,000,000

Melville (AGOR-14) (*Courtesy UCSD*)

OCONOSTOTA

Leading particulars

Length, overall	30·49 m (100 ft)
Length, waterline	29·57 m (97 ft)
Navigational draught (full load)	3·20 m (10·5 ft)
Beam	7·62 m (25 ft)
Displacement (full load)	327·7 long tons
Operating crew:	
Officers	5
Crew	3
Scientific accommodation	6
Maximum speed	11 knots
Cruising speed	10 knots
Range at cruising speed	4,400 miles
Minimum sustained speed	½ knot
Maximum endurance	16 days
Maximum sea state at cruise speed	Sea State 5
Enclosed laboratory area	8·93 m² (96 ft²)
Propellers	One, fixed pitch
Shaft horse power	900 s.h.p.
Fuel capacity	16,580 US gallons
Potable water capacity	5,500 US gallons

Cranes: Articulated Paybrook PL-6-J. Capacity 3,150 kg (7,000 lb) at min. radius 2·13 m (7 ft); 1,125 kg (2,500 lb) at max. radius 5·49 m (18 ft)

Oconostota (*Courtesy UCSD*)

Potable water capacity 5,500 US gallons
Cranes: Articulated Paybrook PL-6-J. Capacity 3,150 kg (7,000 lb) at min. radius 2·13 m (7 ft); 1,125 kg (2,500 lb) at max. radius 5·49 m (18 ft)

ELLEN B. SCRIPPS

The *Ellen B. Scripps* was converted during the course of construction from an offshore drilling rig supply vessel to an oceanographic research vessel. Construction started in 1964 and she was launched in 1965 for use in hydrographic, oceanographic, geological, seismic and biological work.

Leading particulars

Length, overall	28·96 m (95 ft)
Length, waterline	28·05 m (92 ft)
Navigational draught (full load)	1·83 m (6 ft)
Depth to main deck	3·05 m (10 ft)
Beam	7·32 m (24 ft)
Displacement (full load)	233·8 long tons
Operating crew:	
Officers	4
Crew	1
Scientific accommodation	8

Ellen B. Scripps (*Courtesy UCSD*)

Maximum speed	9·5 knots
Cruise speed	9·0 knots
Range at cruise speed	6,480 miles
Maximum endurance	30 days
Sustained min. speed	2·0 knots
Maximum sea state at cruising speed	Sea State 4
Enclosed laboratory area	10·04 m² (108 ft²)
Propellers	Two fixed pitch
General motors	Two 12V-71, 340 bhp each, 1,200 rpm (Diesel)
Fuel capacity	21,250 US gallons
Potable water capacity	7,500 US gallons

Crane: Daybrook PL6-J: Capacity 3,150 kg (7,000 lb) at min. radius of 2·13 m (7 ft); 900 kg (2,000 lb) at max. radius of 6·10 m (20 ft)

Capital cost	$US 250,000 (1965)

Ellen B. Scripps (*Courtesy UCSD*)

THOMAS WASHINGTON

Leading particulars

Length, overall	63·72 m (209 ft)
Navigational draught (full load)	4·98 m (16 ft 4 in)
Displacement (full load)	1,915 long tons
Beam, overall	14·02 m (46 ft)
Operating crew:	
Officers	10
Crew	15
Scientific accommodation	25
Maximum speed	13·2 knots
Cruise speed	11·5 knots
Range at cruise speed	8,700 miles
Maximum endurance	29 days
Sustained minimum speed	0–1·0 knots
Maximum sea state at cruising speed	Sea State 6
Enclosed laboratory area	(Total 95·38 m² (1,025·6 ft²) includes Computer room 29·70 m² (319·4 ft²))
Propellers	one, 5 blade, stainless, fixed pitch
Shaft horse power	1,000 s.h.p.
Fuel capacity	51,000 US gallons
Potable water capacity	2,800 US gallons

Cranes: Bucyrus Erie M-5: Capacity 5,400 kg (12,000 lb) at 5·49 m (18 ft) radius (min.) 1,260 kg (2,800 lb) at 16·46 m (54 ft) (max. radius)

Daybrook PL 6-J Articulated: Capacity 3,150 kg (7,000 lb) at 2·13 m (7 ft) radius (min.) 900 kg (2,000 lb) at 6·10 m (20 ft) radius (max.)

Thomas Washington (*Courtesy UCSD*)

Thomas Washington (*Courtesy UCSD*)

DUKE UNIVERSITY, MARINE LABORATORY

R/V EASTWORD

Leading particulars

Length, overall	35·82 m (117 ft 6 in)
Length, waterline	31·40 m (103 ft)
Navigational draught (full load)	3·96 m (13 ft)
Depth to main deck	4·23 m (13·9 ft)
Beam	8·68 m (28 ft 6 in)
Displacement (full load)	610 long tons
Operating crew:	
Officers	6
Crew	9
Scientific accommodation	15
Maximum speed	11·5 knots
Cruise speed	10·5 knots
Range at cruise speed	4,500 miles
Maximum endurance	21 days
Maximum sea state at cruise speed	Sea State 3
Enclosed laboratory area	42·91 m² (461·5 ft)
Propellers	One controllable pitch
Engine	One Fairbanks-Morse diesel, 640 s.h.p.
Fuel capacity	17,500 US gallons
Potable water capacity	8,500 US gallons

Cranes: Articulated Boom: Capacity 675 kg (1,500 lb) at minimum radius 1·49 m (5 ft), 675 kg (1,500 lb) at maximum radius 2·44 m (8 ft). Stedt Hydraulic Telescoping Boom: Capacity 3,600 kg (8,000 lb) at minimum radius of 2·44 m (8 ft), 1,350 kg (3,000 lb) at maximum radius of 5·18 m (17 ft)

FLORIDA STATE UNIVERSITY

TURSIOPS

Tursiops is a 19·8 m (65 ft) research vessel. She is powered by a Caterpiller 375 diesel engine which drives a model 130, 116·8 cm (46 in) diameter Wilkinson controllable-pitch propeller. She was built in 1953 and converted in 1955. Displacement is 95 tons. Crew 3, scientists 5.

UNIVERSITY OF GEORGIA SKIDAWAY INSTITUTE OF GEORGIA

KIT JONES

A 19·5 m (64 ft) research vessel, built in 1938 and converted in 1958. Displacement 90 tons. Crew 3. Scientists 4.

UNIVERSITY OF HAWAII

KANA KEOKI

This vessel is used by the Hawaii Institute of Geophysics and owned by the University of Hawaii. She was built as an offshore supply vessel in 1967 and converted in 1970.

Leading particulars

Length, overall	47·56 m (156 ft)
Length, waterline	43·59 m (143 ft)
Draught, full load	3·35 m (11 ft)
Beam	10·97 m (36 ft)
Crew	15
Scientists	15
Enclosed laboratory area	131·13 m² (1,410 ft²)
Speed:	
Cruise	11 knots
Maximum	13·5 knots
Range:	
at cruise speed	12,500 miles
at max. speed	10,000 miles
Endurance constraint	refrigerated stores
Maximum endurance	42 days
Sustained minimum speed	1 knot
Sea-state capability	Cruise speed can be sustained in Sea State 4
Propellers	Two fixed pitch
Rudders	Two
Engines	Two 900 s.h.p. diesels
Fuel capacity	80,000 US gallons
Potable water capacity	30,000 US gallons
Manoeuvring device	Tunnel thruster, forward, 130 s.h.p., 900 kg (2,000 lb) thrust

Kana Keoki (Courtesy University of Hawaii)

MOANA WAVE (AGOR-22)

Built 1973 by Halter Marine Services, New Orleans

Leading particulars

Length, overall	53·0 m (174 ft)
Depth to main deck	3·05 m (10 ft)
Beam, overall	10·98 m (36 ft)
Displacement (full load)	950 long tons
Operating crew:	
Officers, Crew	9–11
Scientific accommodation	9–11
Maximum speed	13·5 knots
Cruise speed	12 knots
Range at cruise speed	8,000 miles
Minimum sustained speed	1 knot
Maximum endurance	40 days
Propellers	variable pitch
Bow thruster	150 h.p.
Engines	Two, diesel—800 h.p. each

9,000 gallon holding tank

Cost: $US 3·9 million (1973) total, including $US 0·83 million instrumentation

The vessel will be operated on research and development projects for the US Navy.

THE JOHNS HOPKINS UNIVERSITY CHESAPEAKE BAY INSTITUTE

RIDGELY WAREFIELD (Catamaran)

Built 1967, Bethlehem Steel Ship Building Corporation, Baltimore

Leading particulars

Length, overall	32·32 m (106 ft)
Length, water line	30·18 m (99 ft)
Navigational draught (full load)	2·67 m (8 ft 9 in)
Depth to main deck	3·35 m (11 ft)
Beam	10·37 m (34 ft)
Displacement (full load)	162 long tons
Operating crew	
Officers	3
Crew	5
Scientific accommodation	10 (8 male, 2 female)

Maximum speed	18 knots
Cruise speed	16 knots
Minimum sustained speed	2 knots
Maximum endurance	30 days
Maximum sea state at cruise speed	Sea State 4
Enclosed laboratory area	41·85 m^2 (450 ft^2)
Propellers	Two, fixed pitch
Engines	Four 600 h.p. Cummins VT-12-825M 2,350 s.h.p.
Fuel capacity	12,000 US gallons
Potable water capacity	750 US gallons

Cranes: Telescoping boom has winch with 915 m (3,000 ft) by 1·27 cm ($\frac{1}{2}$ in) SS wire. Capacity: 1,125 kg (2,500 lb) at 5·18 m (17 ft) minimum radius, 551·25 kg (1,225 kg) at maximum radius 7·62 m (25 ft)

UNIVERSITY OF MIAMI ROSENSTIEL SCHOOL OF MARINE AND ATMOSPHERIC SCIENCE

CALANUS

This vessel was commissioned in September 1971. She was designed by Rudolph F. Matzer and Associates, Inc., and bulit by Atlantic Marine, Inc., Florida. Construction was financed partially by the National Science Foundation, with additional funds from the University of Miami and donations from private individuals. The vessel is specially designed and built for shallow-draught, in-shore and off-shore work.

Leading particulars

Length, overall	19·13 m (62 ft 9 in)
Length, waterline	17·5 m (57 ft 6 in)
Beam, moulded	6·1 m (20 ft)
Depth, moulded	2·64 m (8 ft 8 in)
Draught (full load)	1·47 m (4 ft 10 in)
Gross tons	83
Net tons	56
Displacement (Long tons)	80 full load 60 light load
Speed, maximum sustained	9 knots
Endurance	14 days
Crew	2
Scientific personnel	6
Water (P & S tanks)	1,710 US gals
Fuel (P & S tanks)	4,818 US gals (95%)
Main engines	Two GM-6-71, 165 s.h.p., 1,800 rpm
Ship's service generators	Two 20 kw, 208 volt, 3 phase, 60 cycle, 1,200 rpm. Prime movers GM 2-71
Propellers	Two 5 blade, bronze 76·2 cm (30 in) diameter, 55·9 cm (22 in) pitch
Hydraulic system	Main engine PTO, 17 gpm, 1,410 kg cm: (2,000 lb/in)
Workboat	3·96 m (13 ft) C-Craft inflatable rubber boat, O/B motor 18 h.p.
Life saving equipment	25-Man, Encapsulated rubber boat, auto-release

JAMES M. GILLISS (AGOR-4)

The R/V *James M. Gilliss* was one of the first US Navy oceanographic research ships, and is named after Captain James Melville Gilliss (1811–1865) who founded the Naval Observatory and served as its director until 1865. She was built in Sturgeon Bay, Wisconsin, in 1961–62 and was delivered to the US Navy in late 1962. Before assuming her role as an oceanographic research ship, *Gilliss* took part in the search for the submarine *Thresher.*

On November 22, 1970, *Gilliss* was formally assigned to the University of Miami by the Oceanographer of the Navy. Since then she has undergone extensive modification to fit her for her future work, with the financial help of both the US Naval Oceanographic Office and the National Science Foundation. The vessel is used, with the support of these and other agencies, for oceanographic research programmes by the University's Rosenstiel School of Marine and Atmospheric Science.

The *Gilliss* is equipped with conventional oceanographic equipment: Niskin bottles, reversing thermometers, corers, trawls, and a variety of winches. Electronic instruments include sound velocity-temperature-pressure probes and recorders, precision echo sounders and a sub-bottom acoustic profiler.

Scientific personnel, using available deck machinery, are capable of handling a wide range of physical, chemical, biological and geological programmes. The versatile craft permits the School to fulfil many presently neglected areas of research and to increase its scientific complement at sea. The School is able to provide cruise time on *Gilliss* for visiting investigators from other universities that lack a major oceanographic vessel and wish to participate in tropical marine research.

Gilliss is based at the Port of Miami Oceanographic Centre. During her first year of use by the University of Miami, R/V *Gilliss* will carry out research expeditions in the tropical Atlantic Ocean, the Caribbean Sea, the Gulf of Mexico, the Gulf of Panama, and the south-eastern Pacific Ocean.

USNS *Gilliss* (AGOR-4) (*Courtesy U.S. Navy*)

Leading particulars

Length, overall	63·41 m (208 ft)
Length, water line	57·93 m (190 ft)
Navigational draught (full load)	4·66 m (15 ft 3$\frac{1}{2}$ in)
Depth to main deck	1·91 m (6 ft 3 in)
Beam	11·41 m (37 ft 5 in)
Displacement (full load)	1,422 long tons
Operating crew:	
Officers	7
Crew	15
Scientific accommodation	19
Maximum speed	13·8 knots
Cruise speed	12·0 knots
Range at cruise speed	11,500 miles
Minimum sustained speed	4 knots
Maximum endurance	40 days
Maximum sea state at cruise speed	Sea State 5
Enclosed laboratory area	109·55 m^2 (1,178 ft^2)
Propellers	one, 5-blade
Engine:	
Caterpillar D-397	1,000 s.h.p.

Fuel capacity	5,000 US gallons
Potable water capacity	20,000 US gallons
Cranes:	
Telescopic Boom	Capacity: 9,000 kg (20,000 lb) at min. radius 3·66 m (12 ft) 1,800 kg (4,000 lb) at max. radius 14·63 m (48 ft)
Articulated Boom	Capacity: 6,750 kg (15,000 lb) at min. radius 0·61 m (2 ft) 225 kg (500 lb) at max. radius 5·49 m (18 ft)
Moveable 'A'-frame	Capacity: 13,500 kg (30,000 lb) at min. radius 1·83 m (6 ft) 13,500 kg (30,000 lb) at max. radius 3·66 m (12 ft)
Moveable 'A'-frame	Capacity: 9,000 kg (20,000 lb) at min. radius 0·61 m (2 ft) 9,000 kg (20,000 lb) at max. radius 1·83 m (6 ft)
Boom	Capacity: 9,000 kg (20,000 lb) at min. radius 0 m (0 ft) 9,000 kg (20,000 lb) at max. radius 1·52 m (5 ft)

COLUMBUS ISELIN

Iselin is a relatively recent addition to the UNOLS fleet. She was launched on 1 March 1972. The vessel is 52 m (170 ft) long, and is in use by the University of Miami School of Marine and Atmospheric Science.

Designers: Rudolph F. Matzer and Associates, Inc., Jacksonville, Florida

Builders: J. Bellinger Shipyards, Inc., Jacksonville, Florida

Leading particulars

Length, overall	51·82 m (170 ft)
Length, water line	48·78 m (160 ft)
Navigational draught (full load)	3·20 m (10 ft 6 in)
Depth to main deck	4·57 m (15 ft)
Beam	10·97 m (36 ft)
Displacement (full load)	830 long tons
Operative crew:	
Officers	5
Crew	7
Scientific accommodation	13
Maximum speed	13·5 knots
Cruise speed	13·5 knots
Range at cruise speed	7,000 miles
Minimum sustained speed	1·0 knots
Maximum endurance	35 days
Maximum sea state at cruise speed	Sea State 5
Enclosed laboratory area	117·92 m² (1,268 ft²)
Propellers	Two, controllable pitch, Liaaen-CG 50
Shaft horse power, (two Caterpillar D398A diesels)	1,450 s.h.p.
Fuel capacity	30,000 US gallons
Potable water capacity	750 US gallons
Cranes:	
Telescopic Boom	Capacity 10,800 kg (24 000 lb) at min. radius of 3·04 m (10 ft) 1,800 kg (4,000 lb) at max. radius of 14·63 m (48 ft)
Articulated Boom	Capacity 675 kg (1.500 lb) at min. radius of 0·60 m (2 ft) 225 kg (500 lb) at max. radius of 5·48 m (18 ft)
'A'-frame	Capacity 13,500 kg (30,000 lb) at min. radius of 1·82 m (6 ft) 13,500 kg (30,000 lb) at max. radius of 2·43 m (8 ft)
'A'-frame	Capacity 13,500 kg (30,000 lb) at min. radius of 1·21 m (4 ft) 13,500 kg (30,000 lb) at max. radius of 1·82 m (6 ft)

UNIVERSITY OF MICHIGAN

INLAND SEAS

A 1943-built vessel of 34·8 m (114 ft) length converted in 1962.

MYSIS

A 15·0 m (50 ft) length vessel built in 1963.

NOVA UNIVERSITY

GULF STREAM

A 16·8 m (55 ft) length vessel built in 1963.

OREGON STATE UNIVERSITY DEPARTMENT OF OCEANOGRAPHY

CAYUSE

R/V *Cayuse* is a steel hull purse seiner-type motor vessel built in 1968. She is designed for research in the area of the continental shelf and slope off the Pacific Northwest coast, and she fills the gap between the 54·88 m (180 ft) deep sea vessel *Yaquina* and the 10·06 m (33 ft) coastal water vessel *Paiute*. From home port in Newport, *Cayuse* sails on cruises ranging from 1 to 60 days. She has made extended cruises to the Gulf of California and to the Auke Bay area in Alaska. Much of her schedule calls for servicing oceanographic and meteorological buoys in the offshore buoy programme.

Part of the time *Cayuse* is used for coastal oceanography and marine fisheries research under the Sea Grant College programme. Oceanographic technicians from other institutions in Oregon are trained on this vessel as a part of the Sea Grant programme.

Cayuse has a hydrographic winch with over a mile of 4·76 mm ($\frac{3}{16}$ in) wire rope and a deep-sea winch with 9·53 mm ($\frac{3}{8}$ in) wire designed for obtaining samples with dredges, trawls, nets, coring tubes, and water samplers down to depths of approximately four miles. The main boom has a capacity of three tons for work with the deep-sea winch. An auxiliary boom has a 1½-ton capacity and is used to handle ship supplies and scientific equipment.

There are quarters for seven crew members and for eight scientists, plus a wet laboratory for hydrographic work, a dry laboratory, and an electronic laboratory. Graduate students in oceanography receive much of their training on *Cayuse*.

Leading particulars

Length, overall	24·33 m (79·8 ft)
Navigational draught (full load)	3·05 m (10·0 ft)
Freeboard to weather deck (in way of work area at full load)	1·05 m (3·5 ft)
Beam	7·32 m (24 ft)
Displacement (full load)	173 long tons
Gross tonnage	165
Operating crew	
Officers	4
Crew	3
Scientific accommodation	8
Enclosed laboratory area	31·06 m² (334 ft²)
Cruise speed	8 knots
Maximum speed	10 knots
Range at cruise speed	5,000 miles
Range at maximum speed	3,000 miles
Maximum endurance	20 days
Maximum sea state at cruising speed	Sea State 4
Propellers	one, fixed pitch
Shaft horse power	380, diesel
Fuel capacity	10,500 US gallons
Potable water capacity	1,550 US gallons
Cranes:	
main boom, capacity	2,700 kg (6,000 lb) at 9·15 m (30 ft) max.
auxiliary boom capacity	1,350 kg (3,000 lb) at 5·18 m (17 ft) max.
movable A frame	900 kg (2,000 lb) at 2·13 m (7 ft) max.

YAQUINA

R/V *Yaquina*, a deep-sea vessel, is equipped to carry out all types of marine research. At sea about 70 per cent of the time, *Yaquina* is usually on cruises lasting up to 30 days. However, occasionally she sails on extended cruises of up to three months to conduct oceanographic explorations in other parts of the ocean world . In the summer of 1966, *Yaquina* carried scientists to Honolulu, Hawaii, and to Adak, Alaska, on her first long cruise which lasted 100 days. In the winter and spring of 1969 she travelled as far south as Peru during a research cruise lasting 114 days. Her longest cruise returned her to the south-east Pacific area again commencing August 1971, not to return to home port until May 1972.

Yaquina is equipped with three large electric winches for lowering equipment into the ocean. The deep-sea coring winch has almost six miles of 1·25 cm ($\frac{1}{2}$ in) wire rope for coring operations and for lowering heavy dredges to collect geological and biological material from the deep ocean floor. The 4·75 mm (3/16 in) wire on the hydrographic winch is used only for light-weight gear, this winch is used mainly for bringing up water samples from as deep as 9,150 m (30,000 ft). The trawl winch, with 9·53 mm ($\frac{3}{8}$ in) wire, is used to tow nets and recording instruments from the stern of the ship.

Quarters are provided for 19 crew members and for 17 scientists. In addition, eight oceanographic laboratories are available to the scientists for sample and data processing.

Although primarily a research vessel, *Yaquina* is also used to train graduate students in oceanography.

Leading particulars

Length, overall	54·83 m (179·83 ft)
Length, waterline	51·31 m (168·3 ft)
Navigational draught (full load)	3·51 m (11·5 ft)
Displacement (full load)	865 long tons
Beam	9·76 m (32 ft)
Operating crew	
Officers	6
Crew	11
Scientific accommodation	17
Maximum speed	12 knots
Cruise speed	10 knots
Range at cruise speed	5,300 miles
Maximum endurance	28 days
Sustained minimum speed	1 knot
Maximum sea state at cruising speed	Sea State 5
Enclosed laboratory area	121·18 m² (1,303 ft²)
Propellers	Two, fixed pitch
Shaft horse power	1,000 s.h.p.
Fuel capacity	31,500 US gallons
Potable water capacity	20,500 US gallons
Cranes	
Gantry (stern)	5,400 kg (12,000 lb) at 3·35 m (11 ft)
Side Tripod boom	5,400 kg (12,000 lb) at 2·74 m (9 ft)
Side Hydro boom	900 kg (2,000 lb) at 1·83 m (6 ft)

UNIVERSITY OF RHODE ISLAND

TRIDENT

R/V *Trident* is a steel hull, double chined vessel built by Higgins Industries. Many of these ships were built as well deck vessels. This particular vessel had a continuous upper deck and was fitted out as a floating machine shop.

The ship has conventional aft and forepeak tanks which are used for fuel. A double bottom is continuous from the forepeak to the mid portion of the engine room. This is divided into 5 tank spaces fore and aft. Each tank space is divided into Port/Stb'd tanks by a continuous centre vertical keel. The No. 1 tank is solid ballast. The No's. 2, 3 and 5 tanks are used for fuel. An antiroll tank, common with No. 3 fuel tank is also used for fuel. The No. 4 tank is used for fresh water and is separated from fuel tanks by cofferdams. Below the main deck the original No. 1 cargo hold is still used as such. The No. 2 hold has been converted into a winch room for the deep sea winch. The auxiliary engine room is next aft with the engine room immediately aft of it. The space above the after peak tank contains one room of crew quarters for six persons.

Between the main deck and upper deck, the space above the forepeak tank is fitted with paint locker, bos'n locker and laundry. Aft of the forepeak bulkhead the space is divided into main laboratory, darkroom, electronics laboratory, scientific office/computer room, galley, messroom, hospital and quarters. Women and technicians are quartered in two rooms for four persons.

Leading particulars

Length, overall	54·83 m (179 ft 10 in)
Length, waterline	50·76 m (166 ft 5$\frac{3}{4}$ in)
Navigational draught (full load)	3·19 m (10 ft 9$\frac{3}{16}$ in)
Depth to main deck	4·34 m (14 ft 2$\frac{13}{16}$ in)
Beam	9·76 m (32 ft)
Displacement (full load)	1,021 long tons
Operating crew	
Officers	7
Crew	11
Scientific accommodation	13
Maximum speed	10·5 knots
Cruise speed	10·0 knots
Range at cruise speed	7,000 miles
Minimum sustained speed	4 knots
Maximum endurance	30 days
Maximum sea state at cruise speed	Sea State 5
Enclosed laboratory area	170·19 m² (1,830 ft²)
Propellers	Two, fixed pitch, manganese bronze, four-blade, 1·83 m (6 ft) diameter, 2·02 m (6·62 ft) pitch
Engines	Two, GM 6-278A, 750 rpm diesel, 1,000 s.h.p. total
Fuel capacity	38,000 US gallons
Potable water capacity	9,000 US gallons
Cranes:	
Cargo Boom, capacity	2,016 kg (4,480 lb) at min. radius 9·15 m (30 ft)
	2,016 kg (4,480 lb) at max. radius 16·77 m (55 ft)
'A'-frame, capacity	9,000 kg (20,000 lb) at min. radius 0 m (0 ft)
	9,000 kg (20,000 lb) at max. radius 3·05 m (10 ft)

UNIVERSITY OF SOUTHERN CALIFORNIA ALLAN HANCOCK FOUNDATION

VELERO IV

The *Velero IV* is of the 'San Diego' type live bait tuna clipper design, with a welded steel hull, the main engine forward and the hull divided into ten holds. Renovation of the ship in 1972 converted previously unusable hold space into two large below deck laboratories and two large cargo holds. A previously open wet laboratory on the main deck has also been converted into an enclosed laboratory. Accommodations are available for 11 crew and 12 scientists on extended cruises.

Two 500 US gallon sewage holding tanks store sewage. Usually these are emptied into the open sea at the end of a cruise.

Leading particulars

Length, overall	33·54 m (110 ft)
Navigational draught (full load)	3·96 m (13 ft)
Beam	8·23 m (27 ft)
Displacement (full load)	540 long tons
Operating crew:	
Officers	5
Crew	6
Scientific accommodation	9
Maximum speed	9 knots
Cruise speed	9 knots
Range at cruise speed	8,000 miles
Maximum endurance	60 days
Sustained minimum speed	4 knots
Maximum sea state at cruising speed	Sea State 4
Enclosed laboratory area	15·07 m² (162 ft²)
Propellers	One, fixed pitch
Engine	Atlas Imperial diesel, 600 s.h.p.
Fuel capacity	27,000 US gallons
Potable water capacity	14,000 US gallons
Crane:	
Movable 'A' Frame, capacity	3,150 kg (7,000 lb)

TEXAS A & M UNIVERSITY

GYRE (AGOR 21)

Leading particulars

Built during 1973 by Halter Marine Shipyards, New Orleans. Research operations were due to commence at the beginning of 1974.

Length, overall	50·30 m (165 ft)
Navigational draught (full load)	3·35 m (11 ft)
Beam	10·98 m (36 ft)
Displacement (full load)	950 long tons
Operating crew: Officers, Crew	13
Scientific accommodation	18
Maximum speed	14 knots
Cruise speed	12 knots
Range at cruise speed	8,000 miles
Maximum endurance	42 days
Propellers	Two, variable pitch
Engines	Two, Caterpillar diesels 1,880 h.p.
Cranes	Equipment-handling cranes; 2 U-frames on stern and starboard side aft

UNIVERSITY OF WASHINGTON DEPARTMENT OF OCEANOGRAPHY

HOH

Hoh is a 19·8 m (65 ft) research vessel, originally built in 1943 and converted for its present use in 1962. Displacement 91 tons. Crew 2, scientists 6.

ONAR

The same length as *Hoh*, this vessel was built in 1954 and converted in 1963. Displacement 95 tons.

THOMAS G. THOMPSON (AGOR-9)

Leading particulars

Length, overall	63·66 m (208 ft 10 in)
Navigational draught (full load)	4·57 m (15 ft)
Beam	11·89 m (39 ft)
Displacement (full load)	1,362 long tons
Operating crew	
Officers	7
Crew	15
Scientific accommodation	20
Maximum speed	14 knots
Cruising speed	12·5 knots
Range at cruise speed	10,000 miles
Minimum sustained speed	½ knot
Maximum endurance	40 days
Maximum sea state at cruise speed	Sea State 4
Enclosed laboratory area	139·5 m² (1,500 ft²)
Propellers	One, fixed pitch
Shaft horse power	1,125 s.h.p.
Fuel capacity	80,000 US gallons
Potable water capacity	3,000 US gallons
Cranes:	
Articulated Boom, capacity	5,400 kg (12,000 lb) at min. radius of 3·04 m (10 ft); 1,125 kg (2,500 lb) at max. radius of 16·46 m (54 ft)
Moveable A-frame capacity	Raise load 27,000 kg (60,000 lb) at min. radius 0·30 m (1 ft); Raise load 27,000 kg (60,000 lb) at max. radius 4·57 m (15 ft)
Capital cost	$US 3,000,000 (1965)

WOODS HOLE OCEANOGRAPHIC INSTITUTION

ATLANTIS II

This vessel was built by the Maryland Shipbuilding and Drydock Co. under a grant from the National Science Foundation. She was delivered to the owners in January 1963.

Atlantis II (Courtesy Woods Hole Oceanographic Institution)

Leading particulars

Certification authority clearance: U.S.C.G.-ABS

Length, overall	64 m (210 ft)
Beam, overall	13·4 m (44 ft)
Draught	4·9 m (16 ft)
Displacement	2,300 tons
Gross tonnage	1,529
Full speed	12 knots
Cruising speed	11 knots
Range at cruise	6,000 miles
Endurance	60 days
Maximum anchoring depth	7,620 m (25,000 ft)
Main engines	Reciprocating Steam, Skinner Uniflow 1,400 s.h.p., 160 rpm
Auxiliary engines	Steam turbine, 450 s.h.p., 720 rpm
Propellers	Twin screw
Controls	Bridge controlled. Bow thruster position holding

On-board equipment

Craneage facilities: One crane 15-ton maximum, 9·1 m (30 ft) above main deck

Winching facilities: Trawl winch, 150 h.p., 9,100 m (30,000 ft); two hydrographic winches, 35 h.p., 7,620 m (25,000 ft); BT winches, portable

Navigational equipment: Two S-band radars, Satellite receiver, Loran A, RDF, one gyro-compass, one fathometer

Survey and search equipment: Two 12 kc echo sounders

Crew

Sea-going complement, officers, men: 11 officers—20 crew

Scientific complement: 25

Crew safety facilities and equipment: U.S.C.G. Approved

CHAIN (AGOR 17)

A US Navy Rescue and Salvage Tug converted into an oceanographic research vessel. Conversion by Savannah Machine and Foundry in 1959, having been built in 1944 by Basalt Rock Co., Napa, California.

Leading particulars

Length, overall	64·9 m (213 ft)
Beam, overall	12·5 m (41 ft)
Draught	4·9 m (16 ft)
Displacement	2,100 tons
Gross tonnage	1,328

Performance

Full speed	14 knots
Cruising speed	12·5 knots
Range at cruise	10,500 miles
Maximum anchoring depth	7,620 m (25,000 ft)

Main machinery description

Main engines	Diesel electric, 3,000 s.h.p., 160 rpm
Auxiliary engines	Diesel, 400 s.h.p., 800 rpm
Propellers	Twin screw

On-board equipment

Craneage facilities: 15-ton maximum, 6·1 m (20 ft) above deck

Winching facilities: Trawl winch, 150 h.p., 9,100 m (30,000 ft); One hydrographic winch, 35 h.p., 7,620 m (25,000 ft); BT winches portable

Navigational equipment: One S-band radar, satellite receiver, Loran A and Loran C, RDF, one gyrocompass, one fathometer

Survey and search equipment: 12 kc echo sounder

Pumps, compressors and ancillary power equipment and services: portable in vans

Crew

11 officers 22 crew

Scientists 25

Crew safety facilities and equipment: Public Vessel subject to Inspection by the US Navy

Chain (AGOR-17) (*Courtesy Woods Hole Oceanographic Institution*)

Chain (AGOR-17) (*Courtesy Woods Hole Oceanographic Institution*)

GOSNOLD

A US Army Small Freighter, built in 1943, and converted in 1962 for use as an oceanographic research vessel.

Leading particulars

Length, overall	30·2 m (99 ft)
Width, overall	7·9 m (26 ft)
Draught	2·7 m (9 ft)
Gross tonnage	166

Performance

Full speed	7 knots
Cruising speed	6 knots
Range at cruise	1,500 miles
Maximum anchoring depth	7,620 m (25,000 ft)

Main machinery description

Main engine	Kahlenburg Diesel, 250 s.h.p., 250 rpm
Auxiliary engine	Diesel, 75 s.h.p., 1,800 rpm
Propeller	Single screw, controllable pitch

On-board equipment

Craneage facilities: One crane, 1½ ton maximum, 3 m (10 ft) above main deck

Winching facilities: One trawl winch, 50 h.p., 4,570 m (15,000 ft), one hydrographic winch, 25 h.p., 6,100 m (20,000 ft), BT winches portable

Navigational equipment: One S-Band radar; Loran A; one gyrocompass, one fathometer

Survey and search equipment: One 12 kc echo sounder

Pumps, compressors and ancillary power equipment and services: portable in vans

Crew

4 officers—4 crew

Scientists: 6

Crew safety facilities and equipment: Two inflatable ocean life rafts, automatic CO2 fire extinguishing system, salt water fire mains

Gosnold

KNORR

An oceanographic research vessel built in 1970, owned by the US Navy and operated by Woods Hole Oceanographic Institution under contract with the Office of Naval Research.

Knorr (*Courtesy Woods Hole Oceanographic Institution*)

Leading particulars

Length, overall	74·64 m (244 ft 10 in)
Navigational draught (full load)	4·57 m (15 ft)
Beam	14·02 m (46 ft)
Displacement (full load)	1,915 long tons
Operating crew: Officers, Crew	25
Scientific accommodation	25
Maximum speed	11·5 knots
Cruising speed	10 knots
Range at cruising speed	10,000 miles
Minimum sustained speed	0·1 knots
Maximum endurance	60 days
Enclosed laboratory area	316·20 m² (3,400 ft²)
Engines	One Enterprise, Mod. DMR-38, 2,500 h.p. at 360 rpm
Craneage facilities	One 30 ton max. and one 25 ton max. at 15 m (50 ft) and 9 m (30 ft) respectively
Propellers	Two, cycloidal—one forward, one aft (J. M. Voith Mod. 32E and 24E)
Winches	Two hydro winches—9,100 m (30,000 ft) 0·48 cm (3/16 in) wire rope; One coring winch—9,100 m (30,000 ft) 1·6 cm ($\frac{5}{8}$ in) wire rope
Maximum anchoring depth	7,620 m (25,000 ft)
Navigational equipment	One S-band radar, one X-band radar, Satellite receiver, Loran A, RDF, two gyro compasses, one fathometer
Survey and search equipment	Two 12 kc echo sounders

U.S. NAVAL OCEANOGRAPHIC OFFICE

BARTLETT (T-AGOR 13)

Leading particulars

Year built	1969
Length	63·4 m (208 ft)
Beam	11·89 m (39 ft)
Draught	45·73 m (15 ft)
Displacement	1,340 tons
Crew	26
Scientific complement	15
Home port	San Francisco, Calif.

USNS *Bartlett* (T-AGOR-13)

BOWDITCH (T-AGS 21)

Leading particulars

Year built	1945
Year converted	1958
Length	138·7 m (455 ft)
Beam	18·9 m (62 ft)
Draught	9·45 m (31 ft)
Displacement	13,050 tons
Crew	86
Scientific complement	17
Home port	New York, N.Y.

USNS *Bowditch* (T-AGS-21)

CHAUVENET (T-AGS 29)
(see also *Harkness*)

Leading particulars

Year built	1969
Length	118·9 m (390 ft)
Beam	16·46 m (54 ft)
Draught	4·88 m (16 ft)
Displacement	4,200 tons
Crew	69
Scientific complement	12
Home port	San Francisco, Calif.

An Artist's concept of how *Harkness* and *Chauvenet* will look when built (T-AGS-29)

DAVIS (T-AGOR 5)

Leading particulars

Year built	1963
Length	63·4 m (208 ft)
Beam	11·28 m (37 ft)
Draught	4·57 m (15 ft)
Displacement	1,034 tons
Crew	26
Scientific complement	15
Home port	San Francisco, Calif.

USNS *Davies* (AGOR-5)

DUTTON (T-AGS 22)

Leading particulars

Year built	1944
Year converted	1958
Length	138·7 m (455 ft)
Beam	18·9 m (62 ft)
Draught	9·45 m (31 ft)
Displacement	13,050 tons
Crew	86
Scientific complement	17
Home port	New York, N.Y.

USNS *Dutton* (T-AGS 22)

DE STEIGUER (T-AGOR 12)

Leading particulars

Year built	1969
Length	63·4 m (208 ft)
Beam	11·89 m (39 ft)
Draught	4·57 m (15 ft)
Displacement	1,340 tons
Crew	26
Scientific complement	15
Home port	San Francisco, Calif.

USNS *De Steiguer* (T-AGOR 12)

ELISHA KANE (T-AGS 27)

Leading particulars

Year built	1965
Length	86·89 m (285 ft)
Beam	14·63 m (48 ft)
Draught	4·57 m (15 ft)
Displacement	2,560 tons
Crew	48
Scientific complement	30
Home port	New York, N.Y.

USNS *Elisha Kane* (T-AGS 27)

HARKNESS (T-AGS 32)
(see also *Chauvenet*)

Leading particulars

Year built	1970
Length	118·9 m (390 ft)
Beam	16·46 m (54 ft)
Draught	4·88 m (16 ft)
Displacement	4,200 tons
Crew	69
Scientific complement	12
Home port	New York, N.Y.

KELLAR (T-AGS 25)

Leading particulars

Year built	1969
Length	63·7 m (209 ft)
Beam	11·89 m (39 ft)
Draught	4·27 m (14 ft)
Displacement	1,300 tons
Crew	29
Scientific complement	13
Home port	New York, N.Y.

USNS *Kellar* (T-AGS 25)

LYNCH (T-AGOR 7)

Leading particulars

Year built	1965
Length	63·4 m (208 ft)
Beam	11·28 m (37 ft)
Draught	4·57 m (15 ft)
Displacement	1,034 tons
Crew	26
Scientific complement	15
Home port	New York, N.Y.

USNS *Lynch* (T-AGOR 7)

MICHELSON (T-AGS 23)

Leading particulars

Year built	1945
Year converted	1958
Length	138·7 m (455 ft)
Beam	18·9 m (62 ft)
Draught	9·45 m (31 ft)
Displacement	13,050 tons
Crew	86
Scientific complement	17
Home port	San Francisco, Calif.

USNS *Michelson* (T-AGS 23) (*Courtesy U.S. Navy*)

SANDS (T-AGOR 6)

Leading particulars

Year built	1965
Length	63·4 m (208 ft)
Beam	11·28 m (37 ft)
Draught	4·57 m (15 ft)
Displacement	1,034 tons
Crew	26
Scientific complement	15
Home port	New York, N.Y.

USNS *Sands* (T-AGOR 6)

SILAS BENT (T-AGS 26)

Leading particulars

Year built	1965
Length	86·89 m (285 ft)
Beam	14·63 m (48 ft)
Draught	5·18 m (17 ft)
Displacement	2,560 tons
Crew	44
Scientific complement	34
Home port	San Francisco, Calif

USNS *Silas Bent* (T-AGS 26)

U.S. NAVAL RESEARCH LABORATORY

GIBBS (T-AGOR 1)

Leading particulars

Year built	1942
Year converted	1958
Length	94·74 m (310 ft 9 in)
Beam	12·53 m (41 ft 1 in)
Draught	3·6 m (12 ft)
Displacement (full load)	2,800 tons
Crew	48
Scientific complement	26

USNS *Gibbs* (T-AGOR 1)

HAYES (T-AGOR 16)

Leading particulars

Built by Todd Shipyards

Length, overall	75·1 m (246 ft 5 in)
Length, waterline	67·07 m (220 ft)
Beam, max.	24·39 m (80 ft)
Beam, each hull	7·32 m (24 ft)
Draught, full load	5·74 m (18 ft 10 in)
Displacement (full load)	3,080 tons
Speed, cruise	15 knots
Endurance	6,000 miles at 13·5 knots
Accommodation	
Crew	44
Scientists	25
Engine	2,400 h.p. diesel
Propellers	Controllable pitch
Traction winch	2,273 kg (5,000 lb) pull, 24·4 m (80 ft) per min. in depths to 6,100 m (2 000 ft)

Vessel cost (1971) $14 m. and $2 m for scientific equipment

USNS *Hayes* (T-AGOR 16)

MISSION CAPISTRANO (T-AG 162)

Leading particulars

Year built	1944
Year converted	1960
Length	159·6 m (523 ft 5 in)
Beam	20·7 m (68 ft)
Draught	9·38 m (30 ft 9 in)
Displacement	17,000 tons
Crew	60
Scientific complement	15
Home port	Bayonne, New Jersey

USNS *Mission Capistrano* (T-AG 162)

MIZAR (T-AGOR 11)

Leading particulars

Year built	1957
Year converted	1964
Length	81·15 m (266 ft 2 in)
Beam	15·70 m (51 ft 6 in)
Draught	5·34 m (17 ft 6 in)
Displacement (full load)	3,886 tons
Crew	40
Scientific complement	19
Home port	Washington, D.C.

USNS *Mizar* (T-AGOR 11)

UNIVERSITY OF CALIFORNIA SCRIPPS INSTITUTION OF OCEANOGRAPHY

ARGO (AGOR 18)

Leading particulars

Type	Rescue and salvage
Hull	Steel
Year built	1944
Year acquired by SIO	1959
From whom acquired	USN
Owner	USN
Length	64·94 m (213 ft)
Beam	12·20 m (40 ft)
Draught	4·60 m (15 ft 1 in)
Displacement (full load)	2,079 tons
Cruising speed	13 knots
Maximum speed	14 knots
Minimum speed	1/2 knots
Range	8,000 miles
Endurance	60 days
Crew	32
Scientific party	24

Argo (AGOR 18) (*Courtesy Scripps Institute of Oceanography*)

DOLPHIN

Construction of R/V *Dolphin* started in October 1967 and she was launched in April 1968. The yacht is intended to be used mainly for biological and chemical research.

RV *Dolphin*

Leading particulars

Length, overall	29·27 m (96 ft)
Beam, overall	6·71 m (22 ft)
Height, overall	10·67 m (35 ft)
Draught	2·13 m (7 ft)
Gross tonnage	96·09
Full speed	14·5 knots
Cruising speed	12 knots
Range at cruise	1,800 nautical miles
Maximum anchoring depth	36·6 m (120 ft) 20 fathoms at a scope of three to one
Main engines	Two diesel reduction at 400 s.h.p. each
Auxiliary engines	Two diesel 115 s.h.p. at 1,200 rpm
Propellers	Two fixed pitch
Pumps	Two self-priming bait tank/bilge pumps
Facilities for diving	Scuba
Sampling equipment	Fishing gear and nets, has two trolling booms
	Vessel has stem-mounted 'A' frame for light duty net hauls
Winching facilities	Hydraulic 0·64 cm ($\frac{1}{4}$ in) wire 150 m (492 ft)
Navigational equipment	Radar, Loran A and ADF
Survey and search equipment	Radar and limited sonar

Pumps, compressors and ancillary power equipment and services: one air compressor at 100 lb, two 37·5 KVA 3 phase 120/208 volts AC generators

Organic and inorganic laboratory facilities are provided

Crew	One officer, three seamen
Scientists	7
Capital cost	\$US 1,200,000 (1968)

FLIP

The Floating Instrument Platform (FLIP) was built for the US Navy in 1962 by Gunderson Brothers Shipbuilding of Portland, Oregon. The vessel was developed by the Marine Physical Laboratory at Scripps Institution of Oceanography, University of California. FLIP has no motive power of her own and is towed to the operating site in the horizontal position. When up-ended on site by flooding ballast tanks the vessel has facilities both above and below the surface for oceanographic research. It is claimed that when in the vertical position, movements vertically of less than 0·30 m (1 ft) occur in waves 12·2 m (40 ft) high. FLIP normally carries a crew of 5 and a scientific party of up to 11.

FLIP (*Courtesy UCSD*)

Leading particulars

Length	108·23 m (355 ft)
Beam	6·1 m (20 ft), 3·66 m (12 ft) horizontally and vertically, respectively
Draught	3·05 m (10 ft), 91·46 m (300 ft) horizontally and vertically, respectively
Displacement	2,100 tons (vertical)

ST-908

This vessel is the smallest research vessel belonging to the Scripps Institution of Oceanography and is used for limited off-shore work. She is 13·72 m (45 ft) long.

ST-908 (*Courtesy UCSD*)

UNIVERSITY OF DELAWARE COLLEGE OF MARINE STUDIES
JOANIE

Specific purpose of vessel

Light dredging, trawling, crab scrapes, shallow-water work, water sampling

Leading particulars

Built	1966
Acquired	1970
Length, overall	9·76 m (32 ft)
Beam, overall	3·35 m (11 ft)
Height, overall	3·66 m (12 ft)
Draught	0·61 m (2 ft)
Displacement	7 tons
Net tonnage	5

Performance

Full speed	25 knots
Cruising speed	15 knots
Range at cruise	40 miles maximum
Endurance at low speed	60 miles at 7 knots
Position holding capability	Sea state 2; ability—fair
Maximum anchoring depth	23·44 m (75 ft)
Towing capacity	Only for dredging and trawling

Main machinery description

Main engines	One Chrysler 225 I-O engine
Propeller type	One 3-blade aluminium on Volvo outboard drive
Pumps	Two electric bilge pumps

UNIVERSITY OF DELAWARE COLLEGE OF MARINE STUDIES
SKIMMER

Specific purpose of vessel

Hydro surveys, Nansen casts, small coring operations, seismic profiling, plankton trawls, small bottom grabs and trawling.

Leading particulars

Built	1970
Length, overall	12·8 m (42 ft)
Beam, overall	3·96 m (13 ft)
Height, overall	8·54 m (28 ft) plus radio antennas
Displacement	15 tons

Performance

Full speed	13 knots
Cruising speed	11 knots
Range at cruise	150 mile steaming radius
Endurance at low speed	300 mile steaming radius at 6 knots
Position holding capability	Sea State 4; ability—fair
Maximum anchoring depth	60·98 m (200 ft)
Towing capacity	Limited to trawling and dredging

Main machinery description

Main engines	Two Detroit diesel 6V53N engines; 197

On-board equipment

Facilities for carrying major surface equipment: 9·3 m² (100 ft²) of deck space

Sampling equipment: Trawls grabs, dredges, plankton samplers, etc. owned by College of Marine Studies and available with vessel

Craneage facilities: One boom 225 kg (500 lb) capacity

Winching facilities, capacities: One electric winch, 450 kg capacity, 100 m, 0·64 cm cable (1,000 lb capacity, 330 ft, ¼ in cable)

Navigational equipment: One magnetic compass, one sextant

Survey and search equipment: Ray Jefferson Depthmeter 400

Pumps, compressors and ancillary power equipment and services: 12 volt DC power available from engine, One VHF Radio—25 w

Laboratory and technical facilities

None

	s.h.p. each at 2,800 rpm
Propeller type	Two Columbia 4-bladed

Pumps: One electric bilge pump, one manual bilge pump, one bilge pump with clutch driven off of main engine pulley

On-board equipment

Deck space: 20·93 m² (225 ft²)

Sampling equipment: Four Nansen bottles with thermometres plus trawls, grabs, dredges, plankton samplers, water samplers, etc. owned by College of Marine Studies and available with vessel

Craneage facilities: Two booms from mast; capacity 450 kg (1,000 lb); 2·44 m (8 ft) clearance

Winching facilities: One double drum hydraulic winch—1st drum 183 m, 0·79 cm (600 ft, 5/16 in) cable, 2nd drum 183 m, 0·41 cm (600 ft, 5/32 in) cable

Navigational equipment: One Nelco Autofix 500 Loran A & C with cycle matching, one ENAC DX Navigator Loran A/C, Decca 101 Radar, magnetic compass, sextant

Survey and search equipment, radar, sonar, TV, seismic: Furuno FG-200 Mark 3 fathometre, Benmar DI-XA fathometer

Pumps, compressors and ancillary power equipment and services: One fresh water pump in laboratory, one sea water pump in laboratory, one 10 kw generator 120 volt AC,

Crew

Sea-going complement: One captain

Scientific complement: Six scientists (including technicians and students)

Crew safety facilities and equipment: One life-ring, flares, life-jacket with light for every person on board

Availability

Lease, charges: $US 27/hour

Charter availability: Chartered on time available basis

10, 60 Hz, 12 volt DC available from engines, one Konel AM-DSB Radio—85 w, one VHF Radio—25 w

Laboratory and technical facilities

Laboratory spaces include approximately 5·58 m² (60 ft²), counter space with a 'wet' bench and a 'dry' bench; pressurized salt and fresh water; 0·11 m³ (4 ft³) freezer; 0·11 m³ (4 ft³) refrigerator

Crew

Sea-going complement: One captain, one engineer

Scientific complement: 4–6 scientists (including technicians and students) or 12–14 students for instructional cruises

Accommodation: Berthing for captain and engineer only; head equipped with one manual toilet

Crew safety facilities and equipment: One Beaufort 8 man ocean life-raft; two horseshoe life-rings with lights, day and night flares; life-jackets with lights for every person on board

Availability

Lease, charges: $US 45/hour, $US 540/12 hour

Charter availability: Chartered on time available basis

UNIVERSITY OF DELAWARE COLLEGE OF MARINE STUDIES

WOLVERINE

Specific purpose of vessel

Stem trawling, hydro casts, bottom grabs, dredging, bottom coring, plankton sampling.

Conversion: Originally built in 1947 as a ketch-rigged motor sailer, converted in 1963–64.

Leading particulars

Length, overall	14·33 m (47 ft)
Beam, overall	3·96 m (13 ft)
Height, overall	9·76 m (32 ft)
Draught	1·68 m (5·5 ft)
Displacement	15 tons
Gross tonnage	35

Performance

Full speed	9 knots
Cruising speed	8 knots
Range at cruise	150 miles steaming radius, 300 miles maximum range
Position holding capability	Sea State 3; ability—fair
Maximum anchoring depth	45·73 m (150 ft)
Towing capacity	Limited to trawling, dredging

Main machinery description

Main engines: Two Detroit Diesel 4I-53N engines; 100 s.h.p. at 2,400 rpm

Propeller(s) tyoe: Two Federal 3-bladed

Pumps: One electric bilge pump, one manual bilge pump, one bilge pump with clutch, driven off of main engine pulley

On-board equipment

Facilities for carrying major surface equipment: Deck space 13·95 m² (150 ft²)

Sampling equipment: Three Nansen bottles with thermometers plus trawls, grabs, dredges, plankton samplers, water samplers, etc. owned by College of Marine Studies and available with vessel

Craneage facilities: Variable angle A-frame, 2 ton capacity, 2·44 m (8 ft) clearance

Winching facilities: One double drum hydraulic winch—each drum 228·66 m, 0·79 cm, 675 kg (750 ft, 5/16 in, cable; 1,500 lb) capacity; one hydrographic electric winch—106·71 m, 0·41 cm, 450 kg (350 ft, 5/32 in cable; 1,000 lb) capacity

Navigational equipment: One Nelco Autofix 500 Loran A & C w/cycle matching, one ENAC D-X Navigator Loran C, Decca D202 Radar, magnetic compass, sextant

Survey and search equipment: Raytheon DE-122 Depth Sounder, Ferrograph 'Offshore' 500 Echo Sounder

Pumps, compressors and ancillary power equipment and services: One 7 kw generator 120 volt AC 1 0 60 Hz, one 3 kw generator 120 volt AC 1 0 60 Hz, one electric fresh water pump in laboratory, one manual fresh water pump in galley head, one Apelco AM-DSB Radio—50 w, one VHF Radio—25 w

Laboratory and technical facilities

Laboratory spaces include approximately 6·51 m² (70 ft²) of 'wet' and 'dry' bench space; pressurized fresh water supply; 0·11 m³ (4 ft³) refrigerator; 0·11 m³ (4 ft³) freezer

Crew

Sea-going complement: One captain, one engineer

Scientific complement: 10–12 scientists (including technicians and students)

Accommodation: Galley with 4 burner gas range and oven, 0·11 m³ (4 ft³) refrigerator; head equipped with sink, manual toilet; berthing for 2 crew plus 4 scientists

Crew safety facilities and equipment: One Eliot 6 Man Limited Service life raft; 2 horseshoe liferings with lights; day and night flares; lifejackets with lights for every person on board

Availability

Lease, charges: $ US 45/hour, $ US 540//12 hours

Charter availability: Chartered on time available basis

THE JOHNS HOPKINS UNIVERSITY APPLIED PHYSICS LABORATORY

This laboratory operates two oceanographic research vessels.

COVE

A modified US Navy MSI class inshore minesweeper.

BONNIE DUNDEE

An 25·9 m (85 ft) Luders-built Ketch-rigged motor sailer.

NATIONAL SCIENCE FOUNDATION OFFICE OF POLAR PROGRAMS

HERO

A converted trawler.

Hero serves as a floating laboratory for investigations of both manning the terrestrial subjects primarily in the Antarctic Peninsula area.

Design history

Start of construction: 1967
Date of launching: March 1968
Date trials were completed: July 1968

Basic dimensions

Length, overall	38·11 m (125 ft)
Width, overall	9·25 m (30 ft, 4 in)
Draught	5·03 m (16·5 ft)

Displacement	640 tons
Crew	12
Cruise speed	10 knots
Range at cruise speed	5,000 nautical miles
Engines	Two 380 h.p. diesel
Propeller	One three-bladed
Controls	Manual and automatic

Facilities for diving: Compression and ancillary equipment for support of scuba divers

Sampling equipment- Normal over-the-side biological collection gear

Winching facilities: Electric Deep Sea trawl winch with three drums; hydrographic winch

Navigational equipment: Decca RM 326 and RMS 14 radars; Bendix ADF 168 Radio Director Finder

Survey and search equipment: Simrod Sonar; Kelvin Hughes Depth Sounder

Laboratory and technical facilities

Hydrographic Lab.; Analytical Lab.; Dissecting Lab.; Microbiology Lab.

Crew

Sea-going complement	12 Officers and men
Scientific complement	8
Crew safety facilities	Two 25-man inflatables

Operators

Holmes & Narver, Inc.,
400 East Orangethorpe Avenue,
Anaheim, California 92801

During a research trip off the coast of Argentina in 1971–73 by the R/V *Hero*, thirty-five Southern Right Whales were sighted, an animal on the endangered species list.

OCEAN SEARCH, INC.

ALCOA SEAPROBE

A deep ocean search and recovery ship designed and built by the Aluminum Company of America. For search, identification and recovery, the ship employs an unmanned search vehicle on the end of a drill pipe string. The drill pipe with its sensors and recovery tools can be deployed using conventional seagoing drilling-ship techniques through a centre well down to a depth of 2,440 m (8,000 ft). For recovery, the pipe system is capable of heavy lifts up to 200 tons at 1,829 m (6,000 ft). The vessel has an open centre well, 3·66 m × 10·98 m (12ft × 36 ft) as the work area.

Design history

Built by Peterson Builders, Inc., Sturgeon Bay, Wisconsin

Keel laid, February 1969

Launched: 17 November 1970

Owner and operator: Ocean Search, Inc., accepted vessel on 1 September 1971

Basic dimensions

Length, overall	74·09 m (243 ft)
Beam	15·24 m (50 ft)
Hull draught	2·44 m (8 ft)
Total draught (over vertical axis Voith-Schneider blades)	4·27 m (14 ft)
Displacement	1,700 tons

Performance

Speed	10 knots
Range	6,600 miles to 9,000 miles
Endurance	45 to 60 days

Other data

Main power: Two 800 kw diesel-electric generators

Auxiliary power: Two 250 kw diesel-electric generators

Propulsion: Two Voith-Schneider cycloidal omni-directional propulsion units

Auxiliary deck equipment: Two 5-ton cranes; Oceanographic winch-interchangeable drums

Ship control: Decca ship control consoles on bridge and in search/recovery control centre

Primary ship construction material: 5456—H117 aluminium plate; 5456—H111 aluminium extrusions

Derrick—Height: 40·24 m (132 ft) (above waterline); capacity: 250-ton hook load with safety factor of 2; material: 6061—T6 aluminium tubing; 5456—H321 aluminium plate

Drawworks: EMSCO 800 with 600 h.p. motor generator power supply for DC control

Pipe: 0·12 m ($4\frac{1}{2}$ in) external upset-internal flush; Sections 18·29 m (60 ft) in length

Pipe handling: Semi-automatic pipe handling system

Berthing: All spaces air conditioned; crew 30, scientific party 19

US Flag—Coast Guard inspected and ABS certified.

Alcoa Seaprobe

Deep ocean floor work capabilities

The business end of the *Alcoa Seaprobe* system is located at the end of a pipe string made of 18·29 m (60 ft) segments of drill pipe threaded together to reach the ocean depths required. Depending upon the task at hand, sensor systems are deployed for fine grain bottom search and examination, target identification and marking, or oceanographic sampling and measurement. Recovery devices ranging from multi-purpose grappling claws to coring tools are available for lifting bottom objects or cores to the surface. A cable affixed to the exterior of the pipe provides the necessary electrical power, telemetry control signals, and data transmission circuits between the shipboard control consoles and the sensor systems probing the deep ocean.

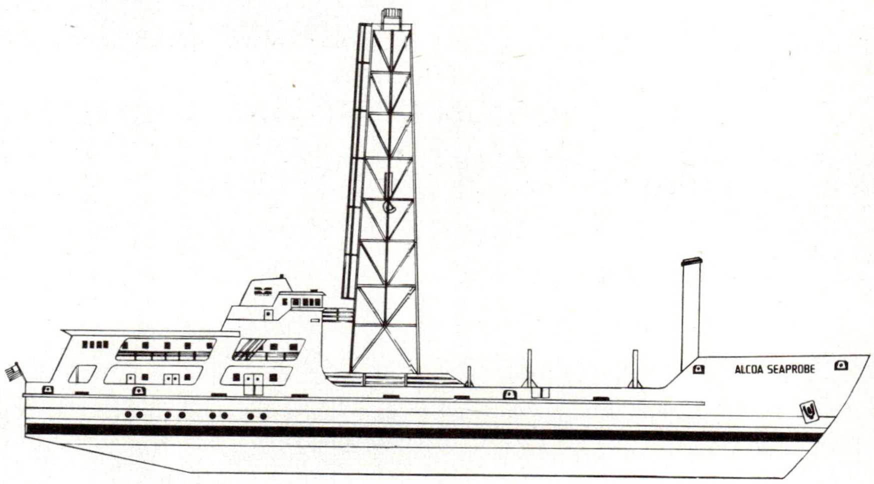

Alcoa Seaprobe, elevation diagram

Search

The basic search 'pod' deploys sidescan sonar to sweep a 730 m (2,400 ft) path along the sea floor. The pod is arranged with forward looking sonar, television, still camera, target illumination systems, and a dropable acoustic beacon to use in marking specific target locations. Additional sensors, detector and other special devices may be added to the pod to meet the requirements of specific missions.

Recovery

Heavy object recovery devices utilize rugged electro-mechanical and hydraulic systems for closure and holding control. Precise positioning of the recovery device with respect to the target is effected by sonar and transponder sensing devices in concert with remotely monitored television and target illumination systems.

Sampling

In concert with television examination, bottom samples may be selectively recovered for the purpose of magnetic, chemical, biological, geological and paleontological studies. Shipboard precision navigation systems and a thruster system to control the end of the drill string permit finite positioning of devices and sensors with regard to the ocean floor.

Coring

Alcoa Seaprobe can be outfitted to core in water depths to 5,488 m (18,000 ft). A bottom-oriented acoustic reference system is used to maintain position in the open ocean and permits wire line recovery of cores for paleontologic and magnetic stratigraphy.

Cores can also be recovered for studies concerned with the mechanism of sediment supply, transport, deposition, and to confirm seismic survey data for scientific or other exploratory purposes.

Relevant technical papers: 'Operational Evaluation of a Deep Ocean Search and Recovery Ship' by Wilbur G. Sherwood, Marine Technology Society, 8th Annual Conference.

Surface Support Vessel for Subsea Operations by Wilbur G. Sherwood. Paper No. OTC 1837 given at the Fifth Annual Offshore Technology Conference, Houston, Texas, 29 April–2 May 1973.

WOODS HOLE OCEANOGRAPHIC INSTITUTION.

(see also under UNOLS fleet)

ASTERIAS

An auxiliary Dragger-type research vessel built in 1931.

Leading particulars

Length, overall	12·5 m (41 ft)
Beam, overall	3·96 m (13 ft)
Draught	1·22 m (4 ft)
Gross tonnage	30
Full speed	8 knots
Cruising speed	$7\frac{1}{4}$ knots
Range at cruise	525 miles
Maximum anchoring depth	61 m (200 ft)
Main engines	One Diesel, 65 s.h.p., 1,800 rpm
Auxiliary engines	One Diesel, 10 s.h.p., 1,800 rpm
Propeller type	Single screw
Winching facilities	Small trawl winch 61 m (200 ft), portable BT winch
Navigational equipment	Radar, RDF, Fathometer
Crew	1
Scientists	3

LULU

See under Submersible Support Ships.

Asterias

UNION OF SOVIET SOCIALIST REPUBLICS

USSR ACADEMY OF SCIENCES BAIKAL LIMNOLOGY STATION

ALBATROS
A research expedition vessel.

Albatros ***(Courtesy Novosti Press)***

USSR ACADEMY OF SCIENCES

COSMONAUT VLADIMIR KOMAROV

The worlds largest research vessel, launched at the Baltic Shipyards in 1967.

Length	156 m (511·7 ft)
Accommodation	240

This vessel has been principally used, it is believed, in connection with earth satellite work and interplanetary probes.

Cosmonaut Vladmir Komarov ***(Courtesy Novosti Press)***

USSR ACADEMY OF SCIENCES INSTITUTE OF OCEANOLOGY

ACADEMICIAN KURCHATOV

Built in the German Democratic Republic at Wisman in 1966 for composite oceanographic research.

Leading particulars

Displacement	6,680 tons
Length	124·2 m (407·38 ft)
Beam	17·0 m (55·76 ft)
Height of sides	8·2 m (26·90 ft)
Draught	6 m (19·68 ft)
Main engines	Two 4,000 h.p.
Speed	18 knots
Range	14,900 miles
Electric power plant	2,425 kw

The ship has 26 laboratories with a total area of 417 m^2 (4,500 ft^2). There are 16 oceanological winches, 7 of which are wire hawser and 9 cable. The ship is fitted with stabilizers and is air-conditioned throughout.

In January 1973 the *Akademik Kurchatov* was reported as having left the Baltic port of Kaliningrad for the Caribbean and the Gulf of Mexico where research would be carried out under the international 'Seacar' programme. The ship would operate for three and a half months in the area of the Puerto-Rican, Venezuelan and other sea depressions with depths up to 10,000 m (32,800 ft). Attempts would be made to collect samples of the soil and specimens of animal life from the seabed.

The long-term 'Seacar' programme, sponsored by Unesco, involves scientists from Britain, Denmark, the USA and certain Latin American countries. They have succeeded in obtaining quantitative characteristics of the inter-connection between thermal processes of the ocean surface with the atmosphere, which is of great importance for weather forecasting.

Academician Kurchatov (*Courtesy Novosti Press*)

USSR ACADEMY OF SCIENCES INSTITUTE OF LIMNOLOGY SIBERIAN BRANCH

G. YU VERESCHAGHIN

The first research ship to be used on Lake Baikal.

Yu Vereschaghin (*Courtesy Novosti Press*)

USSR ARCTIC MARINE BIOLOGY INSTITUTE

DIANA

No details available, see photograph.

Diana (*Courtesy Novosti Press*)

MARITIME HYDROMETEOROLOGICAL SERVICE ADMINISTRATION

ACADEMICIAN SHIRSHOV

Built in the German Democratic Republic and delivered to Vladivostock in early 1968. This vessel is the flagship of the Maritime Hydrometeorological Service fleet. She is equipped with roll stabilizers and fitted out with 29 laboratories and a computer. Her sister ship is the *Akademik Korolyov*.

***Akademik Shirshov* (Courtesy Novosti Press)**

MOSCOW STATE UNIVERSITY

MOSKOVSKY UNIVERSITET

Launched in 1971, this vessel will be used by more than thirty chairs of the Moscow State University to conduct research. Three-D colour television is used down to a depth of 100 m (328 ft) for studying the animal life in the oceans.

For studying deep-water fishing, the ship is equipped with a special trawl capable of operating at a depth of 3 km (1·86 miles or 9,820 ft).

***Moskovsky Universitet* (Courtesy Novosti Press)**

PACIFIC RESEARCH INSTITUTE OF FISHERIES AND OCEANOGRAPHY (TINRO)

LAKHTAK

This vessel has been used by the Magadan branch of the Institute in the study of marine mammals in Northern waters.

***Lakhtak* (Courtesy Novosti Press)**

SOVIET ANTARCTIC EXPEDITION

OB

Flagship of the Antarctic fleet, by September 1973 she had concluded eighteen runs to Antarctica and during her last trip she was trapped in ice for over three months.

Main propulsion machinery is diesel electric. A helicopter landing deck is provided from which an Mil-8 helicopter has operated.

Ob (*Courtesy Novosti Press*)

SOVIET FAR EAST RESEARCH FLOTILLA

PROFESSOR DERYUGIN

A new vessel, 1968, for use by the Pacific Research Institute of Fisheries and Oceanography.

She was built in the Ukraine.

Professor Deryugin (*Courtesy Novosti Press*)

UKRAINIAN ACADEMY OF SCIENCES

AKADEMIK VERNADSKY

This vessel joined the Soviet expeditionary fleet in 1969 and is designed to study hydrophysical processes in the oceans and adjacent atmosphere.

Leading particulars

Length	124 m (406·7 ft)
Beam	17 m (55·8 ft)
Draught	6 m (19·7 ft)
Speed	18 knots

In December 1973 the *Akademik Vernadsky* was reported as having set out for the equatorial zone of the Atlantic, its eighth scientific expedition. During a four-month cruise the vessel, which has 60 Soviet hydrophysicists, biologists and other scientists on board, would carry out investigations under the international programme for the study of the seas and oceans and the use of their resources.

The expedition's leader, oceanographer Pavel Ganson, says that the expedition will chiefly study the tropical zone of the Atlantic between the 9° and 12°N, which has always attracted oceanographers by its higher biological productivity.

It will also be studying the ocean floor relief, magnetic and electrical fields, horizontal and vertical structure of water salinity and changes in the current as well as carry out geological and seismological investigations in the tropical zone.

Water exchanges between the Atlantic and the Mediterranean through the Straits of Gibraltar are also being studied together with commercial fishery research, mineral prospecting, anti-seismic protection, oceanic pollution and weather forecasting.

Akademik Vernadsky (*Courtesy Novosti Press*)

ADDITIONAL RESEARCH VESSELS

ACADEMICIAN OBRUCHEV

Built in 1953 in the German Democratic Republic and refitted in 1954 for oceanographic research in the littoral zone.

Leading particulars

Displacement	78·6 tons
Length	18·25 m (59·86 ft)
Beam	5·22 m (17·12 ft)
Height of sides	2·5 m (8·2 ft)
Draught	2 m (6·56 ft)
Main engine	150 h.p.
Speed	8 knots
Range	1,000 miles

Akademician Obruchev (*Courtesy Novosti Press*)

ANDROMEDA

This Soviet research ship was used in 1973 in an expedition formed to conduct wide-scale research on pollution of the sea due to industrial waste in the Norwegian and North Seas.

ARGUS

No details available, see photograph.

Argus (*Courtesy Novosti Press*)

PROFESSOR DOBRYNIN

Built in the Soviet Union in 1963 for oceanographic research in the littoral zone.

Leading particulars

Length	25·23 m (82·7 ft)
Beam	6·6 m (22·1 ft)
Height of sides	2·8 m (9·2 ft)
Draught	2·54 m (8·33 ft)
Displacement	104·5 tons
Speed	9 knots
Main engine	150 h.p.
Range	1,350 miles

The ship has a well-equipped laboratory and an oceanological winch.

Professor Dobrynin (*Courtesy Novosti Press*)

COSMONAUT YURI GAGARIN

A research vessel in use with the Russian space programme.

Cosmonaut Yuri Gagarin (*Courtesy Novosti Press*)

Cosmonaut Yuri Gagarin (*Courtesy Novosti Press*)

MIKHAIL LOMONOSOV

This vessel is equipped with sixteen laboratories and is used for general scientific research in the oceans.

Mikhail Lomonosov (*Courtesy Novosti Press*)

DMITRY MENDELEYEV

One of seven research vessels built at the Wismark shipyards in the German Democratic Republic. The vessel is fitted out with 27 laboratories.

In mid-1973 the *Dmitry Mendeleyev* concluded geochemical explorations of the Pacific Ocean.

Apart from Soviet researchers, the expedition included scientists from the USA and Sweden. The vessel covered about 25,000 miles in the Pacific, over the route Nakhodka–Mexico–Gulf of California–Funafuti atol–West Somoa–Fiji Islands–New Hebrides–Tokyo–Vladivostok.

PRIBOI

A 4,000-ton survey vessel fitted out with twenty-two laboratories. In March 1973 she completed a joint Soviet–American research programme in the Bering Straits on sea and ice conditions.

YU M SHOKALSKY

A research expedition ship.

Yu M. Shokalsky (*Courtesy Novosti Press*)

SEVERNOYE SIYANIE
This ship is referred to as a floating power plant.

Severnoye Siyanie (*Northern Lights*) (*Courtesy Novosti Press*)

SERGEI VAVILOV
Soviet 'Lebedev' class research vessel, built in 1954 and comprehensively equipped.

Sergei Vavilov (*Courtesy Royal Navy*)

VITYAZ

Flagship of the Soviet research fleet for nearly 20 years until 1967, having been built in Germany in 1939. In 1947–48 she was refitted for composite oceanographic studies. She has engaged in 54 research voyages covering some 600,000 miles.

Leading particulars

Length	109·4 m (358·8 ft)
Beam	14·56 m (477·6 ft)
Height of sides	8·75 m (28·7 ft)
Draught	5·86 m (19·2 ft)
Displacement	5,710 tons
Main engines	3,000 h.p.
Speed	12 knots
Range	21,000 miles
Laboratories	20, area 267 m² (2,900 ft²)
Electric power plant	554 kw
Oceanological winches	13 (2 cable, 11 wire rope)

PROFESSOR ZUBOV
A Soviet polar research ship.

Vityaz (*Courtesy Novosti Press*)

OCEAN-DATA AND MAJOR NAVIGATIONAL BUOYS

FRANCE

ATELIERS ET CHANTIERS DE BRETAGNE DEPARTEMENT TECHNIQUES NOUVELLES

SURFACE BUOY/SELF CONTAINED OCEANOGRAPHIC STATION

This buoy is designed to measure the following parameters:

air temperatures;
wind speed and direction;
air humidity;
atmospheric pressure and variation of same;
water temperatures;
current speed and direction;
wave motion;
dissolved oxygen content in water;
water salinity;
sound speed through the water.
No further details are available.

The A.C.B. self-contained oceanographic station

COMEX EQUIPMENT

BOREM BUOY

The Spar buoy structure was chosen for its great stability. The buoy is composed of an immersed large float onto which an emerging small secondary float is attached. The mast supporting the meteorological instruments is located above the secondary float. A kentledge is attached to the lower part of the main float by a lattice, and the ballast is located under the kentledge.

The structure is moored by means of a float. This technique accommodates a certain swing and a lack of interaction from the anchoring on the buoy stability.

The structure is protected against corrosion and fouling by nine coats of paint and a cathodic protection, and can thus remain in water for more than two years without maintenance.

When installing, the buoy is lowered horizontally into the water and towed in this position to its mooring site. It is then vertically tilted, with diver intervention, by water ballasting in less than 15 minutes.

The structure is equipped with sensors designed for operation in rough conditions.

(a) A cup type anemometer ensures a 1% linear measurement.

(b) A vane coupled to a Magnetic North compass delivers a signal directly proportional to the wind direction.

(c) Swell is measured by a Teflon coated steel wire Wave Staff, vertically stretched along the mast. The electronic circuitry is compact and accurate due to the use of integrated circuits.

The electronic and transmitter container is located at the highest part of the main float. A sequencer switches on the transmitter during the daytime only. The signals are received and recorded onshore with an I.R.I.G. standard receiver.

The power supply container is immersed, and because of its zero weight in water can easily be replaced by divers. The batteries have a three to six months life.

The buoy is also equipped with a passive radar reflector and a flashing beacon which operates automatically at night.

Special markings on the structure permit measurement with a cine theodolite.

Specifications

General characteristics

Overall height	28·5 m (94 ft)
Above surface	10 m (33 ft)
Weight	8·5 tons
Pitch	Less than 2° up to 15 sec. periods negligible over 20 sec.

Wind speed sensor module

Sensor	Cup type
Speed range	0–100 knots
Speed accuracy	±5% (sensor is 1%)
Information pick up	Magnetic switch
Pulse rate	Four per revolution

Wind direction sensor module

Direction vane	Compass mounted neutrally buoyant in oil filled — shock proof housing
Direction range	0–360°
Direction accuracy	±7° (sensor accuracy 5°)
Direction linearity	±1%
Information pick up	Linear potentiometer

Swell sensor—Wave staff

Swell amplitude	±5·2 m (17 ft) Precision of 5% for periods of 0–7 seconds. Precision of 8% for periods of 7–15 seconds
Swell periods	0–15 seconds. Precision 1%

The above specifications refer to the complete system with accelerometer.

Transmitter

Frequency range	215 to 260 MHZ
Power	300 mW

Power supply

Replaceable standard 1·5 volt batteries ensuring a 3 to 6 month autonomy.

Electronic

Integrated micro circuits. Modules can be added if more parameters are needed.

On request

Atmospheric pressure sensor
Current meter
Temperature sensor
Accelerometer

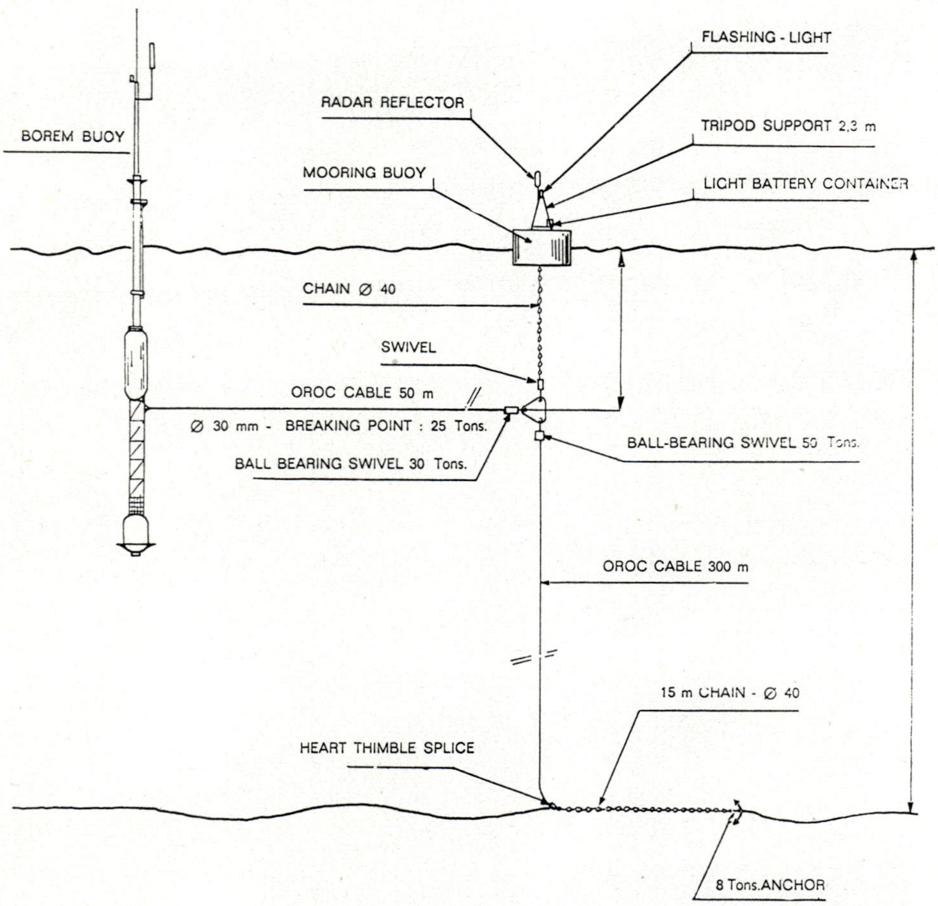

Anchoring arrangement of the Borem buoy

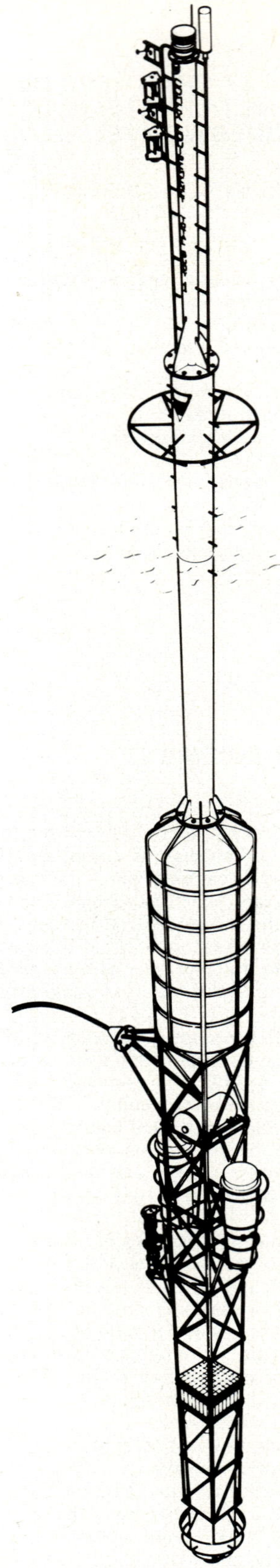

Comex Equipment's Borem buoy

GERMANY

DORNIER SYSTEM GmbH

DEEP-SEA ROLL - STABILIZED BUOY

A buoy design for the North Sea—Baltic Measuring Network, Project VEMNO. This measuring network is to be realised in several stages:

(1) Development and testing of the first prototype buoy by mid-1975.

(2) Construction of a prototype measuring network with approximately five buoys in the Heligoland Bight as from 1976.

(3) Construction of an operational measuring network in the North Sea and the Baltic Sea as from 1978.

Phases 1 and 2 are planned as independent national programmes, in contrast to the operational measuring network which is intended to operate as part of an overall European and/or world-wide network. Programmes of this nature are under preparation at the present time within the framework of the European Communities (Action COST 43), the IOC and WMO (IGOSS = Integrated Global Ocean Station System).

The measuring station will consist of a buoy serving as a support for the meteorological sensors. The oceanographical sensors will be concentrated in so-called sensor packages and attached to a cable at various depths below the buoy. The planned prototype will be a buoy adapted to the special wave conditions of the Heligoland Bight and displaying quasi-stable behaviour in the waves. The buoy will be walkable and equipped with a diesel generator for energy supply as well as a winch which will serve to raise and lower the sensor packages so that both measuring at fixed depth horizons and profile measuring will be possible in the trial phase.

The equipment for this buoy is being developed and tested within the framework of the VEMNO project. It comprises the sensor data acquisition sub-system with at first four sensors as well as the data acquisition and cable transmission, data processing, radio transmission and energy supply subsystems. The work also involves the integration and testing of the equipment and sea trials with the prototype buoy described above. The trials will take place in a sea test field to the north-west of Heligoland where, besides the prototype buoy, a stationary, manned platform will be established and equipped with the VEMNO devices.

Basic dimensions

Diameters, hull section	3·4 m (11·2 ft)
stabilizing torus	5·3 m (17·3 ft)
Height, overall	12 m (40 ft)
Draught	4·3 m (1·41 ft)

Weights

Displacement	8 tons
Payload	4 tons

Structure

Steel torus and hull, aluminium mast section

Production

Five North Sea Types are planned to be built

Operational aspects

Limiting wave conditions:	
height	5·0 m (16·4 ft)
period	8 s
Limiting current conditions	2·0 m/s (6·7 ft/s)
Mooring	single point mooring for deep sea

Measurement capability

Oceanographic parameters: currents, temperature, with a separate mooring line and sub-surface float: temperature, depth, currents, conductivity, wave height in various depths (up to 1,200 m (4,000 ft)

Meteorological parameters

Temperature, pressure, wind

Data system

Capacity	up to 14 sensor packages with up to 8 different sensors for oceanographic parameters and up to 8 meteorological sensors
Sampling frequency	variable, max 10 Hz.
Presentation	magnetic tape on a shore station or on the buoy
Transmission	HF- or VHF-radio link to a shore station (later on UHF with a geostationary satellite as a relay station)

Handling

Lifting points	three
Towing points	one, in the middle of the hull section
Launching methods	by crane of supply vessel

Power installation

Power source	Diesel engine and batteries
Duration of unattended operation	one year
Maintenance and overhaul requirements	one year for service, three to five years for overhaul

Application

Northern North Sea	German buoy network, and another buoy network as navigation buoys with specials

Availability

Prototype 1974

Test model of Dornier's Deep Sea Roll-Stabilised Buoy in a wave-tank

MASCHINENBAU GABLER GmbH

HOVERING BUOY

Designed as an instrument carrier for the collecting of seismic data, employing hydrophone and tape recorder

Availability

Prototype 1974

Basic dimensions

Length	2·3 m (7·5 ft)
Beam	1·8 m (5·9 ft), with antenna, 2·9 m (9·5 ft)
Height	2·6 m (8·5 ft), with antenna, 4·6 m (15·0 ft)
Draught	normally fully submerged, but surfaced draught is 2·3 m (7·54 ft)

Weights

Displacement	2·2 metric tons
Payload	80 kg (176 lb)

Structure: In mild steel. There are two pressure vessels for instrumentation, two for batteries and five for high pressure air

Production

One prototype was constructed in 1973

Operational aspects

Operating depth and time (12 h max) are set before launching, depth control and hovering are achieved by high pressure air—controlled flooding and freezing system. Surfacing after operating time is achieved by automatic blowing of ballast tank. Maximum submerged depth: 200 m (656 ft).

Measurement capability

Physical and chemical properties, mainly accoustical, collecting and storing of data as required.

Handling

One pad eye for lifting and towing. Launching: by ship's crane or loading gear.

Power installation

Power source: lead-acid battery, 24 v, 240 Ah; High-pressure air for hovering and surfacing

Cost (1973): Approximately DM200,000.

The Maschinenbau Gabler Hovering Buoy

UNITED KINGDOM

HAWKER SIDDELEY DYNAMICS LIMITED

(Member company of SEATEK which also includes EMI Electronics Ltd., and R. H. Green & Silley Weir.)

LANBY

A large automatic navigational aid buoy.

The equipment carried in this buoy is designed using modular concepts and therefore provides considerable flexibility to meet particular customer requirements.

Designs are already available covering power generation and monitoring/control systems.

Diameter	12·2 m (40 ft)
Height, overall	12·2 m (40 ft)
Draught	1·2 m (4 ft)
Displacement	84 tons
Pay load	26 tons

Operational aspects (Nominal)

Towing speed	7 knots
Wind limits	100 mph
Wave limits	12·2 m (40 ft) high
Mooring arrangements for deep and shallow water	Single point with concrete sinker and Admiralty type anchor. Depth 5 to 50 fathoms
Current limits	7 knots

Data system

Capacity: (30 monitoring channels—buoy to shore (provision for further 20) (20 control channels—shore to buoy (provision for analogue channel)

Sampling frequency: once every ½ hour or 'on demand'

Presentation: visual instrumentation and audible alarm

Transmission: UHF duplex—FM modulation. Duplicated transmitters with buoy automatic changeover. Parallel receivers

Handling requirements

Lifting points: Four twin deck bollards

Towing points: Lifting bollards

Launching methods: Crane lift or slipway

The SEATEK—LANBY

Power installation

Power source: 12½ b.h.p. Petters PH2W diesel engine driving a 5½ kw alternator. This unit is repeated three times for sequential operation to a pre-determined programme

Duration of unattended operation: 6 months

Maintenance and overhaul requirements: Standard diesel engine practice

Positioning

By ship tow to station location

Availability

18 months from order date

Costs

Purchase, capital costs—By negotiation dependent on customer requirements. In region of £150,000 to £200,000 per buoy. Shore station extra.

D.B.1 (pending allocation of ODAS identification)

An oceanographic and meteorological data collection buoy, to be designed, built and commissioned under contract to the UK Department of Trade and Industry. This 7·6 m (25 ft) diameter buoy is the first large oceanographic and meteorological sensor carrier of British origin and will form the UK contribution to the proposed European network of such devices, and eventually to a world-wide buoy network. The equipment carried in this buoy is designed using modular concepts and therefore provides considerable flexibility to meet particular customer requirements.

Diameter	7·6 m (25 ft)
Height, overall	9·8 m (32 ft)
Draught	1·06 m (3 ft 6 in)
Displacement	40 tons

Structure

7·6 m (25 ft) diameter by 2·1 m (7 ft) deep discus hull supporting a main mast tube of ·45 m (18 in) diameter on which is constructed a sensor platform. A spar retracts into this main mast. Lloyds Grade A and D mild steel plate and tube for the hull, with upperworks of N8 alloy plate and HE9 alloy tube.

Operational aspects (Nominal)

Towing speed	7 knots
Wind limits	100 mph
Wave limits	18·29 m (60 ft) high
Mooring arrangements for deep and shallow water	Three point mooring with concrete sinker and Admiralty type anchor. Depth 10 to 100 fathoms
Current limits	5 knots

Measurement capability

Sensors covering:

(A) *Oceanographic*
1. Surface currents
2. Mid water and near bottom currents
3. Sea surface temperature
4. Sea bed temperature and pressure
5. Wave data
6. Salinity at various depths
7. Dissolved oxygen at various depths
8. Density at various depths

(B) *Meteorological*
1. Air temperature
2. Air/sea temperature difference
3. Barometric pressure
4. Relative humidity
5. Rainfall
6. Wind speed direction
7. Visibility
8. Radiation

(C) *General*
1. Compass
2. Pollution

Data system

Capacity: 50 data channels accepting analogue, FM or BCD (Serial or parallel) provision for further 50 channels

Sampling frequency: Once every hour plus transmission of selected channels for 20 minutes every three hours

Presentation: By print out

Transmission: HF 1 out of 10 tone modulation buoy to shore only multi frequency shift keying system (Piccolo)

Handling requirements

Lifting points: 4 twin deck bollards

Towing points: Lifting bollards

Launching methods: Crane lift or slipway

The second buoy design from Hawker Siddeley Dynamics Ltd: D.B.1

Power installation

Power source: Air depolarisation dry cells giving 15 watts mean

Duration of unattended operation: 6 months

Maintenance and overhaul requirements: Replacement as required

Positioning

By ship tow to station

Availability

18 months from order date

Costs

By negotiation dependent upon customer requirement. Price in region of £130,000 Shore station extra.

UNITED STATES OF AMERICA

GENERAL DYNAMICS, ELECTRONICS DIVISION

ESB

An environmental service buoy (ESB) for environmental data acquisition and telemetry.

Basic dimensions

Diameter, overall	12 m (40 ft)
Height, overall	10 m (32·8 ft)
Draught	1·1 m (3·7 ft)

Weights

Displacement	87,000 kg (192,000 lb)
Payload	2,450 kg (5,500 lb) (ballast of 42,500 kg (94,000 lb)

Structure

Main components: Sensor system, data processing system, command/communication system, power system

Principal structure: Thick discus hull with superstructure, both of welded steel construction

Production

Ten built (1973)

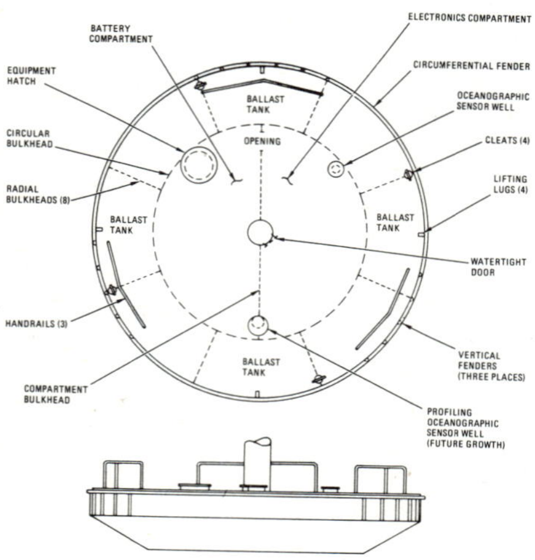

Environmental Service Buoy (ESB/12)

Operational aspects

Towing speed 10 knots

Wind limits To 175 knots

Wave limits: Breakers 12 to 18 m (40 to 60 ft) high, waves (not breaking) of any height

Mooring arrangements: Single point deep water or shallow water moorings, synthetic line, taut or slack

Current limits: To 10 knots

Measurement capability

Meteorological, oceanographic, and pollution monitoring sensors. One shore data collection station or interface to user facilities can command/control/record buoy operations. One station can function with multiple buoys in a network over very long ranges—4,000-4,800 km (2,500-3,000 miles).

Data system

Small digital computer or hard-wired controller.

Capacity: Up to 200 sensors

Sampling frequency: Variable, can be in real-time

Presentation: Digital formatting aboard buoy

Transmission: Single sideband PSK or FSK on narrowband HF channels allocated by WARC, range to 4,800 km (3,000 miles)

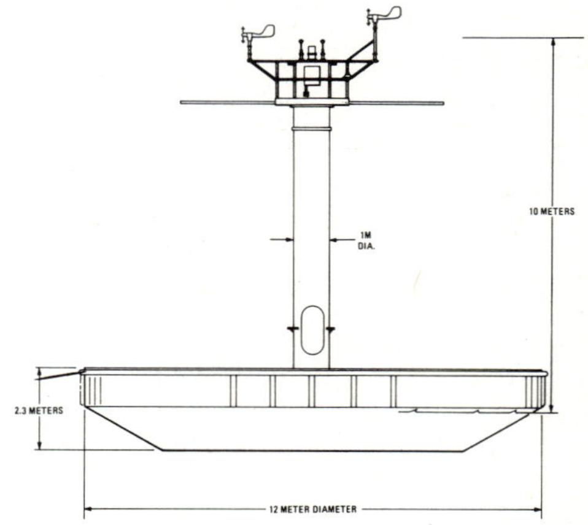

Environmental Service Buoy (ESB/12)

Handling requirements

Lifting points: Four

Towing points: Four on deck, one on chine

Launching methods: Crane lift into water, or over land by mobile transporter

Power installation

Power source: All battery system capable of a one-year operation

Duration of unattended operation: one year

Maintenance: one year. Overhaul: three years

Positioning

Can be moored anywhere in ice-free waters.

LNB/40 AND LANBY/40

A large navigational buoy (LNB) and a similar version produced for Trinity House in the UK is called a large automatic navigational buoy (*LANBY*), both designed as major aids to navigation and light-ship replacement.

These automatic navigational buoys provide the highly reliable service of manned lightships and fixed platforms at greatly reduced costs and in extreme environmental conditions: winds of 150 to 175 knots, currents to 10 knots, breakers 12 to 18 m (40 to 60 ft) high, waves (not breaking) of any height. They can also perform other primary aids-to-navigation missions (such as sea lane marking) in the open ocean in waters of unlimited depth for long periods without service of any kind.

For its principal mission, the LNB is equipped with beacon lights, fog signal, radio beacon, and racon (radar transponder). Beacon lights are available with intensities to one million effective candelas; fog signal range is 3 miles; radio beacon range varies from 10 to 50 miles; racon range is consistently better than 10 miles. The operational status of all aids is continuously monitored ashore. One shore command station can control up to six buoys.

In its secondary role, the LNB can automatically report to the shore station meteorological, oceanographic, and pollution data via the same monitor link that commands the aids to navigation. At the shore station, in addition to the environmental data, all buoy operational functions are displayed: beacon light primary or secondary, radio beacon transmitter primary or secondary, sound signal power, radar transponder on or off, command link transmitter main or standby, engine generator main or standby, etc.

A LANBY buoy on station in UK coastal waters

Audible alarms supplement the visual display if an operating parameter exceeds pre-set limits.

Power aboard the LNB is provided at 120 V, 60 Hz, by continuously-operating diesel generators. All systems can operate for 6 months without service or maintenance of any kind. Ample facilities are provided for ease of service on station: electric power, ventilation, lighting, tools, and work space. The LNB can operate for 2 years before refueling, 3 years before overhaul.

Basic dimensions

Diameter, overall	12 m (40 ft)
Height, overall	13·7 m (45 ft) (24·2 m (80 ft) with whip antenna atop superstructure)
Draught	1·12 m (3 ft 9 in)

Weights

Displacement	100 tons
Payload	50 tons

Structure

Main components: Beacon lights, sound signals, radio beacon, racon, control/communications system, power system

Principal structure: Thick discus hull with superstructure, both of welded steel construction. Rated 100A1 by Lloyds

Production

Seven LNBs now on station in US coastal waters (1973). Four LANBYs in use by Trinity House of UK (built by Camper and Nicholsons, Gosport, Hampshire, UK). Two additional LANBY being constructed for 1974 use.

Operational aspects

Towing speed	To 10 knots
Wind limits	150 to 175 knots

Wave limits: Breakers 12 to 18 m (40 to 60 ft) high, waves (not breaking) of any height

Mooring arrangements: Shallow water to depths of 91·5 m (300 ft): single-point or three-point moorings. Deep water to depths of 1,220 m (4,000 fathoms), synthetic line

Current limits: To 10 knots

A US Coast Guard LANBY buoy (General Dynamics) on Station

Measurement capability

Meteorological, oceanographic, and/or pollution monitoring systems.

Data system: Optional

Capacity: Up to 200 sensors

Sampling frequency: variable, usually once per hour

Presentation: Digital conversion aboard buoy

Transmission: Uses same monitor link used to command aids-to-navigation from shore station

Handling requirements

Lifting points: Four

Towing points: Three on chine, four on deck

Launching methods: Crane lift into water or over land by mobile transporter

Power installation

Power source: Two diesel engine-generators

Duration of unattended operation: 6 months

Maintenance and overhaul requirements: Maintenance: 1 year. Overhaul: 3 years

Positioning

Can be moored anywhere in ice-free waters.

Purchase

Capital costs: Depends on options selected but UK LANBY was reported to cost £195,000

Applications

Aids to navigation, lightship replacement, environmental data gathering stations. One shore station can monitor/control up to five buoys. Distances depend on radio range and/or relay station location. The United States LNB network is shown in the accompanying map and the UK *LANBY*'s will replace lightships at Morecambe, Cork, the Shambles, the Owers and the Lynn Well lighthouse. The LNB's are classed 100A1 by Lloyds Registry of Shipping. Lifetime of the hull and operating systems is in excess of 20 years.

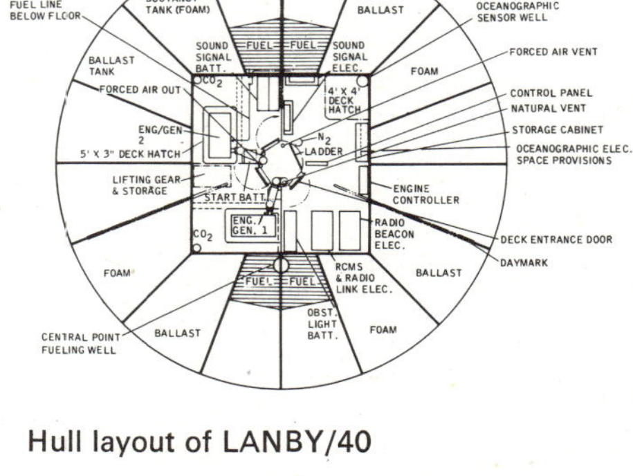

Hull layout of LANBY/40

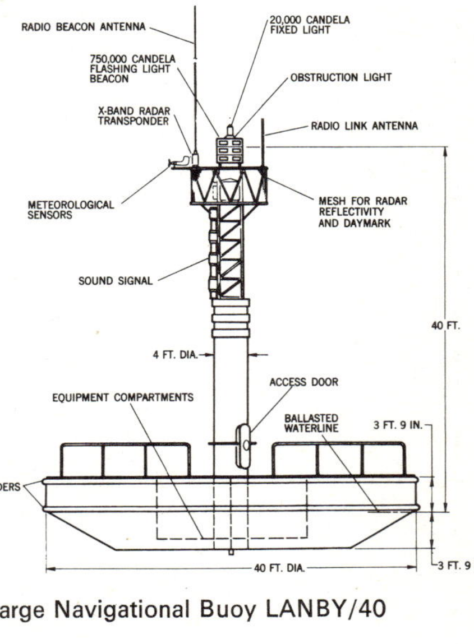

Large Navigational Buoy LANBY/40

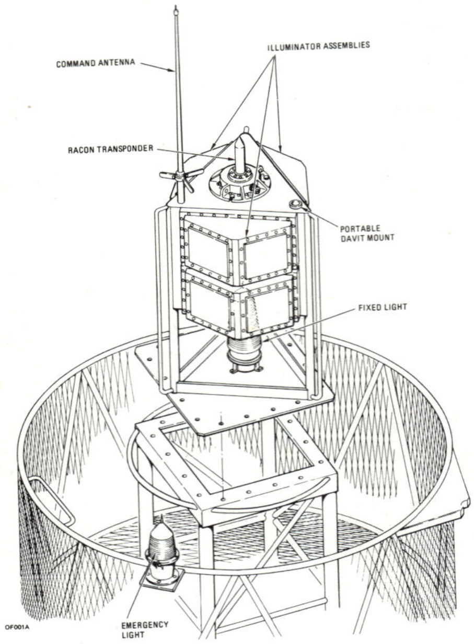

Arrangement of xenon light assemblies on LANBY buoys

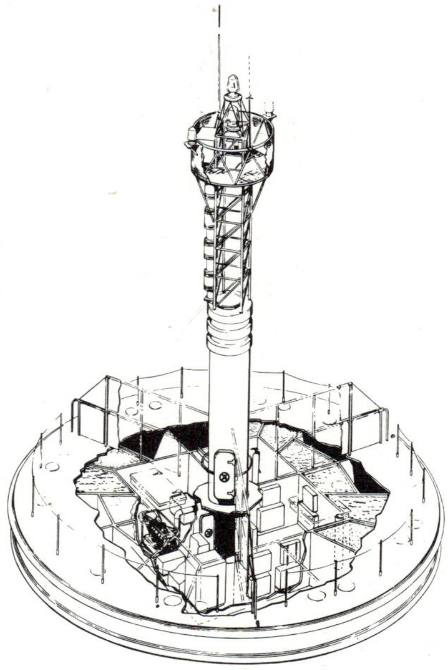

Sectional perspective of LANBY/40

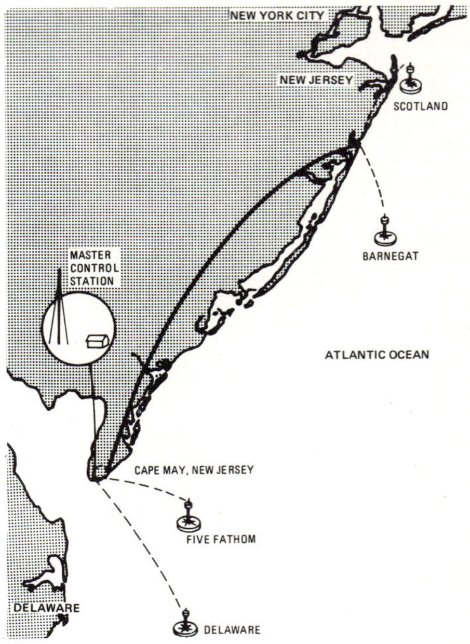

Map shows LANBY network on US East coast

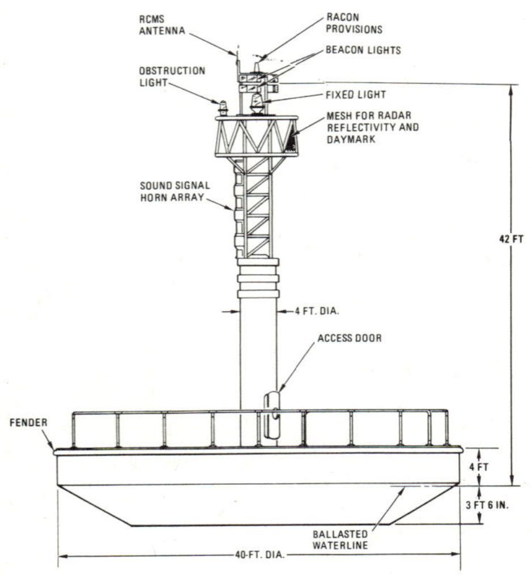

Elevation diagram of a Large Automatic Navigational Buoy (LANBY/40)

LANBY/40

MODS/30

A medium ocean data station (MODS) for environmental data acquisition and telemetry. Two features distinguish MODS/30 as compared to the large Ocean Data Stations also produced by General Dynamics: the hull diameter is 9·1 m (30 ft) rather than 12·2 m (40 ft) and the entire mast assembly is removable for ease of shipboard handling and transportation. In all other respects, its capabilities approach those of its larger counterpart.

Basic dimensions

Diameter, overall	9·15 m (30 ft)
Height, overall	10 m (30 ft)
Draught	0·94 m (3 ft 1 in)

Weights

Displacement	46 tons
Payload	23 tons (including fuel)

Structure

Main components: Sensor system, data processing system, command/communication system, power system

Principal structure: Thick discus hull with superstructure, both of welded steel construction

Production

One built (1973).

Operational aspects

Towing speed	To 10 knots
Wind limits	125 to 150 knots

Wave limits: Breakers 9·15 to 13·5 m (30 to 45 ft) high, waves (not breaking) of any height

Mooring arrangements: Single-point deep water or shallow water moorings. Synthetic line, taught or slack

Current limits To 9 knots

Measurement capability

Meteorological, oceanographic, and pollution monitoring sensors.

One shore data collection station or interface to user facilities can command/control/record buoy operations. One station can function with multiple buoys in a network over long ranges 3,200 to 4,000 km (2,000 to 2,500 miles).

Data system: Small digital computer or hard-wired controller

Capacity: Up to 200 sensors
Sampling frequency: Variable, can be in real-time
Presentation: Digital formatting aboard buoy
Transmission: Single-sideband PSK or FSK on narrowband HF channels allocated by WARC, range to 4,000 km (2,500 miles)

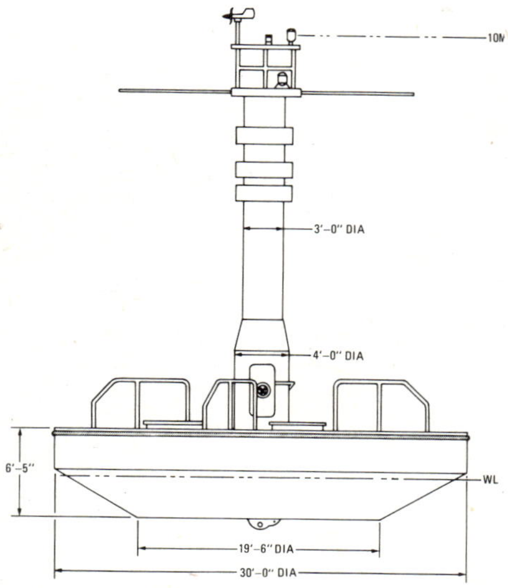

Elevation diagram of a Medium Ocean Data

Handling requirements
Lifting points: Four
Towing points: Four on deck, one on chine
Launching methods: Crane lift into water, or over land by mobile transporter

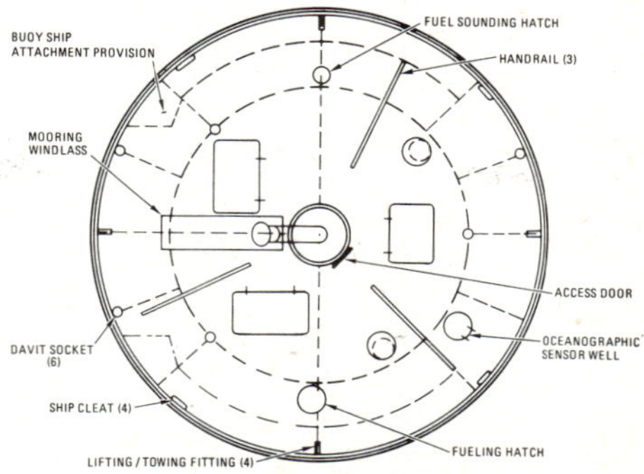

General hull layout of ODS/40

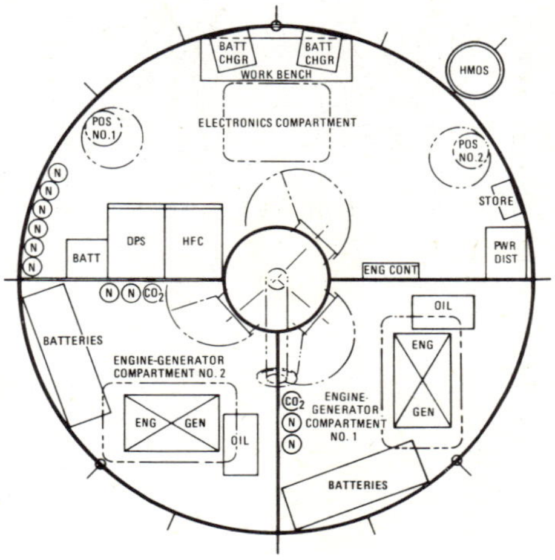

Equipment location on MODS/30

General Dynamics buoy, MODS/30 on station

Power installation
Power source: Two diesel engine-generators
Duration of unattended operation: 1 year
Maintenance: 1 year. Overhaul: 3 years

Positioning
Can be moored anywhere in ice-free waters

Applications
Weather monitoring networks, storm warning, scientific data accumulation, pollution monitoring

ODS/40
An ocean data station (ODS) for environmental data acquisition and telemetry.
The ODS/40 is the largest meteorological and data recording buoy produced by General Dynamics, having a diameter of 12·2 m (40 ft) and a loaded displacement of 100 tons.
The steel hull of the ODS/40 is classed 100A1 by Lloyds Registry of Shipping. It is constructed to high-quality ship-building standards, and incorporates ship-type fittings. Rated towing speed is 10 knots. Tows as long as 3,000 miles are routine, including transits of the Panama Canal.
Like the other Ocean Data Stations in its family, ODS/40 can be deployed in any depth of water. Single-point deep-water moors are customary. Moorings vary from taut (synthetic line, nominally shorter than water depth) to moderately slack, depending upon environmental conditions.
ODS/40 can mount quantities of meteorological sensors at two levels: 5 and 10 m (16·4 and 32·8 ft). Mounting provisions in and under the hull, and on the mooring assembly, can accommodate a variety of oceanographic sensors of several types. Redundant sensors for reliability and data verification are typical.

General Dynamics ODS/40 buoy

An ODS/40 buoy on one of General Dynamics' transporters

Basic dimensions

Diameter, overall	12 m (40 ft)
Height, overall	11 m (36 ft)
Draught	1·12 m (3 ft 9 in)

Weights

Displacement	100 tons
Payload	50 tons

Structure
Main components: Sensor system, data processing system, command/communication system, power system
Principal structure: thick discus hull with superstructure, both of welded steel construction

Production
Eight have been deployed in the Atlantic, Pacific Oceans, and in the Gulf of Mexico. Two additional buoys are under construction (July 1973)

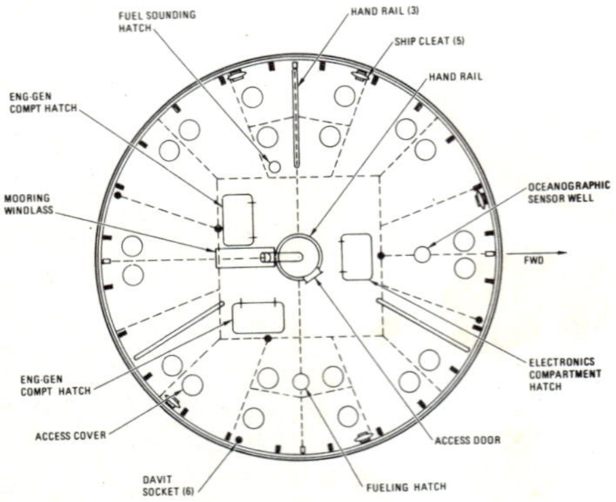

General hull layout of ODS/40

Operational aspects

Towing speed To 10 knots
Wind limits 150 to 175 knots
Wave limits: Breakers 12 to 18 m (40 to 60 ft) high, waves (not breaking) of any height
Mooring arrangements: Single-point deep water or shallow water moors. Synthetic line, taut or slack
Current limits: To 10 knots

Measurement capability

Meteorological, oceanographic, and pollution monitoring sensors.

One shore data collection station or interface to user facilities can command/control/record buoy operations. One station can function with multiple buoys in a network over very long ranges 4,800 to 6,400 km (3,000 to 4,000 miles)

Data system: Small digital computer or hard-wired controller
Capacity: Up to 200 sensors
Sampling frequency: Variable, can be in real time
Presentation: Digital formatting aboard buoy
Transmission: Single-sideband PSK or FSK on narrowband HF channels allocated by WARC, range to 5,650 km (3,500 miles)

Handling requirements

Lifting points: Four
Towing points: Four on deck, one on chine
Launching methods: Crane lift into water, or over land by mobile transporter

Power installation

Power source: Two diesel engine-generators, 120 V, 60 Hz, up to 7 kW available
Duration of unattended operation: 1 year
Maintenance: 1 year. Overhaul: 3 years

Positioning

Can be moored anywhere in ice-free waters but can withstand substantial accumulations including up to 12·5 cm (5 in) on all superstructure surfaces.

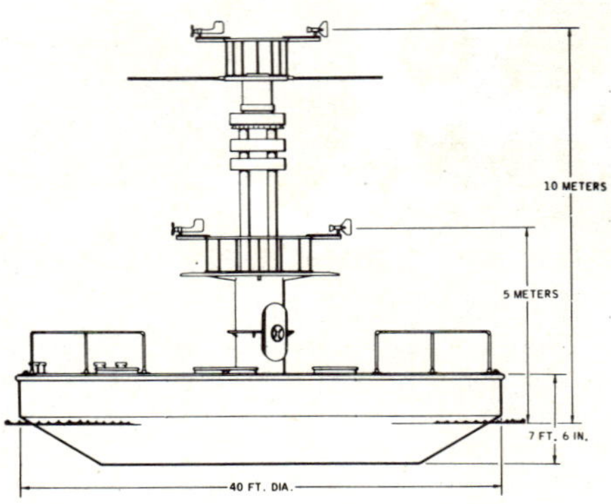

Ocean Data Station (ODS/40) buoy

A view of an ODS/40 and a SODS/16 at the General Dynamics Marine Test Centre in San Diego Bay

General Dynamics' Large Ocean Data System (ODS/40)

Applications

Weather monitoring networks, storm warning, scientific data accumulation, pollution monitoring

SODS/12

A small ocean data station (SODS) for pollution monitoring and environmental data acquisition

Basic dimensions

Diameter, overall	3·65 m (12 ft)
Height, overall	3 m (10 ft)
Draught	0·355 m (1 ft 2 in)

Weights

Displacement	2,350 kg (5,250 lb)
Payload	1,135 kg (2,500 lb)

Structure

Main components: Sensor system, data processing system, command/communication system, power system

Principal structure: Thick discus hull with mast. Welded steel construction

Production

Six built (1973)

Operational aspects

Towing speed:	6 knots
Wind limits:	80 to 100 knots
Wave limits:	Breakers 3·65 to 5·48 m (12 to 18 ft) high

Mooring arrangements: Single point or three point mooring, deep or shallow water, synthetic lines

Current limits

To 6 knots

Measurement capability

Pollution monitoring, meteorological, and oceanographic sensors

One shore station or interface to user facilities can command/control/record buoy operations. One station can function with multiple buoys in a network over medium ranges 1,200 to 1,600 km (750 to 1,000 miles).

Data system: Hardwired controller

Capacity: Up to 80 sensors

Sampling frequency: Variable, can be in real time

Presentation: Digital formatting aboard buoy

Transmission: Single-sideband PSK or FSK on narrowband HF channels allocated by WARC, range to 1,609 km (1,000 miles)

Handling requirements

Lifting points: Three

Towing points: Three on deck three on chine

Launching methods: Crane lift into water, launch ramp from dockside or ship, over land by mobile transporter

Power installation

Power source: All battery system capable of 1 year operation

Duration of unattended operation: 1 year

Maintenance: 1 year, Overhaul: 3 years

Positioning

Lakes, estuaries, and ice-free coastal waters

Applications

Pollution monitoring, scientific data accumulation, weather monitoring

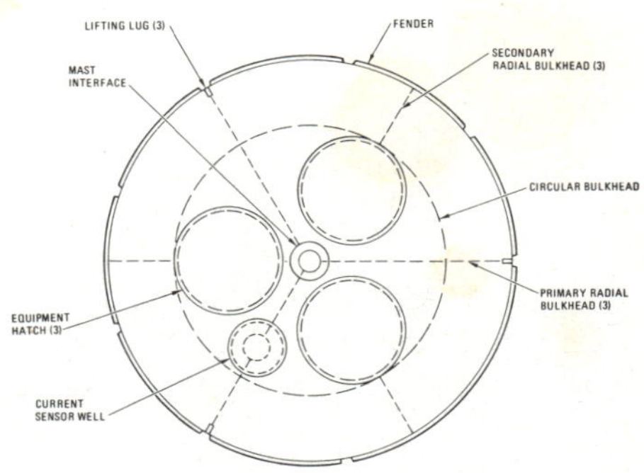

General hull arrangement of SODS/12

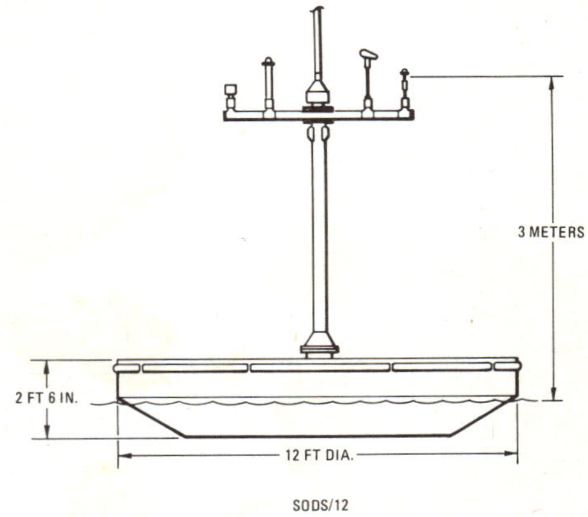

Elevation diagram of SODS/12

General Dynamics SODS/12

SODS/16

A small ocean data station (SODS) for environmental data acquisition and telemetry

Basic dimensions

Diameter, overall:	4·87 m (16 ft)
Height, overall:	5 m (16·4 ft)
Draught:	4·55 m (1·5 ft)

Weights

Displacement:	5,700 kg (12,600 lb)
Payload:	1,810 kg (4,000 lb)

Structure

Main components: Sensor system, data processing system, command/communication system, power system

Materials or principal structure: Thick discus hull with mast. Welded steel construction

Production

Thirty built.

Operational aspects

Towing speed:	8 knots
Wind limits:	90 to 110 knots
Wave limits:	Breakers 4·87 to 7·3 m (16 to 24 ft) high

Mooring arrangements: Single point or three point, deep or shallow water, synthetic lines

A view of two SODS/16s on tow in the Pacific Ocean

Current limits

To 7 knots.

Measurement capability

Meteorological, oceanographic, and pollution monitoring sensors

One shore data collection station or interface to user facilities can command/control/record buoy operations. One station can function with multiple buoys in a network over medium ranges 1,600 to 2,400 km (1,000 to 1,500 miles)

Data system: Hardwired controller
Capacity: Up to 100 sensors
Sampling frequency: Variable, can be in real time
Presentation: Digital formatting aboard buoy
Transmission: Single-sideband PSK or FSK on narrowband HF channels allocated by WARC, range to 2,430 km (1,500 miles)

Handling requirements
Lifting points: Three
Towing points: Three on deck, three on chine
Launching methods: Crane lift into water, launch ramp from dockside or ship, over land by mobile transporter

Power installation
Power source: All battery system capable of 1 year operation

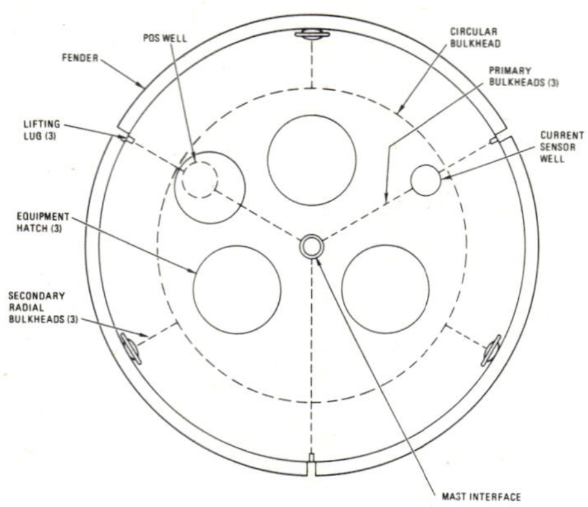

Plan diagram of SODS/16

Duration of unattended operation: 1 year
Maintenance: 1 year. Overhaul: 3 years

Positioning
Lakes, estuaries, and ice-free coastal waters

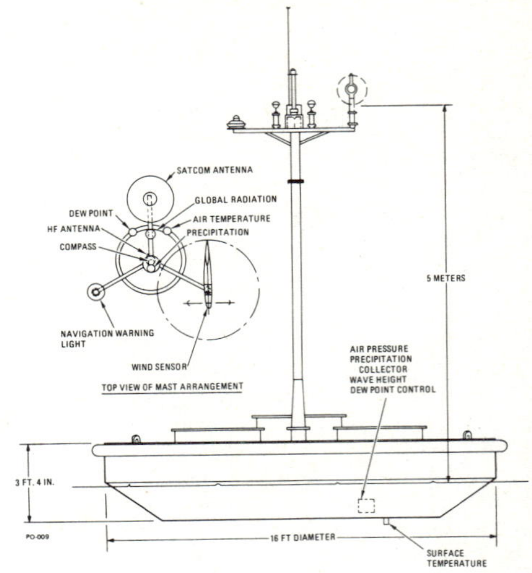

Elevation diagram of SODS/16 buoy

Applications
Weather monitoring networks, scientific data accumulation, pollution monitoring

GENERAL ELECTRIC RE-ENTRY AND ENVIRONMENTAL SYSTEMS DIVISION

NOAA DRIFTING LIMITED CAPABILITY BUOY (DLCB)

The Ocean Systems Programs Department of General Electric was selected by the National Data Buoy Center of the National Oceanic and Atmospheric Administration, Department of Commerce, in the summer of 1971 to design both a drifting and moored meteorological oceanographical environmental monitoring buoy system. This system is commonly referred to as the Limited Capability Buoy (LCB) System.

The communications, data acquisition, and buoy experience developed by General Electric over the last decade were directly applicable to design, build and demonstrate this buoy system. In order to achieve the goal of a cost effective operational system, the design was so conceived as to provide a maximum number of common subsystems between the drifting and moored systems. These common subsystems are contained in a surface float which is interchangeable between the Drifting LCB and Moored LCB Systems.

The attachment of a mooring line data line and a mooring system to the drifting buoy will transform it into a moored limited capability buoy. The mooring system consists of a subsurface float attached to the anchor by a taut mooring line and to the surface float by the slack data mooring line. This two stage buoy system features automatic deployment and eliminates the critical problem of initial mooring system damage and failure which is incident to the deployment of single stage moorings, especially in deep water.

The initial buoys are equipped to make the measurements shown in the accompanying table. Reserve power and buoyancy in the surface float provide the capability to add other sensors for specific applications.

In the cutaway drawing, the LCB subsystems are shown in a schematic configuration. The principal subsystems are the HF communication system, the meteorological sensor system, and the power system. Space is provided in the buoy for the installation of either the Navy Navigation Satellite Position Fixing System or the OMEGA Position Fixing System. When operated in a drifting buoy, these systems will furnish position information about the buoy so that the buoy's position may be plotted to indicate the movement of water masses.

General Electric's Drifting Limited Capability Buoy (DLCB)

Basic dimensions
Diameter 1·16 m (3 ft 10 in)
Height (overall) 3·05 m (10 ft)
Draught 1·22 m (4 ft)

Weight
Displacement 727 kg (1,600 lb)
Payload 182 kg (400 lb)

Structure
Main components Spherical and Cylindrical
Material Aluminum

Production
Number built (1973) 2
Wind limit 100 kts
Wave limit 45 ft

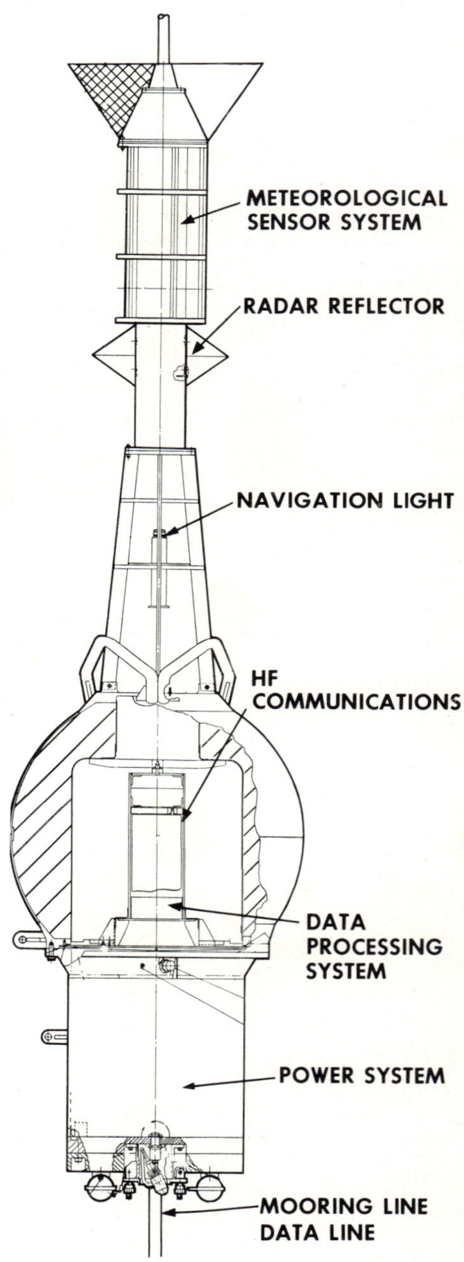

General arrangement of a DLCB

MEASUREMENT CAPABILITIES OF LCB

Quantity	*Range*	*Allowable Error*
Air Pressure	900 to 1,000 MB	1·0%
Air Temperature	—10° to 40°C	0·5°C
Wind Speed—North	1 to 50 m/sec	1 m/sec or 5%
Wind Direction	0 to 360°	5°
Water Temperature (2 to 200 meters depth, (6·5 to 650 ft))	—2 to 40°C	0·1°C
Water Pressure (at 200 meters, (650 ft))	15 to 23 Kg/cm² 217 to 330 lb/in²	0·5 Kg/cm² 7·13 lb/in²

SEA ROBIN

An experimental buoy which has been used as a development platform to evaluate systems designs, equipment and procedures required for future ocean based data acquisition systems.

Basic dimensions

Diameter	1·22 m (4 ft)
Height (overall)	3·95 m (13 ft)
Draught	3·35 m (11 ft)

Weights

Displacement	1,363 kg (3,000 lb)
Payload	225 kg (500 lb)

Structure

Main components	Conic and cylindrical sections
Material	Aluminum

Production

No. built (1973)	1

Operational aspects

Towing speed	2 kts
Wind limits	64 kts
Wave limits	14 m (45 ft)
Mooring (deep and shallow)	Pre-cut line
Current limits	3 kts

Handling requirements

Lift points	3
Tow points	2
Launch method	Over-the-side

Power installation

Power source	Battery
Duration (unattended)	30 days

Positioning

Navigation light
Colour code
Watch circle 91 m (300 yds)

Applications

Environmental monitoring and communications relay

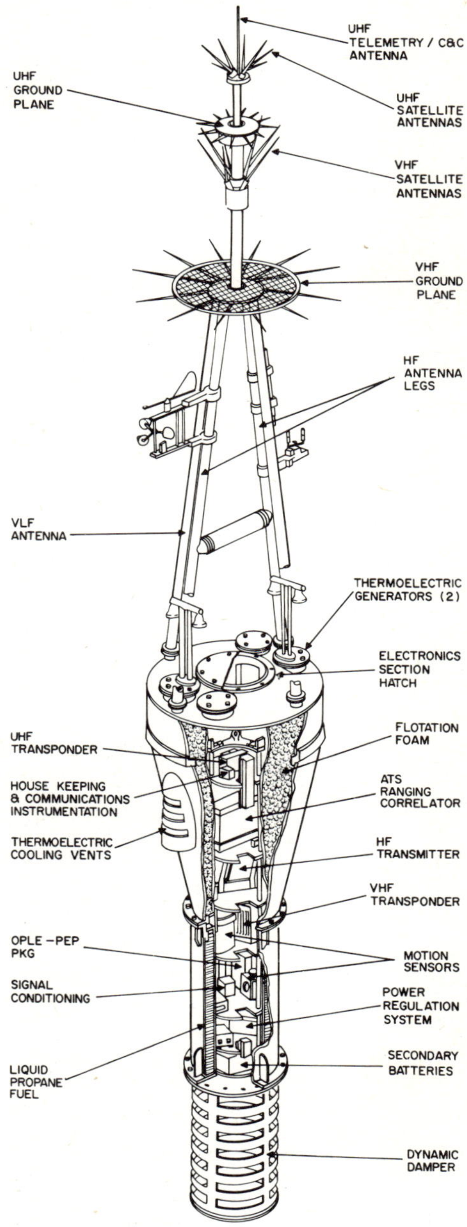

Equipment layout of Sea Robin buoy

LOCKHEED MISSILES AND SPACE COMPANY

ECOBUOY

The Ecobuoy is a small version of a meteorological and oceanographic data buoy designed and built for the National Oceanic and Atmospheric Administration. The buoy has been designed for ease of deployment and maintenance, and to withstand the most severe environmental conditions encountered in coastal, estuarine and inland waters.

Data processing

The Ecobuoy's data-processing system acquires data hourly from the sensor suite, and the data are instantaneously converted to digital form and transmitted to shore. Simple modifications permit data to be acquired and transmitted either more or less frequently than hourly. The use of low-power-consumption COSMOS solid-state logic elements enables a service life on-station of up to 6 months.

An alternative configuration of the Ecobuoy provides for recording data onboard the buoy, for later recovery, instead of transmitting the data periodically to a shore station.

Sensors

The sensors are off-the-shelf items, selected for the proven accuracy and ruggedness essential to extended operation in a marine environment. The parameters that the buoy monitors, in its basic configuration, are: surface water temperature, conductivity (salinity), dissolved oxygen, and pH. Sensors can be added to monitor other parameters, such as current, turbidity, wind speed, and air temperature, and provision can be made for collecting data at subsurface levels. The buoy's reserve buoyancy 773 kg (1,700 lb) under zero current conditions, provides high flexibility to meet additional needs or unusual environments.

Communications

The VHF communications system is the buoy-to-shore link for transmission of data. Data received on shore are recorded on tape for subsequent processing. An HF system can be provided for increased communication range.

Mooring

For shallow depths, the mooring system consists of a Danforth and a deadweight clump, attached to the buoy by a chain of sufficient length to give a mooring scope of 3:1. At greater depths, or where accurate positioning of the buoy is necessary, a taut nylon line is used instead of the chain. Mooring systems can be tailor-made to fit special requirements such as large tidal variations or high currents.

Handling

The size and weight of the Ecobuoy make handling very economical. It can be transported over land on a boat trailer and launched from a small boat ramp; it can be launched from a dock by means of a simple pipe A-frame and chainfall on the rear of a pickup truck; and it can be towed to and from a monitoring station by a small power boat.

Safety

The Ecobuoy is fitted with an identification/obstruction light on the mast to provide night time warning. During daylight, its red and yellow striping (international buoy colours) makes it readily visible.

Leading particulars

Overall length	2·24 m (7ft 4 in)
Diameter	0·91 m (3 ft)
Draft	1·3 m (4 ft 3 in)
Mast height above waterline	1·45 m (4 ft 9 in)
Displacement	590 kg (1,300 lb)

Sensor capability

Sensor	*Range*	*Accuracy*
Water temperature	0 to 40°C	±0·5°C
Conductivity	0 to 1,000 micromhos/cm	±2% full scale
Dissolved oxygen	0 to 20 ppm	±0·2 ppm
pH	0 to 12 pH	±0·1 pH

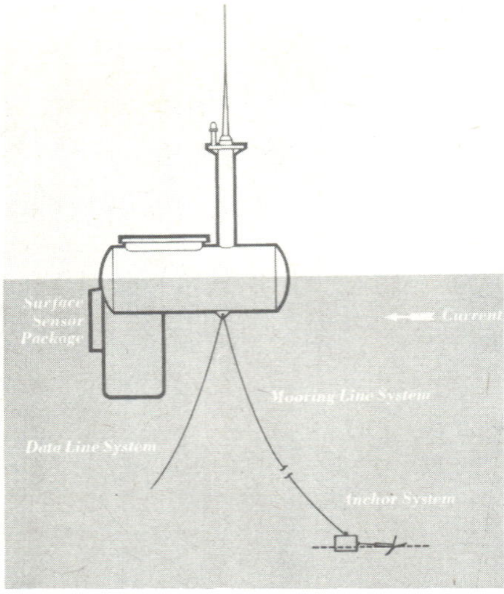

Mooring arrangement of the Lockheed Ecobuoy

LOCKHEED MISSILES AND SPACE COMPANY

HIGH CAPABILITY BUOY

This buoy was developed under a $1·4 million contract to the National Oceanic and Atmospheric Administration (NOAA). The buoy has a boat-shaped hull and a deep keel extending 6·1 m (20 ft) beneath the surface. This form of hull, together with horizontal stabilizers at the top rear of the keel, is designed to greatly reduce drag compared to conventional buoy hulls, increase chances of survival in severe storms and also ease handling for maintenance crews. The design permits reduced size to achieve an acceptably stable ocean platform.

Length	8·8 m (29 ft)
Width	3·05 m (10 ft)
Displacement	21 tons

The buoy is equipped with sensors that measure water temperatures, pressures, current speed and salinity at various depths and collects meteorological data from instruments mounted on its mast, automatically transmitting such data to shore stations.

The Lockheed High Capability buoy

MOORED LIMITED CAPABILITY BUOY (MLCB)

Lockheed's Moored Limited Capability Buoy (MLCB) was conceived as a highly reliable, yet relatively low cost, device for gathering on-station meteorological and oceanographic data. The MLCB will survive a 4-knot current, 100-knot wind, and 14 m (45-ft) waves. Onboard zinc-air batteries provide a total installed energy of 36 kilowatt-hours, sufficient to power the buoy for up to 12 months. The buoy was designed and built for The National Oceanic and Atmospheric Administration. By October 1973, three of these buoys had been built and more were planned.

Onboard data processing

The MLCB's data processing system acquires data hourly from meteorological and oceanographic sensors. During the last 10 minutes of each hour, 128 samples are collected. These data are processed onboard the buoy, averaged, and stored. Every six hours, a synoptic data report is transmitted to shore. Various alternative modes can be dictated by command from shore—for example, acquisition and processing of real-time data, transmission of the last hour of stored data, or transmission of the last six hours of stored data.

Sensors

The sensors are off-the-shelf items, selected for the proven accuracy and ruggedness essential to extended operation in an ocean environment. The parameters that the buoy monitors, in its present configuration, are wind speed and direction, air temperature and pressure water temperature to depths of 200 m (656 ft) and water pressure at 200 m (656 ft). A planned modification will enable sensing of current and sensors can be provided to monitor such parameters as wave height, conductivity, and dew point. Pollution indicators—for example, pH and dissolved oxygen—can also be added.

Communications

The high-frequency communications system comprises a buoy-to-shore link for transmission of data to a shore station and a shore-to-buoy link for interrogation of the buoy sensors, if desired. The latter also permits alternative operating modes to be put into effect, such as sequencing the receiver to a 25-percent duty cycle in order to extend buoy on-station time. Capability to convert to a UHF-satellite radio link is also provided.

Self-mooring

The entire mooring system is self-deploying—to depths as great as 7,300 m (24,000 ft) with a taut moor at any depth. The support ship merely lowers the buoy and mooring canister overboard and continues on its course.

Three subsystems comprise the MLCB's mooring system: anchor, mooring line, and data line.

Two anchors stabilize the buoy. A clump anchor controls the vertical pull of the buoy. A Danforth anchor resists horizontal pull.

The mooring line is a 1·59 cm ($\frac{5}{8}$-in) diameter nylon line with a 1·6 mm ($\frac{1}{16}$-in) thick polyurethane sheath on the upper 1,500 m (4,920 ft) to protect against fish-bite.

The 200-m (656 ft) data line is separate from the mooring line so that retrieving it will not influence the mooring. The nucleus of the data line is torque-free stranded steel. Imbedded in the line are 6 thermistors—at 2, 10, 20, 50, 100, and 200 m (6·5, 32·8, 65, 164, 328, and 656 ft) and, at the bottom of the line, a water-pressure transducer. Surrounding the stranded steel are 16 copper conductors. The entire assemblage is protected by a polyurethane sheath.

Navigation aid

An identification and obstruction light, visible for five nautical miles, is mounted on the mast. At night, the light flashes for 3 seconds, then turns off for 17 seconds. A photosensitive cell turns the light off in response to adequate daylight.

Leading particulars

Overall length	3·43 m (11 ft 3 in)
Diameter	1·37 m (4 ft 6 in)
Draught	2·26 m (7 ft 5 in)
Mast height above waterline	3·53 m (11 ft 7 in)
Displacement	2,500 kg (5,500 lb)

Purchase cost (1973) $ U.S. 100,000

The Lockheed Moored Limited Capability buoy designed and built for the NOAA

Sensor capabilities

Sensor	*Range*	*Accuracy*
Wind Speed and Direction	0 to 80 m/sec	5% or 1 m/sec, whichever is greater
Air Temperature	−10 to +40°C	0·5°C
Air Pressure	900 to 1,100 mb	1 mb
Water Temperature	−2° to +40°C	0·1°C
Water Pressure	15 to 23 kg/cm² (220 to 330 lb/in²)	0·5 kg/cm² (7·15 lb/in²)

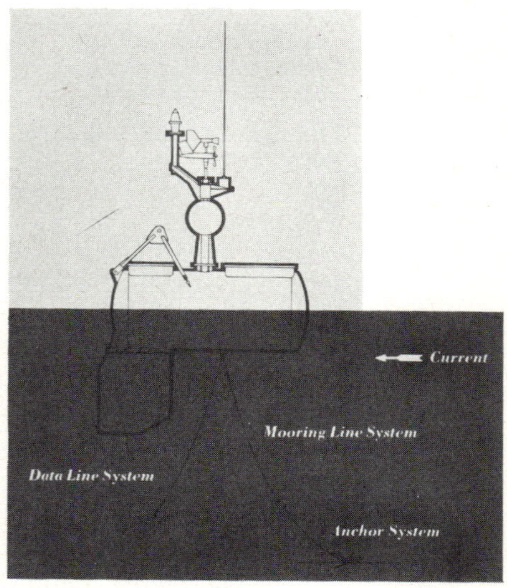

Mooring arrangement of the Lockheed Moored Limited Capability Buoy

SCRIPPS INSTITUTION OF OCEANOGRAPHY

BUMBLEBEE BUOY

This buoy, of the moored ocean instrument platform surface type is designed for long-term monitoring of air/sea interaction in the North Pacific storm belt

Basic dimensions

Length	6·1 m (20 ft)
Beam	2·4 m (8 ft)
Height	1·98 m (6 ft 6 in) above water line, exclusive of antenna arrays
Draught	0·456 m (1 ft 6 in) exclusive of underwater probes

Weight

Displacement	840 kg (1,850 lb) empty hull
Payload	1,750 kg (4,000 lb)

Scripps Institute's 6·1 m (20 ft) *Bumblebee* buoy, showing radio antenna and the wind-driven

Structure

Horizontal floating cylinder with V slot cut out along the floating area to make a self-righting catamaran hull

Plywood cylinder 0·95 cm ($\frac{3}{8}$ in) thick with internal bulkheads and stringers. Foam molded nose over bow structure, and foam in hulls provides absolute floatation. Weight of foam in hulls provides the weight bias to make cylinder self-righting. Plywood and urethane foam are all covered with fibreglass and colour impregnated resin for protection and visibility

Production

16 have been built (1973 July)

Operational aspects

10 knots towing limit on long tow rope, unlimited tow speed on short, bow lifting tow rope

No wind limit

No wave limit

Chain moor for shallow waters, taut nylon rope for 182 m (100 fathoms) or more

Measurements

Primarily: wind speed and direction, air temperature, sunlight and sea temperatures at 10 points to 300 m (1,000 ft) depth

Specialty: since buoy is non-metallic, radar, radio, magnetism problems have to be measured from special instrumentation

Data system

Capacity—recording up to 6 months normal Longer if needed

Sampling frequency—once an hour

Dials of all instruments are photographed

Transmission—used only when wanted

Handling

Lifting bar on top integral with buoy for crane launch

Bow towing eye

Launching done by crane, by davit, by pushing off the stern, or by sliding down a ramp

Power

Batteries with windmill generator recharge system

6 months to 1 year duration of unattended operation

6 months to 1 year maintenance and overhaul requirements

Positioning

Anchored buoy, sight fixed by navigation satellite

Availability

Designed as a do-it-yourself system—hull available from local boat builder

Costs

Buoy hull $4,300 (US) (1972 cost), less in quantity. Instrument suite $25,000 (US), depends on requirements

Applications

Designed as a low-cost, home-built system to operate in any ocean environment free of ice.

With windmill generator and battery, it can telemeter information if that is desired. Profiling system (BT-STD, etc.) provides much greater detail of water column than thermister chain, but is limited by available power. Profiling system is ideal for shallow water areas where tidal depths change greatly.

Additional information

Buoys have lasted as long as two years on station in the storm belt of the North Pacific Ocean. Exposed instruments are waterproof due to gale-driven spray, rain and threat of capsize. Buoys are self-righting and designed to be boarded at sea under rough conditions if necessary and are basically unsinkable. All systems are above the water line so that instruments or batteries can never be flooded by leakage from damage or human intervention. The plywood hull will crush easily if rammed by a ship thus limiting its navigational hazard if adrift or dead.

The Scripps Institute North Pacific study 4·87 m (16 ft) Bumblebee buoy

UNIVERSITY OF WASHINGTON APPLIED PHYSICS LABORATORY

ARCTIC DATA BUOY

An environmental data buoy for polar ocean area applications, eight having been built by early 1974.

Hull body length	7 m (22·9 ft)
Mast height	3 m (9·8 ft)
Hull diameter	20 cm (7·9 in)
Mast diameter	8 cm (3·1 in)
Draught	6 m (19·7 ft)
Displacement	170 kg (374 lb)
Payload	120 kg (316·8 lb)

Operational aspects

Wind limits	70 mph
Wave limits	0·9 m (3 ft)
Current limits	NA-Drifting buoy

Measurement capability:
- Atmospheric pressure;
- Atmospheric temperature;
- Water temperature;
- Water velocity;
- Wind velocity

Data system	Polar orbiting satellite communications
Capacity	Sample of 28 quantities per satellite orbit
Sampling frequency	Satellite orbital period is 110 minutes
Presentation	Print out of raw data telemetered to satellite
Transmission	400 MHz, 6 watts

Handling requirements

Launching methods: Installed in a 22 cm (8·7 in) diameter hole cut in ice through to ocean

Power installation

Power source: 45 kg (99 lb) of low temperature mercury cells

Duration of unattended operation: 2 years

Maintenance and overhaul requirements: none (expendable)

Positioning

Accuracy ±2 km by satellite

University of Washington Arctic Data buoy

Availability

Developmental

Costs

Depends upon sensor suite, etc. About $5,000 US

MAJOR DIVING FACILITIES

FRANCE

COMEX PRO

DIVING BELLS

Three levels of capability are provided for as shown in the following table.

Design operating depth	100 m (330 ft)	200 m (660 ft)	400 m (1,320 ft)
Gas storage capability	90 m³ (3,150 ft³)	120 m³ (4,200 ft³)	120 m³ (4,200 ft³)
Cylinders, 9·96 m³ (350 ft³) each	Seven bottom mixture One decompression mixture, one O_2	Ten bottom mixture One decompression mixture, one O_2	Ten bottom mixture One decompression mixture, one O_2
Gas manifold	Yes	Yes	Yes
Decompression mixture and O_2 manifold	Yes	Yes	Yes
External manifold for cylinder filling	Yes	Yes	Yes
Bell weight	3,000 kg (6,600 lb)	3,000 kg (6,600 lb)	3,500 kg (7,700 lb)

Comex 1,000 ft Saturation Diving Bell on board the ODECO rig *Ocean Rover*

200 m (660 ft) DIVING BELL

Accommodation	One bell man, two divers
Weight	3,000 kg (6,600 lb)
Internal diameter	1·47 m (4 ft 10 in)
Internal volume	3·87 m³ (136 ft³)
Tunnel diameter	0·61 m (2 ft)
Working pressure	20 kg/cm² (286 lb/in²)
Maximum external pressure	12 kg/cm² (172 lb/in²)
Buoyancy	250 kg (550 lb)
Outside diameter	2·18 m (7 ft 2 in)
Height, legs folded	2·16 m (7 ft 1 in)
Height, legs extended	3·05 m (10 ft)

Equipment

One hub for clamping on a SDC
Twelve portholes 14 cm (5·5 in) diameter
Numerous feedthroughs
Twelve cylinders 9·96 m³ (350 ft³) each for mixed gas and oxygen storage
Gas manifold
Internal and external pressures monitoring panel
External manifold for cylinders refill
Internal and external lighting
Telephone and intercommunication system
Carbon dioxide scrubber
Two lests 500 kg (1,100 lb each) droppable from the inside
Two circular bumpers
Three collapsible legs
Electromechanical cable (40 tons capability)
Power supply 440 volts 60 cycles 10 KVA or 380 volts 50 cycles 8 KVA

Options

Hydraulic outside door closing system
Horizontal propulsion units
Heating system
Complete electrical assembly for powering heated suits and gas heaters

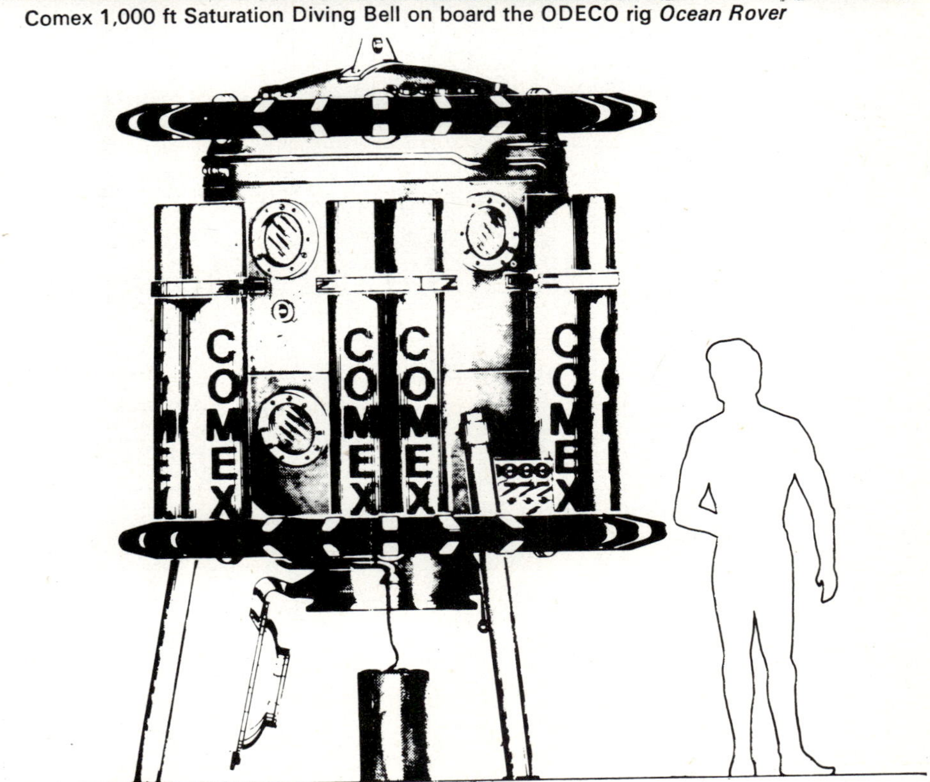

Comex 660 ft Diving Bell

RDO 1500 DECK DECOMPRESSION CHAMBER

General features

RDO 1500 is the standard chamber for deep diving; it enables the execution of decompression programmes under optimum comfort and safety conditions.

Technical data

External diameter	1·53 m (6 ft)
Overall length	4·70 m (15 ft 5 in)
Overall width	2·40 m (8 ft)
Overall height	2·30 m (7 ft 8 in)
Chamber volume	4·2 m³ (148 ft³)
Lock volume	2·2 m³ (78 ft³)
Working pressure	12 bars (172 lb/in²)
Weight	7,000 kg (15,400 lb)

Equipment

One chamber, 1 lock
Removable skids
Four pressure locking doors
Guiding device for clamping the diving bell
Five portholes
One medical lock, diameter 200 mm (8 in) length 600 mm (2 ft)
Hydraulic balancing system of the upper door
Two banks
Two seats
Internal manifold for O_2 and breathing mixtures distribution
One safety valve
Monitoring panel for the control of gas storage, lock and chamber pressure
Protected by metallisation and epoxyline paint
Intercommunication
Flash proof lights
Powered by 24 volt, AC or DC

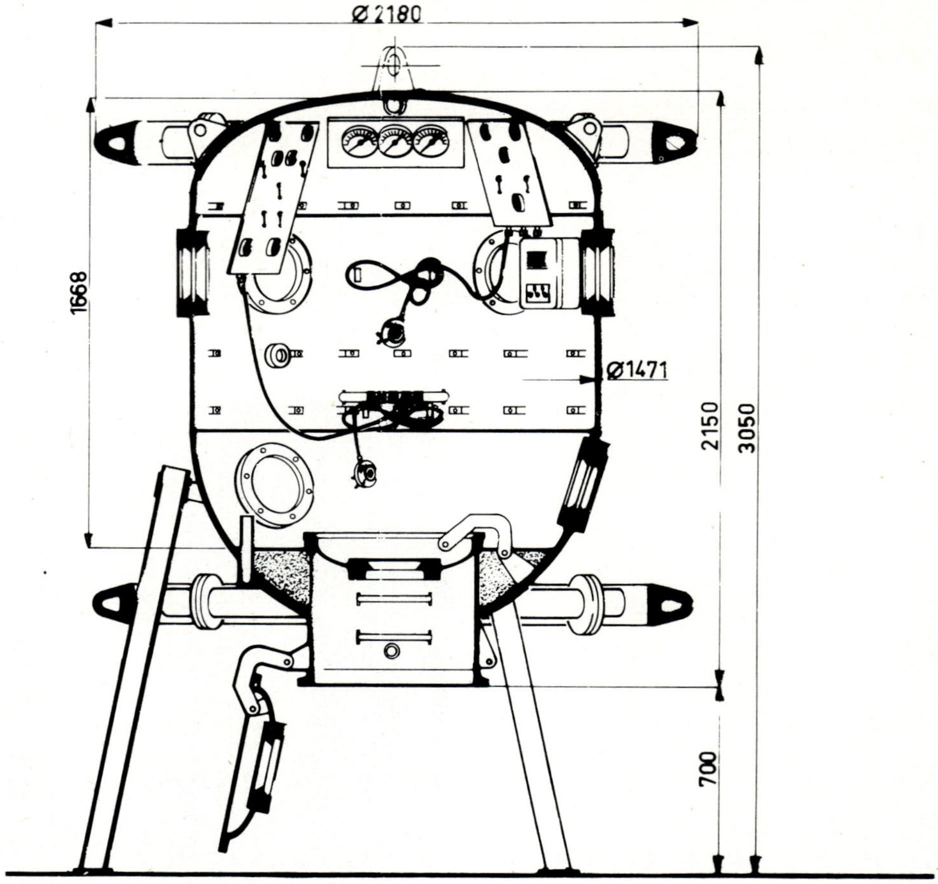

Comex 660 ft Diving Bell

Options

Closed circuit TV
Extra portholes
Ring adaptater for different chambers
Thermal insulation
Upper deck
CO_2 scrubber
Atmosphere regeneration system
Air conditioning system
Expiration regulator assembly
Working pressure 20 bars (285 lb/in²)

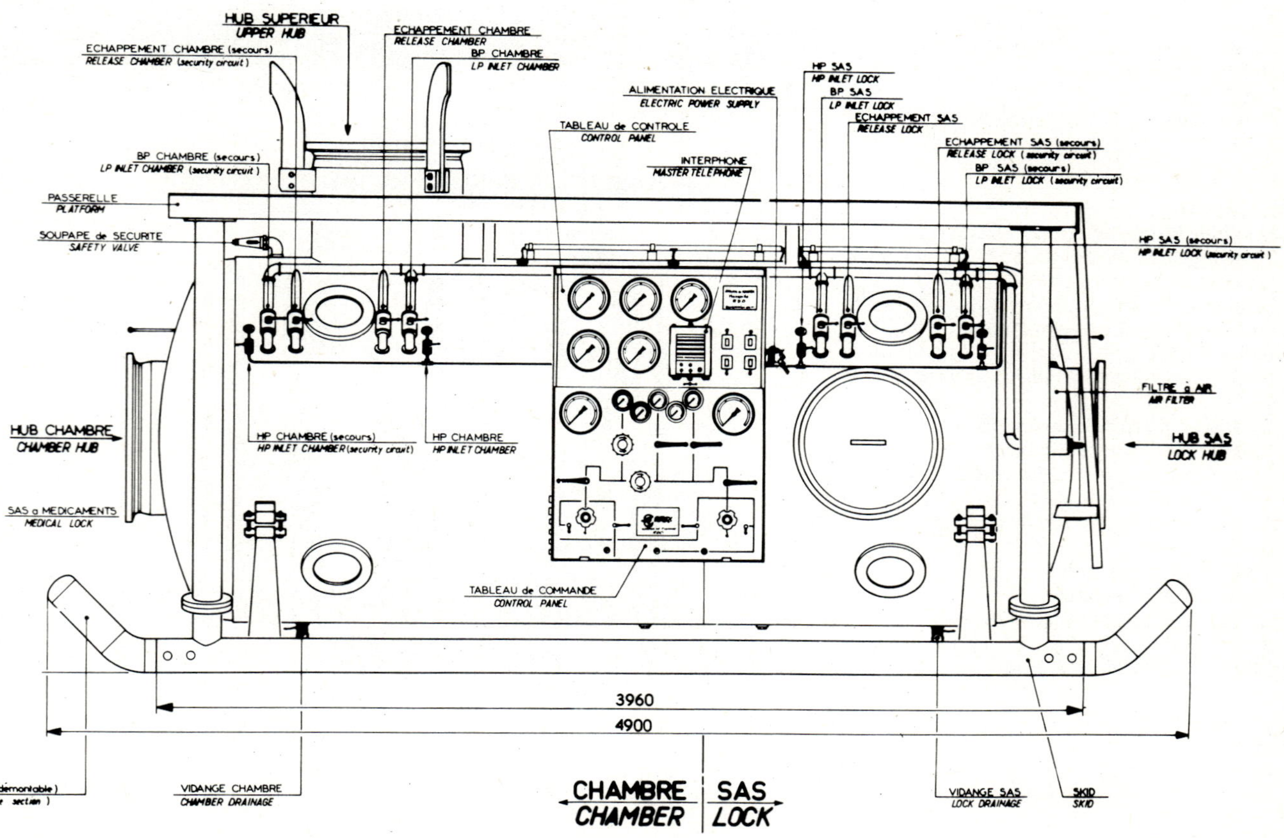

Comex Deck Decompression Chamber RDO 1500: (R) Recompression; (D) Decompression;

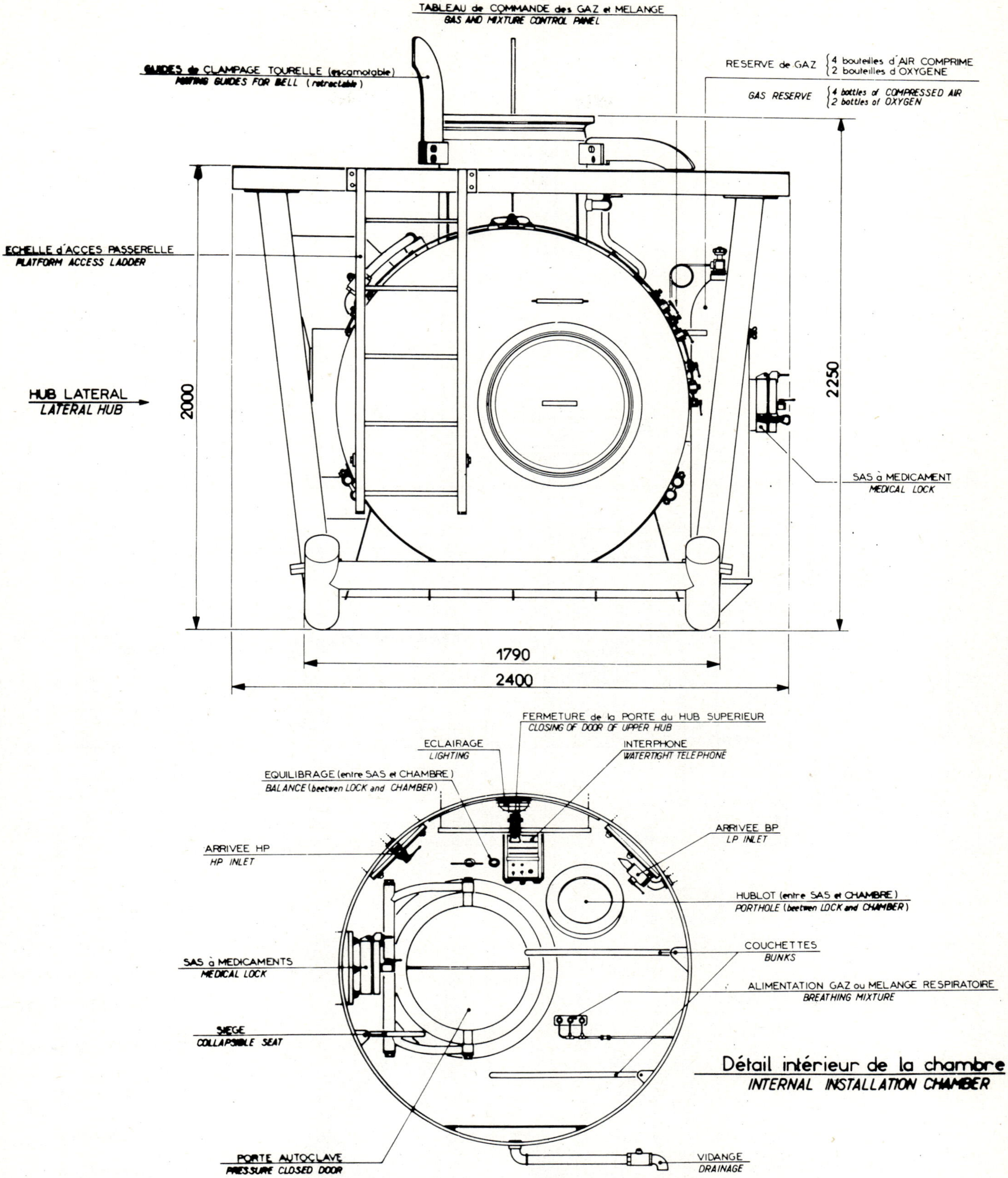

Comex Deck Decompression Chamber RDO 1500: (R) Recompression; (D) Decompression; (O) Oxygen therapy

RDO 1800 DECK DECOMPRESSION CHAMBER

General features

RDO 1800 is specially designed for use on deep diving worksites, when saturation diving is planned, because of its size, working pressure and sophisticated equipment. The designation RDO is derived from:

R: Recompression
D: Decompression
O: Oxygen therapy

Technical data

External diameter	1·80 m (6 ft)
Overall length	5 m (16 ft 5 in)
Overall width	2·40 m (6 ft 6 in)
Overall height	2·60 m (7 ft 6 in)
Chamber volume	5·60 m³ (197 ft³)
Lock volume	3·30 m³ (116 ft³)
Working pressure	20 bars (286 lb/in²)
Weight	9,000 kg (19,800 lb)

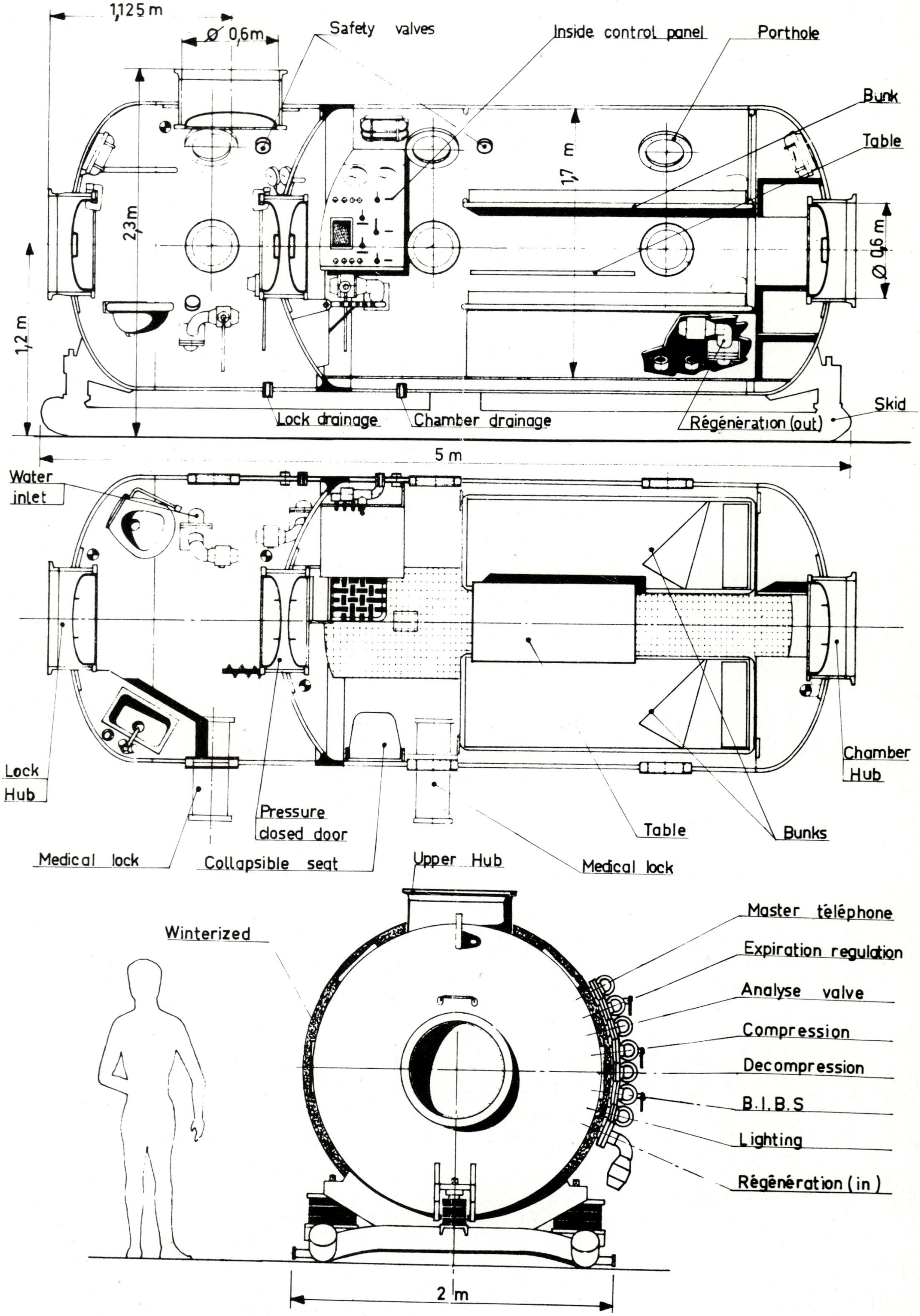

Comex Deck Decompression Chamber RDO 1800

Equipment

One chamber, one lock
Removable skids
Five pressure locking doors
Guiding device for clamping the diving bell
Twelve portholes 140 mm (5½ in) diameter
Two medical locks: diameter 200 mm (8 in), length 600 mm (2 ft)
Hydraulic balancing system of the upper door
Four banks
Two seats
Internal manifold for O_2 and breathing mixtures distribution
Two safety valves
Monitoring panel for the control of gas storage, lock and chamber pressures
Protected by metallisation and epoxyline paint
Intercommunication
Flash proof lights
Powered by 24 volts, AC or DC

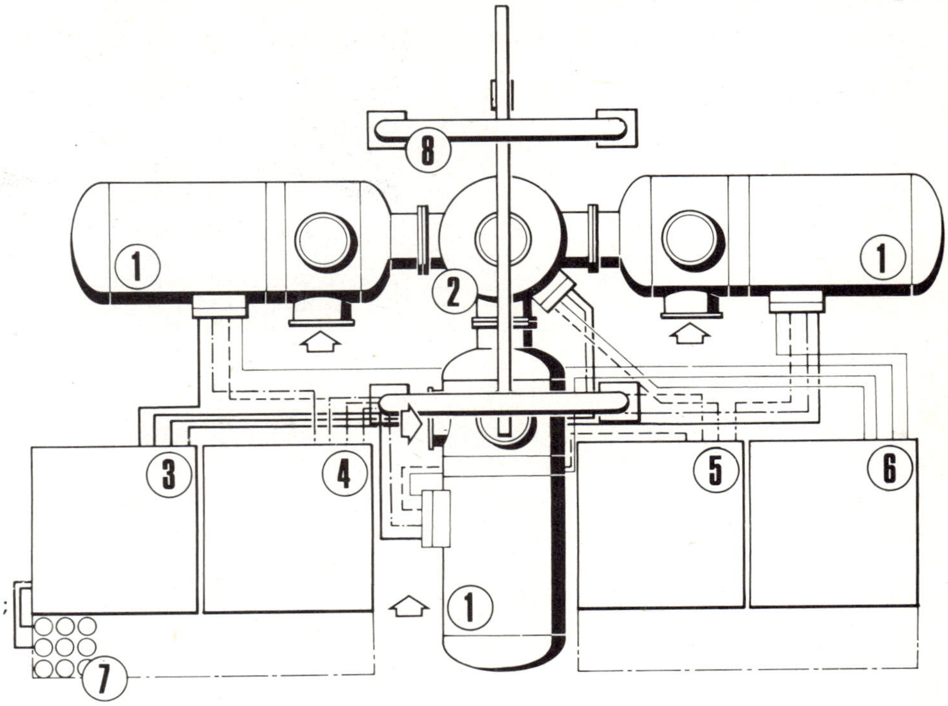

Installation of Saturation Systems: (1) RDO 1800; (2) lock; (3) gas mixtures; (4) regeneration; (5) control; (6) analysis; (7) reserve gas; (8) control consol

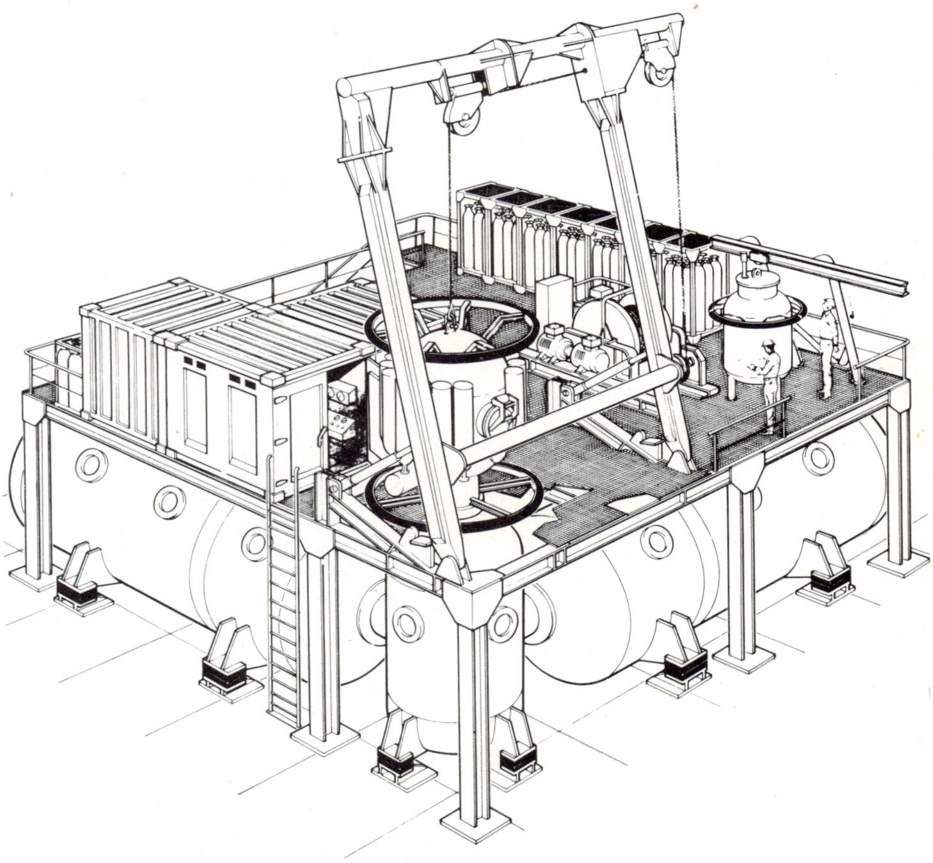

Saturation System

Comex 1,000 ft Saturation System on board the rig *Ocean Rover* in the North Sea

Options

- Closed circuit TV
- Extra portholes
- Ring adaptater for different chambers
- Sanitary equipment in case of saturation
- Thermal insulation
- Upper deck
- CO_2 scrubber
- Atmosphere regeneration system
- Expiration regulator assembly

Comex Deep Diving System on board the rig *Sea Quest* in the North Sea

GERMANY

DEUTSCHE BABCOCK AND WILCOX

DEEP DIVING SYSTEM

The Babcock deep diving system serves for industrial underwater work, especially in connection with oil extraction, mineral extraction, pipeline construction, other civil engineering work, geological explorations and salvage operations.

Without losing its efficiency the system may be applied to difficult missions in the field of ocean engineering and to simple work as well.

The modular construction principle allows the application of individual systems each of which can be used appropriately as a single unit or in combination, when difficult and complex problems are to be solved and when the conventional means available have proved incapable of performing the task.

The system components

Control Chamber (CC)

The control chamber is a diving sphere having a lower and lateral door with a coupling mechanism; it may be coupled at the bottom with a personnel transfer chamber (PTC) and laterally with a deck decompression chamber (DDC). In the whole system the CC serves as the pilot cockpit for the PTC, as lock chamber or as shuttling lift for observers and divers between the PTC installed on sea floor and the DDC located on deck. The CC may be applied both at zero pressure and at overpressure. Individual application as observation sphere or divers' chamber is also possible.

Personnel Transfer Chamber (PTC)

The PTC is an upright cylindrical chamber having an upper and a lateral door with a coupling mechanism; it may be coupled at the top with a CC and laterally with a DDC. In diving application the chamber will be flooded at one third for exit and entry. The divers use the lateral door. Advantages: safety against blocking by obstacles or contact with the ground; convenient entry and exit by swimming; possibility of rescuing a diver who has met with an accident. The CC and PTC may be used as a double chamber.

The PTC may be applied individually as a divers' chamber or also as an observation chamber at zero pressure.

Deck Decompression Chamber (DDC)

The DDC is a normal decompression chamber for living under pressure (also helium saturation), with lock chamber, air conditioning plants and closed gas system. The chamber is provided with a lateral coupling for a CC or a PTC and with a separate door for the lock chamber.

Crane system

The crane completely prevents swinging and oscillation of the load being heaved over board. The load is held by the crane from the moment of lifting from the deck to the moment of complete immersion into the water and

The Babcock and Wilcox Deep Diving System showing a model of the Deck Decompression Chamber (DDC), Personnel Transfer Chamber (PTC) and Control Chamber (CC)

The control chamber and personnel transfer chamber (model) prior to lowering into the water

back. This operation has so far limited the industrial application of deep diving. The crane may handle a CC and a PTC individually and in combination, but is also designed for any other loads to be handled in ocean engineering. It consists of two lateral masts each with three arms which can be swivelled by 90°, and are operated hydraulically and synchronised electronically (instead of a mechanical coupling). The upper cross-beam is equipped with a hydraulically rotating catch. Erection aids as well as bases with self-aligning bearings for coping with vibrations and deformations without producing forces are located on ship deck.

Guide rope system

The CC and the PTC are normally taken to the ground anchor by two lateral guide ropes which are free from twisting. In the water the ropes remain motionless. On deck they are led through oscillation and shock absorbers to constant-pullwinches. Only one short rope sling is in motion in order to make up for the vibrations.

Communication system

All stations of the system (CC, PTC, DDC, control desk, doctor, winch operator, command bridge, superintendent of work, etc.) are interconnected by a television cable. By a multiplex system any number of voice channels, TV signals, measured values, control signals, etc. may be transmitted into the cable. Furthermore the electric supply power is transmitted.

Each signal has another frequency; small transmitters/receivers modulate and demodulate the frequencies. Any radio sets (walkie-talkies, etc.), TV sets and cameras may be connected direct.

The system eliminates all cable problems which have been up to this point a nuisance; furthermore it may be extended at will. Furthermore, radio stations may be coupled to the network both as transmitters and receivers.

Life support system

By means of flexible tubes, all noted systems may be connected. The communication system allows ideal remote supervision and control (gas analyses, physiological control values, etc.). Extension is possible if desired.

Logistics

In addition to the system there are air plants, etc. The helium plant with cryogenic helium recovery (purification) may also produce oxygen.

CONTROL CHAMBER (CC)

Purpose

Diving chamber, especially as observation and pilot chamber as well as lock for persons, but also suitable for divers' missions.

Design

Sphere of approximately 1·8 m (5 ft 11 in) in diameter. Lateral gate with coupling for connection to DDC and as lock entry. Lower gate with coupling for connection with PTC and as divers' exit.

At top attachment for carrying rope. Annular flooding tank for autonomous manoeuvring under water, especially on guide rope system.

Laterally guides and brake for guide rope.

Numerous windows, at top, laterally and at bottom.

Equipment

Two bayonet doors with coupling
Ten windows, six searchlights
Flooding tank with valves
Hose connection
Two oxygen cylinders, 10 litres/200 kp/cm² (2,830 lb/in²) gauge
One oxygen analyser
Four compressed air cylinders, 20 litres/200 kp/cm² (2,830 lb/in²) gauge (possibly helium)
Pressure control and valves
Discharge ballast
Rope guide and brake
Benches to lie on
Sonic depth finder
Communication system
Warm water heating

The unit positioned for working

Further expansion

Horizontal propeller
Vertical propeller
Manipulator

Possibilities of operation

(*a*) *CC individually, without carrying cable and without propeller*

The CC may control its ascent and descent by flooding tanks and discharging ballast. This allows using it without carrying cable. It is possible to employ communication electric cables, etc., safety cables and hoses. (engineless submarine boat)

Use as observation chamber or for exit of divers

(*b*) *CC individually, without carrying cable, but with propellers*

With or without current supply cable etc.

Use: the same as for the submarine boat. Controllable motors with propellers are attached to the chamber.

(*c*) *CC individually, without carrying cable, but with guide ropes*

The CC is placed into and lifted out of the water by the crane. Descent and ascent along the guide ropes are controlled from the CC direct by means of flooding tanks, ballast and guide rope brake.

(*d*) *CC individually, with electric carrying rope and horizontal propellers*

The Control chamber may be employed on a carrying rope which, additionally, functions as electric supply cable. The CC is suspended from the cable, but may swing out laterally under precise control by motors and propellers; thus it may approach an object. This allows particularly good work with manipulators and good observation. The wave movements of the ship may be absorbed by current stops when using elastic cables, while this function is performed by shock and wave absorbers when using normal cables.

(*e*) *CC individually, with carrying cable and guide ropes*

The CC may slide upwards and downwards on the guide ropes; upon so doing it is controlled by a carrying cable.

(*f*) *CC and PTC in combination*

All the operations described may be performed by CC and PTC in combination.

(*g*) *CC as lock to PTC, with guide rope*

When CC and PTC operate with the guide rope, they may separate and unite under water and above water. In this way CC may serve as a link between surface and PTC and may transport personnel as well as material upwards and downwards.

Likewise it is possible to connect CC and DDC, i.e. to perform again the function of a lock.

(*h*) *CC as a link to the underwater station*

The CC may establish the connection from the surface to underwater stations or sunken submarine craft with or without guide rope, with or without motors. Coupling is possible laterally or below the CC, and the possibility exists of positioning by means of propellers.

PERSONNEL TRANSFER CHAMBER (PTC)

Purpose

Diving chamber for the exit of divers at great depth. Base for industrial work with semi-permanent installations. Also suitable as observation chamber.

Design

Vertical cylinder with a height of approximately 2·2 m (7 ft 2·6 in) and a diameter of approximately 1·4 m (4 ft 7·1 in). Lateral door for exit under water as well as for coupling to stations, such as DDC on ship or underwater habitats or submarine craft under water. At top with coupling for connection to CC.

At top attachment for carrying cable. Annular flooding tank for autonomous manoeuvring under water, especially on guide rope system with simultaneous use of the CC.

Laterally guides and brakes for guide rope.

Lateral windows.

Supply system for diving apparatus and working systems.

Equipment

Two bayonet doors with coupling
Four windows, four searchlights
Flooding tank with control valves
Hose connection
Sixteen cylinders, 50 litres/200 kp/cm^2 (2,380 lb/in^2) gauge, for oxygen, air and helium
One oxygen analyser
Pressure control and valves
Water level control
Hose connections
Rope guide and brake
Seats
Sonic depth finder
Communication system
Warm water heating

Further expansion

Working systems and facilities
Horizontal propeller
Vertical propeller
Manipulator

Possibilities of operation

See (a) to (h) for Control Chamber (CC) for the same functions which may be performed by the PTC.

DECK DECOMPRESSION CHAMBER (DDC)

Purpose

The first purpose is the use as recompression and treatment chamber for divers who have met with an accident. Then the chamber may also be used as accommodation under pressure for divers in diving chamber operations with coupling under pressure and with decompression.

Design

The normal medical compression chamber according to the BABCOCK standard system is further expanded for special use. Particularly suitable for deep diving with helium saturation (long times): chamber with a diameter of 2·1 m (6·89 ft), length approximately 7 m (23 ft) with manlock, pressure 30 kg/cm^2 (425 lb/in^2) gauge. The door with the coupling mechanism is fitted laterally to the main chamber.

Equipment

As for medical compression chamber

Additionally:

One bayonet door with coupling
Six beds
WC
Shower bath
Dehumidifying
Insulation with protection from radiation
Helium control plant
Oxygen analyser
CO_2 analyser

Resilience and shock absorption of the platform for placing on uneven ship decks and for eliminating deformation forces on the coupling connection of PTC/CC

Further expansion

Decompression computer and control
Locks and larger dimensions
Internal equipment
Expansion to underwater station

Possibilities of operation

(*a*) *Recompression, in the lock, of divers who have met with an accident*

When the main chamber is already used by divers, recompression of divers who have met with an accident and who are transported to the chamber from outside is possible in the lock. Temporarily they may also stay in the main chamber, depending upon the pressure prevailing there. Then it is not absolutely necessary to interrupt the work.

(*b*) *Recompression, in the main chamber, of divers who have met with an accident*

This corresponds to the normal use as medical compression chamber or as divers' compression chamber, respectively

(*c*) *As a living room and base*

Several teams of divers can live in the DDC under pressure (also with helium atmosphere) and alternatingly go to work in the depth of the sea by the PTC and/or the CC. Especially the PTC may be stationed at the bottom. By means of the CC the divers go from the DDC on deck to the PTC at bottom and work from there. The tools may remain at bottom. Observers, etc. may likewise go downwards and upwards by means of the CC.

(*d*) *As an underwater station*

It is possible to expand the DDC further so as to be used as an underwater station. A wet lock room would be added for entry and exit. The normal lock door would likewise be developed to become a bayonet door also capable of withstanding outside pressure. The dashboard would be transferred to the inside and the connecting flanges will not be subjected to any modification; the chamber body (steelwork) is consequently identical to the normal chambers. A flooding equipment and the throw-off ballast would be added to the chamber and insulation would be carried out so as to be pressure proof.

DIVING BELL

Design operating depth: 300 m (1,000 ft)
No other details available.

Babcock and Wilcox Diving Bell

UNITED STATES OF AMERICA

FMC CORPORATION
US NAVY MARK I DEEP DIVE SYSTEM

Delivered to the US Navy in March 1969. The Mark I Deep Dive System is the most recent hardware development in support of saturated diving. It provides US Naval Salvage with the capability of performing observation, bounce or saturated diving to a depth of 243 m (850 ft). It has a life-support system with capacity to support two teams of divers for 14 days. The Mark I components are air transportable and can be used on a variety of ships such as the ASR, ARS, ATS, ATA, and HLC.

The Personnel Transfer Capsule (PTC) in its principal mode is used to carry divers from the Deck Decompression Chamber (DDC) complex aboard the ship to the work site, or to the spot from which diver excursions will be made. In this mode, the PTC maintains the divers in an artificial atmosphere which has a gas pressure equal to the ambient sea water pressure at their destination depth. When used on working dives, the PTC can carry two or three divers at internal saturation pressures equivalent to 243 m (850 ft) depths. At final equipment depth, a diver may leave the PTC through the lower lock and be sustained on 'hookah' lines at distances up to 30 m (100 ft). The PTC can also be used as a 'diving bell', with atmospheric air at about 1·05 kg/cm^2 (15 lb/in^2) absolute. In this mode, the PTC is used only for observation and, of course, the occupants remain inside the vessel. Divers can also go down in the PTC at atmospheric pressure and pressurise at depth.

The PTC is equipped to provide the following services:

Heated diving suits;
Internal lighting;
Communication with the surface, including closed-circuit television;
Exterior searchlights;
Atmospheric purification and control;
Breathing and pressurisation gas supply.

The PTC is handled from the fantail of the ATS or other ships by means of a multifunction Strength-Power-Communication Cable (SPCC). It is raised from its stowage facility and swung over the side by the aft crane with the aid of several small tugger winches.

The Deck Decompression Chamber complex consists of two DDCs and an Entry Lock (EL). The DDCs are vessels each made from a 1·95 m (78 in) cylindrical centre section and two ellipsoidal heads. A flanged entry trunk, with hatch, is fitted on each end and a medical lock, which serves as a pass-through for food and small items, is provided near one end. Several viewports are provided. These permit illumination of the interior of the DDC from exterior light sources and surveillance of the inhabitants' activities by means of an exterior closed-circuit television system (CCTV). An helium-oxygen atmosphere system is provided. It is capable of maintaining correct CO_2 and oxygen levels at gas pressures equivalent to seawater pressure at 300 m (1,000 ft) depth. Amenities inside the DDCs include bunks, stowage, bath and toilet facilities and air conditioning.

The EL is spherical with four flanged entry trunks for connection with the DDCs, PTC, and for external entry. The EL has its own atmospheric system which is similar to that in the DDC chambers (it can be used as a DDC in an emergency). It permits access between the DDCs and either the deck of the ship or the PTC.

The DDC complex elements are monitored and controlled from the Main Control Console (MCC). The MCC is also connected by an umbilical cord running through the slip rings on the Deck Winch and then into the PTC. The DDC elements are further connected to the Life Support System (LSS) unit that provides the CO_2 gas scrubber system, sanitary system, and deoderisation unit for DDC hotel functions.

The Life Support Package (LSP) contains the deck equipment life support system and the sanitary system. The life support system

The MARK I Deep Dive System

provides metabolic gas conditioning for the atmosphere of the DDCs and the EL, atmospheric carbon-dioxide removal, humidity control and temperature control. The sanitary system provides hot and cold water to the DDCs and collects waste from the waste lines in the DDCs. The packages are mounted on steel frames, with the sides and top enclosed in a steel housing.

The Power Control Console (PCC) contains all power-control components, including circuit breakers, emergency batteries, battery charger, inverter, transformer and also the MCC air-conditioner unit. Primary power for the LSP and the Deck Winch are also controlled from the Power Console.

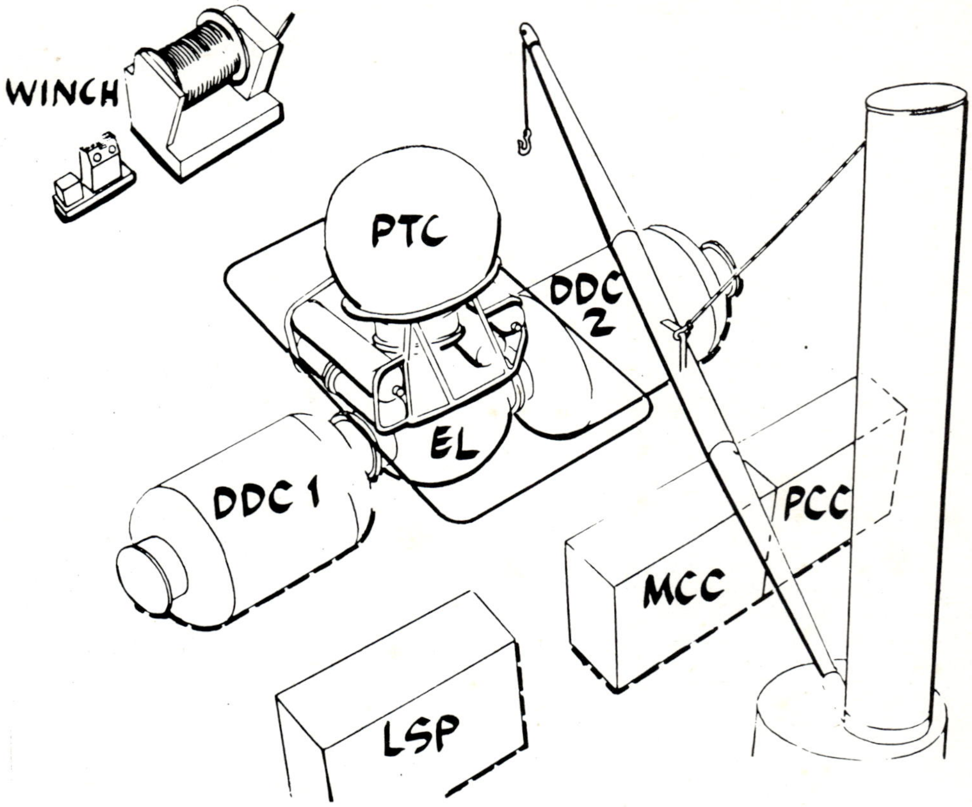

FMC Corporation MARK I Deep Dive System for the US Navy (major components)

Technical data

Personnel Transfer Capsule

Material	HY80 Steel
Construction	Welded

Dimensions

Outside diameter at equator	2·0 m (79 in)
Total height with support frame	3·66 m (144 in)
Inside height (including trunk)	2·44 m (96 in)
Trunk diameter	7·62 m (30 in)
Trunk length	0·58 m (23 in)

Gas flasks

Helium (He):	
Quantity	2
Volume (cubic feet water volume)	0·28 m³ (10 ft³)
Pressure	208 kg/cm² (3,000 lb/in² g)
Helium–oxygen mix ($He–O_2$):	
Quantity	1
Volume (cubic feet water volume)	0·28 m³ (10 ft³)
Pressure	208 kg/cm² (3,000 lb/in² g)
Oxygen (O_2):	
Quantity	1
Volume (cubic feet water volume)	0·112 m³ (4 ft³)
Pressure	208 kg/cm² (3,000 lb/in² g)
Operating temperatures:	
In air	20°F to 120°F
In water	29°F to 90°F
Emergency oxygen (O_2):	
Quantity	1
Volume (cubic feet water volume)	0·002 m³ (0·072 ft³)
Pressure	208 kg/cm² (3,000 lb/in² g)
Anchor:	
Type	Lead
Weight	454 to 945 kg (1,000 to 2,080 lb)
Maximum operating pressure:	
Internal	31 kg/cm² (444 lb/in² g)
External (1,000 ft depth)	31 kg/cm² 444 lb/in² g)
Electrical power:	
Surface supplied (through SPCC)	440 volts, 3 ø, 60 hz
Emergency (on-board batteries)	24 volt, DC
CO_2 scrubber:	
Type	Baralyme
Capacity	30 man-hours at 5% surface-equivalent concentration of carbon-dioxide
Hookah system (UBA):	
Quantity	2
Type	Mark IX
Hose length	30·5 m (100 ft)

Emergency Mask Breathing System (MBS):

Quantity	2
Type	Mark IX
Helium-oxygen duration	2 man-hours

Downhaul winch

Manufacturer	Electro Kinetics Corp.
Type	2-speed, electric, reversible
Overall dimensions	0·76 × 0·56 × 0·86 m (30 × 22 × 34 in)
Maximum weight in air (w/cable installed)	665 kg (1,463 lb)
Power requirements	440 volt, 3 ø, 60 hz
Drum capacity	381 m (1,250 ft)
Cable (CRS 6X37 IWRC):	
Diameter	9·5 mm (⅜ in)
Length	381 m (1,250 ft)
Break strength	5,450 kg (12,000 lb)
Performance:	
Max. rated load	1,136 kg (2,500 lb) at 40 fpm
Min. rated speed	1,136 kg (2,500 lb) at 12 to 15 fpm
Water temp.	20°F to 90°F
Water pressures	518 m (1,700 ft)
Brake	Automatic

Strength-Power-Communication-Cable

Manufacturer	Simplex Corp.
Type construction	Double layer plow shear steel
Dimensions:	
Overall dia.	5·08 cm (2 in)
Overall length	425 m (1,400 ft)
Breaking strength	64,091 kg (141,000 lb)
Fatigue strength	11,360 kg (25,000 lb) at 60,000 cycles
Electrical and communications conductors:	
No. 10, stranded-wire conductors	9
Multistrand coaxial cables—type RG-59/U	3
No. 18, shielded single conductors	12
No. 18, shielded conductor pairs	12

SPCC Deck Winch

Type	Electro-hydraulic
Drive	Fixed displacement hydraulic motor
Control	Electro-mechanical, variable speed
Drum diameter	1·5 m (60 in)
Drum capacity (2·12-in cable)	442 m (1,450 ft)
Motor:	
Manufacturer	General Electric
Type	30 h.p., 3 ø, 440 vac AC, 60 hz
Dimensions (overall):	
Length	3·91 m (154 in)
Height	2·44 m (96 in)
Width	2·79 m (110 in)
Total weight	11,360 kg (25,000 lb)
Manual brakes:	
Type	Flat band
Automatic brakes:	
Type	Hydraulic, disc
Level wind	Chain-driven, two-sheave
Slip ring assembly:	
Coaxial for 3 RG 59/U coaxial cables	6
Red brass for 440 vac AC through copper graphite brushes	6
Coin silver for 9 circuits of AWG 18 through silver graphite brushes	18
Coin silver for 12 circuits of AWG 18 through silver graphite brushes	36
Indicators:	
Cable tension	0 to 13,630 kg (0 to 30,000 lb)
Cable speed (payout or reel-in)	0 to 18 m/min (0 to 60 ft/min)
Total footage	resettable 455 m (1,500 ft)

Performance:
Rated load (raise or lower): 4,090 kg (9,000 lb) at 40 ft/min)
Maximum rated load (raise): 13,630 kg (30,000 lb) at 10 ft/min)
Maximum rated speed (raise or lower): 2,270 kg (5,000 lb) at 60 ft/min)
Environmental requirements:
Operating: 20°F to 120°F
Stowage: 20°F to 120°F

Deck Decompression Chamber
Dimensions (overall):
Length 4·11 m (162 in)
Height 2·34 m (92 in)
Width 2·34 m (92 in)

Entrance Lock
Dimensions (overall):
Length 2·34 m (92 in)
Height 2·75 m (108·5 in)
Width 3 m (118 in)

Main Control Console
Dimensions (overall):
Length 4·27 m (168 in)
Height 2·29 m (90 in)
Width 1·75 m (69 in)

Power Control Console
Dimensions (overall):
Length 2·29 m (90 in)
Height 2·29 m (90 in)
Width 1·75 m (69 in)

Life Support Package
Dimensions (overall):
Length 4·32 m (170 in)
Height 2·26 m (89 in)
Width 1·27 m (50 in)

Development programme

The Mark I was developed as a military version of the commercial ADS IV System with increased performance capability. As such, it was to be built to military specifications and fully certifiable in conformance with NAV-SHIP-090-028-2010. Because of the need for the system for potential salvage operations, it was developed in a minimum of time consistent with the certification requirements. The major items of hardware were delivered 21 months after contract award.

The design analysis phase involved a technical evaluation of the contract plans to determine feasibility from a technical as well as certification point of view. As a result of this evaluation, sixty major changes were made, of which the following are typical:

The PTC was reconfigured to provide for increased buoyancy at a lesser weight;

The DDCs were lengthened to permit a separation of the wet and dry areas;

PTC capacity was increased from two to three men;

HY80 was selected as the structural material on all pressure chambers due to certification requirements (Category I material);

PTC on-board gas was increased to permit self-pressurisation on bottom. Military specification flasks were substituted for commercial to permit certification;

The main console was separated into two packages (the PCC and MCC) since the console as configured would not fit through the ATS hatch.

Because of the urgency of the programme, the various phases were performed concurrently. Initially, it was felt that the critical path for the programme involved the procurement of the HY80 heads and forgings. However, because of the excellent performance of these suppliers, this was not so. The actual critical path involved electrical components primarily because of the failure of the suppliers to appreciate the stringent quality control requirements, such as material traceability, required as part of certification.

OCEAN SYSTEMS INC.

ADVANCED DIVING SYSTEMS MK IV (ADS IV)

The system consists of a Submersible Decompression Chamber (SDC), a Deck Decompression Chamber (DDC), a Control Van (CV), and related life support and handling equipment. The use of additional DDC's and a saturation living chamber extends the system's basic capabilities.

The SDC transports a two-man diving team from the surface to the undersea work site at surface pressure, minimising the divers' decompression obligation. Once the scope of the task is determined by visual observation and the operation planned, the divers pressurise the chamber to ambient sea pressure and leave the vehicle. Breathing mix is supplied to the divers through an umbilical hose and constant communication is maintained with the surface. The SDC remains on station throughout the work phase providing the divers with a safe and comfortable base.

Upon completion of their task the divers return to the SDC and seal the hatches to maintain their environment. During ascent the divers may initiate decompression or they may remain at pressure should additionl work be necessary. Once returned to the surface the SDC is mated to a DDC which continues diver decompression or maintains the divers at work site pressure in relative comfort. During their stay in the DDC the divers are maintained in a carefully controlled and monitored environment. Ready access to the divers is provided throughout the decompression stage.

ADS IV offers the added capability of being used as an observation bell where non-divers (engineers and supervisors) may descend to the work site while maintained at a one atmosphere environment to observe and survey a work site or equipment. Virtually unrestricted visibility is provided through 15 viewports of unique design (patent pending)

Modular in concept, virtually all components of all systems are interchangeable and compatible. All chambers bolt together with O-ring seals. The small size and inherent simplicity of ADS IV provides 'fly-away' portability, use aboard ships and barges of opportunity, a greater margin of safety during operations and a minimum of maintenance requirements.

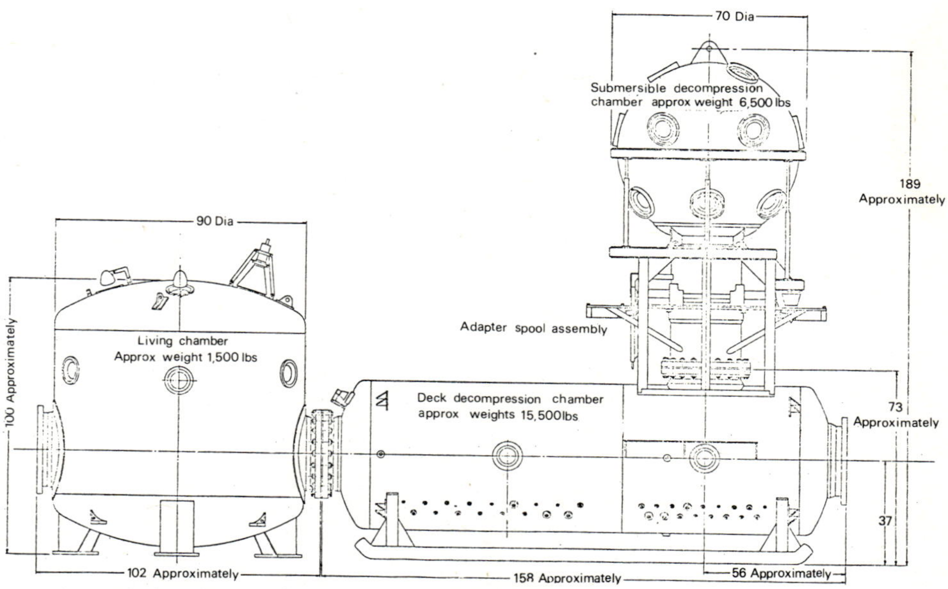

Ocean Systems, Submersible Decompression Chamber, Deck Decompression Chamber and Living Chamber

Operational specifications

SDC

Operational pressure, internal/external: 35 kg/cm² (500 lb/in²), 344 m (1,125 ft) seawater 31·1 kg/cm² (445 lb/in²) 300 m (1,000 ft)

Crew capacity: 2

Buoyancy: 225 kg (500 lb) positive with anchor jettisoned

Anchor weight: 910 kg (2,000 lb)

Weight in air: 2,930 kg (6,500 lb)

External diameter: 1·8 m (72 in) at bumper ring

Height (overall): 3·2 m (10 ft 6 in)

Communications: Two independent 2-wire systems

Life support: Self-contained or surface supplied breathing mix-CO_2 removal units

Emergency ascent winch: Modified hydraulic winch with 300 m (1,000 ft) of wire rope

DDC

Internal design pressure: 18·5 kg/cm² (267 lb/in²) 181 m (600 ft) seawater

Communications: 2 wire system in each lock

Life support: CO_2 removal units, O_2 mask breathing. Complete environmental monitoring

Crew capacity: Two 2-man crews

Length: Average 4·87 m (16 ft)

Height (mating flange to deck): 2·32 m (92 in)

Weight in air: 5,000 kg (11,000 lb)

External diameter: 1·36 m (54 in)

The pressure vessels have been proportioned to meet the requirements of the 1968 ASME boiler and pressure vessel code, Section VIII, Division I and the marine engineering requirements of the USCG, as published in the Federal Register, Vol. 33, No. 245, Part II, 18 December 1968.

First prototype date
1964

Ocean Systems, Submersible Decompression Chamber

Development history
1964—ADS I & II
1965—ADS III
1966—ADS IV

Commercial data
Capital cost (1973): $US 250,000

Availability
18 Systems—ADS IV—In Operation
3 Systems—Under construction (1973 August)

Applications and operations to date
Phillips Petroleum, North Sea: 1970 to present.
Esso, Australia, Bass Strait: 1969–1971.
Esso Exploration, Inc., Africa/Europe: 1966 to present.
Numerous operations on all continents in support of offshore oil field drilling and construction activities.

Future developments
Increase in Diver Saturation/Depth capability.

WILSON MARINE SYSTEMS INC.

MARK IV DEEP DIVING SYSTEM

The Mark IV Diving System is a completely unitised, skid-mounted system consisting of a diving bell, a double-lock decompression chamber, a gas control console, a pneumatic and hydraulic handling system, a 213 m (700 ft) unsheathed umbilical to carry gas and power to the bell, and a diesel-driven, air-cooled compressor.

The system is built to ASME and U.S. Coast Guard standards. American Bureau of Shipping code is available at extra cost.

Diving Bell
Hull

The diving bell portion of the Mark IV Diving System is a 1·68 m (66 in) inside diameter sphere, with an operational depth capability of 198 m (650 ft).

Hatches

There are two through-hull coamings with double doors. One is at the bottom for diver egress and the other one is at the side for transfer-under-pressure to the skid-mounted decompression chamber. Each 0·61 m (24 in) inside diameter hatch is O-ring sealed with a simple dog retainer to effect initial seal.

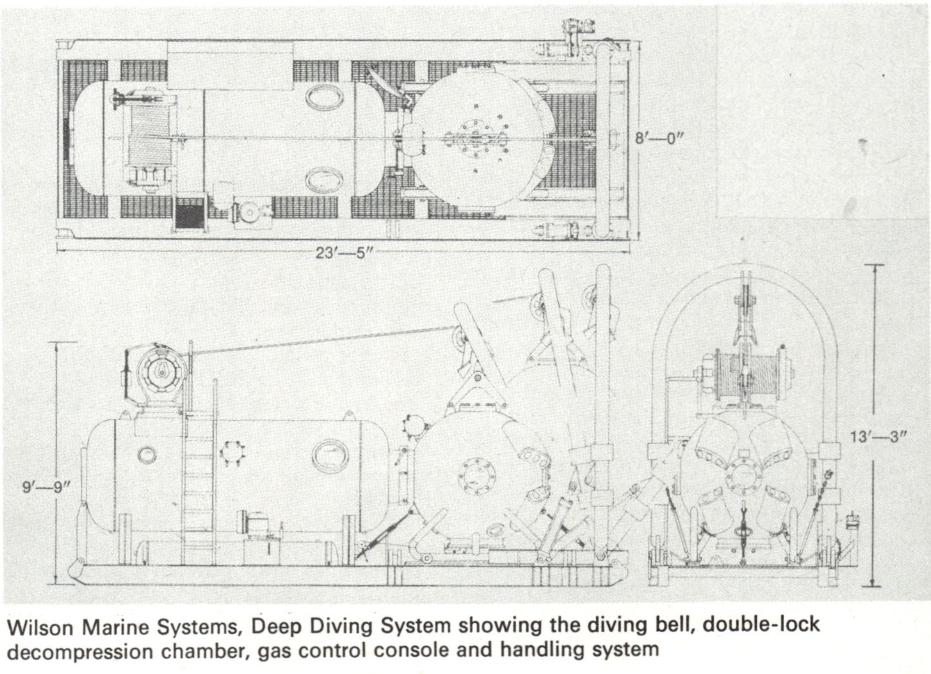

Wilson Marine Systems, Deep Diving System showing the diving bell, double-lock decompression chamber, gas control console and handling system

Viewports

There are three 0·305 m (12 in) inside diameter side viewports and a total of six 15·2 cm (6 in) inside diameter top and bottom viewports, arranged for omnidirectional visibility. The double acrylic, 90-degree conical frustum ports are O-ring sealed to withstand pressure in both directions. Ports have been tested on a quick rupture basis to a pressure of 160 kg/cm² (2,300 lb/in² g). Long term tests (10 days) at 32 kg/cm² (450 lb/in² g), room temperature, indicated no tendency to creep under continuous load.

Life support

A dive control panel in the bell provides for blowing down or venting the bell through the umbilical and delivering umbilical gas to two divers. A back-up system will sustain the divers automatically from onboard gas storage, in the event of umbilical pressure drop. A regulator, with two masks, delivers stored breathing gas to the divers in the bell.

Gas storage

Standard onboard high pressure bottle storage consists of four racks of three 2·05 m³ (72 ft³) scuba bottles attached to the side of the bell. One rack contains pure oxygen, which is regulated with metered flow to replace oxygen consumed metabolically by the divers in the bell. Additional onboard gas supply may be requested to offer a maximum storage of twelve scuba bottles, attached to the hull of the bell, and six 7·0 m³ (244 ft³) cylinders, in two racks, attached to the bell legs.

Gauging

Gauging in the bell monitors oxygen and breathing gas supply pressures, umbilical, and diver's supply pressures. Two caisson gauges read bell depth and internal bell pressure. Bell depth and diver depths are monitored at the surface.

Diver umbilical

Two 22·9 m (75 ft) diver umbilicals, consisting of four No. 16 AWG conductors for communications, 0·95 cm ($\frac{3}{8}$ in) inside diameter air supply hoses, and 0·47 cm (3/16 in) Kluge lines, are standard.

CO_2 scrubber

Carbon dioxide removal is accomplished by a low profile, wall-mounted chemical absorbent scrubber. A 12-volt electric blower circulates bell air through a replaceable scrubber canister. Each canister holds 8·2 kg (18 lb) of sodasorb, rated for 36 manhours of one atmosphere service.

Emergency power

A 12-volt storage battery, externally mounted in its own pressure vessel, serves as a backup power source for the scrubber and interior lights. The communications equipment has its own 12-volt dry cell.

Communications

A telephone amplifier in the bell will maintain diver communications, in the event of failure of the topside unit.

Ballast

A 910 kg (2,000 lb) cylindrical ballast, suspended beneath the bell legs, may be released by the diver inside the bell to give the bell positive buoyancy for emergency ascent.

Protective ring

A ring of 7·6 cm (3 in) pipe on top of the bell offers protection for gas and electrical penetrators.

Decompression Chamber

Hull

The double-lock decompression chamber has a 1·83 m (6 ft) long main chamber and a 1·22 m (4 ft) long entry lock. The inside diameter of both locks is 1·37 m (54 in). This standard chamber is designed to accept divers directly from a depth of 1·37 m (450 ft). There are twenty through-hull penetrations for gas and electrical feedthroughs. Chambers can be built for greater pressures, as may be required for hyperbaric facilities and saturation diving systems.

The Wilson diving bell being lowered into the water

Hatches

There are three hatches in the decompression chamber.

The diving bell mates to the decompression chamber with a 0·66 m (26 in) Tube-Turn hub and yoke assembly, equipped with a hand-cranked chain drive. This Tube-Turn assembly offers transfer-under-pressure capability up to 14 kg/in² (200 lb/in² g) 137 m (450 ft) of sea water.

The entrance hatch on the Tube-Turn end of the main lock is a 0·61 m (24 in) inside diameter, round door and seals with a 1·27 m ($\frac{1}{2}$ in) cross section diameter O-ring. Doors are faced and O-ring grooves are machined for positive sealing.

The hatch separating the main and entry locks and the hatch on the end of the entry lock are 0·66 m × 0·74 m (26 in × 29 in) inside diameter, ob-round. The ob-round doors facilitate passage of personnel, reduce the swing radius of opening hatch doors, and permit the removal of doors from the chamber.

Viewports

There are four side viewports in the chamber. The two in the main chamber have a 0·305 m (12 in) viewing diameter, and the two in the entry lock have a 0·152 m (6 in) viewing diameter. These are also double-paned for safety. A 0·152 m (6 in) diameter port is located on top of each chamber lock for externally mounted chamber lights.

Life support

The chamber locks are equipped with oral/nasal masks for oxygen and mixed gas decompression. Three masks are in the main chamber and two in the small chamber. The main lock has two bunks and an 0·20 m (8 in) diameter medical lock. Larger diameter medical locks are available at extra cost.

Transfer-under-pressure

The Wilson transfer-under-pressure system offers complete interchangeability of any standard Wilson chamber or bell; or the use of multiple chambers with a single bell.

Handling System

The skid

The steel skid serves as the base for two air receivers, handling davits, double-lock decompression chamber, hydraulic power package, air winch, winch and hydraulic control station, and the 'heart' of the system, the gas control console. All components are installed on the skid to effect bell chamber mating.

Handling davits

This assembly lifts the bell from its tracks. The ballast weight is attached and the bell, with ballast, is lowered into the water.

Hydraulic cylinders move the gantry inboard or outboard.

Hydraulic power unit

A 220/440 volt, 3-phase, explosion proof electric motor drives a hydraulic pump for operating the two 12·7 cm (5 in) diameter, 0·56 m (22 in) stroke, cylinders which move the bell gantry. This hydraulic unit includes a pair of high pressure accumulators, which serve as shock absorbers. Also included is a motor controller and a remote control station.

Air winch

The air winch holds 366 m (1,200 ft) of 1·9 cm (¾ in) diameter non-twist cable, or 213 m (700 ft) of 2·22 cm (⅞ in) cable. It has a maximum lift, in water, of 30·5 m (100 ft) per minute when the inlet pressure is at least 7·05 kg/in² (100 lb/in²). The controls for the winch and the controls for the hydraulic package are located together, at the front of the skid. The recommended operating air pressure is 5·6 to 7·05 kg/cm² (80 to 100 lb/in²). Short duration operation, at 10·5 kg/cm² (150 lb/in²) is permissible but not recommended.

Main strength line

The main strength line (bell lifting wire) consists of 229 m (750 ft) of 1·91 cm (¾ in) diameter, non-rotating improved plow steel cable with a poured 2·54 cm (1 in), closed spelker socket attached. The spelker socket attaches to the bell wishbone lifting assembly. The wire passes through a 0·35 cm (14 in) open sheave on the gantry and is spooled on the drum of the air winch, mounted atop the chamber. The breaking strength of this line is 24 tons. (The bell, with divers and equipment, is approximately 4·5 tons.)

Console

The gas control console is composed of three panels. The left panel has controls for the bell and diver gas lines. The centre panel has controls for the main chamber gas supply, lighting, and communications. The panel on the right has controls for the entry lock gas supply. The console has gauges to monitor all gas and air supplies and regulators to reduce high pressure gas to the pressure desired for breathing masks. A pneumofathometer system, also in the console, enables the diving supervisor to be aware of bell and diver depth at all times. Valving on the console permits the routing of air or mixed gas throughout the diving system.

Air compressor unit

The compressor has a capacity of 3·02 m³ (107 ft³) displacement, at 17·6 kg/cm² (250 lb/in² g) and is skid-mounted with a two cylinder, air-cooled, diesel engine. The following features are standard:

Hand cranking is standard, electric or air starting at extra cost;
Belt drive;
Variable speed controls;
Fuel and lube oil filters, oil pressure gauge;
Fuel pump with automatic venting;
Sixty US gallon, ASME approved surge tank;
Stainless steel flex line between compressor and surge tank;
Extra heavy duty piping and fittings;
ASME approved safety valves;
Vibration and shock-free mounting;
Removable safety guards;
Industrial grade primer and paint;
Unit receives 4-hour test run prior to final inspection and shipment;
Size: 1·02 m (40 in) wide × 2·01 m (79 in) long × 1·22 m (48 in) high, 3·4 m³ (120 ft³);
Weight: 1,340 kg (2,950 lb) (ready for shipment);
Extra options: Deluxe air filter, carbon monoxide filter, five-hose outlet manifold.

Umbilical description

The standard umbilical is 213 m (700 ft) in length, and consists of the following components:

(1) One 1·91 cm (¾ in) inside diameter breathing gas hose, with a minimum burst pressure of 350 kg/cm² (5,000 lb/in² g), serves as bell blowdown and divers supply hose.

(2) Two 0·64 cm (¼ in) inside diameter hoses, 840 kg/cm² (12,000 lb/in² g) minimum burst pressure, transmit pressures to the surface console, giving readouts of diver and bell depths in feet of seawater.

(3) Two shielded pairs of No. 16 AWG wire are provided for communications.

(4) Internal bell power is supplied by four conductors of No. 10 AWG wire. Two conductors supply 12-volt DC power for the interior bell lights and CO_2 scrubber motor. Two additional conductors are provided to supply power for additional instrumentation. While 110 volts do not exceed the insulation rating of this cable, the use of 110 volt AC is discouraged for hyperbaric environment.

(5) External bell power is supplied by four conductors of No. 6 AWG wire with copper stranding, rubber insulation, and neoprene jacket. This cable has a 600-volt insulation rating and is intended primarily for underwater lights on the bell.

(6) The standard umbilical is a taped bundle with stainless steel, anti-kink springs, and 37-degree J.I.C. swivels on all hose ends. The internal bell power and communication cables terminate in a single, right angle, multi-pin connector and through-hull penetrator assembly of stainless steel. External bell power cables are terminated with wet-pluggable type, underwater connectors.

General information

The Wilson Marine Systems Mark IV Deep Diving System consists of five packages when shipped. (1) The skid unit, which includes bell, chamber, hydraulic unit, gantry, volume tanks, and ballast weight, has an overall dimension of 2·4 (8 ft) wide × 6·7 m (22 ft) long × 3·05 m (10 ft) high and weighs approximately 10,910 kg (24,000 lb). (2) The compressor unit is 1·22 m (48 in) wide × 1·63 m (64 in) long × 1·117 m (44 in) high, and shipping weight is 1,363 kg (3,000 lb). (3) The air winch is removed from the top of the decompression chamber and shipped on a pallet. It is 0·61 m (2 ft) wide × 1·22 m (4 ft) long × 1·22 m (4 ft) high, and weight 818 kg (1,800 lb) with 213 m (700 ft) of 1·9 cm (¾ in) wire on the drum. (4) The unsheathed umbilical is 213 m (700 ft) long with a diameter of 5·08 cm (2 in), and weighs approximately 523 kg (1,150 lb). It may be spooled or coiled on a pallet. Approximate size is 1·98 m³ (70 ft³). (5) A box of miscellaneous equipment and spares is 0·92 m (3 ft) wide × 0·92 m (3 ft) long × 0·92 m (3 ft) high. Weight is approximately 182 kg (400 lb) and size is 1·0 m³ (36 ft³).

When at sea, the system would require 18·5 m² (200 ft²) of deck space for the skid, 2·8 m² (30 ft²) of deck space for the compressor, and working space in the vicinity of both. For operational diving, the system uses 220 or 440 volt, 3-phase, AC power for the hydraulic unit, 110 volt for the gas control console, a DC power supply for the CO_2 scrubber, and 13·8 kg/in² (200 lb/in²) air, if available, to augment the system compressor. Six men can operationally dive the system. Two divers would be in the bell, a diving supervisor at the console, a winch and hydraulic control man at the bell handling station, and two line handlers—one of whom may be a standby diver.

The system is sand-blasted, primed with zinc-rich primer, and painted with epoxy marine paint. Colour may be selected by customer. All exposed metal items are plated, iridited, anodized, or of stainless steel or monel to resist the salty environment.

To facilitate operation and familiarisation of the system, a 135-page 'Operation, Maintenance, and Lubrication' manual for the Mark IV Diving System is included.

A training programme is available, for the use of the system, and requires from two to six weeks for completion. This familiarisation programme, when requested by the customer, is priced on a day rate basis plus transportation, food, and lodging for the representative of Wilson Marine Systems.

Optional accessory equipment

A 516-70-A 300 Class I steel in all pressure-resisting structural components, for operating temperatures to: Approx. Minus (25°F).

American Bureau of Shipping certification on entire diving system.

Additional umbilical hoses:
(a) Additional gas supply hose;
(b) Separate bell venting hose;
(c) Hot water from surface to divers.

Additional umbilical cables and electrical equipment:
(a) Television inside bell;
(b) Television on exterior of bell;
(c) Electric cable for heated diver suits;
(d) Electric cables and instrumentation for life support.

Protective sheathing of polyurethane tubing for umbilical.

Variations in number or location of viewports on bell and chamber.

Enlarged chamber dimensions.

Divers breathing apparatus and related equipment:
(a) Gas hats and recirculators;
(b) Air hats;
(c) Full face mask with oral/nasal unit;
(d) Emergency 'come home' bottles;
(e) Weight belts;
(f) Chaffing pants/jock combination.

A 10·2 cm (4 in) quick-sealing medical lock for the bell.

A spare parts package consisting of: fittings, O-rings, acrylic viewports, valves and miscellaneous support equipment.

Television camera (underwater).

Oxygen sensors and read-out instruments.

Carbon dioxide sensors and read-out instruments.

Enclosed dive control booth for personnel comfort in adverse weather or noisy conditions.

OFFSHORE DRILLING RIGS
WORLD FLEET 1974

Name OWNER/OPERATOR	Type	Built	Limiting Water Depth (ft)	Dimensions of Platform/Hull (ft) and shape Length×breadth ×depth (draught)	Tank Capacities (bbl=barrel) Fuel	Freshwater	Mud (dry/liquid)	Cement
Abu Dhabi Marine Areas Ltd., Das Island, Iran								
ADMA Constructor	platform	UK 1962	100	195×90×14·5 (rectangular)	300 s/tons	740 s/tons	177 s/tons	—
Arabian American Oil Company, P.O. Box 5000, Dhahran, Saudi Arabia								
Aramco MDP2	self-elevating	USA 1966	200	190×169×25 (hexagonal)	1,500 bbl	4,000 bbl	6,700 cft	—
Arabian Drilling Company, P.O. Box 932, Riyaah, Saudi Arabi								
Arabdrill 1	tender/fixed platform	France 1966(C)	200	50×70 (rectangular)	2,800 bbl	1,400 bbl	1,000 bbl	—
Arabdrill 2	tender/fixed platform	France 1966(C)	200	60×80 (rectangular)	2,500 bbl	4,600 bbl	700 bbl	—
Atlantic Drilling Company								
Under construction	semi-submersible	Finland (1975)	600	355×221 (rectangular)	16,900 bbl	14,400 bbl	18,000 cft	6,000 bags 1,450 bbl
Under construction	semi-submersible	Norway (1975)	600	355×221 (rectangular)				
Associated Marine Services								
La Ciencia	drill ship	USA 1944		136×24·5×(11·5)	178 bbl			
Atwood Oceanics Inc., Town and Country Office Park, Suite 301, 10555 Katy Freeway, Houston, Texas 77024								
Big John	drill barge	USA 1970	600	352×76×19 (ship shape)	—	8,000 bbl	10,000 cft 1,700 bbl	10,000 bags
Chikamauga	self-elevating	Singapore 1974	300	230×200×26	—	—	5,000 cft 1,400 bbl	10,000 bags
Gettysburg	drill barge	Singapore 1973	600	348×70×22 (ship shape)	—	11,000 bbl	6,400 cft 2,000 bbl	15,000 bags
Lauritzen	drill barge	Reported as being converted in Australia						
Margie	semi-submersible	Singapore 1973	600	202×182×110	—	10,715 bbl	11,200 cft 2,000 bbl	12,000 bags
Shiloh	drill ship	USA 1968	600	352×76×19	—	(distilling equip)	6,240 cft	—
Shenandoah	self-elevating	USA (1975)						
Vicksburg	self-elevating	USA (1976)					1,467 bbl	
BP Clyde Tanker Co. Ltd., BP House, Ropemaker Street, London EC2								
Sea Quest	semi-submersible	UK 1966	600	342×66 (triangular)	766 tons	10,500 bbl	18,000 cft	—
Blandford Shipping Company								
Under construction	semi-submersible	Norway (1974)	600	355×221 (rectangular)				

Name OWNER/OPERATOR	Type	Built	Limiting Water Depth (ft)	Dimensions of Platform/Hull (ft) and shape Length×breadth ×depth (draught)	Tank Capacities (bbl=barrel) Fuel	Freshwater	Mud (dry/liquid)	Cement
Canam Offshore Ltd., P.O. Box 7074, Shreveport, Louisiana 71107								
Gulftide	self-elevating	UK 1967	250	179×150×20 (rectangular)	3,650 bbl	5,400 bbl	10,000 cft	—
Celtic Drilling Company								
Under construction	semi-submersible	Finland (1975)	1,000	355×221×130 (rectangular)	16,900 bbl	14,400 bbl	8,500 cft 1,530 bbl	6,000 bags 1,530 bbl
Chevron Oil Company, P.O. Box 128, Harvey, Louisiana, USA								
S45 Rowan 19	submersible	USA 1956	25	156×90×11 (rectangular)	155 tons	490 tons	660 bbl	3,000 bags
S55 Noble 20	submersible	USA 1956	63	190×178×12 (rectangular)	206 tons	900 tons	1,190 bbl	3,000 bags
Compania Shell de Venezuela, Apartado 809, Caracas, Venezuela								
GP 9	tender/fixed platform	Neth. 1958	150	200×100×10 (rectangular)	372 m^3	1,422 m^3	412 m^3	—
GP 10	tender/fixed platform	Neth. 1958	150	168×80×10 (rectangular)	696 m^3	1,422 m^3	412 m^3	—
Consumers Gas Company, 19 Toronto Street, Toronto, Canada								
Mr Neil	self-elevating/ self-propelled	USA 1967	80	118×52 (catamaran)	280 bbl	280 bbl	500 bags 240 bbl	500 bags
Creole Petroleum Corporation, Apartado 889, Edificio Creole, Caracas, Venezuela								
Creole C 201 *Creole C 202* *Creole C 203* *Creole C 225* *Creole C 226* *Creole C 334*	drill barges	converted 1963–67	110	180×70×11·25 (rectangular)	1,600 bbl	800 bbl	2,980 bbl	—
	drill barges	converted 1967–71	110	180×70×11 (rectangular)	730 bbl	850 bbl	1,700 bbl	—
Crestwave Offshore Services Inc., 500 Veterans' Memorial Bvd., P.O. Drawer J, Metairie, La. 70005, USA								
Topper I	self-elevating	USA 1967	120	154×128×18 (triangular)	600 bbl	1,700 bbl	1,640 cft 700 bbl	1,640 cft
Topper II	self-elevating	USA 1968	125	165×142×19 (triangular)	750 bbl	2,500 bbl	4,000 cft 1,300 bbl	4,000 cft
Topper III	self-elevating	USA 1972	300	230×200×25 (triangular)	2,000 bbl	5,000 bbl	4,200 cft 1,500 bbl	2,200 cft

Name OWNER/OPERATOR	Type	Built	Limiting Water Depth (ft)	Dimensions of Platform/Hull (ft) and shape Length×breadth ×depth (draught)	Tank Capacities (bbl=barrel) Fuel	Freshwater	Mud (dry/liquid)	Cement
Deep Sea Drilling Company, P.O. Box 25, 5032 Minde, Norway								
Bergendrill 1	semi-submersible	Norway 1974	1,000	355×220×130	16,326 bbl	12,500 bbl	8,500 cft 1,530 bbl	8,500 cft
Bergendrill 2	semi-submersible	Norway (1975)	1,000	355×220×130	16,326 bbl	12,500 bbl	8,500 cft 1,530 bbl	8,500 cft
Diamond M. Drilling Company, P.O. Box 22738, Houston, Texas 77026, USA								
Century	semi-submersible	USA 1973	600	270×173×85	5,100 bbl	10,200 bbl	9,120 cft 1,575 bbl	3,500 bags
Diamond M 99	self-elevating	USA 1972	300	208×178×22 (triangular)	4,000 bbl	6,300 bbl	6,200 cft	3,500 bags
Under construction	semi-submersible	USA (1974)	600– 1,000	290×200×95			9,150 cft 1,575 bbl	3,500 bags
On order	self-elevating	USA (1975)						
On order	self-elevating	USA (1976)						
On order	semi-submersible	USA (1975)						
On order	semi-submersible	USA (1976)						
Dixilyn Corporation, 1012 First City Nat. Bank Bldg., Box 14067, Houston, Texas 77021, USA								
Dixilyn 150	self-elevating	USA 1966	185	193·6×182×23·5 (triangular)	1,000 bbl	6,000 bbl	4,400 cft 1,250 bbl	3,080 cft
Dixilyn 370	self-elevating	USA modified 1972	370	195×200 (triangular)	1,500 bbl	4,000 bbl	6,820 cft 1,500 bbl	1,320 cft
Pentagone 85	semi-submersible	USA 1974	660	323×336×22·4			14,480 cft 2,600 bbl	6,600 bags
Pentagone 86	semi-submersible	USA 1974	660	323×336×22·4			14,480 cft 2,600 bbl	6,600 bags
Dolphin International/Fred Olsen								
Bideford Dolphin	semi-submersible	Norway 1974	1,000	365×220×120			17,000 cft 1,450 bbl	600 bags
Borgyn Dolphin	semi-submersible	Norway 1974	1,000	365×220×120			17,000 cft 1,450 bbl	6,000 bags
Dresser Offshore Services Inc., 1360 Oil & Gas Bldg., New Orleans, La. 70112, USA								
Dresser I	self-elevating	USA 1970	70	50×96 (rectangular)	150 bbl	360 bbl	3,000 cft 190 bbl	1,500 bags
Dresser III	self-elevating	USA 1968	70	65×114 (rectangular)	180 bbl	620 bbl	1,000 bags 200 bbl	500 bags
Dresser IV	self-elevating	USA 1968	70	65×112 (rectangular)	180 bbl	620 bbl	1,000 bags 200 bbl	500 bags
Dresser V	self-elevating	USA 1969	70	65×112 (rectangular)	180 bbl	620 bbl	1,000 bags 200 bbl	500 bags
Dresser VI	self-elevating	USA 1969	70	65×112 (rectangular)	180 bbl	620 bbl	1,000 bags 200 bbl	500 bags
Dresser VII	self-elevating	USA 1969	60	50×96 (rectangular)	150 bbl	360 bbl	1,500 bags 190 bbl	1,500 bags

Name OWNER/OPERATOR	Type	Built	Limiting Water Depth (ft)	Dimensions of Platform/Hull (ft) and shape Length×breadth ×depth (draught)	Tank Capacities (bbl=barrel) Fuel	Freshwater	Mud (dry/liquid)	Cement

Field International Corporation,
930 Milan Bldg., San Antonio, Texas 78205
USA

Name	Type	Built	Limiting Water Depth (ft)	Dimensions	Fuel	Freshwater	Mud (dry/liquid)	Cement
Earl Rowe— San Antonio	self-elevating	Singapore 1971	250	203×168×22	2,515 bbl	6,120 bbl	26,000 cft 2,600 bbl	16,000 cft
Bob Bushman	self-elevating	Singapore 1974	250	203×168×22	2,515 bbl	6,120 bbl	26,000 cft 2,600 bbl	16,000 cft
Pat Rutherford	semi-submersible	USA 1974	600	330×205×126			10,250 cft 2,000 bbl	5,000 bags

Drillships Ltd. (Atwood Oceanics)

Regional Endeavor, *Regional Explorer* — drill ship — These vessels are reported to be under conversion in Australia for completion in 1974. They are understood to have limiting depth of 1,000 ft.

Fearnley & Eger
Oslo

Name	Type	Built	Limiting Water Depth (ft)
On order	semi-submersible	Finland (1976)	600+
On order	semi-submersible	Finland (1977)	600+

Fluor Corporation Ltd.,
2500 S. Atlantic Blvd., Box 7030,
East La, Branch, Los Angeles, California
90022 USA

Name	Type	Built	Limiting Water Depth (ft)	Dimensions	Fuel	Freshwater	Mud (dry/liquid)	Cement
Mr Arthur	submersible	USA 1957	80	181×151×16	1,800 bbl	4,000 bbl	7,000 bags 1,200 bbl	1,000 bags
Mr Gus II	self-elevating	USA 1957	180	173·3×104×22·5	1,800 bbl	8,475 bbl	4,400 bags 1,660 bbl	1,000 bags
Mr Mel	self-elevating	USA 1972	300	230×200×36 (rectangular)	2,000 bbl	6,000 bbl	4,000 cft 1,200 bbl	3,000 cft
Mr Sam	self-elevating	USA 1967	155	162×108×14	1,250 bbl	6,120 bbl	7,600 bags 1,244 bbl	2,050 cft
Mr Si	self-elevating	USA (1974)	350					
Caldrill I	drill ship	USA 1964	600	176×33 (ship shape)	476 bbl	476 bbl	300 bbl	300 bags
Wodeco I	drillbarge	USA 1958	600	252×68×(17·5) (ship shape)	1,835 bbl	9,400 bbl	1,050 bbl	2,100 cft
Wodeco III	drillbarge	USA 1964	1,000	365×75×(32) (ship shape)	7,806 bbl	20,575 bbl	6,400 cft 2,700 bbl	3,200 cft
Wodeco IV	drillbarge	USA 1966	1,500	360×82×(28·4) (ship shape)	10,000 bbl	19,238 bbl		
Wodeco V	drillbarge	USA 1965	1,000	365×75×(32) (ship shape)		20,575 bbl	6,400 cft 2,770 bbl	
Wodeco VI	drillbarge	converted 1967	600	342×76×(19)	6,000 bbl	14,550 bbl	4,200 cft 1,500 bbl	2,100 cft
Wodeco VII	drillbarge	converted 1967	600	446×80×(29)	12,700 bbl	19,500 bbl	11,500 cft 3,270 bbl	11,500 cft
Wodeco VIII	drillbarge	Taiwan 1974	600	487×82·5×42·5	23,000 bbl	42,000 bbl	15,400 cft 6,120 bbl	

Name OWNER/OPERATOR	Type	Built	Limiting Water Depth (ft)	Dimensions of Platform/Hull (ft) and shape Length×breadth ×depth (draught)	Tank Capacities (bbl=barrel) Fuel	Freshwater	Mud (dry/liquid)	Cement
Foramer SA, 7 Avenue de L'Europe 92310 Sévres, France								
Aramco No 1	self-elevating	USA 1958	90	93×104×10 (rectangular)				
Ile de France	self-elevating	Neth. 1966	200	181×176×24 (pentagonal)	2,200 bbl	8,900 bbl	10,000 cft	2,200 bags
Forex Neptune, 20 Avenue Rapp, Paris 7e, France								
Neptune 1	self-elevating	France 1965	250	200×166×23 (triangular)	2,800 bbl	1,400 bbl	9,700 cft 1,250 bbl	
Neptune Gascogne	self-elevating	France 1965	290	164×174×23 (triangular)	3,500 bbl	1,600 bbl	10,500 cft 1,250 bbl	
Neptune 7	semi-submersible	France 1969	660	332×318×35	3,600 bbl	6,430 bbl	14,150 cft 1,800 bbl	
Pentagone 82	semi-submersible	USA 1973	660	338×325×317 (pentagonal)	4,830 bbl	8,475 bbl	13,165 cft 2,228 bbl	
Pentagone 83	semi-submersible	France 1974	660	338×325×317 (pentagonal)	4,830 bbl	8,475 bbl	13,165 cft 2,228 bbl	
Pentagone 84	semi-submersible	Finland 1974	660	338×325×317 (pentagonal)	4,763 bbl	6,900 bbl	13,165 cft 2,228 bbl	
Global Marine Inc., Global Marine House, 811 W Seventh Street, Los Angeles, California 90051, USA								
Cuss I	drillbarge	Converted Navy craft	12,000	260×58×(10) (ship shape)	1,430 bbl	1,500 bbl	850 cft 4,722 bbl	
Glomar II	drillship	USA 1962	600	267·8×58×(16·4) (ship shape)	2,000 bbl	11,200 bbl	5,000 cft 500 bbl	2,740 cft
Glomar III	drillship	USA 1963	600	267·8×58×(16·4) (ship shape)	2,000 bbl	11,200 bbl	5,000 cft 500 bbl	2,300 cft
Glomar IV	drillship	USA 1963	600	267·8×58×(16·4) (ship shape)	2,450 bbl	14,900 bbl	5,000 cft 500 bbl	
Glomar V	drillship	USA 1963	600	267·8×58×(16·4) (ship shape)	2,450 bbl	14,900 bbl	5,000 cft 500 bbl	2,300 cft
Glomar XII	drillship	USA 1972	600	400×65×(20) (ship shape)	8,150 bbl	15,630 bbl	11,840 cft 3,000 bbl	4,540 cft
Glomar Challenger	drillship	USA 1968	600	400×65×(20) (ship shape)	10,873 bbl	15,634 bbl	6,150 cft 3,000 bbl	6,150 cft
Glomar Conception	drillship	USA 1967	600	400×65×(21) (ship shape)	8,150 bbl	15,630 bbl	6,150 cft 3,000 bbl	6,150 cft
Glomar Grand Banks	drillship	USA 1972	600	380×65×27 (ship shape)	8,150 bbl	15,712 bbl	13,900 cft 3,100 bbl	4,540 cft
Glomar Grand Isle	drillship	USA 1967	600	400×65×(21) (ship shape)	8,150 bbl	15,634 bbl	6,150 cft 3,000 bbl	4,540 cft
Glomar North Sea	drillship	USA 1965	600	380×64×(19) (ship shape)	1,778 bbl	13,570 bbl	7,200 cft 3,700 bbl	1,820 cft
Glomar Sirte	drillship	USA 1965	600	380×64×(19) (ship shape)	1,333 bbl	13,570 bbl	7,640 cft 3,700 bbl	1,820 cft
Glomar Tasman	drillship	USA 1965	600	380×64×(19) (ship shape)	1,778 bbl	13,570 bbl	7,200 cft 3,700 bbl	1,820 cft
Glomar Coral Sea	drillship	USA 1974	600	410×65×(21) (ship shape)			6,590 cft 3,000 bbl	
Under construction	drillship	USA (1975)	600	450×72×35 (ship shape)			6,590 cft 3,000 bbl	
Poly-Glomar Driller	semi-submersible	Norway (1975)	1,000	355×220×120	2,180 bbl	2,170 bbl	17,000 cft 1,450 bbl	6,000 bags

Name OWNER/OPERATOR	Type	Built	Limiting Water Depth (ft)	Dimensions of Platform/Hull (ft) and shape Length×breadth ×depth (draught)	Tank Capacities (bbl=barrel) Fuel	Freshwater	Mud (dry/liquid)	Cement
Golden Lane Drilling Company, 1011 Tremont Street, Galveston, Texas 77550, USA								
Goldrill IV	drillship	USA 1966	600	205×44×19 (ship shape)	852 bbl	5,500 bbl	5,000 cft	3,000 cft
Gotaas Larsen-Rowan Exploration Company, 122 E 42nd Street, New York, NY 10017 USA								
Norjald	semi-submersible	Norway 1974	600	320×266×128	6,900 bbl	14,580 bbl	4,800 cft 1,500 bbl	4,800 cft
Norskald	semi-submersible	Norway 1974	600	355×221×120	6,900 bbl	14,580 bbl	4,800 cft 1,500 bbl	4,800 cft
Gowart/Olsen								
Pentagone 88	semi-submersible	Finland (1975)						
Pentagone 89	semi-submersible	Finland (1976)						
Indian Oil and Natural Gas Commission, Directorate of Drilling, Dehran Dun, Uttar Pradesh, India								
Sagar Samrat	self-elevating	Japan 1972	250	181×176×24 (pentagonal)	10,000 bbl	3,700 bbl	6,000 cft 1,600 bbl	
Iran Pan American Oil Company (IPAC) Box 1466, Teheran, Iran								
Pan II			170	200×104×15 (rectangular)	2,400 bbl	4,400 bbl	4,200 bags 1,500 bbl	2,800 bags
Impexmin								
Under construction	self-elevating	Romania (197 ?)	300					
Japan Drilling Co. Ltd., Mori Bldg 11, Akefune-cho, Shiba-Nishikubo, Minatu-ku, Tokyo, Japan								
Fuji	self-elevating	Japan 1969	175	156×105×18 (rectangular)	1,500 bbl	3,940 bbl	2,700 cft 3,200 bbl	2,700 cft
Hakurgo Go	self-elevating	Japan 1958	110	203×151×20 (triangular)	850 bbl	6,000 bbl	3,900 cft 6,150 bbl	1,900 cft
Hakuryo II	semi-submersible	Japan 1971	656	275·5×200 (rectangular)	5,742 bbl	7,181 bbl	11,537 cft 13,414 cft	
Hakuryo III	semi-submersible	Japan 1974	820	275·5×200 (rectangular)	5,742 bbl	7,181 bbl	11,537 cft 13,414 cft	
Hakuryo IV	semi-submersible	Japan (1975)	850	275·5×200 (rectangular)				

Name OWNER/OPERATOR	Type	Built	Limiting Water Depth (ft)	Dimensions of Platform/Hull (ft) and shape Length×breadth ×depth (draught)	Tank Capacities (bbl=barrel) Fuel	Freshwater	Mud (dry/liquid)	Cement
Key International Drilling Company, P.O. Box 268, Hamilton, Bermuda								
Key Biscayne	self-elevating	Singapore 1972	250	203×168×22 (triangular)	2,000 bbl	5,500 bbl	4,000 cft 1,300 bbl	4,000 cft
Key West	self-elevating	USA 1973	300	230×200×36 (triangular)	2,000 bbl	6,000 bbl	4,000 cft 1,000 bbl	5,000 cft
Key Victoria	self-elevating	UK 1974	250					
Loffland Brothers Company, P.O. Box 4567, Tulsa, Oklahoma 74114 USA								
Ocean Master II	self-elevating	USA 1966	300	167·3×180·1 (hexagonal)	2,425 bbl	6,630 bbl	1,850 bbl	4,000 bags
Marine Drilling Company, 900 Corpus Christi State National Bldg., Corpus Christi, Texas 78401 USA								
J Storm I	self-elevating	USA 1969	225	170×210×16		5,250 bbl	9,500 cft 1,500 bbl	
J Storm III	self-elevating	USA 1974	225	170×210×16			9,500 cft 1,500 bbl	
Marlin Drilling Co. Inc., P.O. Box 51887, OCS, Lafayette, La. 70501, USA								
Marlin 3	self-elevating	USA 1966	265	164×86×20	1,500 bbl	4,650 bbl	6,300 cft 1,200 bbl	6,300 cft
Marlin 4	self-elevating	USA 1966	173	168×86×20	1,500 bbl	5,490 bbl	6,300 cft 1,2000 bbl	6,300 cft
Marlin 6	self-elevating	USA 1971	300	208×178×22	4,100 bbl	5,430 bbl	5,000 cft 1,200 bbl	3,000 cft
Marlin 7	semi-submersible	USA (1975)	1,000	222×175			11,550 cft 2,000 bbl	
Mission Drilling & Exploration Corporation, 736 Poydras Street, New Orleans La. 70130, USA								
Mission Exploration	drillship	USA 1973	600	523×68×39	18,000 bbl	16,056 bbl	9,000 cft 2,000 bbl	12,000 bags
Mitchell, George & Associates, 3900 The Shell Plaza, Houston, Texas 77002, USA								
Nordrill	drillbarge	converted 1961	200	245×45×17				
Movible Offshore Inc.,								
Movible Rig 3	submersible	USA 1955	45	200×146×11	970 bbl	7,064 bbl	1,209 bbl	5,900 bags
Movible Rig 16	self-elevating	Singapore 1973	250	109×166×16 (rectangular)	1,796 bbl	4,722 bbl	3,000 cft	3,150 cft
Movible Rig 17	self-elevating	Singapore 1974	250	109×166×16 (rectangular)	1,796 bbl	4,722 bbl	1,671 bbl 3,000 cft 1,671 bbl	3,150 cft

Name OWNER/OPERATOR	Type	Built	Limiting Water Depth (ft)	Dimensions of Platform/Hull (ft) and shape Length×breadth ×depth (draught)	Tank Capacities (bbl=barrel) Fuel	Freshwater	Mud (dry/liquid)	Cement
Narvaz S.A.R.L., 7, Avenue de l'Europe, 92310 Sévres, France								
Astragale	drillship	France 1967(C)	1,800	197×42·5	500 bbl	3,660 bbl	225 bbl	
Noble Drilling Corporation, 2200 Fourth National Bank Bldg., Tulsa, Oklahoma 74119, USA								
Noble S 55	submersible	USA 1955	60	190×178×12	1,680 bbl	5,455 bbl	5,550 cft 1,190 bbl	2,475 cft
Nordic Offshore Drilling Company, Oslo, Norway								
Havdrill	drillship	Neth. 1973	1,000	490×70×(24)	3,725 tons	280 tons		
A/S Norsedrill,								
Drillmaster	semi-submersible	France 1973	660	338×325×317	4,830 bbl	8,275 bbl	13,160 cft 2,440 bbl	
Ocean Drilling & Exploration Company (ODECO) 1600 Canal Street, P.O. Box 61780, New Orleans, La. 70112, USA								
Barge A	submersible	USA 1957	75	190×160×13	1,233 bbl	4,850 bbl	1,200 bbl	
El Dorado	submersible	USA 1957	70	300×200×16	3,500 bbl	4,500 bbl	1,200 bbl 7,000 bags	2,000 cft
John Hayward	submersible	USA 1955	30	180×120×14	1,000 bbl	4,000 bbl	1,200 bbl	
Margaret	submersible	USA 1957	65	300×200×16 (rectangular)	3,500 bbl	4,500 bbl	1,200 bbl	
Mr Charlie	submersible	USA 1954	80	220×134×14 (rectangular)	1,200 bbl	3,500 bbl	1,200 bbl	
ODECO Seven	submersible	USA 1957	35	180×120×15	1,050 bbl	3,750 bbl	3,500 cft 1,600 bbl	2,500 cft
Ocean Tide	drillship (self-elevating)	UK 1971	250	276×130×22 (rectangular)	3,000 bbl	5,000 bbl	4,500 cft 1,600 bbl	4,500 cft
Ocean King	self-elevating	USA 1973	300	197×242× (triangular)	2,000 bbl	6,000 bbl	4,000 cft 1,400 bbl	
Ocean Star	self-elevating	USA 1969	173	168×86×20	1,500 bbl	4,800 bbl	7,340 cft 1,200 bbl	2,000 bags
Ocean 66	self-elevating	USA 1958 (overhauled 73)	100	217·5×117×14	1,144 bbl	6,700 bbl	6,800 cft 1,300 bbl	
Ocean Digger	semi-submersible	Australia 1967	600	365·5×217·5 (rectangular)	4,400 bbl	10,230 bbl	900 bbl 2,100 bags	5,000 bags
Ocean Driller	semi-submersible	USA 1963	600	377 each side (triangular)	4,750 bbl	6,800 bbl	970 bbl 7,000 bags	3,000 bags
Ocean Endeavour	semi-submersible	Australia 1974	600	320×262 (octagonal)		14,670 bbl	9,600 cft 1,500 bbl	3,000 bags
Ocean Explorer	semi-submersible	USA 1964	600	377 each side (triangular)	3,170 bbl	6,800 bbl	970 bbl 2,000 bags	5,000 bags
Ocean Koknei	semi-submersible	Japan 1973	600	319×266×128 (octagonal)		14,670 bbl	9,600 cft 1,500 bags	3,000 bags
Ocean Prospector	semi-submersible	Japan 1972	600	263×340×126	4,000 bbl	13,700 bbl	4,500 bags	4,500 bags
Ocean Queen	semi-submersible	USA 1965	600	365×217 (rectangular)	4,460 bbl	9,430 bbl	2,100 bags 900 bbl	4,000 bags
Ocean Rover	semi-submersible	USA 1973	600	320×262 (octagonal)		14,670 bbl	9,600 cft 1,500 bbl	3,000 bags
Ocean Ranger	semi-submersible (self-propelled)	Japan (1975)	1,500					
Ocean Scout	semi-submersible	USA 1974	600	202×182 (rectangular)		10,380 bbl	7,500 cft 1,150 bbl	3,000 bags
Ocean Traveler	semi-submersible	USA 1966	500	362×217 (rectangular)	4,460 bbl	9,430 bbl	2,100 bags 900 bbl	4,000 bags
Ocean Victory	semi-submersible (self-propelled)	USA 1972	600	320×362 (octagonal)	6,980 bbl	14,670 bbl	9,320 cft 1,370 bbl	
Ocean Viking	semi-submersible	Norway 1967	600	369·5×300×136 (rectangular)	4,460 bbl	9,430 bbl	2,100 bags 900 bbl	4,000 bags
Ocean Voyager	semi-submersible (self-propelled)	Norway 1974	600	320×362 (octagonal)	6,980 bbl	14,670 bbl	2,100 bags 900 bbl	4,000 bags

Name OWNER/OPERATOR	Type	Built	Limiting Water Depth (ft)	Dimensions of Platform/Hull (ft) and shape Length × breadth × depth (draught)	Tank Capacities (bbl=barrel) Fuel	Freshwater	Mud (dry/liquid)	Cement
Offshore Company, 3411 Richmond Avenue, P.O. Box 2765 Houston, Texas 77001, USA								
Discoverer I	drillship	USA 1963	1,200	320×70×25 (ship shape)	3,148 bbl	7,961 bbl	1,617 bbl	
Discoverer II	drillship	Japan 1966	1,200	362×70×26 (ship shape)	5,170 bbl	7,960 bbl	1,700 bbl	
Discoverer III	drillship	Japan 1970	1,000	380·5×70×26 (ship shape)	5,170 bbl	7,960 bbl	1,700 bbl	
Discoverer 534	drillship	USA 1974	3,000	524×80×32 (ship shape)	7,750 bbl	24,570 bbl	4,980 cft 1,200 bbl	
On order	drillship							
Antares	self-elevating	Singapore 1973	110	150×120	1,600 bbl	2,000 bbl	2,000 cft 990 bbl	
Enterprise	self-elevating	Germany 1967	118	190×106×15 (rectangular)	3,000 bbl	5,801 bbl	9,800 cft 1,500 bbl	9,800 cft
Estrellita	self-elevating (tender assisted)	USA 1969	150	132×108×15 (rectangular)				
Gemini	self-elevating (tender assisted)	Singapore 1973	200	112×112×16 (square)				
Hustler	self-elevating	USA 1965	90	122×108×12·3	700 bbl	1,360 bbl	600 bbl 500 cft	
Jubilee	self-elevating	USA 1965	200	170×130×17	3,000 bbl	3,700 bbl	1,400 bbl	
Mercury	self-elevating (self-propelled)	UK 1969	250	276×130×22 (ship shape)	10,000 bbl	3,700 bbl	1,600 bbl	
North Star	self-elevating	UK 1965	250	170×130×17·3	3,000 bbl	3,700 bbl	1,400 bbl	5,000 bags
No 52	self-elevating	USA 1955	100	220×117×13·5	2,124 bbl	6,955 bbl	1,448 bbl	2,100 bags
No 54	self-elevating	USA 1956	125	200×104×15	3,970 bbl	9,250 bbl	210 tons 1,070 bbl	50 tons
No 55	self-elevating	USA 1956	120	200×105×15	3,000 bbl	3,590 bbl	1,000 cft 1,070 bbl	50 tons
No 60	self-elevating (tender assisted)	USA 1961	120	108×101×12				
Orion	self-elevating	UK 1966	275	179×150×17 (rectangular)	2,608 bbl	4,700 bbl	10,000 cft 1,000 bbl	
Regasus	self-elevating	Germany 1957	125	188×106×13 (rectangular)	1,000 bbl	1,438 bbl	716 bbl	
Polar Cub	self-elevating	USA 1964	110	108×101×12 (triangular)				
On order	self-elevating	Singapore (1975)						
Chris Chenery	semi-submersible	Germany 1974	1,000	270×233×20	8,720 bbl	8,455 bbl	10,500 cft 1,700 bbl	
Colonel Drake	semi-submersible	USA 1974	600	214×178×20	7,670 bbl	5,650 bbl		
SCP III Mark 2	semi-submersible	Spain 1975	1,000	270×233×20	8,720 bbl	8,455 bbl	10,500 cft 1,700 bbl	
Pentagone 87	semi-submersible	France 1975	660	338×325×317			13,160 cft 2,440 bbl	
Oslo Drilling A/S, Kangensgate 6, Oslo 1, Norway								
Odin	semi-submersible	Norway 1974	600	326×283	6,942 bbl	14,600 bbl	12,000 cft 1,600 bbl	
Penrod Drilling Company, 3333 First National Bank Bldg., Dallas, Texas 75202, USA								
Penrod 50	submersible	USA 1958	50	196×134 (rectangular)	1,482 bbl	7,960 bbl	8,166 bbl	
Penrod 51	submersible	USA 1957	60	190×150 (rectangular)	980 bbl	3,923 bbl	1,384 bbl	
Penrod 53	self-elevating	USA 1965	300	167·25×180·1 (hexagonal)	1,000 bbl	6,000 bbl	825 bbl	
Penrod 54	self-elevating	USA 1966	300	167·25×180·1 (hexagonal)	1,000 bbl	6,000 bbl	825 bbl	
Penrod 55	self-elevating	USA 1966	150	200×182 (triangular)	1,000 bbl	6,000 bbl	825 bbl	
Penrod 56	self-elevating	USA 1966	150	200×182 (triangular)	1,000 bbl	6,000 bbl	825 bbl	
Penrod 57	self-elevating	USA 1966	150	200×182 (triangular)	1,000 bbl	6,000 bbl	825 bbl	
Penrod 58	self-elevating	USA 1968	220	242×197×22 (triangular)	2,100 bbl	3,985 bbl	6,000 cft 1,275 bbl	6,000 cft
Penrod 59	self-elevating	USA 1968	225	242×197×22 (triangular)	5,000 bbl	3,985 bbl	6,000 cft 1,275 bbl	6,000 cft

Name OWNER/OPERATOR	Type	Built	Limiting Water Depth (ft)	Dimensions of Platform/Hull (ft) and shape Length×breadth ×depth (draught)	Tank Capacities (bbl=barrel) Fuel	Freshwater	Mud (dry/liquid)	Cement
Penrod 60	self-elevating (propulsion asstd)	USA 1971	300	230×200×26 (pentagonal)	5,886 bbl	2,836 bbl	6,500 cft 1,700 bbl	
Penrod 61	self-elevating (propulsion asstd)	USA 1972	350	230×200×26 (pentagonal)	5,886 bbl	2,836 bbl	6,500 cft 1,700 bbl	
Penrod 62	self-elevating (propulsion asstd)	USA 1972	350	230×200×26 (pentagonal)	5,886 bbl	2,836 bbl	6,400 cft 1,700 bbl	
Penrod 63	self-elevating (propulsion asstd)	USA 1973	330	209×220×25 (pentagonal)	5,400 bbl	3,050 bbl	6,000 cft 1,250 bbl	
Penrod 64	self-elevating (propulsion asstd)	USA 1973	300	230×200×26 (pentagonal)		4,352 bbl	6,500 cft 1,400 bbl	
Penrod 70	semi-submersible (propulsion asstd)	Spain 1973	800	245×185×111 (rectangular)		5,150 bbl	4,500 cft 1,700 bbl	
Penrod 71	semi-submersible (propulsion asstd)	USA 1973	1,000	288×216×138 (rectangular)	7,200 bbl	7,200 bbl	8,000 cft 1,700 bbl	
Penrod 72	semi-submersible (propulsion asstd)	USA 1974	1,000	288×216×138 (rectangular)	7,200 bbl	7,200 bbl	8,000 cft 1,700 bbl	
Penrod 73	semi-submersible (propulsion asstd)	USA 1974	1,000	255×191×128 (rectangular)	12,000 bbl	5,620 bbl	8,000 cft 1,700 bbl	
Penrod 74	semi-submersible (propulsion asstd)	Singapore 1974	1,000	255×191×128 (rectangular)	12,000 bbl	5,620 bbl	8,000 cft 1,700 bbl	
Penrod 75	semi-submersible (propulsion asstd)	USA 1974	1,000	288×216×138 (rectangular)	7,200 bbl	7,200 bbl	8,000 cft 1,700 bbl	
Perforadora Mexico SA, Av. Insurgentes Sur 432, 90 Piso, Mexico 7								
Reforma	drillbarge	USA 1966	600	381×75×21 (ship shape)	7,750 bbl	24,567 bbl	4,980 cft 1,200 bbl	4,980 cft
Revolucion	drillbarge	USA 1966	600	381×75×27 (ship shape)	7,750 bbl	24,567 bbl	4,980 cft 1,200 bbl	4,980 cft
Petroleo Brasileiro SA, Praca 22 de April, 36, Rio De Janiero, Brazil								
Petrobras I	self-elevating	Portugal 1968	100	145×108×15 (rectangular)	1,480 bbl	400 bbl	1,300 bbl	
Petrobras II	drillship	Japan 1973	984	362·6×70×26 (ship shape)	5,170 bbl	7,960 bbl	7,060 cft 1,700 bbl	
Petrobras III	self-elevating	USA 1974	320	234×210×26	7,000 bbl	4,000 bbl	7,200 cft 1,600 bbl	
Petroleos Mexicanos, Ave Marina Nacional 329, Mexico 17								
Independencia	drillbrage	USA 1966	600	381×75×27 (ship shape)	7,750 bbl	24,567 bbl	1,200 bbl	4,980 cft
Reading & Bates								
Douglas Carver	drillship	UK 1974	600	475×72×23 (ship shape)			15,000 cft 2,000 bbl	
E. W. Thornton	drillship	USA 1965	1,000	278×105×34 (catamaran)	5,200 bbl	11,440 bbl	6,000 cft 2,600 bbl	
J. W. Bates	drillship	Denmark 1967	600	638×77×57·25 (ship shape)	20,000 bbl	90,224 bbl	2,750 cft 1,950 bbl	
Chris Seger	self-elevating	USA 1970	115	176×162×20	1,500 bbl	5,500 bbl	10,000 bags 1,000 bbl	5,000 bags
Milton G. Hulme	self-elevating	Singapore 1974	250	180×132×16			6,000 cft 1,500 bbl	3,000 bags
Mr Jack	self-elevating	USA 1965	300	205×204×23·5 (triangular)	1,480 bbl	5,200 bbl	5,000 cft 1,200 bbl	2,000 bags
W. D. Kent	self-elevating	USA 1969	175	148×140 (triangular)	800 bbl	13,000 bbl	4,000 cft 2,300 bbl	2,200 cft
Mr Louie	self-elevating	USA 1958	150	241×133× (rectangular)	1,778 bbl	4,326 bbl	1,692 bbl	
J. W. McLean	self-elevating	USA 1970	225	180×132×16 (square)	894 bbl	4,790 bbl	6,000 cft 1,500 bbl	
D. M. Saunders	self-elevating	USA 1970 (modified)	150	272×60×27 (triangular)	3,079 bbl	10,614 bbl	5,500 cft 2,300 bbl	3,300 cft
C. E. Thornton	self-elevating	USA 1958	110	153×139×15	500 bbl	4,000 bbl	1,000 bbl	4,000 bags
Demaga	self-elevating	USA 1974	300	230×200·5×26			7,280 cft 1,200 bbl	5,000 bags
F. G. McLintock	self-elevating	USA (1975)	300	230×200×26			7,280 cft 1,200 bbl	5,000 bags

Name OWNER/OPERATOR	Type	Built	Limiting Water Depth (ft)	Dimensions of Platform/Hull (ft) and shape Length×breadth ×depth (draught)	Tank Capacities (bbl=barrel) Fuel	Freshwater	Mud (dry/liquid)	Cement
A/S Rosshavet & A/S Vesthold								
Under construction	semi-submersible	Norway (1975)	1,000	355×220×120	2,180 bbl	2,170 bbl	1,700 cft 2,170 bbl	
Robray Offshore Drilling Company, 21 Robin Home, Sherton Way, Singapore 1, Rep. of Singapore								
Robray 300	self-elevating	Singapore 1974	300	236×212				
Rowan Companies Inc., 1900 Postoak Tower Bldg., 5051 Westheimer, Houston, Texas 77027, USA								
Rig 4	submersible	USA 1970	20	210×54×48	1,814 bbl	6,500 bbl	1,250 cft 1,450 bbl	1,250 cft
R-M City	submersible	USA 1973	20	200×70	2,800 bbl	7,650 bbl	2,250 cft 2,800 bbl	
Rowan Alaska	self-elevating	USA (1975)	300					
Rowan Anchorage	self-elevating (propulsion asstd)	Singapore 1972	250	203×168×22 (triangular)	3,400 bbl	5,740 bbl	3,000 cft 1,300 bbl	2,000 cft
Rowan Houston	self-elevating (propulsion asstd)	USA 1969	200	203×168×22	3,400 bbl	5,470 bbl	3,000 cft 1,300 bbl	2,000 cft
Rowan Louisiana	self-elevating	USA (1975)	350					
Rowan New Orleans	self-elevating (propulsion asstd)	USA 1970	200	203×168×22	3,400 bbl	5,470 bbl	3,000 cft 1,300 bbl	2,000 cft
Rowan Texas	self-elevating (propulsion asstd)	USA 1972	225	203×168×22	3,400 bbl	5,470 bbl	3,000 cft 1,300 bbl	2,000 cft
Rowan Midland	semi-submersible	USA (1975)	600					
Saipem SpA, CP 4159, Milan, Italy								
Saipem Due	drillship	Italy 1973	—	431×72×40 (ship shape)	110,500 bbl	2,300 tons	10,000 cft 3,600 bbl	5,000 cft
Gatto Selvatico	self-elevating	Italy 1961	150	186×152×22 (triangular)	16,200 bbl	19,500 bbl	14,300 bbl	
Rerro Negro	self-elevating	Italy 1961	120	186×152×22 (triangular)	16,200 bbl	19,500 bbl	14,300 bbl	
Scarabeo II	semi-submersible	Italy 1968	600	348×300×156 (triangular)	5,000 bbl	10,925 bbl	1,415 bbl	9,600 cft
Scarabeo III	semi-submersible (propulsion asstd)	Germany 1974	1,000	358×329×146 (triangular)	4,800 bbl	10,750 bbl	10,000 cft 1,700 bbl	
Scarabeo IV	semi-submersible (propulsion asstd)	Germany 1975	1,000	358×329×146 (triangular)	4,810 bbl	10,750 bbl	10,000 cft 1,700 bbl	
Salvesen Offshore Drilling Company Ltd., 50 East Fettes Avenue, Edinburgh EH4 1DN, UK								
Dalmahoy	drillship	USA 1973	100	329×50×29 (ship shape)	5,593 bbl	11,026 bbl	4,800 cft	4,800 cft

Name OWNER/OPERATOR	Type	Built	Limiting Water Depth (ft)	Dimensions of Platform/Hull (ft) and shape Length×breadth ×depth (draught)	Tank Capacities (bbl=barrel) Fuel	Freshwater	Mud (dry/liquid)	Cement
Santa Fe International Corporation 2222, One Wilshire Bldg., Los Angeles, California 80017, USA								
Reina del Lago	drillbarge	USA 1971	—	180×70×11				
Rey del Lago	drillbarge	USA 1971	—	180×70×11				
Swamp Master	drillbarge	USA 1966	22	200×70 (rectangular)	2,800 bbl	7,650 bbl	2,400 bbl	—
Swamp Queen	drillbarge	USA 1970	22	200×70 (rectangular)	2,800 bbl	7,650 bbl	2,400 bbl	—
Britannia	self-elevating	UK 1968	230	224×200×25 (triangular)	1,070 bbl	4,800 bbl	8,000 cft 1,850 bbl	8,000 cft
Explorer	self-elevating	1959	160	190×106				
Rig 103	self-elevating	Singapore 1974	250	203×168×22	1,500 bbl	6,000 bb	1,200 bbl	
Rig 105	self-elevating	Singapore 1975	250	203×168×22	1,500 bbl	6,000 bbl	1,200 bbl	
Blue Water 2	semi-submersible	USA 1964	1,000	204·6×204·6×83 (square)	1,350 bbl	6,000 bbl	1,200 bbl	
Blue Water 3	semi-submersible	USA 1966	1,000	220×198×83 (rectangular)	1,350 bbl	6,000 bbl	1,200 bbl	
Blue Water 4	semi-submersible	USA 1974	1,500	220×198×83 (rectangular)		6,300 bbl		
Mariner I	semi-submersible	USA 1968	600	106×270×42 (catamaran)	3,360 bbl	13,830 bbl	5,600 cft	
Mariner II	semi-submersible	USA 1973	600	106×270×42 (catamaran)	3,360 bbl	13,830 bbl	5,600 cft 1,250 bbl	
Southern Cross	semi-submersible	Australia (1975)	1,500					
Seadrilling Netherlands BV, Wagenaarweg 13, The Hague, Netherlands (25 percent owned by Sedco Inc.)								
Sedneth 1	semi-submersible	Neth. 1967	600	270×239×106 (rectangular)	5,080 bbl	10,627 bbl	1,650 bbl 7,500 bags	3,250 bags
Sedneth 2	semi-submersible	Neth. 1968	200	189×183×24 (pentagonal)	2,380 bbl	2,000 bbl	1,200 bbl 7,500 bags	3,250 bags
Sedco/Sedneth 701	semi-submersible	Canada 1973	800	295×245×130	5,000 bbl	8,700 bbl	2,000 bbl	17,325 cft
Seadrilling Corporation, 307 Maryland Casualty Bldg., New Orleans, La. 70112, USA								
Seadrill 2	submersible	USA 1958	35	180×136 (rectangular)	940 bbl	2,000 bbl	1,200 bbl	
Seadrill 3	submersible	USA 1958	35	192×125 (rectangular)	1,485 bbl	2,700 bbl	1,400 bbl	
Shell International Petroleum Maatschappij BV, Box 162, Carel van Bylandlaan, The Hague, Netherlands **Shell UK Exploration and Production Ltd.,** Shell Centre, London SE1 **Shell BP Petroleum Development Company of Nigeria Ltd.,**								
Cowrie One	self-elevating	Neth. 1968	—	164×75·5×12·75	60 tons	50 tons	1,200 bbl	
Seashell	self-elevating	Neth. 1959	120	209×105×17·5	300 tons	6,800 bbl	1,200 bbl	190 tons
Staflo	semi-submersible	UK 1967	600	264×220·5×113 (rectangular)	428 tons	119 tons	1,976 bbl	
Skinner Drilling Company Ltd., P.O. Box 46, San Fernando, Trinidad, W1								
Skillco 145	self-elevating	USA 1959	100	196×80×13 (rectangular)	2,784 bbl	1,392 bbl	1,148 bbl	

Name OWNER/OPERATOR	Type	Built	Limiting Water Depth (ft)	Dimensions of Platform/Hull (ft) and shape Length×breadth ×depth (draught)	Tank Capacities (bbl=barrel) Fuel	Freshwater	Mud (dry/liquid)	Cement
Sedco Incoroporated, Cumberland Hill, 1901 North Akard, Dallas, Texas 75201, USA								
Sedco 445	drillship	Japan 1971	—	445×70×32 (ship shape)	2,368 tons	3,080 tons	3,360 bbl	
Sedco Gusto	self-elevating	Neth. 1971	180	230×91×15	2,000 bbl	2,300 bbl	2,000 bbl	
Sedco 135 *Sedco 135A* *Sedco 135D* *Sedco 135E* *Sedco 135F* *Sedco 135G*	semi-submersible	USA 1965 Japan 1965 Neth. 1966 Japan 1966 Canada 1967 Japan 1968	600	342·5×340×166 (triangular)	5,080 bbl	10,627 bbl	8,105 cft 1,650 bbl	3,250 bags
Sedco H *Sedco I*	semi-submersible	Canada 1970	1,000	360×360×360 ×146 (triangular)	5,080 bbl	10,636 bbl	7,500 bags 1,650 bbl	3,250 bags
Sedco J *Sedco K*	semi-submersible	Canada 1972 USA 1973	800	342·5×340×166 (triangular)	5,080 bbl	10,627 bbl	8,105 cft 1,650 bbl	3,250 bags
Sedco 700	semi-submersible (propulsion asstd)	USA 1973	800	195×225 (rectangular)	5,000 bbl	8,700 bbl	6,000 cft 1,500 bbl	6,000 cft
Sedco 702	semi-submersible (propulsion asstd)	USA 1973	1,000	231×197×130 (rectangular)	5,000 bbl	8,700 bbl	6,000 cft 1,500 bbl	6,000 cft
Sedco 703	semi-submersible (propulsion asstd)	USA 1974	1,000	231×197×130 (rectangular)	5,000 bbl	8,700 bbl	6,000 cft 1,500 bbl	6,000 cft
Sedco 704	semi-submersible (propulsion asstd)	Canada 1974	1,000	195×225 (rectangular)	5,000 bbl	8,700 bbl	6,000 cft 1,500 bbl	6,000 cft
Sedco 705	semi-submersible	Canada (1975)	1,000					
Sedco 706	semi-submersible	USA (1975)	1,000					
Smedvig Tankrederi, P.O. Box 201, 4001 Stavanger, Norway								
West Venture (*Norrig 5*)	semi-submersible	France 1973	600	335×349×153 (pentagonal)	2,500 bbl	3,130 bbl	4,500 cft 2,000 bbl	5,000 cft
Societé Maritime de Service, 162 rue de Faubourg, St Honoré, Paris 8, France								
Pelican	drillship	Neth. 1972	900	490·25×70·25 ×41 (ship shape)	3,500 m³	8,200 bbl	120 m³	
Totalmar 1	drillship	Neth. 1974	1,000	492×70×41	7,000 bbl	8,200 bbl	1,700 bbl	
Southern Marine Drilling Company, Corpus Christi State National Bldg., Room 808, Corpus Christi, Texas 78401, USA								
Stormdrill V	self-elevating	USA 1967	125	145·5×205 ('A' shaped)	737 bbl	3,775 bbl	340 bbl	3,000 bags
Storm Drilling Company, 4141 Southwest Freeway, P.O. Box 22791, Houston, Texas 77027, USA								
Cyclone	drillship	USA 1971	600	379×70×29 (ship shape)	5,136 bbl	18,567 bbl	2,000 bbl	4,870 cft
Typhoon	drillship	USA 1969	600	380×70×29 (ship shape)	5,497 bbl	18,560 bbl	2,000 bbl	11,000 cft
Hurricane	drillship	USA 1974	600	378×70×28 (ship shape)	—		1,200 bbl	9,600 cft
Stormdrill I	self-elevating	USA 1964	70	142·5×94 ('A' shape)	560 bbl	1,990 bbl	1,000 bags	—
Stormdrill II	self-elevating	USA 1965	152	164×224 ('A' shape)	1,550 bbl	4,340 bbl	3,000 bags	—
Stormdrill III	self-elevating	USA 1965	152	164×224	1,550 bbl	4,340 bbl	3,000 bags	—
Stormdrill IV	self-elevating	USA 1966	175	164×224	1,550 bbl	4,340 bbl	3,000 bags	—
Stormdrill VI	self-elevating	USA 1969	224	226×132×16	1,550 bbl	4,340 bbl	1,500 bbl	—
Stormdrill VII	self-elevating	USA 1971	250	166×109×16	1,550 bbl	4,340 bbl	1,500 bbl	4,050 cft
Stormdrill VIII	self-elevating	Singapore (1975)	250					
Zephyr I	semi-submersible	USA 1973	600	202×182 (square)	4,500 bbl	10,400 bbl	9,600 cft 1,600 bbl	
Zephyr II	semi-submersible	USA 1974	600	202×182 (square)	4,500 bbl	10,400 bbl	9,600 cft 1,600 bbl	

Name OWNER/OPERATOR	Type	Built	Limiting Water Depth (ft)	Dimensions of Platform/Hull (ft) and shape Length×breadth ×depth (draught)	Tank Capacities (bbl=barrel) Fuel	Freshwater	Mud (dry/liquid)	Cement
Sun Marine Drilling Company, Raymond International J.V., 4250 Wilshire Blvd., Los Angeles, California 90010, USA								
George F. Ferris	self-elevating	USA 1963	200	190×122·5×17 (barge shape)			1,000 bbl	2,000 bags
John C. Marthons	self-elevating	USA 1963 (mod. 1971)	125	195×102×14 (rectangular)			1,000 bbl	4,500 cft
Transocean Drilling Company, Transworld Drilling Company Ltd., Ken McGee Bldg., Oklahoma City, Oklahoma 73102, USA								
Transocean I	self-elevating	German 1965	195	225·75×140×18 (barge shape)	1,285 bbl	2,400 bbl	6,800 cft 2,000 bbl	7,000 bags
Transocean II	self-elevating	UK 1966	217	225·75×140×18 (barge shape)	1,285 bbl	2,400 bbl	6,800 cft 2,000 bbl	7,000 bags
Transocean II	semi-submersible	German 1974	600	400×58×23	900 bbl	4,260 bbl	4,500 cft 1,500 bbl	4,500 cft
Transworld Rig 44	submersible	USA 1954	40	140×72×13·5 (rectangular)	942 bbl	3,900 bbl	930 bbl	—
Transworld Rig 45	submersible	USA 1956	35	142×69×13·5 (rectangular)	1,064 bbl	3,838 bbl	630 bbl	—
Transworld Rig 46	submersible	USA 1956	70	242×202	1,092 bbl	3,950 bbl	800 bbl	—
Transworld Rig 47	submersible	USA 1957	70	252×202	1,380 bbl	8,780 bbl	800 bbl	
Transworld Rig 54	submersible	USA 1963	175	337×387×20 (triangular)	2,015 bbl	8,780 bbl	1,000 bbl	
Transworld Rig 50	self-elevating	USA 1965	70	170×105 (rectangular)	750 bbl	2,140 bbl	700 bbl	
Transworld Rig 59	self-elevating	USA 1967	125	180×136 (rectangular)	750 bbl	2,500 bbl	1,100 bbl	
Transworld Rig 62	self-elevating	USA 1974	300	174×146×20 (rectangular)	2,000 bbl	5,900 bbl	1,500 bbl	6,750 cft
Transworld Rig 63	self-elevating	Singapore (1975)	250					
Transworld Rig 58	semi-submersible	Neth. 1966	600	248·5×222×130 (rectangular)	600 bbl	4,800 bbl	1,200 bbl	
Transworld Rig 61	semi-submersible	Japan 1970	600	400×58×23 (ship shape)	786 bbl	2,425 bbl	1,600 bbl	
Taiheiyo Tankai Kogyo Co., Mori Bldg., No 10, 28 Shiba-Nishikubo, Sakuro Gawa-cho, Minato-ku, Toyko, Japan								
Tankaigo No 1	drillship		300	180×28×14 (ship shape)		100 tons	115 tons	20 tons
Underwater Gas Developers Ltd., 19 Toronto Street, Toronto 1, Canada								
Timesaver II	self-elevating	Canada 1965	100	92×71×10 (rectangular)			80 bbl	
Translake 1	self-elevating	Canada 1958	45	48×46×3 (triangular)			60 bbl	
Translake 2	self-elevating	Canada 1958	50	49×46×5 (triangular)			60 bbl	
USSR, Ministry of Oil and Gas, Moscow								
Apsheron	self-elevating	USSR 1966	50					
Azerbaijan	self-elevating	USSR	65	88×167×18				
Chazar	self-elevating	Neth. 1967	200	164×148×24			300 tons	150 tons
Kaspy	self-elevating	USSR 1974						
K.S Viking Offshore A.S., (In association with Field International) Norway								
Under construction	semi-submersible	Norway (1975)	600	355×221×120				
Waage Drilling A.S., Kongenst 6, Oslo, Norway								
Waage Drill I	semi-submersible	USA 1973	600	320×293 (rectangular)	6,942 bbl	14,600 bbl	12,000 cft 1,600 bbl	
Waage Drill II	semi-submersible	USA 1974	600	320×293 (rectangular)	6,942 bbl	14,600 bbl	12,000 cft 1,600 bbl	

Name OWNER/OPERATOR	Type	Built	Limiting Water Depth (ft)	Dimensions of Platform/Hull (ft) and shape Length×breadth ×depth (draught)	Tank Capacities (bbl=barrel) Fuel	Freshwater	Mud (dry/liquid)	Cement
Walker Huthnance Offshore Company, 514 Southwest Tower, Houston, Texas 77002, USA								
Ranger I	self-elevating (propulsion asstd)	USA 1969	70	115×74×8 (rectangular)	700 bbl	700 bbl	350 bbl	820 cft
Ranger II	self-elevating (propulsion asstd)	USA 1970	70	115×74×8 (rectangular)	700 bbl	700 bbl	350 bbl	820 cft
Ranger III	self-elevating (propulsion asstd)	USA 1974	70	115×74×8 (rectangular)	700 bbl	700 bbl	350 bbl	820 cft
Gulf Commander	self-elevating	USA —	100	113	500 bbl	1,450 bbl	600 bbl	1,640 cft
West Burne International Drilling Ltd., 535 Seventh Ave., SW Calgary 2, Alberta, Canada								
Westdrill I	self-elevating	Canada 1969	150	136×131×14 (rectangular)	1,000 bbl	1,000 bbl	650 bbl	
Western Oceanics Inc., 1212 Main Bldg., Room 484, Houston, Texas 77002, USA								
Western Delta	self-elevating	USA 1971	120	133×129×15 (rectangular)	800 bbl	4,500 bbl	2,400 cft 600 bbl	—
Western Star	self-elevating	USA 1969	110	130×119×16 (triangular)	525 bbl	2,330 bbl	2,460 bbl 600 cft	—
On order	self-elevating	Singapore (1975)	250					
On order	self-elevating	Singapore (1975)	250					
On order	self-elevating	Singapore (1976)	250					
On order	self-elevating	Singapore (1976)	250					
Western Pacesetter I	semi-submersible	USA 1974						
Western Pacesetter II	semi-submersible	USA 1975	600	200×260×20 (rectangular)	6,650 bbl	13,300 bbl	9,000 cft 1,700 bbl	900 cft
Western Pacesetter III	semi-submersible	USA 1974						
Wilh. Wilhelmsen Oslo, Norway								
On order	semi-submersible	Norway (1975)	600+					
On order	semi-submersible	Finland (1976)	600+					
Zapata Corporation, 1701 Houston Club Bldg., Houston, Texas								
Zapata Investigator	drillship	Australia 1967	600	260×70×(15) (ship shape)	312 tons	2,333 tons	540 tons	—
Zapata Navigator	drillship	USA 1970	600	340×65×29·5 (ship shape)	1,000 bbl	5,000 bbl	1,500 bbl	—
Zapata Trader Rig 21	drillship	USA 1974	600	430×77×32 (ship shape)				
Chaparral	self-elevating	USA 1966	300	205×205×23 (triangular)	1,387 bbl	5,499 bbl	1,600 bbl	—
Endeavour	self-elevating	USA 1965	250	190×190×22 (triangular)	1,300 bbl	5,230 bbl	1,600 bbl	—
Heron	self-elevating	USA 1966	300	205×205×23 (triangular)	1,387 bbl	5,499 bbl	1,600 bbl	—
Intrepid	self-elevating	USA 1965	300	167·3×180·2 (hexagonal)	1,326 bbl	7,777 bbl	1,860 bbl	—
Vinegarroon	self-elevating	USA 1957	165	164×150 (triangular)	980 bbl	5,000 bbl	1,500 bbl	1,700 bags
Zapata Explorer	self-elevating	USA 1967	225	237×236 (triangular)	1,387 bbl	7,400 bbl	1,600 bbl	2,850 cft
Zapata Nordic	self-elevating	USA 1972	300	235×236 (triangular)	1,566 bbl	6,992 bbl	1,500 bbl	
Louisiana	semi-submersible	USA 1968	600	180×150×12 (square)	850 bbl	4,950 bbl	1,200 bbl 7,000 bags	
Zapata Ugland	semi-submersible (propulsion asstd)	USA 1974	1,000	350×210×130 (rectangular)	10,000 bbl	10,700 bbl	1,700 bbl 10,000 cft	10,000 cft
On order	semi-submersible	USA (1975)	1,000					
On order	semi-submersible	USA (1975)	1,000					
On order	semi-submersible	USA (1976)	1,000					

OFFSHORE DRILLING RIGS

AUSTRALIA

EVANS DEAKIN INDUSTRIES LTD SHIPBUILDING DIVISION

Name	Owners	Type	Design	Delivery date
Southern Cross	Santa Fe-Pomeroy, Inc	semi-submersible	Santa Fe & Lloyds	May 1975

SOUTHERN CROSS

Type

Compact semi-submerisble rig, twin-hulled, column-stabilised, designed for deep water, high seas and low temperature operation

Designers

Santa Fe Engineering Services Division in co-operation with the Ocean Engineering Section of Lloyd's Register of Shipping

Delivery

May 1975

Classification

Lloyd's Register of Shipping Class OU 100 A1 Mobile Drilling Platform

Principal features

Minimum motion in operating conditions coupled with long natural periods in heave, pitch and roll

Positive stability throughout the range of draughts with full deck load

Drilling operations possible in 6·1 m (20 ft) seas with survival capability in 33·5 m (110 ft) seas

Ability to submerge and to deballast in any loading condition

Total structural steel weight of 5,500 short tons combining the most economical initial capital outlay commensurate with structural and motion characteristics which match those of much larger vessels

Speed of up to 8 knots during location moves

Dimensions

Vessel:	
Length overall	79·3 m (260 ft 0 in)
Beam	48·2 m (158 ft 0 in)
Each hull:	
Beam	12·2 m (40 ft 0 in)
Depth	6·6 m (21 ft 9 in)
Bilge and gunnel radii	0·91 m (3 ft 0 in)
Columns	
Ends, Length	9·1 m (30 ft 0 in)
Breadth	7·32 m (24 ft 0 in)
Centre, Length	12·2 m (40 ft 0 in)
Breadth	6·7 m (22 ft 0 in)
Corner radius	1·8 m (6 ft 0 in)
Height (to machinery deck)	24·9 m (81 ft 9 in)
Upper hull:	
Length	65·8 m (216 ft 0 in)
Beam	42·7 m (140 ft 0 in)
Height above keel to:	
Machinery deck	24·9 m (81 ft 9 in)
Texas deck	29·2 m (95 ft 9 in)
Drill floor	35·4 m (116 ft 9 in)
Spider deck	22·8 m (74 ft 9 in)

Design capabilities

Drilling depth capability	6,097 m (20,000 ft)
Water depth capability	610 m (2,000 ft)
Design seas:	
At maximum draught	12 m (40 ft) waves
Survival	33 m (110 ft) waves
Transit speed (calm)	8 knots
Ambient temperature range	
Accommodations	0° to 120°F
Structural	−4° to 120°F

Hydrostatic data

Towing draught	5·9 m (19 ft 3 in)
Towing displacement	9,330 short tons
Drilling draught	12·8 m (42 ft)
Drilling displacement	15,740 short tons

Capacities

Drilling water	6,730 bbls
Potable water	750 bbls
Fuel oil	5,030 bbls
Drill water/fuel oil (combination tank)	2,300 bbls
Liquid mud	1,400 bbls
Bulk cement	274 m^3(9,000 ft^3)
Bulk mud	
Sack mud storage	4,500 sacks
Drill pipe storage	6,097 m (20,000 ft)

Accommodation

80 men

Safety equipment

Three Brucker 28-man capsules are provided as well as four 25-man inflatable life rafts.

An artist's impression of *Southern Cross* (*Courtesy Santa Fe-Pomeroy Inc.*)

Mooring system
Eight 13,630 kg (30,000 lb) Danforth anchors. Anchor cables include eight lengths of 1,463 m (3,000 ft) 6·3 cm ($2\frac{1}{2}$ in) galvanised mooring cable

Thruster units
Four 1,500 h.p. Baylor Thruster units fitted with double right angle drive/reduction train, full 360° horizontal movement, Kort nozzle propellers

Drilling equipment
Oilwell E2000 draw works, rated 2,000 h.p. for 6,097 m (20,000 ft) depth, driven by two GE 752 motors.
Derrick: Lee C. Moore, mast 46·3 m (152 ft) clear height, capacity, 454,545 kg (1,000,000 lb)
Rotary table: Oilwell 147·6 cm ($37\frac{1}{2}$ in)

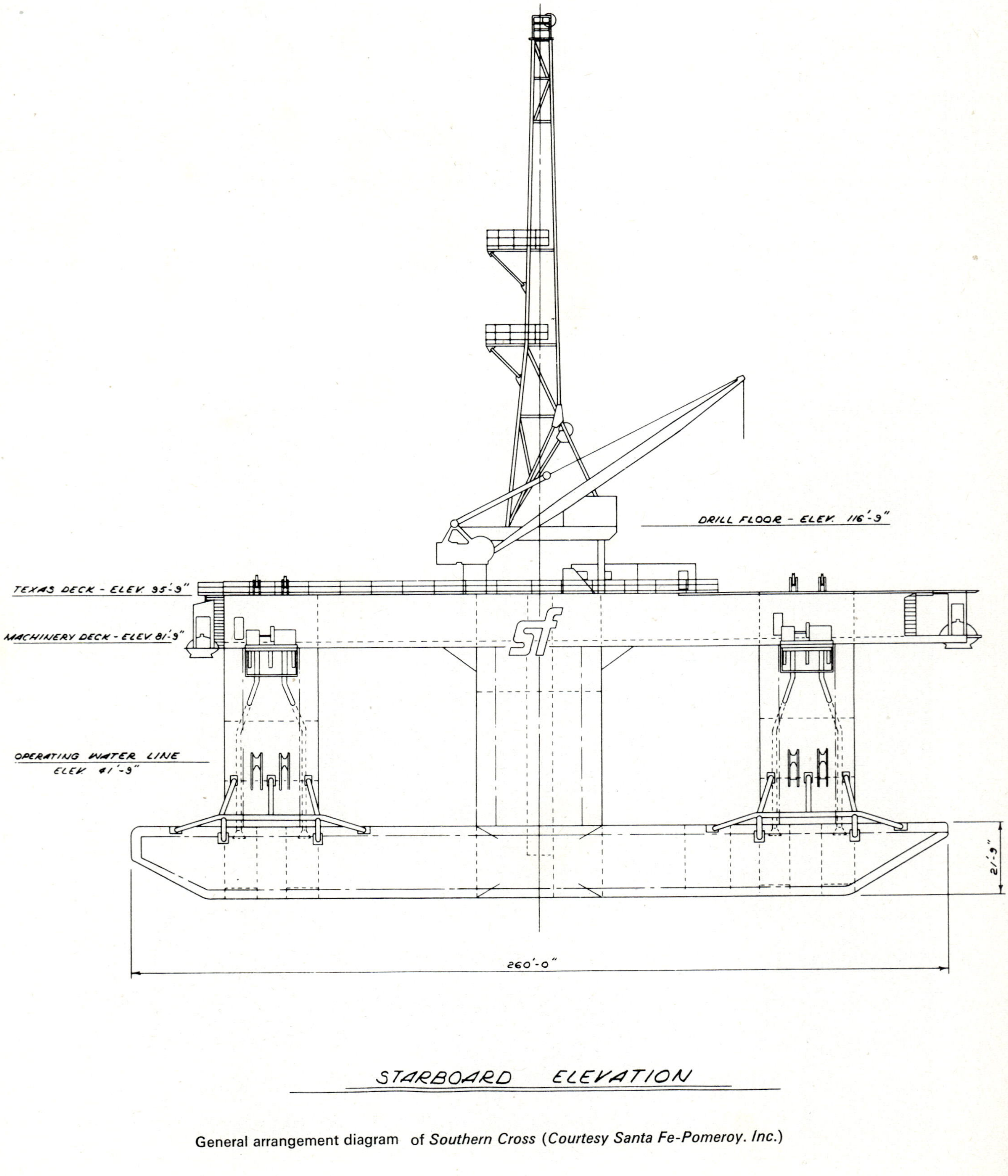

General arrangement diagram of *Southern Cross* (*Courtesy Santa Fe-Pomeroy. Inc.*)

TRANSFIELD PTY. LTD

Name	Owners	Type	Design	Delivery date
Ocean Endeavour	Offshore Drilling and Exploration Co. (ODECO)	semi-submersible		

OCEAN ENDEAVOUR

Specifications

Type	Semi-submersible
Length	104·9 m (344 ft)
Width	80·3 m (263·5 ft)
Displacement	9,123 tons
Max. drilling depth capability	7621 m (25,000 ft)
Cost	$9·5 million

CANADA

HAWKER SIDDELEY CANADA LTD HALIFAX SHIPYARDS DIVISION

Name	Owners	Type	Design	Delivery date
Sedneth 701	Sea Drilling Netherlands (Sedneth)	semi-submersible		1973
Sedco 704	Sedco International, SA	semi-submersible		1974
Sedco J	Southeastern Commonwealth	semi-submersible		1972

SEDCO J

Specifications

Completed in the autumn of 1972, the *Sedco J* was the third rig of this type to be constructed by Hawker Siddeley Canada.

Type	Semi-submersible
Length	104·4 m (342 ft 6 in)
Breadth	103·7 m (340 ft)
Accommodation	58
Displacement	19,700 tons
Maximum drilling depth capability	7,620 m (25,000 ft)
Maximum water depth capability	243·9 m (800 ft)
Propulsion	Two 3,350 h.p. DC electric motors, twin Kort nozzles
Survival capability	Sea state 9

Storage capacities

Bulk mud	7,500 sacks
Liquid mud	1,650 bbl
Cement	3,250 sacks
Fuel	5,080 bbl
Drilling water	10,160 bbl
Potable water	476 bbl

Cranes

Two Manitowoc	82·45 tons at 6·7 m (22 ft)
One Clyde	72 tons at 7·6 m (25 ft)

Drilling equipment

Draw works	OWS E-3000
Pumps	OWS 1700 P
Power	Four EMD 645-12
Derrick	43·3 m (142 ft)

Mooring

Nine 1,363·6 kg (30,000 lb) anchors
Nine 853.7 m (2,800 ft) chains, 7·6 cm (3 in) diameter

Thrusters

Three 2,000 h.p.

Cost

$20 million

When *Sedco J* was positioned for its first drilling probe, 215 miles east of St. Johns in 114·3 m (375 ft) of water, it set a world's distance record for a drillling probe offshore.

Sedco 1 sister rig of *Sedco J*, leaving Halifax Shipyard (*Courtesy Hawker Siddeley Canada, Ltd.*)

SEDNETH 701

Specifications

Type	Semi-submersible
Length, overall	89·9 m (295 ft)
Breadth, overall	74·7 m (245 ft)
Accommodation	95
Maximum drilling depth capability	7,620 m (25,000 ft)
Maximum water depth capability	244 m (800 ft)
Survival capability	18·3 m (60 ft) seas plus

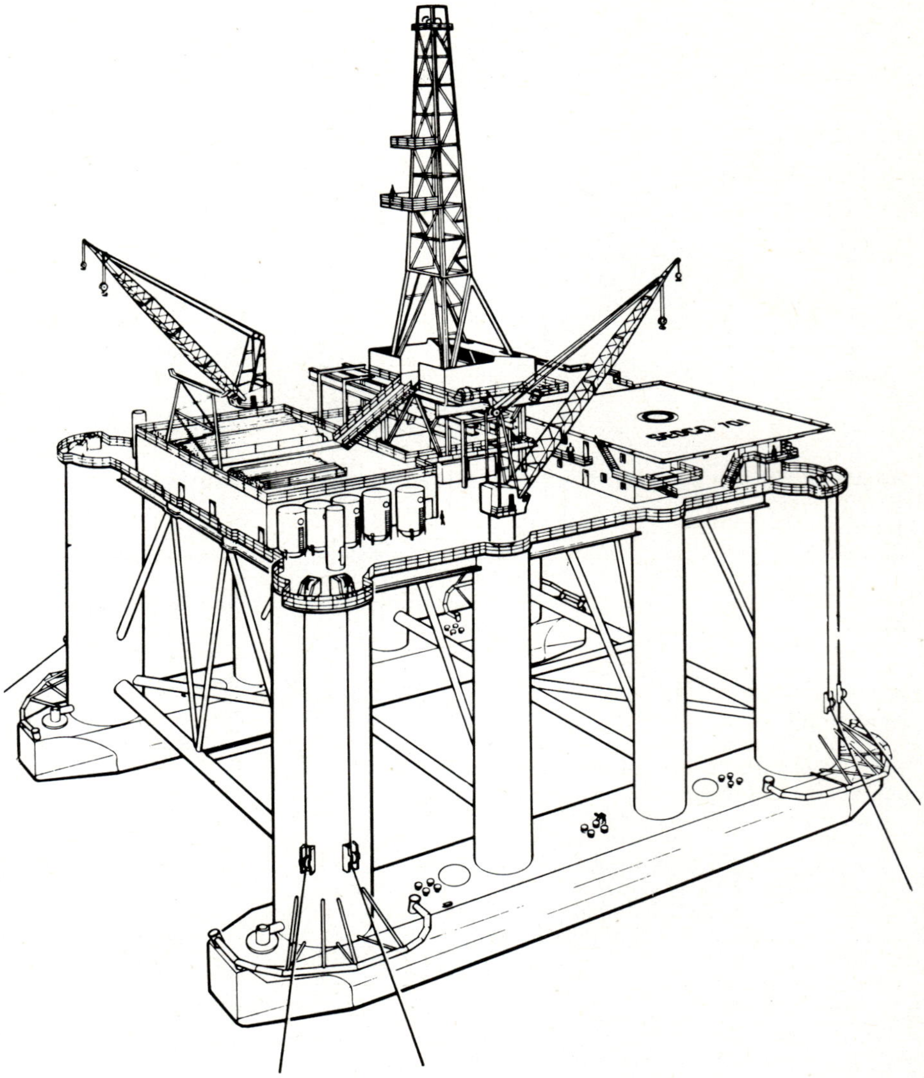

Sedneth 701 (*Courtesy Hawker Siddeley Canada, Ltd.*)

FINLAND

RAUMA REPOLA OY, MANTYLUOTO
(For Aker H-3 rig details, see Aker group)

Name	Owners	Type	Design	Delivery date
N/A	Atlantic Drilling Co., England	semi-submersible	Aker H-3	September 1975
N/A	Celtic Drilling Co., England	semi-submersible	Aker H-3	April 1975
N/A	Unaphalt Roads Ltd, England (Kingsnorth Marine Drilling Ltd)	semi-submersible	Aker H-3	October 1974
N/A	Unaphalt Roads Ltd, England (Kingsnorth Marine Drilling Ltd)	semi-submersible	Aker H-3	December 1974
Pentagone 84 (see *Pentagone 82*)	Forex Neptune SA & International Schlumberger Group	semi-submersible	Forex-Neptune Pentagone	November 1973
*Pentagone 85** (*Pen Lyn 1*)	Dixilyn with Panhandle Eastern Pipe Line Co subsidiary Pen-Lyn Co.	semi-submersible	Forex-Neptune Pentagone	April 1974
N/A	Wilh Wilhelmsen, Oslo, Norway	semi-submersible	Aker H-3	July 1976
N/A	Stolt-Nielsen Group, Oslo, Norway	semi-submersible	Aker H-3	April 1976
N/A	Fearnley & Eger, Oslo, Norway	semi-submersible	Aker H-3	September 1976

* Building at Pori works.

PENTAGONE 85

Type
Semi-submersible—Pentagone design, built under licence from Neptune-Forex, Paris

Fabrication start date
February 1973

Commissioning date
May 1974

Areas of intended use
Worldwide

Dimensions, weights, draughts

Length, overall	99·1 m (325 ft)
Width, overall	103 m (338 ft)
Height	94·2 m (309 ft)
Main deck length	82 m (269 ft)
Main deck width	46 m (151 ft)
Platform clearance at drilling draught	14·8 m (48 ft 6 in) with 55·9 cm (22 in) draught
Drilling draught	21·9 m (72 ft)
Displacement at drilling draught	17,600 long tons
Displacement at transit conditions	12,010 long tons
Draught of pontoons	6·7 m (22 ft) transit condition, 21·9 m (72 ft) drilling (normal)

Operating draughts in various wave conditions: (Design)
21·9 m (72 ft) draught, 14·9 m (49 ft) wave, 9 sec period
19·8 m (65 ft) draught, 20·7 m (68 ft) wave, 12 sec period
15·8 m (52 ft) draught, 29·9 m (98 ft) wave, 17 sec period

Capabilities
Maximum water depth for drilling: 201·2 m (660 ft)
Maximum drilling depth capability 9,146 m (30,000 ft)
Limiting wind and current conditions: 138 mph wind, 2·5 knot current

Accommodation
74

Tug requirements
13,500 h.p.

Design towing speed
6·5 knots

Power plant
Main engines—type D-399 Cat. Engine Seven 1,250 h.p. engines, 8,750 h.p. total
Generators: GE 8000 Gen, 930 kw and 1,162 KVA each
Pumping systems for ballast, drill water, fuel, mud: Ballast, Ten 50 h.p. 300 m³/hour, Mud (drilling), Two 1,600 h.p. Triplex, Drill water, Two 36 h.p. 100 m³ (3,500 ft³)/hour, Mixing, Two 100 h.p. centrifugal 6 × 8, Fuel, Two 10 h.p. 25 m³ (885 ft³)/hour, Supercharging, Two 75 h.p. centrifugal 6 × 8

Thrusters
None at present. However, pontoons are designed and built so that thrusters may be added at a later date without any changes. Necessary piping and wiring are included in columns. Switch gear is included in the SCR unit.

Storage capacities

Drill water	4,700 bbls
Fuel oil	4,700 bbls
Potable water	3,645 bbls
Wash water	1,962 bbls
Liquid mud	Five tanks, 2,140 bbls
Bulk mud	Four 51·3 m³ (1,810 ft³) P-tanks, 205·2 m³ (7,240 ft³) total
Other	Four 51·3 m³ (1,810 ft³) P-tanks, bulk cement, 205·2 m³ (7,240 ft³) total
Total supply capacity	4,160 metric tons (4,085 l.tons)
Deck storage capacity	2,100 metric tons (2,062 l.tons) in drilling position

Mooring system
Ten Brissoneau & Lotz anchor winches, capacity 1,740 m (5,700 ft) of 7 cm (2¾ in) wire rope, Ten 13,640 kg (30,000 lb) LWT anchors

Drilling equipment

Derrick:	
height	48·8 m (160 ft)
hook load capacity	522,727 kg (1,150,000 lb) Nominal
Drill pipe diameter	4,573·3 m (15,000 ft 5 in) 4,573 m 8·9 cm (15,000 ft 3½ in)
BOP stack:	
bore	47·6 cm (18¾ in)
working pressure	4,545 kg (10,000 lb)
Rotary table	95·2 cm (37½ in)

Other features: Four 36,360 kg (80,0000 lb) Riser Tensioners. Four 7,270 kg (16,000 lb) Guideline Tensioners

Motion characteristics

Natural period in pitch	32 to 40 seconds
Natural period in roll	32 to 40 seconds
Natural period in heave	20 seconds
Ballasting facilities	Ten 300 m³ (10,600 ft³) 50 h.p. electric

Helicopter deck
Super Frelon or S-61 design facility

FRANCE

FRANCE OFFSHORE GROUP: CIE FRANCAISE D'ENTREPRISES METALLIQUES (C.F.E.M.), LE HAVRE. L'UNION INDUSTRIELLE D'ENTREPRISE (UIE) CHANTIERS NORMANDIE

Name	Owners	Type	Design	Delivery date
Norrig V renamed *West Venture*	Peder Smedvig Tankrederi A/S, Stavanger with Scheepsbouwbelangen, Holland		Marcon NV, Den Haag	27 July 1973
Pentagone 83 *Drill Master*	Norsedrill & Co. Sigurd Herlofson Group Oslo		I.F.P.-Neptune	End of 1973
Pentagone 87	Offshore Drilling Co. (Kulukhundis Group (UK))		Forex Neptune	1974

WEST VENTURE
West Venture is a semi-submersible five column stabilised offshore unit for drilling wells up to 30,000 ft in both floating and sit-on-bottom positions.
The rig is designed for extreme operational conditions allowing for worldwide operation including all year drilling in North Sea and North Atlantic areas up to the iceline.
West Venture is registered with Det Norske Veritas under class:
✠ OP 1 strength and stability for column stabilised platform.

Drilling performance

Operational areas	Worldwide North Sea and North Atlantic up to the iceline	
Drilling depth	up to 9,146 m (30,000 ft)	
Operational water depths		
position	floating	sit on bottom
range, m	30 to 300	10 to 40
range, ft	100 to 1,000	33 to 131
Operational variable loads		
position	floating	sit on bottom
metric tons	2,500	up to 7,000

Dimensions

	m	ft
Radius of circle through centres of stabilisation columns	46·00	151
Length, moulded	102·22	335
Breadth, moulded	106·50	349
Depth main deck, moulded	46·60	153
Depth cellar deck, moulded	42·60	140
Depth of floaters, moulded	14·00	46
Diameter of floaters, I.D.	19·00	62
Diameter of stabilisation columns, I.D.	9·30	30·5
Diameter of circular bracings, O.D.	2·00	6·6
Depth underside of lower bracings, moulded	14·20	47

Position of rotary table

	m	ft
Top rotary beams above base line	57·00	187
Centre of rotary table aft of unit's centre	15·00	49

Freeboard

	m	ft
At drilling draught (design conditions)	25·00	82
deck openings top of columns above sea level	21·60	71
main deck above sea level	21·60	71
cellar deck above sea level (airgap)	17·60	58
At towing draught	11·90	39
underside lower bracings above sea level	2·30	7·5
At light draught	9·50	31
underside lower bracings above sea level	4·70	15·4

Maximum ballast capacity
Total number of ballast tanks, 31
Total ballast capacity, 17,230 m³ (600,000 ft³)

Living quarters
The living quarters are arranged in a three level accommodation package placed on top of the space frame structure.

The living quarters are of high standard and provide accommodation for 77 men divided over:

6 × 1 berth cabins
22 × 2 berth cabins
6 × 4 berth cabins
1 × 3 berth cabin

The accommodation package is fully air-conditioned on basis of the following design conditions:

Condition	Temperature centigrades inside	outside	Relative humidity % inside	outside
Summer	+29	+35	45	70
Winter	+18	−20	abt. 45	—

Helicopter deck
The helicopter deck located on top of the accommodation package is designed for Sikorsky S-61 N helicopters.

Its effective landing area measures 25·2 × 25·2 m² = approx 83 × 83 ft.²

It is provided with 1·50 m × approx 5 ft wide safety nets.

Ballast pump rooms
Five ballast pump rooms are provided for, one at the lower part of each stabilisation column. The pump rooms are accessible by means of an elevator.

Main diesel engines
Six Caterpillars D 399 TA, 1,100 h.p. each at 1,200 rpm, total 6,600 h.p. continuous.

Generators
Six Alsthom Generators PA 100 G, 1,150 KVA each, 600 volts.

Thyristors (SCR's)
Six Alsthom Thyristor converters for DC motor's feeding. Rated current 970 Amp corresponding to 800 h.p. power on the motor shafts. Total capacity 4,800 h.p. continuous.

Electric motors
Three Draw works motors of 800 h.p. each.
Four Pump motors of 800 h.p. each.
One Rotary table motor, 500 h.p.
Two Cement pump motors of 500 h.p. each.

West Venture, March 1973 (*Courtesy C.F.E.M.*)

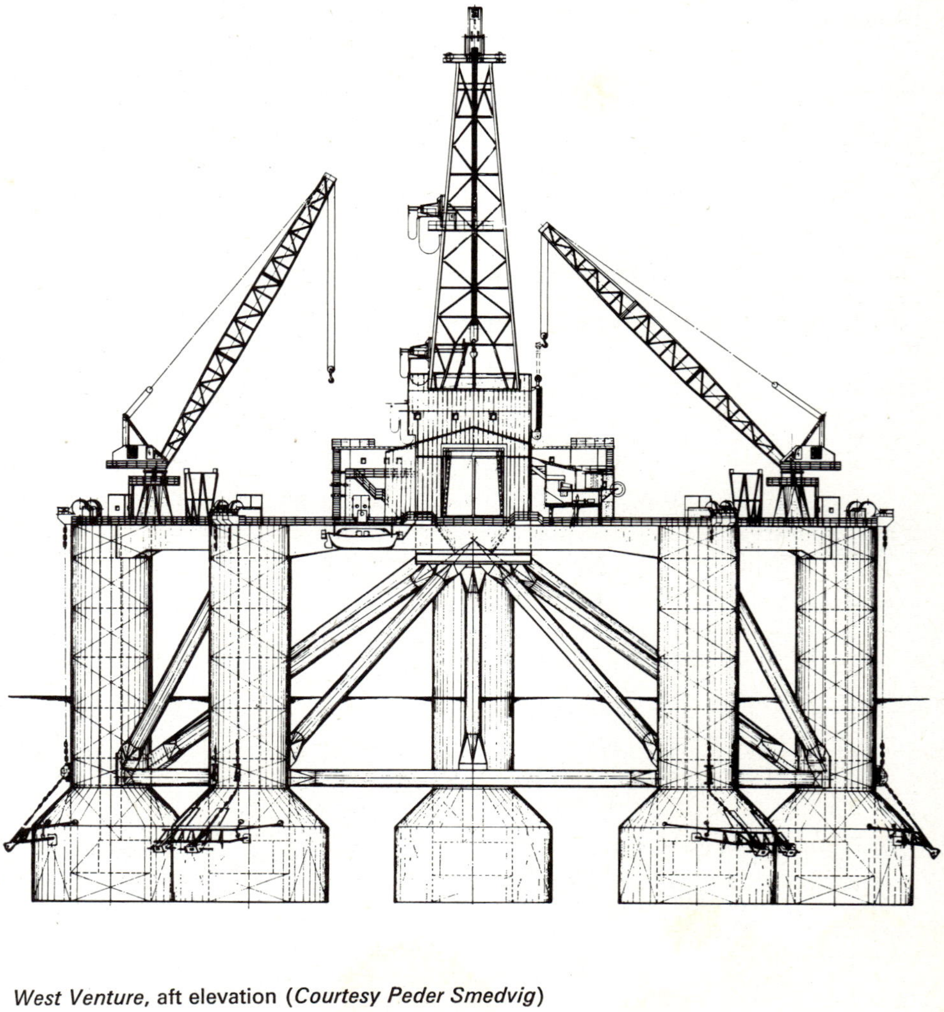

West Venture, aft elevation (*Courtesy Peder Smedvig*)

AC power
3 × 750 KVA transformers, 3 phase, 440 volts, 60 Hz.

Emergency generator
One Caterpillar D-343TA diesel engine with 250 kw, 3 phase, 440 volts, 60 Hz, AC generator.

The emergency system is independent of the main system and powers the following functions:

Emergency lights and sounds throughout the unit including quarters, engine room, derrick floor and navigation lights.
Instrumentation and remote controls in Control Room.
Salt water cooling pumps, fire pumps and other fire fighting equipment.
Radio and Communication systems.
Ballast Pumps.

Natural frequencies

Motion	Natural frequency rad/sec	natural period secs
Heave	0·349	18
Pitch	0·126	50
Roll	0·143	44

Mooring system

Anchors:	
number	10
weight	13,600 kg (30,000 lb)
Anchor chains:	
number	10
length	1,100 m (600 fathoms)
min. breaking strength	438,000 kg (960,000 lb)
Windlasses:	
number	10
pulling force	
at 2 m/min	182,000 kg (400,000 lb)
at 10 m/min	35,000 kg (77,000 lb)
brake capacity	340,000 kg (750,000 lb)
chain stopper capacity	500,000 kg (1,100,000 lb)

Towing

Loading condition:	
Light weight	14,000 metric tons
Variable load	1,200 metric tons
Displacement (salt water)	15,200 metric tons
Draught	10·50 m

Assumptions:			
Underwater hull	clean		
Sea condition	smooth		
Windspeed	zero		
Class of tug(s) (h.p.)	8,000	10,000	2×10,000
Towing speed (knots)	3·0	3·3	4·5

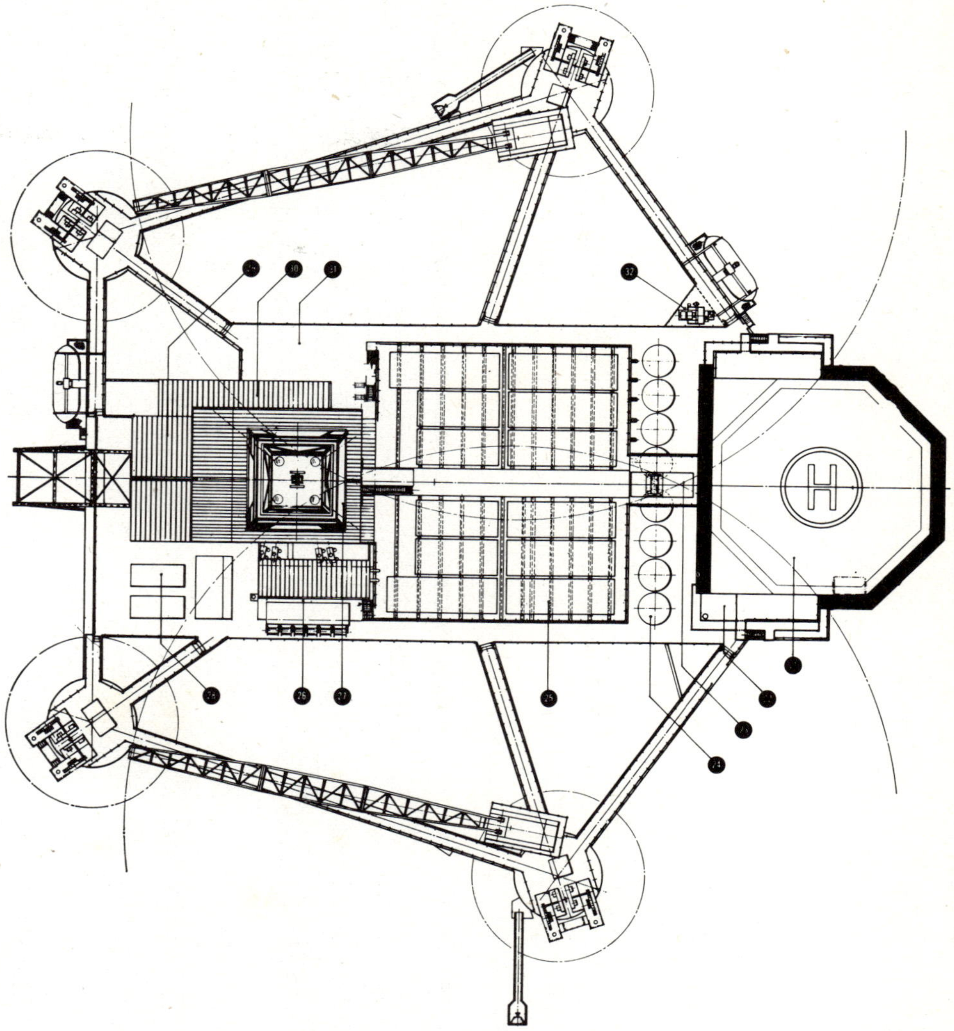

West Venture, top elevation (*Courtesy Peder Smedvig*)

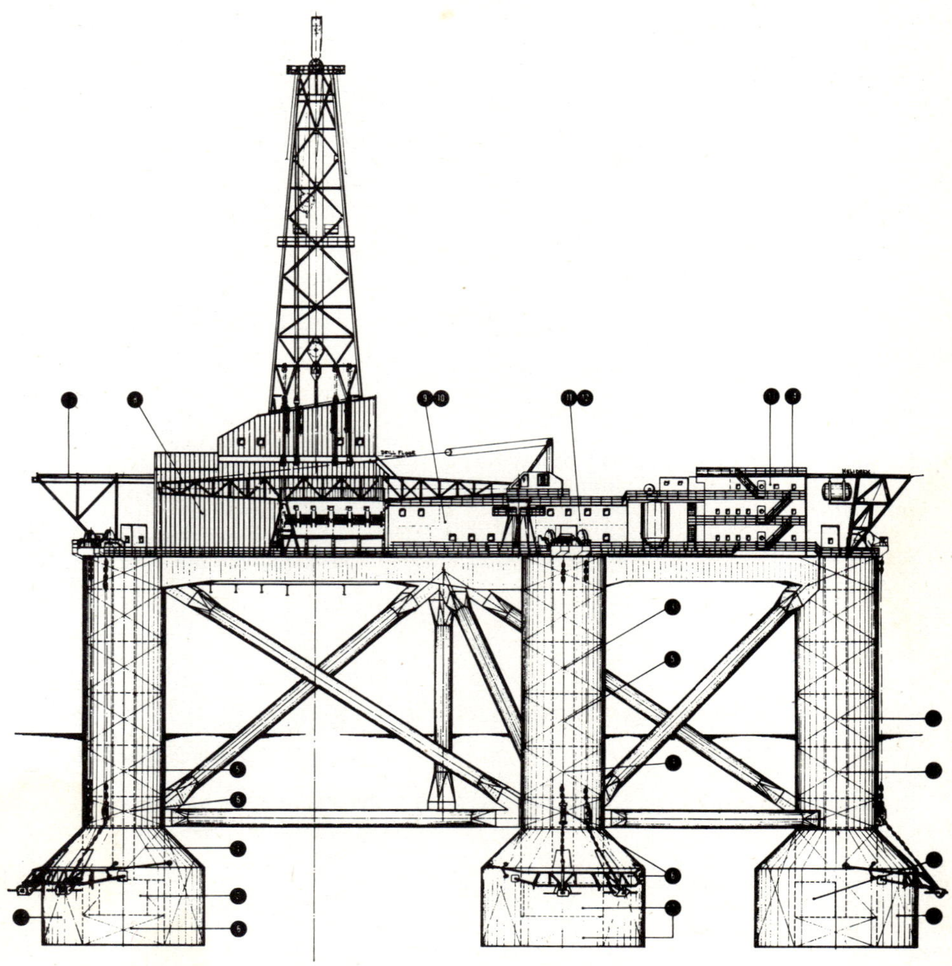

West Venture, side elevation (*Courtesy Peder Smedvig*)

DRILL MASTER
PENTAGONE 83

Type
Semi-submersible

Classification
Lloyd's Register of Shipping

Dimensions

diameter of the circle on which the five columns are placed	85 m (279 ft)
length	99·4 m (326 ft)
width	103 m (338 ft)
main deck area	2,300 m² (25,030 ft²)
height between base of pontoons and upper deck	40·8 m (134 ft)
pontoons	
diameter	21·9 m (72 ft)
height	7·5 m (24 ft 7 in)
diameter of columns	9·5 m (31 ft 3 in)
weight	9,130 metric tons

Draught

towing	7 m (23 ft)
drilling	21·9 m (72 ft)

Displacement

(drilling position)	17,600 metric tons

Storage capacity

sea water	60,000 bbls
drilling water	4,700 bbls
fresh water	3,700 bbls
fuel oil	4,000 bbls

Design criteria

wind	approx. 230 km/h (138 mph)
waves:	
period	17 s
height	30 m (98 ft 5 in)

Mooring
10 lines each with:
40,000 bbls 00 anchor
152·4 m 6·9 cm (5,000 ft 2¾ in) anchor wiring
1 anchor winch
1 anchor wire tensioner:
Martin Decker constant tension indicating system on all anchor wires

The erection of *Pentagone 83* at Le Havre with *West Venture* (*Courtesy C.F.E.M.*)

Deck cranes
2 electric powered Richier G 90
boom revolving cranes, max. capacity 40 mt. at 9·1 m (30 ft)
2 electric powered Richier G 80
boom revolving cranes, max. capacity 12 mt. at 9·1 m (30 ft)
one 100 mt. capacity bridge crane for handling BOP

Draw works
EMSCO C 3 type 2 driven by three 1,000 h.p. DC electric motors

GERMANY (FEDERAL REPUBLIC OF)

BLOHM & VOSS AG, OFFSHORE DIVISION

Name	Owners	Type	Design	Delivery date
SCP III Mk 2	The Offshore Co., Houston, Texas	semi-submersible	The Offshore Co. & Blohm & Voss AG	1974
Scarabeo III (Hull No. 884)	Saipem, S.p.A.	semi-submersible	The Offshore Co. & Blohm & Voss AG	1974
Scarabeo IV (Hull No. 886)	Saipem, S.p.A.	semi-submersible	The Offshore Co. & Blohm & Voss AG	1974

CHRIS CHENERY

Type
Self-propelled, semi-submersible, column-stabilised drilling rig

Class
A.B.S. +A1 (E)(M) +AMS for unrestricted ocean use

The rig consists of an upper hull, two parallel lower hulls with six columns and eight diagonal bracing members. In the aft part of each of the lower hulls are installed two 2,250 h.p. electric motors acting over a reduction gear on a Kort nozzle propeller with 4,000 s.h.p.

A centreline drilling well with derrick is arranged amidships and the heliport is located at the aft deck.

Fully air-conditioned living quarters can accommodate 72 persons.

The unit will be capable of working in water depths of up to 305 m (1,000 ft) and will have a rated drilling capacity of 7,622 m (25,000 ft).

The lower hulls are round to improve the motion characteristics and to reduce the resistance in transition condition as minimum wetted surface is obtained.

The upper hull consists of a twin-deck design having all machinery and accommodation incorporated inside the hull, thus giving a large free area on the main deck and providing ample space for storage and working purposes.

The unit is scheduled to operate initially in Western European waters and is designed to operate in the most extreme conditions of the North Sea.

Extensive model basin tests have been made and indicate that the vessel will have unprecedented motion characteristics in drilling condition.

Principal dimensions

Lower hulls:	
Length, overall	115·37 m (378 ft)
Beam, overall	73·76 m (242 ft)
Diameter	10·67 m (35 ft)
Stabilising columns:	
6 columns, each	8·23 m (27 ft)
Upper hull:	
Length, overall	82·30 m (270 ft)
Width, overall	67·97 m (223 ft)
Depth	6·10 m (20 ft)
Height:	
Base line to main deck	42·67 m (139 ft)
Draught:	
Transit	8·53 m (22 ft 6 in)
Drilling	21·34 m (70 ft)
Still water speed	8 knots
Propulsion	Two 4,000 h.p. 3·6 m (12 ft) diameter screws
Mooring	Eight 0·83 × 1,097 m (2 ft 6 in × 3,600 ft) long high-strength chains, each with 20·43 t anchors

Liquid storage capacities

Fuel oil	1,118 tons
Wash water	419 tons
Drill water	1,256 tons
Potable water	102 tons
Water distillation	11,200–15,200 litres/h
Cranes	Two 75 tons, two 30 tons
Quarters	72 persons
Helicopter deck	24·4 × 27·4 m (80 ft × 90 ft) for Sikorsky S-61

Drilling equipment

Rated drilling depth, 7,622 m (25,000 ft)

Draw works: National 1625 DE, powered by two Westinghouse 370 DC motors

Engine: Two EMD Model S-20E-4GW-20 power units with one 2,500 kw AC generator each, two EMD Model S-12EI-GW-20 power units with one 1,050 kw AC generator each, stand-by electric set Caterpillar D 398 with Kato AC generator

Pumps: Two EMSCO F 1600, 1,600 h.p. Triplex pumps each powered by two Westinghouse 370 DC motors rated at 900 h.p. each continuous

Derrick: EMSCO 48·8 × 12·2 × 12·2 m (160 ft × 40 ft × 40 ft) with travelling block guide, 500 ton Crown Block, nominal gross capacity of 454,545 kg (1,000,000 lb)

Substructure: The Offshore Company design, 10·7 m (35 ft) high

Rotary: National Model C-375 driven by two speed transmission powered by Westinghouse Model 370 D3 motor rated 900 h.p.

Travelling block: EMSCO Model RA-60-6 with six sheaves

Swivel: National type P 650 with standard bail and link bumper support

Shale shaker: Solids separator and dual shaker

Drill pipe: 4,573·3 (15,000 ft 5 in) o.d. Grade S 135, range 2

Accumulator: Koomey Model 26320-3S, 230 gallon, 211 kg/cm² (3,000 lb/in²) WP

A scale model of the Blohm & Voss SCP III Mk. 2 self-propelled, semi-submersible drilling rig *Chris Chenery* (*Courtesy Blohm & Voss AG.*)

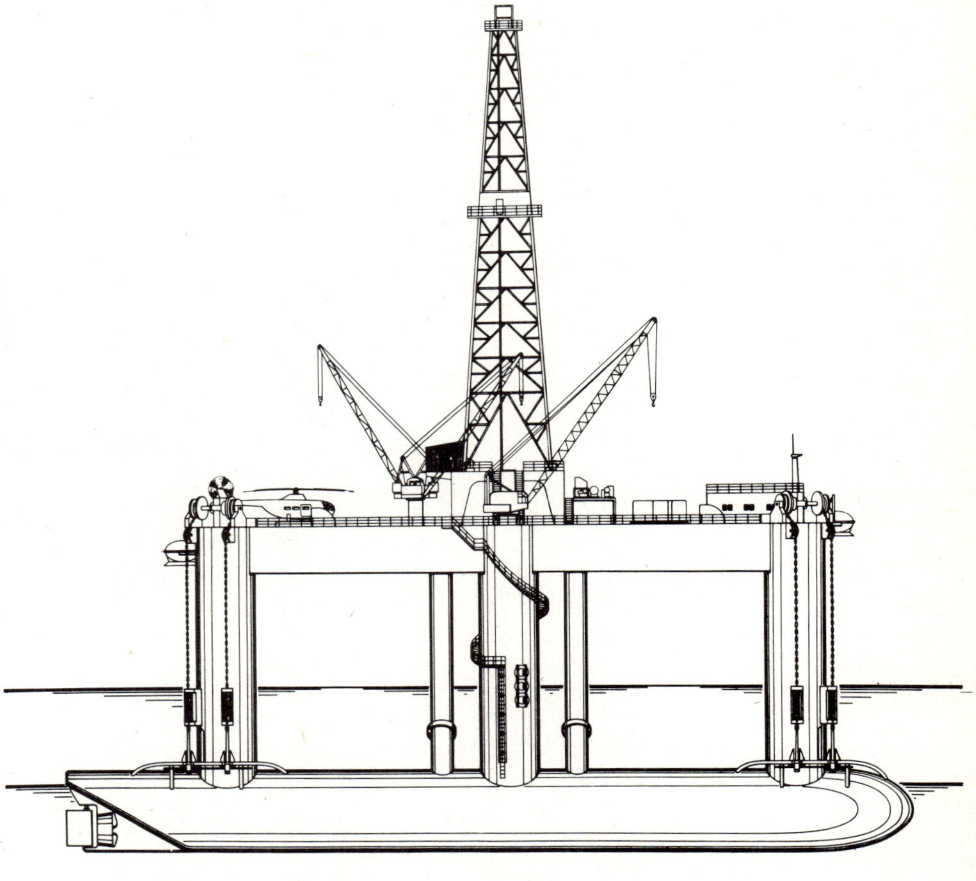

Side elevation of *Chris Chenery* (*Courtesy Blohm & Voss AG*)

SCARABEO III & IV

Type
Self-propelled, semi-submersible drilling platform

Class
A.B.S. A1 M Column-Stabilised Drilling Unit RINA 100 AS-1.1 Nav. G 'pf'

The *Scarabeo III* is triangular in shape, has a total height of 115 m (377 ft), including the derrick, and width of about 110 m (361 ft).

Supporting legs consist of three caissons with special horseshoe-shaped footings. Each footing measures 40 × 20 × 9 m (131 × 65 ft 6 in × 29 ft 6 in) and is divided into watertight compartments which may be filled for ballasting and for fuel and fresh water storage. The three caissons are 12 m (39 ft) in diameter and are connected by a frame of steel pipes ranging from 1·80 to 2·80 m (5 ft 9 in to 9 ft) in diameter.

The main deck with the drilling rig, auxiliary equipment and accommodation is supported on the upper part of the caissons. The platform is fitted with Blohm & Voss M 1000 Accommodation System accommodating 72 persons in air-conditioned cabins with one, two or four berths.

The unit will be capable of working in water depths of up to 304·9 m (1,000 ft) and will have a rated drilling depth of 9,146·3 m (30,000 ft).

The design of the rig provides excellent motion characteristics and meets survival criteria of 30·5 m (100 ft) high waves with 17 to 18 second periods.

The rig is propelled by three thruster units installed in the footings, developing 2,000 h.p. each.

Principal dimensions

Footings:	
Height	39·62 m (130 ft)
Width	19·81 m (65 ft)
Depth	9·15 m (30 ft)
Stabilising columns	Three columns each 12·00 m (39 ft) dia.
Platform:	
Length	109·34 m (358 ft 6 in)
Width	100·31 m (329 ft)
Height from base line to main deck	44·50 m (146 ft)
Transit draught	8 m (26 ft)
Still water speed	approx 6 knots

Propulsion: Three Schottel thruster units, each rated at about 2,000 h.p.

Mooring: Nine steel anchor ropes, reach of 2,020 m (6,625 ft 6 in) length and 75 mm (3 in) dia. with nine 13·6 ton anchors. Brissoneau & Lotz Marine winches

Liquid storage capacities:	
Fuel oil	1,500 tons
Drilling water	3,000 tons
Potable water	136 tons
Quarters	72 persons
Helicopter deck	25 m (82 ft) dia. for Sikorsky S-61
Water distillation	Two MECO type evaporators, each 40 t/24 h
Cranes	Three fully revolving ARMCO cranes of 30·4 to 36·6 m (100 to 120 ft) boom length, about 50 ton capacity each

A scale model of the Blohm & Voss *Scarabeo III* and *IV* type semi-submersibles (*Courtesy Blohm & Voss AG*)

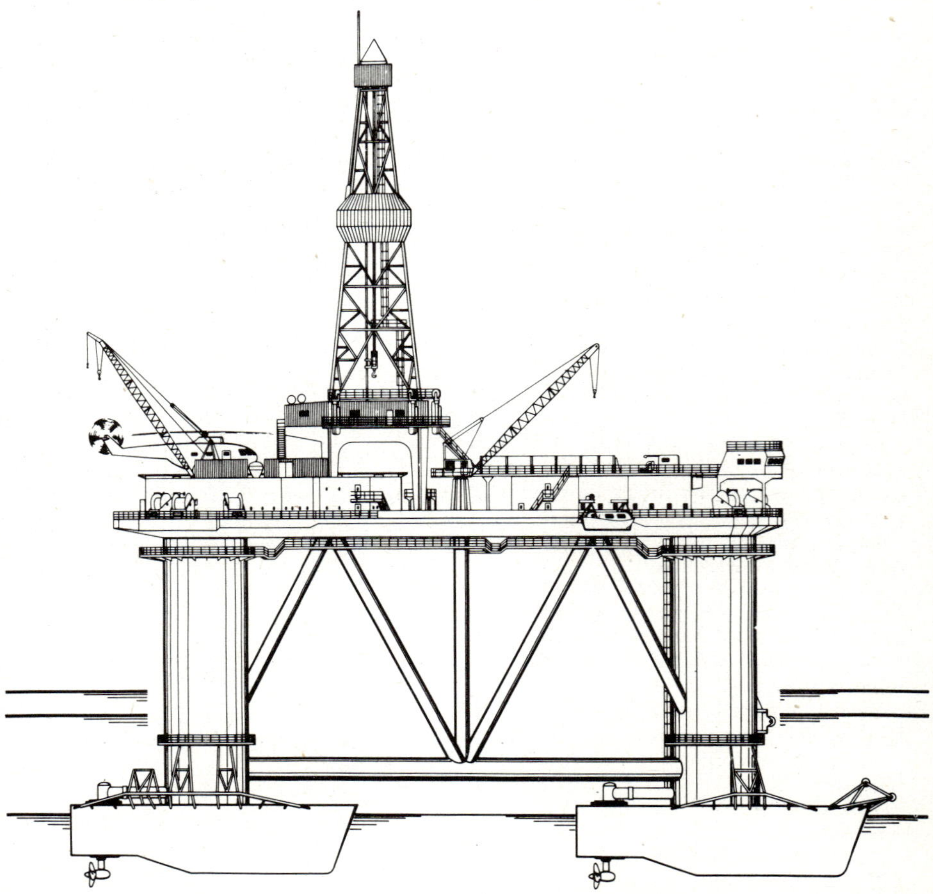

Side elevation of *Scarabeo III* and *IV* (*Courtesy Blohm & Voss AG*)

Drilling equipment
Rated drilling depth 9,150 m (30,000 ft)
Draw works: ARMCO, type National 1625-DF, driven by three Siemens electric motors
Engines: Seven 800 h.p. Caterpillar diesel engines, type D 399, one emergency Caterpillar diesel engine of same type, each driving one Siemens generator

Pumps: Three ARMCO slush pumps, each driven by two Siemens 800 h.p. electric motors
Derrick: 12·2 × 12·2 m (40 ft × 40 ft) base, 48·8 m (160 ft) high mast, designed to accommodate motion compensator
Substructure: Friede & Goldman, 12 m (39 ft) high

Rotary: ARMCO, type National C-495, driven by one 800 h.p. Siemens electric motor
Travelling block: IDECO-Pignone UTB 480-6-55 with six 1·4 m (4 ft 7 in) sheaves and unitised hook
Swivel: IDECO-Pignone TL-500

JAPAN

MITSUBISHI HEAVY INDUSTRIES, LTD, HIROSHIMA

Name	Owners	Type	Design	Delivery date
Ocean Ranger	ODECO with Fearnley & Eger (Oslo)	semi-submersible		
Ocean Kokvei	ODECO (Nihon, S.A.)	semi-submersible		1973
Sagar Samarit	Goot of India	jack-up		Nov. 1972
Hakuryu III	Japan Drilling Co.	semi-submersible		May 1974

HAKURYU III

Type
Semi-submersible, sister rig to *Hakuryu II*

Specifications

Length, overall	101·02 m (331 ft 4 in)
Breadth	66·97 m (219 ft 8 in)
Height of main deck	34·96 m (114 ft 8 in)
Draught, operating	19·97 m (65 ft 6 in)
Max. water depth capability	304·88 m (1,000 ft)
Max. drilling depth	9,146 m (30,000 ft)
Limiting wind speed	120 knots
Self-propulsion speed	8·5 knots
Accommodation	100
Drill water	8,302 bbls
Active mud	1,132 bbls
Storage mud	2,139 bbls
Fuel oil	6,775 bbls
Potable water	2,528 bbls
Bulk mud	181 m³ (6,400 ft³)
Bulk cement	181 m³ (6,400 ft³)

Hakuryu II, sister rig to *Hakaryu* (*Courtesy Mitsubishi Heavy Industries, Ltd.*).

OCEAN RANGER

Specifications

Length	119·21 m (391 ft)
Width	79·88 m (262 ft)
Propulsion power	14,000 h.p.
Accommodation	100
Helicopter pad	for S-65 helicopter
Operating depth:	
standard	460 m (1,500 ft)
with extra equipment	1,070 m (3,500 ft)

OCEAN KOKUEI
Type
Semi-submersible, self-propelled
See *Ocean Victory* for principal details.

Specifications
Hull: Octagonal shape, width at operating level 79·88 m (262 ft). Lower hull consists of four 8·54 m (28 ft) diameter cylindrical tubes parallel to each other framed together to form a base of 80·98 m × 97·5 m (266 ft × 320 ft)
Working deck: Main deck 2,057 m² (22,136 ft²) Cellar deck and drill floor 686 m² (7,381 ft²)
Operating depth: Afloat 183 m (600 ft) with 8 anchors and chain supplied (additional chain may be added for deeper water)
Propulsion: 7,000 h.p. twin screw diesel electric—Kort nozzle
Quarters: For 81 men
Radios: Marine radio telephone

Heliport: Overall size 26·22 m × 25·3 m (86 ft × 83 ft)—Heavy wheel load area 25·3 m × 25·3 m (83 ft × 83 ft)

Equipment
One Lee C. Moore 12·2 × 12·2 × 43·29 m (40 ft × 40 ft × 142 ft) cantilever mast 453,600 kg (1,000,000 lb) API static hook load capacity.
One Continental EMSCO C-3 type 2 draw works grooved for 3·5 cm (1⅜ in) line, V-200 Parmac hydromatic brake, EMSCO cat heads, sandreel assembly mounted on draw works, driven by three 1,000 h.p. DC motors.
One Continental EMSCO 95·3 cm (37½ in) rotary driven by 1,000 h.p. DC motor with two speed transmission.
One 650 ton swivel Continental EMSCO L-650.
One Continental EMSCO MA-60-6-1 travelling block—650 ton.

One Byron-Jackson 500 ton hook.
One Lee C. Moore 2·1–18·3 m (7–60 ft) sheave crown.
One Koomey accumulator—320 gallon 211 kg/cm² (3,000 lb/in²) W.P. with electric master and remote panels.
One 34·6 cm (13⅝ in) 5,000 lb/in² BOP system: 2 Shaffer 'LWS' double BOP ram units with 1 shear, blind ram, one 353 kg/cm² (5,000 lb/in²) Hydril and 183 m 40·6 cm (600 ft 16 in) Vetco MR-4 marine riser.
One 20 in, 2,000 lb/in² BOP system: 1 ram, 1 Hydril and 183 m (600 ft) of 61 cm (24 in) Vetco MR-4 marine riser.
One Riser tensioner system—109,000 kg (240,000 lb).
Two Fairbanks-Morse, 12 cylinder Model 38D8 ⅛ opposed piston diesel engines, each rated 2,040 h.p. at 720 rpm driving two 1,600 kw traction type generators and one 1,000 KVA AC generator.

Two Fairbanks-Morse, 12 cylinder, Model 38D8 $\frac{1}{8}$ opposed piston diesel engines, each rated 2,040 h.p. at 720 rpm driving two 1,600 kw traction type generators.

Two EMSCO F-1300 triplex pumps 7 × 12 driven by 1,300 h.p. DC motor each supercharged with 5 × 6 Mission centrifugal pump.

Two Mission 6 × 8R, H-30 centrifugal mud mix pumps with 26·7 cm (10½ in) impellers and 100 h.p. AC motors.

Two Rumba Model 4860 DU dual shale shakers.

10,000 ft of 12·7 cm (5 in) OD 8·86 kg (19·50 lb) Grade E, R-3 drill pipe with 12·7 cm (5 in) X-Hole tool joints.

5,000 ft of 12·7 cm (5 in) OD 8·86 kg (19·50 lb) X-105, R-3 IEU drill pipe with 12·7 cm (5 in) Hole tool joints.

Twenty-one 19·7 cm (7¾ in) OD × 7·14 cm (2 13/16 in) ID drill collars with 16·83 cm (6⅝ in) API Reg. Conn.

One taper tap and overshot for rig drill pipe and drill collars.

One link belt speeder Model ABS 218 crane with 36·6 m (120 ft) boom.

One Clyde special Model H 27103 heavy duty stiff leg derrick with 41·16 m (135 ft) boom rated 40 tons at 9·15 m (30 ft) radius and 15 ton at 39·63 m (130 ft) radius utilising Manitowoc Model 290 two drum shaft hoist.

One Halliburton HT-400 cement unit, Pioneer T-16-4 Desilter; Pioneer T-10-6 Desander, Pit-O-Graph.

Two 40-ton Houston Systems overhead cranes.

Eight 1 drum electric anchor windlasses, each with one 1,000 h.p. DC motor.

Eight anchors with 1,006 m (3,300 ft) of 7 cm (2¾ in) steel stud link anchor chain.

Storage capacity

Fuel	6,972 bbls
Drill water	14,320 bbls
Potable water	384 bbls
Dry mud	140 s. tons
Bulk mud and cement	9,600 cu. ft
Liquid mud	1,625 bbls

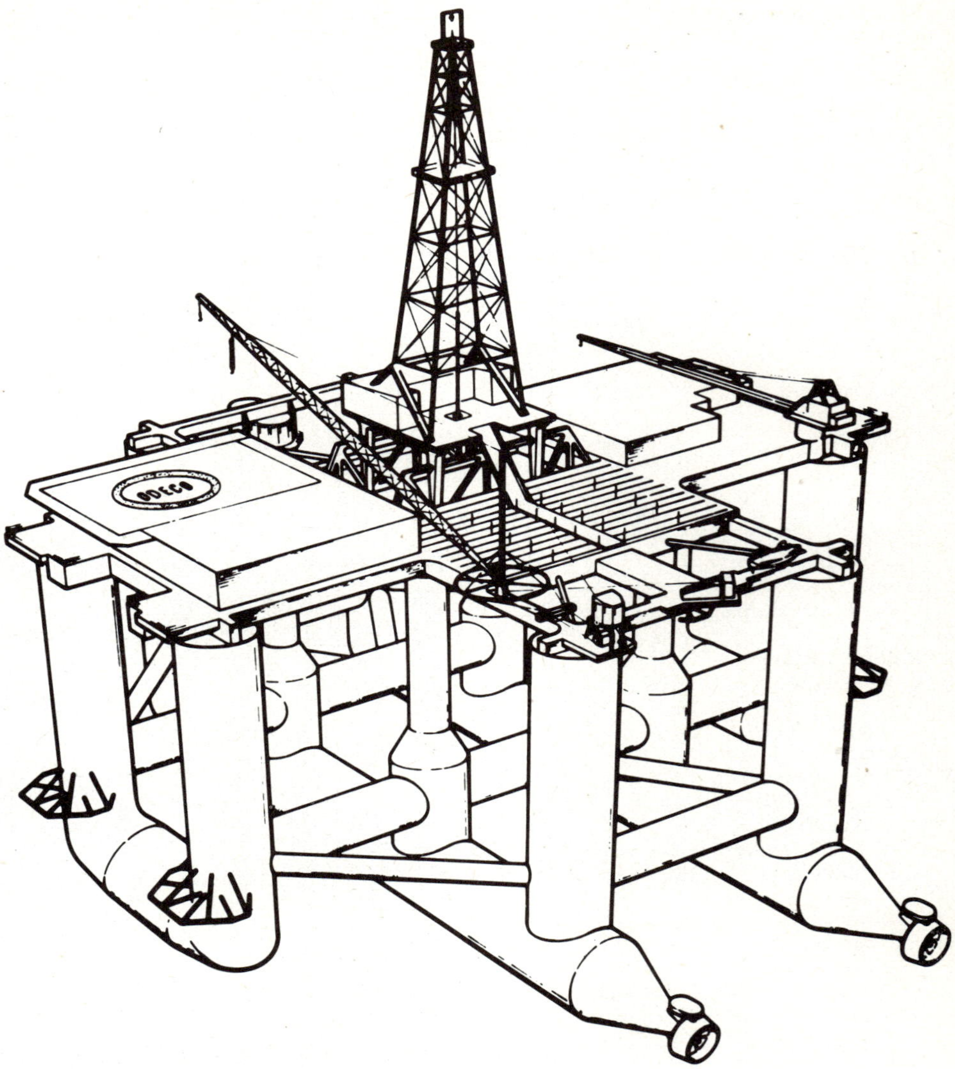

Ocean Kokuei (*Courtesy ODECO*)

NORWAY

AKER GROUP (NYLANDS VERKSTED, OSLO)

Name	Owners	Type	Design	Delivery date	Hull No.
Ocean Voyager	Canam Offshore Ltd (ODECO), Panama	semi-submersible	ODECO	November 1973	693
Odin Drill	Oslo Drilling A/S & Co., Oslo, Norway (Hagb. Waage/Wilh. Wilhelmsen together with the Canadian companies Home Oil and Bow Valley)	semi-submersible	ODECO	January 1974	694
Deep Sea Driller (Bergen Rig 1)	Deep Sea Drilling Company A/S, Bergen, Norway (The rig is chartered to the French company ELF)	semi-submersible	Aker H-3	January 1974	695
Norjald (NOR 102)	Gotaas-Larsen Shipping Corporation, New York, USA/Fearnley & Eger, Oslo, Norway with the American company Rowan International, Inc as partners	semi-submersible	Aker H-3	May 1974	697
Borgny Dolphin	Fred. Olsen & Co., Oslo, Norway—with the American company Dolphin International as partners	semi-submersible	Aker H-3	July 1974	700
Bideforce Dolphin	Blandford Shipping Co. Ltd, London, England—with the American company Dolphin International as partners	semi-submersible	Aker H-3	October 1974	701
Bergen Drill	Deep Sea Drilling Company A/S, Bergen, Norway	semi-submersible	Aker H-3	4th quarter 1974	702

Name	Owners	Type	Design	Delivery date	Hull No.
Haakon Magnus	K/S Viking Offshore, Kristiansand, Norway	semi-submersible	Aker H-3	4th quarter 1974 Christening Oct. 31	703
	The Shipping Company Einar Rasmussen, Kristiansand, Norway—with the American company Global Marine as partners	semi-submersible	Aker H-3	1st quarter 1975	704
	Atlantic Drilling Co., Ltd, England	semi-submersible	Aker H-3	May 1975	705
	Gotaas-Larsen		Aker H-3	March 1976	
	Gotaas-Larsen		Aker H-3	2nd quarter 1976	
	New Company, with Arne Blystads Rederi and A/S Det Dansk Fransk Dampskibsselskab		Aker H-3	2nd quarter 1976	
	Svene Ditlev-Simonsen & Co., Oslo, Norway		Aker H-3	3rd quarter 1976	

OCEAN VOYAGER (see above table)

Delivered in November 1973, *Ocean Voyager* was in use by Shell in January 1974 in block 205/21, 75 miles north west of Kirkwall, Orkney.

AKER H-3

The self-propelled semi-submersible drilling rig, designated 'Aker H-3', is now selling in numbers to the offshore drilling industry. The first H-3 platform was completed in 1973. The Aker Group's experience with semi-submersibles goes back to 1965 when the Group received a contract for building of the ODECO rig *Ocean Viking*. This was the first rig to make a commercial oil find in the Norwegian part of the North Sea (the Ekofisk field). The engineering department of the Aker Group has considered a number of alternative proposals for semi-submersible drilling platforms. Numerous proposals had been analysed before the decision was taken to prepare detailed design specifications of the H-3 type. The comparison study of the alternative designs was partly carried out using special computer programmes developed by the Aker Group for this purpose.

The main objectives in designing the 'Aker H-3' platform were to achieve a drilling unit with the following key features:

Ability to operate in rough, deep sea conditions as encountered in the North Sea area;

Favourable wave motion response characteristics;

High degree of mobility (self-propulsion);

Large supply capacity.

At the same time the size of the platform itself was kept at a minimum to reduce capital costs. This means not only optimising the design arrangement and dimensions, but also optimising the steel structural design. This has been done, utilising the Aker Group's shipbuilding experience, and the extensive com-

On the left, *Ocean Voyager* and on the right, *Odin*, at the Akers Oslo yard (*Courtesy Aker Group*)

puter programmes which Det norske Veritas has prepared for structural design.

The pontoons are square to improve the motion characteristics, and this makes it possible to achieve better connection between pontoons and columns. Resistance in travelling condition is also reduced as less wetted surface is obtained.

The design of the main deck incorporates the deckhouse etc. into the steel structure. This, together with extensive use of high tensile steel, has made the H-3 platform very light and has improved the storage capacity.

A new feature is also the extensive use of the pontoons and columns for supplies. This improves stability and makes it possible to reduce the diameter of the columns, which again improves the motion characteristics. Pontoons and columns are divided by watertight bulkheads to the extent that possible leakage resulting from a collision etc. will not hazard the stability of the platform (two compartment standard).

A model of the platform was tested in the ship model experiment tank at Wageningen. Very good motion characteristics in drilling conditions were achieved, and the survival test proved that the 'Aker H-3' behaved exceedingly well even in waves of 30·5 m (100 ft).

Special care was taken to check every part of the design. Close co-operation has been established with Det norske Veritas in this respect employing advanced computer programmes, etc. In addition to this, an American consultant with long experience of semi-submersible platforms was engaged to verify the main items. The basic data was also checked by Offshore Dynamics, London.

By July 1973, twelve rigs of the Aker H-3 design had been contracted.

This was the largest number of semi-submersibles ever to be built of a single design, and at the time of going to press this number stood at eighteen.

Licensed production of the Aker H-3 design in the USSR is contemplated.

Principal dimensions

Length, overall	108·2 m (355 ft)
Width, overall	67·4 m (221 ft)
Main deck length	69·2 m (227 ft)
Main deck width	61·0 m (200 ft)
Depth to main deck	36·6 m (120 ft)
Large column diameter	7·9 m (26 ft)
Pontoon height	6·7 m (22 ft)
Pontoon width	11 m (36 ft)

Operation characteristics

Water depth	183 m* (600 ft)
Drilling depth	7,620 m (25,000 ft)
Draught of platform in drilling condition	21·3 m (70 ft)
Platform clearance at drilling draught	15·2 m (50 ft)

* Based on 915 m (3,000 ft) of chain. Water depth can be extended by increasing anchor chain lengths.

Aker H-3 type hull in Netherlands ship model basin tank (*Courtesy Aker Group*)

A scale model of the Aker H-3 type rig (*Courtesy Aker Group*)

Supply capacities

	bbls	long tons
Bulk, mud and cement (tanks in columns) 18,000 ft^3	3,205	900
Sack material area—6,000 sacks		300
Pipe rack—5,000 ft^2		900
Liquid mud	1,450	400
Fuel oil	16,900	2,270
Drill water	14,400	2,255
Total supply capacity		7,025
Max. variable deck load (incl. 600 L–tons bulk in columns)		2,350

Displacements

	long tons
Pontoons total	13,700
Columns total below drilling draught	5,300
Displacement at drilling draught	19,000
In transit condition	12,500

Motion characteristics

Natural heave period	22 seconds
Natural roll period	34 seconds
Natural pitch period	30 seconds

Equipment and outfitting

According to normal practice, machinery and equipment are decided by the customer. The total weight of this equipment is estimated to be 2,800 long tons.

The platform is foreseen equipped as itemised below, but modifications can be made to meet customer's special requirements:

Draw works: Rated for 7,620 m (25,000 ft) drilling depth, equipped with three 800 h.p. motors

Derrick: 12·2 × 12·2 m (40 × 40 ft) base, 48·78 m (160 ft) high mast or derrick, designed to accommodate motion compensator

Mud pumps: Two Triplex pumps, each equipped with two 800 h.p. top-mounted motors

BOP system: Moon well area design to accommodate any commonly used BOP system

Power plant: Four 2,200 h.p. Bergen Diesels, each driving one AC generator. Thyristor plant for AC/DC conversion

Propulsion: Two DC propulsion units, each rated at about 2,500 h.p. Propeller steering through Kort nozzle rudders. Remote control from wheel house.

Windlasses: Four double anchor windlasses with quick release chain stoppers and tension indicators. Remote control. Low pressure hydraulic winches of NORWINCH make are foreseen.

Anchors: Eight 13,640 kg (30,000 lb) anchors

Chain: Eight 915 m (3,000 ft) stud link chains

Lower fairleads: Eight single roller wildcat anchor chain fairleads. W/180° turning

Cranes: AKERS revolving cranes, 100–120 ft boom length. 40 tons main hoist; 15 tons ship hoist

Accommodation: Air conditioned cabins for 75 persons, arranged on two decks, in 1- 2- and 4-berth cabins

Control rooms: Wheel house, radio room and combined ballast mooring control room

Heliport: 25·3 × 25·3 m (83 × 83 ft) heliport installed above drilling quarters, designed for Sikorsky S-61 or equivalent

AKER H-4

Design work is to be completed in 1974, on a larger version of the H-3 to drill in water depths down to 457·2 m (1,500 ft). Displacement will be approximately 33,000 tons, and cost about N.Kr. 300 million.

Construction of the first Aker H-3 oil drilling rig at the Aker Group yard at Verdal (*Courtesy Aker Group*)

Outfitting of the first Aker H-3 rig at Bergen (*Courtesy Aker Group*).

A/S FRAMNAES MEK. VAERSTAD/TROSVIKGRUPPEN

Name	Owners	Type	Design	Delivery date
N/A	K/S Dyvi Drilling A/S, Oslo, Norway	semi-submersible	Aker H-3	November 1974
N/A	A/S Rosshavet A/S Vestfold, Sandefjord, Norway	semi-submersible	Aker H-3	May 1975

NORMARIG A/S, LANGESUND

Name	Owner	Type	Design	Delivery date
N/A	Gotaas-Larsen/Rowan Company	semi-submersible	Aker H-3	November 1973

SINGAPORE

BETHLEHEM SINGAPORE PRIVATE LTD

Name	Owners	Type	Design	Delivery date
Teledyne Rig 16	Teledyne Movable Offshore	jack-up	Bethlehem	January 1973
Teledyne Rig 17	Teledyne Movable Offshore	jack-up		1973
N/A	Storm Drilling Co.	jack-up (mat) (max. water depth 76 m (250 ft)) cost $14.5 m		By October 1975
Transworld Rig 63	Transworld Drilling Co, Ltd	jack-up		November 1974
Milton G. Hulme	Reading & Bates	jack-up		1974

TRANSWORLD RIG 63

Type
Jack-up, three leg mat supported

Completion
November 1974

Platform

Length, overall	54·8 m (180 ft)
Breadth, overall	40·2 m (132 ft)
Depth	4·9 m (16 ft)
Drilling slot	4·79 m² (50 ft²)

Mat
64 m × 51·8 m × 3·04 m (210 ft × 170 ft × 10 ft)

Columns
Three, 95·1 m (312 ft) long, 3·6 m (12 ft) diameter

Maximum drilling depth capability
7,621 m (25,000 ft)

Maximum water depth capability
76·2 m (250 ft)

Accommodation
78 men

Application
Anticipated to be in South East Asia area

MILTON G. HULME

Type
Jack-up

Hull

Length, overall	54·88 m (180 ft)
Breadth, overall	40·24 m (132 ft)

Water depth, maximum
76 m (250 ft)

Drilling depth, maximum
6,100 m (20,000 ft)

Storage
Mud and cement bulk: 175·39 m³ (6,000 ft³) plus 3,000 sacks
Mud and cement, liquid: 1,500 bbl

Cranes
Two Le Tourneau 25·4 tonnes (25 tons) at 9·15 m (30 ft) radius

RIG 17

Under construction 1973 fabrication having started 1 January and planned commissioning date, 15 January 1974. Rig 16 is of the same design.

Specifications

Length, overall	68·9 m (226 ft)
Width, overall	51·8 m (170 ft)
Height	95·1 m (312 ft)
Main deck length	50·6 m (166 ft)
Main deck width	33·2 m (109 ft)
Platform clearance at drilling draught	7·6 m (25 ft)
Drilling draught	76·2 m (250 ft)

Capabilities
Maximum water depth for drilling: 76·2 m (250 ft)
Drilling depth capability: 7,621·9 m (25,000 ft)
during hurricane season 60·9 m (200 ft)

Accommodation
82 men

Tug requirements
Design towing speed: 3 knots with 6,000 combined tug horsepower
Power services: Main engines—type: Two 2,200 h.p. EMD 16-645E1; Generators: Four EMD D-79-6B and two EMD D-32K

Drilling power system—Electric
Pumping systems for ballast, drill water, fuel, mud: Mission centrifugal pumps for mud changing, mud mixing, salt water, ballast, and fuel service. Two Continental EMSCO F-1600 triplex mud pumps with Mission supreme fluid ends.

Liquid storage capacities

Drill water	4,386 bbls
Fuel oil	1,796 bbls
Potable water	472 bbls
Wash water	2,372 bbls
Liquid mud	1,637 bbls
Bulk mud	93 m³ (3,280 ft³)
Other:	
Bulk cement	89 m³ (3,150 ft³)

Jacking system
Bethlehem hydraulic jack

Drilling equipment

Derrick:	
height	Continental-EMSCO 44·8 m (147 ft)
hook load capacity	475 tons
Drill pipe diameter	12·7 cm (5 in)
BOP stack:	
bore	50·8 cm (20 in)
working pressure	141 kg/cm² (2,000 lb/in²)
Rotary table	Continental-EMSCO T-3750

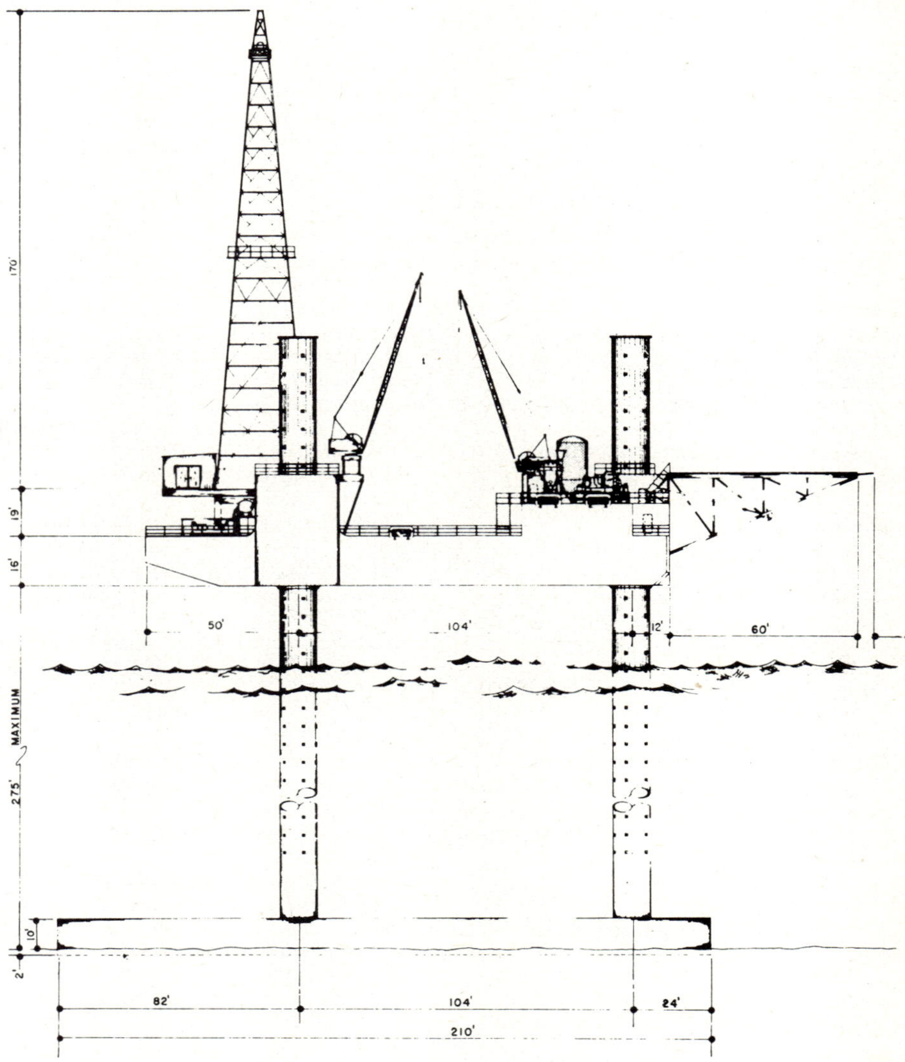

Side elevation of Teledyne *Rig 16* and *17* (*Courtesy Teledyne Movible Offshore Inc.*)

Other features
Draw works: Continental-EMSCO Model C-3 Type II
Helicopter deck: 18·2 m × 21·3 m (60 ft × 70 ft)

FAR EAST LEVINGSTON SHIPBUILDING LIMITED

Name	Owner	Type	Design	Delivery date
Penrod 74	Penrod Drilling Co.	semi-submersible		1974

PENROD 74

Type
Semi-submersible rig, twin hulled, column-stabilised

Delivery
Early 1974

Operations area
New Zealand, initially

Dimensions

Platform:	
Length	77·7 m (255 ft)
Breadth	58·2 m (191 ft)
Columns	6
Total height, base of hull to upper deck	39 m (128 ft)

MARATHON LE TOURNEAU

Name	Owners	Type	Design	Delivery date
Rig 103	Santa Fe Drilling Co.	jack-up		1974
Margie	Atwood Oceanics, Inc	semi-submersible		1973
Rowan-Anchorage	Rowan International, Inc	jack-up		July 1972
Key Biscayne	Key International Drilling Co.	jack-up		July 1972
Rig 105	Santa Fe Drilling Co.	jack-up	(Rig 103 type)	1975
F. G. McClintock	Reading & Bates	jack-up		January 1975
Chickamauga	Atwood Oceanics, Inc	jack-up		1974

RIG 103

Specifications

Type	Jack-up
Length	61·8 m (203 ft)
Width	51·2 m (168 ft)
Depth	6·7 m (22 ft)

Maximum water depth: 76·2 m (250 ft) with zero leg penetration and 7·6 m (25 ft) air gap between water and bottom of platform. 10·7 m (35 ft) air gap at 45·7 m (150 ft) water depth

Drilling cantilever: 6·1 m (20 ft) wide and 12·2 m (40 ft) long working grid

Accommodation: 78

Sewage treating unit: Demco WT-5000-C (chemical)

Helicopter deck: diameter 21·3 m (70 ft) can handle Sikorsky S-61

Jacking system

Mfg/design	Le Tourneau
Number of legs	3
Maximum length	108·8 m (357 ft)
Jacking speed at maximum load	0·304 m/min (1 ft/min)
Leg bottom support	
Total support area (3 cans)	464·5 m² (5,000 ft²)

Maximum operating water depth

76·2 m (250 ft) with zero leg penetration and 7·6 m (25 ft) air gap between water and bottom of platform. 10·6 m (35 ft) air gap at 45·7 m (150 ft) water depth

Propulsion

Has provision for two thrusters of 1,000 h.p. each

Towing speed in calm sea

3·6 knots with 2,400 (est.) h.p. sea tug

Deck cranes

Capacity: 45 tons at 7·6 m (25 ft) radius. Boom length 30·5 m (100 ft)

Capacity: 45 tons at 7·6 m (25 ft) radius. Boom length 30·5 m (100 ft)

Capacity: 45 tons at 7·6 m (25 ft) radius. Boom length 30·5 m (100 ft)

Make and model: Le Tourneau RD-120 AS

Drilling equipment

Rated drilling depth: 6,079 m (20,000 ft)

Draw works and drive: Continental EMSCO C-2 with two GE-752 motors

Main power plant: Four Cat D-399 engines, each driving one GE 930 kw 600 volt generator/SCR system

Main mud pumps: Two National N-1300 Duplex rated at 1,300+ h.p.

Mud mixing pumps: Two Mission GX8R at 100 h.p. each

BOP equipment: SIPM furnished/bare boat charter

Rotary table: T-3750/53¼

Maximum support load: 340,909 kg (750,000 lb)

Substructure: Skidded cantilever mount/ maximum variation at 795,454 kg (1,750,000 lb)

Maximum setback load: 181,818 kg (400,000 lb)

Mast/Derrick: Continental EMSCO Model 20-R, 9·1 m × 9·1 m × 42·6 m (30 ft × 30 ft × 140 ft) clear

Maximum hook load: 453,600 kg (1,000,000 lb)

Cementing unit: Haliburton HT-400 D

Logging unit: SIPM furnished/bare boat charter

Drill pipe and drill collars: SIPM furnished/ bare boat charter

Storage capacities

Bulk mud	142 m³ (5,000 ft³)
Bulk cement	85 m³ (3,000 ft³)
Liquid mud	1,200 bbls
Fuel oil	1,500 bbls
Drilling water	1,500 bbls
Potable water	1,000 bbls
Desalinisation unit	Two Meco, capacity 660 US gph each

Other equipment

Shell mud/gas separator, dual tandem and dual conventional shale shakers

Personnel survival equipment

Standard plus two watercraft 50-man-each life boats

CHICKAMAUGA

To work in Singapore area.

Building

1974

Dimensions

Length	70·1 m (230 ft)
Breadth	60·9 m (200 ft)
Max. water depth	91·4 m (300 ft)
Max. drilling depth	6,079 m (20,000 ft)

Maximum operating sea state

Operating	9·1 m (30 ft)
Survival	15·2 m (50 ft)

Storage

Mud and cement bulk	142 m³ (5,000 ft³) plus 10,000 sacks
liquid	1,400 bbls

Power

Five Caterpillar D-399, 1,325 h.p. each

Cranes

Three Le Tourneau 45 tons at 7·6 m (25 ft) radius

Accommodation

78 men

SPAIN

HIJOS DE J. BARRERAS, VIGO

Name	Owners	Type	Design	Delivery date
Penrod 70	Penrod Drilling Co.	semi-submersible		1973
	The Offshore Co. with I. M. Skaugen & Co. (Norway) plus others	semi-submersible	SCP III *Mk. 2*	1974

SCP III—MARK II

Type

Semi-submersible

Lower hulls

Length, overall	121 m (397 ft)
Diameter	10·7 m (35 ft)

Top hull

Length	82·3 m (270 ft)
Width	68·0 m (223 ft)

Accommodation

90

Maximum water depth capability

183 m (600 ft)

Maximum drilling depth capability

7,620 m (25,000 ft)

Cost

$24 million

UNITED KINGDOM

MARATHON SHIPBUILDING COMPANY (U.K.) LIMITED

Name	Owners	Type	Design	Delivery date
Penrod 64	Penrod Drilling Company			Sept. 29, 1973
N/A	Keydril, subsidiary of Gulf Oil	N/A (reported cost £5.8 m)	N/A	N/A

PENROD 64

Specifications

Length, overall	70·1 m (230 ft)
Beam, overall	60·9 m (200 ft)
Depth	7·9 m (26 ft)
Maximum water depth capability	71·6 m (235 ft)
Maximum drilling depth capability	9,146 m (30,000 ft)
Maximum sea state	15·3 m (50 ft) waves
Maximum winds	125 mph

Accommodation 78 men

This rig will be operated in the North Sea. It was launched in only 13 months from Marathon taking over the Clydebank yard.

Storage capacities

Bulk mud	184 m³ (6,500 ft³)
Drilling water	3,273 bbls
Fresh water	1,080 bbls

Cranes

Three, at 45 tons at 7·6 m (25 ft)

UNITED STATES OF AMERICA

ALABAMA DRYDOCK AND SHIPBUILDING CO., MOBILE, ALA.

Name	Owners	Type	Design	Delivery date
Century	Diamond M. Drilling	semi-submersible	Korkut Engineers, Inc, New Orleans, Louisiana	1972
N/A	Diamond M. Drilling	semi-submersible	Breit Engineering Co., New Orleans, Louisiana	1974

CENTURY

Delivered towards the end of 1972.

Hulls designed for minimising heave response in short period waves.

Specifications

Length, overall	82·32 m (270 ft)
Width, overall	52·74 m (173 ft)
Lower hulls:	
Lengths	82·32 m (270 ft)
Widths	10·67 m (35 ft)
Depths	7·62 m (25 ft)
Height (drill floor to keel)	35·98 m (118 ft)
Draught:	
operating	15·24 m (50 ft) (displacement 15,000 l. tons)
transit	6·1–7·01 m (20–23 ft)
Air gap between upper deck underside and top of lower hulls	16·46 m (54 ft)
Max. water depth:	
normal	183 m (600 ft)
with additional chain	305 m (1,000 ft)
Max. drilling depth	9,150 m (30,000 ft)

Propulsion assist: Four 750 h.p. DC motors driving two 2·44 m (8 ft) diameter stainless steel propellers, Kort nozzles

Anchoring	Eight 1,160 m (3,800 ft) chains, 13,636 kg (30,000 lb) anchors
Constructional steel weight	5,200 tons
Draw works	Two 1,000 h.p. motors
Derrick	48·8 m (160 ft)
Drill water	1,575 bbls

Accommodation 64

Cost $ US 21 million

Storage

Mud and cement:	
Bulk	258·6 m³ (9,120 ft³)
Sacks	3,500
Liquid	1,575 bbls
Fuel	5,100 bbls
Water:	
Drilling	9,800 bbls
Potable	400 bbls

NAME N/A

Type

Semi-submersible, designed by Breit Engineering Co.

Specifications

Hull:	
Length, overall	88·41 m (290 ft)
Breadth, overall	61 m (200 ft)
Rated water depth	183 m–305 m (600 ft–1,000 ft) allows operation over 95% of the world's oceans
Rated drilling depth	9,150 m (30,000 ft)
Accommodation	77
Operating draught	16·8 m (55 ft)
Moorings	Eight anchors, each 15 s. tons

Storage

Mud and cement:	
Bulk	258 m³ (9,120 ft³)
Sacks	3,500
Liquid	1,575 bbls

AVONDALE SHIPYARDS, INC, BELLE CHASSE, NEW ORLEANS, LA.

Name	Owners	Type	Design	Delivery date
Sedco 702	Sedco, Inc		Earl & Wright	March 17 1973 commissioned
N/A	Zapata Offshore Co.		SS-2000	Spring 1975
N/A	Zapata Offshore Co.		SS-2000	mid 1976
N/A	Zapata Offshore Co.		SS-2000	mid 1976
Waage Drill I	K/S Waage Drilling A/S Co. with Santa Fe International Services	semi-submersible	ODECO (*Ocean Voyager* class)	August 1973
Waage Drill II	K/S Waage Drilling A/S Co. with Santa Fe International Services	semi-submersible	ODECO (*Ocean Voyager* class)	1974
Pacesetter III	Western Oceanics Division of Western Company of North America	semi-submersible		1973
Sedco 703	Sedco			
Ocean Rover	ODECO (Canam Offshore Ltd, Hamilton, Bermuda)	semi-submersible	ODECO	1973
Ocean Victory	ODECO (Canam Offshore Ltd)	semi-submersible	ODECO	1972

SEDCO 702

Designed by Earl & Wright, *Sedco 702* was the first twin-hull mobile unit to be completed.

Type

Rectangular, twin hull, column-stabilised mobile offshore drilling unit

Design objectives

To operate in deep, rough water (and speed)

Fabrication start date
January 1972

Commissioning date
April 1973

Capital cost
$ US 27,000,000 (1973)

Areas of intended use
North Sea

Customer
Conoco

Unique features
Largest BOP, rise pipe and rotary table in world. Tallest drilling mast, automatic anchor line back-off and take-up system

Dimensions, weights, draughts

Length, overall	91·5 m (300 ft)
Width, overall	83·8 m (275 ft)
Height:	
keel to mast top	103·7 m (340 ft)
keel to main deck	39·6 m (130 ft)
Main deck length	70·4 m (231 ft)
Main deck width	60·1 m (197 ft)
Platform clearance at drilling draught	11·1 m (36 ft 6 in) when at the 25·9 m (85 ft) draught
Drilling draught	24 m to 27 m (80 ft to 90 ft)
Displacement at drilling draught	24,980 short tons at (80 ft) draught
Displacement at transit conditions	5·2 m (17 ft) draught, 14,025 short tons
Displacement of pontoons	6·4 m (21 ft) draught, 16,955 short tons
Displacement of columns below water at drilling draught	18,025 short tons

Operating draughts in various wave conditions: 5–8 m (16–26 ft) Moving–moderate seas. 15–18 m (50–60 ft) Excessive motions at shallow draught (therefore move at deep draught). 24–27 m (80–90 ft) Drilling and station keeping

Capabilities
Maximum water depth for drilling: 915 m (3,000 ft)
Drilling depth capability: Draw works for 7,622 m (25,000 ft) depth
Sea state, current and wind conditions, including maxima, with corresponding drilling depths: 5,183 m (17,000 ft) 3½ degree rolls, 4·6 m (15 ft) heave with 55 knot wind

Dynamic positioning capability
Consists of a position Ref. system and four azimuthing thrusters. Position information is displayed on a scope in barge control room, also a console can control the thrust and direction of 4 thrusters

Self-propulsion
Eight knots ahead and astern

Range with corresponding speed
Forty days or 5,000 miles—8 knots under tow

Accommodation
102

Design towing speed
Ten knots + with assistance of 4 thrusters

Power services
Main engines—type: Four 1,600 h.p. DC electric motors (thrusters)
Generators: Three 2,625-KVA generators (3,070 b.h.p.)
Drilling power system: DC Thyristor drilling system
Propulsion power system: Four thrusters
Pumping systems for ballast, drill water, fuel mud: Four 2,500 gpm ballast pumps, Two 1,700-h.p. mud pumps
Drill water: One 350-gpm pump for each hull and 1 for main deck
Fuel: One 100-gpm pump for each hull
Thrusters—thrusts: 8,000 kg (17,600 lb) at zero speed fully azimuthing with full thrust in any direction. Number of screws: Four; screw diameter: 3·05 m (10 ft)

Liquid storage capacities

Draw works brake cooling water	2,200 US gallons
Drill water	492,000 US gallons
Fuel oil	300,000 US gallons
Potable water	56,000 US gallons
Liquid mud	4,000 bbls or 167,900 US gallons
Bulk mud	108 m³ (3,850 ft³)
Other:	
Lube oil, waste oil, cooling water etc.	9,000 US gallons

Deck storage capacity
2,500 short tons

Mooring system
Eight 13,636 kg (30,000 lb) LWT Anchors

Drilling equipment

Derrick:	
height	Mast height = 48·8 m (160 ft)
hook load capacity	454,545kg (1,000,000 lb)
Drill pipe diameter	12·7 cm (5 in)
BOP stack:	
bore	54 cm (21¼ in)
working pressure	700 kg/cm² (10,000 lb/in²)
height	12·5 m (41 ft)
weight	180 tons
Rotary table	125·7 cm (49½ in) Oilwell
Riser pipe:	
length	305 m (1,000 ft)
diameter	61 cm (24 in)
weight/ft	387 kg/m (260 lb/ft)

Other features
Four external lines: Two 12·7 cm (5 in) and two 7·6 cm (3 in)

Motion characteristics

Natural period in pitch	26·0 seconds
Natural period in roll	35·0 seconds
Natural period in heave	22·5 seconds

Ballasting facilities: 2,673,000 US gallons capacity
Diving system: 305 m (1,000 ft) depth capability
Helicopter deck: Approximately 647 m² (6,950 ft²), 25·3 m (83 ft) diameter for Sikorsky S-61 helicopter

Developments possible
Future complete dynamic stationing without mooring lines.

SS-2000
Orders for three semi-submersible rigs of a new class, SS-2000 series, were placed at the end of 1973 by the Zapata Corporation, together with an option on a fourth. Deliveries of these rigs will extend from spring 1975 to the middle of 1976. They are of twin catamaran hull form with six stabilising columns.

Specifications

Maximum water depth capability	610 m (2,000 ft)
Maximum drilling depth capability	7,620 m (2,500 ft)
Length, overall	79 m (260 ft)
Beam, overall	61 m (200 ft)
Depth to drilling deck	24 m (80 ft)
Displacement	16,750 long tons at drilling draught of 14 m (45 ft)

Accommodation 90 men

Cost
About $17 m (building)

Design
ABS and US Coast Guard standards

WAAGE DRILL I
Built: 1973, christened 30 August 1973. This drilling unit is operated jointly by Santa Fe International Services, Inc. Orange, Calif., and K/S WAAGE Drilling A/S and Co.

Specifications
Designers: Ocean Drilling and Exploration Co. (ODECO)

Length, overall	91·46 m (300 ft)
Beam, overall	89·33 m (293 ft)
Maximum water depth capability	183 m (600 ft)
Maximum drilling depth capability	7,620 m (25,000 ft)

WAAGE DRILL II
Delivery 1974, a sister rig identical to *Waage Drill I.*

PACESETTER III
See *Pacesetter I* (Bethlehem Steel Corporation).

OCEAN ROVER
Hull
Octagonal form, tubular construction

Length, overall	97·56 m (320 ft)
Breadth lower hulls	81·1 m (266 ft)
Lower hull dia.	8·54 m (28 ft)
Height of upper deck above bottom of hull	39·02 m (128 ft)
Main deck area	2,122·34 m² (22,136 ft²)
Transit draught, approx.	7·62 m (25 ft)
Drilling draught	21·34 m (70 ft)
Displacement at drilling draught	18,900 tonnes (20,800 short tons)
Displacement at transit draught	13,270 tonnes (14,600 short tons)
Operating water depth (with standard chain—additional may be carried)	183 m (600 ft)
Drilling depth rated	9,150 m (30,000 ft)
Drilling water capacity	14,318 bbls
Fuel oil capacity	6,972 bbls
Potable water capacity	384 bbls
Liquid mud capacity	1,161 bbls
Bulk mud and cement capacity	272 m³ (9,600 ft³)
Dry mud capacity	140 short tons

Propulsion
Two diesel electric 7,000 h.p. total, driving two 1·52 m (5 ft) diameter 5-bladed propellers in Kort nozzles

Accommodation
78

Cranes
One Pyramid 49·39 m (162 ft), 500 short ton capacity
One Continental EMSCO swival 650 short ton capacity
One Clyde Model CP-96 with 36·59 m (120 ft) boom
One Clyde Model H 27103 derrick, with 36·59 m (120 ft) boom, rated 40 short tons at 9·15 m (30 ft) radius and 16·5 short tons at 36·59 m(120 ft) radius
One Houston Systems 40 short tons

ODECO's *Ocean Rover* in transit (*Courtesy ODECO*)

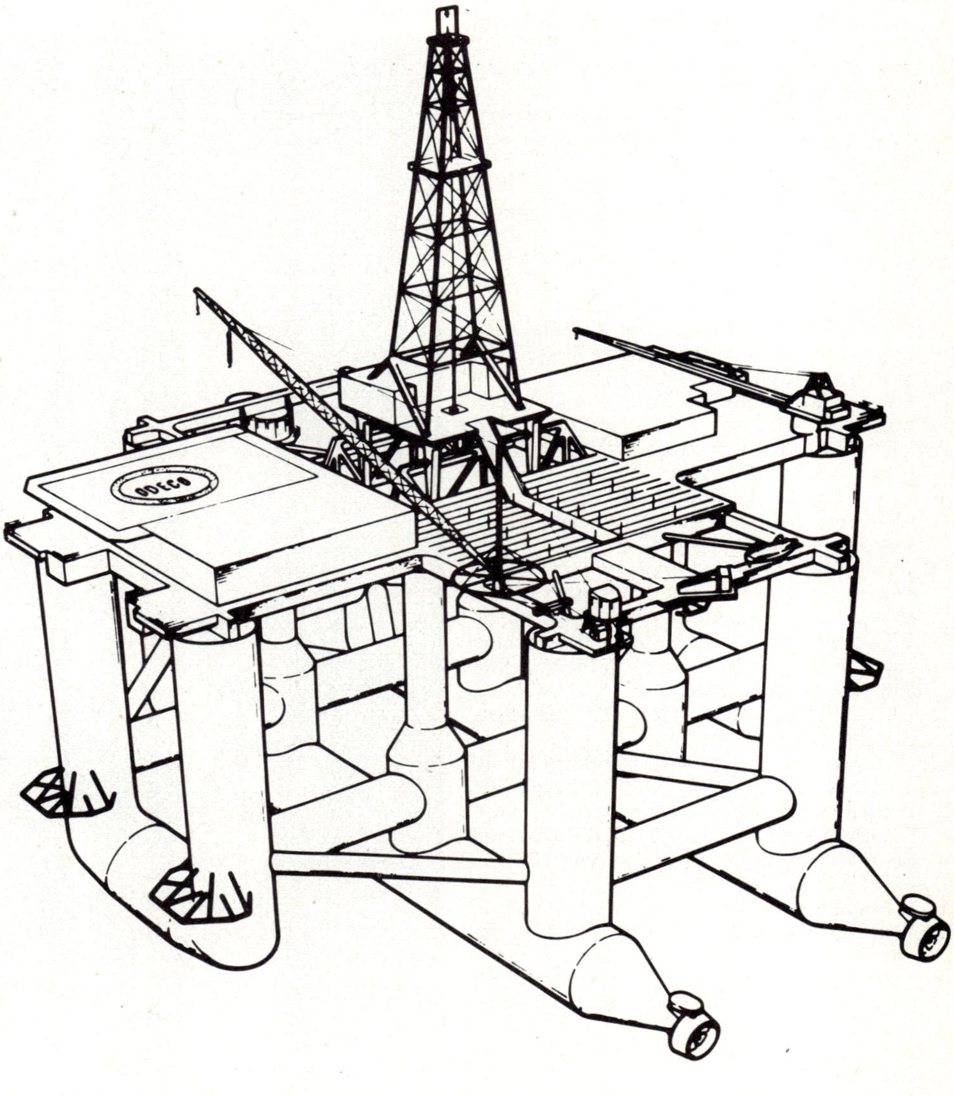

A sketch of *Ocean Rover* showing the two Kort propulsion units (*Courtesy ODECO*)

OCEAN VICTORY

Launched in the autumn of 1972, *Ocean Victory* left the Avondale Shipyards on 7 October 1972 for a 5,200 mile self-propelled crossing of the Atlantic, to take up position for drilling in North Sea UK Block 15/17. She was the first self-propelled, semi-submersible.

The *Ocean Victory* is sister rig to the Mitsubishi-built *Ocean Prospector*.

Specifications

Length	97·56 m (320 ft)
Width	68·90 m (226 ft)
Height (main deck to base line)	36·6 m (120 ft)
Max. drilling depth	7,620 m (25,000 ft)
Max. water depth	183 m (600 ft)
Displacement	10,714 l. tons
Main deck area	2,057 m² (22,136 ft²)

Propulsion: Two 3,500 h.p. electric motors, each driving 2·74 m (9 ft) diameter Kort propellers

Speed: 7 knots forward, 3 knots astern

Accommodation: 81

Equipment

One Pyramid 12·2 × 12·2 × 49·4 m (40 ft × 40 ft × 162 ft) cantilever mast 454,500 kg (1,000,000 lb) API static hook load capacity.

One Continental EMSCO C-3 type 2 draw works grooved for 3·50 cm (1⅜ in) line, V-200 Parmac hydromatic brake, EMSCO cat heads, sandreel assembly mounted on draw works, driven by three 1,000 h.p. DC motors.

One Continental EMSCO 95.25 cm (37½ in) rotary driven by 1,000 h.p. DC motor with two speed transmission.

One 650-ton swivel Continental EMSCO L-650.

One Continental EMSCO MA-60-6-1 3/8 travelling block—650-ton.

One Byron-Jackson hydrahook—500-ton.

One Pyramid 15·2 cm–152 cm (6–60 in) sheave crown, 2·54 cm–152 cm (1–60 in) fast line sheave.

One Koomey accumulator—320-gallon, 3,000 lb/in² W.P. with electric master and remote panels.

One 34·61 cm (13⅝ in) 705 kg/cm² (10,000 lb/in²) BOP system: 2 Shaffer 'LWS' double BOP ram units with 1 shear blind ram, 70·5–353 kg/cm² (1–5,000 lb/in²) W.P. Hydril and 183 m 40·64 cm (600 ft 16 in) marine riser.

One 50·8 cm, 907·2 kg (20 in, 2,000 lb) BOP system: 1 ram and 1 Hydril and 183 m (600 ft) of 61 cm (24 in) marine riser.

Two EMD S-16-645 diesel engines, each ABS rated 1,950 h.p. at 900 rpm driving two 1,600 kw traction type generators and one 1,000 KVA AC generator.

Two EMD S-16-645 diesel engines, each ABS rated 1,950 h.p. at 900 rpm driving two 1,600 kw traction type generators.

Two EMSCO F-1300 triplex pumps 7 × 12 driven by 1,300 h.p. DC motor each supercharged with 5 × 6 Mission centrifugal pump.

Two Mission 6 × 8R, H-30 centrifugal mud mix pumps with 26·7 cm (10½ in) impellers and 100 h.p. AC motors.

Two Rumba Model 4860 DU dual shale shakers.

10,000 ft 12·7 cm (5 in) OD 8·86 kg (19·50 lb) Grade E. R-3 drill pipe with 12·7 cm (5 in) X-Hole tool joints.

5,000 ft 12·7 cm (5 in) OD 8·86 kg (19·50 lb) X-105, R-3 I.E.U. drill pipe with 12·7 cm (5 in) X-Hole tool joints.

Twenty-four 16·5 cm (6½ in) OD × 2¾ in ID drill collars with 12·7 cm (5 in) API Reg. Conn.

Thirty 20·32 cm (8 in) OD × 7·6 cm (3 in) ID drill collars with 16·83 cm (6⅝ in) API Reg. Conn.

One taper tap and overshot for rig drill pipe and drill collars.

ODECO'S *Ocean Victory* self-propelled, semi-submersible, typical of ODECO'S *Ocean Prospector, Ocean Rover, Ocean Voyager* and *Ocean Endeavour* (*Courtesy ODECO*)

Ocean Victory on location in the North Sea (*Courtesy ODECO*)

One Clyde Model CP-96 crane with 36·6 m (120 ft) boom.

One Clyde special Model H 27103 heavy duty stiff leg derrick with 36·6 m (120 ft) boom rated 40 tons at 9·15 m (30 ft) radius and 16·5 tons at 36·6 m (120 ft) radius utilising Clyde frame 7—three-drum shaft hoist.

One Haliburton HT-400 cement unit, Pioneer T-16-4 Desilter; Pioneer T-10-6 Desander, Pit-O-Graph.

One 40-Ton Houston Systems overhead crane.

Eight Bendix Skagit 1 drum electric anchor windlasses, each with one 1,000 h.p. DC motor.

Eight anchors with 1,100 m (3,600 ft) of 7·6 cm (3 in) steel stud link anchor chain.

Storage capacity

Fuel	6,972 bbls
Drill water	14,320 bbls
Potable water	384 bbls
Dry mud	140 short tons
Bulk mud and cement	2·72 cm³ (9,600 ft³)
Liquid mud	1,344 bbls

BETHLEHEM STEEL CORPORATION FORT McHENRY SHIPYARD, BALTIMORE

Name	Owner	Type	Design	Delivery date
Ocean Scout	Ocean Drilling & Exploration Co. (ODECO)	semi-submersible	Breit Engineering, Inc, New Orleans	10 December 1973

OCEAN SCOUT

Designed for unrestricted ocean service.

Delivered

10 December 1973

Specifications

Lower hulls:	
length	61·54 m (202 ft)
width, each	9·76 m (32 ft)
depth	8·54 m (28 ft)
Width, overall	55·49 m (182 ft)
Four corner columns, dia.	9·76 m (32 ft)
Four interior columns, dia.	3·05 m (10 ft)
Max. drilling depth	6,100 m+ (20,000 ft+)
Max. water depth	183 m (600 ft)
Draught, operational	14·63 m (48 ft)
Accommodation	59
Derrick	44·82 m (147 ft)
Anchoring	Eight 13,63·64 kg (30,000 lb) anchors
Chain on each	762·20 m (2,500 ft) of 6·35 cm ($2\frac{1}{2}$ in) stud link
Draw works	3,000 h.p.
Mud pumps	Two 1,300 h.p.
Primary power plant	5,100 h.p.
Constructional steel weight	approx. 7,000 tons
Helicopter pad	20·43 × 17·07 m (67 ft × 56 ft)

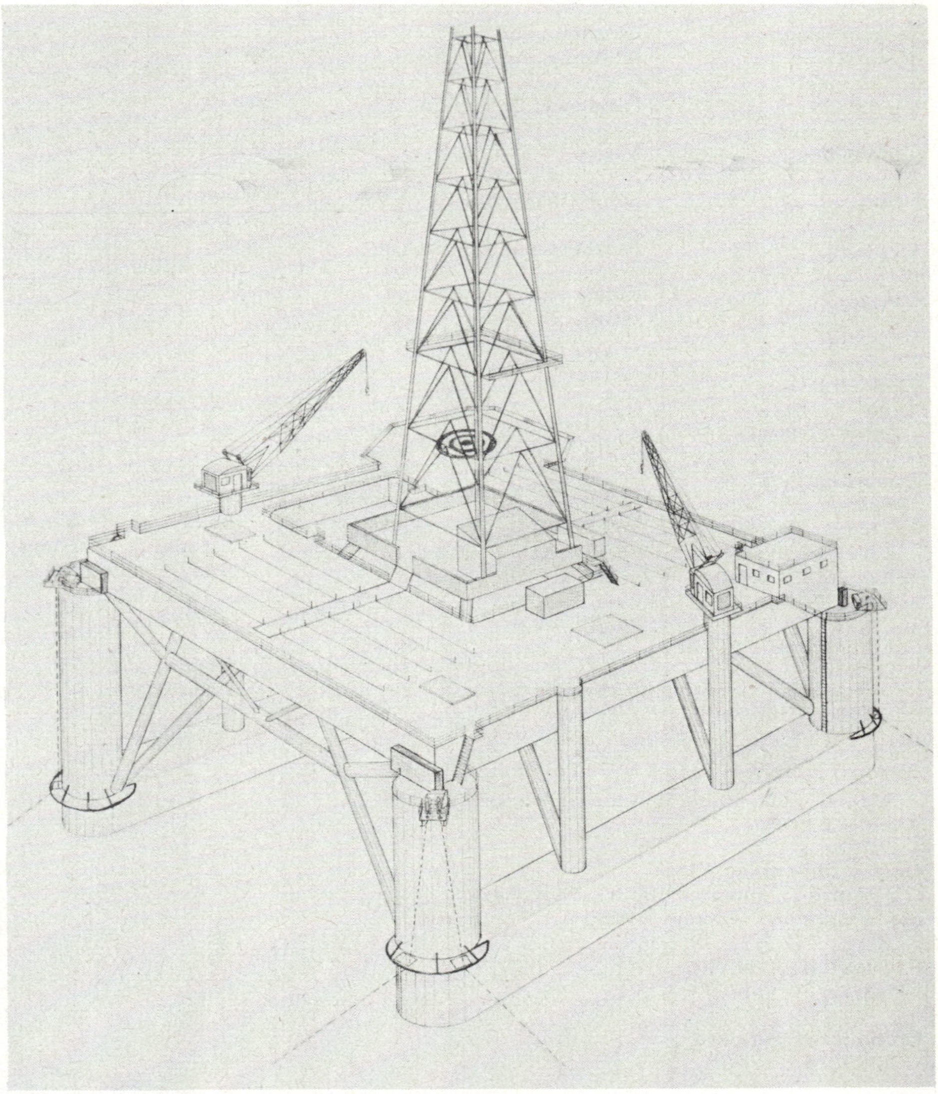

ODECO's *Ocean Scout*, designed by Breit Engineering, Inc. (*Courtesy ODECO*)

Equipment

One Pyramid 12·2 × 12·2 × 44·82 m (40 ft × 40 ft × 147 ft) cantilever mast 453,600 kg (1,000,000 lb) API static hook load capacity.

One Continental EMSCO C-3 type 2 draw works grooved for 3·81 cm ($1\frac{1}{2}$ in) line, V-200 Parmac hydromatic brake, EMSCO cat heads, sandreel assembly mounted on draw works, driven by three 1,000 h.p. DC motors.

One Oilwell 95·25 cm ($37\frac{1}{2}$ in) rotary driven by 1,000 h.p. DC motor with two speed transmission.

One 650-ton swivel Continental EMSCO L-650.

One Continental EMSCO MA-60-6-1 3/8 travelling block—650-ton.

One Byron-Jackson Hydrahook—500-ton.

One Pyramid 15·2–152 cm (6–60 in) sheave crown, 2·54–152 cm (1–60 in) fast line sheave.

One Koomey accumulator—320-gallon 211 kg/cm² (3,000 lb/in²) W.P. with electric master and remote panels.

One 34·61 cm ($13\frac{5}{8}$ in) 353 kg/cm² (5,000 lb/in²) WP BOP system, 70·5–353 kg/cm² (1–5,000 lb/in²) WP Hydril and 183 m 40·54 cm (600 ft 16 in) marine riser.

One 50·8 cm (20 in) 140 kg/cm² (2,000 lb/in²) WP BOP system: 1 ram, 1 Hydril and 183 m (600 ft) of 61 cm (24 in) marine riser.

Two EMD S-16-645 diesel engines, each ABS rated 1,950 h.p. at 900 rpm driving two 1,600 kw traction type generators and one 1,000 KVA AC generator.

One EMD S-16-645 diesel engine ABS rated 1,950 h.p. at 900 rpm driving two 1,600 kw traction type generators.

Two EMSCO F-1300 triplex pumps 7 × 12 driven by 1,300 h.p. DC motor each supercharged with 5 × 6 Mission centrifugal pump.

Three Mission 6 × 8R, H-30 centrifugal mud mix pumps with 26·7 cm ($10\frac{1}{2}$ in) impellers w/100 h.p. AC motors.

One Rumba Model 201 DU dual shale shaker.

10,000 ft 12·7 cm (5 in) OD 8·86 kg (19·50 lb) Grade E, R-3 drill pipe with 12·7 cm (5 in) X-hole tool joints.

5,000 ft 12·7 cm (5 in) OD 8·86 kg (19·50 lb) X-105, R-3 I.E.U. drill pipe with 12·7 cm (5 in) X-hole tool joints.

Twenty-one 19·7 cm ($7\frac{3}{4}$ in) OD × 7·6 cm (3 in) ID drill collars with 16·83 cm ($6\frac{5}{8}$ in) API Reg. Conn.

One taper tap and overshot for rig drill pipe and drill collars.

Two Link-Belt ABS 238 cranes with 36·59 m (120 ft) booms.

One Halliburton HT-400 cement unit, Pioneer T-10-4 desilter; Pioneer S-3-12 desander.

One 40-ton Kranco type TRP-BG overhead crane.

Eight Bendix-Skagit 1 drum electric anchor windlasses, each with one 1,000 h.p. DC motor.

Eight anchors with 762·2 m (2,500 ft) of 6·35 cm ($2\frac{1}{2}$ in) steel stud link anchor chain.

Storage capacity

Fuel	4,500 bbls
Drill water	4,500 bbls
Potable water	1,600 bbls
Bulk mud and cement	204 m³ (7,200 ft³)
Liquid mud	1,200 bbls

BETHLEHEM STEEL CORPORATION, BEAUMONT

Name	Owners	Type	Design	Delivery date
Zephyr I	Maersk A. P. Moller & Dearborn-Storm Corp.	semi-submersible	Storm Drilling Co.	1973/1974
Zephyr II	Storm Drilling Co.			1973
Pat Rutherford Snr.	Viking Offshore A/S with Field International Drilling Co. (FIDC)	semi-submersible	Aker H-3	1974
Western Pacesetter I	Western Oceanics Division of Western Co. of North America	semi-submersible	Friede & Goldman, Inc.	1973
Western Pacesetter II	Western Oceanics Division of Western Co. of North America	semi-submersible	Friede & Goldman, Inc. (same basic design as for *Pacesetter I*)	1974
Zapata Ugland	Zapata Corporation			1973
Marlin No. 7	Marlin Drilling Co., Inc	semi-submersible		March 1975
Rig 62	Transworld Drilling Co. Ltd	jack-up		Early 1974
J Storm III	Marine Drilling	jack-up		June 1974
N/A	Storm Drilling Co.	jack-up (mat) (cost $14.5m)		By Oct. 1975
SS 3000	Zapata Marine Drilling, Inc	semi-submersible	Bethlehem Beth Drill	1974
Ranger III	Walker-Huthnance Offshore Co.	jack-up		1974

ZEPHYR I

Type
Semi-submersible

Specifications

Hull:	
length, overall	61·59 m (202 ft)
breadth, overall	55·49 m (182 ft)
Rated water depth	183 m (600 ft)
Rated drilling depth	7,620 m (25,000 ft)
Accommodation	82
Storage:	
Mud, bulk	272 m³ (9,600 ft³)
Mud, liquid	1,600 bbls
Fuel	4,500 bbls
Water:	
drilling	9,000 bbls
potable	1,400 bbls
Derrick	44·82 m (147 ft), capacity (500 tons)
Crane	One Manitowoc 70 tons at 3·66 m (12 ft) radius
Sea state:	
operating	Be-8(c)
survival	Unrestricted
Power	Three GD-EMD MD 16 E8GW, 2,200 h.p. each
Anchors	Eight, each 15 tons
Anchor chains, each	854 m (2,800 ft) long

ZEPHYR II
Sister rig of *Zephyr I*

PAT RUTHERFORD, SNR.
This semi-submersible is of the Norwegian AKER H-3 design.

Design objectives
Designed to withstand 100 ft, 15 seconds period wave and 100 knot wind

Fabrication start date
July 1973

Commissioning date
July 1974

Capital cost
$20 US million (1973)

Areas of intended use
Anywhere between ice caps north and south.

Unique features
Extensive use of low temperature steel in high stress areas permitting year-around operation above the 60° latitude.

Dimensions, weights, draughts

Length, overall	100·61 m (330 ft)
Width, overall	62·50 m (205 ft)
Height	37·84 m (126 ft) bottom of lower hull to main deck
Main deck length	67·7 m (222 ft)
Main deck width	53·35 m (175 ft)
Platform clearance at drilling draught	12·20 m (40 ft) to 15·24 m (50 ft)

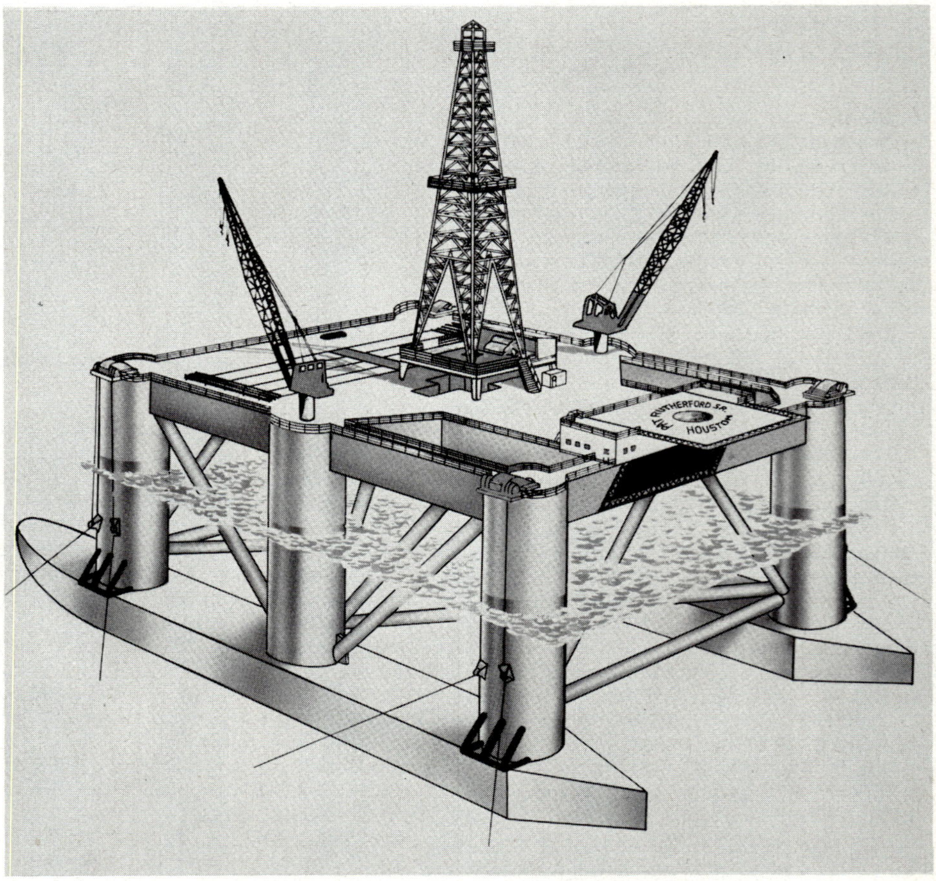

Pat Rutherford, Snr., on Aker H-3 type rig for Field International Drilling Co

Drilling draught 18·29 m (60 ft) to 21·34 m (70 ft)
Displacement at drilling draught 23,500 l. tons
Displacement at transit conditions 11,000 l. tons
Displacement of pontoons 17,000 s. tons
Displacement of columns below water at drilling draught: 9,500 s tons.
Operating draughts in various wave conditions: 21·34 m (70 ft) draught up to 18·29 m (60 ft) wave, 16·77 m (55 ft) draught in 18·60 m+ (61 ft+) waves.

Capabilities
Maximum water depth for drilling: 183 m (600 ft)
Drilling depth capability: 7,620 m (25,000 ft)

Crew accommodation
Quarters for 76 men and a six-berth hospital

Tug requirements
8,000 IHP
Design towing speed: 8 knots

Power services
Main engines—type: 5 cat. D 399—horsepower (actual and ABS rating) per engine: Act.—1,325 h.p., ABS—1,250 h.p.
Generators: 2 GE 1,850 kw, 1 GE 930 kw
Drilling power system: GE SCR system and GE Model 752 drilling motors
Pumping systems for ballast, drill water, fuel, mud: Centrifugal pumps

Liquid storage capacities

Brake cooling water	15 bbls
Drill water	10,000 bbls
Fuel oil	4,000 bbls
Potable water	1,500 bbls
Wash water	from drill water
Liquid mud	2,000 bbls
Bulk mud and cement	297 m³ (10,500 ft³)
Other	Sack storage 139 m² (1,500 ft²) for mud, chemicals and cement

Deck storage capacity
2,400 Kips

Mooring system
Eight 13,636 kg (30,000 lb) anchors, each attached to 1,006 m (3,300 ft) of 7·62 cm (3 in) chain

Drilling equipment
Derrick:
height 48·8 m (160 ft) Lee C. Moore standard derrick
hook load capacity 445,500 kg (1,000,000 lb)
Drill pipe dia. 12·7 cm (5 in)
BOP stack:
bore
(*A*) 54 cm (21¼ in)
(*B*) 34·61 cm (13⅝ in)
working pressure
(*A*) 141 kg/cm² (2,000 lb/in²)
(*B*) 705 kg/cm² (10,000 lb/in²)
height
(*A*) 7·93 m (26 ft)
(*B*) 11·59 m (38 ft)
weight
(*A*) 63,036 kg (140,000 lb)
(*B*) 100,000 kg (220,000 lb)
Rotary table National 37½ in
Riser pipe:
length
(*A*) 183 m (600 ft)
(*B*) 183 m (600 ft)
diameter
(*A*) 40·64 cm (16 in)
(*B*) 60·96 cm (24 in)

Motion characteristics

Natural period in pitch	33·3 seconds
Natural period in roll	38·0 seconds
Natural period in heave	20·4 seconds
Ballasting facilities	Yes
Diving system	Yes

Helicopter deck: 25·30 m × 25·30 m (83 ft × 83 ft)
Developments of the rig possible: to drill in 305 m (1,000 ft) of water

PACESETTER I
This self-propelled semi-submersible rig was designed by Friede and Goldman and is capable of drilling anywhere in the world having a speed capability of 8 knots.

Specifications

Length, overall	79·27 m (260 ft)
Breadth, overall	60·98 m (200 ft)
Two hulls, each	15·24 m (50 ft) beam
Six vertical columns, each	9·76 m (32 ft) diameter
Upper deck:	
length	62·14 m (203 ft 10 in)
breadth, max.	58·23 m (191 ft)
Accommodation	79 men, including hospital
Draught	
transit	5·18 m (17 ft)
displacement:	10,500 tons
normal drilling	18·29 m (60 ft)
displacement	19,288 tons
severe storm	13·72 m (45 ft)
displacement:	17,080 tons
Design:	
max. wind speed	100 knots
max. wave height	30·49 m (100 ft)
max. current speed	5 knots
Natural periods, normal drilling condition:	
heave	19·4 seconds
pitch	34·0 seconds
roll	34·0 seconds

WESTERN PACESETTER I
Design water depth
183 m (600 ft)

Moorings
Eight 7·62 cm (3 in) chains each 915 m (3,000 ft) long

Anchors
Eight Baldt each 13,636 kg (30,000 lb)

Propulsion
Eight electric motors, four to each shaft, 6,000 h.p. total power
Two propellers 3·05 m (10 ft) diameter, in fixed Kort nozzles
Static thrust 79,545 kg (175,000 lb)
Speed in calm water 10 knots

Storage

Upper compartments	
Bulk cement	263 m³ (9,000 ft³)
Sacks	6,000
Drill mud	1,700 bbls
Fuel oil	190 bbls
Potable water	100 bbls
Lower hulls:	
Fuel oil	6,694 bbls
Drill water	12,230 bbls
Ballast	46,525 bbls
Potable water	1,227 bbls

Cranes

One National	86 tons at 6·1 m (20 ft) radius
One National	60 tons at 9·15 m (30 ft) radius
One Manitowoc	38 tons at 5·79 m (19 ft) radius
One Derrick	46·34 m (152 ft)

RANGER III
Type
Jack-up

Specifications

Length	35·06 m (115 ft)
Width	22·56 m (74 ft)
Max. water depth capability	21·34 m (70 ft)

Accommodation 30

Storage capacities

Bulk mud and cement	23 m³ (820 ft³) plus 1,000 sacks
Liquid	350 bbls
Water, drilling	380 bbls
Water, potable	400 bbls

MARLIN No. 7
Type
Semi-submersible

Specifications

Length	67·68 m (222 ft)
Width	53·35 m (175 ft)
Max. water depth capability	30·49 m (1,000 ft)
Max. drilling depth capability	914·63 m (30,000 ft)

Accommodation 72

Storage capacities

Bulk mud and cement	327 m³ (11,550 ft³)
Liquid mud	2,000 bbls

GULFPORT SHIPBUILDING CORPORATION, PORT ARTHUR, TEXAS
(subsidiary of Levingston Shipbuilding Co.)

Name	**Owner**	**Type**	**Design**	**Delivery date**
Santa Fe Mariner 2	Santa Fe International Corporation	semi-submersible		1973

SANTA FE MARINER 2
Type
Semi-submersible

Specifications

Length	82·32 m (270 ft)
Width	32·32 m (106 ft)
Height (to main deck)	12·8 m (42 ft)
Number of stabilising columns	6
Displacement	9,300 short tons (at summer loadline)
Operating draught	8·23 m (27 ft)
In transit draught	3·81 m (12 ft 6 in)
Air gap at operating draught	4·57 m (15 ft) (to bottom main deck structure)
Helicopter deck	21·34 m (70 ft) × 24·09 m (79 ft)

Largest type helicopter usable S-61
Quarters to accommodate 80 men + 6-man hospital
Sewage treating unit: Demco unit to handle all waste from living quarters
Heave response: 35% of wave height for 7 seconds wave and/or natural heave period 16·3 seconds
Natural period: 14·7 seconds for pitch; 22·8 seconds for roll
Towing speed: 8 knots in a calm sea

Mooring equipment
Number of anchors: 8
Weight each 10,000 kg (22,000 lb)
Number anchor lines: 8; length of each 915 m (3,000 ft)
Make-up of each anchor line: 915 m (3,000 ft) of 5·40 cm (2⅛ in) wire rope
Tensile strength each anchor line: 169,100 kg (372,000 lb)
Line pull capacity of each anchor winch: at low speed w/250% motor rating 169,100 kg (372,000 lb)

Maximum operating water depth
183 m (600 ft)

Riser pipe equipment
192 m of 47·31 cm (630 ft of 18⅝ in) OD × 45·40 cm (17⅞ in) ID. Maximum connector OD 50·80 cm (20 in). Slip joints: 13·72 m (45 ft) stroke. Type Riser connectors 50·80 cm (20 in) Riserlock (National) Riser tensioning system: 4 Vetco Model RT-66-40 hydraulic ram tensioners

Deck cranes
Capacity: 40 tons at 5·49 m (18 ft) radius. Boom length 30·49 m (100 ft).
Capacity: 10 tons at 28·96 m (95 ft) radius. Boom length 30·44 m (100 ft).
Make and model: Manitowoc Sea Crane (2)

Drilling equipment
Rated drilling depth: 6,100 m (20,000 ft)
Draw works and Drive: National 1320 UE
Main power plant: General Motors EMD, three 16-645 E, each with one EMD 1500 kw DC generator
Main mud pumps (number, type and h.p.): Two National N-1300, each with two EMD D79 traction motors
Mud mixing pumps: (2) Mission 6 × 8R, with 75 h.p. motors
BOP equipment: National stack with two 16 3/4 GK 5000 Hydrils and three Shaffer LW's 16 3/4 5000 ram preventers with a Koomey control unit
Rotary table: National C-375 electric
Substructure: Pyramid 12·20 m × 12·20 m × 12·80 m (40 ft × 40 ft × 42 ft) high
Maximum setback load: 227,000 kg (500,000 lb)
Mast/Derrick: Pyramid dynamic derrick 149·39 m, 12·20 m (162 ft, 40 ft) square base

Santa Fe *Mariner 2* (*Courtesy Santa Fe International Corp.*)

Maximum hook load: 632,000 kg (1,392,000 lb)
Heave compensation: One Vetco Model 400-20S single cylinder; 181,400 kg (400,000 lb) maximum dynamic capacity; 6·10 m (20 ft) stroke
Cementing unit: One Haliburton HT-400
Logging unit: Schlumberger
Drill pipe and drill collars: 3,350 m 12·70 cm (11,000 ft 5 in) Grade E with 16·2 cm (6⅜ in) OD tool joints, and 1219·2 m 12·70 cm (4,000 ft 5 in) Grade P 110 with 16·2 cm (6⅜ in) OD tool joints

Storage capacities

Bulk mud	80·76 cm³ (2,850 ft³)
Bulk cement	80·76 m³ (2,850 ft³)
Liquid mud	1,200 bbls
Fuel oil	3,368 bbls
Drilling water	6,976 bbls
Potable water	2,100 gallons
Desalinisation unit	2-Aqua Chem
Capacity	300 gph each

Other covered storage: 5,600 sacks bulk storage 177 m² (1,900 ft²) storage area

Personnel survival equipment
Two Brucker survival capsules

IHC HOLLAND MARINE CORPORATION

Name	Owner	Type	Design	Delivery date
Penrod 63	Penrod Drilling Co.	jack-up		Early 1974

LEVINGSTON SHIPBUILDING CO., ORANGE, TEXAS

Name	Owners	Type	Design	Delivery date
Sedco 700	Sedco Maritime, Inc.	semi-submersible	Sedco	1973
Sedco K	Sedco Maritime, Inc. Dallas, Texas	semi-submersible		1973
Blue Water No. 4	Santa Fe	semi-submersible		1973

SEDCO 700

Type
Semi-submersible

Delivery
1973

Specifications
Main deck:

length, overall	68·60 m (225 ft)
width, overall	59·45 m (195 ft)
depth, overall	1·83 m (6 ft)

Lower hulls:

length	85·37 m (280 ft)
width	15·24 m (50 ft)
depth	6·40 m (21 ft)
Four stabilising columns, dia.	9·15 m (30 ft)

Four intermediate columns, dia.	5·49 m (18 ft)
Height, base line to:	
main deck	39·63 m (130 ft)
under deck	38·11 m (125 ft)
Draught:	
operating, max.	24·39 m (80 ft)
towing, max.	5·18 m to 5·79 m (17 ft 19 ft)

Maximum sea state capability; waves 30·49 m to 36·59 m (100 ft to 120 ft)

Propulsion assist enables the rig to be towed at 10 knots

Anchoring: eight chains, 1,220 m (4,000 ft) of 7·62 cm (3 in)

Maximum drilling depth: 7,620 m (25,000 ft)

Storage

Lower hulls:	
drill water	8,000 bbls
fuel oil	5,000 bbls
potable water	700 bbls
bulk mud	170 m³ (6,000 ft³)
Main deck:	
liquid mud	1,500 bbls
bulk mud and cement	166 m³ (5,850 ft³)

BLUE WATER No. 4

Type

Column stabilised semi-submersible

Specifications

Length, overall	74·7 m (245 ft)
Beam, overall	71·65 m (235 ft)
Depth to Texas deck	35·06 m (115 ft)
Draught, drilling	18·29 m (60 ft)
Draught, towing	4·16 m (13 ft 8 in)
Allowable deck load	2,000 short tons
Light ship	10,350 short tons
Displacement at 18·29 m (60 ft)	19,500 short tons
TPI, drilling draught	5·75 short tons
Number and size of columns	Four 7·62 m (25 ft) diameter each
Normal operating draught	Approx. 18·29 m (60 ft)
Number and size of anchors	Eight 13,600 kg (30,000 lb) modified LWT with mud hooks
Mooring lines, two from each corner	1,220 m (4,000 ft) of 7·62 cm (3 in) dia. wire
Mooring chain	As required
Water depth limitations	30·49 m (100 ft) to 457·32 m (1,500 ft)
Towing speed	Approximately 7 knots
Helicopter deck capacity	Designed to support Sikorsky S61
Number, capacity and make of cranes	Two 30 ton at 9·76 m (32 ft) Manitowac
Ship service water	2,500 bbls
Fuel capacity	3,000 bbls
Accommodation	80 men
Hospital (Solas)	7 men

Rig data

Size and type of mast	Lee C. Moore Dynamic 43·29 m (142 ft) × 12·20 m (40 ft) base cantilever mast rated 591,000 kg (1,300,000 lb)
Dynamic rating	12 degrees any direction, 8 second period, with 100 knot wind
Draw works	National Type 1625-DE with elmagco 7838 brake, sandline, micromatic driller and electric drive

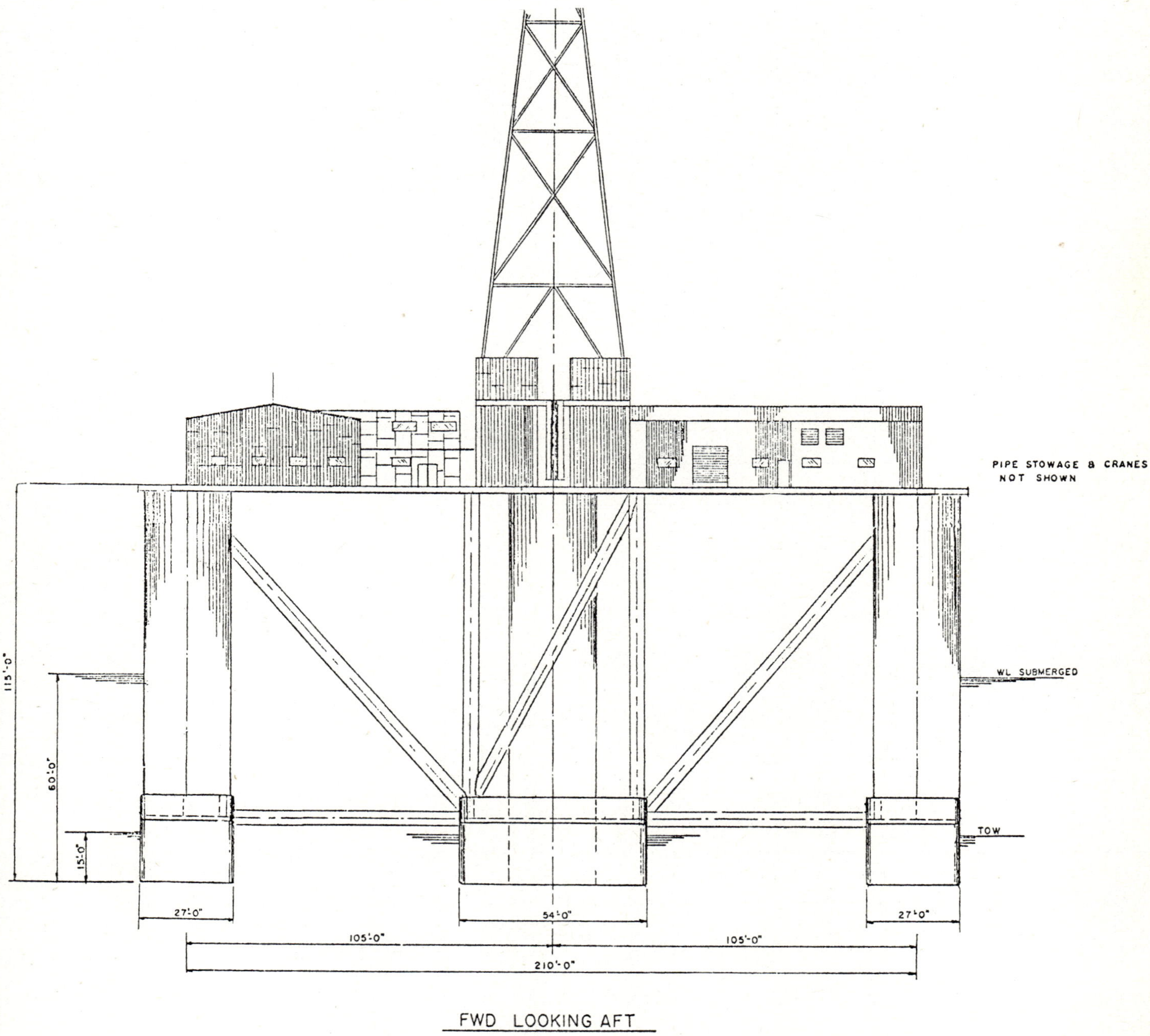

Elevation diagram of the Santa Fe *Blue Water 4* (*Courtesy Santa Fe International Corp.*)

Pumps	Two National Type N-1600 with Electric Drive
Liquid mud capacity	Total 1,200 bbls
Bulk storage	Total 210 m³ (7,200 ft³) bulk storage capacity plus sack storage
Drill water capacity	7,000 bbls
BOP equipment available	As required
BOP package clearance under substructure	7·93 m (26 ft)
Spider beam capacity (retractable)	273,000 kg (600,000 lb)
Bridge crane capacity	70 tons
Opening through deck	7·62 m (25 ft) dia.
Hull well clearance	7·62 m (25 ft) dia.
Riser system and capacity	Six hydraulic ram tensioners. Maximum total capacity 164,000 kg (360,000 lb)
Re-entry guideline system	Four wire capacity 17,100 kg (56,000 lb)
Diving equipment	As required

Machinery and equipment

(1) National Type 1625-DE draw works, complete with cat heads, sandline drum, micromatic driller, electric drive and guards.

(2) Two National N-1600 mud pumps, complete with strainers, pulsation dampeners, crosses, electric drives with guards, stroke counter and jib crane.

(3) National C-375 rotary, with independent electric drive.

(4) National 660-G-500 Hook Block, grooved for 3·81 cm ($1\frac{1}{2}$ in) wire line.

(5) National Type N-1324 Swivel, complete with sub, gooseneck and bumper.

(6) Lee C. Moore Dynamic 43·29 m × 12·20 m (142 ft × 40 ft) base cantilever mast.

(7) General Motors 'EMD' drilling rig power package consisting of four SR-16-W Skid Units keel cooled. Each unit drives one 1,500 KVA, 600 volt, 3 phase, 60 cycle, AC generator.

(8) One Emergency Generator Set, EMD SR-10, driving a 700 KVA, 600 volt, 3 phase, 60 cycle, AC generator.

(9) One SCR Unit and AC Distribution Board for DC and AC equipment.

(10) Four double drum mooring winches.

(11) Four Murray & Tregurtha, or equivalent 1,200 h.p. thrusters.

(12) One Swaco degasser.

(13) One Demco V-60 desander.

(14) 1·52 m (5 ft) drill pipe with 15·88 cm ($6\frac{1}{4}$ in) and 20·32 cm (8 in) collars.

(15) One Vetco heave compensator.

(16) Navigation and communication equipment.

(17) Fork-lift truck.

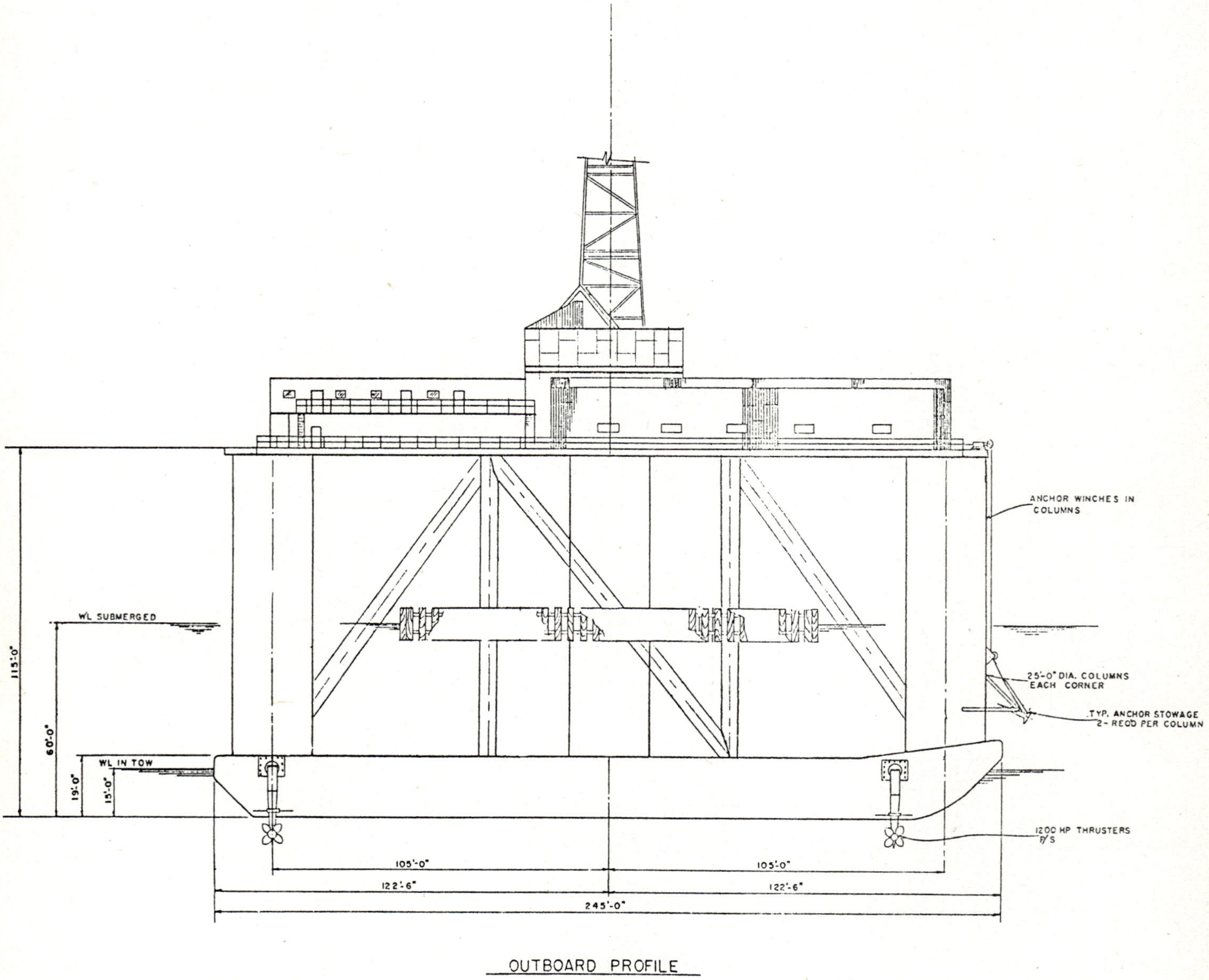

Elevation diagram of the Santa Fe *Blue Water 4* (*Courtesy Santa Fe International Corp.*)

MARATHON LE TOURNEAU, GULF MARINE DIVISION, BROWNSVILLE, TEXAS

Name	Owners	Type	Design	Delivery date
Topper III	Crestwave Offshore Services, Inc.	jack-up		1972
Penrod 71	Penrod Drilling Co.	semi-submersible	Penrod Drilling Co.	1973
Penrod 72	Penrod Drilling Co.	semi-submersible	Penrod Drilling Co.	1974
Penrod 73	Penrod Drilling Co.	semi-submersible	Penrod Drilling Co.	1975
Pentagone 82	Schlumberger Group Sea & Land Contractors, Inc.	semi-submersible	Forex Neptune	1973
Penrod 75	Penrod Drilling Co.	semi-submersible	Penrod Drilling Co.	1974
Colonel Drake	The Offshore Co. with Amoco Amoshore Drilling Co.	semi-submersible	(New design)	early 1974
Discoverer 534	The Offshore Co.	semi-submersible		March 1974
Bob Buschman	Field International Drilling Co.	jack-up		1974
Zapata Nordic	Zapata Corporation	jack-up		
Demaga	Reading & Bates	jack-up		Jan. 1974

TOPPER III

Working in the Gulf of Mexico.

Dimensions

Length	70·12 m (230 ft)
Breadth	60·98 m (200 ft)
Max. water depth	91·5 m (300 ft)

Storage

Mud and cement bulk: 178·5 m³ (6,300 ft³) plus 6,300 sacks, liquid 1,450 bbls, fuel 2,000 bbls, drilling water 5,000 bbls

Power

Caterpillar D-399 TA

Derrick

44·82 m (147 ft)

Cranes

1 × Le Tourneau 60 ton at 17·68 m (58 ft) radius

1 × Le Tourneau 45 ton at 7·62 m (25 ft) radius

Accommodation

71 persons

PENROD 71, 72, 75

Launched in 1973, the *Penrod 71* with a displacement of 42,000 long tons was the largest displacement rig constructed. She was the second semi-submersible rig ordered by the Penrod Drilling Company.

Specifications

Lower hull:	
length	90·85 m (298 ft)
width, overall	65·85 m (216 ft)
cross section	9·15 m × 21·34 m (30 ft × 70 ft)
Upper hull:	
length	71·04 m (233 ft)
breadth	59·76 m (196 ft)
depth	5·49 m (18 ft)
Height of upper deck above bottom of lower hull	42·07 m (138 ft)
Towing draught	4·57 m (15 ft)
Drilling draught	21·34 m (70 ft)
Drilling water capacity	4,200 bbls
Fuel oil capacity	7,200 bbls
Liquid mud capacity	1,700 bbls
Bulk cement capacity	113·4 m³ (4,000 ft³)
Designed deck load	2,036 tonne (2,000 ton)
Displacement at drilling draught	42,000 long tons
Mooring system	Eight × 13,600 kg (30,000 lb) anchors and 1,500 m (5,000 ft) of 7·94 cm ($3\frac{1}{8}$ in) dia. chain per anchor
Max. water depth	304·88 m (1,000 ft)
Max. sea-state	15·24 m (50 ft) significant with 138 mph wind speed (approx. 230 kmph)

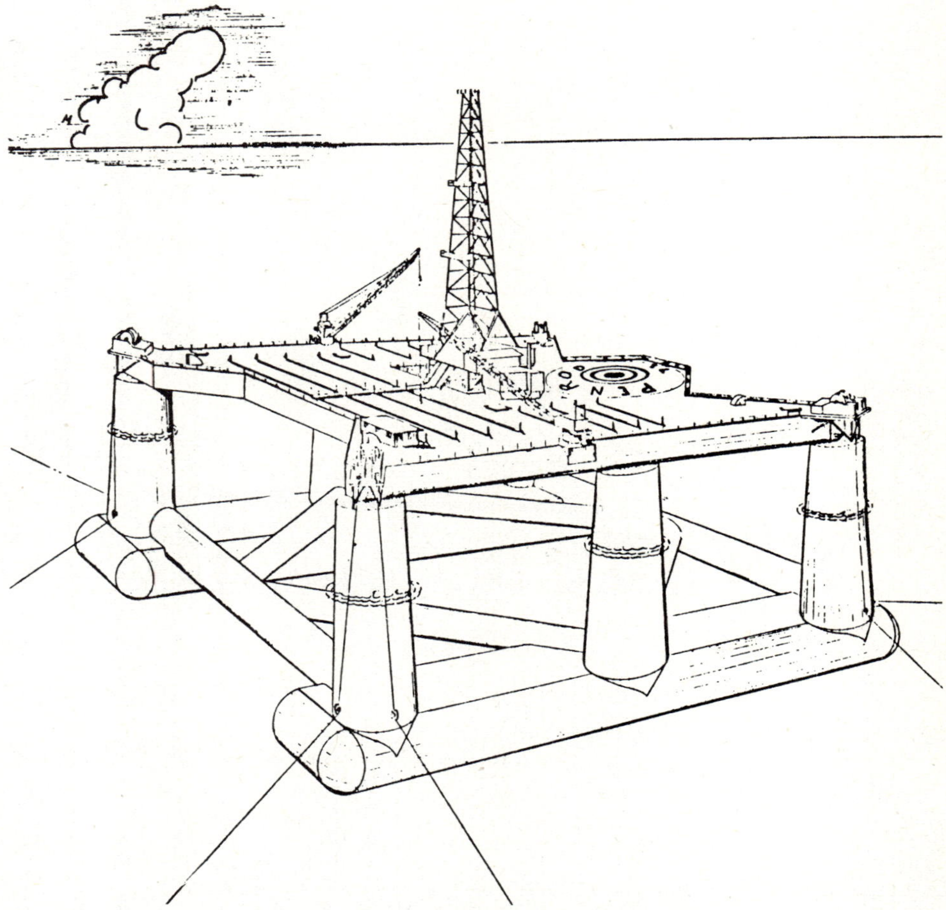

Sketch of *Penrod 71* semi-submersible rig (*Courtesy of Penrod Drilling Co.*)

Natural periods at 21·34 m (70 ft) draught

roll	35 seconds
pitch	37 seconds
heave	38 seconds

Power: Four diesel giving total 13,000 h.p. Baylar Solid State conversion from AC to DC

Cranes: Two 110 ton

Maximum drilling depth: in excess of 9,150 m (30,000 ft)

Accommodation for 86 with air conditioning and heating suitable for North Sea and tropics

Cost: Penrod 71 was reported to have cost over $20 million, in 1971

PENTAGONE 82

Commissioned on August 25 1973, *Pentagone 82* was the first offshore drilling rig to be commissioned from the Gulf Marine Division of the Marathon Le Tourneau Company. This rig type was designed by the Institut Français du Petrole and Societé de Forages en Mer Neptune, Paris.

The *Pentagone 82* is a column-stabilised semi-submersible, pentagonal shape, and propulsion assisted.

Areas of intended use

North Sea—Hudson Bay (Canada)

Work capability

Worldwide

The rig is owned by Sea and Land contractors, Inc and their first customer is Conoco-Aquitaine Company of Canada.

Dimensions, weights, draughts

Length, overall	99·09 m (325 ft)
Width, overall	103·05 m (338 ft)
Height	40·85 m (134 ft) (to upper deck), 99·06 m (317 ft) overall height
Main deck	709 m² (25,030 ft²)

Drilling draught: Operating 21·95 m (72 ft), Hurricane 15·8 m (52 ft)

Displacement at drilling draught: 19,000 tons 21·95 m (72 ft) draught

Displacement at transit conditions: 11,900 tons

Capabilities

Maximum water depth for drilling: 304·88 m (1,000 ft)

Drilling depth capability: 6,100 m (20,000 ft)

Rig design criteria: Wave 30·49 m (100 ft) period 17 seconds, current 2·5 knots, Wind 120 knots

Propulsion assisted: Two thrusters 2 000 h.p. each. One steerable thruster 750 h.p.

Crew accomodation

Quarters to accommodate 74 men

Tug requirements

13,500 h.p.

Design towing speed: 5 knots with 13,500 h.p. tug

Power services

Main engines. type: Four S.12 E 1-GW-GM diesel engines, horsepower per engine: 1,650

Generators: Four GE AC generators 1,500 h.p./1,050 kw each at 900 rpm, 480 volts, 60 hz, ABS continuous rating

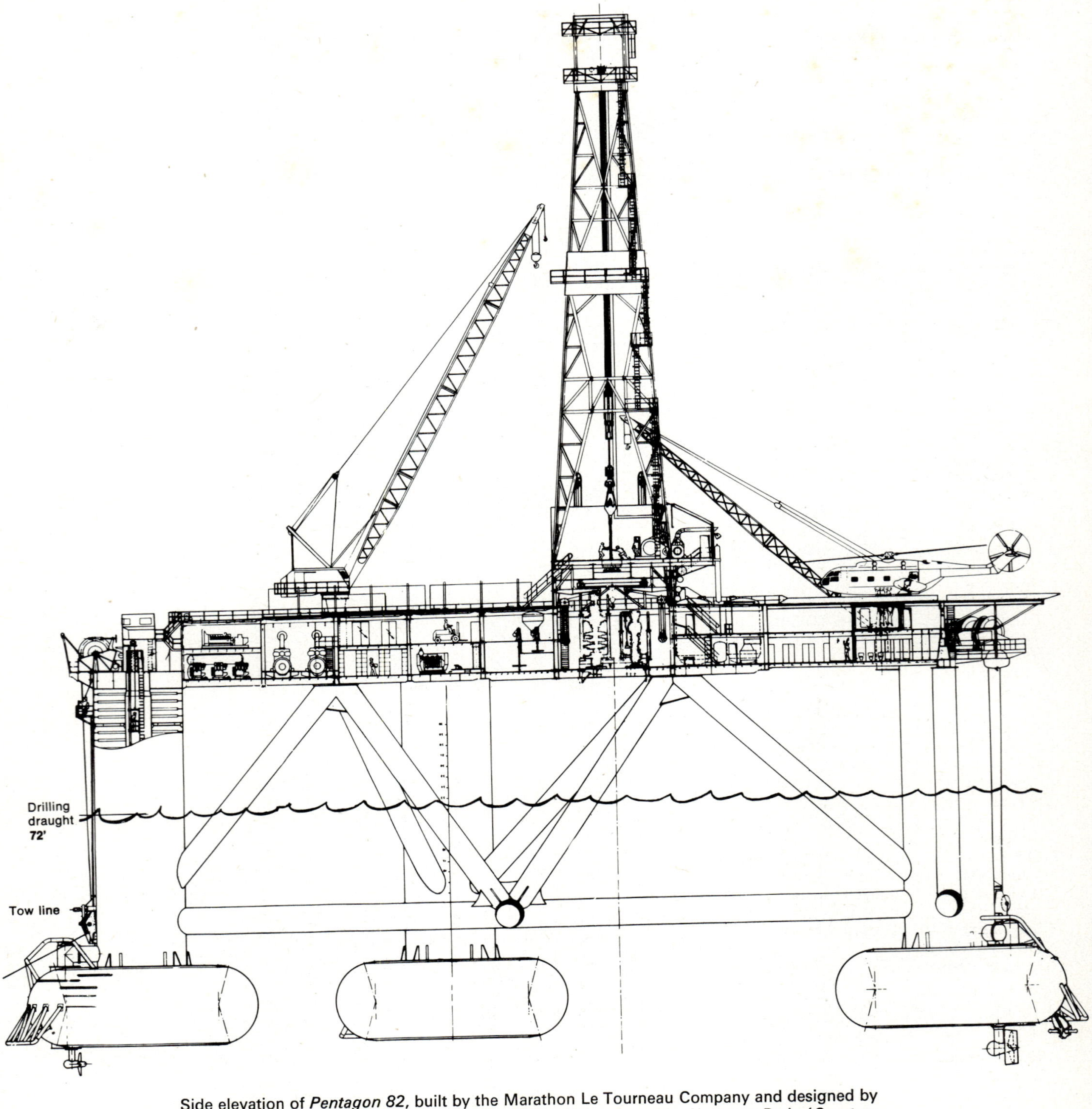

Side elevation of *Pentagon 82*, built by the Marathon Le Tourneau Company and designed by the Institut Francais due Petrole and the Societe de Forages en Mer Neptune, Paris (*Courtesy Forex Neptune*).

Pumping systems for ballast, drill water, fuel, mud: Mud (drilling), Two EMSCO triplex F 1600 (1,600 h.p. input) main mud pumps. Mud (mixing), One EMSCO duplex D 375 pump

Thrusters, motor types: GE GT 608 and GE 752 DC motors (for assistance to propulsion); number of screws: Two fixed direction, one steerable; screw diameter: Two 3·05 m (120 in) (fixed), one 2·08 m (80 in) (steerable)

Liquid storage capacities

Drill water	4,830 bbls
Fuel oil	4,830 bbls
Potable water	3,645 bbls
Liquid mud	2,540 bbls
Two bulk mud	bentonite 66·9 m³ (2,360 ft³)
Other	Six bulk cement and/or baryte 306 m³ (10,800 ft³)

Deck dead load capacity

2,100 tons at drilling draught, 1,500 tons at transfer draught

Mooring system

Ten anchors 30 kips each, six back anchors 20 kips each. Ten anchor lines, 1,700 m (5,576 ft) of 7 cm ($2\frac{3}{4}$ in) cable + 245 m (803·6 ft) of 6·35 cm ($2\frac{1}{2}$ in) chain

Drilling equipment

Derrick:	
height	44·82 m (147 ft), 523,000 kg (1,150,000 lb) API capacity
hook load capacity	860 kips (dynamic factor included)
Drill pipe dia.	12·70 cm (5 in)
BOP stack:	
bore	52·71 cm ($20\frac{3}{4}$ in), 34·61 cm ($13\frac{5}{8}$ in)
working pressure	141 kg/cm² (2,000 lb/in²), 352–705 kg/cm² (5,000–10,000 lb/in²)
Rotary table	EMSCO T 3750

Riser pipe: VETCO integrated type, dia.: 55·88 cm (22 in), 40·64 cm (16 in)—wall: 1·27 cm (0·500 in), 1·59 cm (0·625 in)

Other features

Two hole position indicators (Taut wire line and Honeywell acoustic). One angle riser indicator—Honeywell acoustic. One motion compensator—VETCO

Diving system

One diving bell with one main winch 10,000 kg (22,000 lb) capacity. Two decompression chambers, compressor units, Helium vessels and auxiliary equipment

Helicopter deck

619 m² (6,461 ft²) unobstructed area, Sikorsky S.61 or Puma SA 330 helicopters

Motion characteristics

Natural period in pitch	Over 30 seconds
Natural period in roll	Over 30 seconds
Natural period in heave	19·5 seconds

Ballasting facilities

Ten ballast pumps 1,320 gallons/mn each at 2·54 kg/cm² (36 lb/in²)

COLONEL DRAKE

Specifications

Length, overall upper hull	65·24 m (214 ft)
Two lower hulls, overall length	101·52 m (333 ft)
Six stabilising columns, dia.	8·54 m (28 ft)

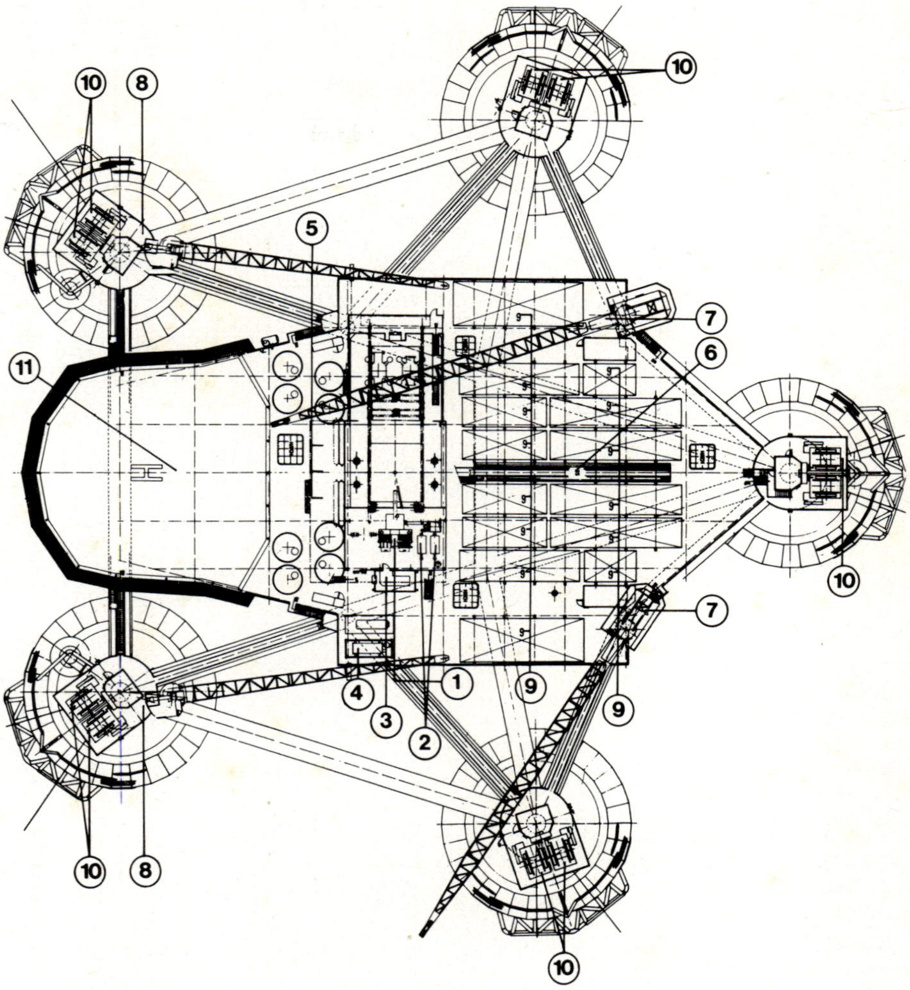

The main deck of *Pentagone 82*. Key: (1) Shale skakers; (2) Desilting unit; (3) Geology lab; (4) Production unit; (5) Accumulator unit for motion compensator; (6) Walk way; (7) Revolving crane, 40 T; (8) Revolving crane, 12 T; (9) Pipe rack; (10) Anchor winches; (11) Heli deck

Max. water depth capability	183 m (600 ft)
Max. drilling depth capability	7,620 m (25,000 ft)
Powerplant	6,000 h.p.
Self-propulsion speed	6 knots
Cost	Approx. $ US 20 m

Work area, initially North Sea.

BOB BUSCHMAN

Type

Jack-up

Design objectives

Operate in 76·22 m (250 ft) of water with 80 mph winds and simultaneous 9·15 m (30 ft) waves

Commissioning date

July 1974

Capital cost

$10,300,000 (1973)

Areas of intended use

Anywhere between 60° Latitudes

Dimensions, weights, draughts

Length, overall	61·89 m (203 ft)
Width, overall	51·22 m (168 ft)
Depth of platform	6·71 m (22 ft)
Main deck length	61·89 m (203 ft)
Main deck width	51·22 m (168 ft)
Platform clearance at drilling draught	7·62 m (25 ft) air gap

Capabilities

Maximum water depth for drilling	76·22 m (250 ft)
Drilling depth capability	6,100 m (20,000 ft)

Crew accommodation

Quarters for 48 Nationals and 30 expatriates and a 5-man hospital

Tug requirements

6,000 IHP

Design towing speed: 5 knots

Power services

Main engines:	
type	Five Cat D-399
horsepower per engine	Actual 1,325 h.p. ABS 1,250 h.p.
Generators	Six GE DC 930 kw Four Le Tourneau AC 600 kw
Drilling power system	GE Model 752 DC motors

Liquid storage capacities

Brake cooling water	15 bbls
Drill water	4,000 bbls
Fuel oil	2,000 bbls
Potable water	800 bbls
Wash water	From drill water
Liquid mud	1,400 bbls
Bulk mud	88 m³ (3,000 ft³)
Cement	117 m³ (4,000 ft³)

Mooring system

Four 4,550 kg (10,000 lb) anchors and four 3·49 cm ($1\frac{3}{8}$ in) × 760 m (2,500 ft) cables

Jacking system
Le Tourneau

Drilling equipment
Derrick:
height | 44·8 m (147 ft) standard
hook load capacity | 480,000 kg (1,056,000 lb)
Drill pipe dia. | 12·70 cm (5 in)
BOP stack:
bore
(*A*) | 50·8 cm (20 in)
(*B*) | 34·61 cm ($13\frac{5}{8}$ in)
working pressure
(*A*) | 141 kg/cm² (2,000 lb/in²)
(*B*) | 705 kg/cm² (10,000 lb/in²)
Rotary table | 95·25 cm ($37\frac{1}{2}$ in)

Helicopter deck
21·34 m × 21·34 m (70 ft × 70 ft)

ZAPATA NORDIC
Type
Jack-up

Specifications

Length	72·26 m (237 ft)
Width	71·95 m (236 ft)
Accommodation	78

Storage

Bulk mud	121·1 m³ (4,275 ft³)
Liquid mud	1,600 bbls
Cement	121·1 m³ (4,275 ft³)
Fuel	1,550 bbls
Drilling water	5,700 bbls
Potable water	1,200 bbls

MARATHON LE TOURNEAU, VICKSBURG, MISS.

Name	Owners	Type	Design	Delivery date
Key West	Key International Drilling Co.	Jack-up		1972
Mr. Mel	Fluor Drilling Services, Inc.	Jack-up	Jack-up	Sept. 1972
Rig 64	Penrod Drilling Co.	Jack-up		1973
Ocean King	Odeco	Jack-up		1973
Rowan Texas	Rowan International, Inc.	Jack-up		1972
Penrod 62	Penrod Drilling Co.	Jack-up		1972
Rowan Louisiana				

OCEAN KING
Type
Jack-up

Specifications

Length, overall	70·1 m (230 ft)
Beam, overall	60·9 m (200 ft)
Leg length	125 m (410 ft)
Maximum water depth capability	103·6 m (340 ft)
Maximum drilling depth capability	7,621·9 m (25,000 ft)
Power plant	5,000 h.p. diesel electric
Accommodation	59 men
Helicopter pad	for Sikorsky S61 N

The Ocean King commenced operations in the Gulf of Mexico and is the twenty-fourth rig belonging to ODECO.

ODECO'S *Ocean King* jack-up rig (*Courtesy ODECO*)

Equipment
One Continental EMSCO 12·2 × 12·2 × 44·82 m (40 × 40 × 147 ft) derrick 453,600 kg (1,000,000 lb) API static hook load capacity

One Continental EMSCO C-3 type 2 draw works grooved for 3·49 cm ($1\frac{3}{8}$ in) line, V-200 Parmac hydromatic brake, EMSCO cat heads, sandreel assembly mounted on draw works, driven by three 1,000 h.p. DC motors

One Continental EMSCO 92·25 cm ($37\frac{1}{2}$ in) rotary driven by 1,000 h.p. DC motor with two speed transmission

One 650 ton swivel-Continental EMSCO L-650

One Continental EMSCO MA-60-6-1 3/8 travelling block—650 ton

One Byron-Jackson 500 ton hook

One Continental EMSCO 17·8–152·4 cm (7–60 in) sheave crown

One Koomey accumulator—150 gallon 211 kg/cm² (3,000 lb/in²) W.P. with electric master and remote panel

One 34·61 cm ($13\frac{5}{8}$ in) 352 kg/cm² (5,000 lb/in²) BOP system: double and single Shaffer 'LWS' BOP ram units, 70·5–352 kg/cm² (1–5,000 lb/in²) Hydril GK

Three Fairbanks-Morse, 10 cylinder Model 38D8 1/8 opposed piston diesel engines, each rated 1,600 h.p. at 720 rpm driving two 1,600 kw traction type generators and one 1,000 kw AC generator

Two EMSCO F-1300 triplex pumps 7 × 12 driven by 1,300 h.p. DC motors each supercharged with 5 × 6 Mission centrifugal pump

Three Mission 6 × 8R, H-30 centrifugal mud mix pumps with 25·4 cm (10 in) impellers w/100 h.p. AC motors

One Rumba Model 102 dual shale shaker

10,000 ft 12·7 cm (5 in) OD 8·86 kg (19·50 lb) Grade E, R-3 drill pipe with 12·7 cm (5 in) X-Hole tool joints

5,000 ft 12·7 cm (5 in) OD 8·86 kg (19·50 lb) X-105, R-3 I.E.U. drill pipe with 12·7 cm (5 in) X-Hole tool joints

Twenty-one 19·69 cm ($7\frac{3}{4}$ in) OD 7·14 cm (2 13/16 in) ID drill collars with $6\frac{5}{8}$ API Reg. Conn.

One taper tap and overshot for rig drill pipe and drill collars

Two Le Tourneau Series PCM-120 AS with 30·49 m (100 ft) boom—rated 25 tons at 15·24 m (50 ft) and 40 tons at 7·93 m (26 ft)

One Haliburton HT-400 cement unit, Pioneer T-16-4 desilter, Pioneer T-10-4 desilter, Pioneer S-3-R desander, Schlumberger Unit, Pit-O-Graph

Four series W-1500 TS electric anchor winches with 915 m (3,000 ft) of 3·18 cm ($1\frac{1}{4}$ in) diameter wire rope and anchors

Storage capacity

Fuel	2,000 bbls
Drill water	5,000 bbls
Potable water	1,000 bbls
Dry mud	140 short tons
Bulk mud and cement	176·82 m³ (6,240 ft³)
Liquid mud	1,400 bbls

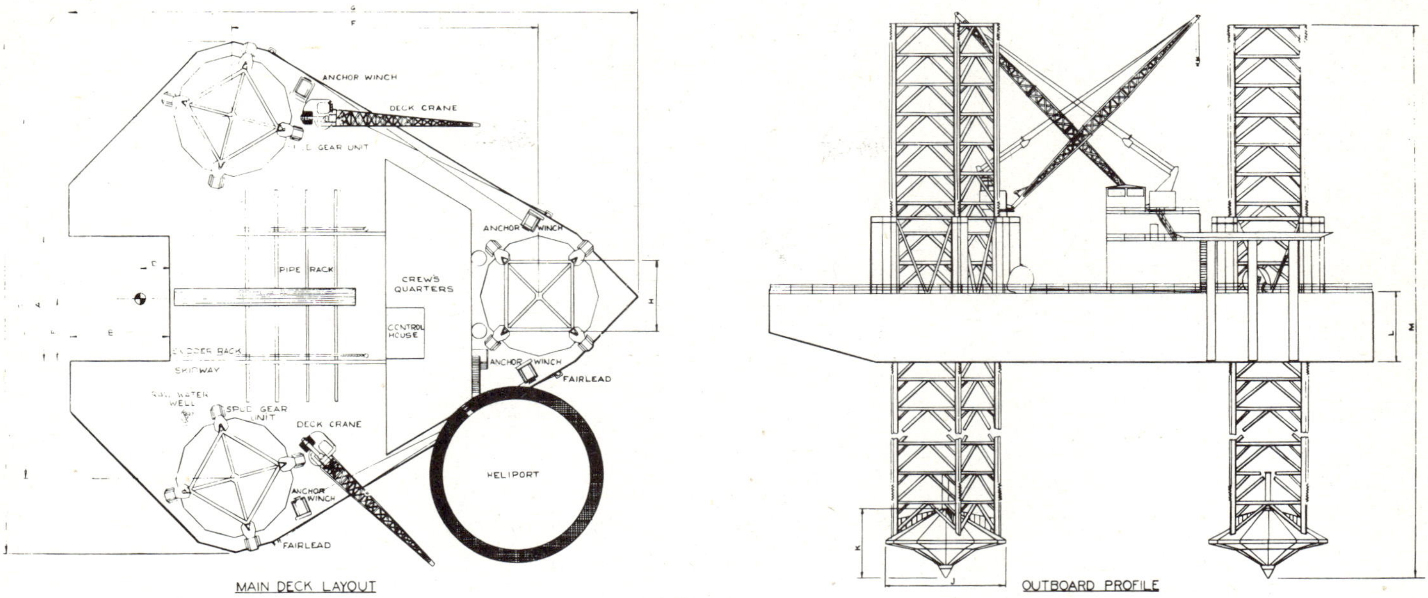

Elevation and plan diagrams of *Ocean King* (*Courtesy ODECO*)

ROWAN LOUISIANA

Type
Jack-up

Delivery
Second quarter of 1975

Specifications

Leg length	111·59 m (366 ft)
Max. water depth capability	106·71 m (350 ft)
Max. drilling capability	9,150 m (30,000 ft)

Cost
$ US 15 million

MR. MEL

The most powerful rig of its type in the world (August 1973) *Mr. Mel* can also drill in deeper water than most jack-ups. Fabrication started on 28 March 1972 and the rig was commissioned on 12 September 1972.

Specifications

Length, overall	70·1 m (230 ft)
Width, overall	60·9 m (200 ft)
Hull depth	7·9 m (26 ft)
Main deck length	70·1 m (230 ft)
Main deck width	60·9 m (200 ft)

Platform clearance above water in drilling conditions: 13·7 m (45 ft)

Maximum water depth for drilling: 106·7 m (350 ft) (non-hurricane season), 91·5 m (300 ft) (hurricane season)

Drilling depth capability: 7,621·9 m (25,000 ft)

Sea state, current and wind conditions, including maxima, with corresponding drilling depths: 100 mph winds, 10·7 m (35 ft) waves, 106·7 m (350 ft) water

Accommodation
74 men

Design towing speed
5 knots

Power services
Total of 8,800 h.p.

Prime movers type	three EMD SR 16 diesel
	two EMD SR 8 diesel
Generators	three 1,500 kw alternators
	two 700 kw AC generators

Drilling power system: Hutchison-Hayes SCR system

Mr. Mel, a jack-up rig, built by Marathon Le Tourneau, Vicksburg, Miss., for Fluor Drilling Services, Inc. (*Courtesy Fluor Drilling Services, Inc.*).

Pumping systems for ballast, drill water, fuel, mud: Two National 12P-160-7 1·27 cm × 30·5 cm ($\frac{1}{2}$ in × 12 in) triplex mud pumps. One National 9P-100-6 0·635 cm × 23·5 cm ($\frac{1}{4}$ in × $9\frac{1}{4}$ in) triplex mud pump. Two 6 × 8 centrifugal (100 h.p. each)

Storage capacities

Brake cooling water	2,300 bbls
Drill water	5,700 bbls
Fuel oil	3,170 bbls
Potable water	900 barrels
Liquid mud	1,300 bbls
Bulk mud	Six 29·7 m³ (1,050 ft³) 'P' tanks
Other	Sacked material, 5,000 sacks

Mooring System
Four Le Tourneau W-1500TS winches

Jacking System
Le Tourneau Rack & Pinion

Drilling equipment

Derrick:	
height	44·8 m (147 ft)
hook load capacity	600,000 kg (1,320,000 lb)
Drill pipe diameter	12·7 cm (5 in) OD
BOP stack:	
bore	50·8 cm (20 in) Hydril
working pressure	34·6 cm ($13\frac{5}{8}$ in) Cameron Type A 5000
height	34·6 cm ($13\frac{5}{8}$ in) Cameron U Double 5000
weight	34·6 cm ($13\frac{5}{8}$ in) Cameron U Single 5000
Rotary table	National C-375

Helicopter deck

21·6 m (71 ft) diameter

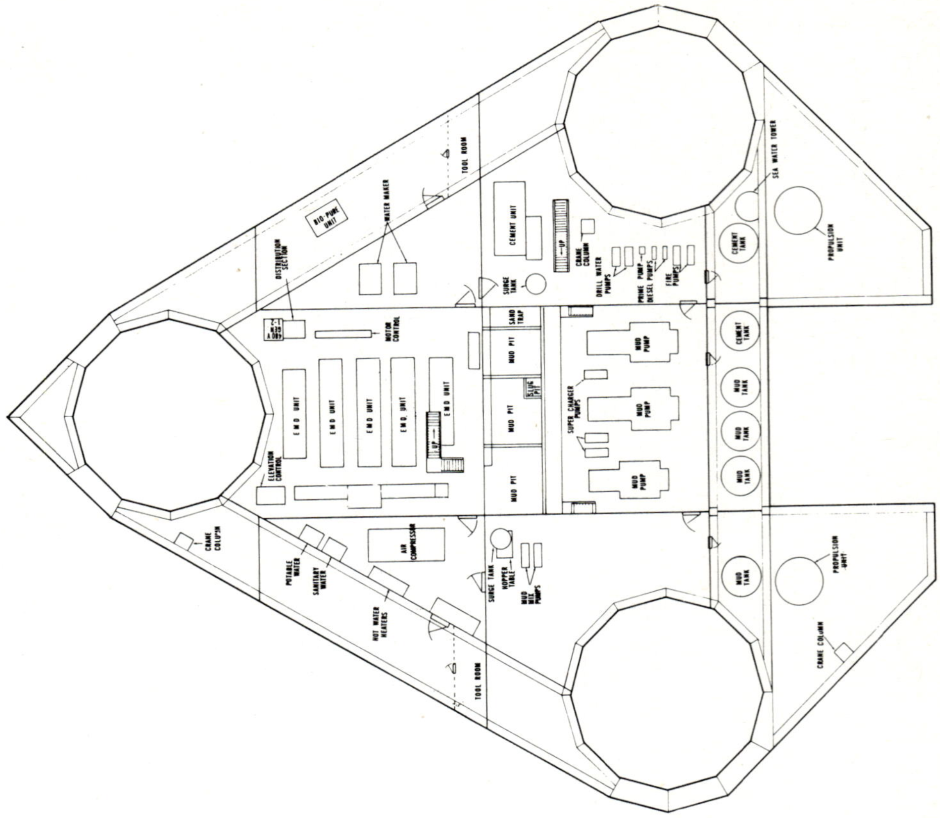

Layout of *Mr. Mel* (*Courtesy Fluor Drilling Services, Inc.*)

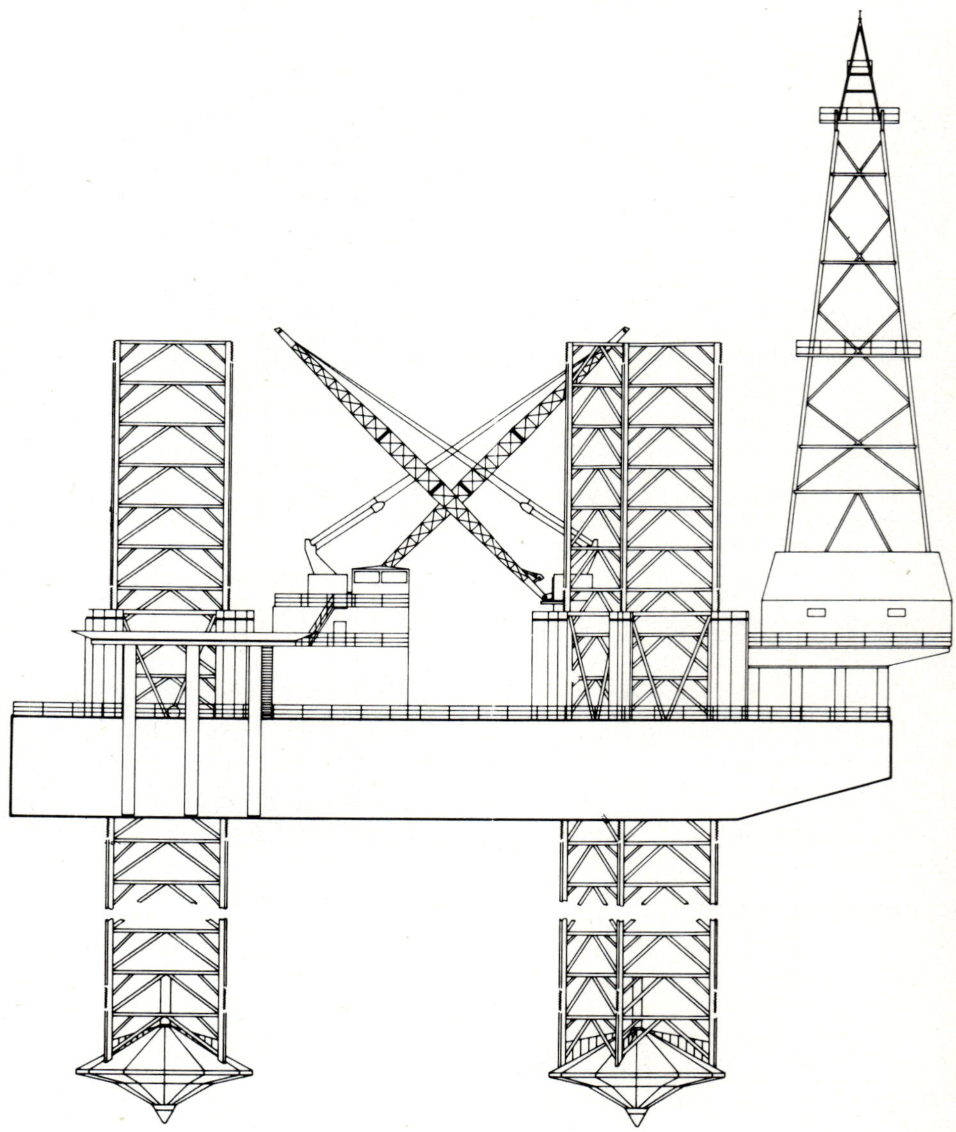

Layout of *Mr. Mel* (*Courtesy Fluor Drilling Services, Inc.*)

METAL ARTS CO., HOUSTON

Name	Owners	Type	Design	Delivery date
Pentagone 86 (*Pel Lyn 2*)	Dixylyn International, Inc.	Semi-submersible, self-propelled	IFP Forex Neptune	September 1974

PENTAGONE 86
(for further details, see *Pentagone 85*)

Type
Semi-submersible, self-propelled

Specifications

Hull:

length, overall	98·48 m (323 ft)
breadth, overall	102·44 m (336 ft)
Water depth, max.	198 m (650 ft)
Water depth, min.	26 m (85 ft)
Rated drilling depth	9,150 m (30,000 ft)
Draught, transit	7·01 m (23 ft)
Displacement, transit	12,000 short tons
Draught, drilling	18·3 m to 21·3 m (60 ft to 70 ft)
Displacement, drilling	18,000 short tons
Storage, mud and cement	411 m³ (14,500 ft³) plus 6,600 sacks
Power units	Seven Caterpillar D399

Propulsion by two main propellers and one bow propeller

Accommodation	74 men
Mooring	10 anchors, each 14·5 tons

Sea states:

survival	22 m (72 ft) waves
operating	18 m (59 ft) waves

SURVIVAL SYSTEMS

UNITED KINGDOM

WATERCRAFT LTD.

WATERCRAFT-SCHAT SURVIVAL SYSTEM

This system has been designed for application in drilling rigs, oil production platforms and semi-submersible drilling units.

Approvals

The system has been approved by major National Authorities and exhaustive on-board tests have been held by the British Department of Trade and Industry. At its inception the craft was fire-tested under rigorous conditions in the presence of I.M.C.O., the Fire Research Establishment and the Department of Trade and Industry, together with representatives of Oil Companies and other Classification Societies. The result of the tests proved that the temperature inside the craft did not exceed a maximum of 120°F (54°C) reducing to 109°F (42·8°C) amidships, all temperatures taken at head level.

Operation

The system enables the crew to evacuate a rig and to be afloat totally enclosed and independent of the outside atmosphere. The craft is fully protected by means of a water-spray system which enables it to go through an oil fire up to a distance of one mile. The craft is extremely manoeuvrable with a speed of over 6 knots, which is the speed prescribed by the 1960 SOLAS Convention and is fuelled for at least 24 hours operation. Inherent in the low silhouette design is versatility of operation, with the ability to make headway against wind and tides, and to maintain station or position. To fully cover a platform unit against all weathers it is preferred by some owners to fit from two to four systems per unit, depending on conditions encountered in different parts of the world.

Embarcation of the craft is through two large watertight openings which permit up to 50 persons to be seated, with doors closed ready for lowering in 30 seconds (during which time the main engine can be hydraulically started for immediate getaway).

Immediate lowering is effected by pulling and maintaining tension on the control wire which passes through the canopy top adjacent to the control position from which the craft is driven and the engine controlled. The craft descends at a controlled speed and being suspended from two points will not spin. The descent can be halted at any point by releasing the control wire. The boat release system is operated by a single lever which releases both hooks simultaneously when the craft is waterborne. It is impossible to release one hook without the other. One of the advantages of this system is that the craft is very suited for emergency rescue purposes and easily recoverable.

Watercraft survival craft

The Watercraft Survival Craft is constructed of glass fibre reinforced plastic (with fire resistant additives). The outside of the craft is coloured International Orange for easy recognition.

A compressed air-system is fitted which operates a continuous water shed for 10 minutes covering the whole of the craft, with an air supply sufficient for the personnel as well as the engine running at full throttle. A positive pressure is built up within the craft preventing the entry of fumes and gases. Air is not re-circulated.

The engine fitted is a Lister HRW2GMR which has a closed cooling system and heat exchanger. It is capable of being hand or hydraulically started and is fully approved for this type of craft, including re-starting after capsizing. The engine develops approximately 29·5 h.p. at 2,200 rpm and will give the craft a speed of over 6 knots, and has fuel for a minimum of 24 hours at this speed.

Controls are sited at the helmsman's position, including boat lowering, hydraulic engine starting, craft release, speed and steering. There is also an emergency steering system.

Among other features of the craft are: Double lifelines all round craft for emergency boarding; all side decks have a non-skid surface; all fittings are non-ferrous or hot dipped galvanised and rust treated with white enamelled finish; each person has an individual seat belt; there are built-in toilet facilities; there are special openings for natural ventilation and automatic closing ventilators in the event of capsizing; large opening doors permit the loading of the craft to take place in a very short period of time.

The craft is self-righting and is complete with built-in polyurethane foam buoyancy.

Schat launching system

Outriggers, support struts and cross beam between outriggers.

Lead sheaves on system with rust-resistant shafts.

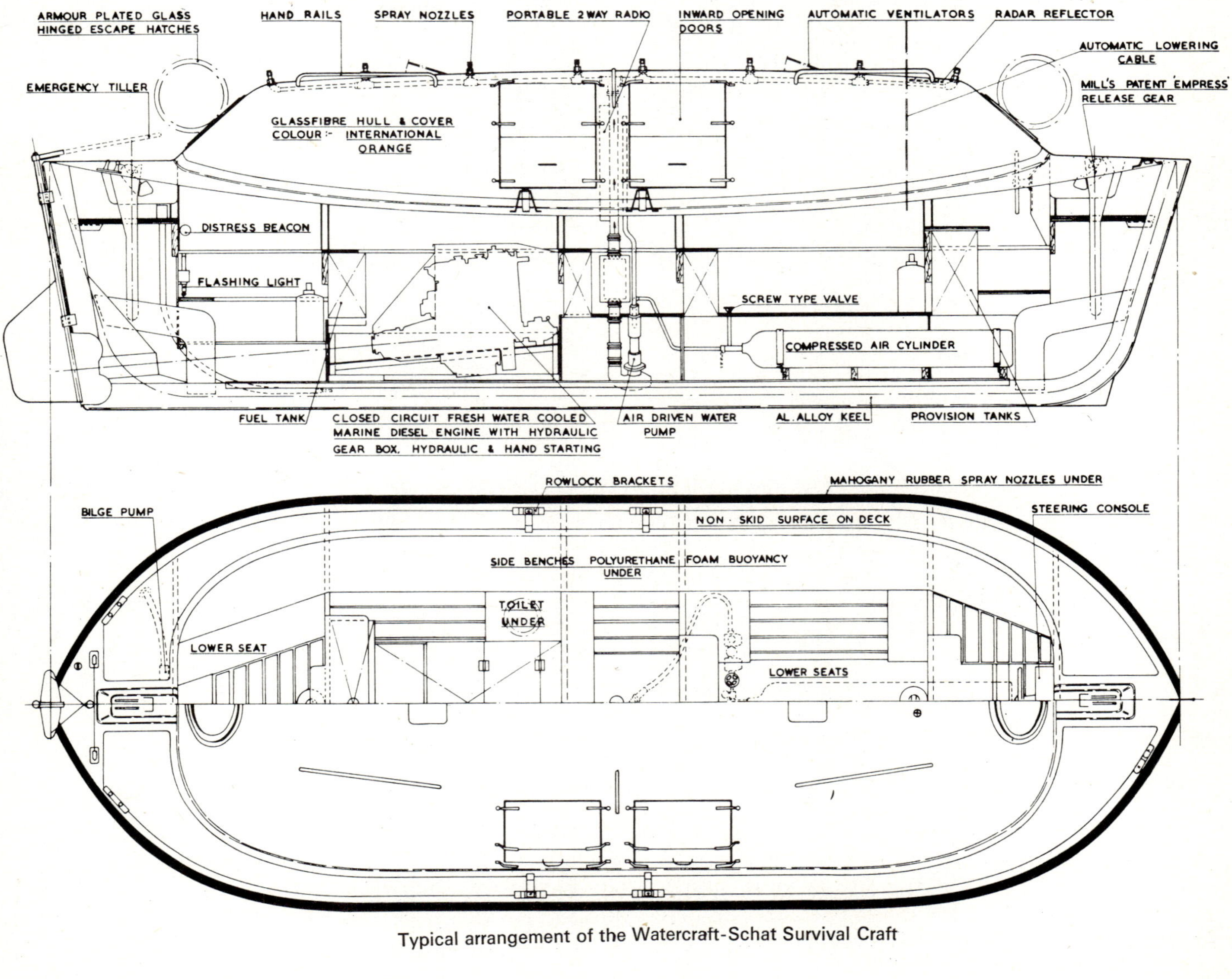

Typical arrangement of the Watercraft-Schat Survival Craft

Galvanised steel rope for lowering up to 39·6 m (130 ft).

Galvanised securing wires and fittings for holding craft when stowed.

Suspension links between wire rope fall and craft.

Tecalemit grease nipples and one grease gun per rig.

Compensating device to take up stretch in wire rope.

Schat electric boat winch mounted on outrigger with all mechanism totally enclosed and all gearing in an oil bath.

The winch allows the craft to be lowered under gravity without power at a controlled speed of 36 m (120 ft) per minute.

The holding brake can be operated from the winch position or from within the craft. Hoisting normally by electric motor but in the event of power failure by hand crank handle.

The crank handle is not rotated when the motor is running or the craft is being lowered by gravity.

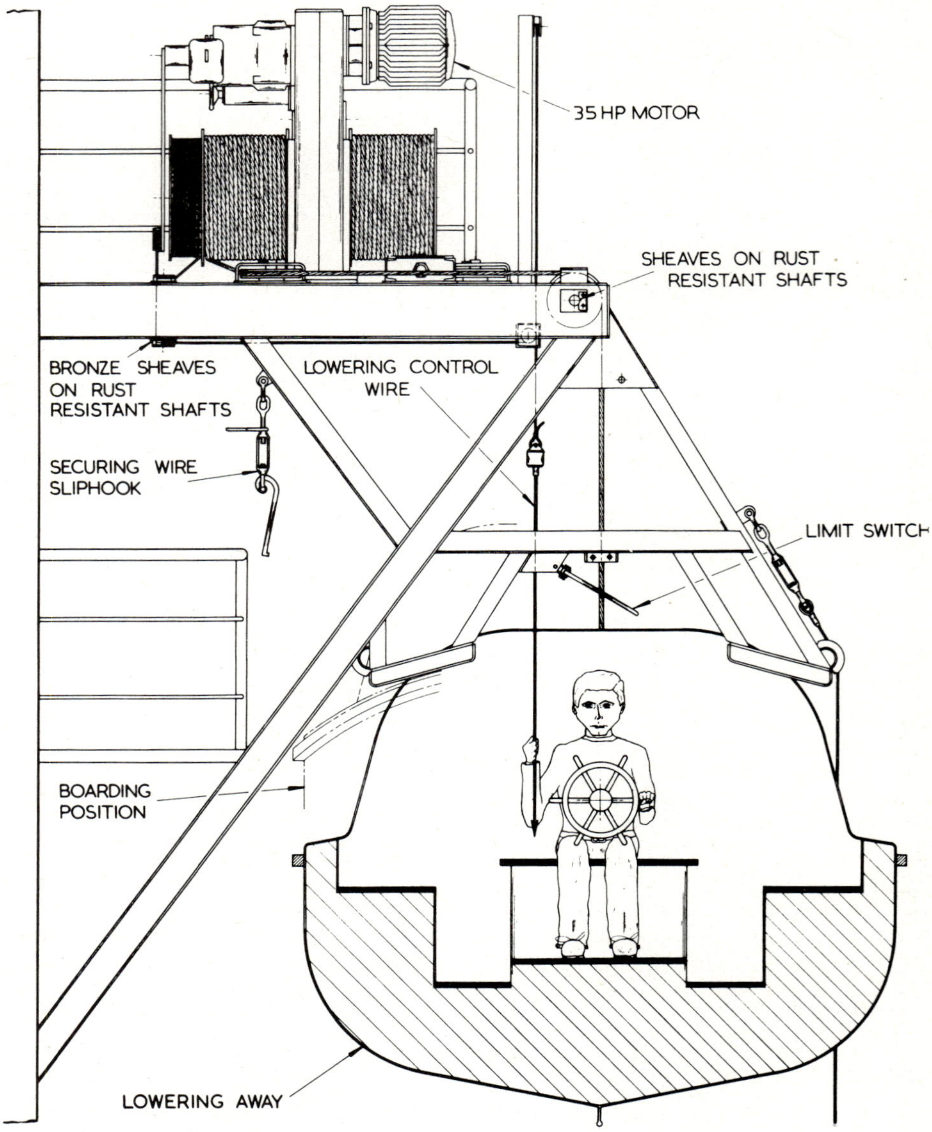

Launching system of the Watercraft-Schat Survival Craft

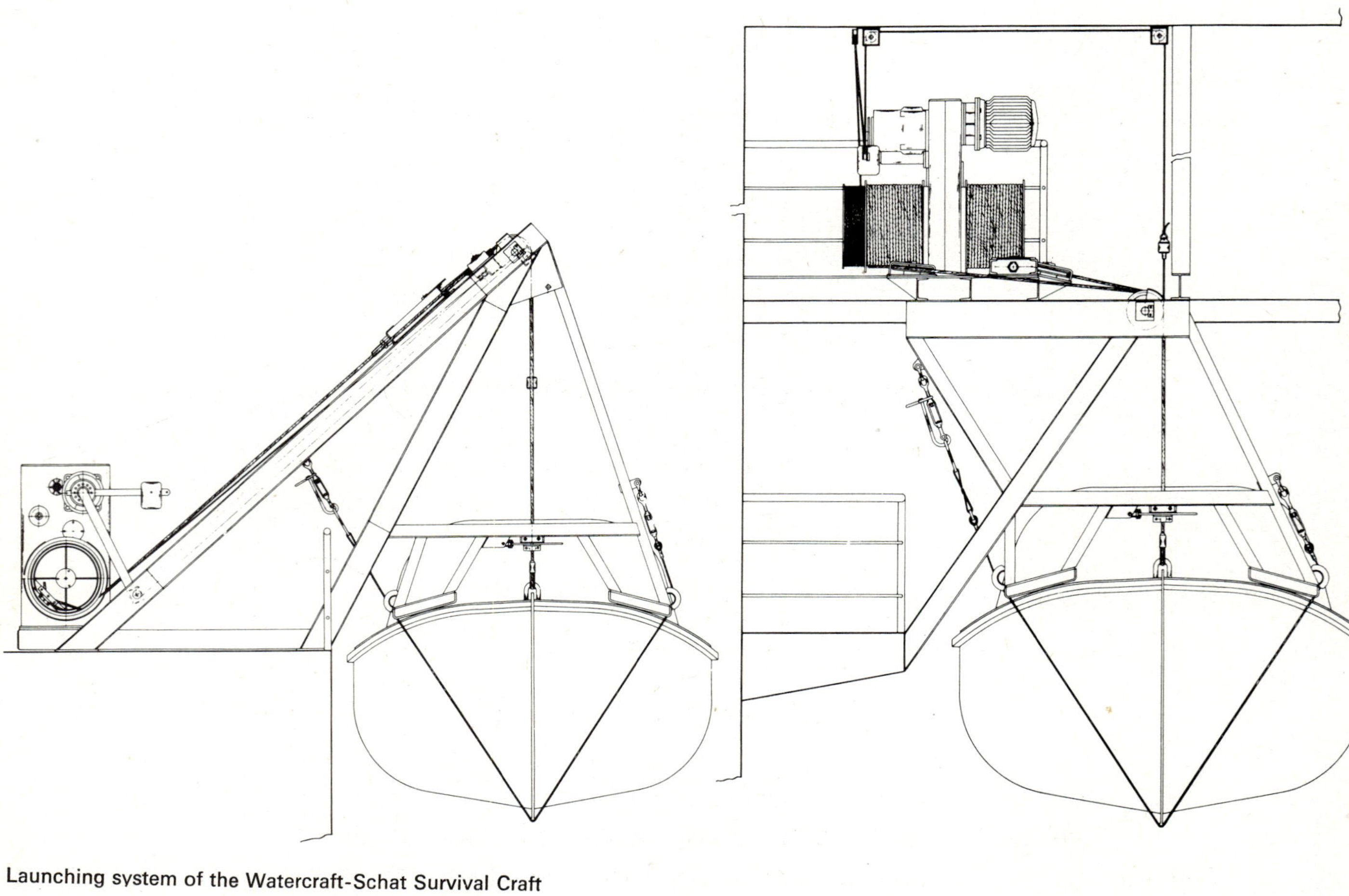

Launching system of the Watercraft-Schat Survival Craft

UNITED STATES OF AMERICA

B.F. GOODRICH COMPANY

ESCAPE SLIDE SYSTEMS

On offshore oil platforms, fire or other emergency situations demand immediate evacuation of personnel to a safe distance.

The B. F. Goodrich Company has experience in the successful development of escape slide systems for Boeing 747 jet transports. The same principle of quick evacuation has now been applied to offshore drilling operations.

System components

Inflatable Escape Slide

The inflatable multi-tubular slide is constructed of high-strength, durable nylon fabric coated with a neoprene elastomer. Special fabric used in the slide lane provides a smooth surface with uniform sliding characteristics, either wet or dry. At the lower end of the slide, a special 'deceleration surface' slows down the evacuees as they prepare to enter the attached life raft. High sides on the slide lane add safety during descent. The under side of the slide is fitted with a unique inflated truss which increases stiffening of the entire unit and prevents bending and sagging under large loads.

Inflatable Life Raft

Each escape system incorporates an inflatable life raft joined to the lower end of the slide by a quick release restraint device. Release is accomplished instantaneously by pulling a lanyard at the end of the slide after evacuees are safely in the raft.

The raft is compartmentalised into identical flotation chambers placed one on top of the other. This construction provides not only a safety factor in the event of one chamber being damaged, but it allows the raft to be used either side up, should the unit capsize.

Rafts are equipped with two sets of paddles, a knife, and a flashing locator light. When specified, additional survival equipment or rafts with a canopy for protection in severe weather conditions are available.

Escape Slide Inflation System

Inflation of the escape slide is accomplished by gaseous nitrogen stored in a reservoir at 137·5 kg/cm² (2,000 lb/in²). When a valve is tripped, the gas passes through a pressure regulator and into six aspirator pumps, located in the wall of the inflatable slide. The high pressure air passing through the aspirator pumps creates a low pressure area in the pump barrel and causes ambient air to be drawn into the inflatable. A set of check doors in each aspirator close after inflation is complete to prevent the air from escaping.

Inflation system components are treated for corrosion resistance by cadmium plating, anodising, or painting with an epoxy paint. Inflation systems are also available in stainless steel.

Life Raft Inflation System

The life raft automatically inflates when the escape slide is deployed by means of two aspirator pumps located in the wall of the raft, one in each chamber. A check valve in each air line maintains the inflation of the two chambers independently of each other.

Mounting Apron Assembly

The mounting apron is made of carbon steel, welded or bolted onto the platform. On the horizontal surface of the apron are located two 90° cradle segments on which the storage container rests. Outrigger arms project from each side of the apron. Double guy lines are attached to each side of the apron and run to the tips of the outrigger arms. These guy lines secure the inflatable in its proper position, even in high winds, both during and after deployment.

Storage Container

A drum shaped container, split lengthwise, holds the inflatable parts in a single compact package on the apron assembly. When released, the container rolls off the cradles with one half falling to the left side and the other to the right side. A tether line attached to one half is connected to the main air supply valve. The impact of the falling container half actuates the air valve to initiate the inflation sequence automatically. A tether line is also attached to the other container half to prevent loss of the container, since the escape slide and raft can be repacked and used again.

Design data

Nominal sill height	**15·2±1·52 m (50±5 ft)**	**19·8±1·52 m (65±5 ft)**
Slide length	22·9 m (75 ft)	29·0 m (95 ft)
Slide volume	42·5 m³ (1,500 ft³)	54·0 m³ (1,900 ft³)
Deployment time	10 seconds	20 seconds
Approx. system weight:		
Slide sub assembly	193 kg (425 lb)	295 kg (650 lb)
Raft	18 kg (40 lb)	18 kg (40 lb)
Slide Inf. system	454 kg (1,000 lb)	500 kg (1,100 lb)
Raft Inf. system	20 kg (45 lb)	20 kg (45 lb)
Appron assembly	545 kg (1,200 lb)	545 kg (1,200 lb)
	826 kg (2,710 lb)	1,380 kg (3,035 lb)
Inflation medium: gaseous Nitrogen		
Slide	147,420 cm³ at 137·5 kg/cm² (9,000 in³ at 2,000 lb/in²)	19,656 cm³ at 137·5 kg/cm² (12,000 in³ at 2,000 lb/in² g)
Raft	5,405 cm² at 207 kg/cm² (330 in³ at 3,000 lb/in² g)	5,405 cm² at 207 kg/cm² (330 in³ at 3,000 lb/in² g)
Slide type		0·61 m (2 ft) wide lane
Nominal slide angle		45°
Raft size overall		2·13 m × 3·66 m (7 ft × 12 ft)
Raft volume		2·7 m³ (95 ft³)
Raft capacity:		
Seated		24 persons
Clinging to lifeline		26 persons
Total		50 persons
Slide and raft operating pressure		0·102 kg/cm² (1·5 lb/in² g)
Max. evacuation rate		60 persons/min

Deployment of the Goodrich Escape Slide System: one, six and eight seconds after release

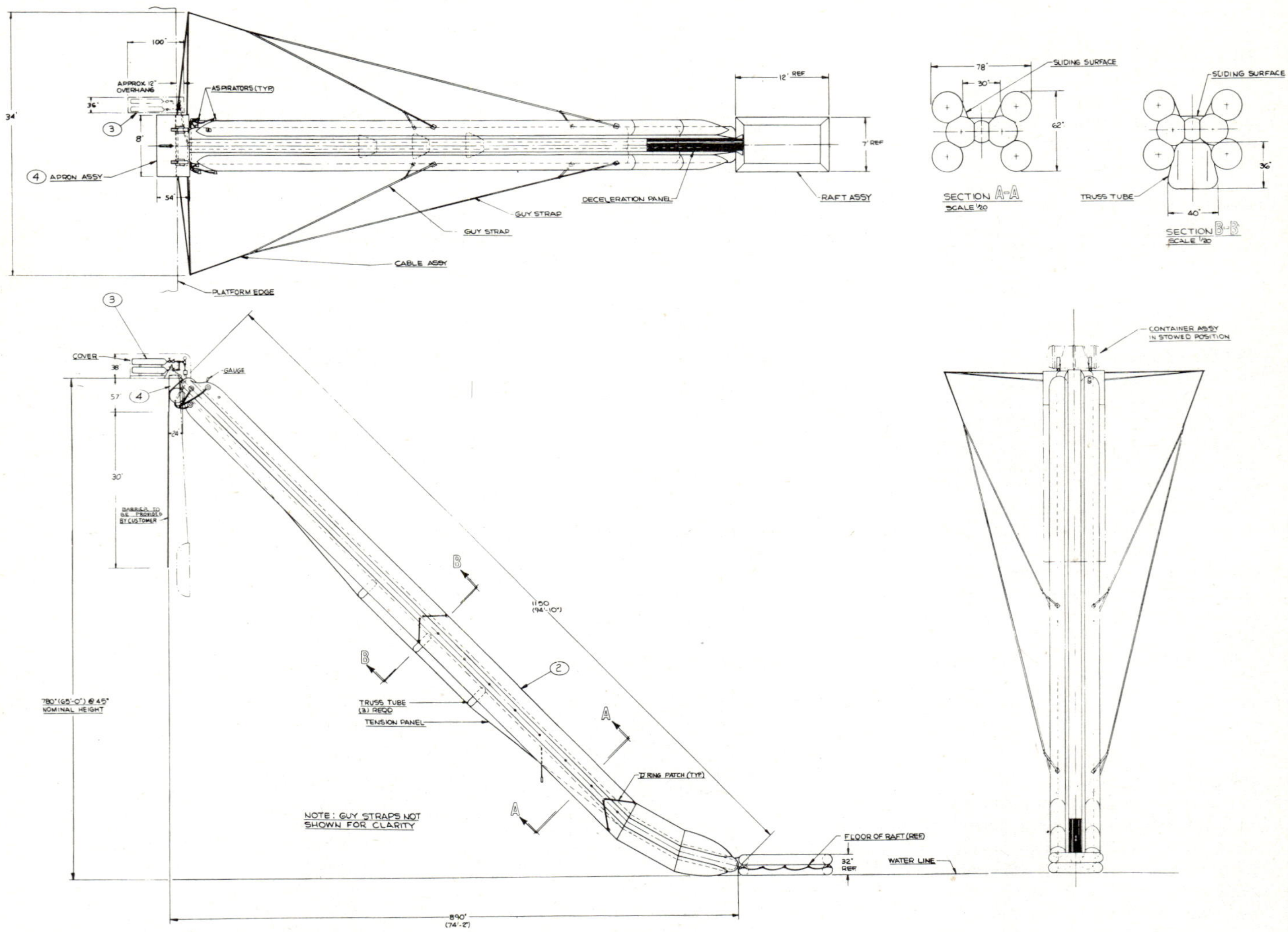

General arrangement of the 20 m (65 ft) level Goodrich Escape Slide System for offshore drilling platforms

WHITTAKER CORPORATION

WHITTAKER SURVIVAL SYSTEM

This survival system has been developed specifically for use with oil drilling rigs and offshore production platforms. The system consists of a survival capsule and a launching facility.

The capsule, with its 40 h.p. engine running, can be winched into the water without power within 60 seconds. It can then be manoeuvred on the spot to ensure a quick getaway in any direction at a speed of 5 knots. This is especially important whilst the capsule is still in close proximity to oil rig legs. There is no difficulty in launching in heavy seas or where there is a list.

The capsule is made from glass fibre reinforced plastics and cannot be capsized. It has an international orange coloured surface with a radar reflective dome and two watertight and fireproof hinged doors.

28 passengers can be accommodated and the capsule is fully equipped to meet their needs for a five-day period. There is a minimum of one hour's breathing time; compressed air is not used, thus eliminating any danger of explosion.

Unlike most other systems, the Whittaker capsule can readily be used for training purposes as it is easy to recover.

The Whittaker System is fully approved by international authorities including the Department of Trade and Industry, UK, and the United States Government Authority; it is also accepted by Sjøsartsdirektoratet, Norway, and the French Ministry of Transport.

The Whittaker 28-man Survival Capsule propelled by a 40 h.p. Westerbeke diesel engine

The capsule

Included in the dome of the capsule are ventilating manifolds, a fire sprinkler system, handrails, hoisting arrangements and windows. The capsule is approximately 4·27 m (14 ft) in diameter, 2·74 m (9 ft) high, and with equipment aboard weights approximately 2,270 kg (5,000 lb).

The internal structure provides seating accommodation for 28 passengers, including a toilet, water-tight food and water storage compartments, and compartments for tools, survival equipment, battery, engine, bilge pump, fuel tank, lighting, instrumentation, air purification and exhaust systems.

Permanent equipment supplied with each life sphere engine

Each capsule is supplied with a 40 h.p. Model 4·108 LS Westerbeke Marine Engine; liquid cooled using a sea water/fresh water heat exchanger. A fifty ampere alternator, controlled by a solid state regulator is provided, along with a pre-wired instrument panel containing a tachometer with hour meter, an oil pressure gauge, an ammeter and a temperature gauge.

Engine starting system

This hydraulic system includes:
Hand pump with hydraulic reservoir combined
Control Valve
Cranking Motor (supplied with Engine)
Engine Driven Pump
High-Pressure Filter
Oil Pressure Gauge
Hydraulic Accumulator

Additional equipment

Xenon Blinker Light with Battery
Air Purification System
Motion Sickness Bags
Twenty-nine seat belts

Equipment carried in the capsule

Plastic bailer	1
Boat hook	2
Bucket	2
Compass and mounting	1
Ditty bag	1
Drinking cup	1
Fire extinguisher	2
First-aid kit	1
Flashlight	1
Hatchet	2
Heaving line	2
Jack-knife	1
Lantern	1
Life preservers	2
Matches (boxes)	2
Milk, condensed (pounds per person)	1
Mirror, signalling	2
Oil, illuminating (quarts)	1
Painter (100·ft—$\frac{5}{8}$ in diameter nylon rope)	1
Provisions (pounds per person)	2
Signals, distress, Floating orange smoke	2
Signals, distress, Red hand flare	12
Signals, distress, Red parachute flare	12
Tool kit	1
Water (quarts per person)	3
Whistle, signalling	1
Fishing kit	1
Bilge pump (permanently attached)	1
Life-saving signals	
Instruction plaque	1
Lifeline becketed to outside of capsule	1
Rudder and tiller arm	1

The launching system
Electric winch

The winch consists of a 7·5 h.p. 208–220/440 volt, 3-phase, 60 cycle, 280 rpm output gearhead electric motor with motor starter-transformer-main line disconnect and raise, push-button control switch, which will drive through a mechanical clutch, and an overrunning clutch on the disc brake. From this point, the system consists of a blower, 33 h.p. at 1,800 rpm, a two-stage planetary, a spur gear mesh, and the wire rope drum. A manual crank can be used for final positioning of the capsule. A hand wheel is available for use or paying out cable if required during capsule retrieval cycles. The winch is rated for 5,000 kg (11,000 lb) working load. Winch weight is 1,430 kg (3,150 lb).

The Whittaker Survival Capsule

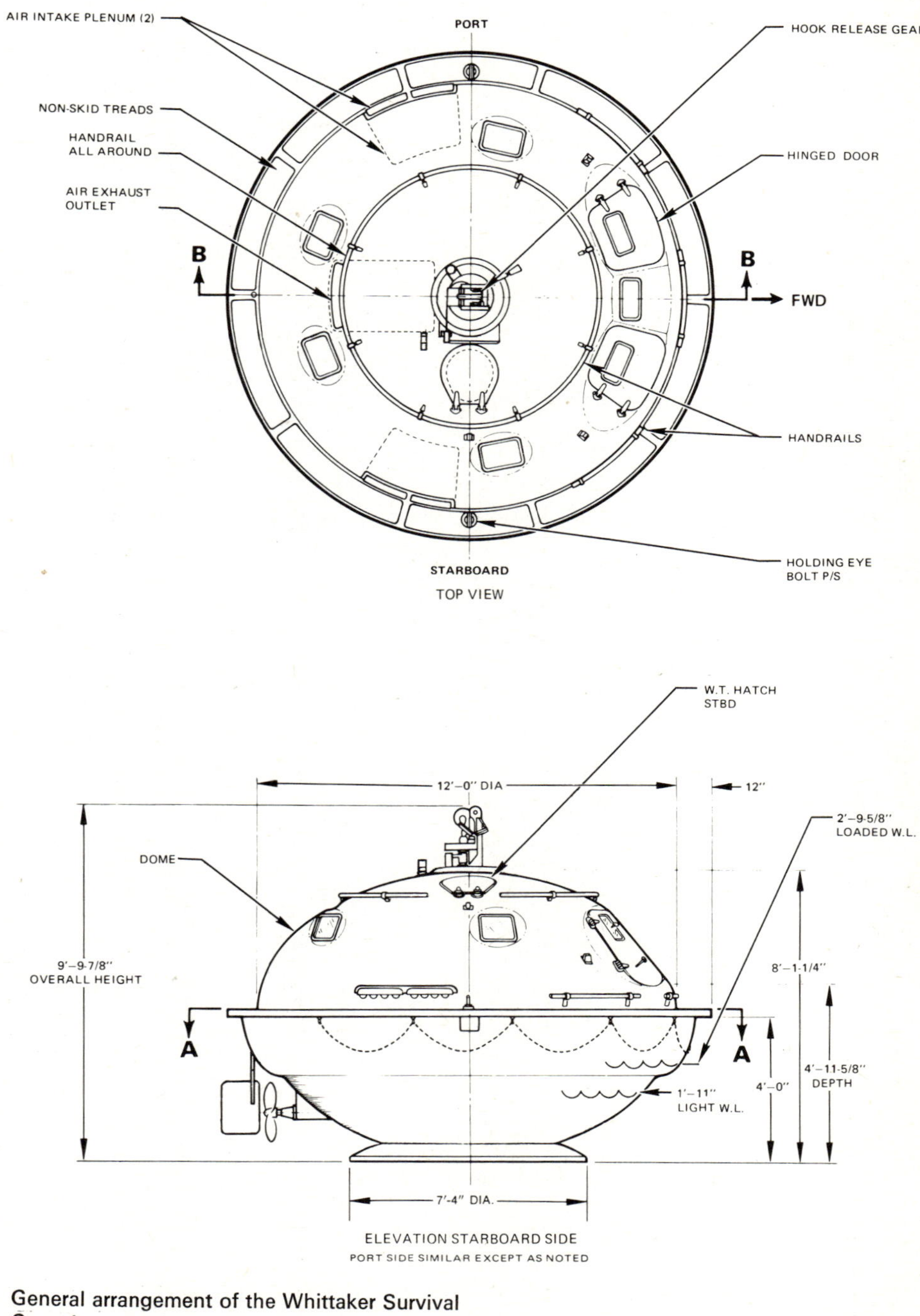

General arrangement of the Whittaker Survival Capsule

Holding and launch platform

(1) Survey and assistance in location of capsules for rig.

(2) Re-engineering if necessary for holding and launching platform.

(3) Consultation on reinforcement of rig if necessary.

(4) Holding and launching platform in subassembly to be welded on rig under our supervision.

(5) Platform weight 2,136 kg (4,700 lb).

(6) Complete set of platform drawings.

(7) Swivel block.

All electrical installations to capsule and welding equipment as well as crane assistance to be supplied by rig owner or contractor.

Applications

Several hundred installations of the Whittaker Survival Capsule system are now in being on a wide variety of drilling platforms. throughout the world.

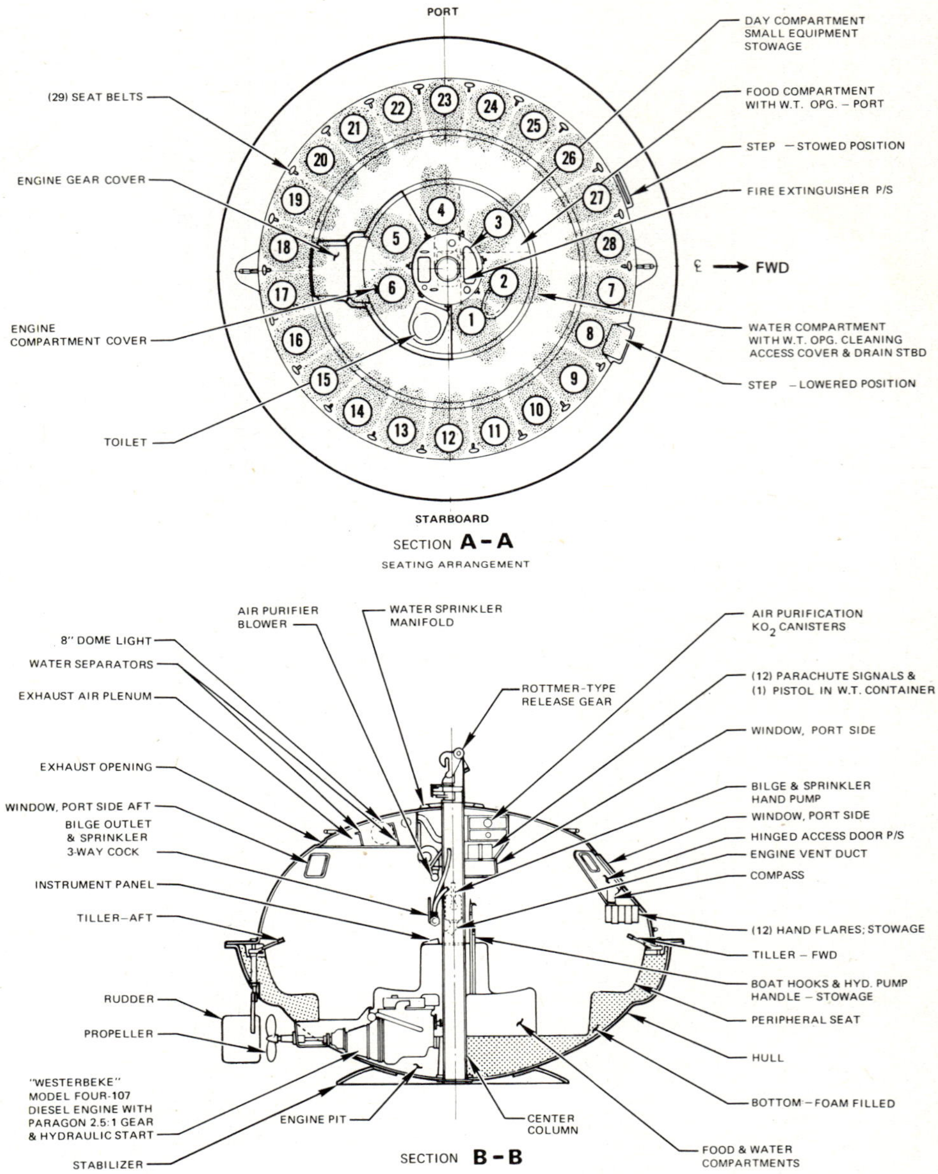

General arrangement of the Whittaker Survival Capsule

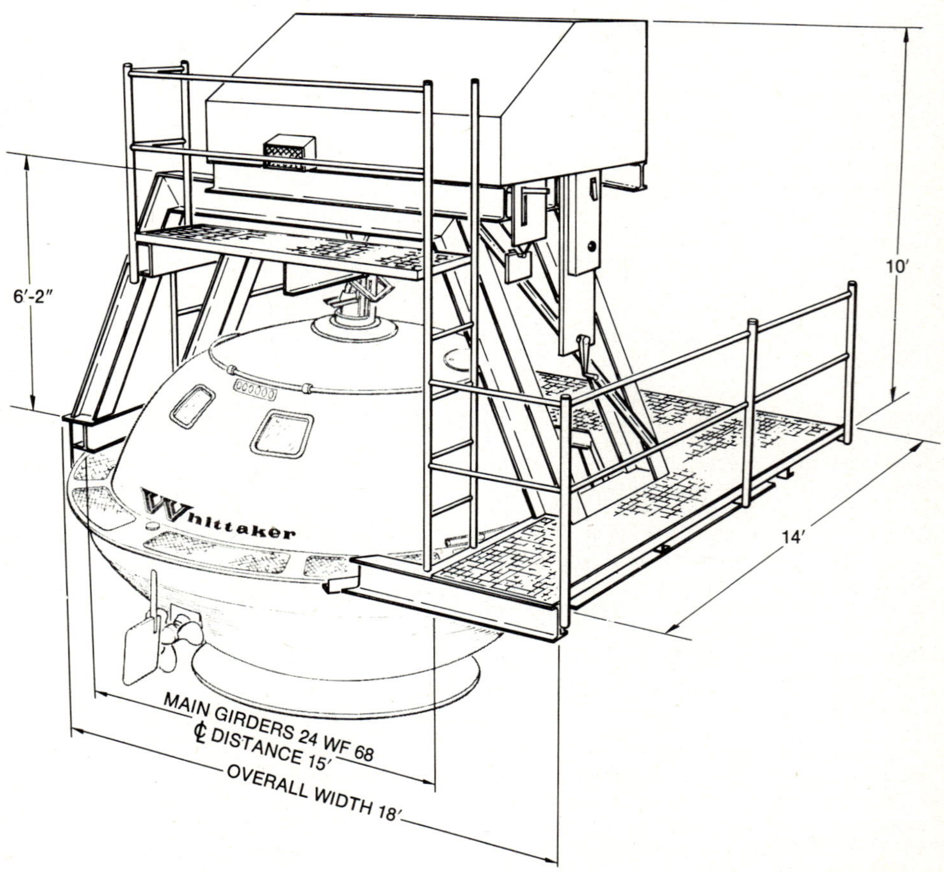

Launching system for the Whittaker Capsule

TUGS AND SUPPLY VESSELS BUILDERS

ARGENTINA

ASTILLEROS ALIANZA S.A.,
San Martin 50,
Buenos Aires.

ASTILLEROS ARGENTINOS RIO DE LA PLATA S.A.,
Tucuman 1438,
Buenos Aires.

ASTILLEROS CORRIENTES,
Corrientes 545,
Buenos Aires.

ASTILLEROS MESTRINA S.A.,
Pedro de Mendoza 2233,
Buenos Aires.

ASTILLEROS PRINCIPE MENGHI Y PENCO S.A.,
Ribera Sud del Riachuelo,
Isla Maciel—Avellaneda,
Provincia de Buenos Aires.

ASTILLEROS Y FABRICAS NAVALES DEL ESTADO S.A.,
Corrientes 672,
Buenos Aires.

ASTILLEROS VICENTE FORTE DE FORTE Y CACCIOLA,
25 de Mayo 611,
Buenos Aires.

CANESTRARI HERMANOS S.A.C.I.F.,
Zuviria 620,
Tigre,
Provincia de Buenos Aires.

EMPRESA LINEAS MARITIMAS ARGENTINAS,
Corrientes 389,
Buenos Aires.

RYAN ASTILLEROS ARGENTINOS S.A.I.C.,
Pinzon 121/123,
Buenos Aires.

TALLERES DE REPARACIONES NAVALES,
Avenida Gaspar Melchor de Jovellanos,
Buenos Aires.

TALLERES NAVALES MARINO, S.R.L.,
Montana 42,
Isla Maciel—Avellaneda,
Provincia de Buenos Aires.

YACIMIENTOS PETROLIFEROS FISCALES,
Avenida Roque Saenz Pena 777,
Buenos Aires.

AUSTRALIA

CARRINGTON SLIPWAYS PTY. LTD.,
Old Punt Road,
Tomago 2322,
New South Wales.

LADY SARAH
For P&O Australian Offshore Services.

Length, overall	57·9 m (190 ft)
Beam, extreme	13·2 m (43 ft 5 in)
Draught, max.	4·37 m (14 ft $4\frac{1}{4}$ in)

Total power: 4,000 h.p. Four 8 cylinder Daihatsu diesel, 2 shafts

LADY VILMA
For P&O Australian Offshore Services.

Length, overall	57·9 m (189 ft 9 in)
Beam, extreme	13·2 m (43 ft 5 in)
Draught, max.	4·37 m (14 ft $4\frac{1}{4}$ in)

Total power: as for Lady Sarah

LADY LAURIE
Built 1970 for P&O Australian Offshore Services.

Length, overall	49·27 m (161 ft 8 in)
Beam, extreme	10·39 m (34 ft 1 in)
Draught, max.	3·35 m (11 ft)
Total power	3,200 h.p., two English Electric diesel
Deadweight	687 tons
Speed	12·5 knots

LADY LORNA
Same details as for *Lady Laurie*

ADELAIDE SHIP CONSTRUCTION,
Birkenhead,
South Australia.

BASS SHORE
Built 1973 for Osmarine (Australia) Pty. Ltd.

Length, overall	56·40 m (185 ft)
Beam, extreme	11·89 m (39 ft)
Draught, max.	4·50 m (14 ft 9 in)
Deadweight	963 tons
Total power	4,000 h.p., two 16 cyl. Mirrlees-Blackstone
Speed	14 knots
Bollard pull	47 tons
Propellers	Fixed pitch, Kort Bow-thruster

SOUTHERN TIDE
(a sister ship will be built)
Built 1973 for Tidewater Port Jackson Marine
Designers: Burness Corlett & Partners Ltd Basingstoke, UK

Length, overall	54·15 m (177 ft 7 in)
Beam, extreme	12·19 m (39 ft $11\frac{3}{4}$ in)
Draught, max.	4·12 m (13 ft 6 in)
Deadweight	800 tons
G.R.T.	787
Total power	3,900 h.p., two 16 cyl. General Motors, EMD 16-645 E2 diesels
Speed	14 knots
Bollard pull	62·4 tons
Propellers	Nissho-Iwai A.B.S. ✠A1

SMIT-LLOYD 31
Built 1968 for Smit-Lloyd (Australia) Pty. Ltd.
Total power 3,000 h.p.

SMIT-LLOYD 32
Built 1968 for Smit-Lloyd (Australia) Pty. Ltd.
Total power 3,000 h.p.

SMIT-LLOYD 33
Built 1969 for Smit-Lloyd (Australia) Pty. Ltd.
Total power 3,000 h.p.

SMIT-LLOYD 34
Built 1969 for Smit-Lloyd (Australia) Pty. Ltd.
Total power 3,000 h.p.

WALKERS LTD.
SMIT-LLOYD built 1973
Owner: Smit-Lloyd (Australia)

Length, overall	54·81 m (179 ft 10 in)
Beam, extreme	12·29 m (40 ft 4 in)
Draught, max.	4·40 m (14 ft $5\frac{1}{4}$ in)
Deadweight	860 tons
Total power	2,000 h.p., two 8 cyl. De Industrie
Speed	13 knots

BELGIUM

BOELWERF N.V.,
Dijk 6,
2690 Temse.

CANADA

ALLIED SHIPBUILDERS LTD.,
1870 Harbour Road,
North Vancouver,
British Columbia.
LADY PAMELA
Built 1971 for I.O.S.

Length, overall	55·77 m (182 ft 11 in)
Beam, extreme	12·17 m (39 ft 11 in)
Draught, max.	4·68 m (15 ft $4\frac{1}{4}$ in)
Deadweight	1,033 tons
Total power	4,200 h.p., four 6 cyl. Hydquist & Holm geared to two shafts
Speed	14 knots
Bollard pull	70 tons

Gibraltar Straits

B.C. MARINE SHIPBUILDERS, LTD.,
101 N. Victoria Drive,
Vancouver
GIBRALTAR STRAITS

BEL-AIRE SHIPYARDS LTD.,
1667 Columbia Street,
North Vancouver,
British Columbia.
CATHY-B
Built 1972 for Nordic Offshore Services

Length, overall	56·10 m (184 ft)
Beam, extreme	14·02 m (46 ft)
Total power	5,280 h.p., two 12 cyl. English Electric
Speed	14 knots

Haida Brave, one of the most powerful tugs to be constructed in Canada by the Halifax Shipyard Division of Hawker Siddeley

FERGUSONS
Halifax Shipyards, Division of
Hawker Siddeley.
HAIDA BRAVE

RIVTOW STRAITS LTD.,
see B.C. Marine Shipbuilders, Ltd.

STAR SHIPYARDS LTD.,
61 Duncan Street,
New Westminster,
British Columbia.
MARY B VI
Built 1972 for Nordic Offshore Services

Length, overall	52·13 m (171 ft)
Beam, extreme	11·89 m (39 ft)
Total power	3,280 h.p., two 16 cyl. General Motors
Speed	12 knots

YARROWS LTD.,
Canteen Road,
Esquimalt Harbour
Victoria,
British Columbia.

DENMARK

ARHUS FLYDEDOK A/S,
P.O. Box 23,
DK.8100 Århus C,
(Århus Floating Dock Ltd.)
Tel: 06-134000
Telex: 64462
(Subsidiary of Weco-Shipping, Copenhagen)
Weco-Shipping I/S
113 Rungsted Strandvej DK
2960 Rungsted Kyst
Tel: 01-86 6500
Telex: 15237 weco dk
MAERSK FIGHTER (built 1967)
MAERSK FEEDER (built 1967)

Two vessels built in 1967 to German Lloyd's Class +100 A 4 'K' Eisklasse 1 and specially designed for servicing oil drilling platforms. Vessels have twin screws and rudders, bow thruster and engines of 2,400 h.p. giving them a cruising speed of 13 knots.

Principal dimensions

Length	52·66 m (172 ft 8 in)
Breadth	11 m (36 ft 1 in)
Depth	4 m (13 ft 1 in)
Draught loaded	3·38 m (11 ft 2 in)

Tonnage

Gross register tonnage	499 tons
Net register tonnage	235 tons

Cargo capacity

Total cargo capacity	800 tons
Cargo deck capacity	300 tons
Clear deck space	248 m² (100 ft × 26 ft = 2,670 ft²)

2,400 HP Class Supply Vessel 1

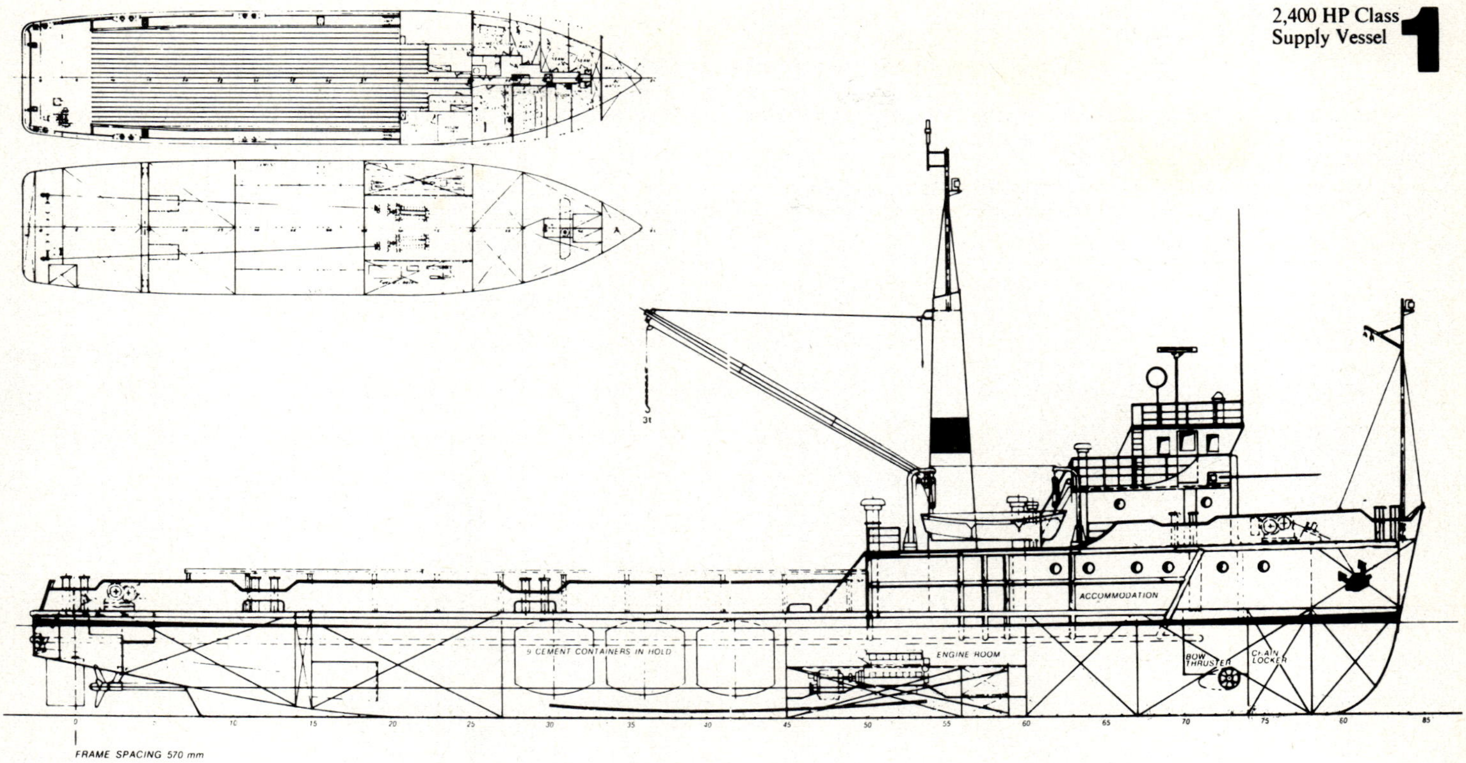

General layout diagrams of the 2,400 h.p. Class supply vessel

Bulk pressure tanks below deck: Nine of 15·3 m³ (540 ft³) = 137·7 m³ (4,860 ft³)
Rig fuel: 238·7 m³ (8,450 ft³) = 198 tons
Water: 440·9 m³ (15,500 ft³) = 441 tons
Bunker fuel: 70·8 m³ (246·5 ft³) = 59 tons
Refrigerated storage: 9·95 m³ (350 ft³) at −18°C/0°F; 9·95 m³ (350 ft³) at +4°C/39°F

Accommodation
Officers: Four in 4 single-berth cabins
Crew: Six in 3 double-berth cabins
Passengers: Eight in 2 four-berth cabins

Wheelhouse
Of special design with good visibility all around and complete remote control of main engines and bow thruster.

Navigation equipment etc.
Radar: Terma Pilot 7 T 36
Decca: Navigator Mark 12
Gyro compass: Sperry SR 120
Echo sounder: Atlas Nereus
MF radio telephone 100 watt: M.P.P.
VHF radio telephone 20 watt: Storno

Anchors/mooring gear
4,070 kg (8,960 lb) hydraulic windlass with two fixed capstan barrels for 1,109 kg (2,440 lb) anchor on 284 m (930 ft) of chain to starboard, and 1,109 kg (2,440 lb) anchor on 207 m (680 ft) of chain to port. 1,530 kg (3,360 lb) hydraulic windlass placed aft with capstan barrel for 509 kg (1,120 lb) stern anchor on 180 m (590 ft) of chain.

Derrick and derrick gear
3,050 kg (6,720 lb) SWL derrick 12 m (39 ft) long with corresponding cargo winch.

Main engines
Two M.A.N. V 8 V 16/18 T sixteen-cylinder marine diesel engines, with a total maximum output of 2,400 i.h.p. and a continuous output of 1,932 b.h.p. Cruising speed of 13 knots at a fuel consumption of approx. 8 tons/day of gas oil fully loaded.

Auxiliaries
Three M.A.N. D 2148 M eight-cylinder diesel engines each of 160 b.h.p. at 1,500 r.p.m. driving three 135 KVA (108 kw) 220/380 volt AC 50 cycles generators.

Cargo pumps
Three pumps of each 80 m³ (2,810 ft³) per hour capacity against 60 m (197 ft) head.

Compressor for pressure tanks
One high pressure screw compressor of 1,000 m³ (35,200 ft³) air per hour at 3·5 kg/cm² (50 lb/in²). Discharging rates of approx. 60 tons cement or 55 tons barytes per hour against 60 m (197 ft) head.

Bow thruster
One 130 b.h.p. Tornado bow thruster with a thrust of 1·6 tons.

Fifteen 499 g.r.t., 3,800 b.h.p. single deck, twin-screw tug supply vessels have been built (or are on order) by Århus Flydedok A/S.
Sister vessels:
MAERSK SERVER, 1971
MAERSK SUPPORTER, built October 1971
MAERSK SUPPLIER, February 1972
MAERSK SHIPPER, May 1972
MAERSK HELPER, June 1972
MAERSK HANDLER, September 1972
MAERSK HAULER, December 1972
Owner: Maersk Supply Service, division of A. P. Moller, Denmark.
TBN, on order, Yard No. 155, January 1974
TBN 156, February 1974
TBN 157, August 1974
TBN 158, October 1974
TBN 160, February 1975
TBN 161, April 1975
TBN 162, August 1975
TBN 163, October 1975
Owner: China National Machinery Import and Export Corporation, Peking.
Owner: Rederiaktieselskabet Danneborg
New building (two)

Classification society

Registered tonnage	499 g.r.t.
Net registered tonnage	230
Cargo capacity	600 tons
Main engine	3,800 b.h.p.
Speed loaded	13 knots
Bow thruster	130 b.h.p.
Bollard pull	40 tons max.
Deck area	29·5 × 7·5 m (96·7 × 24·6 ft)
Fuel oil (rig)	238 m³ (8,400 ft³)
Fresh water	11 m³ (390 ft³)
Crew	10 man
Consumption	about 7·25 tons diesel oil daily

Lloyds + 100 A1, Tug/supply vessel, Ice Class 3

Main particulars

Length, overall	53·10 m (175 ft)
Length, b.p.	50 m (164 ft ½ in)
Breadth, moulded	11 m (36 ft 1 in)
Depth	4 m (13 ft ½ in)
Draught, loaded	3·45 m (11 ft 4 in)
Deadweight	730 tonnes
Deck cargo up to max.	370 tonnes
Cement containers (8)	119 m³ (4,250 ft³)
Rig water	372 m³ (13,100 ft³)
Ballast water	41 m³ (1,445 ft³)

Technical description
Twin-screw supply vessel, with Kort-nozzles and twin-rudders. Singledeck construction with all accommodation and wheelhouse forward, intended for oil rig supply and towing.
Deck area strengthened for heavy cargo, 2·5 t/m² (0·23 t/ft²), fitted with Stern Roller, made by Hitzler and towage and anchor handling winch, made by Thrige Titan, 100 t.
Radio telephone, radio direction finder, VHF 26 channels, Emergency radio intercommunication system, 15 stations

Depth Sounder with recorder, Rudder Angle Indicator, Radar, Gyrocompass and Automatic Pilot

Accommodation for 11 crew members in single and twin bed cabins and for 12 passengers.

Machinery

Main engines: Two non-reversible, single acting, four stroke diesel type 8 MU452 AK, make M.A.K., 1,900 b.h.p. at 425 rpm

Propellers: Two four bladed C/P propellers in stainless steel, type PL 63/180, make Liaaen, fitted in Kort-nozzles

Bow thruster: One type Tornado with 1·5 ton thrust at 130 h.p. 1,450 rev/min

Auxiliary diesel engines

Three single acting, four stroke diesel engines with turbocharging and fresh water cooling: Type GASI 11-05, made by Scania Vabis

Cement system

Eight Cement Containers, compressor made by Guterhoffnungshytte, type SK 12, for discharge of cement. Capacity about 1,000 m^3/hour (35,100 ft^3/hour) air suction (at 1 ato), Motor 125 h.p., 380 volt, 3,000 rev/min

Electric power supply

Three diesel driven, directly coupled, constant voltage synchronous AC generators with self-exitation, each of 160 KVA, 3 × 380 volt, 50 Hz cos $\theta = 0{\cdot}8$, arranged for parallel running

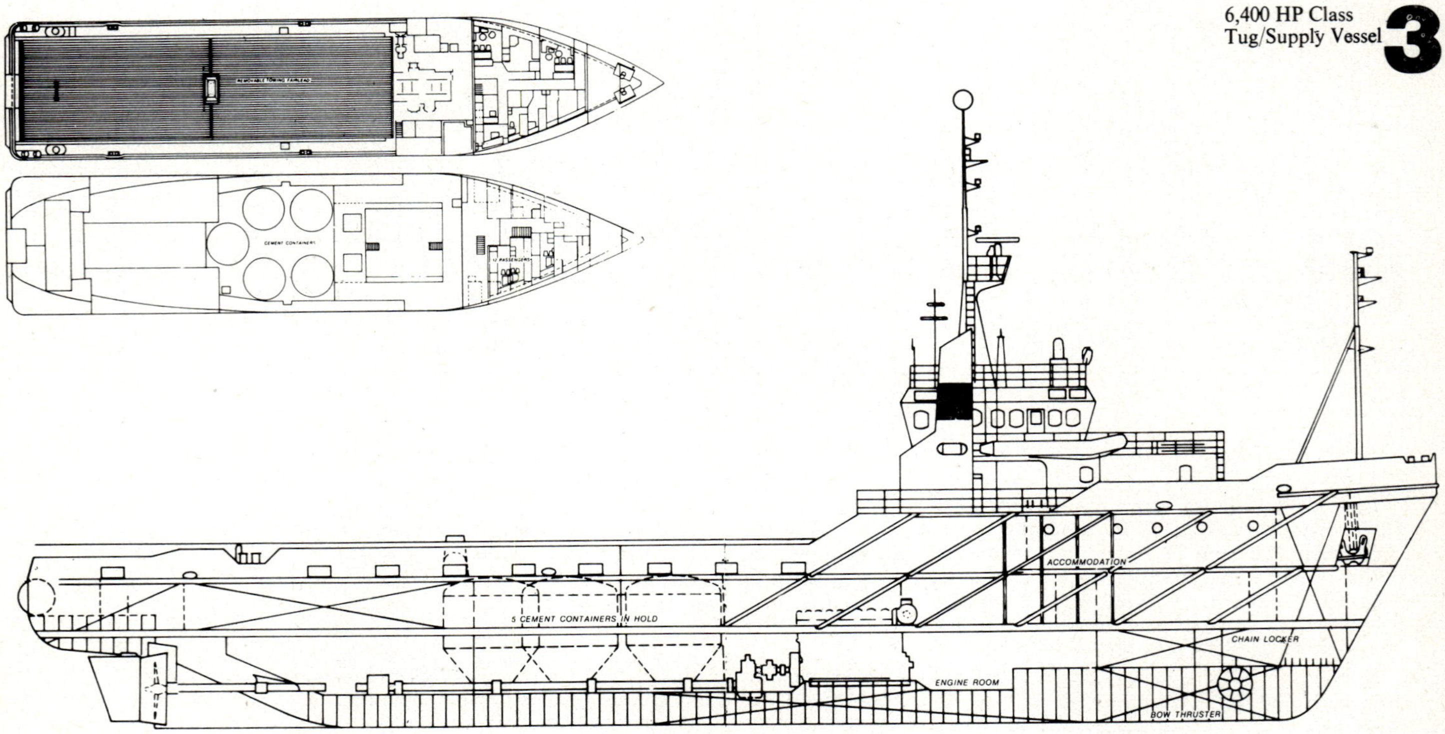

General layout diagram of the 6,400 h.p. Class tug / supply vessel

MERSELISBOREG

Built 1973, lauched August. No details.

6,400 H.P. CLASS TUG/SUPPLY VESSEL

Seven identical supply-, towing- and anchor-handling vessels are being built to Norwegian Veritas Class +1A1 Ice C* Tug/Supply EO. The vessels are specially designed for duties in connection with operation of large semi-submersible platforms at great water depths and in extreme working conditions world-wide. The vessels will have excellent seakeeping qualities, with high bow forms and with very large free board and cargo capacity. They will be fitted with variable pitch propellers in nozzles, and engines of 6,400 h.p. giving a bollard pull of 68 metric tons, and powerful bow thruster of 5 tons capacity. The vessels are also equipped with combined heavy-duty anchor handling and towing winch with a pull of 100 tons and a large stern roller.

Principal dimensions

Length	58·5 m (191 ft 11 in)
Breadth	12·2 m (40 ft)
Depth	6·1 m (20 ft 1 in)
Draught:	
660 DWT	3·9 m (12 ft 9 in)
1,000 DWT	4·4 m (14 ft 5 in)
1,400 DWT	5·3 m (17 ft 3 in)

Tonnage

Gross register tonnage	499/850 tons

Cargo capacity

Total cargo capacity	1,400 tons
Cargo deck capacity	500 tons

Clear deck space: 33·5 m × 9·14 m (110 ft × 30 ft) = 307 m^2 (3,300 ft^2)

Bulk pressure tanks below deck: Five of 34 m^3 (1,200 ft^3) = 170 m^3 (6,000 ft^3)

Rig fuel: 350 m^3 (12,400 ft^3) = 300 tons

Water: 510 m^3 (18,000 ft^3) = 510 tons

(Fuel and water tanks partly interchangeable according to requirements)

Ballast water: 300 m^3 (10,600 ft^3) = 300 tons

Bunker fuel: 150 m^3 (5,300 ft^3) = 130 tons

Refrigerated storage: Total 15 m^3 (600 ft^3) in 2 movable refrigerated containers $-21°C$

Accommodation

Officers and crew: Eleven in 11 single cabins

Passengers: Twelve in 3 four-berth cabins

The accommodation is fully airconditioned

Wheelhouse

Large windows enabling view all around. Dual remote control of main engines, propeller pitch and bow thruster

Navigation equipment etc.

Radar: Terma Pilot 7T36
Decca: Navigator
Gyrocompass: Sperry Mark 37
Auto pilot
Echo sounder: Simrad ES2BN
Radio direction finder: Taiyo
HF radio telephone: SSB 400 watt
VHF radio telephone: 26 channels

Anchors, mooring and derrick gear

The anchoring system will enable the vessel to anchor in up to 305 m (1,000 ft) of water. Two 6-ton power capstans. 5-ton cargo winch. Derrick can be fitted

Anchor-handling and towing facilities

One 100-ton heavy duty 'Brattvaag' winch (double drum waterfall type) with a capacity of 1,082 m (3,550 ft) of 5·08 cm (2 in) dia. wire on each drum, and a static load capacity of 250 tons. Vessel fitted for rig moves and ocean towage.

Stern roller

Length: 3·05 m (10 ft). Diameter: 1·525 m (5 ft)

Main engines

Two M.A.K. 8 M 453 AK diesel engines with a total maximum output of 6,400 i.h.p. Continuous output of 5,300 b.h.p. Variable pitch propellers in nozzles

Speed and consumption

Maximum speed loaded $14\frac{1}{2}$ knots at approximately 19·5 tons/day. At 13 knots approximately 12 tons/day

Bollard pull

65 metric tons at continuous output and 68 metric tons at maximum output

Auxiliaries

Three GM diesel engines driving three generators with total capacity of 617 KVA, 440/220 volt AC 60 cycles

Cargo pumps

Three pumps of each 100 m³ (3,520 ft³) per hour against 70 metres head. One pump of 50 m³ (1,760 ft³) per hour against 70 m (230 ft) head

Compressor for pressure tanks

One Atlas Copco BE 63A compressor of 890 m³ air per hour at 3·5 kg/cm² (50 lb/in²). Discharging rates of approximately 60 tons cement or 55 tons barytes per hour against 70 m (230 ft) head

Bow thruster

One 5-ton bow thruster driven by a separate diesel engine of 425 b.h.p.

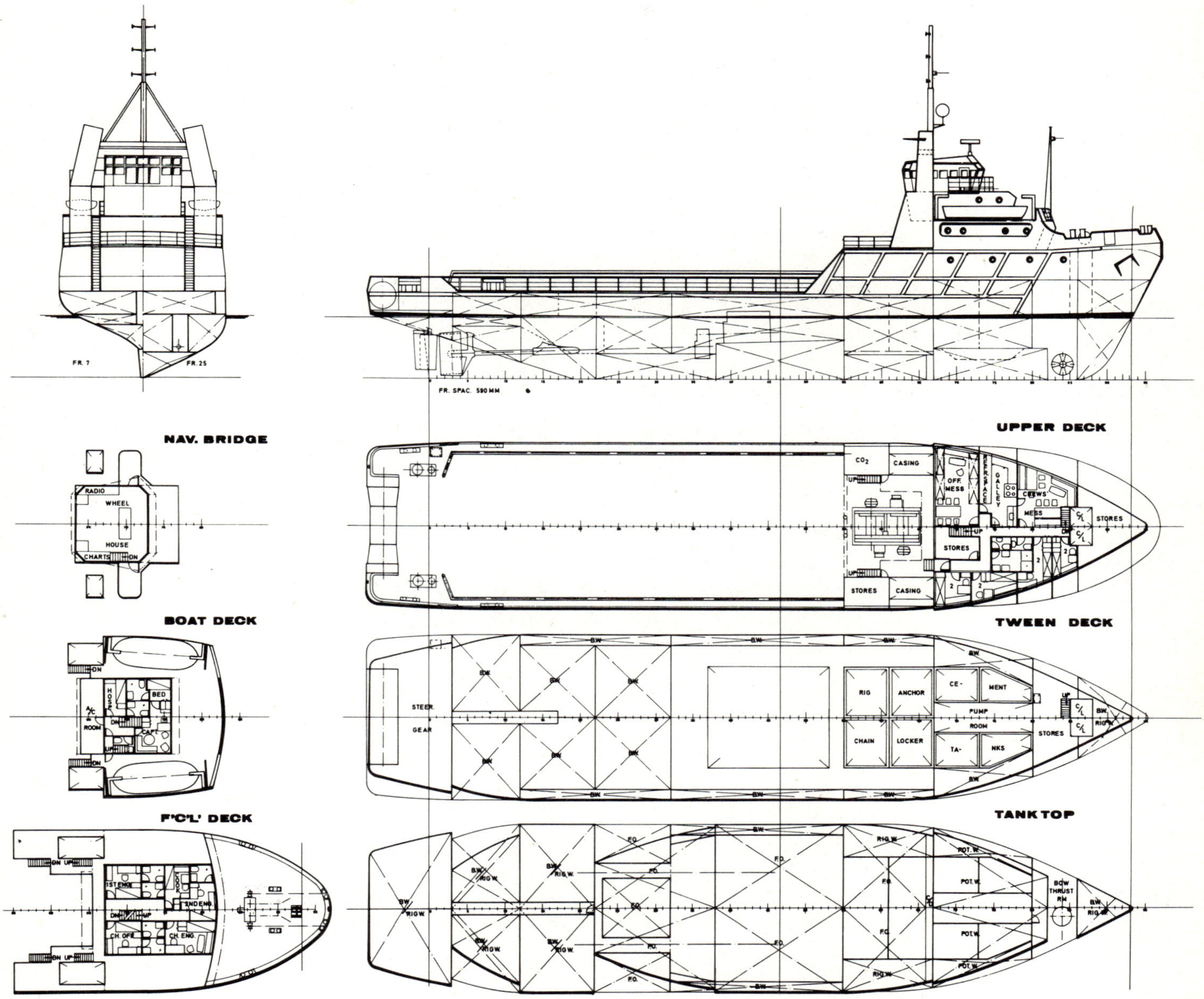

General layout diagrams of Århus Flyedok's 499 g.r.t. supply vessel

NEW DESIGN

Århus Flydedok A/S have a new design of 499/799 g.r.t. open/closed shelterdecker tug/supply vessel with up to 8,000 b.h.p. Deliveries are expected to start in 1975. The design incorporates increased freeboard, better manoeuvrability and greater flexibility for cargo carrying, than previous designs.

Classification: N. V. ✠ 1A1 Tug/Supply Vessel, Ice C. S.F.

Open/Closed Shelterdecker, twin-screw supply vessel, double-skin-hull construction, no fuel tanks exposed to seawater. Intended mainly for rig anchor handling, towing and oil rig supply, suitable for world-wide operations.

Principal particulars

Length, overall	appr. 62·30 m (190 ft)
Length between perpendiculars	appr. 55 m (167 ft 8 in)
Breadth moulded	appr. 13 m (39 ft 7 in)
Depth to upper deck	appr. 6·70 m (20 ft 5 in)
Depth to tween deck	4·90 m (14 ft 11 in)
DWAT open/closed	appr. 940/1,740 tons
Corresponding draught	appr. 4·83/6·05 m (15·8/19·8 ft)
Corresponding freeboard	appr. 2·12/0·90 m (6·95/2·95 ft)

Cargo capacities

Deck cargo	appr. 600/750 tons
Deck area, 36 × 11·2 m (118 × 36·7 ft)	appr. 400 m² (4,300 ft²)
Cement in 4 tanks	appr. 6·5/4·5 t/m² (0·60/0·42 t/ft²)
(Claudius Peters)	appr. 170 m³ (6,000 ft³)
Fuel oil tanks	appr. 700 m³ (24,650 ft³)
Rig water tanks	appr 500 m³ (17,600 ft³)
*Potable water tanks	appr. 165 m³ (5,900 ft³)
*Ballast water tanks	1,060 m³ (38,000 ft³)
Rig chain locker (Dual purpose with*)	appr. 520 m³ (18,350 ft³)

Propulsion

Four or two main engines (M.A.K.)	up to 8,000 b.h.p.
Corresponding speed	appr. 17 knots
Corresponding bollard pull	appr. 86 tons (metric)

Propellers

Twin-screw C/P	
Propellers in nozzle	appr. 2·85 m dia. (9·35 ft)
Becker rudders	

Bow thruster

Brunvoll with CP propellers	up to 550 b.h.p.

Anchor winch
Brattwaag, double drum type — appr. 135 tons

Stern roller
Hitzler, 7 ft dia., shock load — appr. 250 tons

Accommodation
Fourteen single cabins for crew and one hospital
Eight pullman beds/hinged cots for passengers

Cargo handling
All liquid cargo and cement can be handled through hoses on aft deck and from f'c'l', the latter enabling discharge over the bow in bad weather

Cargo pumps
One fire- and deck wash pump Cap: 100 m^3/h at 60 m head (3,520 ft^3/h at 195 ft)
One fresh water pump Cap: 100 m^3/h at 60 m head (3,520 ft^3/h at 195 ft)
One fuel discharge pump Cap: 100 m^3/h at 60 m head (3,520 ft^3/h at 195 ft)
Cement handling by compressed air system Claudius Peters

Deck cargo
A movable deck MacGregor system, can be fitted as an option, facilitating discharge of all cargo from stern area

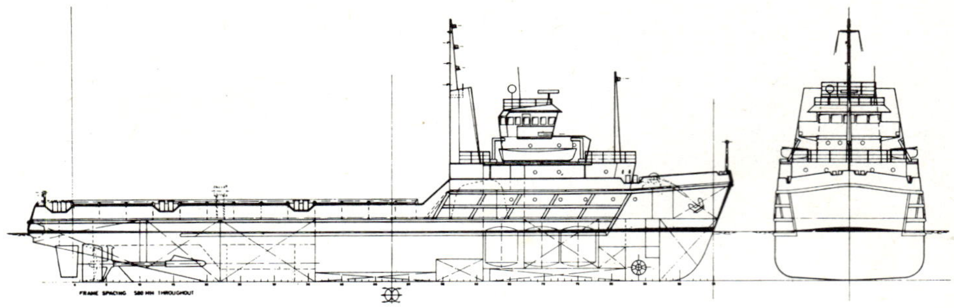

A new design by Webco-Shipping for a 6,000/10,000 b.h.p. 'Combo-tug, supply and anchor handling vessel'

Wheel house
Centralised one-man operated manoeuvring panels and perfect view for all operations. Vessel equipped with all modern electronic aids

Further new designs
Designs have been prepared for more powerful supply vessels among them the 'Combo-Tug-Supply and Anchor Handling Vessel' of about 1,000 dwt, shelterdecker of 6,000/10,000 b.h.p., bollard pull 70/160 tons.

AALBORG VAERFT A/S,
(Aalborg Shipyard, Ltd.)
P.O. Box 661,
DK-9100 Aalborg,
Tel: 08-12 44 33
Telex: 69705 yard dk
New building: Tugs

SVENDBORG SKIBSVAERFT,
(Svendborg Shipyard Ltd.)
DK-5700 Svendborg.
Tel: 09-21 22 21
Telex: 58119 svyard dk
New building: Tugs and supply vessels.

FINLAND

HOLLMING OY,
P.O. Box 14,
26101 Rauma 10.

LAIVATEOLLISUUS OY,
P.O. Box 152,
20101 Turku 10.
(Purchased by Valmet Oy in 1973)

NAVIRE OY,
Parainen.

SAVONLINNA SHIPYARD,
Savonlinna.

VALMET OY,
Pansio Works,
Turku 15.

WARTSILA OY,
Turku Shipyard,
Turku.

FRANCE

ATELIERS ET CHANTIERS C. AUROUX,
Arcachon.

LADY MIRIAM
Built 1971 for International Offshore Services.

Length, overall	55·75 m (182 ft 11 in)
Beam, extreme	12·17 m (39 ft 11 in)
Draught, max.	4·68 m (15 ft 4¼ in)
Deadweight	1,050 tons
Total power	2 shafts, 4 engines British Polar

SOCIÉTÉ PROVENÇALE DES ATELIERS TERRIN,
287, chemin de la Madrague-Ville,
13 Marseille 15e,
BP 819.

GREEN FISH
Built 1970 for Feronia International Shipping

Length, overall	40·77 m (133 ft 9 in)
Beam, extreme	7·80 m (25 ft 7 in)
Draught, max.	3·20 m (10 ft 6 in)
Deadweight	208 tons
Total power	2,100 h.p. two 6 cyl. Polar
Speed	14½ knots

BROWN FISH
Same details as for Green Fish

GERMANY (FEDERAL REPUBLIC OF)

J. G. HITZLER,
Lauenburg.

ANSGARITURM
Built 1969 for Offshore Supply Association
Owner: Hansa, Bremen.

Length, overall	53·68 m (176 ft 1 in)
Beam, extreme	11·25 m (36 ft 11 in)
Draught, max.	3·35 m (11 ft)
Total power	1,900 h.p., two 12 cyl. diesel MWM
Speed	12½ knots
Deadweight	763 tons

HOLSTENTOR (ex NVG 67)
Built 1966

Length, overall	52·87 m (173 ft 5 in)
Beam, extreme	11·25 m (36 ft 11 in)
Draught, max.	3·35 m (11 ft)
Deadweight	806 tons
Total power	1,900 h.p., two 12 cyl. MWM
Speed	12½ knots

WILDHADITURM
Built 1973 for Offshore Supply Association

Length, overall	57·52 m (188 ft 8 in)
Beam, extreme	12·02 m (39 ft 5 in)
Draught, max.	4·50 m (14 ft 9 in)
Deadweight	936 tons
Total power	4,600 h.p., two 16 cyl. MWM
Speed	15½ knots
Bollard pull	65 tons

WASSERTOR
Built 1971

Length, overall	56·51 m (185 ft 5 in)
Beam, extreme	12·04 m (39 ft 6 in)
Draught, max.	4·40 m (14 ft 5¼ in)
Total power	4,600 h.p., two 16 cyl. diesel, M.A.N.
Speed	13½ knots
Deadweight	903 tons

PAGENTURM
Built 1967 for Offshore Supply Association

MUHLENTOR
Built 1972

Length, overall	52·87 m (173 ft 5 in)
Beam, extreme	11·33 m (37 ft 2 in)
Draught, max.	3·44 m (11 ft 3½ in)
Total power	1,900 h.p., two 8 cyl. MWM
Deadweight	745 tons
Speed	12½ knots

MARIENTOR
Built 1971 for Offshore Supply Association

Length, overall	54·50 m (178 ft 9 in)
Beam, extreme	11·33 m (37 ft 2 in)
Draught, max.	3·80 m (12 ft 5½ in)
Deadweight	768 tons
Total power	3,000 h.p., two 12 cyl. MAN
Speed	13½ knots

KLOSTERTOR
Built 1970 for Offshore Supply Association

Length, overall	54·52 m (178 ft 10 in)
Beam, extreme	11·33 m (37 ft 2 in)
Draught, max.	3·79 m (12 ft 5¼ in)
Deadweight	658 tons
Total power	4,000 h.p., two 12 cyl. MAN
Speed	13½ knots

JOHANNISTURM
Built 1973 for Offshore Supply Association

Length, overall	52·87 m (173 ft 5 in)
Beam, extreme	11·33 m (37 ft 2 in)
Draught, max.	3·44 m (11 ft 3½ in)
Deadweight	813 tons
Total power	1,900 h.p., two 8 cyl. MWM
Speed	12½ knots

JAKOBITURM
Built 1972

Length, overall	52·87 m (173 ft 5 in)
Beam, extreme	11·33 m (37 ft 2 in)
Draught, max.	3·44 m (11 ft 3½ in)
Deadweight	846 tons
Total power	1,900 h.p., two 8 cyl. MWM
Speed	12½ knots

IMKENTURM
Built 1967 for Offshore Supply Association

Length, overall	52·87 m (173 ft 5 in)
Beam, extreme	11·28 m (37 ft)
Draught, max.	3·35 m (11 ft)
Total power	1,932 h.p., two 16 cyl. diesel MAN
Speed	12½ knots
Deadweight	792 tons

ROLAND WERF,
Bremen.

BERLINERTOR (ex NVG 2-67)
Built 1965 for Offshore Supply Association (V.T.G.)

Length, overall	52·58 m (172 ft 6 in)
Beam, extreme	11·35 m (37 ft 3 in)
Draught, max.	3·28 m (10 ft 9 in)
Deadweight	738 tons
Total power	1,932 h.p., two 16 cyl. MAN
Speed	12¾ knots

HONG KONG

HONG KONG & WHAMPOA DOCK CO. LTD.,
Kowloon Docks.

KAU LUNG
Built 1972 for Hong Kong Salvage & Towing Co. Ltd.

Length, overall	32·08 m (105 ft 3 in)
Beam, extreme	9·35 m (30 ft 8 in)
Draught, max.	4·07 m (13 ft 4¼ in)
Total power	2,100 h.p., two 6 cyl. Mirrlees-Blackstone
Speed	12 knots
Deadweight	224 tons
Bollard pull	33 tons

ISRAEL

ISRAEL SHIPYARDS LTD.,
P.O. Box 1282,
Haifa.

JAPAN

MITSUI SHIPBUILDING AND ENGINEERING CO. LTD.,
6-4 Tsukiji 5-chome, Chuo-ku,
Tokyo.

ALEGRATE
Supply and anchoring boat.

MEXICO

TRANSPORTACION MARITIMA MEXICANA S.A.,
Insurgentes Sur 421,
Edificio B. 805–806,
Mexico, D.F.

NETHERLANDS

BARKEMEYER STROOBOS M.V.,
Groningen.
(member of Cono ship group)

IBIS ONE
Built 1973 for A/S Lom and Co. (J. Larsen) (supply vessel)

Length, overall	53·15 m (174 ft 4 in)
Beam, extreme	11·5 m (37 ft 8½ in)
Draught, max.	3·77 m (12 ft 4½ in)
Deadweight	725 tons
Speed	13 knots
Total power	5,600 h.p., two Normo KUMB-16
Bollard pull	54 tons

IBIS TWO
Built 1973 for A/S Lom and Co. (J. Larsen)
Details as for *Ibis One*

J. H. BODEWES,
Millingen.

SMITHBANK
Built 1970
Owner: N.V. Nieuwe Rotterdamse Sleepdienst, Netherlands

Length, overall	39·67 m (130 ft 2 in)
Beam, extreme	13·28 m (43 ft 7 in)
Draught, max.	4·91 m (16 ft 1 in)
Deadweight	160 tons
Power	Two 12 cyl. N.V. Werkspoor
Speed	13 knots

INDUSBANK
Same details as for *Smithbank*

BARRACUDA
Built 1970 for Smit-Tak

Length, overall	45 m (147 ft 7 in)
Beam, extreme	11·81 m (38 ft 9 in)
Draught, max.	3·23 m (10 ft 7¼ in)
Deadweight	428 tons
Total power	1,440 h.p., two 6 cyl. diesel
Speed	11 knots

EDDA SALVATOR
Built 1972 for Salvator Marine Services

Length, overall	53·63 m (175 ft 11 in)
Beam, extreme	12·24 m (40 ft 2 in)
Draught, max.	4·25 m (13 ft 11⅜ in)
Deadweight	1,060 tons
Total power	6,160 h.p., two 16 cyl. Hydquist Holm
Speed	13 knots
Range	14,000 miles
Bollard pull	73 tons

SMIT PIONEER
Built 1972 for Smit-International (Antilles) N.V.

Length, overall	43·85 m (143 ft 10 in)
Beam, extreme	10·65 m (34 ft 11 in)
Draught, max.	4·55 m (14 ft 11¼ in)
Deadweight	376 tons
Total power	Two 9 cyl. W. H. Allen
Speed	14 knots

SMIT ENTERPRISE
Built 1972 for Smit-International (Antilles) N.V.
Details as for *Smit Pioneer*

SMIT-LLOYD 12
Built 1967

SMIT-LLOYD 42
Built 1972 for Smit-Lloyd

Length, overall	53·18 m (174 ft 5 in)
Beam, extreme	12·30 m (40 ft 4 in)
Draught, max.	4·43 m (14 ft 6½ in)
Deadweight	851 tons
Total power	Two 8 cyl. D&J Boot De Industrie
Speed	13 knots

SMIT-LLOYD 43
Built 1972
Same details as for *Smit-Lloyd* 42.

SMIT-LLOYD 45
Built 1972
Same details as for *Smit-Lloyd* 42.

SMIT-LLOYD 47
Built 1973, owner: Smit-Lloyd

Length, overall	54·81 m (179 ft 10 in)
Beam, extreme	12·29 m (40 ft 4 in)
Draught, max.	4·44 m (14 ft 6·75 in)

Deadweight 860 tons
Power Two 8 cyl. D&J Boot De Industrie
Speed 13 knots

SMIT-LLOYD 105
Built 1973 for Smit-Lloyd

Length, overall	63·90 m (209 ft 7 in)
Beam, extreme	13·31 m (43 ft 8 in)
Draught, max.	4·75 m (15 ft 7 in)
Deadweight	1,115 tons
GRT	1,295
Speed, max.	15 knots
Bollard pull	90,909 kg (200,000 lb)
Total power	7,500 b.h.p., two Stork-Werkspoor TM 410 diesels, at 550 rpm

DE MERWEDE,
Hardinxveld.

RODE ZEE
Built 1968
Owner: Smit International Ocean Towage & Salvage Co., Netherlands

Length, overall	68·50 m (224 ft 9 in)
Beam, extreme	12·60 m (41 ft 4 in)
Draught, max.	5·70 m (18 ft 8½ in)
Deadweight	950 tons
Total power	11,000 i.h.p., two Stork-Werkspoor TM 410 geared to single shaft
Speed	17 knots

NOORDZEE
Built 1970
Same details as for *Rode Zee*

POOLZEE
Built 1971
Same details as for *Rode Zee*

DE WAAL.

SMIT-LLOYD 104
Built 1973
Owner: Smit-Lloyd

Length, overall	63·93 m (209 ft 7 in)
Beam, extreme	13 m (43 ft 8 in)
Draught, max.	4·75m (15 ft 7in)
Deadweight	1,100 tons
Total power	7,500 h.p., two 6 cyl. Stork-Werkspoor
Speed	15 knots
Bollard pull	90 tons

SMIT-LLOYD 41
Built 1971
Owner: Smit-Lloyd

Length, overall	54·81 m (179ft 10 in)
Beam, extreme	12·29 m (40 ft 4 in)
Draught, max.	4·71 m (15 ft 5¼ in)
Deadweight	860 tons
Total power	4,000 h.p., two 8 cyl. D&J Boot 'De Industrie'
Speed	13½ knots
Bollard pull	40,909 kg (90,000 lb)

SMIT-LLOYD 44
Built 1972
Owner: Smit-Lloyd
Details as for *S-L 41*

SMIT-LLOYD 46
Built 1972
Owner: Smit-Lloyd
Details as for *S-L 41*

SMIT-LLOYD 48
Built 1973
Owner: Smit-Lloyd
Details as for *S-L 41*

JONKER

OIL EXPLORER and OIL PROSPECTOR
Built 1972 for Ocean Inchcape Ltd., UK

Length, overall	56·41 m (185 ft 1 in)
Beam, extreme	12·70 m (41 ft 8 in)
Draught, max.	4·25 m (13 ft 11½ in)
Deadweight	1,036 tons
Total power	4,130 h.p., two 12 cyl. W. H. Allen
Bollard pull	40 tons
Speed	13 knots

Designed by Ysselwerf

VAN DE GIESSEN

SMIT-LLOYD 103
Built 1970 for Smit-Lloyd

Length, overall	61·82 m (202 ft 10 in)
Beam, extreme	12·06 m (39 ft 7 in)
Draught, max.	4·71 m (15 ft 5½ in)
Deadweight	669 tons
Power	Two 8 cyl. De Industrie
Speed	14 knots

OIL PRODUCER and OIL SUPPLIER
Built 1972 for Ocean Inchcape Ltd., UK
Designed by Ysselwerf

Length, overall	56·51 m (185 ft 5 in)
Beam, extreme	12·75 m (41 ft 10 in)
Draught, max.	4·27 m (14 ft)
Deadweight	1,075 tons
Total power	4,130 h.p., two 12 cyl. W. H. Allen
Speed	14 knots

VEROLME SCHEEPSWERF,
Heusden.

LADY MARIANN
Built 1970
Owner: International Offshore Services

Length, overall	48·59 m (159 ft 5 in)
Beam, extreme	10·06 m (33 ft)
Draught, max.	3·35 m (11 ft)
Deadweight	750 tons
Total power	2,800 h.p., two 8 cyl. British Polar
Speed	11½ knots

YSSELWERF N.V.,
Rotterdam.

CHANNEL SHORE
Built 1972 for Offshore Marine Ltd.

Length, overall	55·66 m (182 ft 7 in)
Beam, extreme	11·43 m (37 ft 6 in)
Draught, max.	4·28 m (14 ft ½ in)
Deadweight	837 tons
Total power	4,000 h.p., two Lister Blackstone
Speed	14·3 knots
Bollard pull	57 tons
Endurance	35 days at full speed, 42 days at 85% max. power

DOGGER SHORE
Due 1974 for Offshore Marine Ltd

Length, overall	59·45 m (195 ft)
Beam, extreme	12·80 m (42 ft)
Draught, max.	4·45 m (14 ft 7 in)
Deadweight	1,057 tons
Total power	5,600 h.p.
Speed	14½ knots
Bollard pull	66 tons

OCEAN SHORE
Built 1972 for Offshore Marine Ltd
Details as for *Channel Shore*

VIKING SHORE
Built 1973 for Offshore Marine Ltd
Details as for *Dogger Shore*

OIL DISCOVERER
Built 1973, owner Ocean Inchcape Ltd (Oil, United Kingdom)

Length, overall	59·50 m (195 ft)
Beam, extreme	13 m (42 ft 7 in)
Draught, max.	4·35 m (14 ft 3¼ in)
Deadweight	1,090 tons
Total power	5,600 b.h.p., two 16 cyl. PVBCS12-F Allen (Amalgamated Power Engineering Ltd, (APE)), 750 rpm
Bollard pull	73 tons
Speed	15¼ knots
Bow thruster	Gill jet type, thrust 3·7 tons

Cargo carrying:

potable water	200 tons
drilling water	385 tons
fuel oil	385 tons
cement or bulk cargo	173 m³ (6,070 ft³)
Deck cargo	550 tons
Refrigerated cargo	8·5 m³ (300 ft³)
Accommodation	12 passengers

Anchor handling and towing winch: hydraulically-powered, waterfall type, 143 tons static pull

OIL VENTURER
Built 1973 for Ocean Inchcape Ltd

Length, overall	59·50 m (195 ft)
Beam, extreme	13 m (42 ft 7 in)
Draught, max.	4·35 m (14 ft 3¼ in)
Deadweight	1,090 tons
Total power	5,600 h.p., two 16 cyl. Allen PVBCS12-F, 750 rpm
Speed	15¼ knots
Bollard pull	73 tons

NEW ZEALAND

WHANGAREI ENGINEERING & CONSTRUCTION CO. LTD.,
Whangarei,
North Island.

NORWAY

P. HOIVOLDS MEK VERSTED A/S,
Kristiansand.

AILOS
Built 1972 for Nordsk Bjergningskompagni A/S, Bergen.

Length, overall	45·45 m (149 ft 1 in)
Beam, extreme	11·03 m (36 ft 2 in)
Draught, max.	5·80 m (19 ft ½ in)
Total power	Two 16 cyl. Normo Gruppen
Speed	16 knots
Bollard pull	70 tons

BAUS
Built 1971 for Bukser

Length, overall	36·28 m (119 ft)
Beam, extreme	10·04 m (32 ft 11 in)
Draught, max.	5·56 m (18 ft 3 in)
Deadweight	416 tons
Total power	One 8 cyl. Atlas-Mak-Masch
Speed	13½ knots

BJONN
Built 1969 for Bukser
Length, overall 25·66 m (84 ft 2 in)
Beam, extreme 7·52 m (24 ft 8 in)
Draught, max. 34·68 m (11 ft 4½ in)
Total power 1,200 h.p., one 6 cyl. diesel

HERKULES
Built 1960

AUKRA BRUK A/S

BATSERVICE VERFT A/S,
Arendal.
ELLERNTOR
Built 1972 for Offshore Supply Association
Length, overall 56·43 m (185 ft 1 in)
Beam, extreme 12·22 m (40 ft 1 in)
Draught, max. 4·56 m (14 ft 11½ in)
Deadweight 935 tons
Total power Two 12 cyl. MWM
Speed 13½ knots

MICHAELISTURM
Built 1973 for Offshore Supply Association
Length, overall 56·4 m (185 ft)
Beam, extreme 12·22 m (40 ft 1 in)
Draught, max. 4·56 m (14 ft 11½ in)
Deadweight 950 tons
Total power Two 12 cyl. MWM
Speed 14½ knots

BOLSØNES VERFT,
Molde.
RIG MASTER
Built 1973, owner Sandøy Supply A/S & Co
Length, overall 57·61 m (189 ft)
Beam, extreme 10·97 m (36 ft ½ in)
Draught, max. 4·52 m (14 ft 10 in)
Deadweight 820 tons
Total power 4,600 h.p., two 12 cyl. Polar, Nydquist & Holm
Speed 15 knots

ULSTEIN MEK VERKSTED A/S,

VOLDNES A/S,
Fosnavag.
RIG PILOT
Built 1972, owner Sandøy Supply A/S & Co
Length, overall 57·27 m (187 ft 9 in)
Beam, extreme 9·81 m (32 ft 2 in)
Total power 4,000 h.p., two Atlas-Mak-Masch

SINGAPORE

JURONG SHIPYARD LTD.
PSA VALIANT
Built 1973 for Port of Singapore Authority
Length, overall 31·95 m (104 ft 9½ in)
Propulsion Voith Schneider
Speed 12 knots
Cost S $2·9 m (1973)
PSA VICTORY
Built 1973
PSA VIGILANT
Built 1973

VOSPER THORNYCROFT PTE. LTD.
BORNEO GOLD
Built 1970 for Borneo Straits Offshore
Length, overall 49·31 m (161 ft 9 in)
Beam, extreme 11·76 m (38 ft 7 in)
Draught, max. 42·24 m (13 ft 10¼ in)
Deadweight 747 tons
Total power 3,000 h.p., two 12 cyl. Klochner-Humbolt-Deutz

DRAGON
Built 1972
Length, overall 56·69 m (186 ft)
Beam, extreme 13·11 m (43 ft)
Bollard pull 60 tons
Power 4,750 h.p.

SWEDEN

ÅSL-VERKEN,
Åmål,
POSEIDON
Built 1970 for Neptune, Bergnings & Dykeri
Length, overall 44·86 m (147 ft 2 in)
Beam, extreme 10·85 m (35 ft 7 in)
Draught, max. 5·29 m (17 ft 4½ in)
Total power Two 8 cyl. Nydquist & Holm A/B
Speed 15 knots

UNITED KINGDOM

AILSA SHIPBUILDING
Two new tugs for Cory Ship Towage 1974. 700 tons displacement.

APPLEDORE SHIPBUILDERS LTD.
Appledore,
Devon.
SEADOG
Built 1973
Owner: Wimpey Marine Limited
Length, overall 59·15 m (194 ft)
Beam, extreme 12·50 m (41 ft)
Draught, max. 4·88 m (16 ft)
Deadweight 1,000 tons
Total power 2,430 h.p., two B & W Alpha 18V23HU
Bollard pull 63 tons
Propellers Two B & W controllable pitch
Speed 13½ knots
Auxiliary generators Three Dorman 60TZ at 250 KVA each
One Dorman 60AZ at 56 KVA
Fuel capacity 260 tons
Cargo capacity 290 tons
Drilling water 310 tons
Potable water 80 tons
Tanks Four Haliburton, 141·5 m³ (5,000 ft³)

Classed: Lloyds 100 A1, Ice class II, ums and DTI class VII

Engine: Twin B & W Alpha of Denmark Type 18V23HU, 2,810 h.p. at 825 rpm each

SEA LION
Built 1974 for Wimpey Marine Ltd.
SEA WOLF
Built 1974 for Wimpey Marine Ltd.
FASTNET SHORE
Built 1974 for Offshore Marine Ltd.
Owner: Offshore Marine Ltd.
Length, overall 59·45 m (195 ft)
Beam, extreme 12·80 m (42 ft)
Draught, max. 4·45 m (14 ft 7 in)
Deadweight 1,057 tons
Total power 5,600 h.p., two 16 cyl. W. H. Allen
Speed 14¾ knots
Bollard pull 67·5 tons
Endurance 31 days at full speed
37 days at 85% power
Auxiliary generators Two 320 kw, one 120 kw
Propellers Two fixed pitch Kort nozzle 2·29 m (7 ft 6 in) diameter
Bow thruster Gill jet 1·58 m (40 in) dia. Static thrust 3,636 kg (8,000 lb)
Deck area 337 m² (3,630 ft²)
Fuel oil capacity 212,000 gallons US
Drill water 172,500 gallons US
Potable water 40,200 gallons US
Accommodation 23 persons, including crew
Stabilisers Passive type
Cargo pumps Three (21,600 gallons US per hour) at 61 m (200 ft) head

LUNDY SHORE
Built 1974 for Offshore Marine Ltd.
Details as for *Fastnet Shore*

J. BOLSON & SON LTD.,
Poole Shipyard,
West Quay Road,
Poole,
Dorset.
LADY BRIGID
Built 1966 for P&O Offshore Services
Length, overall 48·17 m (158 ft)
Beam, extreme 11·53 m (37 ft 10 in)
Draught, max. 3·67 m (12 ft ½ in)
Deadweight 675 tons
Total power 1,600 h.p., two 8 cyl. Blackstone

BAY SHORE
Built 1971 for Offshore Marine Ltd.
Owner: Offshore Marine Ltd.
Length, overall 56·17 m (184 ft 3 in)
Beam, extreme 11·43 m (37 ft 6 in)
Draught, max. 4·17 m (13 ft 8 in)
Deadweight 845 tons
Total power 5,280 h.p., two English Electric 12 CVS
Propellers Fixed pitch Kort
Bow thruster Gill jet

BROOK MARINE LTD.,
Lowestoft,
Suffolk.

COCHRANE & SONS LTD.,
(Drypool Group)
Selby,
Yorkshire.
SUFFOLK SHORE
Built 1967
Total power 1,600 h.p., two 8 cyl. Mirrlees Blackstone

POLAR SHORE
Built 1971 for Offshore Marine Ltd.
Length, overall 56·28 m (184 ft 7 in)
Beam, extreme 11·76 m (38 ft 7 in)
Draught, max. 4·70 m (15 ft 5 in)
Deadweight 772 tons
Total power 5,280 h.p., two English Electric 12 CVS

Speed	14 knots
Bollard pull	63 tons
Endurance	25 days at full speed, 29 days at 85% power

MONARCH SERVICE
Built 1972 for Zapata Offshore Services Ltd.

Length, overall	56·25 m (184 ft 6 in)
Beam, extreme	11·73 m (39 ft 4 in)
Draught, max.	4·27 m (14 ft 10½ in)
Deadweight	845 tons
Total power	7,040 h.p., two 16 cyl. English Electric 16 CSVM
Speed	16¼ knots
Bollard pull	70 tons

KENT SHORE
Built 1967

Length, overall	52·36 m (171 ft 9 in)
Beam, extreme	11·69 m (38 ft 4 in)
Draught, max.	4·19 m (13 ft 8¾ in)
Total power	1,600 h.p., two 8 cyl. diesel Blackstone
Speed	12 knots
Deadweight	970 tons

ENGLISHMAN
Built 1965

Length, overall	44·72 m (146 ft 8 in)
Beam, extreme	10·59 m (34 ft 9 in)

WELSHMAN
IRISHMAN
SCOTSMAN

SEAFORTH PRINCE
Built 1973 for Seaforth Maritime Ltd.

Length, overall	54·88 m (180 ft)
Beam, extreme	11·80 m (39 ft)
Draught, max.	4·50 m (15 ft)
Total power	5,000 h.p., two Mirrlees Blackstone EZSL16
Speed	14 knots
Bollard pull	60 tons

SEAFORTH CHALLENGER
Built 1973 for Seaforth Maritime Ltd.

Length, b.p.	51·75 m (169 ft 9 in)
Beam , moulded	11·8 m (38 ft 8½ in)
Draught, mean	5·2 m (17 ft)
Total power	5,000 b.h.p., two Mirrless Blackstone EZSL16
Speed	14 knots
Bollard pull	60 tons

SEAFORTH HERO
Built 1973 for Seaforth Maritime Ltd.
Classed to Lloyds + 100A1 and D.T.I. Class VII for world wide operations.

Length, overall	55 m (180 ft 6 in)
Length, b.p.	49 m (161 ft)
Beam moulded	11·80 m (38 ft 8½ in)
Depth moulded	5·20 m (17 ft 1 in)
Loaded draught	4·50 m (14 ft 9½ in)
Deadweight	1,000 tons
Gross tonnage	732 tons
Nett registered tonnage	282·47 tons
Total power	5,000 h.p., two Mirrlees Blackstone EZSL16
Speed	14 knots
Bollard pull	60 tons

Main Propulsion
2 × Mirrlees Blackstone twin bank diesel engines type EZSL16z developing 2,500 b.h.p. each at 1,000 rpm

Connected by Holset Witchita Pneumatic clutch through a Brevo direct reduction gearbox to Kamewa variable pitch propellers which rotate at 242 rpm in Kort Nozzles controlled from the fore and aft bridge consoles

Polar Shore

Scotsman

Electric power
Supplied by three alternators each of 200 kw output driven by Dorman diesel engines type 6QT voltage 440/3Ph with 220 volt single phase transformers

Bow thrust unit
J. S. White Gill Jet Multi directional thruster having a thrust of 3·8 long tons in any direction through 360° to 30° intervals controlled from the forward and aft bridge consoles

Pumps
Cargo oil pump: Hamworthy type pumps 100 ton per hour against a 200 ft head

Cargo fresh water: Hamworthy type pumps 100 ton per hour against a 200 ft head

General service pumps: Hamworthy type pumps 100 ton per hour against a 200 ft head

Bilge ballast pump: Hamworthy type pumps 45 ton per hour against 120 ft head

Seaforth Chieftain

Heavy duty towing/Anchor handling winch

Hydraulically operated from an elevated control console cabin. The winch has two drums each fitted with pneumatic and manually operated brakes. The anchor handling drum has a capacity for 610 m (2,000 ft) of 52 m/m wire and winch can pull 100 tons at 5·8 m (19 ft) per min. The vessel will also be fitted with a separate hydraulic powered reel for storage of a spare tow wire or on which the tow wire can be carried leaving both winch drums free for anchor handling work. The anchor winch will have warping ends with a pull of 20 tons at 30·5 m (100 ft) per min.

There is a removable warping guide fitted to the towing drum for automatic spooling of the tow wire.

Cement tanks

Three Halliburton horizontal tanks each holding 1,230 ft³ of powdered cement having an output of 35/40 tons per hour of cement in ideal conditions. Air is supplied to the tanks from two Ingersoll Rand Air Compressors having a total output of air of 529 ft³ F.A.D. at 40 p.s.i.

General

Vessel's speed	14 knots
Bollard pull	60 long tons (66 short tons)
Bollard pull with bow thrust	63·8 long tons (71·5 short tons)

Passenger accommodation for 12 persons

SEAFORTH CHIEFTAIN
Built 1973 for Seaforth Maritime Ltd.
Details as for *S. Prince* and *S. Hero*

RICHARD DUNSTON (HESSLE) LIMITED,
Haven Shipyard,
Hessle,
Yorkshire.
Tel: 0482 642261
Telex: 527142

POINT CARROLL
Built 1973 for Smit and Cory

Length, overall	38·71 m (127 ft)
Beam, extreme	9·15 m (30 ft 5 in)
Draught, max.	4·50 m (12 ft 8¾ in)
GRT	336
Total power	2,640 h.p., one 12 cyl. English Electric 12CSVM
Speed	14 knots
Bollard pull	37 tons

POINT SPENCER
Same details as for *Point Carroll*

OIL MANAGER
Built 1973 for Ocean Inchcape Ltd.
First of three twin-screw-rudder mooring tugs for Nigerian waters.

RICHARDS (SHIPBUILDERS) LTD.,
Crown Works,
Lowestoft,
Suffolk.

CROMARTY SHORE
Built 1974 for Offshore Marine

Length, overall	59·65 m (195 ft 8½ in)
Beam, extreme	12·80 m (42 ft)
Draught, max.	4·45 m (14 ft 7¼ in)
Deadweight	1,057 tons
Total power	5,600 h.p., two 16 cyl. Allen
Bollard pull	67·5 tons
Speed	14¾ knots
Endurance	31 days at full speed, 37 days at 85% power

FORTIES SHORE
Same details as for *Cromarty Shore*

ORKNEY SHORE
Built 1973 for Offshore Marine

Length, overall	53·58 m (175 ft 9 in)
Beam, extreme	11·40 m (37 ft 6 in)
Draught, max.	4·35 m (14 ft 3 in)
Deadweight	806 tons
Total power	4,000 h.p., two Lister EWSL16M
Speed	14·65 knots
Bollard pull	64 tons

SHETLAND SHORE
Same detailas as for *Orkney Shore*

HALL RUSSELL & CO. LTD.,
P.O. Box 36, York Place,
Aberdeen AB9 8BT, Scotland.

LADY SYLVIA
Built 1973 for International Offshore Services
One 58·23 m (191 ft) tug/supply vessel building, 1974, one 5,000 h.p. 900 g.r.t. building 1974 for International Offshore Services and one 5,750 h.p. vessel for Tidewater Marine Services

CHARLES HILL & SONS LTD.,
Bristol.

MAJESTIC SERVICE
Built 1972 for Zapata Offshore Services Ltd., London

Length, overall	56·25 m (184 ft 6 in)
Draught, max.	45·35 m (14 ft 10½ in)
Deadweight	831 tons
Power	Two 16 cyl. English Electric diesels
Speed	16½ knots

(also building 1973 for Seaforth Maritime, sub-contract from Drypool Group)

NEW BUILDING
1974, 800 g.r.t., 6,160 h.p. sub-contract from Cochrane

Min. of Defence, ocean tug, 900 g.r.t.

CHARLES D. HOLMES & CO. LTD.,
Hull,
Yorkshire.

SCOTT LITHGOW LTD.,
Cartsburn Dockyard,
Greenock, PA 15 5HL
Two tug/supply vessels building 1974 for Seaforth Maritime

ROBB CALEDON SHIPBUILDERS LTD.,
Victoria Shipyards,
Leith,
Edinburgh 6,
Scotland.
Also:
21, Grosvenor Place,
London, SWIX 7JE
Tel: 01-235 5131
Telex: 22797

LLOYDSMAN
Built 1971 for United Towing Ltd., Hull

Length, overall	80·69 m (264 ft 8 in)
Beam, extreme	14·66 m (48 ft 1 in)
Draught, max.	7·41 m (24 ft 3½ in)
Deadweight	1,957 tons
Bollard pull	135 tons
Speed	18/19 knots
Power	Two 10 cyl. Pielstick-Crossley geared to single shaft
Class	Lloyds 100 A1 ice class 3

New Building for South African Marine Corps

Outline specifications

Length, overall	approximately: 95 m (311 ft 7¼ in)
Length, b.p.	85 m (278 ft 9½ in)
Beam, moulded	15·20 m (49 ft 10¾ in)
Depth, moulded	8·60 m (28 ft 2½ in)
Draught, service mean	7·20 m (23 ft 7¼ in)
Deadweight, total	2,020 tons
Bollard pull	150 tons
Service speed	20 knots
Complement	42 persons, plus hospital for 5
Propulsion machinery	19,600 b.h.p.

The ocean salvage tug *Lloydsman*, claimed to be the biggest tug built, or owned, in the UK. This vessel can pull a fully loaded 300,000-ton tanker at a speed of over 7 knots

This vessel will be the world's largest and most powerful tug.

A comprehensive outfit of salvage gear will be carried; the supply will include portable pumps, submersible pumps, sand/mud pumps, portable generators and compressors, underwater tools, diving and associated equipment, salvage anchors, etc.

The fire fighting installation will comprise four monitors capable of operating independently with foam or water. Two salvage/fire pumps each will operate at 600 tons/hour; in addition there are to be ten connections for fire hoses. The pumps will be able to be connected in series to obtain double water pressure. The monitors will be arranged to produce a maximum arc—the throw being some 60 m (197 ft) for foam and 75 m (246 ft) for water with all four monitors working on water, simultaneously. Foam tanks are arranged to carry 40 tonnes of compound.

Approximately 500 tonnes of the tug's ballast tank capacity will be usable for oil-dispersant chemicals. The pump is capable of delivering 8 tonnes/hour.

A spray boom about 5 m (16½ ft) long will be fitted each side of the funnel with a connection to the discharge line. Surface breaker boards will also be towed from the booms to facilitate the mixing of chemicals with the surface oil.

For deep-sea towage, two hydraulically-operated towing winches would be fitted on the upper deck at the break of the bridge super-structure. Details of the winches, both of which are to be of the friction type with hydromotors, planetary gears and friction wheel, are as follows:

Main winch: static load 320 tons pull 150 tons at 9·3 m/min (30 ft/min)

Secondary winch: static load 180 tons pull 80 tons each 8·4 m/min (27·56 ft/min)

Remote controls for all 'winching' operations will be positioned in a control room overlooking the after deck.

Two horizontal sheaves will be arranged immediately abaft each towing winch to ensure fair leads of wires to the winding drums. At the stern two sets of rising fairleads with a horizontal roller will be arranged as an integral unit and will be capable of being lowered below the line of the bulwark by means of hydraulic actuators. Electro-pneumatic release gear is to be provided for this work.

An artist's impression of the tug now being built for South African Marine Corporation.

For handling heavy lifts, a 30 ton 'swinging derrick' crane is to be arranged on the centre-line abaft the funnel—the clear lifting height is 25 m (82 ft) above the summer loadline.

Three control rooms will be provided in the vessel; a conventional wheelhouse/chart-room housing all the navigational, manoeuvring and engine controls; an after house overlooking the main deck for towing winch and manoeuvring operations; and a compartment within the machinery space for control of main engines, auxiliaries, compressors, pumps, etc.

Main propelling machinery for the new tug consists of two Mirrlees KVMR 'Major' 16 cylinder four stroke single acting marine diesel engines, each giving 9,800 b.h.p. at 600 rpm. One unit will be reversible; the other uni-directional. Both machines will drive a controllable pitch propeller, operating within a nozzle, at 135 rpm through a 'twin-input/single-output' reduction gearbox.

Electrical power requirements will be met by two M.A.K. diesel driven alternators each producing 740 kw, 440 volts, 3 phase, 60 cycles/second alternating current. These machines are each designed to operate at 720 rpm. Two 'harbour' sets are also to be installed. M.A.K. engines operating at 600, rpm will each drive an alternator delivering 308 kw at 440 volts, 3 phase, 60 cycles/seconds. A further generating set will be provided, designed to provide power for stricken vessels comprising a gas turbine driven alternator of 1,200 kw.

An 800 h.p. bow thruster unit is fitted forward; the controllable pitch propeller is electrically driven and will give an effective thrust of 9·5 kp.

RYTON MARINE LTD.,
Davy Bank,
Wallsend,
Northumberland.

J. SAMUEL WHITE,
Cowes,
Isle of Wight.
SOUTH SHORE
Built 1965, Offshore Marine Ltd.

UNITED STATES OF AMERICA

AMERICAN BROWN BOVERI ELECTRIC CORP.,
Camden.
CHEROKEE
Built 1927

Length, overall	35·59 m (116 ft 9 in)
Beam, extreme	7·37 m (24 ft 2 in)
Draught, max.	2·88 m (9 ft 5⅜ in)
Power	3,200 h.p., two 8 cyl. diesels

AMERICAN MARINE CORPORATION,
New Orleans,
Louisiana.
STATE VICTORY
Built 1970 for State Boat Corporation

Length, overall	56·71 m (186 ft)
Beam, extreme	12·24 m (40 ft 2 in)
Draught, max.	3·95 m (12 ft 11⅜ in)
Total power	2,250 h.p., two 16 cyl. Caterpillar
Speed	15 knots

ARTHUR LEVY BOAT SERVICE INC.,
Morgan City,
Louisiana.

AVONDALE SHIPYARDS INC.,
Bayou Black,
New Orleans,
Louisiana.
BEN CANDIES
Built 1966 for Otto Candies

Length, overall	38·11 m (125 ft)
Beam, extreme	10·26 m (33 ft 8 in)
Draught, max.	4·54 m (14 ft 10½ in)
Power	Two 16 cyl. General Motors

BURTON SHIPYARDS INC.,
now
BURTON CONSTRUCTION & S.B. CORPORATION
Port Arthur,
Texas.
SIGNAL SERVICE
Built 1965
BUILDING 1973/74 for Aqua Marine Inc., Houston, Texas, four tug/supply vessels

Length	54·84 m (180 ft)
Beam	12·19 m (40 ft)
Draught	4·57 m (15 ft)
Speed, max.	15 knots
Bollard pull	60 tons

Three 16 cyl. GM EMD engines, total h.p. = 5,400

DEARBORN 47
Built for Dearborn Marine Services Inc.,

Length, overall	50·30 m (165 ft)
Beam, extreme	11·69 m (38 ft 4 in)
Draught, max.	3·36 m (11 ft ⅜ in)
Deadweight	600 tons
Total power	2,250 h.p., two 16 cyl. Caterpillar

DEARBORN 46
Built 1971 for Dearborn Marine Services Inc.

Length, overall	50·30 m (165 ft)
Beam, extreme	11·64 m (38 ft 2 in)
Draught, max.	3·65 m (11 ft ¾ in)
Deadweight	600 tons
Total power	2,250 h.p., two 16 cyl. Caterpillar

IDA CANDIES
Built 1973 for Otto Candies

Length, overall	56·40 m (185 ft)
Beam, extreme	12·30 m (40 ft 4 in)

Draught, max.	4·34 m (14 ft 3 in)
Deadweight	600 tons
Total power	3,000 h.p., two 16 cyl. General Motors

NORDIC SERVICE
Built 1973 for Zapata Marine Service Ltd., Houston

Length, overall	56·40 m (185 ft)
Beam, extreme	12·25 m (40 ft 2 in)
Draught, max.	3·66 m (12 ft)
Total power	5,400 h.p., three 16 cyl. General Motors EMB GM16567BC
Propellers	Three Avondale 4-blade (stainless steel) Kort nozzle type
Bow thruster	Brumvoll 4,545 kg (10,000 lb)
Cargo deck capacity	500 tons
Bollard pull	65 tons
Speed	14 knots

THOR SERVICE
Built 1973 for Zapata Marine Service Ltd., Houston
Details as for *Nordic Service*

TITAN SERVICE
Built 1965

Length, overall	48·78 m (160 ft)
Beam, extreme	12·25 m (40 ft 2 in)
Draught, max.	3·26 m (10 ft 8⅜ in)
Total power	900 h.p., two 16 cyl. General Motors

VIKING SERVICE
Built 1973 for Zapata Marine Service Ltd.,

Length, overall	56·40 m (185 ft)
Beam, extreme	12·20 m (40 ft)
Draught, max.	3·92 m (12 ft 10⅜ in)
Total power	1,640 h.p., two 16 cyl. General Motors

DEARBORN MARINE SERVICE CORPORATION,
P.O. Box 1031,
Freeport,
Texas 77541.

EQUITABLE EQUIPMENT CO. INC.,
P.O. Box 8001,
New Orleans,
Louisiana, 70182.
(wholly owned subsidiary of Trinity Industries, Inc.)

THE BERTILE CANDIES

Length, overall	53·35 m (175 ft)
Beam, moulded	11·59 m (38 ft)
Depth, moulded amidships	4·27 m (14 ft)
Draught, DLWL	3·20 m (10 ft 6 in)

The vessel is designed and built to American Bureau of Shipping Class '✠ A-1 ✠ AMS', and is in accordance with the current requirements of the U.S. Coast Guard.

The two propulsion engines in the vessel are General Motors EMD-16-645E2 engines having a total continuous rating of 3,900 b.h.p. Reduction gears are Western Model TSR-21 marine gears. Air-conditioned quarters and electric galley to accommodate 17 men are provided. The vessel is fitted with a bow thruster unit driven by a GM-6-71 engine. The vessel is equipped with a pneumatic bulk mud system.

HALTER MARINE FABRICATORS,
Moss Point,
Mississippi.

TRUE GRIT
Built 1971 for Jackson Marine

Length, overall	55·45 m (182 ft)
Beam, extreme	12·24 m (40 ft 2 in)
Draught, max.	4·49 m (14 ft 8¾ in)
Deadweight	745 tons
Total power	6,000 h.p., two 16 cyl. White Industrial Power

Launching of the *Bertile Candies* at the Madisonville yard of the Equitable Equipment Company

New building

In December 1973, Equitable Equipment Company Inc., were awarded a contract of approximately $10 (US) million by Nolty J. Theriot, Inc. international offshore vessel operator, to build four twin-screw, 8,500 h.p., ocean-going tugs, with fixed propellers and stainless steel Kort propulsion nozzles. They will be built to ABS ✠ A-1, Unlimited Ocean Towing Service, Ice Class C-1, AMS. The estimated total cost, including owner-furnished equipment of the tugs, exceeds $14 (US) million. The contract for these four tugs and the three identical tugs contracted in January 1973 brings to a total of seven such vessels to be built by Equitable for Nolty J. Theriot, Inc. with an estimated total value of approximately $25 (US) million.

These vessels, upon completion, will be put into service with other Theriot tugs presently engaged in the field of oil exploration and construction worldwide.

RED BEARD
Built 1973 for Euro-Pirates International, Inc.

Length, overall	53·66 m (176 ft)
Beam, extreme	11·59 m (38 ft)
Draught, max.	4·27 m (14 ft)
Total power	2,400 h.p., two Stork FBHD
Speed	12 knots

BLACK BEARD
For details see *Red Beard*

KHAIR ED DIN
For details see *Red Beard*

M.T. BETTY G
Built 1970 for De Felice Marine Contractors

HALTER MARINE SERVICES INC.,
P.O. Box 29266,
New Orleans,
Louisiana 70129.
Tel: (504) 254 1222
Telex: 584-200

ESTEFF S. DE FELICE
Built 1971 for De Felice Marine Contractors

Length, overall	42·67 m (140 ft)
Beam, extreme	10·36 m (34 ft)
Draught, max.	4·27 m (14 ft)
GRT	Less than 200
Total power	Two EMD 16-645-E2 diesels
Propellers	Two 274 cm (108 in) four-blade stainless steel
Crew	8
Passengers	8

J. RAY McDERMOTT SHIPYARD,
Morgan City,
Louisiana.
Building eight 10,000 h.p. tug/supply vessels for Offshore Logistics, Inc. 1973. 600 h.p. bow thrusters

MANGONE SHIPBUILDING CO.,
P.O. Box 5446
819, South 80th St.,
Houston,
Texas, 77012.
Tel: (713) 926 9451

OFFSHORE BEAUMONT
Built 1970 for Offshore Co. Ltd., Panama

Length, overall	50·30 m (165 ft)
Beam, extreme	11·03 m (36 ft 2 in)
Draught, max.	3·38 m (11 ft 1 in)
Total power	2,120 h.p., two 16 cyl. General Motors

OFFSHORE BIRMINGHAM
Same details as *Offshore Beaumont*

SEA EXPLORER
Built 1971 for Bugge Supply Ship A/S, Tonsberg, Norway

Length, overall	52·44 m (172 ft)
Beam, extreme	11·61 m (38 ft 1 in)
Draught, max.	3·99 m (13 ft 1 in)
Total power	3,900 h.p., two 16 cyl. General Motors diesel
Deadweight	800 tons
Speed	14½ knots
ABS Classed	

SEA PEARL
Details as for *Sea Explorer*

XAVIER B
Built 1971 for Collins International Inc., Panama
Length, overall 50·30 m (165 ft)
Beam, extreme 11·64 m (38 ft 2 in)
Draught, max. 3·67 m (12 ft ¼ in)
Total power 3,900 h.p., two 16 cyl. General Motors diesel
Deadweight 500 tons

MAMMOTH
Built for Petrel Marine Co.
Length, overall 56·40 m (185 ft)
Beam, extreme 11·59 m (38 ft)
Draught, max. 3·96 m (13 ft)
Total power 5,750 h.p., two EMD 16-645 ES
Speed 15 knots

APOLLO
Built 1967
Length, overall 26·50 m (86 ft 11 in)
Beam, extreme 8·79 m (28 ft 10 in)
Draught, max. 3·20 m (10 ft 6 in)
Total power 2,100 h.p., two Caterpillar D398 B
Speed 11 knots

EASTERN ADVOCATE
Built 1970 for International Offshore Operators Inc.
Length, overall 50·31 m (165 ft)
Beam, extreme 10·98 m (36 ft)
Draught, max. 3·89 m (12 ft 9 in)
Deadweight 930 tons
Total power 2,000 h.p., two EMD 12-567
Speed 14 knots

EASTERN WORKER
All details as for *Eastern Advocate*

RESPLENDOR
Built 1973 for Petroleo Brasileiro S.A.
Length, overall 56·38 m (185 ft)
Beam, extreme 11·59 m (38 ft)
Draught, max. 3·96 m (13 ft)
Total power 5,600 h.p., two 16 cyl. General Motors diesel

RIG ENGINEER
Length, overall 50·31 m (165 ft)
Beam, extreme 10·98 m (36 ft)
Draught, max. 3·96 m (13 ft)
Power Two electro-motive 645

RIG PUSHER
Length, overall 45·72 m (150 ft)
Beam, extreme 10·98 m (36 ft)
Draught, max. 3·35 m (11 ft)
Total power 4,200 h.p., two Caterpillar 398-B
Speed 14 knots

TENDER TARPON
Built 1973 for Wilhelmsen Offshore Supply Services, Oslo, Norway
Length, overall 56·38 m (185 ft)
Beam, extreme 11·59 m (38 ft)
Draught, max. 3·96 m (13 ft)
Deadweight 832 tons
Total power 2,800 h.p., two 16 cyl. General Motors EMD 16-645 EZ
Speed 14½ knots

TENDER TROUT
Built 1972 for Wilhelmsen Offshore Supply Services, Oslo, Norway
Other details as for *Tender Tarpon*

TENDER TURBOT
Details as for *Tender Tarpon*

TENDER TUNA
See Stewart & Stevenson, Marine Services

HERCULES
Length, overall 26·50 m (86 ft 11 in)
Beam, extreme 8·79 m (28 ft 10 in)
Draught, max. 3·20 m (10 ft 6 in)
Total power 2,100 h.p., two Caterpillar D398B
Speed 11 knots

GEMINI
Length, overall 46·41 m (152 ft 3 in)
Beam, extreme 10·98 m (36 ft)
Draught, max. 3·35 m (11 ft)
Total power 2,000 h.p., two 12V-71 tandem General Motors
Speed 13 knots

GWEN
Length, overall 50·31 m (165 ft)
Beam, extreme 11·59 m (38 ft)
Draught, max. 3·96 m (13 ft)
Total power 2,400 h.p., two EMD 12-567 B-A
Speed 14 knots

JUPITER
Length, overall 26·50 m (86 ft 11 in)
Beam, extreme 8·79 m (28 ft 10 in)
Draught, max. 3·20 m (10 ft 6 in)
Total power 2,100 h.p., two Caterpillar D398B
Speed 11 knots

JUPITER
Length, overall 46·41 m (152 ft 3 in)
Beam, extreme 10·98 m (36 ft)
Draught, max. 3·35 m (11 ft)
Total power 2,000 h.p., two 12V-71 tandem General Motors
Speed 13 knots

SOUTHERN SHIPBUILDING CORPORATION,
P.O. Box 1089,
Slidell,
Louisiana 70458.

ACADIAN FREEDOM
Length, overall 51·5 m (169 ft)
Breadth 11·6 m (38 ft)
Depth 4·12 m (13 ft 6 in)
Draught 3·51m (11 ft 6 in)
Speed, max. 15 knots (full load)
Horsepower 3,080 (continuous shaft) b.h.p., 4,000 i.h.p.
Machinery Two main engines Nohab Polar SF18VS
Propellers 4 blade, 2·44 m (8 ft) diameter stainless steel
Bollard pull 40,909 kg (90,000 lb)
Range 9,600 nautical miles
Gross tonnage 199 g.r.t.
Deadweight 700 tons
Under deck bulk capacity 4 tanks 107 m³ (3,800 ft³)
Rig water 2,132 bbls
Fuel oil 2,625 bbls
Potable water 263 bbls
Deck cargo 400 long tons
Main deck 29·88 m × 9·15 m (98 ft × 30 ft) timber decked 15·75 cm (4 in) thick
Refrigerated storage 12·70 m³ (45 ft³) (net)
Electrical capacity Two (2) 75 kw 208/120 volts AC 3 dia. 60 cycle
Bulk cargo air compressor (450 ft³/min at 50 lb/in²) diesel driven
Electronic monitoring 34 point solid state system
Gyro compass w/two repeaters
Auto pilot
Standard magnetic compass w/reflecting binnacle
Loran A/C
ADF 100
Two radars
RF 201 M 30 channel SSB
Collins 55 channel VHF
Decca navigator
Standby receiver
Digital depth recorder
Simrad recording fathometer
Two searchlights, 500 w
Navigation lights as required
Internal sound power telephone
Bow thruster 200 h.p. diesel/hydraulic drive
Towing winch Double Drum Waterfall 113,636 kg (250,000 lb) line pull
Stern roller 1·52 m (5 ft) diameter
Purifier Mechanical centrifuge

Accommodation
Crew: 7 men
Passengers: 12 men, settee berths for 8
All electric galley
Seating for 14
All accommodation spaces electrically air conditioned and heated

Classification
ABS A-1 ✠ AMS: Ocean Towing
ABS Loadline
USCG Certified: Oceans
US Public Health Certificated

Miscellaneous equipment
One 3 ton hydraulic tugger
One 250 amp electric welding machine
One set oxygen acetylene bottles complete with hoses and burning tips
Tool box with necessary tools for anchor handling equipment
One grapnel hook
One chain stripper

ACADIAN VICTORY
All details as for *Acadian Freedom*

STATE BOAT CORPORATION,
3701 Kirby Drive,
Houston,
Texas 77006.
Tel: (713) 528 5359
Telex: 77-5125

STATE POINT
Length 45·73 m (150 ft)
Beam 10·67 m (35 ft)
Draught, max. 3·20 m (10 ft 6 in)
Loaded displ. 900 long tons, gross 196, net 67
Built 1965 New Orleans, Louisiana

Main engines
Caterpillar D398T/A's, total 2,000 h.p.

Generators
General Motors 671's, 3 phase 60 cycle 115/208 volt AC, total 120 kw

Tanks
Fuel: 36,000 gallons. Lube: 300 gallons. Water: 198,750 gallons

Quarters
16 berths, 2 showers 3 toilets, 7 basins

Structural
Outside control station, tube for centre-ship drilling. Distance from rear deckhouse to stern 33·23 m (109 ft). Clear deck space 33·23 m × 8·23 m (109 ft × 27 ft)

Electronic
Radar, radiotelephone, fathometer, Loran
U.S.C.G. certificate when built
Classed: Hull ✠ A1E, Machinery ✠ AMS by American Bureau of Shipping
Fuel consumption 1,800 US gallons/24 hours at continuous cruising speed of 12½ knots. Range 6,500 miles, endurance 45 to 60 days. Speed 5 to 16 mph

STATE VICTORY
Length 56·71 m (186 ft)
Beam 12·20 m (40 ft)
Draught, max. 3·96 m (13 ft)
Loaded disp. 1,650 long tons, gross 199, net 135
Built 1970 New Orleans, Louisiana

Built 1970 New Orleans, Louisiana

Main engines
Caterpillar D399's with 6·6:1 Lufkin gears, total 3,000 h.p.

Generators
General Motors 671N's, 3 phase 60 cycle 115/208 volt AC, total 150 kw

Tanks
Fuel: 74,300 US gallons, Lube: 300 US gallons, Water: 304,200 US gallons

Quarters
Basic: 20 berths, 5 showers, 5 toilets, 6 basins

Features
Two large unassigned rooms, 4·88 m × 6·71 m and 3·35 m × 5·49 m (16 ft × 22 ft and 11 ft × 18 ft) for extra quarters or labs
Clear deck space 36·89 m × 9·75 m (121 ft × 32 ft)
5·49 m × 5·49 m (18 ft × 18 ft) transducer room or through hull drilling well
Walk-in freezer/refrigerator and watermaker

Electronic
Radar, Decca RM-316 (½ to 48 mi)
Radiotelephone, Apelco Model 90C, 200 watt, 8 channel
Radiotelephone, R. F. Communications SSB Model 201M (DN30), 150 watt
Radiotelephone, Apelco, Model AR50MA, 65 watt, 3 channel emergency
Fathometer, Simrad, Model ES2C, (0 to 300 fathoms)
Gyro Compass, Decca with outside azimuth repeater
Auto Pilot, Decca, Arkas
Loran, Konel DX Navigator 'A'
Classed: Hull ✠ A1E, Machinery ✠ AMS by American Bureau of Shipping
Fuel consumption 2,900 US gallons/24 hours at continuous cruising speed of 13 knots. Range 7,800 miles, endurance 45 to 60 days. Speed range 5 to 16 mph

State Point.

Two views (above and below) of the motor vessel *State Victory*.

STATE TREASURE
Length 53·66 m (176 ft)
Beam 11·59 m (38 ft)
Draught, max. 36·60 m (12 ft)
Loaded displ. 1,500 long tons, gross 183 net 124

Main engines
Caterpillar D398's, total 2,000 h.p.

Generators
General Motors 671N's, 3 phase 60 cycle 120/208 volt AC, total 15 kw

Tanks
Fuel: 69,000 US gallons. Lube: 300 US gallons. Water: 266,000 US gallons

Quarters
Basic: 18 expandable to 26 berths, 3 showers, 3 toilets, 4 basins

Structural
Four 14 m³ (500 ft³) in deck mud tanks, wood deck covering
Outside control station
Distance from rear deckhouse to stern 39·63 m (130 ft)
Clear deck space 39·33 m × 8·84 m (129 ft × 29 ft)
Walk-in freezer/refrigerator

Electronic
Radar, SSB radiotelephone, fathometer auto pilot, Loran
U.S.C.G. Certificate when built

State Treasure.

Classed: Hull ✠ A1E, Machinery ✠ AMS by American Bureau of Shipping
Fuel consumption 1,800 US gallons/24 hours at continuous cruising speed of 12 knots. Range 7,000 miles, endurance 50 to 70 days

STATE DIAMOND
All details as for *State Treasure*

STATE QUEEN

Length	50·61 m (166 ft)
Beam	11·58 m (38 ft)
Draught, max.	33·54 m (11 ft)

Loaded displ. 1,400 long tons, gross 194, net 134
Built 1972 Moss Point, Mississippi

Main engines
Caterpillar D398's, total 2,000 h.p.

Generators
General Motors 671N's, 3 phase 60 cycle 120/208 volt AC, total 150 kw

Tanks
Fuel: 43,000 US gallons. Lube: 1,700 US gallons. Water: 233,000 US gallons

Quarters
18 expandable to 24 berths, 3 showers, 3 toilets, 4 basins

Structural
Distance from rear deckhouse to stern 36·59 m (120 ft)
Free deck space 36·28 m × 88·41 m (119 ft × 29 ft)
Walk-in refrigerator/freezer
Outside control station

Electronic
Radar, radiotelephone, fathometer, auto pilot, Loran
U.S.C.G. Certificate when built
Classed: Hull ✠ A1E, Machinery ✠ AMS by American Bureau of Shipping
Fuel consumption 1,800 US gallons/24 hours at continuous cruising speed of 12 knots. Range 7,000 miles, endurance 45 to 60 days. Speed range 5 to 17 mph

State Queen

State Arrow

State Wave

STATE ARROW

Length	50·30 m (165 ft)
Beam	10·98 m (36 ft)
Draught, max.	33·54 m (11 ft)

Loaded displ. 1,300 long tons, gross 188, net 128
Built 1966 New Orleans, Louisiana

Main engines
Caterpillar D398T/A's, total 2,000 h.p.

Generators
General Motors 671's, 3 phase 60 cycle 115/208 volt AC, total 120 kw

Tanks
Fuel: 36,000 US gallons. Lube: 300 US gallons. Water: 246,000 US gallons

Quarters
Basic: 14 berths, 2 showers, 3 toilets, 6 basins. Two unassigned main deck rooms for extra quarters or laboratories.

Structural
Clear deck space aft: 34·45 m (113 ft)
Constructed for McMullen Flume Stabilisation System

Electronic
Radar, Raytheon 1500B ($\frac{1}{2}$ to 32 mi.)
Radiotelephone, Raytheon 90C, 200 watt, 8 channel
Fathometer, Raytheon, recording DE-721 0–366 m (0–1,200 ft)
Loran, Enac/Triton 'A'
Auto pilot, Sperry
Classed: Hull ✠ A1E, Machinery ✠ AMS by American Bureau of Shipping
Fuel consumption 1,800 US gallons/24 hours at continuous cruising speed of $12\frac{1}{2}$ knots. Range 6,500 miles, endurance 45 to 60 days. Speed range 5 to 16 mph

STATE WAVE

Length	38·11 m (125 ft)
Beam	8·54 m (28 ft)
Draught, max.	3·15 m (10 ft 4 in)

Loaded displ. 675 long tons, gross 196, net 133
Built 1965 Houston, Texas

Main engines
Caterpillar D353T/A's, total 1 000 h.p.

Generators
General Motors 671's, 3 phase 60 cycle 115/208 volt AC. total 120 kw

Tanks
Fuel: 40,000 US gallons. Lube: 1,200 US gallons. Water: 50,000 US gallons

Quarters
20 berths, 5 showers, 5 toilets, 7 basins; Laboratory 3·05 m × 7·01 m (10 ft × 23 ft)
Cargo hold 85 m³ (3,000 ft³)
Distance from rear deckhouse to stern 23·17 m (76 ft)
Separated quarters for female personnel
Outside control station aft on deckhouse
McMullen Flume Stabilisation System

Electronic
Radar, Raytheon Model 1500B ($\frac{1}{2}$–32 mi.)
Radiotelephone, Raytheon Model 90C, 200 watt, 8 channel
Fathometer, Raytheon, recording DE-721A 0–488 m (0–1,600 ft)
Auto pilot, Metal Marine, No. 420
All wave receiver, Hallicrafter
Fuel consumption 900 US gallons/24 hours at continuous cruising speed of 11 knots. Range 11,600 miles, endurance 60 days. Speed range 5 to 14 mph

TUGS AND SUPPLY VESSELS
OWNER/OPERATORS

ARGENTINA

ESSO PETROLERA ARGENTINA S.A.,
Avenida Roque Saenz Pena 567,
Buenos Aires.

SHELL CIA. ARGENTINA DE PETROLEO S.A.,
Avenida Roque Saenz Pena 788,
Buenos Aires.

AUSTRALIA

AUSTRALIAN OFFSHORE SERVICES
S/V LADY VILNA,

OSMARINE (AUSTRALIA) PTY. LTD.,
5th Floor, Yorkshire House,
14 Spring Street, Sydney,
New South Wales 2001.
Tel: 241-2701-2
Telex: AA 213171
S/V:
BASS SHORE, Mar. 1973
COOK SHORE, Dec. 1969

P & O AUSTRALIA, LTD.
LADY LAURIE, 1970
LADY LORNA, 1970
LADY SARAH, —
Fourth vessel, 1973

SMIT-LLOYD (AUSTRALIA) PTY. LTD.,
Interocean House, 4th Floor,
261 George Street,
P.O. Box H. 189 (Australia Square),
Sydney 2000, New South Wales.
Tel: 20573
Telex: 20305
SMIT-LLOYD 31, 1968
SMIT-LLOYD 32, 1968
SMIT-LLOYD 33, 1969
SMIT-LLOYD 34, 1969
S/V & T:
SMIT-LLOYD 35, 1973
SMIT-LLOYD 36, 1973

TIDEWATER PORT JACKSON MARINE PTY. LTD.,
No. 2 Jetty, Circular Quay,
Sydney, New South Wales.
Tender/Supply vessels:
AUSTRAL TIDE, 1968
SOUTHERN TIDE 1972
SYDNEY TIDE, 1969
TASMAN TIDE, 1968
VICTORIA TIDE, 1968

BERMUDA

INTERNATIONAL OFFSHORE SERVICES LTD.,
P.O. Box 391, Bank of Bermuda Building,
Hamilton.
S/V:
MASTER BRUNO, 1965
MASTER MARE, 1965

OCEAN INCHCAPE (BERMUDA) LTD.,
Bank of Bermuda Building,
Hamilton.
SV/T OIL DISCOVERER, 1973
S/V OIL EXPLORER, 1972
SV/T OIL PRODUCER, 1972
S/V OIL PROSPECTOR, 1972
SV/T OIL SUPPLIER, 1972

CANADA

ABC PACKING CO. LTD. (see Nordic Offshore Services Ltd.)

CANADIAN OFFSHORE MARINE LTD.,
P.O. Box 9, 2089 Upper Water Street,
Halifax, Nova Scotia.
SCOTIAN SHORE, May 1973
BRETON SHORE, July 1973

INTERNATIONAL OFFSHORE SERVICES,
Calgary.
Three tug/supply vessels.

NORDIC OFFSHORE SERVICES LTD.,
Suite 103, 1199 West Pender Street,
Vancouver, British Columbia.
S/V:
CATHY B, 1972
NORDIC VI, 1972
MARY B VI, 1972

An artist's impression of one of Nordic Offshore Services' supply vessels

Seven vessels building 1973.
Two classes of vessel:
(1) Length 57·9 m (190 ft), beam 13·7 m (45 ft), 5,000 h.p.
(2) Length 53·34 m (175 ft), beam 11·6 m (38 ft), 3,200 h.p.
Both ice-strengthened to ABS 1A standards and more than 1,800 tons deck and liquid cargo capacity.

OFFSHORE MARINE LTD.,
P.O. Box 9, 2089 Upper Water Street,
Halifax, Nova Scotia.
Tel: 902 425-6571/2/3
Telex: 019-226721.

RIVTOW STRAITS LTD.,
2215 Commissioner Street,
Vancouver 15, British Columbia.
Tugs:
NEVA STRAITS, 1962
RIVTOW LION, 1940
RIVTOW ROGUE, 1958
RIVTOW VIKING, 1940

SEASPAN INTERNATIONAL LTD.,
10 Pemberton Avenue,
North Vancouver, British Columbia.
Tugs:
ISLAND NAVIGATOR, 1944
SEASPAN BREEZE, 1944
SEASPAN CHALLENGER, 1944
SEASPAN CHAMPION, 1970
SEASPAN CHIEF, 1966
SEASPAN CHINOOK, 1943
SEASPAN COMMANDER, 1970
SEASPAN DARING, 1944
SEASPAN KING, 1968
SEASPAN MARINER, 1960
SEASPAN MASTER, 1966
SEASPAN MONARCH, 1966
SEASPAN NAVIGATOR, 1970
SEASPAN PACER, 1967
SEASPAN QUEEN, 1964
SEASPAN SOVEREIGN, 1944
SEASPAN VALIANT, 1969
SEASPAN VICTOR, 1970
SEASPAN WARRIOR, 1959
SUDBURY II, 1943
LA POINTE, 1906

CHINA (PEOPLE'S REPUBLIC OF)

CHINA NATIONAL MACHINERY IMPORT AND EXPORT CORPORATION,
Peking.
Eight 499 grt, 3,800 h.p. tug/supply vessels ordered from Aarhus Flydedok for delivery January 1974–October 1975.

DENMARK

MAERSK SUPPLY SERVICES, DIVISION OF A. P. MØLLER,
Kgs. Nytorv 8, 1098 Copenhagen K.
S/V MAERSK FEEDER,* 1967
S/V MAERSK FIGHTER,* 1967
SV & T MAERSK HANDLER, 1972
SV & T MAERSK HAULER, 1972
S/V MAERSK HELPER, 1972
S/V MAERSK SERVER, 1971
SV&T MAERSK SHIPPER, 1972
SV&T MAERSK SUPPLIER, 1972
S/V MAERSK SUPPORTER, 1971
* sister ships

REDERRAKTIESELSKABET DANNEBROG
Two vessels of modified *Maersk Handler* type supplied by Aarhus Flydedok A/S.

WECO SHIPPING I/S
Rungsted.
Designers of (200 ft) supply vessel for North Sea. Two tug/supply vessels and one supply vessel

FRANCE

SOCIETE ENTREPRISE POUR LES TRAVAUX PETROLIERS MARITIMES, E.T.P.M.,
125 rue de Saussure,
Paris 17e.

FERONIA INTERNATIONAL SHIPPING (FISH)
96 rue de la Victoire,
75009 Paris.
Tel: 744 87 49
Telex: 29672
Fleet: 15 tug workboats; 4 landing craft; 1 water taxi.
S/V:
BLACK FISH, 1969
BROWN FISH, 1970
CAT FISH, 1969
DOG FISH, 1944
GREEN FISH, 1970
SAIL FISH, 1972
SILVER FISH, 1971
SOUTH FISH, 1972
STAR FISH, 1973
STEEL FISH, 1973
SUN FISH, 1971
SWORD FISH, 1972
WHITE FISH, 1970

INTERNATIONAL OFFSHORE SERVICES FRANCE,
c/o International Offshore Services (UK) Ltd.
17c Curzon Street, London, W.1.
S/V LADY LAURA, 1965

PROGAMAR INTERNATIONAL,
53 Boulevard des Dames,
13 Marseilles 2e.
Tugs:
ABEILLE No. 1, 1959
ABEILLE No. 2, 1960
ABEILLE No. 3, 1956
ABEILLE No. 7, 1972
ABEILLE No. 8, 1972
ABEILLE No. 9, 1961
ABEILLE No. 14, 1966
ABEILLE No. 15, 1953
ABEILLE No. 16, 1967
ABEILLE No. 10, 1944
ABEILLE No. 17, 1961
ABEILLE No. 18, 1961
ABEILLE No. 19, 1961
ABEILLE No. 20, 1957
ABEILLE No. 21, 1961
ABEILLE No. 23, 1966
ABEILLE No. 24, 1968
ABEILLE No. 25, 1952
ABEILLE No. 26, 1960
ABEILLE No. 27, 1956
ABEILLE No. 28, 1957
ABEILLE No. 30, 1944
PHOCEEN, 1959
PROVENCAL 1, 1972
PROVENCAL 4, 1973
PROVENCAL 12, 1953
PROVENCAL 15, 1964
PROVENCAL 16, 1965

GERMANY (FEDERAL REPUBLIC OF)

BUGSIER-REEDEREI UND BERGUNGS A.G.,
P.O. Box 1583, Johannisbollwerk 10,
Hamburg 11.
(UK agents: Samuel Stewart and Co. (London) Ltd.
Bevis Marks House, Bevis Marks,
London, E.C.3.
Tel: 01-283 2651
Telex: 888435 or 886001
Tugs:
(salvage) ALBATROS, 1965
ARCTIC, 1970
OCEANIC,
(salvage) ATLANTIC, 1959
(salvage) ATLAS, 1960
BALTIC, 1969
BUGSIER 1, 1972
BUGSIER 2, 1972
BUGSIER 6, 1972
BUGSIER 7, 1973
BUGSIER 8, 1954
BUGSIER 16, 1958
BUGSIER 17, 1958
BUGSIER 24, 1959
BUGSIER 25, 1959
(salvage) BUGSIER 26, 1964
(salvage) BUGSIER 27, 1964
BUGSIER 28, 1964
BUGSIER 29, 1964
BUGSIER 30, 1965
BUGSIER 31, 1965
BUGSIER 32, 1969
BUGSIER 33, 1969
GOLIATH, 1941
(salvage) HERMES, 1956

(salvage) HEROS, 1964
(salvage) SEEFALKE, 1970
(salvage) WOTAN, 1972

REEDERI M.S. SEETRANSI,
F.C.H. Stark K.G.,
Hamburg.
Two tugs and two tug/supply vessels.

HONG KONG

HONG KONG AND WHAMPOA DOCK CO.,
Kowloon Docks.
KAU LUNG. 1972

ICELAND

VITA & HAFNAMALASKRIFSTOFEN,
Seljaveg 32,
Reykjavik.
Lighthouse supply vessel:
ARVAKUR, 1962

IRAN

TIDEWATER MIDDLE EAST MARINE SERVICE CO.,
Abadan.
S/V ABADAN, 1956
S/V FARS, 1957

LIBERIA

INTERNATIONAL OFFSHORE OPERATORS INC.,
Monrovia.
S/V
EASTERN ADVOCATE, 1970
EASTERN CONSORT, 1969
EASTERN WORKER, 1970

MALTA

OCEAN INCHCAPE MARSA,
Six tug/supply vessels.

NETHERLANDS

HEEREMA ENGINEERING SERVICE,
64 Scheveningseweg,
The Hague.
One tug, 4,800 h.p.
HUSKY, 1969

INTERNATIONAL OFFSHORE SERVICES (HOLLAND) N.V.,
Calandstraat 49,
Rotterdam.
LADY MARIANN, 1970
LADY MIRIAM, 1971
S/V:
LADY DIANA, 1964
LADY CAROLINE, —
LADY SYLVIA, 1963

NETHERLANDS OFFSHORE CO. N.V. NIEUWE ROTTERDAMSE SLEEPDIENST
Fleet of 15 tugs.
SMITHBANK, 1970
INDURBANK, 1970
BANJAARDSBANK, 1972
SCHOUWENBANK, 1972
STEENBANK, 1972

SMIT-INTERNATIONAL OCEAN TOWAGE & SALVAGE CO.,
5 Westplein,
3002 Rotterdam.
(previously known as L. Smit and Co. International Towing Service)
Tug:
BARENTSZ ZEE, 1957
CLYDE, 1957
ELBE, 1959
HUDSON, 1964

Banjaardsbank, owned by B. V. Nieuwe Rotterdams sleepdienst

MISSISSIPPI, 1960
Firefighting/Salvage tug
NOORDZEE, 1970

Length, overall	68·50 m (224 ft 9 in)
Length, b.p.p.	62 m (203 ft 5 in)
Breadth, overall	12·62 m (41 ft 5 in)
Beam, mld.	12·10 m (39 ft 8 in)
Depth, mld.	6·40 m (21 ft)
Draught	5·50 m (18 ft 1 in)
Gross tonnage	1,325

Tug ORINOCO, 1964
Firefighting/Salvage tug:
POOLZEE, 1971
RODE ZEE, 1968
Tug:
TASMAN ZEE, 1958
THAMES, 1961
Salvage Tug:
WITTE ZEE, 1966
ZWARTE ZEE, 1963

SMIT-LLOYD B.V.,
Westplein 9,
Rotterdam.
S/V&T:
SMIT-LLOYD 1, 1965
SMIT-LLOYD 2, 1965
SMIT-LLOYD 3, 1965

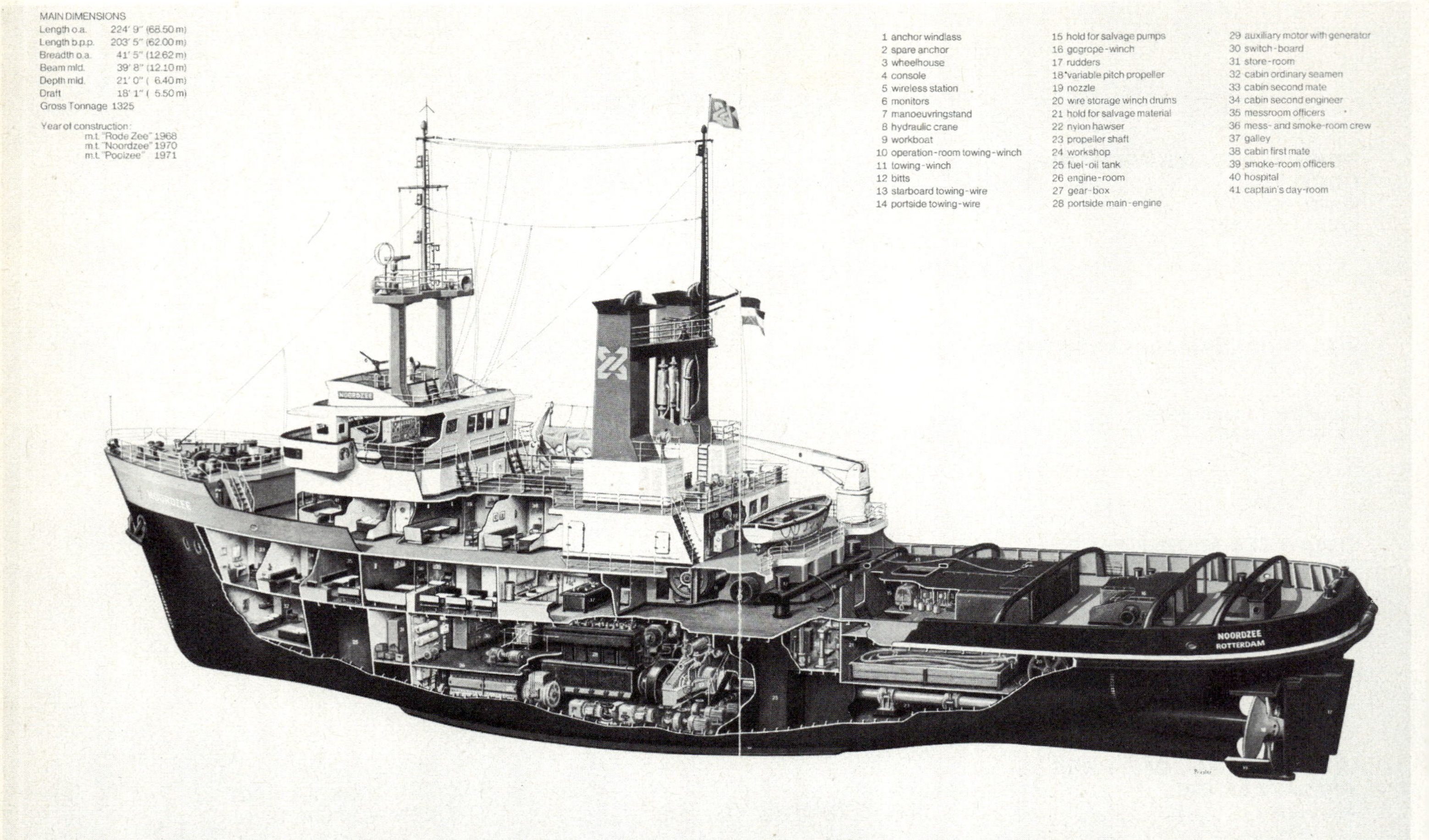

The 11,000 one h.p. ocean going motor tug *Noordzee*, sister ship to *Rode Zee* and *Poolzee*, Key: (1) anchor windlass; (2) spare anchor; (3) wheelhouse; (4) console; (5) wireless station; (6) monitors; (7) manoeuvring; (8) hydraulic crane; (9) workboat; (10) operation-room towing-winch; (11) towing-winch; (12) bitts; (13) starboard towing-wire; (14) portside towing-wire; (15) hold for salvage pumps; (16) gogrope-winch; (17) rudders; (18) variable pitch propeller; (19) nozzle; (20) wire storage winch drums; (21) hold for salvage material; (22) nylon hawser; (23) propeller shaft; (24) workshop; (25) fuel-oil tank; (26) engine-room; (27) gear-box; (28) portside main-engine; (29) auxiliary motor with generator; (30) switch board; (31) store-room; (32) cabin ordinary seamen; (33) cabin second mate; (34) cabin second engineer; (35) messroom officers; (36) mess- and smoke-room crew; (37) galley; (38) cabin first mate; (39) smoke-room officers; (40) hospital; (41) captain's day-room

SMIT-LLOYD 4, 1965
SMIT-LLOYD 5, 1965
SMIT-LLOYD 6, 1965
SMIT-LLOYD 7, 1965
SMIT-LLOYD 8, 1966
SMIT-LLOYD 9, 1966
SMIT-LLOYD 10, 1966

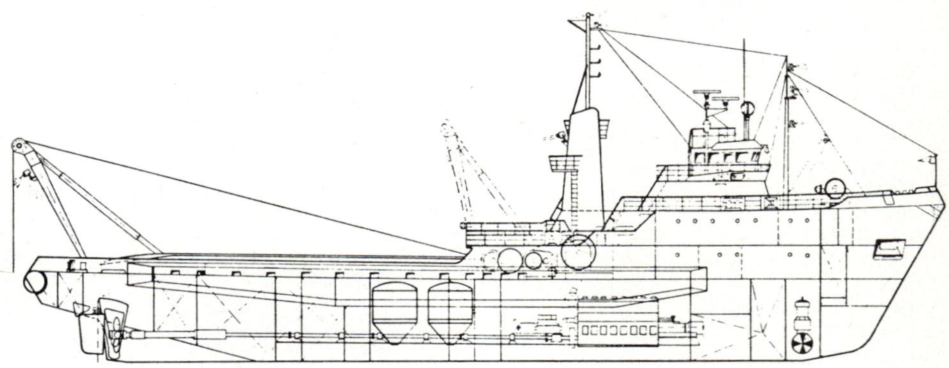

General diagram of the Smit-Lloyd '10' type, 3,000 h.p. tug, anchor handling and supply vessel

General specification of the 3,000 h.p. tug, anchor handling and supply boat

Dimensions

Length, overall	59·7 m (196 ft $\frac{1}{4}$ in)
Beam, overall	11·7 m (38 ft $5\frac{3}{8}$ in)
Depth	4·6 m (16 ft $10\frac{3}{4}$ in)
Draught, max.	4·3 m (14 ft $1\frac{1}{4}$ in)

Tonnage: G.R.T. 798
Class: A.B.S. Ocean Going
Certificates: Dutch Shipping Inspection/A.B.S.
Stability data in accordance with I.M.C.O. regulations

Capacities

Fuel oil for rig and own consumption	280 tons
Drilling water	410 tons
Potable water	95 tons
Deck cargo abt.	400 tons
Deck area	32 m × 6·8 m (105 ft × 29 ft)
Height of cargo bulwark	1·5 m (5 ft)
Bulk cargo abt. (5 vertical tanks underdeck)	4,000 c ft
Deadweight	750 tons

Anchor handling

A two-drum winch, electrically driven, each drum with a heavy brake system operated from a special control platform. The first drum has a single line pull of 160,000 lb max. The second drum has a single line pull of 350,000 lb max. Most vessels are equipped with a 'wildcat' winch; two 15,000 lb capstans; two stern rollers; an A-frame can be installed (latest design for 60 tons lifting capacity); and all special gear such as pelican hooks, wire stoppers, safety links, etc. Floodlights in funnels and A-frame.

Towing

Towing drum on anchor handling winch with 610 m (2,000 ft) of 2 in diameter tow line. Heavy towing bitt, flush railing and bulwark. Three quickly removable towing bars. Bridge control of one capstan to handle tow line. Special shackles, spare wire and pendants, double nylon hawser, etc.

Machinery

Two main engines each rated at 1,500 b.h.p. (continuously) at 300 revs/min. Three bladed controllable pitch propellers reinforced for ice conditions. Blades designed for top efficiency for towing and anchor handling.

Speed

Maximum speed	14 knots
	fuel cons. 12 tons/day
Full speed	13 knots
	fuel cons. 10 tons/day
Cruising speed	12·5 knots
	fuel cons. 8 tons/day
Standby duties	dead slow
	fuel cons. 3 tons/day

Bollard pull: 75,000 lb
Bow thruster with lateral thrust of 7,500 lb

Auxiliary power

For all services 650 h.p. auxiliary power is installed, distributed over four diesel generator sets supplying 220 volts DC to drive winches, pumps, bow thruster, etc.

Cargo transfer capacity

Fuel oil: 125 tons/hour at 40 m (130 ft) head

Drilling water: 2 × 100 tons/hour at 40 m (130 ft) head

Potable water: 30 tons/hour at 61 m (200 ft) head

Bulk cement: five vertical tanks each with 800 cft capacity, working pressure: 44 lb

Discharge stations: Midships on maindeck at port-and starboard side are deck connections fitted with blind flanges

Communications/navigation

Radio telephone combined with direction finder

VHF installation with 36 channels

Talk back communication system

Walkie talkies

Internal telephone system

Decca navigator

Gyro compass and magnetic compass

Automatic pilot

Radar

Echo sounder

Searchlight

Accommodation

7 cabins for crew

Sleeping accommodation for 12 passengers

All accommodation air conditioned and heated

Windlass, anchors, anchor chain

Two bow anchors each 2,500 lb

One spare anchor

Each bow anchor with 274 m (900 ft 1½ in) special studlink anchor chain

Bridge controlled heavy duty windlass

SMIT-LLOYD 11, 1967
SMIT-LLOYD 15, 1968
SMIT-LLOYD 16, 1968
SMIT-LLOYD 17, 1968
SMIT-LLOYD 18, 1969
SMIT-LLOYD 41, 1971

General specification of the 6,000 h.p. tug, anchor handling and supply boat

Dimensions

Length, overall	62·0 m (202 ft 10 in)
Beam, overall	12·0 m (39 ft 6½ in)
Depth	5·3 m (17 ft 6½ in)
Draught	4·7 m (15 ft 5 in)

Tonnage: G.R.T. 870

Class: A.B.S. Ocean Going

Certificates: Dutch Shipping Inspection/ A.B.S. Stability data in accordance with I.M.C.O. regulations

Capacities

Fuel oil for rig and own consumption	325 tons
Alternatively	525 tons
Drilling water	370 tons
Potable water	90 tons
Deck cargo abt.	400 tons
Deck area	30·5 m × 6·8 m (100 ft × 29 ft
Height of cargo bulwark	1·5 m (5 ft)
Bulk cargo abt.	4,000 cft
Deadweight	775 tons

Towing

Single drum towing winch with a capacity of 3,250 ft 6½ in circ. wire and fitted with a spooling gear

Brake designed for 215 tons holding power at 4th layer

Two heavy duty removable towing bars

Heavy towing bitt

Spare towing wire of 990·6 m (3,250 ft)

Nylon hawsers, wire protectors, etc. etc.

Anchor handling

A six drum anchor handling winch with 3 in 'wildcat' driven by a 95/190 h.p. electric motor

One drum for heavy duty service single line pull 325,000 lb

One drum for 3 in 'wildcat' with brake above chain locker 325,000 lb single line pull

One drum for anchor handling service single line pull 120,000 lb

One drum for spare pendants store/cargo derrick wire

One drum for stern anchor wire 610 m (2,000 ft) 5 in circ. or A-frame wire

One drum for spare towing wire 3,250 ft 6½ in circ.

Stern roller of 4·9 m (16 ft) width 1·4 m (4 ft 7 in) diameter

Two capstans of 15,000 lb for handling anchors a self transportable A-frame with SWL of 100 tons is installed

Steel wire pendants, pelican hooks, carpenter wire stoppers

Adjustable floodlights in funnels and A-frame, etc.

Bulk cement: five vertical tanks each with 800 cft capacity, working pressure 44 lb

Discharge stations: Midships on maindeck and at stern are deck connections fitted with blind flanges

Communications/navigation

Radio telephone combined with direction finder

VHF installation with 36 channels

Talkback system

Internal telephone system

Walkie talkies

Decca navigator

Gyro compass and magnetic compass

Automatic pilot

Radar

Echo sounder

Two searchlights on wheelhouse

Accommodation

Single and double cabins for crew

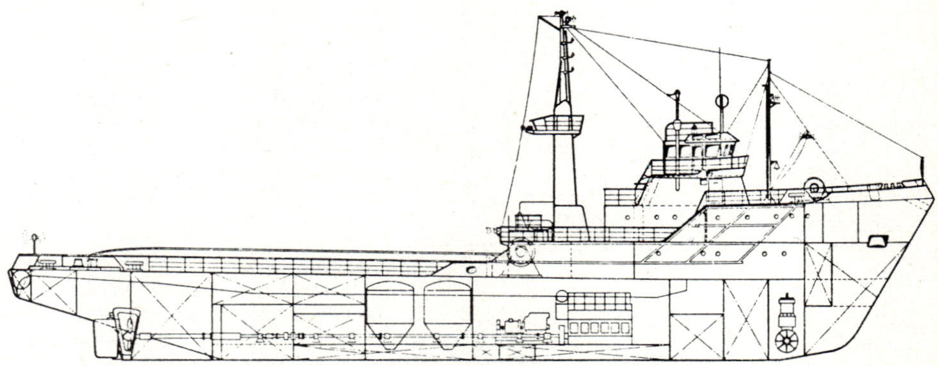

General diagram of the Smit-Lloyd '41' type, 6,000 h.p. tug, anchor handling and supply vessel

Machinery

Two 8 cylinder Industrie engines each rated 3,000 h.p. at 300 revs/min

Two controllable three bladed pitch propellers reinforced for navigation in ice

Blades designed for top efficiency for towing and anchor handling

Propellers bridge-controlled from two positions in wheelhouse

Speed

Maximum speed	14 knots
	fuel cons. 20 tons/day
Full speed	13·5 knots
	fuel cons. 16 tons/day
Cruising speed	13 knots
	fuel cons. 8 tons/day
Standby duties	dead slow
	fuel cons. 4 tons/day

Bollard pull: 125,000 lb

Bow thruster with a lateral thrust of 10,000 lb

Auxiliary power

1 × 16 cylinder 550 h.p. MWM diesel engine driving a 332 kw generator for bow thruster

3 × 5 cylinder Stork diesel engines driving three 100 kw generators

220 volt DC and 1 × 4 cylinder Armstrong Siddeley diesel engine, driving a 20 kw generator

Cargo transfer capacity

Fuel oil: 120 tons/hour at 40 m (130 ft) head

Drilling water: 2 × 100 tons/hour at 40 m (130 ft) head

Potable water: 60 tons/hour at 61·0 m (200 ft head)

Sleeping accommodation for 12–18 passengers

All air conditioned and heated

Windlass, anchors, anchor chain

Two bow anchors each 3,300 lb

One spare anchor

Each bow anchor with 1,800 ft 1¾ in special studlink anchor chain

Each chain locker has a capacity for 1,800 ft 1¾ in anchor chain

SMIT-LLOYD 42, 1972
SMIT-LLOYD 43, 1972
SMIT-LLOYD 44, 1972
SMIT-LLOYD 45, 1972
SMIT-LLOYD 46, 1972
SMIT-LLOYD 47, 1973
SMIT-LLOYD 48, 1973
SMIT-LLOYD 101, 1969

General specification of the 4,000 h.p. tug, anchor handling and supply boat

Dimensions

Length, overall	54·77 m (179 ft 10 in)
Beam, overall	12·3 m (40 ft 4 in)
Depth	5·0 m (16 ft 3½ in)
Draught	4·4 m (14 ft 5¼ in)

Tonnage: G.R.T. 786

Class/A.B.S. Ocean Going

Certificates: Dutch Shipping Inspection/ A.B.S.

Stability data in accordance with I.M.C.O. regulations

Capacities

Fuel oil for rig and 210 tons

own consumption	
Alternatively	360 tons
Drilling water	520 tons
Potable water	146 tons
Deck cargo abt.	400 tons
Deck area	29 m × 9·45 m (95 ft × 31 ft)
Height of cargo bulwark	1·5 m (5 ft)
Bulk cargo abt.	4,000 cft
Deadweight	860 tons

Towing

Towing drum with maximum single line pull of 70,000 lb at a speed of 30 ft/min

Towing wire of 610 m (2,000 ft) 6 in circ.

Heavy duty arrangements with fairleads for towing rope

Removable towing bar

Steelwire pendants, steelwire towing-bridles, nylon hawser, etc. etc.

Towing/anchor handling

Second section of towing drum can be used for anchor handling

Anchor handling

Heavy duty drum with a maximum single line pull of 300,000 lb

Two 'wildcats' on heavy duty drum above chain lockers to store rig anchor chain

Large tapered stern roller of 1·52 m (5 ft) diameter and 5·08 m (16 ft 8 in) width

Two capstans of 15,000 lb

Steel work wires, pelican hook wire and various other equipment

Machinery

Two main engines each rated 2,000 h.p. at 430 revs/min

Two three bladed controllable pitch propellers reinforced for ice conditions

Blades designed for top efficiency for towing and anchor handling

Speed

Maximum speed		13·4 knots
	fuel cons.	12 tons/day
Full speed		13 knots
	fuel cons.	10 tons/day
Cruising speed		12·5 knots
	fuel cons.	8 tons/day
Standby duties		dead slow
	fuel cons.	3 tons/day

Bollard pull: 90,000 lb

Bow thruster with a lateral thrust of 7,500 lb

Auxiliary power

One 12 cyl. 410 h.p. diesel engine driving a 300 h.p. electric motor for the bow thruster

Two 6 cylinder four stroke diesel engines (output of 204 h.p. at 1,200 revs) each with a 220 volt DC generator of 135 kw

One 4 cylinder four stroke diesel engine (output of 96 h.p. at 1,500 revs) with a 220 volt DC generator of 45 kw

Cargo transfer capacity

Fuel oil: 120 tons/hour at 40 m (130 ft head)

Drilling water: 2 × 100 tons/hour at 40 m (130 ft) head

Potable water: 60 tons/hour at 61 m (200 ft) head

Bulk cement: three vertical tanks with a total capacity of +4,000 cft, working pressure 44 lb

Discharge stations:

Midships port and starboard

Aft port and starboard

Communications/navigation

Radio telephone

Direction finder

VHF installation

Talk back communication system

Internal telephone system

Walkie talkies

Decca navigator

Gyro compass and magnetic compass

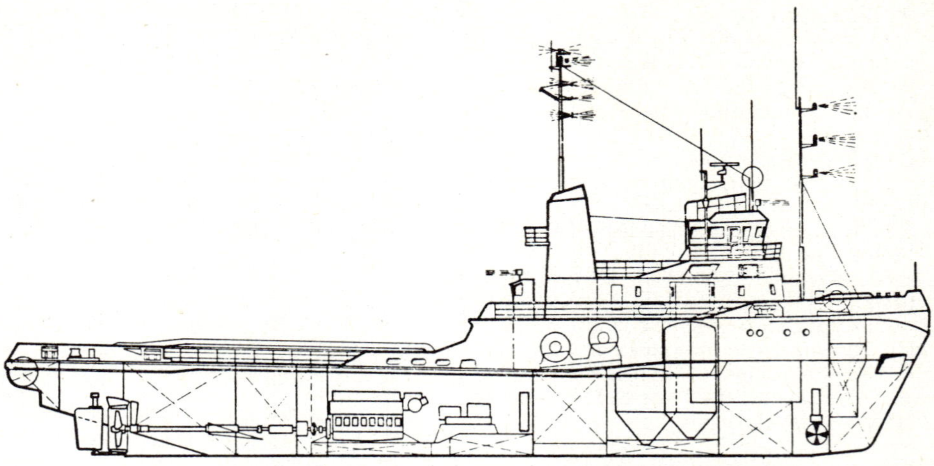

General diagram of the Smit-Lloyd '101' type, 4,000 h.p. tug, anchor handling and supply vessel

Automatic pilot

Radar

Echosounder

Two searchlights

Accommodation

Seven single cabins for crew

Three cabins with sleeping accommodation for twelve passengers

All accommodation air conditioned and heated

Windlass, anchors, anchor chain

Two bow anchors each 2,750 lb

One spare anchor

Each bow anchor with 329·2 m (1,080 ft) 1½ in special stud link anchor chain

Each chain locker has a capacity for 411·5 m (1,350 ft) 1½ in anchor chain

Special features

In the 4,000 h.p. vessels Smit-Lloyd have incorporated all features of earlier vessels and added some extras, resulting from experience gained in working with all type of rigs on a worldwide basis

Some of these special features are:

Controllable pitch propeller in fixed Kort nozzles; large tapered stern roller of 5 ft diameter, width 5·08 m (16 ft 8 in); 'wildcats' for rig anchor chain; chain lockers below 'wildcats, for storage of rig anchor chain

SMIT-LLOYD 103, 1970
SMIT-LLOYD 14, 1967
SMIT-LLOYD 12, 1967
SMIT-LLOYD 104, 1973
SMIT-LLOYD 105, 1973

General specification of the 7,500 h.p. tug, anchor handling and supply boat

Dimensions

Length, overall	63·9 m (209 ft 9 in)
Beam, overall	13·3 m (43 ft 8 in)
Depth	6·3 m (20 ft 8 in)
Draught	4·75 m (15 ft 7 in)

Tonnage: G.R.T. 1,295

Class: A.B.S. Ocean Going

Certificates: Dutch Shipping Inspection/A.B.S.

Stability data in accordance with I.M.C.O. regulations

Capacities

Fuel oil for rig and own consumption (for ocean towage services)	420 tons
Alternatively	620 tons
Drilling water	600 tons
Potable water	110 tons
Deck cargo abt.	500 tons
Deck area	33·5 m × 10·36 m (110 ft × 34 ft)
Height of cargo bulwark	1·5 m (5 ft)
Forward part	2·25 m (7½ ft)
Bulk cargo (5 vertical tanks underdeck) abt.	6,250 cft
Deadweight	1,100 tons

Towing/anchor handling

A three-drum 'waterfall' type towing and anchor handling winch is installed in a winch room forward on the cargo deck

Winch driven by a 125/250 h.p. electric motor with a two-speed gearbox

Winch room provided with fairleads for towing wire, pendant lines, anchor chain etc.

Towing

91·44 cm (36 in) diameter towing drum with a maximum single line pull of 135,000 lb

Holding power of brake system 315 tons and controlled from bridge deck and winch room

Towing winch with 990·6 m (3,250 ft) towing wire of 7¼ in circ.

Wire spooling device installed

Heavy duty towing arrangements with fairleads for towing services under all conditions

Two removable hinged towing bars

Spare towing wire of 3,250 ft 7¼ in circ.

Steel wire pendants, towing bridles, double nylon hawser, towing chain, towing saddles, etc. etc.

Towing/anchor handling

91·44 cm (36 in) diameter towing/anchor handling drum is capable to carry 762 m (2,500 ft) 7¼ in circ. steel wire

Drum will be used for both towing and anchor handling

Maximum single line pull 70,000 lb on the bare drum at a speed of 43 ft/min

In high gear the slack wire speed is 120 ft/min on the bare drum

Holding power pneumatical brake 60 tons

Anchor handling

36 in diameter heavy duty anchor handling drum with a maximum single line pull of 255,000 lb at a speed of 9 ft/min at the bare drum in low gear and 325,000 lb at stall

Capability in high gear 92,500 lb at a speed of 30 ft/min at the bare drum

Pneumatically controlled brake with a holding power of 125 tons

Winch provided with two cantilevered 'wildcats' above chain lockers

Stern roller of 5·0 m (16 ft 6 in) width and 2·13 m to 2·43 m (7 to 8 ft) diameter

Steel work wires, pelican hook wires, buoy catch wires, chain recovery wires, pelican hooks, shackles, chain chasers, etc.

Four adjustable floodlights

Machinery

Two 6 cylinder Werkspoor TM 410 main engines each rated 3,750 b.h.p. continuously at 550 revs/min

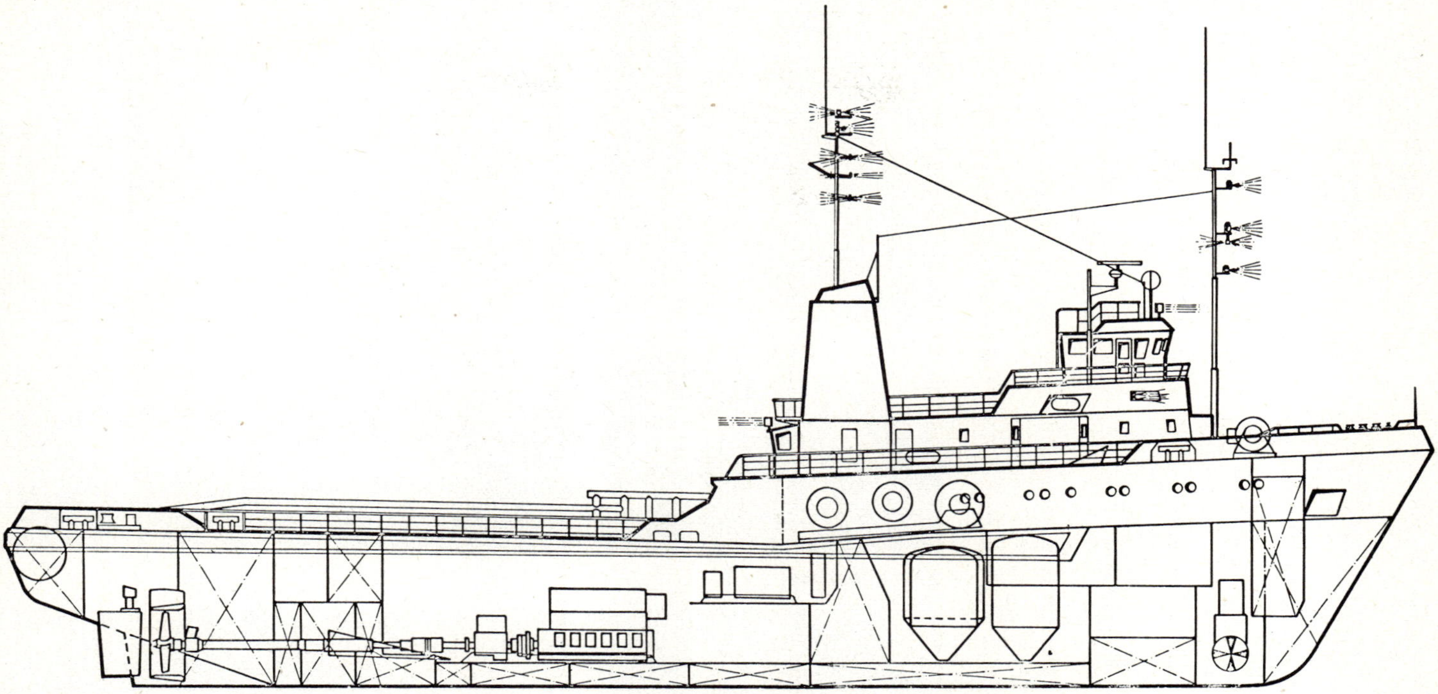

General diagram of the Smit-Lloyd '105' type, 7,500 h.p. tug, anchor handling and supply vessel

Reduction gears between main engines and controllable pitch propellers without power takes off

Two 4 bladed controllable pitch propellers for maximum 205 revs/min

Blades designed for top efficiency for towing and anchor handling

Speed

Maximum speed	15 knots
	fuel cons. 28 tons/day
Full speed	14 knots
	fuel cons. 20 tons/day
Cruising speed	$13\frac{1}{4}$ knots
	fuel cons. 15 tons/day
Standby duties	dead slow
	fuel cons. 5 tons/day

Bollard pull: 200,000 lb

Bow thruster with a lateral thrust of 11,000 lb

Auxiliary power

One 12 cylinder MWM diesel generator for bow thruster

Four diesel generators each with an output of 170 KVA in AC

Cargo transfer capacity

Fuel oil: 90 tons/hour at 61 m (200 ft head)

Drilling water: 2 × 100 tons/hour at 200 ft head

Potable water: 60 tons/hour at 200 ft head

Bulk cement: five vertical tanks with a total capacity of ±6,250 cft, working pressure 44 lb

Two cement compressors are installed

Discharge stations: Midships on main deck at the stern and on forecastle deck are deck connections for cement/barite, fuel oil, drilling water and potable water

Communications/navigation

Radio telephone

Direction finder

VHF installation

'Talkback' system from wheelhouse to capstans, control stand—towing winch, winch room, forecastle

Internal telephone system

'Walkie talkies'

Decca navigator

Gyro compass and magnetic compass

Automatic pilot

Radar

Echo sounder

Two searchlights

Accommodation

Single cabins for crew

Sleeping accommodation for 12 to 18 passengers

All accommodation air conditioned and heated

Windlass, anchors, anchor chain

Two bow anchors each 4,400 lb

One spare anchor

Each bow anchor with 548·64 m (1,800 ft) $1\frac{3}{4}$ in special stud link anchor chain

Each chain locker has a capacity for 975·36 m (3,200 ft) $1\frac{3}{4}$ in anchor chain

Special features

The 7,500 b.h.p. tug/anchor handling/supply vessel is specially designed for the new generation of semi-submersible drilling rigs operating in deeper waters and in remote and rough weather areas

Some special features are: four chain lockers for rig anchor chains, total capacity 7,200 cft; ship's chain lockers with capacity of minimum 975·36 m (3,200 ft) $1\frac{3}{4}$ in anchor chain each; powerful 125/250 h.p. windlass installed for fast hoisting of anchor chain; controllable pitch propellers in fixed Kort nozzles; cargo transfer capacity for delivery of two grades of bulk at the same time; stern roller, diameter 2·13 m to 2·44 m (7 to 8 ft), width 5·08 m (16 ft 8 in); bow thruster, lateral thrust 11,000 lb; remote controlled 3-drum 'waterfall' type towing and heavy duty winch installed in winch room.

SMIT-TAK INTERNATIONAL BERGINGSBEDRIJ N.V.,
Boompjes 90,
Rotterdam.
Salvage vessel:
BARRACUDA, 1970
BEVER, 1956
BUFFEL, 1953
ZEELEEUW
DOLPHIN
BRUINVISCH

TERMINAL INSTALLATIONS INC.,
Rotterdam.

VROON TRADING & SHIPPING CO. LTD.,
P.O. Box 28,
Breskens.
S/V BLUE FISH, 1966
S/V GREY FISH, 1969

VROON N.V.,
Grote Kade 30,
Breskens.

NETHERLANDS ANTILLES

SMIT INTERNATIONAL (ANTILLES) N.V.,
De Ruyterkade 53,
Willemstad,
Curaçao.
Tug:
SCHELDE, 1958
SMIT ENTERPRISE, 1972
SMIT PIONEER, 1972

Smit Enterprise

Smit Pioneer

NEW ZEALAND

BEGG WISEMAN LTD.,
10 Davis Crescent,
Newmarket,
Auckland 1.

FISHER & BLUNDELL LTD.,
P.O. Box 9110,
Auckland 1.

NORWAY

BUGGE SUPPLY SHIPS A/S,
P.O. Box 160,
Tønsberg.
Tel: (Tønsberg) 16 270
Telex: 11556
S/V:
SEA EXPLORER, 1971
SEA PEARL, 1971
SEA GARNET
SEA SUPPLIER
SEA SCOUT
SEA VIKING
SEA GEM, early 1974
SEA WORKER, early 1974
SEA BRUSE, autumn 1974
SEA BURE, autumn 1974

BUKSER OG BJERGNINGSSELSKAPET A/S,
Haunelageret,
Langbryggen,
Oslo.
Icebreaker tug: BAUS, 1971
Salvage tug: BEBE, 1964
Icebreaker tug: BENI, 1967
Icebreaker tug: BJONN, 1969
Salvage tug: BOB, 1963
Icebreaker/salvage tug: BESS, 1960
Icebreaker tug: BRYTEREN, 1955

K/S A/S LOM & CO. (JOHS LARSEN)
Bergen.
IBIS ONE, 1973

NORSK BJERGNINGSKOMPAGNI A/S,
Komediebakken 9,
Bergen,
and
Munkegt 40,
Trondheim.
S/V EDDA SALVATOR, 1972
(see also Salvator Marine Services A/S)
Salvage/icebreaker tug: HERKULES, 1960
Salvage tug: PARAT, 1963
S/V & Tug: STELLA SALVATOR, 1972
Salvage tug: STERKODDER, 1957

Salvage/icebreaker tug: ULA, 1971
Salvage tug: TRAUST, 1958
Salvage tug: SALVATOR, 1948
S/V AILOS, 1972

I/S NORWAY SUPPLY SHIPS INC.,
Stavanger (see Peter Smedvig)

OFFSHORE MARINE LTD.

THE FRED. OLSEN CO.,
Oslo.
Two tug/supply vessels building by Mangone 1973.

PETER SMEDVIG A/S,
P.O. Box 210,
Verksgt 13,
4001 Stavanger.
I/S Norway Supply Ships:
WEST ALBATROSS, 1972
WEST AVOCET, 1971
WEST EAGLE, 1972
WEST OSPREY, 1971

SALVATOR MARINE SERVICES, A/S
Bergen.
S/V EDDA SALVATOR, 1973

K/S SANDØY SUPPLY A/S & CO.
6270 Brattvag,
Alesund.
Four supply vessels under construction
S/V RIG PILOT, 1972
SV&Tug RIG TUGGER, 1973
SV&Tug RIG MASTER, 1973
S/V RIG SAILOR, 1973

TSCHUDI & EITZEN,
Kongensgate 11,
Oslo 1.
Tug/icebreaker SINADER, 1972

VESTERAALENS A/S.,
Stokmarknes.
Two supply vessels building, 1973.

WILHELMSEN OFFSHORE SERVICES,
Postboks 1359,
Roald Amundsens Gate 5,
Oslo 1.
SV & T:
TENDER TURBOT, 1973
TENDER TROUT, 1972
TENDER TUNA, 1972

SINGAPORE

M. J. BATTY & CO. (SINGAPORE) PTY. LTD.,
Fleet of 71 tugs, barges and landing craft.

BORNEO STRAITS OFFSHORE LTD.,
Phoenix Building,
11 Palmer Road.
S/V:
BORNEO GOLD, 1970
GULF GOLD, 1971
JAVA GOLD, 1969
STRAITS GOLD, 1971

GENERAL MARINE INTERNATIONAL SERVICE INC.,
PACIFIC LOGISTICS A/S,
Luyon Stevedoring Corporation (Lusteveco)
2nd Floor, Thongtek Building,
15 Scotts Road.
160 tugs, 5 tug/supply vessels and 4 supply ships.
BATMAN, 1954

SWEDEN

BERGNINGS-OCH DYKERI A/B NEPTUN,
P.O. Box 15054,
S-10465 Stockholm 15.
Salvage tug AJAX, 1960
Icebreaker tug ATLAS, 1968
Icebreaker tug MERCUR, 1964
Icebreaker tug NEPTUN, 1969
Icebreaker tug POSEIDON, 1970
Tug THETIS, 1961

UNITED KINGDOM

BROWN & ROOT—WIMPEY LTD.,
27 Hammersmith Grove,
London, W.6.
S/V WIMBROWN THREE, 1966

CATAMARAN WORKBOATS LTD.,
Newhaven,
Sussex.

CORY SHIP TOWAGE LTD.,
Windsor House,
London Road,
London SW16 4XH.
Salvage tug ANGLEGARTH, 1960
Tug:
AVONGARTH, 1960
BARGARTH, 1966
BUTEGARTH, 1966
Firefighting tug CASHEL, 1959
Salvage tug DALEGARTH, 1960
Tug:
DANEGARTH, 1966
DUNOSPREY, 1968
FALGARTH, 1958
GLENGARTH, 1970
Firefighting tug:
GRAYGARTH, 1970
GREENGARTH, 1970
Tug:
GWENGARTH, 1962
LOWGARTH, 1965
PENGARTH, 1962
PLUMGARTH, 1960
POLGARTH, 1962
RAMSGARTH, 1964
RATHGARTH, 1965
STACKGARTH, 1959
Salvage tug THORNGARTH 1959
Tug:
USKGARTH, 1966
WESTGARTH, 1962
Firefighting tug:
BANTRY BAY, 1968
BRANDON BAY, 1968
DINGLE BAY, 1968
TRALEE BAY, 1968
Tug:
CARRICKFERGUS, 1958
CLANDEBOYE, 1967
Firefighting tug COLERAINE, 1970
Tug:
CRAIGDARRAGH, 1966
CUETRA, 1962
POINT CARROL
POINT GILBERT
POINT JAMES
Firefighting tug POINT MILFORD, 1971
Tug POINT SPENCER

CORY SHIP TOWAGE (CLYDE) LTD.,
76 Carlton Place,
Glasgow, C.5.
Tug:
BRIGADIER, 1961
CAMPAIGNER, 1957
CHIEFTAIN, 1968
Firefighting tug CRUISER, 1959
Tug:
STRONGBOW, 1961
THUNDERER, 1958
VANGUARD, 1964
WARRIOR, 1970
WRESTLER, 1957

GRAY MACKENZIE & CO. LTD.,
40 St. Mary Axe,
London, E.C.3.
Tug:
AL HAMED, 1969
AL JABBER, 1968
AL QADER, 1968
AL QAWI, 1968
AL SALAMA 1957
Landing craft GOWERA, 1957
S/V HAKIM, 1972
Tug:
RABAB, 1969
RAFIK, 1964
RAHIM, 1965
RAMMAH, 1967
RAOOF, 1969
RASHEED, 1969
RECOURSE, 1962
RENCOUNTER, 1962
RIEVER, 1967
RUNNYMEDE, 1969
S/V SAN PEDRO CAPE, 1969

INTERSEAWORK LTD.,
Surrey.
Four tugs.

OCEAN INCHCAPE LIMITED (OIL),
40 St. Mary Axe
London EC3A 8EU.
Tel: 01-283 4680.
Mooring tugs:
OIL DIRECTOR, 1973
OIL SUPERVISOR, 1973
OIL MANAGER, 1973

OIL MARINER, 1973
OIL DRILLER, 1973
OIL PROSPECTOR, May 1972
OIL SUPPLIER, 1973
OIL PRODUCER, September 1973
OIL EXPLORER, November 1973

OFFSHORE INVESTMENTS LTD.,
Edinburgh.

OFFSHORE MARINE LTD.,
Cleveland House,
19 St. James's Square,
London, S.W.1.
Tel: 01-4990 9020
Telex: 9166681
(a Cunard subsidiary, previously part of OSA)

Also:
Block 3, Fish Wharf,
Great Yarmouth, Norfolk.
Tel: 0493-56426
Telex: 97475
Also:
Alexander House,
Abbotswell Road,
Tullos Estate,
Aberdeen.
Tel: 0224-22105
Telex: 73346
Fleet will be 31 strong in 1974.
S/V:
ARCTIC SHORE, 1969
ATLANTIC SHORE, 1968
BAY SHORE, 1971
CAPE SHORE, 1970
*CHANNEL SHORE, 1972
EAST SHORE, 1966
ESSEX SHORE, 1967
ISLAND SHORE, 1970
KENT SHORE, 1967
NORFOLK SHORE, 1967
NORTH SHORE, 1965
NOVA SHORE, 1969
OCEAN SHORE, 1972
ORKNEY SHORE
PACIFIC SHORE, 1969
PETREL SHORE, 1970
POLAR SHORE

* Towed the rig *Glomar IV* from Aberdeen to Korea, 15,477 miles.
Two tug/supply vessels on order from Richards (Shipbuilders) at 31 January 1974 and two from Appledore Shipbuilders, 900 grt, 5,600 h.p.
SV&T POLAR SHORE, 1971
S/V:
SHETLAND SHORE
STRAIT SHORE, 1970
SUFFOLK SHORE, 1967
TROPIC SHORE, 1969

Polar Shore.

OFFSHORE SUPPLY ASSOCIATION, LTD., (OSA)
Carrier House, Warwick Row,
London, S.W.1.
Tel: 01-828 1181
Telex: 27485

By January 1974 OSA had a fleet of 50 supply vessels.
S/V DRAGON, 1972
Landing craft EAGLE
FALCON
S/V:
LION, 1972
TIGER, 1972
WASSERTOR, 1973
ANSGARITURM
ASTERTURM
BINNASTOR
BERLINERTOR, 1965
ELLERNTOR, 1972
IMKENTURM
MICHAELISTURM, 1972
PENGENTURM
NICOLAITURM, 1973
KLOSTERTOR, 1970
MARIENTOR, 1971
WILDHADITURM, 1973
MUHLENTOR, 1972
JAKOBITURM, 1972

SEAFORTH MARITIME LTD.,
Seaforth Centre,
Waterloo Quay,
Aberdeen.
Tel: 573401
Telex: 73387
Two vessels on order from Cochrane and Sons
Two vessels on order from Scott Lithgow at 31 January 1974.
Tug S/V:
SEAFORTH CHALLENGER, 1974
SEAFORTH CHAMPION, 1974
SEAFORTH CHIEFTAIN, 1973
SEAFORTH HERO, 1973
SEAFORTH PRINCE, 1973
SEAFORTH SAGA, 1974
SEAFORTH VICTOR, 1974
SEAFORTH VIKING, 1974

SEA SERVICES SHIPPING CO. AND AGENCY,
100/102 Sandgate Road,
Folkestone,
Kent.
Six vessels on order from Cochrane & Sons, at 31 January 1974
Three 800 grt, 2 × 2,500 h.p. engines.
Three 800 grt, 2 × 3,080 h.p. engines.
S/V:
BORDER BLAZER, 1965
HIGHLAND BLAZER, 1958

UNITED TOWING LTD.,
11 Nelson Street, Hull,
Yorkshire. HU1 1XG.
Thirty-eight tugs.
Tug ENGLISHMAN, 1965
Salvage tug EUROMAN, 1967
Tug HEADMAN, 1963
Salvage tug IRISHMAN, 1967
Tug KROOMAN, 1938
Salvage tug LLOYDSMAN, 1971
Tug:
MOTORMAN, 1965
RIVERMAN, 1963
SCOTSMAN, 1969
SEAMAN, 1967
Salvage tug STATESMAN I, 1966
Tug:
SUPERMAN, 1967
TIDESMAN, 1963
TRADESMAN, 1964
TRAWLERMAN, 1963
TUGMAN, 1964
WATERMAN, 1966
Salvage tug WELSHMAN, 1966
Tug:
WORKMAN, 1963
YORKSHIREMAN, 1967

WIMPEY MARINE LTD.,
28 Hammersmith Grove,
London, W.6.
Two vessels on order from Appledore Shipbuilders at 31 January 1974.
One 900 grt, 4,500 h.p.
One 850 grt, 4,500 h.p.
Tug:
ANDY MITCHELL, 1968
G.W. 180, 1958
S/V:
WIMBROWN ONE, 1965
WIMBROWN TWO, 1965
SEA DOG, 1973

ZAPATA OFFSHORE SERVICES LTD.,
35 Basinghall Street,
London, E.C.2.
SV/Tug:
IMPERIAL SERVICE, 1971
MAJESTIC SERVICE, 1972
MONARCH SERVICE, 1972
PARAMOUNT SERVICE, 1971
CENTURION SERVICE
TROJAN SERVICE

Seaforth Hero (Courtesy Seaforth Maritime)

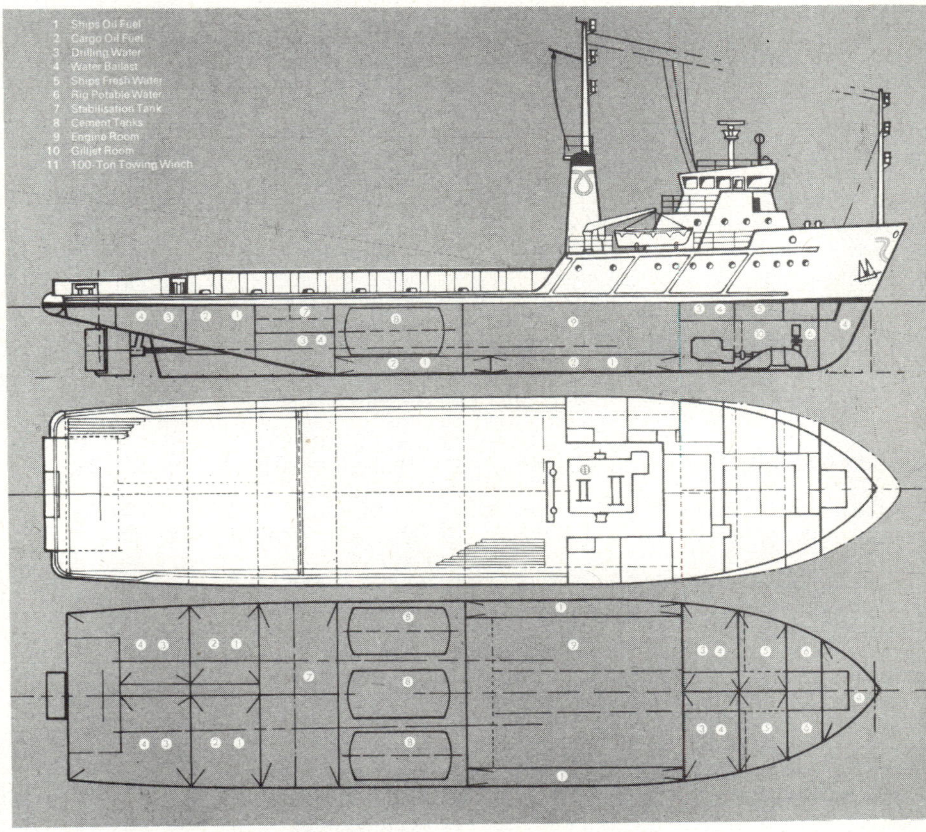

Seaforth Hero (Courtesy Seaforth Maritime)

UNITED STATES AMERICA

ACADIAN MARINE SERVICE INC.,
21st Floor, 225 Baronne Street,
New Orleans, Louisiana 70112.
ACADIAN FREEDOM, 1973
ACADIAN VICTORY, 1973
Plus four others in 1973.

ALLSTATE MARINE & INVESTMENT SERVICE INC.,
Houma, Louisiana.
Seven tugs and three tug/supply vessels.

AMERICAN OFFSHORE FLEET, INC.,
Galliano, Louisiana.
Four tugs.

AMERICAN OFFSHORE INC.,
Houston, Texas.
Eight tug/supply vessels, three supply a bollard pull of 200,000 lb
S/V:
BIGORANGE I, 1972
BIGORANGE II, 1972
CAPRICORN, 1971
DAVID P. GUIDRY 1973
POLAR 901, 1973*
* 8,700 h.p. max. (3 engines) ice rated

Acadian Freedom

AMERICAN TUG BOAT COMPANY,
City Dock, Pier 2,
Everett, Wa.
Tug:
ARTHUR S, 1966
HOWARD H, 1967
THEILINE W, 1952

AQUAMARINE INC.,
10014 Briar Drive,
Houston, Texas 77042.
Tel: (713) 781 3781
Four tug/supply vessels on order in 1973 from Burton Shipyards, Inc.

ASTRO MARINE INC.,
518 First City National Bank Building,
Houston, Texas 77007.
S/V:
MARS. 1965
SOUTHERN CROSS, 1972

ATLAS OFFSHORE BOAT SERVICE INC.,
Belle Chase,
Louisiana.
Seven supply vessels.

B & B TOWING CO. INC.,
P.O. Box 250, Lockport,
Louisiana 70374.
Five tugs.
Tug ETIENNETTE BOLLINGER, 1967

B & L BOAT RENTALS INC.,
New Orleans,
Louisiana.
Two utility supply vessels.

BAROID DIVISION,
N.L. Industries Inc.,
New Orleans, Louisiana.
Five tugs, four supply vessels.

BAY HOUSTON TOWING CO.,
504 National Bank Building,
Galveston, Texas 77550.
Tug:
BARBARA GEORGE, 1957
C. HADEN MASTERTON, 1944
J. HARRIS MASTERTON, 1957

BAYOU MARINE CORPORATION,
Houma, Louisiana.

BELLANGER, INC.,
Golden Meadow,
Louisiana.
Twelve tugs.

BERARD MARINE INC.,
New Iberia, Louisiana.
One tug and four utility supply vessels.

BERWICK BAY OIL CO.,
Berwick,
Louisiana.
Four tugs.

BISSO MARINE CO. INC.,
New Orleans, Louisiana.
Tug W.A.B., 1944

BLUDCO BARGE & TOWING CO.,
P.O. Box 12424,
Houston, Texas.
Three tugs.
Tug BALTIMORE, 1911

BLUE WATER TRANSPORTATION INC.,
Houston, Texas.
Four utility supply vessels.

BROKEE BOAT COMPANY,
Delcambre, Louisiana.
Four tugs.

An artist's impression of the Aqua Marine tug

CENAC TOWING CO. INC.,
End of Palm Street,
Houma, Louisiana.
Twenty-two tugs.
C.T.C.O. 1TT, 1936
Tug CINDY CENAC, 1962

CHERAMI BROS. BOAT CO. INC.,
Golden Meadow, Louisiana.
Fourteen supply vessels.

CRESCENT OFFSHORE TOWING & SALVAGE CO. INC.,
P.O. Box 2690,
New Orleans, Louisiana 70116.
Seven tugs.
Tug:
HUMRICK, 1919
MARIE G. SMITH, 1958
WILLIAM S. SMITH, 1943

CURTIS BAY TOWING CO. OF DELAWARE,
1501 Mercantile Trust Building,
Baltimore 2, Maryland.
Twenty-five tugs.
Tug:
FELLS POINT, 1956
CAPE HENLOPEN, 1973
CAPE MAY, 1973

CURTIS BAY TOWING CO. OF PENNSYLVANIA,
12 South 12th Street,
Philadelphia 7, Pennsylvania.
Tug:
JUSTINE, 1943
REEDY POINT, 1958
NORTH POINT, 1939
QUAKER, 1902

DEARBORN MARINE SERVICE CORPORATION,
P.O. Box 1147, 430 East 2nd,
Freeport, Texas 77541.
Sixteen supply vessels.
S/V:
DEARBORN 46, 1971
DEARBORN 47, 1972
GULF STREAM, 1965
JENNIFER D, 1970
MAXINE D, 1970

DE FELICE MARINE CONTRACTORS INC.,
142 Livingston Place,
Metairie, Louisiana.
Eleven tugs.
Tug NICOLE D. DE FELICE II, 1966
M.T. BETTY G
Ocean tug ESTEFF S. DE FELICE, 1971

DOMAR INC.,
P.O. Box 390,
Morgan City, Louisiana.
One supply vessel.
S/V OBI-I, 1966

DRESSER OFFSHORE SERVICES INC.,
New Orleans, Louisiana.
Two supply vessels.

EASTERN OFFSHORE BOATS INC.,
c/o Arthur Levy Boat Service Inc.,
P.O. Drawer 968,
Morgan City, Louisiana 70380.
S/V:
CLEO STATER, 1969
F.S. WADE, 1969

EASTERN SEA SERVICES,
Stamford, Connecticut.
Four tug/supply vessels.

EURO-PIRATES INTERNATIONAL, INC.,
1 Shell Square, Suite 4324,
New Orleans, Louisiana 70139.
S/V:
RED BEARD, 1973
BLACK BEARD, 1973
KHAIR ED DIN, 1973

FOSS LAUNCH & TUG CO.,
666 West Ewing Street,
Seattle 99, Washington.
Seventy-five tugs, three supply vessels.
S/V ALASKA CONSTRUCTOR, 1944
S/V ALASKA HUSKY, 1966
Tug:
ALASKA ROUGHNECK, 1944
ANDREW FOSS, 1944
BARBARA FOSS, 1944
BRYNN FOSS, 1952
CORAL FOSS, 1957
CATHERINE FOSS, 1967
ELLEN FOSS, 1943
ERIK FOSS, 1908
HENRY FOSS, 1945
JEFFREY FOSS, 1970
JUSTINE FOSS, 1944
LESLIE FOSS, 1970
MARGARET FOSS, 1943
MARY FOSS, 1944
PATRICIA FOSS, 1944
RICHARD FOSS, 1967
SHANNON FOSS, 1957
WEDELL FOSS, 1904

G & H OFFSHORE TOWING CO.,
505 U.S. National Bank Building,
Galveston, Texas.
Twenty-nine tugs.
Tug:
MANTA, 1965
STURGEON, 1957

GATEWAY MARINE INC.
Five tugs, four supply vessels.

GENERAL MARINE TRANSPORT OF SANTA BARBARA,
Santa Barbara, California.
Six utility supply vessels.

GULF COAST TRANSIT CO.,
Tampa, Louisiana.
Five tugs.

GULF MISSISSIPPI MARINE CORPORATION,
P.O. Box 802,
Harvey, Louisiana 70058.
Tel: (504) 347-6231.
Tug:
GULF DUKE, 1967
GULF EXPLORER, 1967
GULF KING, 1967
GULF KNIGHT, 1969
GULF MASTER, 1968
GULF MISS, 1969
GULF SEAS, 1970
GULF SQUIRE, 1966
GULF STAR, 1970
GULF STORM, 1970
GULF STREAM, 1970
GULF WIND, 1970

GULF OVERSEAS MARINE CORP.,
Lafayette, Louisiana.
Tug/S:
JESSE MECHEM, 1973
TITUSVILLE, 1973

GUZZETTA OFFSHORE MARINE SERVICE LTD.,
Berwick, Louisiana.
Three supply vessels.

JACKSON MARINE CORPORATION,
P.O. Box 9631,
Houston, Texas 77015.
and 1 Newark Street,
Hoboken, New Jersey 07030.
Fourteen tugs, one tug/supply, and ten supply vessels.
S/V BEN H. POWELL, 1965
Tug CAPTAIN CARL, 1970
S/V EASTERN MOON, 1970
Tug:
HARBOR STAR, 1943
MISTER CHRIS, 1970
MISTER RICHARD, 1970
S/V TRUE GRIT, 1971

ARTHUR LEVY BOAT CO.,
P.O. Box 1073,
Morgan City, Louisiana 10380.
Fifteen tug/supply and twenty-one supply vessels.
S/V:
JAVA SEAHORSE, 1971
BALTIC SEAHORSE, 1971
AURORA, 1966
P.E. FOOT, 1970
E.R. LEVY, 1966
J.H. LEVY, 1966
BERING SEAHORSE, 1971
DOURADO SEAHORSE, 1971
CORAL SEAHORSE, 1971

LOUISIANA MUD CO. INC.,
Houma, Louisiana.
Three supply vessels.

LUKE BELLANGER GISCLAIR INC.,
Golden Meadow, Louisiana.

MARINE EXPLORATION CO. INC.,
2295 N.W. South River Drive,
Miami, Florida 33125.
Seven tugs.
Tug DENISE, 1944

MOBILE TOWING CO.,
Blakeley Island, P.O. Box 512,
Mobile, Alabama.
Six tugs.
Tug:
AUSTILL PHARR, 1950
JOHN T. WALSH, 1942
PALMER GAILLARD, 1953

MORAN TOWING CORPORATION,
17 Battery Place,
New York, N.Y. 10004.
Thirty-two tugs.
Tug:
E. F. MORAN JNR., 1940
EDMOND J. MORAN, 1940
ESTHER MORAN, 1963
SUSAN A. MORAN, 1941
CATHLEEN E. MORAN, 1949
SEWELL'S POINT, 1942
JULIA C. MORAN, 1949
MARGARET MORAN, 1946
ANNE MORAN, 1938
CHRISTINE MORAN, 1953
MARTHA MORAN, 1936
KERRY MORAN, 1963
SHEILA MORAN, 1941
EUGENE F. MORAN, 1951
CLAIRE MORAN, 1940
LAMBERT POINT, 1949
DIANA L. MORAN, 1956
CAROL MORAN, 1949
CEDAR POINT, 1949
EAGLE POINT, 1942
MARIE MORAN, 1950
HARRIET MORAN, 1938
PATRICIA MORAN, 1962
MICHAEL MORAN, 1913
NANCY MORAN, 1958
MARY MORAN, 1941
CYNTHIA MORAN, 1957
THOMAS POINT, 1941

NAUTICAL MARINE,
Morgan City, Louisiana.
C. CONQUEROR, 1972

NIKISKI MARINE CORP.,
Anchorage,
Alaska.
Fleet of four tugs and four supply vessels planned for 1973.

THE OFFSHORE CO.,
609 Esperson Building,
Houston, Texas 77002.
Six supply vessels.
LION, 1956

OFFSHORE CO. LTD.,
Panama.
S/V:
OFFSHORE BEAUMONT, 1970
OFFSHORE BIRMINGHAM, 1970

OFFSHORE LOGISTICS, INC.,
Lafayette, Louisiana.
Nine supply and nine tug/supply vessels.
S/V MONARCH, 1970

OTTO CANDIES, INC.,
P.O. Box 25, Des Allemands,
Louisiana 70030.
Tel: (504) 722-1311
Eight 10,000 h.p. tug/supply vessels under construction in 1973, see J. Ray McDermott. Total fleet: over 70 vessels, including fifteen tugs, three tug/supply and five supply vessels.
S/V:
PAUL CANDIES
NICK CANDIES
BESSIE CANDIES, 1971
IDA CANDIES, 1973
BERTILE CANDIES, 1973
BEN CANDIES
OTTO CANDIES

PACIFIC INLAND NAVIGATION CO. INC.,
P.O. Box 831,
6305 N.W. Lower River Road,
Vancouver, Washington 98660.
28 tugs.
Tug:
CHEROKEE, 1927
MOHAWK, 1943
OSCEOLA, 1944

PETREL MARINE CO.,
Houma, Louisiana.
Two tug/supply and six supply vessels.

POTT INDUSTRIES, INC.,
St. Louis,
Missouri.
Offshore vessel fleet.
No details.

PUGET SOUND TUG & BARGE CO.,
P.O. Box 3783,
Seattle, Washington 98124.
Fifty tugs and ten supply vessels.
Tug:
ACTIVE, 1944
MARS, 1970
WANDO, 1916
SEA LION, 1965
SEA RACER, 1966
SEA WITCH, 1957
SEA WOLF, 1965

ROBIN BOAT RENTAL SERVICE INC.,
440 Pailet Avenue,
Harvey, Louisiana.
Four tugs.
Tug VINCENT J. ROBIN IV, 1965

ST. BERNARD BOAT RENTAL INC.,
Chalmette, Louisiana.

SEAL CRAFT OPERATORS INC.,
Galveston, Texas.
Nine supply vessels.

SEAL FLEET INC.,
1011, 23rd Street,
Galveston, Texas.
Two supply vessels.
S/V:
CAMPECHE SEAL, 1964
PACIFIC SEAL, 1964

STATE BOAT CORPORATION,
3701 Kirby Drive,
Houston, Texas 77006.
Tel: (713) 528 5359.
Telex: 77-5125.
Also: Morgan City, Louisiana (504) 395 2838
Six supply vessels.
S V:
STATE ARROW, 1966
STATE VICTORY, 1970
STATE BRIGADE, 1973
STATE COMMAND, 1973
STATE QUEEN
STATE DIAMOND

STEWART & STEVENSON SERVICES INC.,
4501 Baldwin Boulevard,
Corpus Christi, Texas 78403.
Thirteen tugs and supply boats building in 1973.
S/V ANDRE B., 1972

STORM DRILLING CO.,
4141 Southwest Freeway,
Houston, Texas 77027.

NOLTY J. THERIOT INC.,
706 Odeco Building,
1600 Canal Street,
New Orleans, Louisiana 70112.
Fourteen tugs. By December 1973, seven 45·6 m (149 ft 6 in) length ocean-going tugs were on order from the Equitable Equipment Company, Inc.
Tug:
PARIS THERIOT
BEA THERIOT, 1967
BILLY B, 1970
CHIPPY T, 1969

ROBERT ALARIO, 1970
RUSTY B, 1970

TIDEWATER AGENCIES INC.,
(same address, New Orleans)
S/V:
LOW TIDE, 1956
SEARCH TIDE, 1967

TIDEWATER BALBOA INC.,
(same address, New Orleans)
S/V:
BONNY TIDE, 1967
EAST TIDE, 1963
NIGER TIDE, 1966
WARRI TIDE, 1967

TIDEWATER BARGE LINES INC.,
2609 N.E. Marine Drive,
Portland 11, Oregon.
Tug:
CAPTAIN BOB, 1945
CAPTAIN LEW S. RUSSELL JR., 1945
INVADER, 1939
MARY GAIL II, 1944
TITAN, 1945

TIDEWATER BOATS INC.,
P.O. Box 19264,
New Orleans, Louisiana 70179.
S/V CARL TIDE II, 1967

TIDEWATER CAMERON INC.,
(same address, New Orleans)
S/V:
ALLEN, 1965
DE SOTO, 1962
FRANKLIN, 1965

TIDEWATER CARIBE, C.A.,
(same address, New Orleans)
S/V:
BRAVE TIDE, 1970
FORCE TIDE, 1970
INDO TIDE, 1971
MALAY TIDE, 1971

TIDEWATER GRAND ISLE INC.,
(same address, New Orleans)
S/V:
CONCORDIA, 1963
GALLANT TIDE, 1972
TENSAS, 1957

TIDEWATER GUARDIANS INC.,
P.O. Box 166,
Berwick, Louisiana.
S/V STACY TIDE, 1961

TIDEWATER INDUSTRIES INC.,
P.O. Box 19264,
New Orleans, Louisiana 70179.
S/V:
ATLANTIC TIDE, 1966
BIG TIDE, 1963

TIDEWATER KEELS INC.,
(address as above, New Orleans)
S/V BRICK TIDE, 1966

TIDEWATER MARINERS INC.,
P.O. Box 2407,
Morgan City, Louisiana.
S/V AMERICAN TIDE, 1965

TIDEWATER MARINE PANAMA, S.A.
(address as above, New Orleans)
S/V:
CAPE TIDE, 1968
CREST TIDE, 1968
GARY TIDE, 1968
STEADY TIDE, 1968

TIDEWATER MARINE SERVICE CORP.,
(address as above) (Berwick)
S/V VEN TIDE, 1964

TIDEWATER MARINE SERVICE INC.,
3308 Tulane Avenue,
New Orleans, Louisiana 70119
Tel: (504) 822 1740.
Telex: 58-4216.
Two vessels on order from Hall Russell at 31 January 1974. Two at 976 g.r.t. 2 × 2,875 h.p.
Tug PARMO, 1935
S/V:
CANADIAN TIDE, 1967
MIN TIDE, 1967

TIDEWATER NAVIGATORS INC.,
P.O. Box 2149,
Morgan City, Louisiana, 70380.

TIDEWATER QUESTS INC.,
(address as above, New Orleans)
S/V:
STOUT TIDE, 1966
STRONG TIDE, 1964

TIDEWATER SHIPS INC.,
(address as above, New Orleans)
S/V BURCH TIDE, 1957

TIDEWATER SUPPLIERS INC.,
(same address, New Orleans)
S/V NORTH TIDE, 1965

TIDEWATER UNDERTAKINGS INC.,
(address as above, New Orleans)
S/V ARCTIC TIDE, 1966

TIDEWATER VENICE INC.,
(address as above, New Orleans)
S/V BOSSIER, 1962

TIDEWATER VESSELS, INC.,
(address as above, New Orleans)
ROUGH TIDE, 1965
VAN TIDE, 1962

WILLIAM-McWILLIAMS CO.,
New Orleans, Louisiana.

ZAPATA MARINE SERVICE LTD., S.A.,
1701 Houston Club Building,
Houston, Texas 77002.
Three tugs, 21 tug/supply vessels, 5 supply vessels.
S/V:
INVINCIBLE SERVICE, 1970
TITAN SERVICE, 1965
TROJAN SERVICE
CENTURION SERVICE
SIGNAL SERVICE
MONARCH SERVICE
MAJESTIC SERVICE
THOR SERVICE, 1973
NORDIC SERVICE, 1973
VIKING SERVICE, 1973
4 more building, 1973

VENEZUELA

TIDEWATER MARINE SERVICE, C.A.,
Calle 76, P.O. Box 151,
Maracaibo.
Tug:
ONOTO, 1949
PEMENO, 1956
RIO BRAVO, 1970

OIL SPILL CONTAINMENT AND RECOVERY SYSTEMS

CANADA

BENNETT POLLUTION CONTROLS LTD.,

SKIMMER

INSHORE/OFFSHORE SKIMMER (in excess of 200 US gallons/minute).

Principle of operation: Boat-carried belt system

Variations of basic type: Mark I 12·2 m× 4·25 m (4·0 ft× 14 ft); Mark II 11·0 m× 3·65 m (36 ft× 12 ft)

Design history

Date of construction of prototype started: January 1969

Date of prototype completed: December 1972

Date trials completed: April 1973

Number of units in service: Two Mark II Skimmer; One Mark I Skimmer

Operation

Under normal operating conditions, only the belt recovery system is utilised. With a recovery capacity in excess of 200 G.P.M. the belt system is adequate to handle the majority of accidental spills.

In operation, oil passes through the perforated, inclined ramp and collects in the trough between the pontoons. The perforated ramp acts as a wave buffer and primary trash barrier. A man standing on the side of the ramp can remove floating trash and place it into the empty boom storage bins. At the end of the trough the floating oil is picked up by an endless belt and then squeezed from the belt as it passes through a pair of neoprene-covered drive rolls. The recovered oil is then collected in two tanks situated directly below the roller assembly and from there overflows into two forward storage tanks.

In the event of a major spill, where large volumes of oil must be processed quickly, the weir system is activated. The weir system consists of an open adjustable sump, located between the pontoons ahead of the belt system. A skimmer mounted rotary vane pump is used to transfer the bulk of oil from the weir to an accompanying oil barge while any remaining oil is simultaneously recovered with the belt system. The Bennett Skimmer is used in conjunction with the Bennett Offshore Boom 1·8 m (6 ft) and/or Bennett Inshore/Offshore Boom 1·02 m (3 ft 4 in), as illustrated and described below.

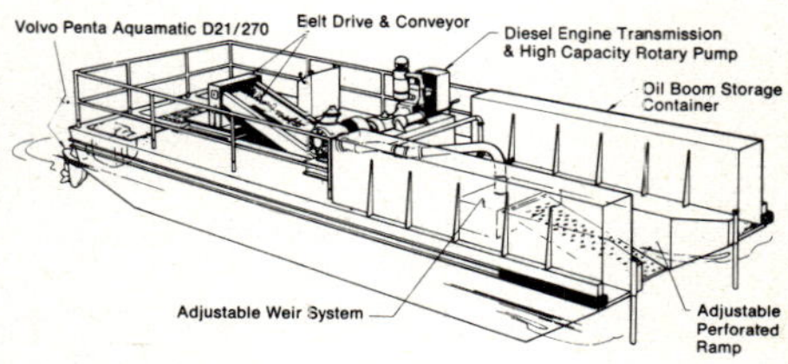

The Bennett inshore/offshore skimmer

Specifications of the Mark I and Mark II Inshore/Offshore Skimmers

	Mark I	Mark II
1·02 m (4 ft) Inshore/Offshore Boom carried onboard	61·0 m (200 ft)	61·0 m (200 ft)
Belt system	2·54 cm (1 in) Polyurethane foam, backed with stainless steel mesh	
Oil storage capacity	80 barrels	30 barrels
Pumping system	15·24 cm (6 in) Rotary vane	10·16 cm (4 in) Rotary vane
Maximum pumping capacity	750 gpm	450 gpm
Prime mover	— — — — Diesel hydraulic — — — —	
Propulsion system (diesel)	2–100 h.p. I/O drive	2–75 h.p. I/O drive
Hull: Length	12·2 m (40 ft)	10·4 m (34 ft)
Beam	4·25 m (14 ft)	3·65 m (12 ft)
Draught	0·69 m (27 in)	0·61 m (24 in)
Weight (dry)	21,400 kg (47,000 lb)	12,700 kg (28,000 lb)

Costs (July 1973)

Approximately $75,000 to $100,000

MINI SKIMMER

The Bennett Mini Skimmer is designed to remove oil and effluent matter from the water surface in confined and restricted areas, e.g. settling basins, as well as rivers, lakes and harbours. The Mini Skimmer is fitted with three adjustable outboard stabilising floats, allowing for precise adjustment of the skimmer's draught.

Drive unit and pump are totally submersible, thereby ensuring continuous performance, even under adverse conditions. Petrol-engine driven pumps can be supplied.

Specification

Weight	Approx. 82 kg (180 lb)
Construction	Aluminium or fibreglass
Saucer size	1·05 m (3 ft 6 in) dia.
Overall dimensions	1·85 m (6 ft 2 in) dia.
Pump	Submersible, electric, 2 h.p., 115/230 volts 60 cycle, single phase
Max. pumping capacity	148 US gpm
Max. head	22·7 m (75 ft) water

The Bennett Mini Skimmer

Booms

Bennett Pollution Controls Ltd. produce oil containment booms of five depths: 0·46 m (18 in), 0·61 m (24 in), 0·915 m (36 in), 1·02 m (40 in) and 1·83 m (72 in).

	0·915 m (36 in)	1·02 m (40 in)
Specifications	*Tanker boom*	*Inshore/Offshore boom*
Freeboard	30·5 cm (12 in)	33·0 cm (13 in)
Draught	61·0 cm (24 in)	68·6 cm (27 in)
Weight/ft run	2·4 kg (5·25 lb)	2·7 kg (6 lb)
Section length	15·24 m (50 ft)	15·24 m (50 ft)
Top strength member	0·96 cm ($\frac{3}{8}$ in) steel cable	0·96 cm ($\frac{3}{8}$ in) steel cable
Bottom strength member	0·96 cm ($\frac{3}{8}$ in) steel cable	0·96 cm ($\frac{3}{8}$ in) steel cable
Total grab tensile	12,750 kg (28,000 lb)	12,750 kg (28,000 lb)
Ballast	Lead weights on bottom cable	Lead weights on bottom cable
Flotation/ETHAFOAM	15·24 cm (6 in) dia. round 5·08 cm × 15·24 cm (2 in × 6 in) oval	20·32 cm (8 in) dia. round 5·08 cm × 15·24 cm (2 in × 6 in) oval
Method of joining boom sections	Fabric slid into PVC joiner, cables shackled	
Costs (July 1973)		
Per foot run	\$ (Can) 18·00	\$ (Can) 20·00
Per metre run	\$ (Can) 59·00	\$ (Can) 65·60

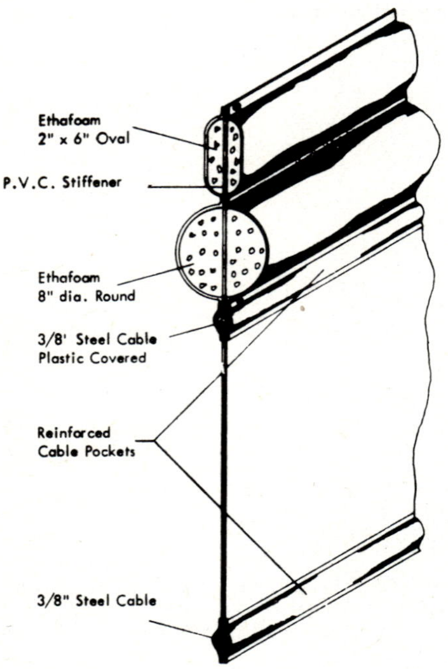

Cutaway diagram of the Bennett inshore/offshore boom

Specifications of the Bennett Inshore Oil Boom

	0·46 m (18 in)	*0·61 m (24 in)*	*0·915 m (36 in)*
Freeboard	0·15 m (6 in)	0·203 m (8 in)	0·305 m (12 in)
Draught	0·305 m (12 in)	0·405 m (16 in)	0·61 m (24 in)
Weight per foot	0·91 kg (2 lb)	1·14 kg (2·5 lb)	1·36 kg (3 lb)
Section length	15·2 m (50 ft)	15·2 m (50 ft)	15·2 m (50 ft)
Top strength member	N/A	5·08 cm (2 in) nylon webbing	0·63 cm ($\frac{1}{4}$ in) steel cable
Bottom strength member	0·63 cm ($\frac{1}{4}$ in) steel cable		
Total grab tensile	3,150 kg (7,000 lb)	5,900 kg (13,000 lb)	6,300 kg (14,000 lb)
Ballast	lead weights crimped to bottom cable		
Flotation—oval ethafoam	one 5·08 cm × 15·2 cm (2 in × 6 in)		
Storage containers	polyethylene coated nylon fabric		
Container size	1·52 × 0·61 × 0·31 m (5 ft × 2 ft × 1 ft)	1·52 × 0·61 × 0·31 m (5ft × 2ft × 1 ft)	1·52 × 0·45 × 0·92 m (5 ft × 1·5 ft × 3 ft)
Method of joining boom sections	fabric slid into pvc joiner, cables shackled		
Costs (July 1973)			
Per foot run	\$ (Can) 7·00	\$ (Can) 8·00	\$ (Can) 12·00
Per metre run	\$ (Can) 23·00	\$ (Can) 26·24	\$ (Can) 39·36

Offshore Oil containment and removal

Equipment required
3—Tugs (600 h.p.)
1—Recovery Barge, approximate size 50·3 m × 13·3 m (165 ft × 44 ft) complete with vacuum pump.
1—Skimmer
2—150 m (500 ft) lengths of **Bennett Offshore Filtration Boom**, complete with towing planes.

The key to a successful operation is the oil boom and the method by which it is deployed.

Boom specification

The Bennett Offshore Boom is 1·8 m (6 ft) high overall, of which approximately 1·2 m (4 ft) is submerged, leaving a freeboard of approximately 0·61 m (2 ft). The boom is completely flexible in a longitudinal direction, which allows it to follow the changing contours of the ocean. The boom is, however, sufficiently stiff vertically, to maintain its stance when subjected to high winds and breaking waves. The submerged section is made of a specially treated filtration material which is oilephilic in nature and permits the passage of water, while the oil is retained. These features together with the inherent structural strength give the Bennett Boom the capability to successfully contain oil even under adverse current and sea conditions.

Deployment

The two 150 m (500 ft) lengths of boom are held to the skimmer (at a distance of not less than 0·91 m (3 ft) by means of two rigid arms, fitted with Universal Joints at both ends. The resulting gap between the skimmer and boom is sealed with a small section of boom. The first 30 m (100 ft) sections of each boom (from the connection at the skimmer) are held together by cross cables at the 10, 30, 50, 70 and 100 ft marks. The attachment to the boom is made by a wire bridle, equally distributing the load across the height of the boom. The cross cables are of a pre-determined length to maintain an angle of maximum 30°.

The towing plane at the end of each 150 m (500 ft) boom section consists of a vertical steel pipe, attached to each of the three longitudinal boom cables, and is headed by a torpedo shaped float.

The angle of tow is determined by the prevailing current in accordance with the table shown above.

No. (3) tug is the control tug, maintaining the position of the recovery barge and skimmer, keeping the tension on the booms and directing tugs (1) and (2).

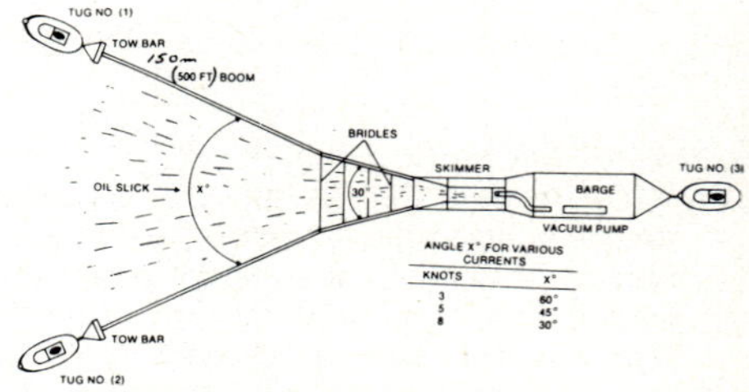

The Bennett oil containment boom

Applications

The above method of offshore oil containment and removal was successfully employed in the Gulf of Mexico, where currents in excess of 2 knots were encountered.

R.B.H. CYBERNETICS

SLICKLICKER

With this system, oil is picked up by a specially designed high capacity continuous endless belt treated to accept oil and reject water. Oil is removed fromthe belt at the upper end of travel by squeeze rollers and deposited in containers by gravity. These containers can be dumped or pumped out, depending upon requirements.

The standard size is with 0·92 m (3 ft) wide belt, 4·57 m (15 ft) long boom, and has a capacity of 50 Imperial gallons per minute of up to 98% oil. Machine size can be scaled up or down. Hydraulic control is provided and the system can be powered electrically or with gasoline or diesel motors.

First conceived in 1968, the first prototype was completed 1969, production prototype completed 1970, and tested on Arrow Oil spill, Chedabucto Bay, Nova Scotia, on Bunker C at freezing temperatures. Some improvements were made at that time and improvement continues.

Forty-seven units are now in service, in Canada, United States, Israel, Japan, Taiwan, England, and present operators (Spring 1974) include Canada's Ministry of Transport, Israel's Eilat-Ashkelon Pipeline Company, Clean Seas (Canada) Ltd, IOCO Refinery of Imperial Oil Enterprises Ltd.

The Slicklicker can be operated from any fairly low, freeboard boat. Oil can be collected in open ended oil drums, in tanks, or collapsible floating elastomeric tanks. The elastomeric tanks provide great flexibility in that they are light to carry when empty, and can be set adrift for later collection when full. Under emergency conditions, the Slicklicker has been set up on two, decked-over, 5·48 m (18 ft) rowboats, and oil collected in floating drums which were set adrift when filled. Its versatility and the fact that it does not clog with debris are two of its greatest assets.

Boat carrying Slicklicker can be dragged behind containment booms that concentrate oil toward Slicklicker; or if guiding floats are built into system, boat can move through slick picking up oil, or boom on Slicklicker can be extended over containment boom to pick up contained oil from outside boom.

The recommended collecting craft is 'Pelikan Cargo Landing Craft' modified to carry Slicklicker and tank, the weight of Slicklicker being 1,136 kg (2,500 lb).

Length	10·92 m (35 ft 10¾ in) (including Slicklicker boom)
Width	3·49 m (11 ft 5¼ in)
Height	1·83 m (6 ft)
Draught:	
Light load	0·25 m (10 in)
Heavy load (4,000 gal. oil)	0·66 m (2 ft 2 in)

Engine: 300 h.p. Volvo diesel with jet drive, for high speed; can be 100 h.p. if speed is not necessary; other options available if desired

Diaphragm pumps to handle all weights of oil and small debris if oil is contaminated with debris

Speed: Twenty to twenty-five knots with 300 h.p. motor

Cost

$43,000 (Can) F.O.B. Victoria, B.C., Canada, with 300 h.p. diesel; less with smaller motor or gasoline motors.

Slicklicker mounted on Rotork Sea Truck during trials at Vancouver

Slicklicker mounted on Rotork Sea Truck during trails at Vancouver

FRANCE

NOUVELLES APPLICATION TECHNOLOGIQUES S.A. (N.A.T.)

VORTEX OIL DRINKER VX150

The Bertin/N.A.T. Vortex Oil Drinker, VX 150

Principle of operation

A vortex is induced in the surface of the water/oil mixture, thereby inducing an inward and downward flow of the upper surface of the water carryir ı the oil.

Design history

Developed by Bertin et Cie.

Prototype manufactured during the second half of 1970.

Trials in calm waters completed in April 1971 and tests on the high seas in November 1971 in the bay of Douarnenez.

Production

Four units presently in service (ordered by the French Merchant Services).

Dimension and weights

Length	7·6 m (24·9 ft)
Width	6·7 m (22·0 ft)
Height	3·0 m (10·0 ft)
Weight	4·7 metric tons

Main components

Floating platform with essentially:

a skirt able to be raised with a bottom (a slit is between the skirt and its bottom); this feature permits drag to be decreased during towing;

a fixed-pitch vertical axis propeller;

a hydraulic station (5 h.p.) set up on the module in order to control the hydraulic impeller (2 h.p.) engine and the various control systems concerning the lifting of the skirt, the suction and discharge valves, the locking of the skirt in up and down positions;

a pumping unit with strainer and different pipes and valves;

a Richier P 492 centrifugal pump actuated by a CEN 160 L 2 B flame-proof electric motor (25 h.p.), 3,000 rpm, being able to operate with 440 volt/60 Hz or 380 volt/50 Hz.

The platform is connected to a ship (or a dock) by a floating pumping pipe and the cables bring energy to the Vortex.

A control cabin, set on the ship (or on the dock) facilitates control of the process.

TheMod of operation

The Vortex recovery system may be raised or lowered from a ship, the skirt being in a raised position, to break free from the weight of water in the skirt. The unit will lie (on its floats) on the ground, the skirt being locked in the up position.

The purpose of the control cabin is to monitor the depth of the oil vortex; the rotation speed of the propeller is adjusted so as to have the strainer inside the oil (pumping is carried out at that time).

When a detection electrode is in the oil, the current is not flowing, whereas the contrary takes place when this electrode is in the water. The control box is protected, waterproof and of marine type, and it includes three galvanometers with adjusting potentiometers. (The current between detection and common electrode will be cut off when the propeller engine is not working.)

Capability of system

The pumping performances are:

Q = 150 m³/hour (5,300 ft³/hour)—suction head: 1 m (3·28 ft) of water—discharge head: 10 m (32·8 ft) of water

Q = 125 m³/hour (4,450 ft³/hour)—suction head: 1 m (3·28 ft) of water—discharge head: 20 m (65·6 ft) of water

Q = 90 m³/hour (3,170 ft³/hour)—suction head: 1 m (3·28 ft) of water—discharge head: 30 m (98·4 ft) of water

Maximum pumping rate	185 m³/hour (6,550 ft³/hour)
Oil recovery efficiency	79%

Limiting conditions

Wind velocity	35 knots (with 45 knot gusts)
Swell	1·5 to 2·0 m (5 to 6·6 ft) wave height 10 to 20 m (33 to 66 ft) wave length

Application

Four units have been ordered by the French Merchant Services and delivered in the following harbours: Le Havre, Cherbourg, Brest and Toulon. They will be operated by the Frency Navy within the framework of the POLMAR plan.

NORWAY

SKUTENG A/S

OIL SKIMMER OS 210

OS 210 underway

Principle of operation

A water and oil layer is sucked into a centre tank by means of a water-jet system. The oil is then separated from the water by floating over a skimming gate to an oil sump, and then it is pumped into four wing-tanks.

The sea water is sucked from the bottom of the centre tank and pumped out by the water jet, which in turn drives the vessel.

All vital functions are hydraulically operated from the pilot house.

Variations of basic type

The OS 210 is made in four versions:

(i) Oil Skimmer

(ii) Oil Skimmer and harbour cleaning vessel, the latter function achieved by a conveyor being lowered into the water in the centre tank

(iii) Oil Skimmer and fire fighting vessel, the latter function achieved by a foam/water mixing unit and a hydraulically operated monitor with a capacity of 1,600 ltr/min (3,531 gals/min) at 8 kp/cm: and a throwing length of 65 m (21·4 ft)

(iv) Oil Skimmer, harbour cleaning vessel and fire fighting vessel, described as above

OS 210 in preparation for an oil skimming operation

Design history

Construction of prototype started June 1970.

Prototype completed July 1971.

Trials completed October 1972.

One has been produced and is in service and two more are now nearing completion.

Weight

Approximately 6,000 kg (13,200 lb).

Method of operation

The vessel is lowered into its working position by filling up the four wing-tanks. The oil/water layer is sucked into the centre tank, where the oil is separated from the water. A mixture of oil and water, dependent on the thickness of the layer of oil on the water, is pumped into the four wing-tanks. Here the oil settles and excess water pumped into the wing tanks is pumped back into the centre tank through the overflow system for a second skimming process. The excess water pumped out of the wing-tanks is being picked up from the bottom of these. This is an automatic/continuous process. When oil is coming out of the 'overflow' system, the wing-tanks are full.

It is also possible to pump the collected oil or oil/water direct from the oil sump and into a barge, a rubber floating tank or other suitable container.

Collecting rate

Dependent on weather conditions, and on thickness of oil layer, than on a smooth water surface the following rates are assured:

Thickness of oil layer	*Capacity*
2 mm (0·08 in)	25 m^3/hour (880 ft^3/hour)
5 mm (1·97 in)	15 to 20 m^3/hour (530 to 710 ft^3/hour)
10 mm (3·94 in)	25 to 30 m^3/hour (880 to 1,060 ft^3/hour)
20 mm (7·88 in)	45 to 55 m^3/hour (1,590 to 1,940 ft^3/hour)
20 mm (7·88 in) and more	Full pump capacity at actual viscosity

Tank (storage) capacity	10,000 litres (2,200 Imp. gals)
Pump capacity	Max. 2,000 litres/min (440 Imp. gals./min)

Dimensions of craft

Length, overall	10·5 m (34·4 ft)
Width, overall	3·8 m (12·4 ft)
Height, overall	4·5 m (14·8 ft)
	2·9 m (9·55 ft) with detached pilot house
Draught, overall	0·6 m (1·97 ft) in ballast
	1·6 m (5·25 ft) in working position

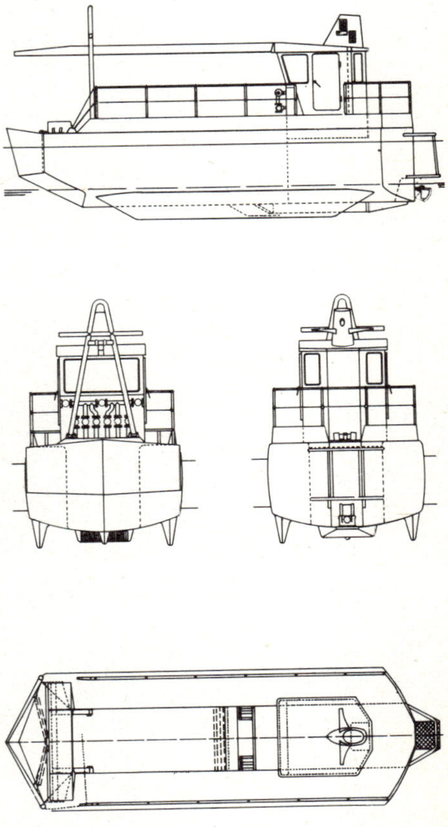

General arrangement drawings of the Skuteng OS 210 oil skimmer

SWEDEN

ATLAS COPCO AB

PNEUMATIC BARRIER

Atlas Copco pneumatic barriers for oil containment have been installed in Antwerp, at Hässelby, Stockholm and also at Emden. Typical costs for the installation of such barriers in harbour areas are around £10,000.

With the Atlas pneumatic barrier a well-defined continuous line of demarcation is established between the oil and the unpolluted water. The volume of air released can be increased if the oil layer becomes greater, if the wind rises, etc.

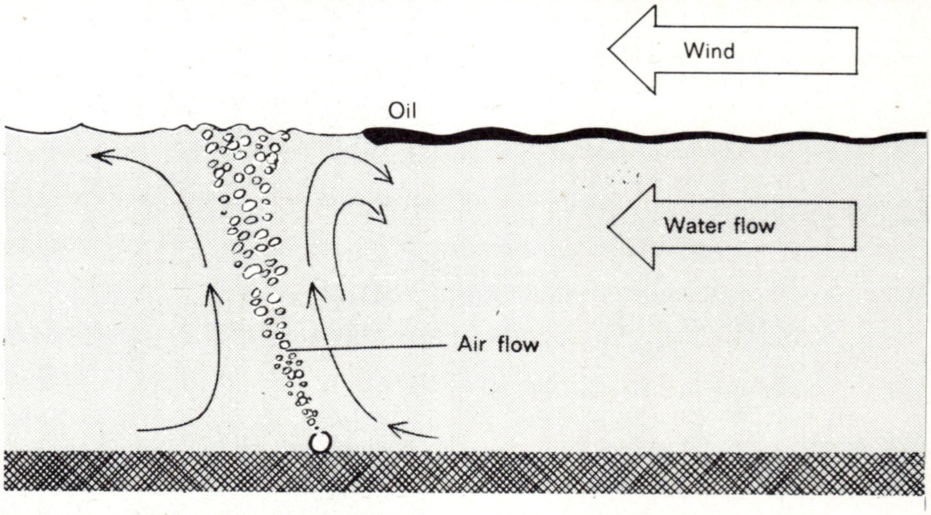

The Atlas Copco pneumatic oil barrier will, under the effects of wind, waves and current stop the advance of the layer of oil by a vertical and horizontal stream of bubbles.

TRELLEBORGS GUMMIFABRIKS AB

RODE ORM BOOM

This disposable boom was the first type developed by Trelleborgs and was designed for use in moderate sea conditions. Experience with the Rode Orm led to the development of the Universal Oil Boom.

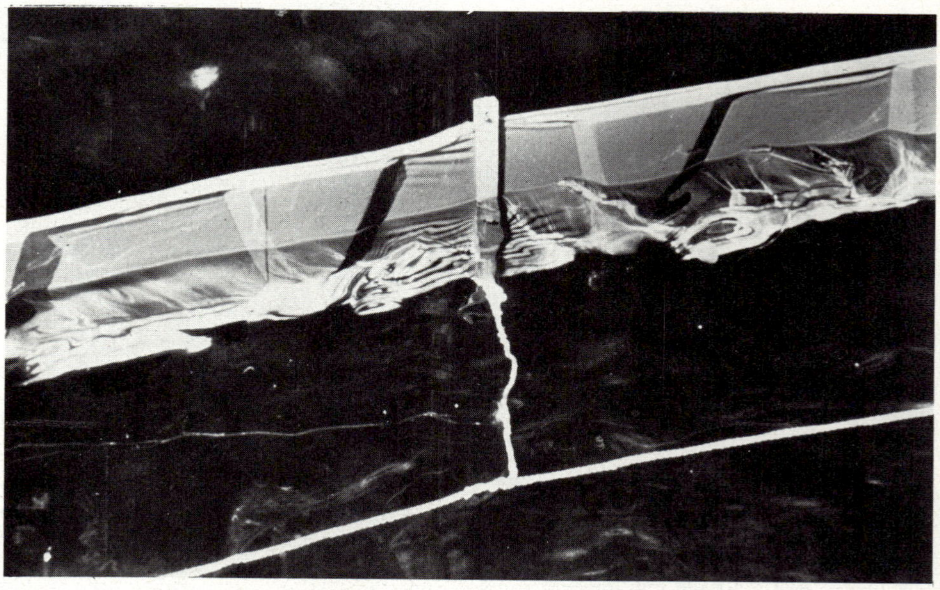

Rode Orm oil boom

Rode Orm oil boom

UNIVERSAL OIL BOOM

The boom is supplied in lengths of 35 m (115 ft), ready-packed on pallets measuring 1·2 m× 1·6 m (3 ft 11 in× 5 ft 3 in). Each section weighs approximately 245 kg (540 lb). Handling is facilitated by hand grips on both sides at about 1 m (3·3 ft) intervals. When stowed on ships or in vehicles, the pallets may be used or the boom sections may be stowed in existing racks, etc., if space is short.

The depth of the boom is 1,050 mm (3 ft 5 in) and it is made of polyurethane-coated woven nylon, with a tensile strength (2 in strip test) of 225 kg (495 lb) and a tear strength of at least 12 kg (25·5 lb). The ageing properties of the thread used are comparable to those of the other components.

Each section incorporates 36 plastic foam floats in protective PVC envelopes. Glass-fibre reinforced plastic panels provide rigidity and reinforcement at securing points. Iron powder in pockets provides the necessary ballast, each sinker pocket containing 5·4 kg (12 lb) of powder.

Tests under wind conditions up to 14 m/s (Beaufort (7 to 8), with currents up to 2·5 knots and waves 2·4 m (8 ft) high have given extremely encouraging results.

Universal Oil boom.

The two illustrations show the deployment of Trelleborgs' Universal Oil Boom, and a close-up view of the boom deployed for a containment operation

Power plant
Volvo TMD 70 B, driving a Castoldi 06 water jet, having a thrust of 600 kp.
Hydraulically driven, infinitely variable screw-spindle pump.

Speed
Approximately 10 knots.

Purchase costs (1973)
In variation: (i) N.kr 510.000
(ii) N.kr 534.000
(iii) N.kr 543.000
(iv) N.kr 563.000

The Skuteng oil skimmer with the bow gates fully open

SKUTENG WSA OIL BOOM

When using the OS 210 in oil spills Skuteng WSA Oil Booms may be attached to the bow gate by a special quick-link connection. These booms are stretched out into a V-form in front of the oil skimmer, making the 'gap' considerably more effective.

The booms are made of polyamide cloth with a PVC layer and can be stored indefinitely. Each length is packed in a quick release storage cover having an outer PVC foil for transport protection. Coupling of the booms can take place whilst laying them out at sea. An iron bar having hoops is placed at one side of the boom's end, the hoops penetrating the eyes of the boom. A rod is then put through the hoops, locking the booms to each other.

Each boom is 25 m (76·2 ft). The 75 cm (29·5 in) long floats have a diameter of 20 cm (0·79 in) and are made of non-hydroscopic hard foam plastic material. The skirt of the boom is 38 cm (15 in) and at the bottom it is equipped with chains, which take the pull when towing.

For better towing three ropes having carabine hooks are connected at one end in a strong towing ring. The weight of one length of 25 m (82 ft) is 65 kg (143 lb). The booms will not tilt by heavy sea and will therefore give good barring under all weather conditions. Maintenance and cleaning after use are very simple.

The Skuteng WSA oil boom

TRYGVE THUNE A/S T-T OIL RECOVERY SYSTEM

The T-T Oil Recovery System has been developed by Mr. Trygve Thune, as a means for mechanical control of oil pollution, to avoid the use of chemicals which destroy not only the composition of the oil, but also the surrounding marine life.

Super T-T Oil Boom

Description

The T-T Oil Boom is made of PVC-coated nylon fabric which according to the International Standard of Measures holds a tensile strength of 332 kg/5 cm (730 lb/2 in) width. The coating is made of PVC which is treated by a special method which makes the coated baffle extremely resistant to oil and sea-water deterioration. The baffle is 0·91 m (3 ft) wide of which 0·6 m (2 ft) are submerged when in water. The T-T Oil Boom comes in standard sections of 50 m (164 ft), which by means of overlapping can be joined together to provide any required length. Each end of the boom is reinforced by aluminium strips. Terylene roping is sewn into the upper and lower edges, and a chain is fixed to the lower edge for holding the stretch, caused by the current, and therefore preventing expansion of the submerged part of the boom baffle. The stabiliser equipment, i.e. the stiffeners, the floats and the balance-weights are spaced at 60 centimetre (2 ft) intervals on the baffle. This makes the T-T Oil Boom very flexible both vertically and sideways, and thereby allows the boom to follow the waves and constantly keep its normal height above the surface under choppy conditions. The buoyancy effect of 300 kg (670 lb) by means of 162 floats, is made of unbreakable expanded PVC, and attached to the baffle by double snap-hooks, together with the force of balance of 80 kg (36 lb) lead weights, makes the T-T Oil Boom extremely stable in the water. The T-T Oil Boom does not lose captured oil from underneath the lower edge, in a current forces in up to 2 knots. The weight of the T-T Oil Boom is low—only 4 kg/m (2·7 lb/ft) and can be stored for immediate use, when folded in zig-zag on a catch-bed.

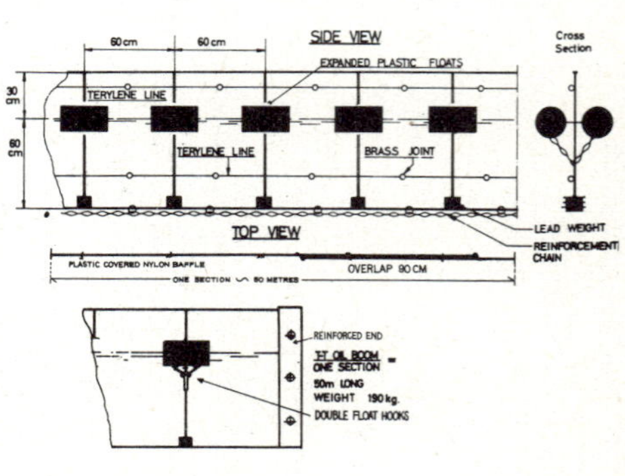

T-T boom construction diagrams

The Trygve Thune Super T-T Oil Boom

Specifications

Freeboard	30 cm (12 in)
Draught	60 cm (24 in)
Standard length	50 m (164 ft)
Weight	4 kg/m (2·7 lb/ft)
Buoyancy	5·53 kg/m (3·75 lb/ft)
Tensile strength	DIN: 53,354 = 310 kg/5 cm (682 lb/2 in)
Strength member	100 kg (220 lb) for chain at the lower edge of the boom
Grab tensile	DIN: 53,354 = 270 kg/5 cm (594 lb/2 in)
Tear strength	DIN: 53,356 = 40 kg/5 cm (88 lb/2 in) for both warp and fill
Storage volume	2·95 m^3 (105 ft^3)

Limiting conditions

Wind velocity	9 knots
Current	1·75 knots
Wave height	0·76 m (2·5 ft)

Trygve Thune have developed magnetic clamps for attaching boom ends quickly to the side of a ship.

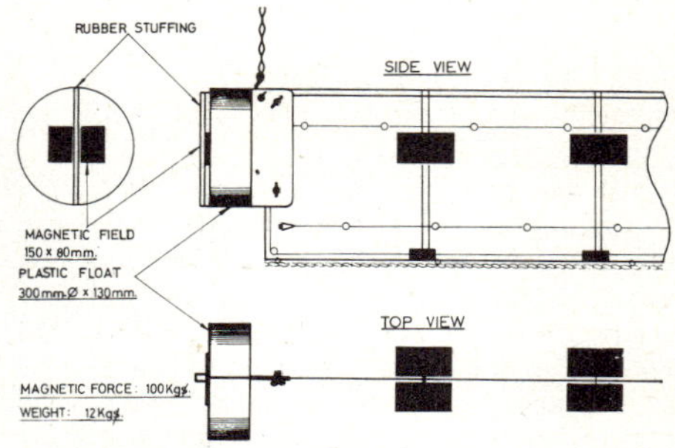

Magnetic boom ends for attachment to ship hulls

Costs (1973)

Super T-T oil boom, standard 50 m (164 ft) length. Norwegian Kr. 7,000 including packing FOB Norwegian Ports.

Paravanes for sweeping purposes, one pair, aluminium, weight 100 kg (220 lb): Norwegian Kr. 4,000 including packing.

Magnet-attachments, one pair, weight 25 kg (55 lb): Norwegian kr. 2,400 including packing.

Applications of the boom

In use by oil companies and harbours in over 45 countries.

T-T OIL RECOVERY VESSEL

This vessel is in the form of a simple pontoon craft on which is mounted a rotating paddle drum and a separating chamber for oil and water. The chamber has a perforated bottom allowing the water to escape whilst the concentrated oil overflows into an oil receiver and is then transferred by a pump to barges or tanks on shore.

Design history

Date construction of prototype started: 1959

Date prototype completed: 1963

Date trials completed: 1965

Production

9 units built.

Weight of system and main components

1,800 kg (818 lb)

Capability of system

Collecting rate: Up to 25 tons of oil in one hour

Capacity: The oil is not stowed on board

Support or collecting craft description

Length, overall	6 m (19·7 ft)
Width, overall	2·5 m (8·3 ft)
Height, overall	1·2 m (4·0 ft)
Draught	Empty tanks—0·2 m (0·66 ft) trimmed down 0·75 m (2·46 ft)
Propulsion—power plant and thrust device	Air-cooled diesel motor twin propellers
Pumping system for oil collection	Diesel-powered centrifugal hydraulic pump
Speed, max. to site	3 knots approx.

Availability

Purchase (1973), capital cost: Norwegian Kr. 190,000 F.O.B. Oslo.

The T-T oil recovery vessel

Operators of the T-T oil recovery system

1968: One complete system to Gulf Oil Terminals (Ireland) Ltd., P.O. Box, Bantry, Co. Cork, Ireland.

1969/70: Three systems to East Coast Services, 343 Washington Street, Braintree, Mass. 02184, USA

1970: One complete system to Merenkulkuhallitus, (Department of Shipping) Laivatoimisto, Helsinki, Finland.

1971: One complete system to Y. Berg, Engineering Consultants & Services Ltd., 16 Matzada St., Ramat-Chen, Ramat-Gan 52235, Israel.

1972: One complete system to Oslo Havnevesenet, (Oslo Harbour Authority) Oslo.

1972: One complete system to the Port Authority and the major oil companies in Reykjavik, Iceland

1973: A complete system is under construction for Petroleo Brasileiro S.A., Rio de Janeiro, Brazil. Delivery of this system took place in April, 1973

UNITED KINGDOM

AMBLER ENGINEERING
SLURP

The kit is centred on the 'Slurp' skimmer. Introduced in 1971, the 'Slurp' has been successfully used on hundreds of oil slicks by Oil Companies, Local Councils, River Boards, Harbour Authorities, British Admiralty, US Navy, Fire Brigades, Government bodies and Industrialists of all types. It is now used around the world with agents in Australia, Japan, USA, Canada, Germany, France, Switzerland, Norway, Sweden and New Zealand.

The 'Slurp' can skim oil from the surface of a river or harbour at up to 2,500 Imp. gallons/hour. It has no moving parts and automatically adjusts itself to the most efficient skimming position after simply being thrown into the water. Oil is drawn in over a curved weir and pumped away through a flexible suction hose. The depth of oil drawn over the weir is automatically controlled by the pumping rate so that adjustments for different thicknesses of oil slick can be remotely made at the pump.

The Slurp skimmer produced by Ambler Engineering.

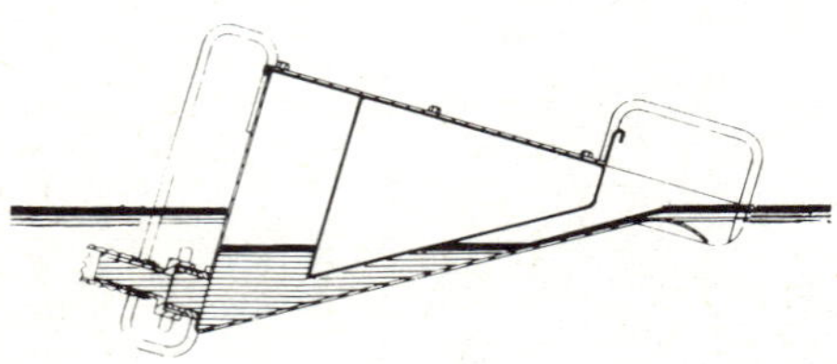
Fast pumping gives a deep weir immersion suitable for the rapid recovery of thick oil slicks

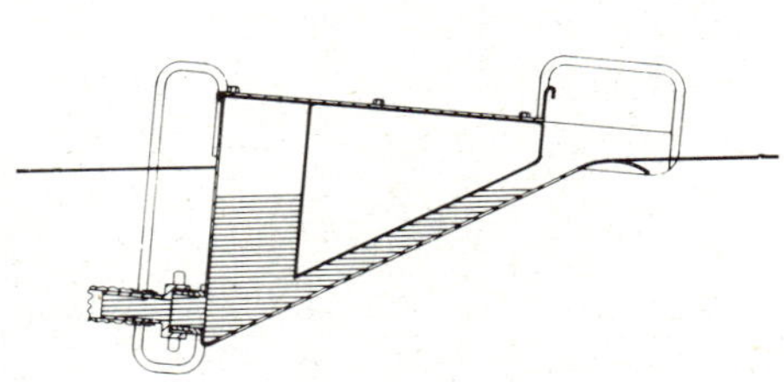
As the slick thins out the pumping rate slows down . The liquid level inside the SLURP rises and the skimmer tilts to reduce the weir immersion

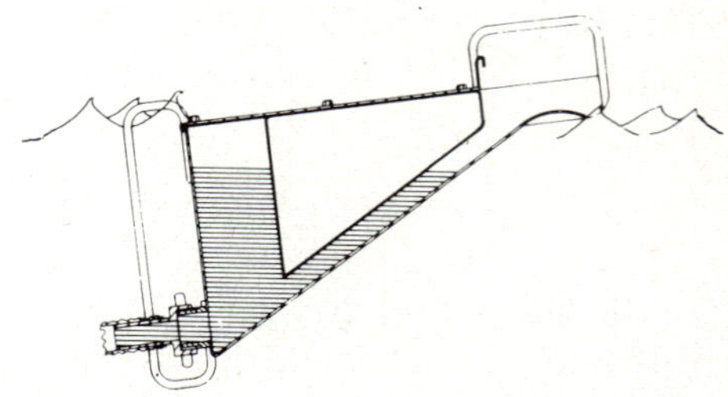
In choppy waves the weir automatically rises until only the wave crests are collected at the rate set by the pump

MEGATOR PUMPS AND COMPRESSORS LTD.

MINI BOOM

The Megator Mini Boom is intended primarily for concentrating oil that has already been contained by a large boom but it can itself be used as a boom under suitable conditions. When the Mini Boom has been positioned to encircle the oil, it is drawn in to reduce the area and so concentrate the oil for more efficient skimming. It also acts as a barrier against flotsam which would choke the skimmer.

The standard boom length is 5 m (16·5 ft) with nylon eyelets at each end enabling two or more lengths to be lashed together and mooring ropes attached. It is 230 mm (9 in) deep and floats with 50 mm (2 in) freeboard. Made from heavy-duty, oil resistant, PVC coated nylon cloth, the Megator Mini Boom has completely enclosed flotation elements of closed cell foam. The mild steel weights at the bottom of the skirt are sealed in and the whole exterior presents a smooth surface which is easy to clean.

The Megator Mini Boom for oil slick containment

The 5 m (16·4 ft) length weighs only 4·8 kg (10·5 lb) and can be folded up for easy transport but is ready for immediate use without having to be inflated or have loose weights added.

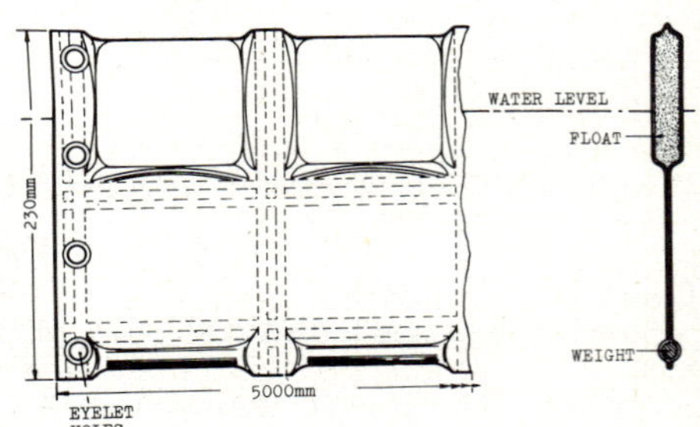

Constructional diagrams of the Megator Mini Boom

THE BRITISH PETROLEUM COMPANY LIMITED

VIKOMA SEAPACK

The Vikoma Seapack is a fully self-contained boom deployment unit which requires only the services of a towing and recovery vessel to complete the operation. The vessel would not need any modification or special equipment other than a simple roller.

The Vikoma Seaboom is a three-tube fabrication in a special Butaclor-coated nylon fabric. The top tube is a (7·5 cm) (3 in) flotation cuff, the intermediate 68·6 cm (27 in) tube is filled with air and the lower 43 cm (17 in) tube is filled with water. The 488 m (1,600 ft) of boom is in a 7·01 m (23 ft) GRP hull and is connected at one end to a diesel driven fan and ducted propeller water pump. These simultaneously fill the air and water chambers of the boom respectively. A compressed air cylinder is also included for inflating the cuff.

No crew is required and no manual attention to the engine compartment is necessary after launching.

The Vikoma Seapack can be transported by road or sea.

Principle of operation: 3-tube configuration boom housed in a self-contained hull requiring only the services of towing and recovery vessel to complete the operation

Design history
Date construction of prototype started: 1970
Date prototype completed: 1971
Date trials completed: System trials still in hand but trials held in Sweden in 1972, UK 1972/1973, Kuwait 1973, Germany 1973

Production
15 produced (at September 1973)

Basic dimensions of system
7·01 m (23 ft) long × 2·44 m (8 ft). Draught 0·56 m (1 ft 10 in), 488 m (1,600 ft) of boom

Weight of system and main components
4 tons

Capability of system
Collecting rate: Depends on current and wind speed
Capacity when deployed: Up to 1,000 tons of oil

Applications
Demonstrations of Boom Deployment and Recovery

Seapack units have been deployed and the boom recovered without damage in all weather conditions in the Solent and adjacent areas on many occasions. Deployment and recovery can now be said to be a routine matter requiring only a towing vessel, preferably fitted with a simple recovery roller, and competent seamen who have seen one or two boom deployments and recoveries. All of these have been done in the main without placing any oil on the water. However, in December 1970, using the original Seaclear deployment technique and in Force 6 winds with short seas 5 to 6 ft (1·5 to 2 m) high, 182 litres (40 gals) of oil were successfully trapped in an 244 m (800 ft) boom formed into a 'U' trap.

In May 1972 the DTI organised a trial in the south west approaches of the English Channel. This trial was bedevilled by extremely bad weather with Force 8 gales and seas in excess of 6 m (20 ft). In the event no boom deployment was attempted because of the extremely severe conditions but during this trial two Seapack units were towed for hundreds of miles at about 7 to 8 knots and were proven to be entirely seaworthy.

In July 1972 the DTI mounted a further trial off Barra in the Outer Hebrides. Again conditions were severe but in the course of the trial a fully deployed boom rode out a Force 7 gale in 2 m (6 ft) high waves. Later in this trial 95% of an 18 ton oil slick was collected in the 'U' trap. The boom closed itself into a tear drop formation and the oil was concentrated to a thickness of approximately 50 mm (2 in).

Handling gear required

Towing vehicle and work boat.

Capital cost

£18,300 ex works (1973).

Licensed manufacturers: Sea clean International Ltd.

BP.'s Vikoma sea-pack assembley after 488m (1600ft) of seaboom has been deployed, inflated and manoeuvred into a U-shaped trap

A bow-view of BP's Vikoma Sea-pack unit under tow at sea

VIKOMA SEASKIMMER

The Vikoma Seaskimmer comprises discs mounted in the form of a toroid allowing oil to enter the pick-up area from all directions. Oil is scraped from the discs and flows into a small collection chamber from which it is pumped to a suitable collection vessel.

The unit is hydraulically operated by remote control from the parent vessel on which is housed the power pack.

Optional extras include a trailer on which the power pack and skimmer are located together with a crane for skimmer deployment.

A mini-skimmer is in the design stage (September 1973).

Design history

Date construction of prototype started: Experimental prototype March 1972.

Date prototype completed: Experimental prototype June 1972; production prototype August 1973

Date trials completed: Still in hand, 1973

BP's Vikoma system for dealing with oil pollution on the open sea. A close-up view of the Seaboom arranged in a U-shape oil trap. The figure-of-eight of the boom, the upper chamber being filled with air and the lower chamber with water, can be seen on the bottom left-hand portion of the photograph

Basic dimensions of systems
3·28 m (10 ft 9 in) dia. × 2·54 m (8 ft 4 in) high

Weight of system and main components
930 kg (2,050 lb)

Method of operation
Disc skimming

Capability of system
Collecting rate: 100 tons of oil per hour

Craft description
Transport by low loader on land and workboat at sea

Power pack
Power Plant Ford 80 bhp diesel engine and hydraulics. Dry weight 1,134 kg (2,495 lb)

Experiments with the VIKOMA SEASKIMMER
Seaskimmer ~~only~~ became available in mid-1972 and its deployment in rough weather requires a fairly large craft equipped with a suitable derrick.

The skimmer has been tested on static tanks at BP's Sunbury Research Centre and the DTI's Warren Spring Laboratory where it removed oil from the water surface at rates of about 100 tons per hour, and it was further demonstrated in calm water in Sweden where it removed the bulk of 545 litres (120 gals) of oil enclosed in a boom within a matter of seconds. During the Swedish trial the skimmer was deployed from one of the Swedish Coastguard workboats under very rough weather conditions, Force 6 gale and waves 2 m (6 ft) high. Unfortunately because of wind direction, it was not possible to spill oil and therefore the skimmer's ability to pick up oil at sea was not demonstrated. However, it did ride the sea well with the pick-up discs appropriately positioned at the air/sea interface.

Further testing of the experimental SEASKIMMER is in hand and a production version will be available before the end of 1973.

A pre-production BP Vikoma Seaskimmer on trial at BP's Sunbury Research Centre

UNITED STATES OF AMERICA

JBF SCIENTIFIC CORPORATION

The DIP concept (dynamic inclined plane) oil skimmer was developed by JBF Scientific Corporation and a demonstration model was built for the US Environmental Protection Agency in 1971. Production equipment is now in use by the US Navy, the Commonwealth of Massachusetts, and the Canadian Government.

JBF manufactures three production models of the DIP skimmer and each of these is described below. In addition, two open ocean prototypes are under construction and these are also described.

DIP 1001

Principle of operation
Oil is forced onto a moving inclined plane, and conveyed below the surface of the water, to a collection well under the skimmer. The oil gravity separates up into the collection well forcing the water out of the bottom. The oil simply rides on the surface of the belt, held on by gravity forces, and then separates up into the well.

Sweeps are available to triple the size of the skimmer mouth while still maintaining a minimum width for transportability over the roads. The sweeps are individually articulated to improve their performance in waves.

Variations of basic type
This skimmer is based upon the DIP concept and all functions are performed pneumatically using an air compressor.

Design history
Date construction of prototype started: June 1970
Date prototype completed: October 1970
Date trials completed: December 1970

Production
Two units produced and in service

Basic dimensions of systems
Width of slick covered: 0·61 m (2 ft) without sweeps, 1·80 m (6 ft) with sweeps

Weight of system and main components
227 kg (500 lb)

Method of operation
The DIP 1001 is designed to operate in congested areas, around piers, between nested ships, and inside booms. The skimmer is pneumatically operated from a compressor. A 7·62 m (25 ft) long wand is used to guide the skimmer through the slick and a remote control box is provided to control the propulsion, pumping, and belt.

Capability of system
(a) Collecting rate: collection rate depends upon the slick thickness and speed at which the skimmer moves through the slick. At 0·5 knots, the following rates apply for all floating oils, irrespective of density or viscosity.

Slick thickness (mm)	*Collection rate (US gals/min) without sweeps*	*Collection rate (US gals/min) with sweeps*
1	2·6	7·8
10	26·0	78·0

(b) Capacity: onboard storage capacity—the system has a 20 US gallon well which is pumped off the skimmer as it collects the oil.

(c) Oil/water percentages: Oil recovery achieved is 90 per cent of all the floating oil presented to the mouth of the skimmer. Free water content in the recovered oil is less than 10 per cent. The skimmer is capable of recovering sorbents and debris compatible with the size of the skimmer mouth and well. Operation is effective in waves to a height of 0·3 m (1 ft).

Support or collection craft description

Length, overall	1·82 m (6 ft)
Width, overall	1·05 m (3 ft 6 in)
Height, overall	0·915 m (36 in)
Draught	0·51 m (20 in)
Propulsion—power plant and thrust device	twin screw air driven
Pumping system for oil collection	diaphragm pump
Speed, maximum to site	transported by trailer or boat and then launched at the spill site

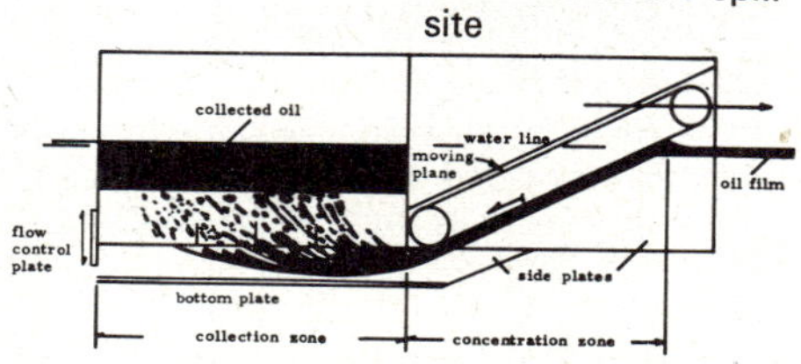

A schematic diagram of the JBF Scientific Corporation's dynamic inclined plane concept for oil skimming and collection

Handling gear required
Boom, davit, or block and tackle capable of lifting 222 kg (500 lb) lifting lugs are provided

Availability
Purchase, capital cost (1973): $19,000 US

This illustration of DIP 1001 shows the device collecting oil by the side of a pier in Boston Naval Shipyard. The cover is removed, showing the pneumatic controls and the diaphragm pump. Recovered oil is pumped up the 7·6 m (25 ft) pipe to a storage tank ashore

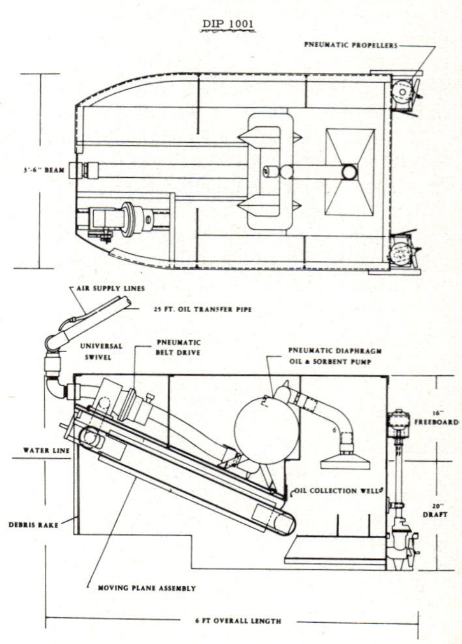

Plan and elevation views of DIP 1001

Operators

Department of the Navy, Naval Facilities Engineering Command, 200 Stovall Street, Alexandria, Virginia 22332

Twelve DIP 1001 units were ordered by the US Navy Facilities Engineering Command in October 1973, as well as six DIP 3001 units, at a total cost of $ (US) 1,416,015.

The DIP 1001 units were provided as a total system: a DIP 1001, two sections of boom, a storage tank and an air compressor, all mounted on a trailer containing a small jib boom for launching the skimmer from a pier. The total recovery system is referred to as the DIP 1002 system.

The units were purchased for river, harbour and coastal oil spills, throughout the United States.

DIP 2001

Principle of operation

Oil is forced onto a moving inclined plane, and conveyed below the surface of the water, to a collection well under the skimmer. The oil gravity separates up into the collection well forcing the water out of the bottom. The oil simply rides on the surface of the belt, held on by gravity forces, and then separates up into the well.

Sweeps are available to triple the size of the skimmer mouth while still maintaining a minimum width for transportability over the roads. The sweeps are individually articulated to improve their performance in waves.

Variations of basic type

This skimmer is based upon the DIP concept and is available in an electrical configuration or a hydraulic configuration.

Design history

Date construction of prototype started: June 1971

Date prototype completed: October 1971

Date trials completed: November 1971

Production

Two produced and in service.

Basic dimensions of systems

Width of slick covered 1·2 m (4 ft) without sweeps, 4·25 m (14 ft) with sweeps.

Weight of system and main components

1,591 kg (3,500 lb)

Method of operation

The skimmer may be moored in a stream using the current to bring oil to the skimmer, or attached to the side of a support boat and used in a sweeping mode to pursue the slick. In either case, the skimmer is unmanned and controlled via a remote control box on the shore or the support boat.

Capability of system

(a) Collecting rate: collection rate depends upon the slick thickness and speed at which the skimmer moves through the slick. At 1 knot, the following rates apply for all floating oils, irrespective of density or viscosity:

Slick thickness (mm)	*Collection rate (US gals min) without sweeps*	*Collection rate (US gals min) with sweeps*
1	9·6	29
10	96·0	290

(b) Capacity: The skimmer has a well capacity of 175 US gallons. Oil is pumped off the skimmer into storage, as the skimmer recovers oil.

(c) Oil/Water percentages:

Oil recovery achieved is 90 per cent of all the floating oil presented to the mouth of the skimmer. Free water content in the recovered oil is less than 10 per cent. The skimmer is capable of recovering sorbents and debris compatible with the size of the skimmer mouth and well. Operation is effective in waves to a height of 0·61 m (2 ft).

Support or collection craft description

Length, overall 3·95 m (13 ft)
Width, overall 2·12 m (7 ft)
Height, overall 1·3 m (52 in)
Draught 0·89 m (35 in)

Propulsion—power plant and thrust device: none (attach to the side of a support boat)

Pumping system for oil collection: two modes are available, an electrically driven positive displacement (Moyno) pump or a direct diesel drive Moyno pump

Speed, maximum to site: typically trailered to the site, or towed to the site with another vessel at 5 knots

Handling gear required

Crane capable of lifting 1,591 (3,500 lb) lifting lugs are provided

Availability

Purchase, capital cost (1973): $47,000 US without sweeps; $50,500 US with sweeps

Operators

Donald L. Corey, The Commonwealth of Massachusetts, Division of Water Pollution Control, 100 Cambridge Street, Boston, Massachusetts 02202

Canadian Government

A test and evaluation programme has been conducted by the Canadian Government on the DIP 2001 oil skimmer and this was probably the first independent acquisition of test data for an oil skimmer system.

The test programme consisted of 14 test runs, over a speed range from 0·88 to 1·37 knots, on both No. 2 fuel oil and crude oil. Their tests produced an arithmetic mean oil recovery of 89·3 per cent of the oil presented to the skimmer (ORF), with a standard deviation of 7·1. Due to scale effects in these results a greater percentage recovery could be expected in service applications.

DIP 3001

Principle of operation

Oil is forced onto a moving inclined plane, and conveyed below the surface of the water, to a collection well under the skimmer. The oil gravity separates up into the collection well forcing the water out of the bottom. The oil simply rides on the surface of the belt, held on by gravity forces, and then separates up into the well.

Sweeps are available to triple the size of the skimmer mouth while still maintaining a minimum width for transportability over the roads. The sweeps are individually articulated to improve their performance in waves.

Variations of basic tyr

This skimmer is based upon the DIP concept and is a fully self contained diesel-hydraulic skimmer boat that operates with a crew of two. The diesel uses a closed cycle, fresh water radiator cooled, diesel engine.

Design history

Date construction of prototype started: January 1972

Date prototype completed: July 1972

Date trials completed: August 1972

DIP 2001 'harvesting' oil flowing from a storm sewer at an international airport. After the oil is recovered, and the water automatically separated, it is pumped ashore for disposal. A timer automatically turns the skimmer off and on periodically so that it may operate unattended

DIP 3001 'harvesting' oil and sorbent materials in Boston Harbour. This unit has been used for eleven months, in the Puget Sound Naval Shipyard, on many spills of different sizes

Production
One produced and in service.

Basic dimensions of systems
Width of slick covered—1·2 m (4 ft) without sweeps, 4·25 m (14 ft) with sweeps.

Weight of system and main components
6,364 kg (14,000 lb)

Method of operation
The skimmer is designed to operate on harbour and coastal spills in waves to a height of 0·915 m (3 ft) and in currents to 3 knots.

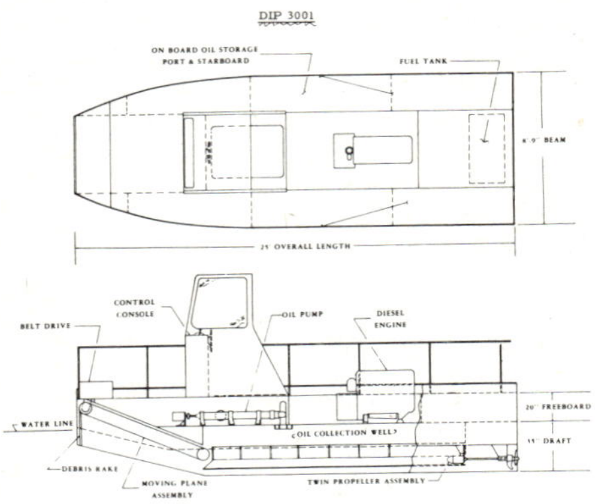

Plan and elevation views of DIP 3001

Capability of system
(a) Collecting rate: collection rate depends upon the slick thickness and speed at which the skimmer moves through the slick. At 2 knots, the following rates apply for all floating oils, irrespective of density or viscosity.

Slick thickness (mm)	*Collection rate (US gals/min) without sweeps*	*Collection rate (US gals min) with sweeps*
1	19	60
10	190	660

(b) Capacity: The skimmer has a total onboard storage capacity of 1,400 US gals (1,000 in the hull and 400 gals in the well).
(c) Oil/water percentages: Oil recovery achieved is 90 per cent of all the floating oil presented to the mouth of the skimmer. Free water content in the recovered oil is less than 10 per cent. The skimmer is capable of recovering sorbents and debris compatible with the size of the skimmer mouth and well. Operation is effective in waves to a heigh of 0·915 m (3 ft).

Support or collection craft description

Length, overall	7·62 m (25 ft)
Width, overall	2·65 m (8 ft 9 in)
Height, overall	1·37 m (55 in)
Draught	0·89 m (35 in)
Propulsion—power plant and thrust device	diesel drives hydraulic system and twin screws are driven hydraulically
Pumping system for oil collection	hydraulically driven Moyno positive displacement pump
Speed, maximum to site	5 knots

Handling gear required
Crane capable of lifting 6,364 kg (14,000 lb), lifting lugs are provided.

Availability
Purchase, capital cost (1973): $120,000 US without sweeps; $127,000 US with sweeps.

Operators
Department of the Navy, Naval Facilities Engineering Command, 200 Stovall Street, Alexandria, Virginia 22332.
Similar unit to DIP 3001 in operation in Puget Sound Naval Shipyard since summer 1972.

DIP 3003

Principle of operation
Oil is forced onto a moving inclined plane, and conveyed below the surface of the water, to a collection well under the skimmer. The oil gravity separates up into the collection well forcing the water out of the bottom. The oil simply rides on the surface of the belt, held on by gravity forces, and then separates up into the well.
Sweeps are available to triple the size of the skimmer mouth while still maintaining a minimum width for transportability over the roads. The sweeps are individually articulated to improve their performance in waves.

Variations of basic type
This skimmer is based upon the DIP principle but the design is integrated into a large hull for skimming in the open ocean.

Design history
Date construction of prototype started: July 1973
Date prototype completed: December 1973
Date trials completed: sea trials January 1974

Production
Prototype on trials

Basic dimensions of systems
Width of slick covered—2·12 m (7 ft) without sweeps, 5·5 m (18 ft) with sweeps

Weight of system and main components
15,000 kg (33,000 lb)

Method of operation
The 3003 is a fully self-contained skimmer boat that may be used in coastal, or offshore, waters to recover oil, debris, and sorbents in either a sweeping mode or a fixed mode.

Capability of system
(a) Collecting rate: collection rate depends upon the slick thickness and speed at which the skimmer moves through the slick. At 2 knots, the following rates apply for all floating oils, irrespective of density or viscosity.

Slick thickness (mm)	*Collection rate (US gals/min) without sweeps*	*Collection rate (US gals/min) with sweeps*
1	30	77
10	300	770

(b) Capacity: the total storage is 4,000 US gals, including 2,000 in the hull and 2,000 in the well
(c) Oil/water percentages: Oil recovery achieved is 90 per cent of all the floating oil presented to the mouth of the skimmer. Free water content in the recovered oil is less than 10 per cent. The skimmer is capable of recovering sorbents and debris compatible with the size of the skimmer mouth and well. Operation is effective in waves to a height of 1·2 m (4 ft).

Support or collection craft description

Length, overall	10·5 m (35 ft)
Width, overall	3·33 m (11 ft)
Height, overall	5·48 m (18 ft) with wheelhouse in place, 3·95 m (13 ft) while being transported
Draught	1·7 m (5 ft 7 in)

Propulsion—power plant and thrust device: propulsion is provided by a diesel driving a Murray and Tregurtha 'Harbourmaster' outdrive. The cooling is provided with a radiator and closed cycle liquid system.
Pumping system for oil collection: diesel driven Moyno positive displacement pump
Speed, maximum to site: 6 knots.

Handling gear required
Crane capable of lifting 15,000 kg (33,000 lb) lifting lugs are provided.

Operators
US Navy will be the operator when the prototype is delivered.

DIP 4001

Principle of operation
Oil is forced onto a moving inclined plane, and conveyed below the surface of the water, to a collection well under the skimmer. The oil gravity separates up into the collection well forcing the water out of the bottom. The oil simply rides on the surface of the belt, held on by gravity forces, and then separates up into the well.
Sweeps are available to triple the size of the skimmer mouth while still maintaining a minimum width for transportability over the roads. The sweeps are individually articulated to improve their performance in waves.

Variations of basic type
This skimmer is based upon the DIP principle but has 2 modes of operation. For low sea states, 0·3 to 0·6 m (1 to 2 ft) wave heights, the skimmer is a fully self contained skimmer boat and is operated with an on-board crew of 2. For high sea states, waves of 0·9 to 1·5 m (3 to 5 ft) and higher, the skimmer is operated unmanned. In this latter mode it is towed astern, and to the side, of a large support vessel. A full remote control capability is provided so that all functions, including propulsion, can be controlled from the bridge of the support vessel. An umbilical between the vessels provides a way of sending commands to the skimmer and allowing oil to be pumped from the skimmer to the support vessel for storage.

Design history
Date construction of prototype started: August 1973
Date prototype completed: October 1973
Date trials completed: November 1973

Production
Prototype under fabrication

Basic dimensions of systems
Width of slick covered—1·5 m (5 ft) without sweeps, 4·55 m (15 ft) with sweeps.

Weight of system and main components
8,182 kg (18,000 lb)

Method of operation
See variations of basic type, above

Capability of system
(a) Collecting rate: collection rate depends upon the slick thickness and speed at which the skimmer moves through the slick. At 2 knots, the following rates apply for all floating oils, irrespective of density or viscosity:

Slick thickness (mm)	*Collection rate (US gals/min) without sweeps*	*Collection rate (US gals/min) with sweeps*
1	21·5	64·5
10	215·0	645·0

(b) Capacity: the total storage (in the well) is 1,200 US gals. Oil is pumped off to the support vessel on large spills.
(c) Oil/water percentages: Oil recovery achieved is 90 per cent of all the floating oil presented to the mouth of the skimmer. Free water content in the recovered oil is less than 10 per cent. The skimmer is capable of recovering sorbents and debris compatible with the size of the skimmer mouth and well. Operation is effective in waves to a height of 1·5 m (5 ft).

Support or collection craft description

Length, overall	7·92 m (26 ft)
Width, overall	3·03 m (10 ft)
Height, overall	3·33 m (11 ft)
Draught	1·4 m (4 ft 8 in)

Propulsion—power plant and thrust device: propulsion is provided by a twin screw diesel-hydraulic system. The diesel is closed cycle radiator cooled.
Pumping system for oil collection: hydraulically driven Moyno positive displacement pump.
Speed, maximum to site: 5 knots under its own power; higher speeds if towed.

Handling gear required
Crane capable of lifting 8,182 kg (18,000 lb) lifting lugs are provided.

LOCKHEED MISSILES & SPACE CO. INC.

CLEAN SWEEP

Principle of operation: Rotating disks connected by vanes with wipers to remove recovered oil from disk surface.

Variations of basic type: Same principle in all. Many packaging configurations.

Design history

Date construction of prototype started: November 1968

Date prototype completed: February 1969

Date trials completed: July 1971

Production

40

Basic dimensions of systems

0·61 m (2 ft) dia. × 1·22 m (4 ft) long; 1·22 m (4 ft) dia. × 2·13 m (7 ft) long

Weight of system and main components

Clean Sweep drum, drive motor and drive components, pump and flotation 450 kg–7,300 kg (1,000 to 16,000 lb)

Method of operation

Oil adheres to rotating disks. Oil is scraped off by stationary wipers and pumped to storage

Capability of system

Collecting rate: Nominal Rating: 0·61 m × 1·22 m (2 ft × 4 ft) 40 US gpm—1·22 m × 2·13 m (4 ft × 7 ft) 200 US gpm

Support or collecting craft description

Length, overall	R2002—7·3 m (24 ft), R2003—7·3 m (24 ft)
Width, overall	R2002—2·9 m (9·5 ft), R2003—5·5 m (18 ft)
Height, overall	R2002—1·79 m (5·8 ft), R2002—4·33 m (14·2 ft)
Draught	R2002—0·305 m (12 in), R2003—0·457 m (18 in)
Propulsion	R2002—50 h.p. outboard, R2003—Twin 40 h.p. hydraulic

Pumping system for oil collection: Positive displacement

Positioning

Helicopter lift or 6 knots surface

Handling gear required

Crane

Availability

Purchase, capital cost (1973): R2002: $ US 5,000–$ US 25,000; R2003: $ US 17,500–$ US 100,000

Operators

Most major US oil companies. Swedish Coast Guard. King Salmon Marine, Eureka, California.

Applications

By August 1973, thirty-six units had been sold in eight countries and many different configurations have been used in conjunction with the basic Clean Sweep rotating drum.

The Lockheed Clean Sweep

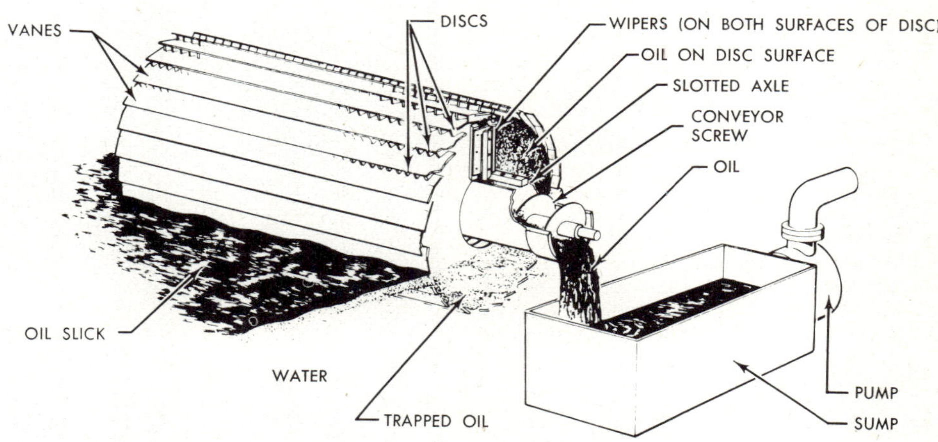

The principle of operation of the Lockheed Clean Sweep Oil recovery device

Robert Yates, the inventor of Clean Sweep, with the rotating drum device. As the Clean Sweep drum rotates in an oil spill oil sticks to the sides of the parallel metal discs and then internal wiper blades wipe the oil from the sides of the discs and direct it to a central channel where it is pumped to storage

The Lockheed Clean Sweep is basically a craft of catamaran form, equipped with a rotating drum of aluminium discs to which the oil will adhere before being wiped off by stationary wipers. The collector system was invented by a Lockheed engineer, Robert Yates. Two sizes of collector drum are available, one of 0·6 m (2 ft) diameter and 1·2 m (4 ft) long which can collect up to 40 US gals per minute, and one which is 1·2 m (4 ft) diameter and 2·13 m (7 ft) long which can collect up to 200 US gals per minute.

First production unit
January 1972

Operators
Ecology Marine Services Co., Martinez, California.
William H. Hutchison and Sons, Wilmington, California.
US Naval Civil Engineering Laboratory, Port Hueneme, California.

The rotating drum of the Lockheed Clean Sweep oil recovering device being prepared for shipment for mounting in two oil clean up boats for the Swedish Coast Guard

HIGH SEAS OIL RECOVERY (HSOR) DEVICE

Principle of operation: Rotating disc-drum
Variations of basic type: Smaller versions for refinery operations (Clean Sweep)
Design history:
Date construction of prototype started: September 1972
Date prototype completed: May 1973
Date trials completed: October 1973
Production:
1
Basic dimensions of Systems:
disc-drum: 1.22 m (4 ft) diameter x 3.96 m (13 ft) long
Weight of System and main components
6820 kg (15,000 lb)
Method of operation:
Discs rotate through oil, picking up oil and shedding water. At vertical point in rotation, oil is scraped off discs into a trough and pumped to collection system.
Capability of System:
Collecting rate: 1000 U.S. gallons per min.
Support craft:
Any Coast Guard cutter can tow system to site. Buoy tenders can carry on deck.
Pumping system for oil collection: hydraulic
Speed, max. to site: if towing—7 knots; if deck cargo—at ship's speed
Handling gear required: 8 ton boom
Availability:
Purchase, capital cost (1973): $U.S.200,000 (HSOR)
Operators:
U.S. Coast Guard Headquarters
400 Seventh Street, S. W.
Washington, D.C. 20590

The Lockheed inflatable Clean Sweep oil recovering device showing the multi-bladed rotating drum

CLEAN SWEEP **(air transportable)**

Built for the United States Coast Guard in 1973. This version is very much larger than the original Clean Sweep and has been designed to operate effectively in sea states with average waves of 1·5 m (5 ft), random waves of up to 3·0 m (10 ft) and in winds of up to 20 knots. The basic structure of the craft consists of an aluminium hull which is supported in the water by four inflatable side bodies. The rotating oil collector drum is 1·2 m (4 ft) in diameter and 4·1 m (13·5 ft) long and in tests carried out in a tank at the Battelle Northwest Laboratory a collecting rate of 750 US gallons per minute has been established, with a water content level of less than 10 per cent.

The inflatable Clean Sweep engaged in tests with a US Coast Guard developed barrier

The Inflatable Clean Sweep oil recovery machine under towing tests in San Francisco Bay.

The US Coast Guard Lockheed Clean Sweep on oil recovery test in a 61 m × 122 m (200 ft × 400 ft) tank at Battelle North-west, Richmond, Washington.

METROPOLITAN PETROLEUM PETROCHEMICALS CO. INC.

MP BOOM

A lightweight flexible boom for oil spill containment.

Design history

Date construction of prototype started: 1968

Date prototype completed: September 1968

Date trials completed: November, 1968

Production

Quantity in service: 113,000 m (375,000 ft) (by July, 1973).

Specification

Fabric	510 gr (18 oz) vinyl-coated nylon, 14 × 14 count, 840 denier. Meets MIL-C-43006 Type 1, Class 1, Heavy Duty
Colour	Yellow
Float fill	6 mm ($\frac{1}{4}$ in) diameter polystyrene foam beads. Float has 15·2 cm (6 in) diameter with 3 compartments
Ballast rod	50 mm × 50 mm (2 in × 2 in) high-density PVC. Will accept up to 7 mm ($\frac{3}{8}$ in) cable
Skirt	30·50 cm (12 in) flexible
Cable	4 mm (5/32 in)—7 × 19 stainless steel type 304 cable
Connector plate	6061-T6 aluminium alloy with black duranodic finish
Boom weight	3 kg/m (2·2 lb/ft)
Storage	2 m × 1 m × 1 m (7 ft × 3 ft × 3 ft) per 30 m section (100 ft)
Shipping weight	115 kg/30 m (250 lb/100 ft) sections
Shipping cube	60·3 cm × 107·3 cm × 182·9 cm ($23\frac{3}{4}$ in × $42\frac{1}{4}$ in × 72 in) 1·19 m^3 (42 ft^3)
Unit	16 1·83 m (6 ft) sections per 30 m (100 ft) nominal length

Availability

Purchase cost (1973): $ US 24·34 metre ($ US 7·85/ft).

Lease charge: $ US 4·10/metre ($ US 1·25/ft) depends on duration.

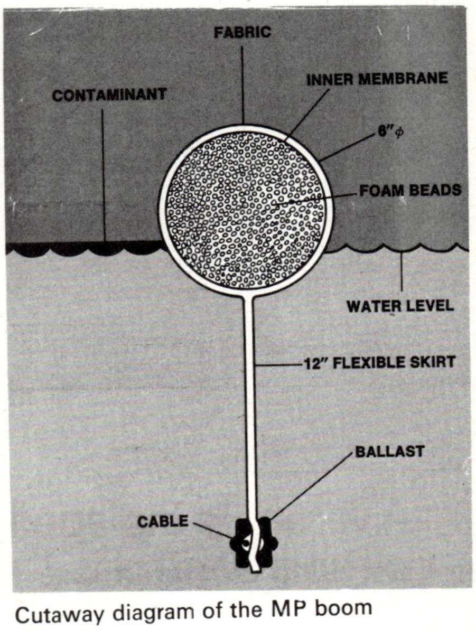

Cutaway diagram of the MP boom

OCEAN SYSTEMS INC

SKIMMER
(125/500/2,000 US GALLONS PER MINUTE)
OIL RECOVERY SYSTEM (ORS-125, ORS-500, ORS-2000)

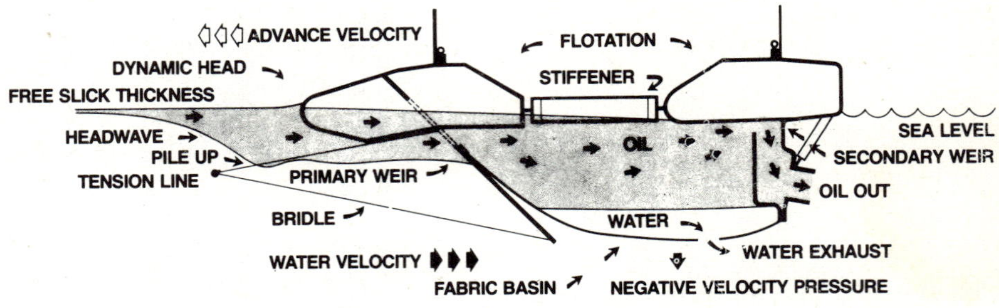

A cross-sectional diagram of the weir section modules

Principle of operation

Buoyant double weir skimmer utilising thickening due to forward motion.

Variations of basic type

Large systems are made-up from modules of the smallest system.

Design history

Originally developed by OSI for the US Coast Guard.

Date construction of prototype started: September, 1972 (ORS-2000)

Date prototype completed: February 1973

Date trials completed: March 1973

These oil recovery systems can be used with a wide variety of vessels, depending primarily on spill conditions.

Handling gear required

ORS-125—manually handled; ORS-500—3-ton boom; ORS-2000—6 to 10 ton boom.

Operation

All models of the ORS collect oil with a minimum of water in the recovery stream under severe environmental conditions 1·5 m (5 ft) sea, 35-knot winds). Since the system acts as an oil barrier there is little loss past the system. Also, there are no moving parts to the system except the pump; as a result maintenance requirements are minimised.

The recovery modules may be used at the apex of commercially available funneling booms to efficiently sweep wide areas in rivers, harbours and offshore regions. The system works for all types of oil, including petrol and up to 90,000 SSU viscosity. Design water speeds range from 0·3 knots to 0·5 knots (current or tow speeds) and the system will work on any oil thickness.

Specification of the ORS-2000 system

Recovery sea state capabilities

2·4 m (8 ft) significant wave height; 3-knot current; 40 knot winds.

Pumping system

Type: Four gear pumps, each driven by two close-coupled hydraulic motors

Capacity: 2,000 gpm

Hose type (suction): Flexible, PVC hose reinforced with PVC spiral fused to wall

Discharge: Light weight collapsible dual hose (similar to MIL-H-82127)

Percentage of oil in discharge stream

80 to 100 per cent when operated as recommended.

Hydraulic power unit output

120 gpm at 1,650 kg/cm^2 (2,400 lb/in^2) differential pressure.

System components

Skimmer unit (9 primary weirs and 16 working secondary weirs); Four oil pumps with hydraulic motors; Four 24·4 m (80 ft) suction hose, 20·3 cm (8 in) diameter; 91·5 m (300 ft) discharge hose, 25·4 cm (10 in) diameter; Hydraulic power unit and hydraulic umbilical 91·5 m (300 ft); Sensor system (oil depth indicator); Handling system; Transportation container and packaging; Option-automatic control system for use with sensor system.

Shipping pallets (dimensions and weights)

Skimmer suction hose: 7·3 m × 2·4 m × 2·73 m (24 ft × 8 ft × 9 ft) 6,330 kg (14,000 lb)

Power unit discharge hose: 4·87 m × 2·4 m × 2·73 m (16 ft × 8 ft × 9 ft) 6,300 kg (14,000 lb)

Deployed dimensions: 1·89 m × 0·61 m × 0·61 m (62 ft wide × 2ft at above × 2 ft below waterline)

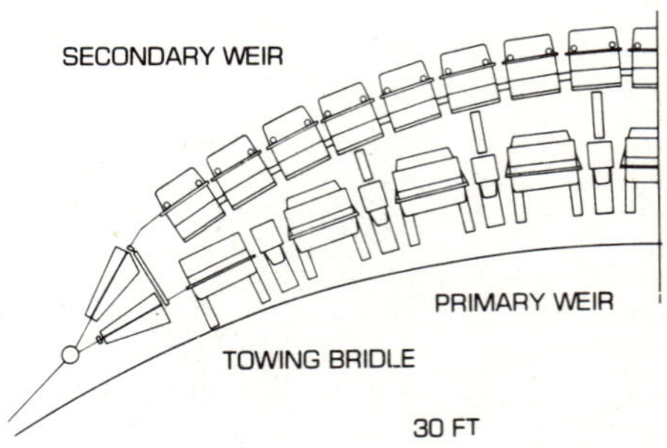

Plan view showing half of the modules of ORS-2000 deployed 0·61 m (2 ft) above water line; 0·61 m (2 ft) below water line

Costs (June 1973)	ORS-125	ORS-500
Purchase, capital cost	$ US 15,200	$ US 98,200

Specifications	ORS-125	ORS-500	ORS-2000
Basic dimensions of systems	3·04 m × 1·68 m × 1·2 m (10 ft × 5 ft 6 in × 4 ft)	9·15 m × 3·04 m × 1·2 m (30 ft × 10 ft × 4 ft)	18·30 m × 3·04 × 1·2 m (60 ft × 10 ft × 4 ft)
Weight of system and main components (skimmer unit, suction hose, etc.)	136 kg (300 lb)	2,040 kg (4,500 lb)	5,460 kg (12,000lb)
Method of operation	Towed or moored	Towed or moored	Towed or moored
Collecting rate	125 (US) gpm	500 (US) gpm	2,000 (US) gpm
Capacity	External storage	External storage	External storage

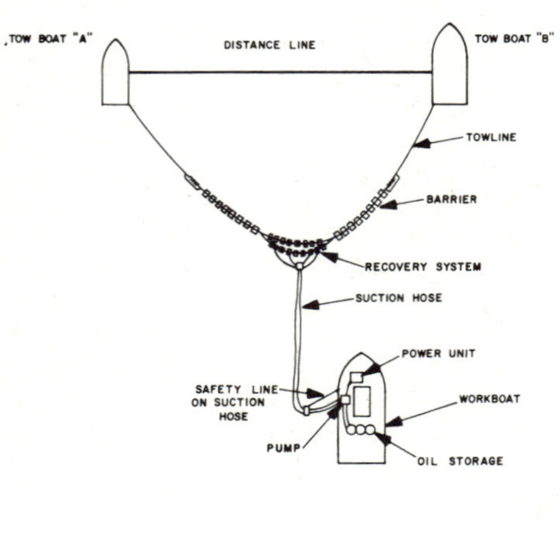

TOWED SYSTEM

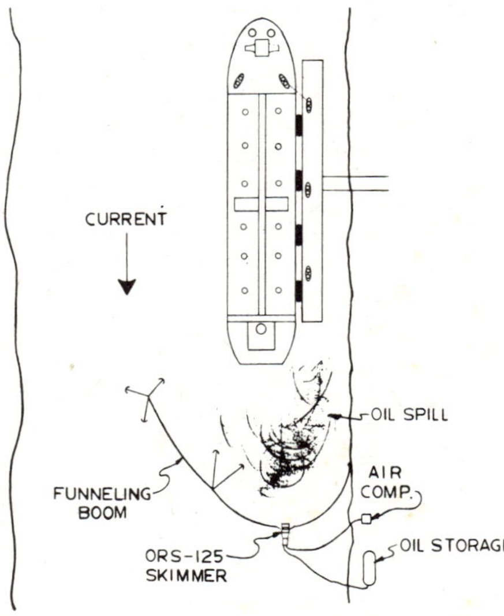

RIVER OR CHANNEL

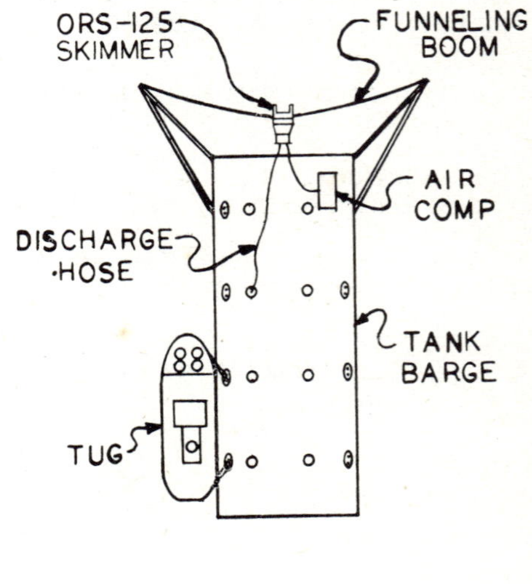

BARGE SYSTEM

Three deployment schemes for Ocean Systems' oil spill control and recovery systems (ORS-125/500/2000). All sizes should use a containment boom to funnel oil to the skimmer and to increase the sweep area. Most commercially available booms may be used

HARBOUR OIL CONTAINMENT BARRIER

Specifications

Typical section length: 15·2 m (50 ft)

Cross section dimensions: 0·061 m× 0·061 m× 1·52 m (2 ft high× 2 ft deep× 5 ft total waterline width (at flotation elements)

Typical section volume: 15·2 m (50 ft) length rolled up), 1·07 m (3½ ft) diameter× 1·22 m (4 ft) high (with floats removed)

Typical section weight: 15·2 m (50 ft) length, 100 kg (220 lb) total

Section weight distribution: 15·2 m (50 ft) length, 77 kg (170 lb) basic barrier weight, 23 kg (50 lb) flotation section weight

Materials

Ethafoam flotation elements 32 kg/m^3 (2 lb/ft^3)

Herculite LR 210 oil resistant nylon reinforced vinyl fabric 0·71 kg/m^2 (21 oz/yd^2), 53·5 kg/cm (300 lb/in) tensile strength

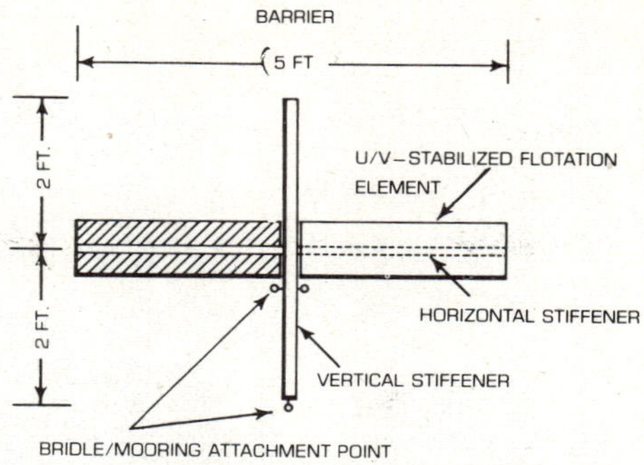

Cross sectional diagram of the Harbour Oil Containment Barrier

Colour

High visibility yellow flotation elements and International orange skirt material

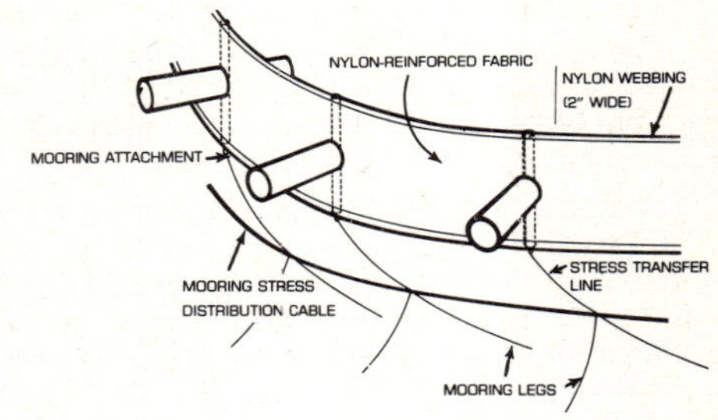

Schematic diagram of the Harbour Oil Containment Barrier

Design operating conditions: 1·2 m (4 ft) maximum wave height

Structural design conditions: 2·4 m (8 ft) wave height; 3 knot currents

OILMOP INC

OIL MOP

The patented Oil Mop process consists of three elements: A Rope Mop which absorbs oil and rejects water; a way to expose the mop to floating oil; and a machine to reclaim the absorbed oil from the mop. The Rope Mop is a configuration of plastic fibres woven into a plastic rope. The resulting thick nap of the Rope Mop is 'oleophilic and hydrophobic'—it accepts oil and rejects water. The Rope Mop is threaded through a Mop Engine which wrings oil from the saturated mop and continuously broadcasts and rebroadcasts the mop out 15 to 30 m or more (50 to 100 ft) around a tail pulley. Secured at a point opposite the Mop Engine, the tail pulley allows the mop to form a continuous looping figure on the surface of the oil slick.

Principle of operation

The rope mop selectively absorbs the oil from the water and reclamation of the absorbed oil is obtained by wringing the oil from the mop; paying the rope mop back out for additional absorbtion in a continuous operation. The operation is similar to a continuous belt with the rope mop floating on the water while absorbing oil. The rope mop consist of 'oleophilic-hydrophobic' fibres interwoven into a polypropylene rope. The rope forms a core to provide flexible towing strength while the fibres protrude out to attach to the oil. The mop floats on the water's surface to give maximum exposure to the oil attracting fibres. The mop has a circular cross section of from 0·10 to 1·0 m (4 in to 36 in) and may be spliced to any length. The rope may be any size from 9·5 mm to 25 mm (⅜ in to 1 in) depending on the strength required for the length. When fully saturated with high viscosity oil, the rope mop weighs up to 45·5 kg/m length (30 lb/ft).

Variation of basic type

* Other than the size as indicated above, the major variation is the type of power plant, i.e. diesel engine, gasoline engine, electric motor, pneumatic motor, or hydraulic unit. The Mark I–4 series mop engine is mounted on an open type drum and when the drum is filled with oil the mop engine is moved to another drum. The Mark II–4 series mop engines are small portable units sometimes used on small pits for permanent installations. All models, except the Mark I–4 series, are available with an integral oil transfer pump. All mop engines are available with automatic controls, trailer mounted, and other ancillaries.

System description

Mop engines	*Min. h.p.*	*Mop size*	*Capacity* *
MK I-4 series	¼	10·2 cm (4 in)	250 bbls/day
MK II-4 series	½	10·2 cm (4 in)	600 bbls/day
MK II-9 series	5	22·9 cm (9 in)	3,000 bbls/day
MK IV-9 series	15	30·5 cm (12 in)	5,000 bbls/day
MK IV-16 series	30	21·0 cm (24 in)	8,000 bbls/day
MK IV-20 series	50	91·5 cm (36 in)	12,000 bbls/day

*

Nominal capacities based on a saturated rope mop. Several factors affect the capacity, of which the most important is the viscosity of the oil to be reclaimed.

Production

Type Mop engine	*Started production*	*Scheduled production*	*No. in service*
MK I-4 series	1973	100	3
MK II-4 series	1971	120	11
MK II-9 series	1971	120	18
MK IV-9 series	1970	–0–	3
MK IV-16 series	1973 (new design)	20	1
MK IV-20 series	1972	2	3

Collecting rate

Type Mop engine	*Basic dimensions* m (in)	*Weight of system* * kg (lb)	*Collection Rate*
MK I-4 series	1·07× 0·38× 0·53 (42× 15× 21)	91 (200)	250 bbls/day
MK II-4 series	1·12× 0·64× 0·84 (44× 25× 33)	405 (900)	600 bbls/day
MK II-9 series	1·83× 1·12× 1·27 (72× 44× 50)	723 (1,600)	3,000 bbls/day
MK IV-9 series	2·24× 1·37× 1·42 (88× 54× 56)	1,363 (3,000)	5,000 bbls/day
MK IV-16 series	2·74× 2·29× 2·16 (108× 90× 85)	4,500 (10,000)	8,000 bbls/day
MK IV-20 series	3·60× 2·74× 2·43 (142× 108× 96)	6,800 (15,000)	12,000 bbls/day

Varies with ancillaries and type power plant.

Design history

Date construction of prototype started: 1966

Date prototype completed: 1970

Date trials completed: The prototype had an unusual trial in January of 1970. Before it was completed, a major oil company had a very large oil spill in the Gulf of Mexico and the still not completed, but operational, unit was mounted on a barge and towed to sea. It's capacity at this time was a little over 200 bbls/day with a 23 cm (9 in) mop, which was satisfactory. Today, this same improved design has reclaimed oil on an average of 3,200 bbls/day over a 27 day period with less than 1 per cent entrained water.

Capacity

The units have very little oil storage capacity with respect to their oil reclamation rate. The storage capacity is designed for a minimum to keep the transfer pump primed with adequate submergence of the suction. Usual field operations provide for a barge storage system, a tank truck relay, plastic lined earthen pits for temporary storage, or in a few cases the oil has been pumped back into the pipe line from where it came.

Support or collection craft description

The table below represents the estimated minimum size craft that the Mop Engines can be mounted on for proper operation. However, this is not always the case, because the system has been mounted on skids with crawler tractors pulling them; 'marsh buggies' (amphibious); trucks; trailers, matted mud lumps; concrete foundations; offshore platforms; tug boats; crew boats; barges; pontoons; lift-boats; and others. This indicates the versatility and portability of the oil mop system of reclaiming oil from the surface of water.

An open sea operation with Oil Mop, using two boats

The Oil Mop recovery system

The Oil Mop system is air transportable; from a telephone call at 4:30 p.m. Friday to New Orleans, La. until 10:30 a.m. Sunday (elapsed time—43 hours) the units have been operational in Anchorage, Alaska. All of this was accomplished via commercial air lines at a very reasonable cost to the customer. A MK IV-16, with rope mop, has been flown from Houston, Texas to Manta, Ecuador on a chartered flight.

Operators

Republic Supply Co., P.O. Box 1995, Oklahoma City, Oklahoma.

Cesco Div. of Browning-Ferris Industries Inc., P.O. Box 15482, Baton Rouge, La. 70821

DiKor, P.O. Box 389, Carmi, Illinois 62821.

Oil Mop Pollution Control Ltd., 80 Commissioners St., Toronto, Ontario.

Several major oil companies, and the United States Navy.

Handling gear required

Mop Engine	*Hook Lift required*
MK I-4 series	Manual—2 men
MK II-4 series	Manual—4 men
MK II-9 series	1-ton hook load
MK IV-9 series	2-ton hook load
MK IV-16 series	5-ton hook load
MK IV-20 series	7½-ton hook load

Availability

Mop Engine	*Purchase* (1973) $ US	*Monthly lease, yearly basis*
MK I-4	$3,500.00	$150.00
MK II-4	8,000.00	350.00
MK II-9	13,000.00	600.00
MK IV-9	16,000.00	850.00
MK IV-16	50,000.00	3,000.00
MK IV-20	60,000.00	3,200.00

Mop engine	*Min. boat size*	*Remarks*
MK I-4 series	3 m (10 ft) skiff	Portable unit
MK II-4 series	4·9 m (16 ft) skiff	Semi-portable
MK II-9 series	7·6 m (25 ft) power boat	
MK IV-9 series	9·15 m (30 ft) power boat	
MK IV-16 series	250 h.p. tug	
MK IV-20 series	600 h.p. tug	

PORT OF PORTLAND SWAN ISLAND SHIP REPAIR YARD AND ENVIRONMENT PROTECTION MACHINES LTD.

BIG SKIMMER

FLOATING WEIR TYPE OIL SKIMMER

Designed by Carl F. Propp this oil skimmer was built in the spring of 1970. The system has a total pumping capacity of 10,000 US gallons/min and is powered by a 238 h.p. diesel engine. The overall length is 10·6 m (35 ft) and beam 5·79 m (19 ft) and the operating draught is 0·92 m (3 ft). The system weighs 22 tons and has been carried on a vessel 8·5 m (28 ft) by 8·5 m (28 ft) with a 3·1 m (10 ft) draught.

Capital cost is $ US 100,000.

SMALL SKIMMER

This version of the oil skimmer has a pumping capacity of 500 US gallons/min. Three units have been built since the first one was built in 1971. The weight of the unit is 4 tons and a collecting capacity of 1,000 US gallons of skimmed oil is provided.

The vessel is 6·24 m (20 ft 6 in) long by 2·13 m (7 ft) wide. A separate free floating adjustable skimming weir is placed within a well 2·89 m (9 ft 6 in) by 1·22 m (4 ft) in the forward section of the vessel and was connected to the main vessel with a flexible joint so as not to impede the free-floating action of the weir. In this way the wave action of the surface water can be accommodated without the intake of a large volume of water. Opening bow doors are fitted to the vessel to create a large skimming path and when closed to function as a bow fairwater. The spillway, which is attached to the floating weir, dumps skimmed material into a pumpwell from which it is pumped into a holding tank where the natural separation of oil and water occurs. This separation is accomplished by decanting water out of the holding tank, thereby allowing the oil or floatable debris to build up on the surface of the contained water. When the holding tank is reasonably full of collected material, the transfer operation is started. The main intake weir is closed, then the skimmed material is directed into the pump sump, and pumped via by-pass piping into a transfer vehicle.

The buoyancy of the vessel is accomplished with two rectangular pontoons held in a catamaran configuration. The forward and aft balance of the vessel is such that the vessel is launched bow in. With the holding tank empty the vessel floats with 0·914 m (3 ft) trim by the bow allowing easy launching and de-hauling. After launching the decant valve is opened filling the holding tank with fresh water. The vessel then floats on an even keel at a draft of 1·06 m (3 ft 6 in).

The propulsion and steering of the vessel is accomplished by a main propulsion unit which rotates 360°. The vessel is also equipped with a bow thruster, which in conjunction with the main propulsion unit allows the vessel to maintain motion in any direction with a minimum of momentum.

The oil and debris pumping system is accomplished by using a sewer-basin clean out pump with a capacity of 500 US gallons per minute.

The power is supplied by one 30 h.p. gasoline engine driving a 2,500 rpm 26 US gallons per minute hydraulic pump which drives the main pump, propulsion and bow thruster. The pump has first call on the flow by using a flow divider. The propulsion and thruster are in series and utilize by-pass oil.

Capital cost is $ US 30,000.

REYNOLDS INTERNATIONAL INC

MEDUSA

(6 ft, 18 ft and 14 ftMK-L)

60 to 500 US gallons per minute oil recovery system

Principle of operation

A vertical impeller pump induces a radial flow towards the centre of the circular unit and the upper areas of this flow, carrying the oil, feed into an annular intake leading the oil suction hose while the lower areas of the water flow are carried downwards and away by the impeller. Thus a two-stage operations is effected.

First construction and prototype kinds were completed in 1971.

Variations of basic type: Three versions of 1·8 m (6 ft) 5·47 m (18 ft) and 4·25 m (14 ft) diameter exist.

Purchase costs (1973): Range from $ US 3,360 to $ US 33,000.

MK-L 1·8 m (6 ft) diameter version (outer weir dia. quoted)

Draught	0·76 m (2·5 ft)
Height	1·35 m (4·5 ft)
Max. diameter	2·32 m (7·67 ft)
Weight	91 kg (200 lb)

Oil plus water total skim rate: up to 1,500 US gals/min

Oil collection up to 60 US gals/min.

Reserve buoyancy of unit: 100 per cent.

Operating conditions: 50 knot winds, waves, 2·4 m (8 ft), currents 4 knots

Method of operation

Oil is induced into the Medusa MK-L from great distances because of the relatively great fluid intake rate created by the vertical impeller pump. This feature hydrodynamically concentrates the incoming oil in the centre of the unit; the water under the oil flows radially inward near the centre and then travels vertically downward into the water pump where it is discharged horizontally in all directions. With a plug valve, the incoming oil is concentrated with the plug valve closed until its thickness reaches 5 to 6·5 cm (2 to 2·5 in). The plug valve is then lifted by pulling a lifting cord in the control rod and the highly concentrated oil (to 100 per cent) is collected over the central weir by the oil pump (driven by the same shaft that drives the water propeller).

MK-L 5·48 m (18 ft) diameter version

Weirs

Outer	Surface following; 12 circularly integrated segments with debris screens.
Inner	Surface following; 6 adjustable vortex producing oil intakes with strainers.

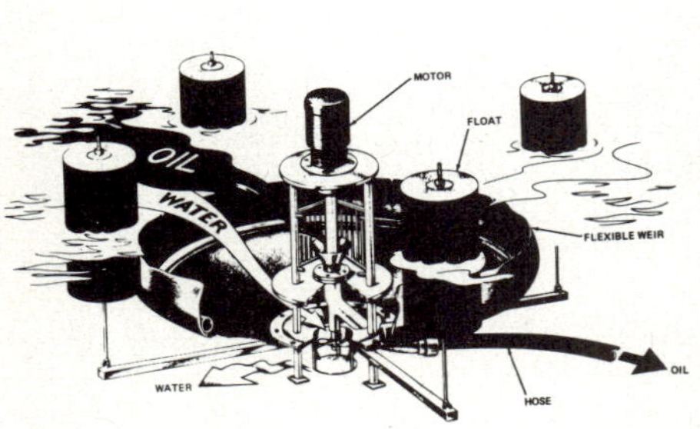

The Medusa MK-L, showing the principles of operation

Capacity

Intake flow	11,360 US gpm
Oil removal flow	287 US gpm at 1·8 m (6 ft) water discharge head
Size	5·48 m (18 ft) diameter
Shape	Circular: 12 flexibly coupled weirs
Weight	2,850 kg (6,400 lb)
Height	2·8 m (9 ft 3 in)
Draught	1·5 m (5 ft)
Primary material	6061 T-6 aluminium alloy
Power source required	30 kw, 440 volt, 3 phase, 60 cycle
Prime movers	
Oil pump motor	Two $7\frac{1}{2}$ h.p., 900 rpm, 220/440 volt, 3 phase, 60 cycle, horizontal, explosion proof, electric motors
Circulator pump motor	One 15 h.p., 600 rpm, 220/440 volt, 3 phase, 60 cycle, vertical, explosion proof, electric motor
Electrical loads	
Oil pump	$8\frac{1}{2}$ amp
Circulator pump	24 amp
Total	41 amp
Oil pumps	Two individually controlled sliding shoe, snore, multi-viscosity, positive displacement pump
Circulator pump	One 3 blade, 66 cm (26 in) diameter impeller
Controls	
Oil pump	Two fused disconnect switch with single stop-start push button control for simultaneous operation
Circulator pump	One fused disconnect switch with single stop-start push button control
Power control cable	Twelve waterproof conductors plus 3·96 cm (5/32 in) galvanised steel cable stress core in flotation housing with waterproof connector one end
Diameter	5·7 cm ($2\frac{1}{4}$ in)
Length	60 m (200 ft)
Oil discharge hose:	
Size	10·15 cm (4 in) inside diameter
Length	30 m (100 ft) with quick disconnect plug one end

Reynolds' 2-stage oil skimmer Medusa (5·48 m (18 ft) diameter)

MK-L 4·25 m (14 ft) diameter version

Weirs

Outer	Surface following; circular multi-segment
Intermediate	Adjustable; circular
Inner	Adjustable; circular with debris screen

Capacity

Intake flow	10,000 US gpm
Oil removal flow	550 US gpm
Size	
Diameter	5·8 m (19 ft)
max.	
Diameter outer weir	4·25 m (14 ft)
Shape	Circular with six radially disposed float support arms
Weight	1,045 kg (2,300 lb)
Height	3·35 m (11 ft 1·75 in)
Draught	1·90 m (6 ft 3 in)
Primary material	6061 T-6 aluminium alloy
Power source required	30 kw, 440 volt, 3 phase, 60 cycle
Prime movers	One 30 h.p. 1,800 rpm 220/440 volt, 3 phase, 60 cycle, explosion-proof, vertical electric motor
Circulator pump motor	Hydraulic motor driven by a variable volume pump through a 3-belt power take off from the 30 h.p. electric motor
Oil pump drive	Inline hydraulic clutch to the 30 h.p. electric motor
Electrical load	35 amp
Oil pump	2-stage 15 cm (6 in) dia. turbine axial flow pump
Circulator pump	Single stage 50·8 cm (20 in) dia. turbine axial flow pump
Controls	
Oil pump	Rotary stop-start switch controlling the clutch
Circulator pump	One fused disconnect switch with stop-start push button control
Power control cable	Twelve waterproof conductors plus 3·96 cm (5/32 in) galvanised steel cable stress core in flotation housing with waterproof connector one end
Diameter	5·7 cm ($2\frac{1}{4}$ in)
Length	61 m (200 ft)
Oil discharge hose	
Size	15·24 cm (6 in) inside dia.
Length	30·5 m (100 ft) with quick disconnect plug one end and socket other end

REYNOLDS VORAXIAL OIL/WATER SEPARATOR

Type	Horizontal voraxial flow; trailer mounted
Trailer	4-wheel with 5·08 cm (2 in) ball hitch assembly, extractable landing leg and highway lighting system constructed of alloy aluminium
Capacity:	
Inlet	700 US gpm
Extractor	73 US gpm
Height	1·43 m (4 ft $8\frac{3}{4}$ in)
Length	5·74 m (18 ft $9\frac{1}{2}$ in)
Width	2·10 m (7 ft)
Weight	2,150 kg (4,850 lb)
Primary material	6061 T-6 aluminium alloy
Power source required	27 kw, 440 volt, 3 phase, 60 cycle
Prime movers	
Separator pump	20 h.p., 1,800 rpm, 220/440 volt, 3 phase, 60 cycle, horizontal electric motor driving a hydrostatic transmission
Extractor pump	$7\frac{1}{2}$ h.p., 900 rpm, 220/440 volt, 3 phase, 60 cycle explosion proof horizontal electric motor
Separator pump	20·4 cm (8 in) open centre axial flow pump. Stainless 30·5 cm (12 in) pitch impeller

Extractor pump	Sliding shoe, snore, multi-viscosity, positive displacement pump
Flow pipe-extractor	20·4 cm (8 in) with inside diameter 1·8 m (6 ft) long heavy wall acrylic view section
Down-stream	5·08 cm (2 in) inside dia. with orifice insert plates and gate valve control
Upstream	2·54 cm (1 in) inside dia. with orifice insert plates and gate valve control
Extractant filtration	Three 5·08 cm (2 in) strainers manifolded in parallel with individual shut off and check valves
Back lighting	Two 75 watt
Controls	
Separator pump	Fused disconnect switch with stop-start push button control
Extractor pump	Fused disconnect switch with stop-start push button control
Back lighting	2 position rotary on-off switch
Power control cable	Four connector S.O. electrical cable with insert one end (for extractor pump connection)
Hose	
Effluent	
Size	20·4 cm (8 in) inside dia. with flange mounting and discharge orifice plates
Length	10·5 m (35 ft)
Extractant	
To ext. pump	
Size	7·62 cm (3 in) inside dia. with quick disconnect plug one end and socket other end
Length	30 m (100 ft)
From ext. pump	
Size	10·1 cm (4 in) inside dia. with quick disconnect plug one end
Length	30 m (100 ft)

Reynolds' MK-L, 4·25 m (14 ft) diameter, 2-stage oil skimmer

The Voraxial oil separator

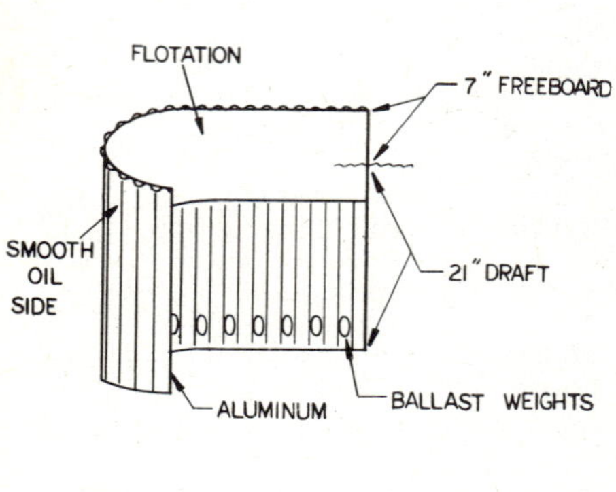

Constructional diagram of the Reynolds oil boom

REYNOLDS ALUMINIUM OIL BOOM
STANDARD TYPE

The oil boom is made of tough resilient sheet aluminium galvanically compatible for long life in a salt-water environment. Tensile strength is well in excess of 910 kg (2,000 lb). A single strip of plastic alloy material provides the flotation to give the 0·71 m (28 in) boom, 0·18 m (7 in) of freeboard, and 0·53 m (21 in) of draught. The bottom of the boom is weighted to provide good metacentric height and stability in a seaway. It provides a reserve buoyancy of approximately 100 per cent.

The side of the boom without flotation is normally the oil containment side. It provides a smooth surface for easy cleaning. The plastic material will not support combustion.

30 m (100 ft) of the boom may be reeled on to a cylindrical spool 1·02 m (40 in) in diameter for ready storage at dockside. This feature, for example, allows an unloading tanker to be completely surrounded by boom in a matter of minutes by one man. It may also be deployed by boat or from shore.

The 0·71 m (28 in) boom weighs 3·87 kg/m (2·6 lb/ft) range. It can be rolled into a compact cylinder and secured by a girth strap for transport by land and sea, or for air drop. Removal of the strap while floating deploys the boom. One hundred foot sections are easily attached with the bolts provided. The boom is made of 3003-H14 aluminium alloy, corrugated vertically.

Inshore type

Lighter and smaller than the standard type, this boom is effective in waters where wind conditions are not severe.

Specifications

Height	38 cm (15 in)
Length	30 m (100 ft) sections
Freeboard	12·7 cm (5 in)
Draught	25·4 cm (10 in)
Weight	1·47 kg/m (1 lb/ft) run

The boom may be stored either palletised or in a spool.

Reynolds' corrugated aluminium oil boom 30·5 m (100 ft)

SHELL PIPELINE CORPORATION PIPELINE RESEARCH AND DEVELOPMENT LABORATORY

OIL SKIMMING BARGE

This barge has some of its components articulated allowing the system to flex with waves. The oil recovery rate is claimed to be 95% at a forward travelling speed of 0·84 m/s (2·75 ft/s). A Vee-shaped 6·1 m (20 ft) wide diverter is employed ahead of the barge. At the collection area at the centre of the vessel a system of flexible articulated baffles are used to reduce the turbulence in the collecting area and so improve the separation of the oil and water.

SLICKBAR INC

MARK 5 AND MARK 6 OIL BOOMS

The Mark 5 is a heavy duty oil boom which projects 0·31 m (12 in) above the water surface and 0·61 m (24 in) below. A heavy duty fabric is used and a 4,500 kg (10,000 lb) tension member to maintain structural integrity. The flotation structure is made of polyethylene foam and covered by a polyethylene skin.

Specification

Swaged cable terminators

The load support cable is attached to the Slickhitch Mark II Connector through a stainless steel swaged cable terminator. This type of construction allows the end sets and cable to carry almost all of the load instead of the skirt.

Fastenings

All fastenings are stainless steel

Skirt material

Woven polyester fabric coated with International Orange PVC with UV inhibitors are provided for longer life. The fabric exceeds USA FTMS 191, required to meet MIL-B-28617 (YD). Temperature range — 20°F to +200° F.

Float strap

Special high tensile PVC coated polyester fabric, 2·54 cm (1 in wide)

SlickhitchMark II connector

The Slickhitch Mark II connector is extruded aluminium, anodised to exceed MIL-A-8625 (Type 2). It is designed to join sections of boom together instantly without any tools. To disconnect sections, the retaining pin is removed and the halves come apart.

Ballast

Ballast is lead, hardened with antimony for longer life. Each weight is securely riveted to the bottom of the skirt. The amount of ballast is varied to meet individual conditions.

The Slickbar oil boom deployed

Tow ropes

15 m (50 ft) braided, 12 mm ($\frac{1}{2}$ in), high tensile floating synthetic rope.

Flotation

10 cm (4 in) or 16·5 cm (6·5 in) high, bright yellow polyethylene foam. The foam is processed to impart a solid polyethylene skin over the entire surface. Ultra violet and oxidation inhibitors are incorporated in the foam for durability and extended life.

Hardened float ends

Float ends are provided with a hard, tapered end during processing for durability and streamlining.

Load support cable

Stainless steel 17·7 × 48 × 0·6 cm (7 × 19 × $\frac{1}{4}$ in) dia. The load support cable supplies tension load capability in excess of 2,272 kg (5,000 lb). During use, loads imposed on the boom are transferred to the cable through numerous attachments along the floats and at every anchor point. Stronger cables are also available.

Anchor points

Anchor points are standard every 30 m (100 ft). Additional anchor points are available.

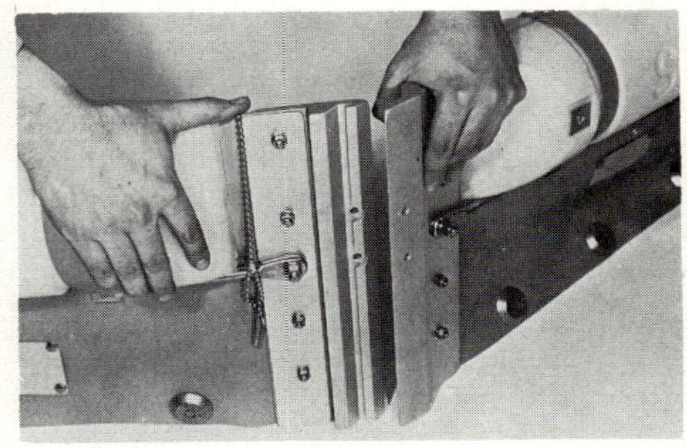

Joint details of the Slickbar oil boom

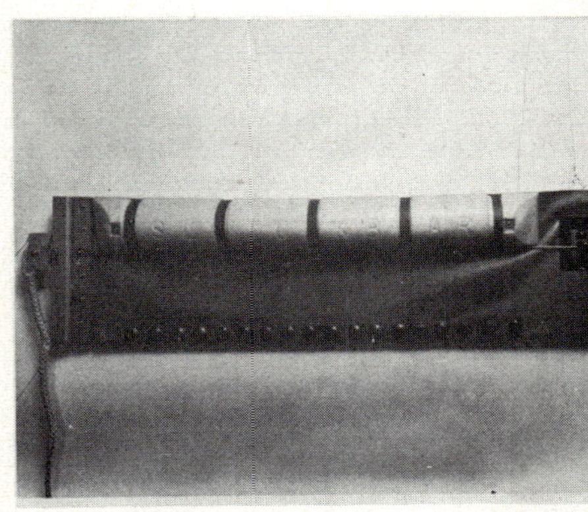

Illustration shows the detailed construction in a segment of the Slickbar oil boom

SKIMMER HEADS

RIGID MANTA RAY SKIMMER HEAD

For use with the Slickbar oil recovery pumping unit.

Diameter (half circle): 1·2 m (4 ft)

Hose coupling: 7·6 cm (3 in) or 10 cm (4 in) available

Capacity: 180 US gals/min

Weight; net: 13·2 kg (29 lb)

Cost, (1973): $ US 266

FLEXIBLE MANTA RAY SKIMMER HEAD

Diameter (half circle): 1·5 m (5 ft)

Hose coupling: 7·6 cm (3 in) or 10 cm (4 in) available

Capacity: 1·27 cm (0·5 in) opening, 80 US gals/min. 2·54 cm (1·00 in) opening, 180 US gals/min

Weight, net: 26 kg (58 lb)

Cost (1973): 1·27 cm (0·5 in) openings, $ US 389; 2·54 cm (1·00 in) openings, $ US 427.50

HIGH CAPACITY ALUMINIUM SKIMMER HEAD

Opening: 5·08 cm (2 in) × 1·2 m (4 ft)

Hose coupling: 10 cm (4 in)

Capacity: 250 US gals/min

Flotation chambers: Two of 0·0328 m^3 (1·16 ft^3)

Cost (1973): $ US 468.50

Three oil recovery skimmer heads made by Slickbar. Left to right: a high capacity aluminium skimmer head; a rigid Manta Ray skimmer head; and a flexible Manta Ray skimmer head

TRANS-VAC 500D PUMPING UNIT

Transvac 500D is a self-contained oil recovery pumping unit. A high capacity vacuum pump is used to move oil, water, debris and air into a receiving vessel from which air and liquid can be continuously evacuated through different outlets, while debris is screened out and retained by a grate inside the tank. Petrol or diesel power may be used to drive the pump equipment.

Design history

Two prototypes built first then production models

Construction of prototype started: April 1970

Prototype completed: April 1970

Trials completed: July 1970

Production

November 1970. Four units in service

Basic dimensions of systems

2·12 m (7 ft) wide, 2·42 m (8 ft) high, 3·64 m (12 ft) long

Weight of system and main components

2,000 kg (4,500 lb)

Method of operation

Engines drive rotary lobe blower and sliding vane discharge pump. Vacuum in receiving vessel is suction.

The Slickbar oil recovery pumping unit (rear and front views)

Capability of system

500 US gpm suction and discharge capacity

Collecting rate: 500 US gpm

Capacity: 1,514 litres (400 US gals) system usually runs empty

Availability

Purchase, capital cost: $ US 22,500 (1973)

SUNSHINE CHEMICAL CORPORATION

SEA BROOM

Sea Broom is a small suction pumping device for use with 'Jansolv-60' Dispersant, developed also by Sunshine Chemical Corporation. The unit is approximately 1·2 m (4 ft) in diameter, 1·0 m (3 ft) high and is powered by a 5 h.p. gasoline engine. Purchase price (1973) is $ US 750.

Sea Broom in operation

UNION OF SOVIET SOCIALIST REPUBLICS

USSR MINISTRY OF THE MERCHANT MARINE

Following the *Torrey Canyon* disaster a contest for the best design of floating oil-trapping device was held in the USSR, and an oil collector device is now in production which can deal with nearly 4 tons of spilt oil an hour, and in addition collect floating refuse.

COMPANY PROFILES (INCLUDING RESEARCH ORGANISATIONS)

ACADIAN MARINE SERVICE INC.

21st floor, 225 Baronne Street, New Orleans, Louisiana 70112, U.S.A.
TEL: (504) 581-7880
TELEX: 58 332
CABLE: Acadian
EXECUTIVE: Captain R. Vorenkamp, Executive Vice-President

ADELAIDE SHIP CONSTRUCTION

Birkenhead, South Australia

A/S AKERS MEK. VERKSTED (AKER GROUP)

Munkedamsveien 45, Oslo 2, Norway

UNIVERSITY OF ALASKA, INSTITUTE OF MARINE SCIENCE

Box 95102, Fairbanks, Alaska 99701, U.S.A.
TEL: (907) 479-7531

APPLEDORE SHIPBUILDERS LIMITED

Appledore, Devon EX39 1LX, England
TEL: Bideford 3281 (Answerback Apship Apldor)
TELEX: 46156
CABLE: Apship Devon
EXECUTIVES: Vice Admiral Sir Peveril William-Powlett, Director; J. Venus, Director; J. R. Young, Director; J. Ball, Director; J. W. Chapman, Director; G. H. Grant, Director; D. W. R. Hickmott, Director; Sir George Briggs, Director; W. R. Downing, Director; M. I. Makin, Director; J. P. Gilfillan, Director.
ORGANISATION: Appledore Shipbuilders Limited is a member of the Court Group.
London Office: 9 Thayer Street, London, W.1
Tel: 01-486 6633. Telex: 27725

ATELIERS ET CHANTIERS DE BRETAGNE ACB

44 Nantes, Prairie-au-Duc, France
TEL: (40) 47.31.32
TELEX/CABLE: Nantac Nante 71960
OFFICE AT: 29 rue du Faubourg St. Honore, 75008 Paris
TEL: 266.51.71 & 265.14.62
TELEX: Nantac Paris 65317

ATLAS COPCO AB

S-105 23 Stockholm, Sweden
TEL: 08-24 30 00
EXECUTIVE: Mr. H. Johnsson, Director of Communications.

BENNETT POLLUTION CONTROLS LTD

119 Charles Street, North Vancouver, British Columbia, Canada
TEL: (604) 929-5451
EXECUTIVES: John A. Bennett, President and Director; Hans R. Banziger, Executive Vice-President and Director; G. N. Munro, Secretary and Director; G. W. Pearmain, Director; Howard Welsh, Operations Superintendent.
ORGANISATION: Formed in early 1969, the company has been engaged in the development of oil spill control and collection systems for both offshore and inshore use.
ASSOCIATED COMPANIES:
Harding Pollution Control Corp,
5719 Kirby Drive, Houston,
Texas 77005, U.S.A.
(wholly owned subsidiary)
REPRESENTATIVES:
Europe
Pechiney Ugine Kuhlmann,
10, Rue du General Foy,
75—Paris 8, France.
South Africa
David Rice & Co (Pty) Ltd,
293A Voortrekker Road,
Maitland, Cape Town,
South Africa
Argentina
Rayo Electronica S.N.L.,
Belgrano 990°—6P,
Buenos Aires, Argentina
New York, New Jersey, Rhode Island and Connecticut
International Underwater Contractors,
264 Fordham Place,
City Island, New York 10464
Eastern Canada
McAllister Towing Ltd,
20 Grey Nuns Street,
Montreal, Quebec
Hong Kong
The China Engineers Ltd,
18th Floor, Sincere Insurance Bldg.
No. 4 Hennessy Road,
Hong Kong

BENTHOS INC.

Edgerton Drive, North Falmouth, Massachusetts 02556, U.S.A.
TEL: (617) 563-5917
EXECUTIVE: Mr. S. O. Raymond, President

BLOHM & VOSS AG

2 Hamburg 1, P.O. Box 720, West Germany
TEL: 3061
TELEX: 02 11047 and 02 11312
CABLE: Blohm d
ORGANISATION: The company has been manufacturing ships and machinery for close on a hundred years. The shareholders in Blohm & Voss are:
August Thyssen-Hutte AG (64·7%)
v. Dietlein (4·9%)
Siemens AG (12·5%)
Blohm Group (17·9%)
ASSOCIATED/SUBSIDIARY COMPANIES: Ottensener Eisenwerk GmbH (wholly owned subsidiary); Barthels & Luders GmbH (wholly owned subsidiary); Hamburgische Schiffbau-Versuchsanstalt GmbH; Marine-Schiffstechnik Planungs-GmbH.

J. BOLSON & SON LTD.

New Quay Road, Poole, Dorset BH15 4AB
TEL: Poole 3201
EXECUTIVE: Mr. M. A. Bolson, Managing Director.

BRITISH AIRCRAFT CORPORATION LIMITED ELECTRONIC AND SPACE SYSTEMS GROUP

P.O. Box No. 77, Filton House, Bristol BS99 7AR
TEL: (0272) 693831
TELEX: 44188
CABLE: Britairgw BSTL
EXECUTIVES: Mr. G. R. Jefferson, Managing Director (Guided Weapons); Mr. D. Rowley, Executive Director, Electronic & Space Systems; Mr. R. G. T. Munday, Product Group Manager, Space Systems & Sounding Rockets; Mr. R. Astbury, Commercial Manager, Electronic & Space Systems; Mr. W. M. Fraser, Marketing Manager, Electronic Systems; Mr. M. Garnett, Project Manager IGS Submersible.
ORGANISATION: B.A.C. interest in the underwater field dates back to 1967 when research was done on the *Kraken* seabed habitat with Imperial College, London, followed by the automated modular habitat—*Bacchus*. Subsequently they designed data handling and processing equipment based on space activities, and mode quantitative mathematical models of underwater mobile systems, mainly towed bodies and cable controlled vehicles.
ASSOCIATED COMPANIES: Vickers Shipbuilding, Vickers Oceanian, GEC, Marconi S & D.S. and Marconi Marina.

THE BRITISH PETROLEUM COMPANY LIMITED

Britannic House, Moor Lane, Moorgate, London, E.C.2, England
TEL: (01) 920-8000
TELEX: 888811
WORKS: BP Research Centre, Chertsey Road, Sunbury-on-Thames, England.
TEL: Sunbury 85533

BROOKE MARINE LIMITED

Lowestoft, Suffolk, England
TEL: Lowestoft 5221
TELEX: 97145
CABLE: Brookcraft Lowestoft
EXECUTIVES: Mr. H. L. Dowsett, Executive Chairman; Mr. A. F. Prouten, Deputy Chairman and Director of Contracts; Mr. R. Straton, Deputy Managing Director; Mr. D. M. Wait, Yard Director; Mr. G. B. Reid, Financial Director; Mr. W. I. Edwards, Technical Director.
ORGANISATION: The company was formed in 1874 and has built ships for 40 different countries.

BRUKER PHYSIK AG

D7501 Forchheim/Karlsruhe, Silberstreifen
TEL: 0721/558070
TELEX: 826 836
EXECUTIVES: Professor Dr. G. Laukien, General Manager.
ORGANISATION: The company formed in 1960, entered the oceanological field in 1967 and its main production lines are: spectrometers, power supplies, magnet systems, and analytical measuring techniques.
ASSOCIATED COMPANIES: In France, Switzerland, England, U.S.A., Italy, Netherands, Canada, Japan.

BURRARD DRY DOCK CO. LTD.

P.O. Box 99, North Vancouver, British Columbia, Canada
TEL: 604-988-2111
TELEX: 04-352652
EXECUTIVE: M. J. Surgenor, Technical Director.
ASSOCIATED COMPANY: Yarrows Limited, Victoria, British Columbia, Canada.

CARRINGTON SLIPWAYS PTY. LTD.

Newcastle, New South Wales. Australia

CENTRE NATIONAL POUR L'EXPLOITATION DES OCEANS (CNEXO)

39 Avenue d'Iena, 75783 Paris Cedex 16, France
TEL: 720.13.69
TELEX: 61775 (OCEANEX F)
WORKS: Section des Engins du CNEXO, Zone Industrielle de Toulon Est, 83130 La Garde. Tel: 41.30.10—41.38.40. Telex: 43037 (OCEANEX LGARD)
EXECUTIVES: Mr. Yves La Prairie, Director General; Mr. Alain Sciard, Head, Service Emploi des Moyens; Mr. Jean Jarry, Head, Section des Engins.
ORGANISATION: CNEXO was formed 1st April 1967. The Section des Engins, formerly Laboratoire du Bathyscaphe, was part of the Centre National de la Recherche Scientifique (CNRS) from 1958 to 1969.

CITY OF LONDON POLYTECHNIC SCHOOL OF NAVIGATION

100 Minories, Tower Hill, London EC3N 1JY, England
TEL: (01) 283 1030
EXECUTIVE: Lt.-Commander K. R. Alger RNR (Retd.), Head of School.

COCHRANE & SONS LTD

Ouse Shipbuilding Yard, Selby, Yorkshire YO8 8AP, England
TEL: Selby 5621
TELEX: 57818
CABLE: Cochrane Selby
EXECUTIVES: P. Curtis, Managing Director; W. M. George, Director; C. R. Shepherd, Director; D. R. Townend, Director; F. F. McGuire, Director.
ORGANISATION: A member of the Drypool Group of Companies.

COCKERILL YARDS HOBOKEN S.A.

B 2710 Hoboken, Belgium
TEL: Antwerp 273880
TELEX: 31175 Cockhob b.
CABLE: Cockerillyards Hoboken
EXECUTIVES: Mr. F. Osterrieth, Chairman; Mr. R. Schenck, Managing Director; Mr. J.-P. Conrad, Manager; Mr. J. Claes, Senior Naval Architect.

CODEVINTEC

27 rue d'Astorg, 75008, Paris, France
TEL: 265-38-43
TELEX: 65.249 (Astrasia)
CABLE: Codevintec-Paris
EXECUTIVE: Mr. H. L. Kyriakidis, Director of Oceanographic Division.

COMEX

13275 Cedex 2 Marseilles, France
TEL: 40.11.70
TELEX: 41.985 Comex MarsL
CABLE: Comex-Marseille
EXECUTIVES: H. G. Delauze, President; J. Lamazou, General Manager; R. Farrington-Wharton, Managing Director, Comex Diving Ltd.
ORGANISATION: The company was formed in 1964 to provide technologically advanced deep diving services to the offshore oil industry. Comex is one of the world's largest diving contractors with subsidiaries all over the world. Through Mr. J. Houlder, Chairman of Comex Diving Ltd, Comex is associated with Houlder Brothers of 53 Leadenhall Street, London.
PRINCIPAL SUBSIDIARY: Comex Diving Ltd, of Great Yarmouth, Norfolk and Dyce, Aberdeen, Scotland.

COMPAGNIE FRANCAISE d'ENTREPRISES METALLIQUES (CFEM)

57 Boulevard de Montmorency, 75016 Paris, France
TEL: 288 49 29
TELEX: 62512
CABLE: Lonfer Paris
WORKS: Blanc-Misseron, Nord; Lauterbourg Bas-Rhin; Maizières-les-Metz, Moselle; Rouen, Seine-Maritime Fos-sur-Mer, Bouches-du-Rhone
EXECUTIVES: Mr. H. Deschenes, Chairman of the Board; Mr. F. P. Goillandeau, Off-shore Manager
ORGANISATION: CFEM, part of Banque de Suez, Pont à Mousson, St. Gobarin Group, was formed in 1967 as a result of the merging of various construction companies, active in the offshore field.

UNIVERSITY OF DELAWARE, COLLEGE OF MARINE STUDIES

Newark, Delaware 19711, U.S.A.
TEL: (302) 738-2841
WORKS: Lewes Field Station, Lewes Delaware 19958, U.S.A. Tel: (302) 645-6674
EXECUTIVES: W. S. Gaither, Dean; K. S. Price, Associate Dean and Director of Field Station; R. B. Biggs, Assistant Dean; J. M. Goodman, Director, Coastal Zone Resource Planning; A. L. Inderbitzen, Director, Marine Operations.
ORGANISATION: The College of Marine Studies was established in 1970 and is a multi-disciplinary, graduate-level institution providing an opportunity for study and research in the fields of marine biology, marine geology, chemical and physical oceanography, ocean engineering and marine affairs.

DEPARTMENT OF AGRICULTURE AND FISHERIES FOR SCOTLAND, MARINE LABORATORY

P.O. Box 101, Victoria Road, Aberdeen AB9 8DB

DEUTSCHE BABCOCK U. WILCOX AG

42 Oberhausen, Germany
TEL: 02132/833810
CABLE/TELEX: 856 951 b dbab
EXECUTIVES: Dipl.-Ing. Schiemichen, Director; Mr. H. L. Ewaldsen, General Manager; Dr. Dreher, Sales Director; Dr. Dickore, Works Manager.
ASSOCIATED COMPANIES:
Brown, Boveri u. Cie
Deutsche Babcock u. Wilcox AG
Dornier System GmbH
Hagenuk GmbH
Babcock, London Ltd
Babcock, U.S.A.
Babcock, Atlantique

DIRECTION DES CONSTRUCTION ET ARMES NAVALES CENTRE d'ETUDES ET DE RECHERCHES TECHNIQUES SOUS MARINES

83800 Toulon Naval, France
TEL: 41.40.20
CABLE: Geninaval Toulon
ADDRESS OF WORKS: As above
SUBSIDIARY/ASSOCIATED ORGANISATIONS:
Centre d'Etudes dependant du Ministère des Armées.

DIRECTION DES CONSTRUCTION ET ARMES NAVALES DE DIEGO-SUAREZ (D.C.A.N.)

DCAN-Diego-Suarez, Madagascar
TEL: Standard Marine No. 221-64
TELEX: Dirnav Diego-Suarez
CABLE: Geninaval Diego-Suarez

DIXILYN INTERNATIONAL INC.

1012 First City National Bank Building, P.O. Box 14067, Houston, Texas 77021, U.S.A.
TEL: (713) 228-9661
TWX: 910-881-1603
EXECUTIVE: Mr. W. N. Plamondon, Marketing Manager.

DORNIER SYSTEMS GmbH

D-7990 Friedrichshafen, Postfach 648, West Germany
TEL: 07545/81
TELEX: 0734359
EXECUTIVES: Dr. Karl Wilhelm Schäfer, (commercial department); Dr. Bernhard Schmidt (technical department); Dr. Helmut Ulke (second in command).
ORGANISATION: Founded in 1962 as part of Dornier Werke GmbH.

DRAGERWERK AG

2400 Lubeck, Postfach 1339
TEL: (0451) 8102-1
TELEX: 026807
WORKS: Moislinger Allee 53155, 2400 Lubeck
EXECUTIVES: Dr. Heinrich Drager; Dr. Christian Drager; Dipl. Kaufm. Theo Drager.

DUKE UNIVERSITY, MARINE LABORATORY

Beaufort, North Carolina 28516, U.S.A.
TEL: (919) 728-2111 and (919) 728-3372
EXECUTIVE: Mr. R. T. Barber, Director of Oceanographic Program; Mr. J. G. Newton, Marine Superintendent.
ORGANISATION: Partially funded by the National Science Foundation.

The Co-operative Program originated in 1961 as a result of National Science Foundation's approval of a proposal submitted by Dr. C. G. Bookhout. The research vessel *Eastward* was designed by Potter, McArthur and Gilbert of Boston, Massachusetts and constructed by the Sturgeon Bay Shipbuilding & Drydock Co, Sturgeon Bay, Wisconsin, in 1963. Research and Training cruises commenced October 1964.

EARL AND WRIGHT

657 Howard Street, San Francisco, California 94105, U.S.A.
TEL: (415) 981-7770
TELEX: 340502
CABLE: Eandw San Francisco
EXECUTIVES: J. G. Wright, Chairman of the Board of Earl and Wright and Director of SEDCO; W. W. Hayes, President; W. M. Martinovich, Senior Vice-President; A. J. Burnell, Supervising Engineer.
ORGANISATION: 1940: Partnership of Earl and Wright; 1955: Earl and Wright became a corporation; 14 August 1968: Earl and Wright became a subsidiary of SEDCO Inc, 1901 North Akard, Dallas, Texas 75201, U.S.A.
ASSOCIATED/SUBSIDIARY COMPANY:
Earl and Wright, c/o CJB House, 10th Floor, Eastbourne Terrace, London, W.2. England.

EQUITABLE EQUIPMENT COMPANY INC.

P.O. Box 8001, New Orleans, Louisiana, U.S.A.
TEL: 504-947-0631
CABLE/TELEX: 058-354
ADDRESS OF WORKS: As above
EXECUTIVES: Mr. C. M. Keeney, President; Mr. A. W. Kersteter, Executive Vice-President; Mr. R. M. Adler, Vice-President Engineering; Mr. J. F. Williams, Vice-President Sales; Mr. F. T. Ventura, Sales Manager Marine Products.
ORGANISATION: Equitable Equipment Company, Inc is a wholly owned subsidiary of Trinity Industries, Inc, Dallas, Texas, U.S.A.

ETUDES ET CONSTRUCTIONS AERONAUTIQUES

17 Avenue du Chateau, Boite Postale No. 16, 92190 Meudon, France
TEL: 027 00-87
TELEX: ECABLVU 20.336 F

FIELD INTERNATIONAL CORPORATION

930 Milam Building, San Antonio, Texas 78205
TEL: (512) 226-4371
TELEX: 767331
CABLE: Fieldco
EXECUTIVES: W. Earl Rowe, Chairman of Board; Robert A. Buschman, President; Reeves L. Smith, Vice-President; Carl Tennis, Secretary; J. D. Boynton, Vice–President, Operations; B. James, Vice-President, Sales; D. E. Edgar, Purchasing Agent; G. Bloodworth, Supervisor, Construction; M. Rutherford, Vice-President; P. Rutherford, Jr., Vice President.
SUBSIDIARY/ASSOCIATED COMPANY:
Viking Offshore A/S.

EXECUTIVES: R. A. Buschman, Chairman; M. G. Rutherford; E.Stalleno; Tore Hansen-Tangen; J. D. Boynton, Vice–President, Operations; B. James, Vice-President, Sales; R. L. Smith, Vice-President and Secretary; D. E. Edgar, Purchasing Agent; G. Bloodworth, Supervisor Construction.

FLUOR DRILLING SERVICES

2500 S. Atlantic Blvd., Los Angeles, California 90040, U.S.A.
TEL: (213) 262-6111
TELEX: 67-4729
CABLE: Wesdril
ADDRESS OF WORKS: As above
TEL: (213) 268-3311
EXECUTIVES: R. A. McClintock, President; J. D. Erlewine, Asst. to President; J. A. Sage, Vice-President; R. F. Woidneck, Vice-President; E. D. Turner, General Manager, Vice-President (Wodeco Division); J. Carpenter, General Manager, Vice-President (Coral Division); N. Haney, Sales Director, Vice-President (Wodeco Division); F. H. Ackema, Vice-President, Marine Design & Construction.
ORGANISATION:
Fluor Corporation (parent company);
Fluor Drilling Services, Inc (subsidiary);
Western Offshore Drilling & Exploration Co (division);
Coral Drilling Co (division).
AREA MANAGERS:
W. Durret, 113 West Sixth Street, Anchorage, Alaska 99501
K. M. Nicholson, P.O. Box 3347, Tehran, Iran
O. B. Persful, P.O. Box 252, Sfax, Tunisia
L. Moore, 101 Ming Center, Clemenceau Av, Singapore 9
B. Harrell, 3 Tung-Hwa South Road, Taipei, Taiwan, Republic of China.

FMC CORPORATION

One Illinois Center, 111 East Wacker Drive, Chicago, Illinois 60601, U.S.A.
EXECUTIVES: Mr. B. C. Carter, Chairman of the Board; Mr. R. H. Malott, President and Chief Executive Officer.

FOREX NEPTUNE

20 Avenue Rapp, Paris 7e, France
TEL: 555-14-00
TELEX: Forex 27920 Paris
Neptune 26692 Paris
EXECUTIVES: Mr. G. Barthe, President; Mr. B. Alpaerts, Executive Vice-President
ORGANISATION: The company is wholly owned by Schlumberger.
ASSOCIATED COMPANIES:
Europe
Forex-Neptune
105, Avenue Raymond-Poincaré, Paris-16e France
ADDRESS: As above
TELEX: Sonarepal 27721
CABLE: Mail address
TEL: 553.97.79
Far East
Forex
Pertamina Oil Center, 55 Djalam Tharim, Djakarta, Indonesia
P.O. Box 2985, Djakarta Indonesia
TELEX: 61-68
CABLE: Petroforex Djakarta
TEL: 54711 to 54716
Middle East
Forex Neptune
Bureau 201—Bloc C. Gefinor Center, Beirut, Lebanon
P.O. Box 8603, Beirut, Lebanon
TELEX: Nepfor 20225
CABLE: Petroforex Beirut
TEL: 346895 & 346896
Western Africa
Forex Company of Nigeria Ltd
Manuwa Street, S.W. Ikoyi, Lagos, Nigeria
MAIL: P.O. Box 6718, Lagos, Nigeria
TELEX: 21293
CABLE: Mail address
TEL: 22303

MASCHINENBAU GABLER GmbH

Niels-Bohr-Ring 5a, P.O. Box 1690, 2400 Lübeck 1, W. Germany
TEL: 0451-3 12 33
TELEX: 26 768
EXECUTIVES: Dipl.-Ing. L. Nohse; Dipl.-Ing. P. Kayser
ORGANISATION:
In 1962, Messrs. Maschinenbau Gabler GmbH was founded as an affiliated firm of Messrs. Ingenieurkontor Lübeck, (Prof. Dipl.-Ing. Ulrich Gabler). In shipbuilding engineering Messrs. Ingenieurkontor Lübeck are one of the most important consulting engineering companies in the world; for many decades their special field has been the design of underwater vehicles. The experience in this area has been of decisive advantage in solving the problems that are dealt with by Messrs. Maschinenbau Gabler GmbH in the field of oceanography.

GENERAL DYNAMICS CORPORATION

Pierre Laclede Center, 7733 Forsyth Boulevard, St. Louis, Missouri 63105, U.S.A.
TEL: (314) 862-2440
TWX: 910-761-2193
TELEX: 44-7630
EXECUTIVES:
BOARD OF DIRECTORS:
David S. Lewis, Chairman, General Dynamics; Gene K. Beare, Executive Vice–President, General Dynamics; Henry Crown, Chairman, Henry Crown & Co, Chicago, Illinois; Nathan Cummings, Chairman, Executive Committee, Consolidated Foods Corp,. New York, New York; Robert F. Ellsworth, President and Chief Executive Officer, Lazard (N.Y.) International, New York, New York; Milton Falkoff, Vice President, Henry Crown & Co, Chicago, Illinois; Albert E. Jenner, Jr., Senior Partner, Jenner & Block, Attorneys, Chicago, Illinois; Earl D. Johnson, Business and Financial Consultant, Greenwich, Connecticut; Gorden E. MacDonald, Executive Vice–President, Finance, General Dynamics; Hilliard W. Paige, President, General Dynamics; Wallace R. Persons, Chairman and Chief Executive Officer, Emerson Electric Company, St. Louis, Missouri; Robert W. Reneker, President and Chief Executive Officer, Swift & Company, Chicago, Illinois
EXECUTIVE COMMITTEE: Henry Crown, Chairman; Nathan Cummings; Albert E. Jenner, Jr.; David S. Lewis; Hilliard W. Paige; Robert W. Reneker
SENIOR EXECUTIVES:
David S. Lewis, Chairman and Chief Executive Officer; Hilliard W. Paige, President; Gene K. Beare, Executive Vice-President and President, GD Commercial Products Co; Gorden E. MacDonald, Executive Vice-President, Finance; Frank W. Davis, Vice-President, General Dynamics, and President, Convair Aerospace Division; Max Golden, Vice-President, Contracts; John T. Hayward, Vice-President, International; M. C. Curtis, Vice-President, General Dynamics, and Deputy General Manager, Electric Boat Division; Algie A. Hendrix, Vice-President, Industrial Relations; Edward J. LeFevre, Vice-President, General Dynamics, and Manager, Washington Office; Edward E. Lynn, Vice-President, General Counsel; Frank Nugent, Vice-President, General Dynamics, and President, Freeman Coal Mining Company and The United Electric Coal Companies; Joseph D. Pierce, Vice-President, General Dyanamics, and General Manager, Electric Boat Division; P. Takis Veliotis, Vice-President, General Dynamics, and President and General Manager, Quincy Shipbuilding Division; Wayne Wells, Vice-President and Treasurer, General Dynamics; John P. Maguire, Secretary, General Dynamics; W. J. Morrow, Vice-President and General Manager, Pomona Division; William W. Oughtred, President, Asbestos Corporation, Limited; Frederick R. Kearns, President, Canadair Limited; Sidney Kirschner, General Manager, Electro Dynamic Division; M. James O'Brien, President, Marblehead Lime Company; Irving Crown, Chairman, Material Service Corporation; Dause L. Bibby, President, Stromberg-Carlson Corporation; Edward T. Keating, Vice-President and General Manager, Stromberg DatagraphiX, Inc; J. Richard Iverson, Vice-President and General Manager, Electronics Division; W. T. Dorrance, Vice-President, Ocean System Programs; J. W. Petre, Manager of Marine Systems.

ORGANISATION:
General Dynamics was incorporated in the State of Delaware in 1952. The principal operating units of the highly diversified corporation are made up of six divisions and eight subsidiary companies. The divisions are the Convair Aerospace Division, the Electric Boat Division, the Electro Dynamic Division, the Electronics Division, the Pomona Division and the Quincy Shipbuilding Division. The subsidiary companies are Canadair Limited, Freeman Coal Mining Company, Marblehead Lime Company, Material Service Corporation, Stromberg-Carlson Corporation, Stromberg DatagraphiX, Inc, The United Electric Coal Companies, and Asbestos Corporation Limited.

Three General Dynamics Divisions are actively engaged in oceanographic and shipbuilding activities (Electric Boat Division, Quincy Shipbuilding Division, and Electronics Division) and, whenever it becomes desirable, the resources of several or all operating units are utilised to accomplish a project. Direct lines of communication are maintained between all operating units and the corporate office in St. Louis, Missouri, to assure the availability and utilisation of these resources. A current example of such utilisation is the application of cryogenic expertise developed by the Convair Aerospace Division in the course of design and production of space launch vehicles to the design of liquefied natural gas tankers now in production at the Quincy Shipbuilding Division.

The Electric Boat Division in Groton, Connecticut, is the world's oldest builder of submarines and is considered the primary nuclear submarine shipyard for the U.S. Navy. Electric Boat also has produced a number of deep research submersibles and has done pioneer work in the field of deep ocean technology.

The Quincy Shipbuilding Division, long a producer of quality surface ships for the US Navy and commercial shipping companies, is presently under contract to build three giant 125,000 cubic metre liquefied natural gas tankers and is negotiating with a number of shipping companies for contracts to build additional LNG ships. The division holds exclusive American manufacturing rights for the Moss-Rosenberg spherical containment design currently under construction in Norway. This design incorporates five 120-foot diameter aluminium spheres containing about 4·4 million cubic feet of liquefied natural gas at –265 degrees F. The Quincy Shipbuilding Division is located at Quincy, Massachusetts.

The Electronics Division, which has units at San Diego, California, and Orlando, Florida, is deeply involved in ocean technology in addition to its other activities (airport surveillance radar, aerospace ground equipment, airborne sonobuoy receivers, large screen displays, communication equipment, range measuring systems, etc.) and, in the case of the division's various Ocean Data Stations and Large Navigational Buoys, the marine engineering background of the corporation at large has been employed. (The hull of the prototype Large Navigational Buoy was built by the Electric Boat Division. Special purpose range instrumentation and space tracking ships produced by Quincy Shipbuilding involved technologies that have been used to advantage in the electronic systems aboard the Ocean Data Stations.)

Despite the extent of previous marine engineering experience, however, it became apparent as the initial buoy programmes entered the development and operational phases that the buoy technique was unique. There was no background of buoy technology as such, and little if any experience with successful deep ocean moorings. Consequently, the buoy development programmes at General Dynamics soon began contributing new and basic knowledge to the field. The accomplishments and contributions of these buoy programmes have been made known to the oceanographic and maritime communities through test and development results that have been widely published nationally and internationally. The net result is that the General Dynamics buoy programmes represent more than 12 years of uniquely successful developmental and operational activity in the relatively new field of buoy technology, covering every aspect of the acquisition and processing of telemetered environmental data from the oceans.

Closely related to the development of Ocean Data Stations (ODSs) and their associated data processing/support systems, was the development of Large Navigational Buoys (LNBs) used as major aids to navigation and for replacement of lightships. Many of the specialised ocean engineering techniques and technical approaches now being used in the data buoy programme have evolved through operational experience acquired with these new ocean navigation systems.

GENERAL ELECTRIC COMPANY OCEAN SYSTEMS PROGRAMS

3198 Chestnut Street, Philadelphia, Pennsylvania 19101, U.S.A.
EXECUTIVES: Mr. E. Ray, Program General Manager; Mr. A. Wait, Habitat Programs Manager; Mr. K. Kelty, Marketing Manager
ORGANISATION: Ocean Systems Programs was formed in 1969 and is the focal point within the Re-Entry and Environmental Systems Division for major ocean oriented research, development and production activities. Activities include 'man-in-the-sea', life support equipment, medical products, buoy systems, ASW surveillance and oil/water separators.
EXECUTIVES: Re-entry and Environmental Systems Division:
O. Klima, Vice-President and General Manager; C. Raber, General Manager, Navy and Advanced Systems Programs Dept.; J. Stewart, Manager Research and Engineering; R. Welsch, Manager Defense Marketing.

GENERAL OCEANOGRAPHICS INC.

2172 DuPont Drive, Suite 18, Newport Beach, California 92664, U.S.A.
SUBMARINE FACILITY: 526 West 182nd Street, Gardena, California 90248, U.S.A.
EXECUTIVES: J. W. Vernon, Chairman; J. R. Curray; D. G. Moore; E. C. Buffington; R. F. Dill; W. B. Huckabay; J. W. Vernon, President, Treasurer, Chief Executive Officer; R. A. Slater, Corporate Secretary; D. N. Privitt, Vice-President, Submersible Designer and Builder; T. S. Crawford, Vice-President, Ship's Master.
ORGANISATION: General Oceanographics, Inc, has been active in offshore industry since 1952. They have designed, built, and used their own submersibles for marine exploration since 1968, making over 1,250 dives off such diverse places as Alaska, British Columbia, Washington, California, Texas, Florida, Virginia, Massachusetts, Michigan, Jamaica, British Honduras, Spain, Morocco and Algeria.

B.F. GOODRICH COMPANY

500 S. Main Street, Dept. 1702, Akron, Ohio 44318, U.S.A.
TEL: 216-379-2049
SUBSIDIARY/ASSOCIATED COMPANY:
B. F. Goodrich GmbH, 6236 Eschborn/Taunus Frankfurter Allee 2-4, West Germany
Tel: 6196-4401

HALL RUSSELL & CO. LTD.

P.O. Box 36, York Place, Aberdeen, AB9 8BT, Scotland
TEL: 0224-29244
TELEX: 73164
CABLE: Halrussel Aberdeen
EXECUTIVE: J. Wright, Managing Director

UNIVERSITY OF HAWAII AT MANOA

2525 Correa Road, Honolulu, Hawaii 96822
CABLE: Unihaw
EXECUTIVE: Mr. J. Frisbee Campbell, Scientific Co-ordinator for Marine Operations.

HAWKER SIDDELEY DYNAMICS LTD.

Manor Road, Hatfield, Herts. AL10 9LL, England
TEL: Hatfield 62300
TELEX: 22324
CABLE: Hawsidyn, Hatf.
EXECUTIVES: Sir John Lidbury, Chairman; Capt. E. D. G. Lewin, RN(Retd.), Managing Director; Dr. G. H. Hough, Deputy Managing Director; J. D. Crane, Commercial Director; R. G. Dancey, Administration Director; P. R. Franks, Director & Divisional Manager Air Weapons; J. McGregor Smith, Financial Director.
J. A. Airey, Executive Director, Space; R. Birch, Executive Director, Finance; L. G. Evans, Executive Director, SRAAM; T. H. Lettice, Executive Director, Contracts; P. C. F. Morgan, Executive Director, Purchasing.
H. R. Leather, Manager, Mechanical Equipment & Systems; C. D. West, General Sales Manager; B. M. Buckley, Sales Manager, Mechanical Equipment & Systems Division.
ORGANISATION: Late in 1966 the Industrial Automation Division of H.S.D. and Trinity House agreed that unmanned buoys could be used to replace manned light vessels. The first buoy was launched in 1969. Since then rationalisation of the Hawker Siddeley Group transferred continuation of work along these lines to the Mechanical Equipment & Systems Division based at Hatfield. Future activities in this field embrace the design, manufacture and development of data buoys for oceanographical, meteorological and pollution applications.
ASSOCIATED COMPANIES: A member company of SEATEK formed with EMI Electronics Ltd and R. & H. Green & Silley Weir Ltd. to undertake research, design, development and manufacturing work in respect of buoys or platforms of any shape or size which deploy sensors for the automatic collection of oceanographic, meteorological or pollution data.

HAWKER SIDDELEY CANADA LTD.

Halifax Shipyards Division, P.O. Box 640, Halifax, Nova Scotia, Canada
TEL: (902) 423-9271
TELEX: 014-422827
CABLE: Halship
DARTMOUTH YARD: P.O. Box 380, Dartmouth, Nova Scotia, Canada
Tel: (902) 469-4311
ORGANISATION: A member of the Hawker Siddeley Group, with the following Divisions and Subsidiaries:

Canadian Car Division
Woodland industries equipment; rapid transit passenger cars (subway, commuter and main line); highway trailers; containers; container chassis; MONOBOX cranes; pollution control equipment

Canadian Car (Pacific) Division
Sawmill and chipping equipment

Canadian Steel Foundries Division
Carbon and alloy steel castings in all sizes

Vancouver Industrial Controls Division
Electric-electronic control equipment; controls for marine applications; motor control centres; solid state battery charging equipment

SUBSIDIARIES

Hawker Industries Limited

Canadian Bridge Division
Galvanized steel transmission towers and poles; sub-stations; guyed masts and self-supporting towers; radar, microwave and scatter antennae

Halifax Shipyards Division
Naval and merchant shipbuilding and repairs; offshore oil drilling unit construction; complete marine services

Trenton Works Division
Railway freight rolling stock; railway axles; industrial mine cars; medium and heavy forgings; pressure vessels; fabricated components; custom machine work

Dosco Overseas Engineering Limited
Mining equipment

Hollybank Engineering Company, Limited
Mine arches; underground roadway supports

Orenda Limited
Aircraft gas turbines; industrial gas turbines for electric power generation, liquid and gas pumping, heating and air conditioning; components for nuclear power applications; computer services; engineering, design, laboratory, testing and graphics services

Canadian General Transit Company, Limited
Lessors of railway tank cars, hopper cars, woodchip cars, box cars, flat cars, etc.; terminals for bulk storage of chemicals, vegetable oils, tallows, etc.

ASSOCIATE COMPANY

Canadian Steel Wheel Limited
Wrought steel wheels for locomotives and railway rolling stock; circular forgings; steel ingots.

HITACHI CONSTRUCTION MACHINERY CO. LTD.

2-10, 1-chome, Uchikanda, Chiyoda-ku, Tokyo, Japan
TEL: (293) 3611
TELEX: 222604
CABLE: Tokhitachitenki

HONGKONG UNITED DOCKYARDS LTD.

Hung Hom, Hong Kong Malaysia
TEL: 3-334111
TELEX: HX 3547
CABLE: Hudrep Hongkong
ORGANISATION: This firm was formed on 1st January 1973 by a merger of the ship repairing interests of Hongkong & Whampoa Dock Co. Ltd. with those of Taikoo Dockyard & Engineering Co. of Hong Kong Ltd.

HYDRO PRODUCTS

P.O. Box 2528, San Diego, California 92112, U.S.A.
TEL: (714) 453-2345

TELEX: 910-322-1133
ORGANISATION: Hydro Products is a Division of Tetra Tech. Inc and a member of the Dillingham group of companies.

IBAK HELMUT HUNGER

D 2300 Kiel 14, Wehdenweg 122, West Germany
TEL: (0431) 707-1
TELEX: 292824
CABLE: Ibak Kiel
EXECUTIVES: Ing.Grad. Helmut Hunger, Owner and General Manager; Dr. Peter Koehler, Head, Research & Development; Ing.Grad. Wilhelm Syassen, Technical Manager; Mr. Rudolf Adler, Sales Manager; Mr. Joachim Bornhoeft, Marketing Manager
ORGANISATION:

The company was founded in 1945 and has been manufacturing electrical marine equipment, such as ship's searchlights, floodlights, electric heaters, etc., since then.

In 1954 they developed the first German underwater TV system for the inspection of underwater structures in the Rhine and other inland waterways.

Developments today include diver-held underwater TV cameras, deep sea television, photographic, and lighting equipment.

About 30 per cent of the production is exported. The company is represented in most countries of the Western world.

IHC HOLLAND LETOURNEAU MARINE CORPORATION

Main Tower, 309 East Main Street, Kilgore, Texas 75662, U.S.A.
TEL: AC214. 984-8691
ORGANISATION:

IHC Holland-LeTourneau Marine Corporation was formed as an initial step in the development of a world-wide enterprise to design and construct a new generation of heavy offshore drilling and production equipment.

The company is a joint venture on the part of IHC Holland, suppliers to the offshore industry, and Richard L. LeTourneau, a Longview, Texas, businessman widely known in the offshore oil industry.

The company meets the needs of oil companies and contractors, to which end the technological skills of American and Dutch engineers and technicians have been pooled.

Arrangements provide for assembly of the large offshore structures to take place in coastal areas all over the world, so as to reduce transportation and insurance costs.

NV INDUSTRIEELE HANDELSCOMBINATE HOLLAND

Marconistraat 2, P.O. Box 6141, Rotterdam
TEL: (010) 25 47 83
TELEX: 23525 ihcc nl
CABLE: Ihcholland
WORKS: IHC Holland Dredger Division, P.O. Box 1, Kinderdijk
Tel: (010) 11 76 00
IHC Holland Offshore Division, P.O. Box 11, Schiedam
Tel: (010) 26 04 20
EXECUTIVES: D. L. H. Smit, President Board of Management; B. Schuil, General Manager, IHC Dredger Division and Executive Vice-President IHC Holland; P. J. M. Verschure, General Manager, IHC Offshore Division and Executive Vice-President IHC Holland; R. Smulders, Managing Director, Offshore Division; W. C. den Hartog, Senior Managing Director, Dredger Division.

SUBSIDIARY/ASSOCIATED COMPANIES:
IHC France, 280–282 Bld. Saint-Germain, 75341 Paris Cedex 07, France
IHC Holland Marine Corporation, P.O. Box 328, Portland, Texas 78374, U.S.A.
IHC Holland (Australia) Pty Ltd, Underwood House, 37/49 Pitt Street, Sydney-2000, Australia
IHC (Malaysia) Sdn. Bhd., 33 Beach Street, Penang, Malaysia

INSTITUTE OF OCEANOGRAPHIC SCIENCES

Wormley, Godalming, Surrey, England
EXECUTIVE: Dr. P. Collar

INSTITUT FRANCAIS DU PETROLE

1 et 4 Avenue de Bois-Preau, 92 Rueil-Malmaison (Hauts-de-Seine), France
TEL: 967.11.10

INTERNATIONAL HYDRODYNAMICS CO. LTD.

145 Riverside Drive, North Vancouver, B.C., Canada
TEL: (604) 929-2391
TELEX: 04-35269 THINKDEEP VCR
EXECUTIVES: R. B. Oldaker, President and Director; W. Forsyth, Director; C. A. Munana, Director; Dr. W. D. Liam Finn, Director; E. R. A. Kelly, Director; Dr. F. J. Lindars, Director; Dr. E. L. Dougherty, Director; B. Tichelman, Vice-President, Administration; J. R. McFarlane, Vice-President, Engineering and Production; R. D. McDonald, Vice-President, Marketing.
ORGANISATION:

The company was formed in 1964 by D. L. Sorte, M. F. Thomson and A. R. Trice, who shared the concept that a small deep diving submersible vehicle could enhance man's ability to perform useful tasks below the surface of the sea. The three partners considered that such a vessel could go much deeper and remain on the work site much longer than a suited diver for research, exploration, salvage and construction. Also, it would permit a non-diving specialist to visit sub-surface work sites and view the task directly under 'shirt sleeve' conditions.

Pisces I was the first of the submersibles, and was completed in 1966. It was soon appreciated that the manned submersible was an ideal platform for the above mentioned uses, including resource investigation, pollution monitoring, etc.

Pisces II was manufactured under contract for Vickers Limited of Barrow-in-Furness, England, which included improved design from experience gained with *Pisces I*. Concurrent with the manufacture of *Pisces II*, *Pisces III* was built to the same design for the company's own use. *Pisces II* and *Pisces III* were certified by American Bureau of Shipping to 730 m (2,400 ft) and 1,060 m (3,500 ft) respectively in 1969.

The next construction was the SDL-I (Submersible Diver Lockout No. 1) which was designed and built during 1970 under contract to the Canadian Department of National Defence and was A.B.S. certified to 610 m (2,000 ft).

In 1971 to 1973 the company designed and built *Pisces IV* and *Pisces V* which are both capable of diving to 2,012 m (6,600 ft) and are certified by A.B.S. *Pisces IV* has been supplied to the Canadian Department of Environment. *Pisces V* is being operated by the company.

The Pisces submersibles have been used on a wide range of activities which include salvage operations, ordnance recovery, geological exploration, oil exploration, pipe line surveys and fisheries research. Operational areas have included the Canadian Arctic Archipelago, Hudson Bay, Lakes Huron and Erie, the west and east coasts of Canada, the United Kingdom and the Bahamas.

In 1973, construction of *Aquarius I* was started. This two-man submersible is considerably smaller than the Pisces type. Further construction of Pisces types and new designs such as diver lockouts and dry transfer are planned for the immediate future.

The company manufactures for lease and sale, deep diving Pisces submersibles with supplementary operational systems, diver lockout submersibles, surface decompression chambers, diving support systems and offshore oil service capabilities. Also, a support vessel, the MV *Hudson Handler* is available, which is a submersible recovery vehicle.
ASSOCIATED COMPANIES:

In 1969 agreement with Vickers Ltd was entered into whereby the company undertook to supply Vickers with Pisces class submersibles. Vickers obtained rights to operate the vessels over a portion of the Eastern Hemisphere.

In July 1973 the Peninsular and Oriental Steam Navigation Company purchased 44 per cent of the company's common stock. P & O and International Hydrodynamics plan to establish operating companies for the purpose of supplying the service of HYCO's submersibles to the offshore industry.

ISRAEL OCEANOGRAPHIC & LIMNOLOGICAL RESEARCH LTD.

21 Hativat Golani Road, P.O. Box 1793, Haifa, Israel
TEL: 04-642345
EXECUTIVE: Mr. A. Golik

ISRAEL SHIPYARDS LTD.

P.O. Box 1282, Haifa, Israel
TEL: 04-724401
TELEX: 4634 (YARD IL)
CABLE: Israyard
EXECUTIVES: Mr. I. Libertovsky, Managing Director; Mr. D. Yallon, Company Secretary; Mr. R. Loewenstein, Comptroller.
ORGANISATION: The company was established in 1961 and has been concerned with the building and repair of Israeli and foreign ships.

JBF SCIENTIFIC CORPORATION

2 Ray Avenue, Burlington, Massachusetts 01803, U.S.A.
TEL: (617) 273-0270
EXECUTIVES: R. A. Bianchi, President; J. H. Farrell, Executive Vice-President; E. E. Johanson, Vice-President; R. W. Neal, Chief Engineer.
ORGANISATION:

JBF Scientific was formed in October 1969 to do research, development and product manufacturing in the field of environmental pollution. The concept of collecting oil by forcing it under the water was conceived and developed by JBF in 1970. Contracts were received from the U.S. Environmental Protection Agency, the Commonwealth of Massachusetts, and the US Navy for units based upon this concept and the first skimmer was launched late in 1971. In 1972 and 1973 production units were delivered to the US Navy and the Canadian Government.

THE JOHNS HOPKINS UNIVERSITY APPLIED PHYSICS LABORATORY

8621 Georgia Avenue, Silver Spring, Maryland 20910, U.S.A.
TEL: (301) 953-7100 and 589-7700
EXECUTIVE: Dr. D. W. Pritchard, Director, Chesapeake Bay Institute and Professor of Earth and Planetary Sciences (Johns Hopkins University).

KAWASAKI HEAVY INDUSTRIES LTD.

Nissei Kawasaki Building, 16-1 Nakamachi-dori 2-chome, Ikuta-ku, Kobe, Japan
TEL: Kobe 341-7731
TELEX: 5622-355 KAWAJU

CABLE: Kawasakihead Kobe
EXECUTIVE:
Mr. Kiyoshi Yotsumoto, President
WORKS:

Harima Works

Land area	444,584 m²
Building area	46,201 m²
Employees	900
Principal products	Plant

Yachiyo Works

Land area	154,146 m²
Building area	33,564 m²
Employees	660

Principal products: Crushers, screens, dryers, mills, cast steel

Noda Works

Land area	151,627 m²
Building area	54,680 m²
Employees	1,110

Principal products: Steel structures

Osaka Works

Land area	201,061 m²
Building area	27,119 m²
Employees	1,600

Principal products: Rolling stock, steel structures, mills and classifiers, automated machines, chemical equipment, boilers, air conditioning machinery, sanitary equipment

Nishi-Kobe Works

Land area	168,065 m²
Building area	25,036 m²
Employees	760

Principal products: Oil hydraulic machinery

Kawasaki Works

Land area	27,320 m²
Building area	17,315 m²
Employees	560

Principal products: Large boilers, sanitary equipment, nuclear equipment

Oihama Works

Land area	64,890 m²
Building area	23,350 m²
Employees	440

Utsunomiya Works

Land area	196,641 m²
Building area	98,985 m²
Employees	190

Principal products: Freight cars, tank cars

Banshu Works

Land area	110,187 m²
Building area	22,180 m²
Employees	440

Principal products: Construction machinery

Kawasaki-Sub Works

Land area	21,281 m²
Building area	9,783 m²
Employees	410

Principal products: Heat exchangers

Shiga Works

Land area	123,314 m²
Building area	11,361 m²
Employees	330

Principal products: Construction machinery, boilers, air conditioning machinery

Gifu Works

Land area	663,184 m²
Building area	177,664 m²
Employees	4,700

Principal products: Aircraft, automobiles, containers

Akashi Works

Land area	427,428 m²
Building area	202,534 m²
Employees	4,900

Principal products: Engines, motorcycles, industrial robots, automated machines

Kakogawa Rolling Stock Works

Land area	47,920 m²
Building area	11,005 m²
Employees	120

Principal products: Freight cars, containers

Sakaide Works

Land area	910,613 m²
Building area	119,828 m²
Employees	3,400

Principal products: Shipbuilding and repairing

Akashi-Minami Works

Land area	46,290 m²
Building area	21,568 m²
Employees	360

Principal products: Turbo machines, couplings

Hyogo Works

Land area	150,799 m²
Building area	101,000 m²
Employees	2,200

Principal products: Rolling stock

Kobe Works

Land area	355,848 m²
Building area	201,346 m²
Employees	9,400

Principal products: Shipbuilding and repairing, machinery

Kakogawa Works

Land area	440,957 m²
Building area	71,429 m²
Employees	1,520

Principal products: Steel structures, package boilers

Hyogo Cast Steel Works

Land area	15,718 m²
Building area	10,909 m²
Employees	200

Principal products: Cast steel

ASSOCIATED COMPANIES:
Kawasaki Kisen Kaisha Ltd, 8 Kaigan-dori, Ikuta-ku, Kobe, Japan
Kawasaki Steel Corporation, 1 Kitahonmachi-dori 1-chome, Fukiai-ku, Kobe, Japan
OFFICES:

Tokyo
World Trade Center Bldg., 4-1 Hamamatsu-cho 2-chome, Minato-ku, Tokyo, Japan
TEL: Tokyo (03) 435-2971
TELEX: 2422851 KAWAJU
CABLE: Kawasakiheavy Tokyo

London
3, St. Helen's Place, London EC3A 6EB, England
TEL: (628) 9915/7
TELEX: 51-886303
CABLE: Kawaju London

New York
29 Broadway, New York, N.Y. 10006, U.S.A.
TEL: White Hall 3-6450
TELEX: 23-420293
CABLE: Nykawasaki NewYork

Düsseldorf
AM Wehrhahn 45, Düsseldorf, F. R. Germany
TEL: 357759, 353122/4
TELEX: 41-8587421
CABLE: Kawaju Düsseldorf

Bangkok
Dusit Thani Bldg., 1-3, Rama IV Road, Lumpini Bangkok, Thailand
TEL: 863951

Taipei
Taihong Bldg., No. 71, Sec. 2, Nanking East Road, Taipei, Taiwan, Republic of China
TEL: 575-141
TELEX: 789
CABLE: Kawasaki Taipei

Hong Kong
Bank of Canton Bldg., No. 6, Des Voeux Road, Central, Hong Kong
TEL: H-223560
TELEX: 802-5690
CABLE: Kawaheavy Hongkong

Sydney
'Kyle House', 31 Macquarie Place, Sydney N.S.W. 2000, Australia
TEL: 241-1123
TELEX: 71-24045

KINGSNORTH SEA CONTRACTORS LTD.

53 Leadenhall Street, London EC3A 2BR, England
TEL: (01) 481-2020
TELEX: 884801/2/3
CABLE: Houlders, London EC3
EXECUTIVE: Mr. P. H. Jones, Chief Executive
ASSOCIATED COMPANY:
Houlder Brothers & Co Ltd

KOMATSU LTD.

No. 3-6, 2-chome, Akasaka, Minato-ku, Tokyo, Japan
TEL: Tokyo 584-7111
TELEX: J22812, J24967
CABLE: Komatsu Ltd Tokyo
ORGANISATION:
The company developed the world's first underwater bulldozer in March 1969 on orders from the Ministry of Construction of Japan.

LABORATOIRE CENTRAL D'HYDRAULIQUE DE FRANCE

10 rue Eugene-Renault, 94700 Maisons-Alfort, France
TEL: 893-28-28
CABLE: Labhydro-Maisons-Alfort

LOCKHEED MISSILES & SPACE CO. INC.

P.O. Box 504, Sunnyvale, California 94088
TEL: 408. 742-6756
TELEX: 346409
CABLE: 9103399221
EXECUTIVES: Mr. S. W. Burriss, President; Mr. R. R. Kearton, Executive Vice-President; Mr. E. P. Wheaton, Vice-President and General Manager, Research and Development Division; Mr. H. P. Kerfoot, Assistant General Manager, Research and Development Division; Mr. J. G. Wenzel, Assistant General Manager and Director, Ocean Systems; Mr. R. W. Kermeen, Assistant Ocean Systems Manager; Mr. R. A. Tyrrell, Acting Manager, San Diego Operations; Mr. R. M. Spath, Manager, Deep Submergence Programs; Mr. R. D. Fuller, Data Buoy Programs and Acting Manager, Program Development; Mr. C. G. Welling, Manager, Ocean Mining Programs; Mr. E. P. Pallange, Ocean Systems Engineering; Mr. J. P. Calder, Assistant Manager, Program Development; Mr. A. S. Dunbar, Ocean Systems Electronics; Mr. E. P. Wheaton, General Manager; Mr. C. E. Rudiger, Sales; Mr. J. S. Hull, Works; Mr. G. W. Phillips, Designer; Mr. M. H. Rosenblum, Project Manager.
ORGANISATION:
Lockheed Missiles & Space Co, Inc, a subsidiary of Lockheed Aircraft Corporation, entered the Ocean Systems field in 1965 with the construction of *Deep Quest*, the research submarine, and entered the buoy field in 1970.
ASSOCIATED COMPANIES:
Lockheed Shipbuilding & Construction Company, Seattle, Washington, U.S.A.
Lockheed Aircraft (Europe), S.A., 7 Old Park Lane, London, W1Y 3LJ, England, and 37 Avenue Pierre Ier de Serbie, 75 Paris 8e, France.

LOCKHEED PETROLEUM SERVICES LTD

(subsidiary of Lockheed Aircraft)
1 Grosvenor Square, New Westminster. British Columbia, Canada
TEL: 604. 524-4451
CABLE: Lops
EXECUTIVE: Mr. G. H. Fahlman
ASSOCIATED COMPANIES:
Lockheed Petroleum Services Ltd.
206–4600 Post Oak Place Drive, Houston, Texas 77027, U.S.A.

TEL: 713/626-9623
7 Old Park Lane, London, W.1, England
TEL: (01) 493-7721

LOM & COMPANY A/S

Bergen , Norway
EXECUTIVE: J. Larsen, Manager
ORGANISATION:
A member of the Conoship Group.

MAERSK SUPPLY SERVICE A.P. MØLLER

8 Kgs. Nytorv, DK-1098 Copenhagen K, Denmark
TEL: (01) 14 15 14
TELEX: 19632
ORGANISATION:
Maersk Supply Service is a division of A. P. Møller. A. P. Møller operates a fleet of almost 100 vessels including VLCCs of approximately 300,000 dwt., modern bulk-carriers, cargo liners, refrigerated vessels and supply vessels.
The fleet of Maersk Supply Service consists of 16 units. 9 of these are working worldwide, 7 are under construction at a Norwegian yard and will be delivered during 1973 and 1974. They have been designed to satisfy all needs connected with offshore supply operations ensuring reliable and efficient service under all conditions. All these vessels are manned by Danish officers and crews with years of experience in the special duties on board supply-, towing-, and anchor-handling vessels in the extreme working conditions of areas like the North Sea.
The vessels are backed up by personnel, towmasters, supervisors, etc. from the head-office in Copenhagen.
SUBSIDIARY/ASSOCIATED
ORGANISATIONS:

Bangkok
Maersk Line, Bangkok Branch, 231/2, Sathorn Road, P.O. Box No. 692, Bangkok, Thailand
TEL: 32926
TELEX: 2651

Hong Kong
Corvetine Shipping Limited
801, Realty Building, 71 Des Voeux Road, Central Hong Kong
TEL: H-24 11 41
TELEX: 3756

Houston
Atlantic Pacific Marine Corporation
2400 Fountain View, Suite No. 322, Houston, Texas 77027, U.S.A.
TEL: (713) 783-0910
TELEX: 762-956

Jakarta
Maersk Line Limited, Jakarta Branch
Jalan Hajam Wuruk 27, P.O. Box 552/DAK, Jakarta, Indonesia
TEL: 42504
TELEX: 4105

London
The Maersk Company Limited
12–20 Camomile Street, London EC3A 7PH, England
TEL: (01) 623 5061
TELEX: 886189

Melbourne
The Maersk Company (Australia) Pty Ltd
185 William Street G.P.O. Box 995 H, Melbourne, Victoria 3001, Australia
TEL: 67-5149
TELEX: 30729

New York
Moller Steamship Company, Inc
One World Trade Center, Suite 3527, New York, N.Y. 10048, U.S.A.
TEL: (212) 466-5300
TELEX: 420372

San Francisco
Maersk Line Agency
1 Maritime Plaza, San Francisco, California 94111, U.S.A.
TEL: (415) 398-1515
TELEX: 372-7432

Tokyo
Maersk Line, Limited, Japan Branch
Palace Building, 1–1 Marunouchi 1 chome, Chiyoda-ku, Central P.O. Box No. 506, Tokyo 100, Japan
TEL: (03) 211-6341
TELEX: 22208

MAKAI OCEAN ENGINEERING CO.

P.O. Box 1194, Kailua, Oahu, Hawaii 96734
TEL: 808. 259-5940
EXECUTIVE: Mr. Henry M. Horn, President

MANGONE SHIPBUILDING COMPANY

Houston, Texas, U.S.A.

MARATHON LeTOURNEAU OFFSHORE COMPANY

Houston, Texas 77002, U.S.A.
ORGANISATION:
Marathon LeTourneau Offshore Company is a subsidiary of Marathon Manufacturing Company, and has yards in Vicksburg, Mississippi; Brownsville, Texas; and the Republic of Singapore.

MEGATOR PUMPS AND COMPRESSORS LTD.

87a Newington Causeway, London SE1 6EQ
TEL: (01) 407-5616
TELEX: 886847
WORKS: Hendon, Sunderland
TEL: Sunderland 5488
EXECUTIVES: Mr. F. W. McCombie, Chairman; Mr. W. A. Lawson, Managing Director; Mr. T. S. Horan, Export Director.
ORGANISATION:
The Company was founded in 1946 to manufacture and market the Megator sliding-shoe pump. Recent developments include a range of oil-spill recovery equipment for dealing with small-scale spills.
SUBSIDIARY/ASSOCIATED
COMPANIES:
The Megator Corporation Ltd, 136 Gamma Drive, Pittsburgh, Pa. 15238, U.S.A. is a wholly owned American subsidiary.

MESSERSCHMITT-BOLKOW-BLOHM GmbH

8 Munich 80, Postfach 80 12 20, West Germany
TEL: (0811) 6000
TELEX: 522279 mbbo

METROPOLITAN PETROLEUM PETROCHEMICALS CO. INC.

25 Caven Point Road, Jersey City, New Jersey 07305
TEL: (201) 434-4451
EXECUTIVES: D. A. Burack, President; T. F. Dalton, General Manager; L. G. Green, Manager Pollution.
ORGANISATION:
Incorporated 1966, New York Corporation. Manufactures and sells specialty chemical products for fuel oil conditioning, cleaners, water treatment products, marine tank cleaners, oil spill boom, pollution products.
ASSOCIATED COMPANIES:
Division of the Pittston Co
Vogelsang-Germany, Gunther Vogelsang, Vogelsang-Schonfeld, 2000 Hamburg 1, Chelehaus C, Germany
Gamlen Chemical, 4 Midland Avenue, Elmwood Park, N.J.
Worthington Corp., 10 Industrial Road, Fairfield, N.J.

MINISTRY OF AGRICULTURE FISHERIES AND FOOD

Fisheries Laboratory, Lowestoft, Suffolk
TEL: Lowestoft 4251
TELEX: 97470
CABLE: Fishlab Lowestoft
EXECUTIVE: Captain T. Sellers, Marine Superintendent

MITSUBISHI HEAVY INDUSTRIES LTD.

5–1 Marunouchi 2-chome, Chiyoda-ku, Tokyo, Japan
TEL: Tokyo 212-3111
TELEX: J22443
CABLE: Hishiju Tokyo
WORKS:
Yokohama Shipyard & Engine Works;
Sagamihara Machinery Works;
Nagoya Aircraft Works;
Nagoya Machinery Works;
Kyoto Machinery Works;
Kobe Shipyard & Engine Works;
Akashi Machinery Works;
Takasago Machinery Works;
Mihara Machinery Works;
Hiroshima Shipyard & Engine Works;
Hiroshima Machine Tool Works;
Shimonoseki Shipyard & Engine Works;
Nagasaki Shipyard & Engine Works
EXECUTIVE: Mr. Gakuji Moriya, President
ORGANISATION:
Mitsubishi Heavy Industries was formed in 1964 and had its origins in the Nagasaki Forge of the Tokugawa Shogunate
SUBSIDIARY/ASSOCIATED
COMPANIES:

Overseas Offices

New York
277 Park Avenue, New York, N.Y. 10017, U.S.A.
TEL: 826-2188, 2189
CABLE: Hishiju NewYork NY
TELEX: NEWYORK 223037 (HISHIJU 223037) 012-7756 (MITSUHEAIND NYK)

London
c/o Mitsubishi Corporation, London Branch, Bow Bells House, Bread Street (Cheapside), London, ECAM 9BQ, England
TEL: 248-8821/8825
CABLE: Hishiju London-EC4
TELEX: LONDON 888251 (MITSUBISHI LDN)

Düsseldorf
4 Düsseldorf, Ludwig Zimmermann-Str-1, Bundesrepublik Deutschland (Federal Rep. of Germany)
TEL: Düsseldorf 17169, 327968
CABLE: Hishiju Düsseldorf
TELEX: DÜSSELDORF 858-7666 (8587666 MHI D)

Oslo
Tollbugata 28, Oslo, Norway
TEL: Oslo 41. 30. 89
CABLE: Hishiju Oslo
TELEX: OSLO 11973 (11973 MHI O)

Hong Kong
6th Floor, St. George's Building, No. 2, Ice House Street, Hong Kong
TEL: 231275, 243765, 231191
CABLE: Hishiju Hong Kong
TELEX: HONGKONG 3468 (HX 3468 HISHIJU)

REPRESENTATIVES:
Los Angeles
Atlantic Richfield Plaza, South Tower, Room 2317, 555 South Flower Street, Los Angeles, CA90017, U.S.A.
TEL: (213) 977-3855
CABLE: c/o Mitsubishi Los angeles Calif
TELEX: LOS ANGELES 067-3139 (MITSUBISHI LSA)

San Francisco
c/o Mitsubishi International Corp., San Francisco Branch, 601 California Street, International Building, San francisco, CA 94108, U.S.A.
TEL: 981-1910 (Mitsubishi International Corp.)
CABLE: c/o Mitsubishi San Francisco Calif
TELEX: 27288 (27288 MICSF UR)

Chicago
c/o Mitsubishi International Corp., Chicago Branch, Suite 4040, John Hancock Center, 875 North Michigan Avenue, Chicago, IL 60611, U.S.A.
TEL: (312) 751-1111 (Mitsubishi International Corp.)
CABLE: c/o Mitsubishi Chicago Ill
TELEX: CHICAGO 0254152 (MITSUBISHI CGO)

Pittsburgh
c/o Westinghouse Nuclear Energy System, Penncenter Building 5, Room 317, P.O. Box 355, Pittsburgh, PA 15230, U.S.A.
TEL: 412-373-4105, 412-256-4384, 412-256-4386, 412-256-4387
CABLE: Hishiju Pittsburgh-Pa
TELEX: 0812443 (HISHIJU PGH)

Philadelphia
c/o Westinghouse Electric Corp., Lester Plant, Licensee Services Lester Branch, P.O. Philadelphia, PA 19113, U.S.A.
TEL: 215-595-3574
CABLE: Mitsubishi Lester-Penn
TELEX: 083-4335 (WESHSINTL LTER)

Windsor
c/o Combustion Engineering Inc, 1000 Prospect Hill Road, Windsor, CT 06095, U.S.A.
TEL: 203-688-1911
CABLE: c/o Combeng Windsor Conn
TELEX: 099297 (COMBENG WSOR)

Paris
c/o Mitsubishi France, S.A., 25, Rue de Berri, Paris (8e), France
TEL: 359-0927 (Mitsubishi Paris)
CABLE: c/o Mitsubishi Paris
TELEX: PARIS-65000 (MITSUBISHI PARIS)

Johannesburg
c/o Mitsubishi Corporation, Johannesburg Branch, 19th Floor, Bank of Lisbon and South Africa Bldg., 37 Sauer Street, Johannesburg, The Republic of South Africa
TEL: 834-1914/8 (Mitsubishi Corporation)
CABLE: c/o Mitsubishi Johannesburg
TELEX: JOHANNESBURG 43-7018 (43-7018 SA) 43-6125 (43-6125 SA)

Taipei
c/o Mitsubishi Corporation, Taipei Branch, 6th Floor, Chung Chen Bldg., No. 79, Sec. 2, Chung Shan North Road, Taipei, Taiwan (P.O. Box 741, Taipei)
TEL: 577211, 577221 (Mitsubishi Corp.)
CABLE: c/o Mitsubishi Taipei
TELEX: TAIPEI 11211 (TP 11211 MITSUBISHI)

Teheran
c/o Ashtad Iran Manufacturing Industries (AIMICO) Co, Ltd, Vesal Shirazi Avenue, Ashtad Building, Teheran, Iran
TEL: 46697
CABLE: Tiller Teheran
TELEX: TEHERAN 2515 (2515 MITSUBISHI TN)

MOBELL MARINE DEVELOPMENT CO. LTD.
23 North Street, Havant, Hampshire, England
TEL: Havant 72092 and 71300
EXECUTIVES: J. Brooks, Director; E. Brooks, Director; E. Bowerman, General Manager; S. Dooley, Operations Manager; M. Terral, Chief Designer; M. Brooks, Sales Director.
ORGANISATION:
The company was formed in 1960.

NATIONAL OCEANIC & ATMOSPHERIC ADMINISTRATION U.S. DEPARTMENT OF COMMERCE
NOAA. Office of Fleet Operations, National Ocean Survey (C7), Rockville, Md. 20852, U.S.A.
TEL: (301) 496-8321
TELEX: 301-496-8539
EXECUTIVES: Dr. Robert M. White, Administrator, NOAA; Rear Admiral A. L. Powell, Director, NOAA National Ocean Survey; Rear Admiral E. A. Taylor, Associate Director, Office of Fleet Operations.

NATIONAL SCIENCE FOUNDATION/OFFICE OF POLAR PROGRAMS
1800 G. Street, N.W., Washington, D.C., U.S.A.
CABLE: Natscifoun, Washington, D.C.
EXECUTIVE: Mary K. Johrde, Head, Office for Oceanographic Facilities and Support, National and International Programs.
ORGANISATION:
Sixteen academic and research institutions in the United States operate a fleet of research vessels which includes twenty major ships capable of open-ocean research as well as a number of smaller ships and boats for coastal and estuarine studies. Although the Foundation is the major source of operating support for these institutions, it does not hold title to the vessels. All of the academic ship operating institutions are members of the University National Oceanographic Laboratory System (UNOLS).

NATURAL ENVIRONMENT RESEARCH COUNCIL
Research Vessel Base, No. 1 Dock, Barry, Glamorgan CF6 6US, Wales
TEL: (0446-2) 77451
TELEX: 497101
CABLE: Nercship Barry
EXECUTIVE: Captain D. M. H. Stobie, Director.
ASSOCIATED/SUBSIDIARY:
Natural Environment Research Council, Alhambra House, 27–33 Charing Cross Road, London WC2, England.

NAVAL UNDERSEA CENTER DEPARTMENT OF THE NAVY
San Diego, California 92132, U.S.A.
EXECUTIVES: Captain R. H. Gautier, U.S.N. Commander; Dr. William B. McLean, Technical Director.
ORGANISATION:
The Naval Undersea Research and Development Center (NUC) was established in July 1967 from related undersea elements of the Naval Weapons Center and the Naval Electronics Laboratory Center with the addition of the Marine Bio-Science Facility in October 1967.
NUC is a major shore installation of the Naval Material Command. The Commander and Technical Director are responsible for the management function and execution of the Center's mission. Six specialised departments, the Arctic Submarine Laboratory and a systems analysis group provide the technical resources for NUC research and development programmes. Technically trained naval officers with extensive operational experience are assigned to major programme areas to provide liaison and advise on 'fleet' requirements. Four other non-technical departments and three staff groups provide general and administrative support to the technical departments.
The Center's laboratories are located in San Diego, Pasadena and Hawaii with support facilities and test sites at San Clemente Island, Long Beach and Morris Dam, California. Other research facilities are situated at Lake Pend Oreille in Idaho and Cape Prince of Wales in Alaska. Small testing and monitoring stations are operated throughout Southern California.
NUC employs 1,400 civilians and has a complement of 370 military personnel. The professional staff includes more than 600 scientists and engineers, 35 per cent with advanced degrees. The civilian technical staff initiates and conducts the research and development programmes.

UNIVERSITY OF NEW HAMPSHIRE ENGINEERING DESIGN AND ANALYSIS LABORATORY
Kingsbury Hall, University of New Hampshire, Durham, New Hampshire 03824, U.S.A.
TEL: (603) 862-1356
EXECUTIVES: Dr. G. H. Savage, Director of the Engineering Design and Analysis Laboratory and Professor of Mechanical Engineering; Dr. Richard S. Davis, Dean, College of Technology.
ORGANISATION:
The Engineering Design and Analysis Laboratory was organised in 1965. Its purpose is to involve both undergraduate and graduate students in 'real' problems of ocean engineering. Additionally it serves as a centre for interdisciplinary ocean engineering research.
The Laboratory establishes working arrangements with other educational institutions, research groups, and engineering firms as required by the nature of the problem.

NORMARIG A/S
Langesund, Norway
ORGANISATION:
A company formed in 1973 by A/S Framnaes, Mek Verksted and Trosvik Verksted and Sverre Munck A/S

NORSE PETROLEUM A/S
P.O. Box 652/653, C. Sundtsgt. 29, 5001, Bergen, Norway
TEL: (05) 23 30 77 and 23 30 78
TELEX: 42051

NOUVELLES APPLICATIONS TECHNOLOGIQUES (NAT)
Boite Postale 3, 78370 Plaisir, France
TEL: 462.25.00
TELEX: 26619
EXECUTIVES: R. H. Levy, President; P. Alba; J. Bertin; J. E. Deltour; J. Douy; P. Marchal; N. Masounabe; B. Salmon; J. Verrien; J. J. Wittmann (Members of Board).
P. Alba, General Manager; J. C. Amande, Technical Development.
ORGANISATION:
NAT is a limited company created in 1971 in order to commercialise the inventions of its shareholders (ELF and BERTIN) particularly the Vortex Oil Drinker. The company has long experience in pollution technology in calm waters (such as storm pits in refineries) and on the high seas.
ASSOCIATED COMPANIES:
Société Bertin et Cie and ELF-ERAP.

REPRESENTATIVES:
U.S.A.
INTEX, 6935 Wisconsin Avenue, Chevy Chase, Maryland 20015

Japan
Toyoda International Sales Inc, 3–17, 3-Chome Kudan Minami, Chiyoda-Ku, Tokyo

South America
Geomet Equipamentos Petroliferos e Eletronicos Ltda av. Rio Branco 156, Sala 1622 Caixa Postal 1765, Rio de Janeiro, Brazil

Benelux
BERAC, 3 Av. Maurice B., 1050 Bruxelles, Belgium
Scandinavia
Dykeri & Underhållskonsult, Liregatan 15, 27100 Ystad, Sweden
Ingeniörsfirma, A. P. Garnier,
65 Artillerigatan, 11445 Stockholm, Sweden

OCEAN DRILLING & EXPLORATION COMPANY (ODECO)

Odeco Building, 1600 Canal Street, P.O. Box 61780, New Orleans, Louisiana 70161, U.S.A.
TEL: (504) 529-2811
TELEX: 58-4124
EXECUTIVES: A. G. Blanke, Jr., Director; J. H. Hayward, Director; C. J. Hoke, Director and Chairman of Executive Committee; A. J. Laborde, President and Director; C. H. Murphy Jr., Chairman of the Board; J. A. O'Connor, Jr., Director; J. T. Hayward, Director Emeritus; H. J. Kelly, Senior Vice-President, Administration and Secretary; J. L. Kilpatrick, Senior Vice-President, Drilling; G. H. Troxell, Jr., Senior Vice-President, Design & Engineering; J. R. Macgregor, Vice-President, Production; L. B. Carpenter, Vice-President, Exploration; O. F. Vaughan, Treasurer; R. E. Roberson, Jr., Controller.
ORGANISATION:

Organised in 1953, ODECO was one of the pioneers in offshore drilling, particularly in the development of the submersible offshore barge. The *Mr. Charlie*, completed by ODECO in early 1954, was an immediate success and set the pattern for subsequent development in the field by ODECO and others.

In May 1963, the Company accepted delivery of its first floating drilling unit, the *Ocean Driller*. The performance of this ODECO-designed unit, including that during the hurricane season and winter weather, confirmed that ODECO had made a substantial advance in offshore drilling technology. In order to obtain increased towing speeds and compete on a world-wide basis, ODECO designed and placed the semi-submersible drilling unit *Ocean Queen* in service in 1965. The performance of this unit led to the construction of four additional semi-submersibles of the same design. In 1971, ODECO placed the *Ocean Prospector* in service as the first self-propelled semi-submersible.

In 1964 ODECO entered the offshore drilling service in the North Sea, and has since expanded into operations around the world, with operations in 1972 being carried on in the North Sea, Japan, Australia and the U.S.A. By 1972 ODECO had 18 rigs in service and six under construction.

ODECO has been the leader in the development of over-water drilling techniques and claims to have made important contributions toward the economic development of the offshore areas. The Company is still looking ahead to continued growth by supplementing its present fleet with similar equipment to explore the deepwater areas of the continental shelves world-wide.
ASSOCIATED/SUBSIDIARY COMPANIES
Canam Offshore Ltd
P.O. Box 1563, Hamilton, Bermuda
Ocean Ventures Pty Ltd
Bryson Centre, 186 Exhibition Street, Melbourne, Victoria 3000, Australia
ODECO (UK) Inc
Alexander House, Abbotswell Base, Craigshaw Road, Tullos, Aberdeen, AB1 4AR, Scotland
Rimrock (UK) Ltd
Site E1, Fish Wharf, Great Yarmouth, Norfolk, England
ODECO Norway Inc
Gjensidiges Hus Kongsgardbakken 6, Stavanger, Norway
ODECO (Australia) Inc
2nd Floor, Park Towers, 517 Hay Street, Perth, 6000, Australia
Ocean Contract Services, S.A.
Akiyama Building, 25 Akefune-cho, Nishikubo Shiba, Minato Ku, Tokyo 105, Japan
Canam Offshore Ltd
Korte Voorhut 20, P.O. Box 4753, The Hague, The Netherlands
ODECO-JILD Offshore Exploration Co, S.A.
Suite 405, Cathay Building, Mount Sophia, Singapore 9

OCEAN SCIENCE AND ENGINEERING INC.

1601 Water Street, Long Beach, California 90802, U.S.A.
TEL: (213) 437-0481
TELEX: 898-406
CABLE: Oceans Rove
WORKS: 5541 Nicholson Lane, Rockville, Maryland 20852
TEL: (301) 881-9320
EXECUTIVES: Mr. K. Warner, General Manager; Mr. R. L. Fagan, Sales Manager; Mr. J. Lovell, Works Manager.
ORGANISATION:

Founded in 1962 by the principal engineers of 'Phase I' of the 'Mohole' project, the company is an engineering design, systems development and environmental study company, concerned with work in the ocean, and particularly concerned with anti oil pollution work.
ASSOCIATED/SUBSIDIARY COMPANIES:
California Shipbuilding & Dry Dock Co, of Long Beach, California, is a wholly-owned subsidiary.

OCEAN SYSTEMS INC.

11440 Isaac Newton Square, Reston Va. 22090, U.S.A.
TEL: (703) 471-1310
TELEX: 899422 OCEANSYS RESTON
EXECUTIVES: C. H. Abbott, Chairman of Board of Directors; R. W. Daniels, Director on Board of Directors; R. A. Greissinger, Director on Board of Directors, Vice-President and General Manager; D. H. Watts, Manager, Diving Division; C. H. Hedgepeth, Operations Manager; J. A. Lawrie, Manager, Engineering Division; H. A. McKenna, Manager, Equipment Division.
ORGANISATION:

Ocean Systems, Inc is a subsidiary of Samson Cordage Works, acquired January 1973. January 1, 1965, OSI was formed to participate in the expanding ocean market in exploration and exploitation of natural resources. The company activities pursue business in ocean diving services, offshore site surveying, marine hardware products, oil pollution abatement, and ocean engineering on an international basis.
SUBSIDIARY/ASSOCIATED COMPANIES:
Ocean Systems, Inc
P.O. Box 1106, 2 Greenway Plaza East, Houston, Texas 77046
TEL: (713) 622-1265

Ocean Systems (UK) Ltd
3 King Street, Aberdeen, Scotland
TEL: (224) 573123 or 124
TELEX: 73264

Ocean Systems Australia Pty, Ltd
5th Floor, 231 Adelaide Terrace, Perth, Western Australia 6000
TEL: 25-4877
TELEX: 93414
CABLES: Oceansyst Perth

Ocean Systems Norge A/S
Stromsteinen, P.O. Box 138, N-4001 Stavanger, Norway
TEL: 27265

OFFSHORE DIVISION

Bunn's Lane, Southtown, Great Yarmouth, Norfolk, England, and Farburn Estate, Dyce, Aberdeen, Scotland
TEL: Great Yarmouth 55680
Dyce 3341/2
TELEX: (Gt. Yarmouth) 97103
(Dyce) 73394
EXECUTIVES: Mr. J. M. Houlder, Chairman; Mr. R. Farrington-Wharton, Managing Director; Mr. H. G. Delauze, Director (Fr.).
ORGANISATION:

Offshore Division is a member of the Comex Group of Companies of Marseilles, France.

OFFSHORE LOGISTICS INC.

Lafayette, Louisiana, U.S.A.

OIL MOP INCORPORATED

Engineers Road, Belle Chasse, Louisiana, 70037, U.S.A.
Mailing Address: P.O. Box 'P', Belle Chasse, Louisiana, 70037, U.S.A.
TEL: (504) 394-6110 (24 hour service)
CABLE: Oil Mop
EXECUTIVES: R. M. Page, Chairman of the Board and Chief Executive Officer; H. M. Rhodes, President and Chief Operating Officer; J. McCloskey, Secretary; J. Little; F. S. Lapeyre; N. DeSoto, Vice-President; C. Horton Smith, Vice-President; A. J. Page, Jr, Treasurer; E. M. Rea III, Field Operations; E. Hepting, Plant Superintendent; T. Schmitz, Chief Engineer.
ORGANISATION:

Oil Mop, Incorporated, both a manufacturer of oil spill pick-up equipment and oil-water separator devices, and an oil spill pick-up service contractor, was formed in March 1969 and began commercial operations in January 1970 with ten employees. The present location includes 5 acres of ground on the intracoastal canal; a fabricating building; an office building; utility buildings; yard storage and work space; and a 175 ft/150 ton loading capacity concrete wharf. Fifteen acres of adjacent property are under option. After $2\frac{1}{2}$ years of operation, Oil Mop, Inc now has over 100 employees.

Oil Mop, Inc operates 24 hours a day, seven days a week, and its equipment and expertise are available on a lease, purchase, or emergency rental basis to businesses or agencies faced with the problem of cleaning up an oil spill. Its customers include most of the major oil companies, pipeline companies, and transportation firms. Two of OMI's biggest customers are the United States Coast Guard and the United States Navy.

The mobility and flexibility of the Oil Mop equipment which is air, truck or boat transportable, has enabled it to operate in many parts of the Western Hemisphere, from Alaska to South America, in open waters or in lakes, rivers, bayous or ponds. The 'Rope Mops' and 'Mop Engines' have operated in the freezing snow and ice conditions of Lake Michigan in the winter, and in the tropical climate of the Panama Canal Zone with equal success.

In addition to the Oil Mop systems and equipment described, OMI manufactures hand mops of the same oleophilic-hydrophobic plastic fibres. These are hand operated and hand wrung and are used to clean up small spills and the small pockets of oil which are almost always present in connection with a large spill.

Oil Mop engineers are completing development of filtering systems for use in ships' bilges and tankers' ballast tanks to ensure that water pumped overboard is free of oil and oil based contaminates. Filters and other methods of *preventing* oil spills are major areas of research at Oil Mop, Inc and will play a large role in OMI's future business.

ASSOCIATED COMPANIES:
Sales agreement with Cesco Inc, Division of Browning-Ferris Industries, Inc, Fannin Bldg., Houston, Texas, U.S.A.
Oil Mop of Texas, Inc
5722 Ransom St., Houston, Texas 77017, U.S.A. Wholly owned subsidiary
TEL: 713-643-4700
DiKor
P.O. Box 389, Carmi, Illinois 62821, U.S.A.
Distributor and Service Contractor
Oil Mop Pollution Control Ltd
80 Commissioners Street, Toronto, Ontario
Export sales agreement with Singer Products Co, Inc
1 World Trade Center, Suite 2365, New York, N.Y. 10048
TEL: 212-432-1400
Arthur Singer, President

OTTO CANDIES INC.

Des Allemands, Louisiana, U.S.A.

PERRY OCEANOGRAPHIC INC.

100 E. 17th Street, Riviera Beach, Florida 33404, U.S.A.
TEL: (305) 842-5261
TELEX: 51-3439
EXECUTIVES: J. H. Perry, Jr., President; B. C. Gilman, Executive Vice-President; J. F. Manuel, Sales Vice-President; W. R. Bryant, Vice-President.
ORGANISATION:
Since 1956, producing commerical submarines, diving systems, underwater laboratories, decompression chambers and other ocean hardware. These products are both sold and leased.
ASSOCIATED COMPANIES:
Perry Ocean Engineering, Inc (design affiliate)
Perry Ocean Services, Inc (leases and operates Perry equipment)
Perry Submarine Builders (construction branch of Perry Oceanographics Inc).

PORT OF PORTLAND

Swan Island Ship Repair Yard, P.O. Box 3529, Portland, Oregon 97208, U.S.A.
TEL: 503-233-8331
CABLE/TELEX: TWX910-464-6151
FAX FDH
EXECUTIVES: C. F. Propp, Ship Yard Manager and Chief Designer.
Licensed Manufacturers of 'Big and Small Skimmers':
Environment Protection Machines, Ltd, Route 2, Box 97, Estacada, Oregon, 97023, U.S.A.

RAUMA REPOLA OY

Mantyluoto Works, Pori, Finland.

R.B.H. CYBERNETICS (1970) LTD.

P.O. Box 4205, Postal Station A, Victoria, British Columbia, Canada
TEL: 604-658-5713
CABLE: Cypapco
WORKS: 859 Devonshire Road, Victoria, British Columbia, Canada
EXECUTIVES: V. N. R. Sewell, President; A. L. Sewell, Secretary; V. N. R. Sewell, General Manager; R. B. H. Sewell, Design Engineer.
ORGANISATION:
The company, founded in 1970, has developed the Slicklicker, an oil spill clean-up device which can pick up oils of viscosities from Bunker C at freezing temperatures to diesel oil, as well as emulsions.

The company has also developed 'Graboil', a reusable oil absorbent primarily developed for use with the Slicklicker, but which can also be used to keep sumps, etc., clean of oil.
ASSOCIATED COMPANIES:
Sub-licencees of R.B.H. Cybernetics (1970) Ltd are:
Dominion Welding Engineering Co Ltd, Keele Street North, Maple, Ontario, Canada
Barbet Engineering Ltd,
62 Glendale Gardens, Leigh-on-Sea, Essex, England
Straits Navigators (Pte) Ltd,
8, Third Lokyang Road, Singapore, 22

RECORDING DESIGNS (EMI) LTD.

Victoria Avenue, Camberley, Surrey, England
TEL: Camberley 26414
CABLE/TELEX: Redel camberley 85272
ADDRESS OF WORKS: As above
EXECUTIVES: Mr. B. P. Collins, Chairman; Mr. P. A. Allaway, Director; Mr. P. I. Hayman, Director; Mr. P. J. Brannigan, Director; Mr. J. M. Kuipers, Director; Mr. D. H. W. R. May, Director; Mr. A. England, General Manager; Mr. E. J. Partridge, Marketing Manager; Mr. R. H. Gardner, Engineering Manager.
ORGANISATION:
The Company was formed in 1959 to specialise in the field of instrumentation magnetic tape recording systems. A rapid expansion of the range of activities followed, initially through underwater weapon instrumentation and has led to the present situation where the company has expertise in all fields of severe environmental instrumentation.
SUBSIDIARY/ASSOCIATED COMPANIES:
RDL is part of the Systems & Weapons Division of EMI Electronics Ltd, and the focal point for underwater instrumentation expertise within this international company. EMI/RDL agents are available in most countries.

REYNOLDS INTERNATIONAL INC.

Box 27002, Richmond, Virginia 23261, U.S.A.
TEL: (804) 282-2301
EXECUTIVES: J. L. Reynolds, Director; A. L. Markel, Director and General Manager; B. B. Gralow, General Manager; P. Fox, Vice-President Sales
ORGANISATION:
The company which is a wholly owned subsidiary of Reynolds Metals Company, entered the oil pollution control equipment and services field in 1969.
ASSOCIATED COMPANY:
Reynolds Submarine Services Corporation (address as above).

UNIVERSITY OF RHODE ISLAND GRADUATE SCHOOL OF OCEANOGRAPHY

Kingston, Rhode Island
TEL: (401) 792-6222 and (401) 792-6203/4
CABLE: Narrmarlab, Kingston, Rhode Island
EXECUTIVES: Dr. J. A. Knauss, Dean and Provost for Marine Affairs; Mr. C. A. Buehrens, Marine Superintendent.
ORGANISATION:
The University was founded in 1895 and the Oceanography Graduate School in 1961.

RIVTOW STRAITS LIMITED

2215 Commissioner Street, Vancouver, B.C. V5L 1A8, Canada
TEL: 255-1133
TELEX: 04-55394
Works and telephone number
B. C. Marine Shipbuilders, Ltd, 101 N. Victoria Drive, Vancouver (255-3121); John Manly Ltd, 2050 Kent Street, Vancouver 16, B.C. (327-6311)
EXECUTIVES: B. Brown, General Manager of B. C. Marine; J. Upex, General Manager of John Manly Ltd; G. A. Newman, Sales.

ORGANISATION:
Some of the many services provided by RivTow Straits Limited are transportation of lumber, logs, wood chips, petroleum products, heavy freight equipment, roll-on roll-off, house moving, lime rock, pebbled lime and sand and gravel.

The company also provides regularly scheduled freight runs and owns a trucking company based in Prince Rupert. Water taxis and bare boat charter services are also available.

RivTow Straits Limited handle over one and a half billion F.B.M. of forest products per annum through phases of log barging, towing, storage and delivery.

Sorting and booming grounds are established at Blind Bay near Nelson Island, in Howe Sound, at Avalon, West Bay, and Plowden Bay. On the Fraser River the sorting grounds are at Pitt River, Mission, and Hope. The Hope sort is utilised for collection of logs that are driven through the Fraser Canyon, considered one of the most unusual fast water log drives anywhere. Annual sorting capacity is over 500,000 F.B.M. The company has leases capable of storing 20,000 log sections.

In the past, many long distance towing contracts have been completed to places such as Japan, Panama, Alaska, Mexico, the US eastern seaboard, and the Great Lakes. In the early part of 1972 we successfully towed a 7,500 ton capacity paper barge from the west coast via the Panama Canal to Searsport, Maine.

The Aggregates Divisions of the company expect to market approximately 650,000 tons of sand and gravel in 1973 in Vancouver, Prince Rupert, Nelson and Kelowna. These communities are also served by R.S.L.'s Ready Mix Division, consisting of Rupert Cement Products (1965) Ltd, Nelson Ready Mix Concrete Ltd and Westbank Ready Mix Concrete Ltd. Tilbury Island Terminals Ltd situated in Delta is now operational as a terminus for both rail and truck coastal barging, primarily to Vancouver Island.

The company will also expect to enjoy a participation in the Mackenzie River and Arctic regions through its newest partially owned subsidiary, Arctic Navigation and Transportation Co Ltd.

The Company's shipyards, B. C. Marine Shipbuilders Ltd and John Manly Ltd construct, refit and repair vessels from boom boats and personnel carriers to vessels of medium size such as the B. C. Auto Ferry presently under construction at B. C. Marine. B. C. Marine Shipbuilders Ltd complements its shipbuilding business with custom-forging and machining. It markets such items as picks, drop-forged hammers, and forks for fork-lift trucks.

John Manly Ltd holds a licensed franchise for the construction and supply of the Kort Nozzle and the Willi Becker Rudder. Its affiliate (Western Propeller Ltd) builds and repairs ship's propellers from small outboard models to large wheels used on modern freighters.
ASSOCIATED COMPANIES:
John Manly Ltd, manufacturing under licence for Willi Becker of Germany the Kort Nozzle, and Sales Agents for Becker Ruder in Canada.
Willi Becker 2 Hamburg 1, Danziger Strasse 22, Germany

SAMSON CORDAGE WORKS

470 Atlantic Avenue, Boston, Massachusetts 02210, U.S.A.
TEL: (617) 426-6550

TELEX: 94-6407
EXECUTIVES: C. H. Abbott, President; W. K. Child, Retired President, Textile Machinery Divisions of North American Rockwell Corp.; R. W. Daniels, Executive Vice-President; W. R. Dewey, Jr., Chairman of the Board; W. R. Driver, Jr., Partner, Brown Bros. Harriman & Co; C. M. Ganson, Partner, Taylor, Ganson & Perrin; F. T. Pratt, Vice-President, Woodstock Corporation
OFFICERS: W. R. Dewey, Jr., Chairman of the Board; C. H. Abbott, President; R. W. Daniels, Executive Vice-President; R. Billings, Vice-President, Sales; S. F. Hazeltine, Treasurer, G. Garland, Assistant Treasurer; C. M. Ganson, Clerk.
ORGANISATION:
The company began developing Braided Rope Systems for the ocean/marine industry in 1960. They have also been major suppliers of single-point mooring systems since 1960.
SUBSIDIARY/ASSOCIATED COMPANIES:
Canada Ropes Limited, Richmond, B.C., Canada;
Metalex Products Ltd, Richmond, B.C., Canada;
Service Bronze and Brass Works, Inc, Portland, Oregon;
Samson Cordage International Corp., Boston Massachusetts;
Ocean Systems, Inc, Reston, Virginia (acquired 29 January 1973)

SANTA FE INTERNATIONAL CORPORATION

Union Bank Square, South Tower, P.O. Box 1401, Orange, California 92668, U.S.A.
TEL: (714) 542-6281
TELEX: 685-525
CABLE: Santafe
EXECUTIVES: E. L. Shannon, Jr., Director President; D. F. West, Director Senior Vice-President; M. A. Kline, Director and Senior Vice-President and General Counsel; G. M. Anderson, Director; S. M. Collester, Director; R. M. Jenney, Director; W. G. White, Director; S. B. McDonald, Director; H. C. Pyle, Director; H. R. Frank, Director; A. B. Henderson, Vice-President and Secretary; R. M. Donaldson, Vice-President, Administration; R. E. Steele, Vice-President; J. J. Syapin, Vice-President; J. E. Warren, Vice-President; A. W. Bramlett, Assistant Secretary, Treasurer; D. C. Reilly, Assistant Secretary, Treasurer; G. L. Womack, Assistant Secretary; S. T. Peeler, Vice-President.

ORGANISATION:
Santa Fe International Corporation, originally known as Santa Fe Drilling Co, was founded in 1946.

Starting in business with eight used rigs, all located in California, Santa Fe has grown into one of the world's largest international drilling contractors with operations in nearly all the major oil-producing countries.

The new *Santa Fe Mariner 2*, the company's fourth semi-submersible-type drilling unit, will bring to 12 the number of offshore rigs owned and operated by Santa Fe. Besides the *Santa Fe Mariner 2*, the company's current $62 million expansion programme also will add eight land rigs, an offshore jackup rig and a third derrick/pipelaying barge to its major equipment list. With completion of units now on order, Santa Fe's Drilling Division will own and operate 58 rigs.

The *Santa Fe Mariner 2* will drill initially in the Gulf of Mexico. Other company-owned rigs currently are under contract in the US and 11 other countries.

Santa Fe's experience as an offshore drilling contractor dates back to the mid-50s when it manned one of the first floating drilling vessels, for the CUSS Group off California. The company became a major factor in offshore drilling operations in 1965 with the acquisition of Blue Water Drilling Corporation, a pioneer in the development of the semi-submersible, column-stabilised drilling unit.

Since then, while building a versatile drilling fleet which includes jackup rigs, shallow-water barges, tender and fixed platform rig, as well as semi-submersibles, Santa Fe has introduced the concept of the semi-submersible vessel in offshore construction operations. Its *Choctaw*, the world's first derrick/pipelaying barge, of that design, will be joined later this year by a similar vessel, *Choctaw II*.

In 1967, through the acquisition of J. H. Pomeroy & Co, Inc, Santa Fe expanded its services to the international oil industry to include engineering and construction. Now known as Santa Fe-Pomeroy, Inc, this division specialises in submarine pipeline installations, erection of offshore platforms and design and construction of oil loading facilities, piers, harbours and other projects for the petroleum industry.

In 1970 Santa Fe acquired Curran & Co, a domestic pipeline contractor. The company now operates as Santa Fe-Curran & Co, a division of Santa Fe International, and specialises in onshore pipeline construction.

ASSOCIATED/SUBSIDIARY COMPANIES:

SANTA FE DRILLING COMPANY

(A Division of Santa Fe International Corporation)
G. M. Anderson, President; C. R. Ball, Jr., Senior Vice-President, Regional Manager, Europe and Africa; S. M. Collester, Senior Vice-President; N. V. Parry, Senior Vice-President, Operations; H. H. Madley, Senior Vice-President, Manager Services; C. F. Dawson, Vice-President, Regional Manager, Middle East; L. M. Jones, Vice-President, Regional Manager, Western Hemisphere and Southeast Asia; I. J. Flowers, Vice-President, Marine Operations; R. C. Parsons, Division Controller; J. M. Dickey, Division Counsel; R. E. Hinson, Personnel Manager; F. Foy, Jr., Drilling Equipment Co-ordinator; R. Hughes, Tax Manager; C. L. Bass, Marine Superintendent.

Alaska
Santa Fe Drilling Company,
P.O. Box 637, Kenai, Alaska 99611
TEL: (907) 283-7580
S. E. Souza, Manager

Houma
Santa Fe Marine, Inc,
1906 Tunnel Blvd., Houma, Louisiana 70360
TEL: (504) 868-2390
S. Orso, Gulf Coast Operations Manager

Houston
Santa Fe Marine, Inc,
4101 North Freeway, Suite 123, P.O. Box 16306, Houston, Texas 77022
TEL: (713) 697-4853
CABLE: Sfemarine
TELEX: 775-218
E. D. Shepherd, President

Egypt
Santa Fe International Corporation,
UAR Operations, P.O. Box 341, Cairo, Egypt UAR
TEL: 800-221
CABLE: Santafe Cairo
TELEX: 2266
M. Baca, Zone Manager

Iran
Santa Fe International Services, Inc,
Ave. Takhte Tavoos, 43 Ave. Darya Noor, P.O. Box 2530, Tehran, Iran
TEL: 623038/631044
TELEX: 2610 'For Santa Fe'
CABLE: Santafe Tehran, Iran
C. F. Dawson, Regional Manager, Middle East; B. G. Parker, Operations Manager, Middle East; D. J. Zimmer, Zone Manager, Iran

Kuwait
Kuwait Drilling Company,
P.O. Box 9066, Ahmadi, State of Kuwait, Arabian Gulf
TEL: 981-598
CABLE: Kutdrill Kuwait Arabian Gulf
H. H. Dean, Drilling Superintendent

Libya
Santa Fe International Services, Inc,
P.O. Box 680, Tripoli, Libyan Arab Republic
TEL: 34110/34111
CABLE: Santafe Tripoli, Libya
G. T. Steadham, General Superintendent

London
Santa Fe (UK) Limited,
Radstock House, No. 5 Eccleston Street, London SW1W 9NA, England
TEL: (01) 730-7151
CABLE: Santafe London SW1 England
TELEX: (851) 264-556
C. R. Ball, Jr., Regional Manager, Europe and Africa; S. D. Carnahan, Manager, Africa; C. K. Orr, Manager, North Sea
Norsmec, Limited,
c/o Santa Fe (UK) Limited, Radstock House, No. 5 Eccleston Street, London SW1W 9NA, England
TEL: (01) 730-7151
CABLE: Santafe London SW1 England
TELEX: (851) 264-556
L. H. Youngblood, Britannia Superintendent

Saudi Arabia
Santa Fe International Services, Inc,
c/o Aramco Dhahran, P.O. Box 219, Abqaiq, Saudi Arabia
TEL: Aramco, Dhahran 42000 (Ask for 5172 in Abqaiq)
CABLE: c/o Aramco, Dammam, Saudi Arabia Attn: Santa Fe, Abqaiq
TELEX: 92811
J. H. Morris, Zone Manager

Trinidad
Santa Fe Drilling Company of S.A.,
No. 5 Queens Park East, Port-of-Spain, Trinidad W.I.
TEL: 36611/38671
CABLE: Santafe Port of Spain
TELEX: 387-249
A. T. Barnardo, Manager

Trucial Coast
Santa Fe Drilling Co (UK) Ltd,
P.O. Box 591, Abu Dhabi, Arabian Gulf
TEL: 61625
CABLE: Santafe Abu Dhabi Arabian Gulf
TELEX: 949-236
C. F. Hammonds, Zone Manager

Venezuela Offices
Santa Fe Drilling Co of Venezuela C.A.,
Apartado 34, Anaco, Estado Anzoategui, Venezuela, S.A.
TEL: 2376
CABLE: Santafe Anaco Venezuela
TELEX: 81191
J. P. Lynch, Zone Manager; I. B. Thrasher, Operations Manager
Santa Fe Drilling Co of Venezuela C.A.,
Apartado 181, Maracaibo, Venezuela, S.A.
R. Hullinger, Operations Manager, Western Venezuela

SANTA FE ENGINEERING & CONSTRUCTION CO.

(A Division of Santa Fe International Corporation)
F. E. Reynolds, President; S. Flournoy, Vice-President and General Manager, Marine and Civil Construction; A. B. Killgore, Vice-

President and General Manager, Marketing and Development; B. J. Warren, Vice-President and General Manager, Offshore Construction; E. T. Van Geem, Vice-President, Contract Administration; P. A. Conrado, Vice-President, Pacific; F. A. Divita, Vice-President, Special Projects; E. K. Ellgaard, Vice-President, Middle East; E. P. Horan, Vice-President, Estimating; D. A. Kerr, Vice-President, Saudi Arabia; P. G. Morken, Vice-President, Domestic Operations; B. W. Passmore, Vice-President, North Sea; C. L. Arrington, Division Counsel; M. Stamos, Controller.
Domestic Operations:
Santa Fe Engineering & Construction Co, 500 Hopper Street Extension, (P.O. Box 837), Petaluma, California 94952
TEL: (707) 763-1918
P. G. Morken, Vice-President, Domestic Operations; M. B. McGowan, Manager, Domestic Construction; A. Brown, Manager, Concrete Products
Gulf Coast Operations:
Santa Fe Engineering & Construction Co, P.O. Box 813, Houma, Louisiana 70360
TEL: (504) 876-7592
L. E. Ruckstuhl, Manager, Gulf Coast Division

Abu Dhabi
CCC-Santa Fe,
P.O. Box 713, Abu Dhabi, Trucial States
CABLE: Consanta
TELEX: CONCO AH 256
J. Mabery, Project Manager

Canada
Bow Valley/Acres/Santa Fe-Pomeroy Arctic Services,
700-630 6th Avenue South West, Calgary, Alberta, Canada T2POT1
TEL: (403) 264-0600
D. T. Jumpsen, General Manager

Indonesia
P. T. Santa Fe-Pomeroy Indonesia,
P.O. Box 2351, Jakarta, Indonesia
TEL: 70730
CABLE: Santafe Jakarta
R. Lewis, President
(Includes Merak Petroleum Base Operations)

Scotland
Santa Fe (UK) Limited,
Shed 4, Victoria Dock, Leith, Edinburgh EH67DZ, Scotland
TEL: 031-554-0363
TELEX: 727-173
B. B. Creel, Vice President, North Sea

London
Santa Fe (UK) Limited,
Radstock House, No. 5 Eccleston Street, London SW1W 9NA, England
TEL: 01-730-7151
CABLE: Santafe London SW1 England
TELEX: (851) 264-556
C. K. Orr, Representative; R. C. Paulson, Manager of Special Projects

Saudi Arabia
Santa Fe-Shuwayer Constructions,
c/o Dhahran Airport Post Office, Dhahran, Saudi Arabia
TELEX: SANTAFE 11
D. A. Kerr, Area Manager

SANTA FE MINERALS, INC.

(A Subsidiary of Santa Fe International Corporation)
14367 Pass Road, Live Oak, California 95953
TEL: (916) 696-2514
J. E. Warren, President and Director; M. G. Cooksey, Vice-President, Assistant Secretary and Director; G. L. Womack, Secretary; D. F. West, Treasurer and Director; A. W. Bramlett, Assistant Secretary-Treasurer and Director; B. Phillips, Jr., Production Engineer.

SANTA FE POMEROY INC.

Box 2432, G.P.O., Brisbane, Queensland 4001, Australia
ORGANISATION:
The company is a subsidiary of Santa Fe International Corporation

SCHAT DAVITS LTD.

Ashley Mead, London Colney, Hertfordshire, England
TEL: Bowmansgreen 22244
TELEX: 21811

SCOTTISH MARINE BIOLOGICAL ASSOCIATION DUNSTAFFNAGE MARINE RESEARCH LABORATORY

P.O. Box 3, Oban, Argyll, PA34 4AD, Scotland
TEL: Oban 2244/6 and 3525
CABLE: Marlab Oban

SCRIPPS INSTITUTION OF OCEANOGRAPHY, UNIVERSITY OF CALIFORNIA, SAN DIEGO

P.O. Box 1529, La Jolla, California 92037, U.S.A.
TEL: (714) 453-2000
EXECUTIVES: Dr. W. A. Nierenberg, Director; Mr. D. M. Brown, Designer, Buoys and Moorings; Mr. M. Sessions, Designer, Instrumentation.

SEAFORTH MARITIME LIMITED

Seaforth Centre, Waterloo Quay, Aberdeen, Scotland
TEL: Aberdeen 573401
EXECUTIVES: I. Noble, Chairman; J. Hann, Chief Executive; L. Kennedy, Chief Marketir.g Executive; M. Ferrier, Chief Financial Executive; N. Smith, Support Services Executive; W. Barrack, Chief Transport Executive; A. Wilkie, Chief Technical Executive; D. Callaghan, Chief Personnel Executive; V. Cook, Chairman Barry Henry & Cook.
ORGANISATION:
Seaforth Maritime Limited was formed in 1972 and is a major Scottish independent offshore oilfield services firm. Principal shareholders of Seaforth are Lyle Shipping, Hogarth Shipping, Sidlaw Industries, North Sea Assets, The Bank of Scotland Finance Company and Viking Resources.
SUBSIDIARY COMPANIES:
J. G. Barrack
Barry Henry & Cook

INGENIOR F. SELMER A/S CIVIL ENGINEERS AND CONTRACTORS

Hausmannsgt. 34, P.O. Box 256, Sentrum, Oslo 1, Norway
TEL: (02) 20 83 30
TELEX: 11246 CONTR
CABLE: Contractor
LOCAL OFFICES: Drammen, Sarpsborg, Skien, Stavanger, Steinkjer, Tromsø
EXECUTIVES: L. Selmer, Managing Director; K. Jacobsen, Technical Director.

SHELL PIPELINE CORPORATION

P.O. Box 2648, Houston, Texas 77001, U.S.A.

SKUTENG A/S

Marstrandgt. 6, Oslo 5
TEL: (02) 21 35 12
CABLE: Skuteng Oslo
TELEX: 11784
EXECUTIVES: K. Engelund, Director; A. Henriksen, Sales Manager; T. Flyen, Chief Engineer; S. S. Wibye, Technical Engineer.
ORGANISATION:
Skuteng A/S has been active in the field of oil recovery for some time, having sold their products to countries in east and west Europe.
ASSOCIATED COMPANIES:
England
Jones & Healy Marine Ltd,
Lloyd Bank Chambers, 18 Broad Street, Wokingham, Berkshire RG 11 I AF

France
Société Française Total Incendie, 19 Rue Lantiez, Paris

Germany
Fa. Total Export GmbH,
Postfach 2430, D-68 Mannheim

Denmark
O. Sommer A/S,
Uplandsgade 70, DK-2300 København S

Sweden
AB Brand & Säkerhetstjänst,
Kungsgatan 30, S-111 35 Stockholm

Greece
Beroo Ltd,
44a Voulis Street, 118 Athens

Australia
United Maritime Agencies Pty Ltd,
P.O. Willoughby North, 2068 N.S.W.

U.S.A.
Arnessen Marine Systems Inc,
55 West 39th Street, New York, N.Y. 10018

Hong Kong
The Jardine Engineering Corporation Ltd.
54, Wong Chuk Hang Road,
P. O. Box 517, Hong Kong

SLICKBAR INC.

Box 295 Saugatuck Sta., Westport, Conn. 06880, U.S.A.
WORKS: 10 Saugatuck Avenue, Westport, Conn., U.S.A.
EXECUTIVES: F. Smith Millard, President; R. M. Blair, Executive Vice-President; J. P. Sullivan, Sales Manager.
ORGANISATION:
Active in oil spill control technology since 1960.
SUBSIDIARY/ASSOCIATED COMPANIES:
Export Department: 85 Washington Street, South Norwalk, Connecticut, U.S.A.
European Office: John Fenn, Slickbar Inc, 4 Market Square, Amersham, Buckinghamshire, England.

SMIT INTERNATIONAL

Westplein 5, Rotterdam 3002, Holland
TEL: 010-36 27 00 (day and night)
TELEX: 22247 (day and night)
CABLE: Smittug Rotterdam
ORGANISATION:
The Smit International Group consists of the following 100 per cent owned Companies
Smit International Ocean Towage and Salvage Company, Rotterdam; Smit International Harbour Towage Company, Rotterdam; Smit Tak International Salvage Company, Rotterdam; Smit Tak Seatransport, Rotterdam; Smit Tak International Salvage & Transport (UK) Ltd, London; Smit International (Antilles) NV, Curaçao; Smit International (Americas) Inc, New York; Smit American Salvage Inc, New York; Smit International (UK) Ltd, London; Smit McLaren Ltd, London; 'Interbergung' Internationale Bergungs & Transport GmbH, Hamburg.

The Smit International Group has an important share holding in the following Companies
Smit & Dirkzwager International Ship Delivery,

Rotterdam; BV Nieuwe Rotterdamse Sleepdienst, Rotterdam; BV Nieuwe Vlissingse Sleepdienst, Flushing; Smit-Lloyd BV, Rotterdam; Smit-Lloyd (Australia) Pty Ltd, Sydney; Smit & Cory International Port Towage Ltd, Maidenhead; Eastern Canada Towing Ltd, Halifax; Smit Overseas Ship Delivery Ltd, London; Smit Moran American Salvage, New York; Sayremar S.A., Madrid.

SMIT-LLOYD B.V.

Westplein 8, P.O. Box 896, Rotterdam-2, Holland
TEL: 010. 36 16 55
TELEX: 23434
CABLE: Smitlloyd, Rotterdam
EXECUTIVES: Mr. R. W. Scheffer, Director; Mr. A. A. W. Pruymboom, General Manager; Dr. J. C. A. Hoogenbosch, Naval Architect; Mr. C. Belder, Technical Superintendent.

ORGANISATION:
The company was formed in November 1964.
ASSOCIATED COMPANIES:
Lijnvaart Group N.S.U.; Holland Amerika Lijn; Smit International
SUBSIDIARIES/ASSOCIATED COMPANIES OVERSEAS:
Smit-Lloyd (Australia) Pty Ltd, Sydney, Australia;
Smit McLaren Ltd, London, England;
Biehl & Co Inc, Houston, Texas, U.S.A.;
Nedlloyd Inc, New York, U.S.A.;
Interocean-E.A.C. Agencies, Singapore, Malaysia.

SMIT TAK INTERNATIONAL SALVAGE & TRANSPORT (UK) LTD.

Baltic Exchange Chambers, 24 St. Mary Axe, London, EC3
TEL: 01-283-4424
TELEX: 88 44 56 'SMITTAK'
EXECUTIVES: P. van den Berg, Director (Dutch); H. Smit, Director (Dutch); F. J. Collis, Director
G. E. Harvey, General Manager.
ORGANISATION:
Formed in 1970 as a subsidiary of Smit-Tak International Salvage, B.V. Rotterdam.
Principal activities: salvage, wreck removal, heavy lifts, ocean going pontoons, diving surveys, well head removal, underwater demolition, cargo recovery, anchor recovery buoy laying, offshore moorings.
REPRESENTATIVES:
Interbergung GmbH, Hamburg;
Sayremar S.A. Madrid.

UNIVERSITY OF SOUTHERN CALIFORNIA, ALLAN HANCOCK FOUNDATION

University Park, Los Angeles, California 90007, U.S.A.
TEL: (213) 746-2929
WORKS: Marine Operations Facility, 100 East Water Street, Wilmington, California 90744, U.S.A.
TEL: (213) 830-4570
EXECUTIVES: Mr. B. C. Abbott, Director, Allan Hancock Foundation; Mr. J. M. Savage, Principal Investigator; Mr. P. F. Irving, Superintendent of Ship Operations.

SUNSHINE CHEMICAL CORPORATION

P.O. Box 17041, West Hartford, Connecticut 06117, U.S.A.
TEL: (203) 232-9227
EXECUTIVES: Mr. A. Flateau, President; Mr. S. Kaufmann, Vice-President; Mr. R. Rydell, Sales Manager.
ORGANISATION:
The company was formed in 1960 and operates in the field of oil spill control.

TELEDYNE MOVIBLE OFFSHORE INC.

Heymann & Travis, P.O. Box 51936, Lafayette, Louisiana, U.S.A.
TEL: (318) 232-5120
TELEX: 58-6613
EXECUTIVES: J. L. Steitle, President and General Manager; C. E. Chism, Controller; J. C. Stanley, Construction Manager; J. H. Brazier, Drilling Manager; T. L. Hull, Chief Engineer.
ORGANISATION:
Teledyne Movible Offshore, a subsidiary of Teledyne, Inc, Los Angeles, California, began operations in 1957 in the offshore contracting business, and presently is operating the following equipment and facilities: (1) Mobile rigs for exploratory and development drilling; (2) Self-contained platform rigs for development drilling and workover; (3) One jack-up operating and another under construction; (4) Engineering for the design of offshore structures; (5) Fabrication yard for the fabrication of offshore structures; and (6) Derrick barges for the installation of offshore structures and pipe lines.
ASSOCIATED COMPANIES:
Teledyne Offshore, Inc,
No. 6 Arcade, Hyatt Hotel, 10–12 Scotts Road, Singapore 9.

TRELLEBORGS GUMMIFABRIKS AB

Fack, S-231 01 Trelleborg, 1, Sweden
TEL: 0410-150 00
TELEX: 329 29, 329 50

ORGANISATION:
Trelleborgs Gummifabriks is situated at Trelleborg, on the south coast of Sweden. The company also has factories in Western Germany (Hamburg), Holland (Hoogezand) and Italy (Turin), and licence agreements with firms in Australia, Japan and U.S.A.
Trelleborg started in 1905 and is today the largest rubber factory in Scandinavia and number ten in Europe.
During the last 20-year period Trelleborg has multiplied its sales six-fold from 94 Mkr to 555. The number of employees has doubled during the same period.
Trelleborg has its own sales companies in Australia, Belgium, Denmark, France, Holland, Italy, Norway, Switzerland, Spain, Great Britain, U.S.A., Western Germany and Austria. The company's products are sold in a hundred or more countries.
In 1972, Trelleborg's sales abroad totalled 250·8 Mkr or 45·2 per cent of total sales. Sales abroad were distributed as follows: Efta 42 per cent, EG 26 per cent, North America 17 per cent and the rest of the world 15 per cent.
The Trelleborg Group, Trelleborgs Gummifabriks AB with subsidiaries in Sweden and abroad, had at the 1972/1973 year-end approx. 6,150 employees. The number of employees at the Parent Company in Trelleborg totalled about 4,000.
Industrial rubber answers for 64 per cent of all sales. The company manufactures hoses, V-belts, conveyor belts, battery containers, roll coverings, fabrics, mats, corrosion protection, rubber sheeting, tread rubber, moulded rubber, rubber/metal (shock-absorbers), containers, gaskets, mouldings and advanced safety products.
Tyres answer for roughly 29 per cent of sales. Production: 30,000 tyres (various kinds) daily. Tyres and tubes for cars, trucks, buses, contract vehicles, agricultural machines and tractors, industrial vehicles, airplanes, motorcycles, mopeds, cycles and wheelbarrows, retreading material, etc., are produced.
Plastic answers for 7 per cent of sales. Products include reinforced plastics and thermoplastics.
The Trelleborg Group pursues a large-scale research and development programme. About 200 persons are employed in laboratories and research departments. More than 15 Mkr is annually invested in research and development.
Investment during the 5-year period 1970–1974 amounted to 375 Mkr and is mainly designated for the continued expansion of the Industrial Rubber Division's resources.
The share capital amounts to 92 Mkr. The number of shareholders is approximately 7,400.
SUBSIDIARY/ASSOCIATED COMPANIES:

Australia
Trelleborg Pty Ltd,
P.O. Box 460, Chatswood N.S.W. 2067, Suite 106, Equity House, 284 Victoria Avenue
TEL: 412 17 62, 41 70 02
TELEX: 243 16

Austria
Trelleborg Gummi GmbH
Zeleborgasse 24, A-1120, Vienna
TEL: 222-83 65 08, 83 65 09
TELEX: 118 68

Belgium
Trelleborg S.A.-N.V.
B-1130 Bruxelles Chaussee de Haecht 1243
TEL: 02-41 6533
TELEX: 233 13

England
Trelleborg Rubber Ltd,
124 Barlby Road, London W10 6BN
TEL: (01) 969 8303
TELEX: 262 477

France
Trelleborg S.A.
Z.I. 'Les Mardelles', 50 Rue Henri Becquerel, F-936 00 Aulnay Sous Bois
TEL: 929-22-91
TELEX: 602 42

Italy
Trelleborg S.p.A.
Via Sansovino 243/55C, I-101 51 Torino
TEL: 011-29 99 52, 21 34 81, 21 14 30
TELEX: 213 79

Netherlands
Trelleborg Rubberfabrieken B.V.,
Postbus 33, Hoogezand, Foxham 52
TEL: 05980-5811
TELEX: 530 55

Spain
Trelleborg S.A.,
16, Diego de León, Madrid 6
TEL: 1-226 99 15, 276 97 92
TELEX: 233 51

Switzerland
Trelleborg Gummi AG,
Postfach 1241, CH-4002 Basle, Gartenstrasse 120
TEL: 061-22 27 55
TELEX: 628 66

U.S.A.
Trelleborg Rubber Company, Inc,
P.O. Box 390, 10 Solon, Ohio 44139
30700 Solon Industrial Parkway
TEL: 216-248-8600
TELEX: 980 282

Germany
Trelleborg GmbH,
D-2000 Hamburg 76, Flotowstrasse
TEL: 0411-220 14 16
TELEX: 212 046

TRYGVE THUNE A/S

Munchsgt 5, Oslo 1, Norway
TEL: 20 45 04

CABLE: Oilboom
EXECUTIVES: Mr. Trygve Thune, Managing Director.
ORGANISATION:
The company was formed in 1954 and has been active wholly in the field of anti-oil pollution equipment since inception.

UNIVERSITY COLLEGE OF NORTH WALES, DEPARTMENT OF MARINE BIOLOGY

Marine Science Laboratories, Menai Bridge, Anglesey LL59 5 EH, Wales
EXECUTIVE: Professor G. E. Fogg

VICKERS OCEANICS LIMITED

P.O. Box 8, Cavendish Park, Barrow-in-Furness, Lancashire, England
TEL: Barrow-in-Furness 25080
TELEX: 65147
CABLE: Oceanics Barrow
EXECUTIVE: Mr. G. G. Mott, Chief Executive and Managing Director.
ORGANISATION:
The company, formed in April 1972, is a member of the Vickers Group.

WARREN SPRING LABORATORY

(Department of Trade and Industry)
P.O. Box 20, Gunnelswood Road, Stevenage, Hertfordshire 2G1 2BX, England
TEL: Stevenage 3388
TELEX: 82250
EXECUTIVE: Dr. P. G. Jeffery, Head of Oil Pollution Division.

UNIVERSITY OF WASHINGTON APPLIED PHYSICS LABORATORY

1013 N.E. 40th Street, Seattle, Washington 98195, U.S.A
TEL: 206 543-1300
TELEX: 910-444-1345
EXECUTIVES: Dr. H. L. Blood, Director; Dr. William R. Davis, Associate Director.

WATERCRAFT LTD.

The Sussex Shipyard, Shoreham-by-Sea, Sussex, England
TEL: Shoreham 4267
TELEX: 877418

WECO SHIPPING I/S

Rungsted Strandvej 113, DK-2960 Rungsted Kyst, Denmark
TEL: (01) 86 65 00
TELEX: 15237
CABLE: Wecoship
EXECUTIVES: E. Baron Wedell-Wedellsborg; Chr. K. Hansen; J. Fleinert; H. Christensen; O. Pedersen; O. Møller.
ORGANISATION:
The company are Managing Owners for Rederiaktieselskabet Dannebrog.
ASSOCIATED COMPANIES:
Arhus Flydedok A/S,
Balticagade, P.O. Box 23, Dk-8100, Arhus, C., Denmark
TEL: (06) 13 40 00
TELEX: 64462
CABLE: Flydedokken
EXECUTIVES: T. Duer; C. Søe; H. J. Henriksen.

WESTINGHOUSE ELECTRIC CORP. OCEANIC DIVISION

P.O. Box 1488, Annapolis Maryland 21404, U.S.A.
TEL: (301) 765-5251
TELEX: 87995
EXECUTIVES: Mr. G. F. Mechlin, Vice-President Ocean Division; Mr. M. S. Macovsky, General Manager, Oceanic Division; Mr. F. T. Brady, Marketing Manager; Mr. P. H. Ockerman, Manager Operations; Mr. B. J. Lowe, Manager, Engineering.

ORGANISATION:
The Oceanic Division of Westinghouse, now a part of the Ocean Research and Engineering Centre, opened at Annapolis, Maryland, in 1967. Since then, it has grown to become one of the world's largest private oceanographic centers. Research, engineering, and hardware development and production are concentrated here in three major technologies: information systems, life-support systems, and submergence systems.
The company's experience with oceanography began more than 30 years ago.
The division is concerned with industries such as agriculture, power generation, oil, gas, chemical, and fishing; government activities such as water and sewage authorities, pollution and environmental control agencies, and recreational planning agencies. The Division can handle any phase of oceanic, fresh-water, and related projects, including feasibility studies, initial planning, development of new concepts and technologies, and complete project management.

WHITTAKER CORPORATION

5159 Baltimore Drive, La Mesa, California 92041, U.S.A
TEL: 714-469-0171
EXECUTIVE: Mr. T. A. Lambert, Manager, International Marketing
SUBSIDIARY/ASSOCIATED COMPANY:
(agent) Gullick Dobson Ltd/Elliott & Garrood Ltd,
P.O. Box 12 Ince, Wigan, Lancashire, England
TEL: 942 41991
Personnel: Captain George Beatty

WIGHT PLASTICS LIMITED

Bembridge Fort, Sandown, Isle of Wight, PO36 8QS, England
TELEX: 86448
CABLE: Micronair Sandown
WORKS: Whitecross Lane, Shanklin, Isle of Wight.
TEL: Shanklin 4351
EXECUTIVES: Mr. J. M. McMahon, Director; Mr. F. H. Mann, Director; Mr. R. P. Gatrell, General Manager.
ORGANISATION:
Formed in 1969, the company specialises in the manufacture of vacuum formed thermo plastics and resin glass laminates for the aircraft and general engineering industries. An underwater transport vehicle is being developed as an extension of the company's activities.
ASSOCIATED COMPANIES:
Crop Culture Group of companies.

WOODS HOLE OCEANOGRAPHIC INSTITUTION

Woods Hole, Massachusetts 02543, U.S.A.
TEL: 617-548-1400
CABLE/TELEX: TWX-710-346-6601
ADDRESS OF WORKS: As above
EXECUTIVES: Dr. P. M. Fye, President of the Corporation, and Director; Mr. C. F. Adams, Chairman of the Board of Trustees; Dr. A. Maxwell, Provost; Dr. R. W. Morse, Associate Director and Dean of Graduate Studies; Dr. F. Webster, Associate Director; Dr. B. Ketchum, Associate Director; Mr. D. D. Scott, Assistant Director for Administration.

ORGANISATION HISTORY:
The Woods Hole Oceanographic Institution was established as a private research organisation in 1930. The Charter granted by the Commonwealth of Massachusetts described the purpose of the Institution: 'To prosecute the study of Oceanography in all its branches; to maintain a laboratory or laboratories, together with boats and equipment and a school for instruction in oceanography and allied subjects. In 1940, the Institution became a year-round research laboratory. Responding to the needs of the US Navy, the work increased dramatically in the next few years. Institution scientists made use of marine data which they had been collecting and computing since 1930, and which now is considered vital to the defence of the United States.
Growth continued after the war, and military projects gave way to basic research. In the past decade, the Institution has become increasingly involved in co-operative national and international cruises, many of which Woods Hole scientists have initiated and organised.
Today, the Institution is the scene of national and international conferences and seminars throughout the year. The published papers of its scientists are circulated to libraries, universities, and laboratories throughout the world; and the staff frequently participates in important symposia in Woods Hole and other countries.
The resident scientific staff is composed of approximately 100 men and women. The Technical Staff includes an additional 100 research specialists and research associates. These 200 resident staff members are supported by approximately 520 additional employees who are technicians, clerical workers, ships' crews, shop personnel and others.
The Institution fleet numbers four ocean-going research vessels: R/V *Atlantis*, R/V *Chain*, R/V *Gosnold*, R/V *Knorr*. The R/V *Lulu* is a catamaran tender that transports the deep submergence vehicle, *Alvin*, to and from dive sites and ports.
Marine research projects are sponsored and financed, for the most part, by Federal agencies. Each Institution scientist or team of scientists devote a part of their time to teaching. The Institution offers graduate and post-doctoral study and training in the marine sciences through degree programmes and through special study and training programmes.

YARROW (SHIPBUILDERS) LIMITED

Scotstoun, Glasgow, Scotland

YARROWS LIMITED, VICTORIA, BRITISH COLUMBIA, CANADA

ASSOCIATED COMPANY:
Burrard Dry Dock Co Ltd,
North Vancouver, B.C., Canada.

YSSELWERF N.V.

Rotterdam, Holland

THE SAFE OPERATION OF MANNED SUBMERSIBLE VEHICLES

by

James W. Dawson

THE SAFE OPERATION OF MANNED SUBMERSIBLE VEHICLES

Paradoxically the greatest setback to the safe operation of manned submersibles has been a long record of safe dive hours, spanning over ten years. Membership of the Marine Technology Society's Advisory Committee for Volume II of their *Safety and Operational Guidelines for Manned Submersibles* gives one a good overall view of the general global scene, together with assistance from high-calibre undersea consultants used by Lloyd's, and the London insurance market. Volume I of this work has without a doubt had a profound effect on both builders and operators of manned and also drone submersibles, in alerting them to hidden dangers, and has dampened the ardour of many backyard builders for impelling themselves into dark and hostile waters.

A particular area causing considerable concern to thinkers in the undersea 'club' is the vulnerability of steel cables, which when constantly changed from air to immersion in sea water and vice versa, tend to fail catastrophically even when they are, or appear to be, in mint condition. This problem is of vital concern to submersibles operated from 'mother ships', as Woods Hole Oceanographic Institution found out, at great cost, in the case of *Alvin's* long sojourn on the seabed. The French 'SP3000', undergoing static pressure tests in 3,200 metres (10,500 ft) of water in the Tyrrhenian Sea, was lost through the rupturing of a cable. It is also of vital importance that equipment, such as shackles used for launch and recovery, is tested and examined at regular intervals, together with the crane and other lifting gear. It could be added that the strain on the drums of winches with a great weight of wire rope overside, can be, and has been, catastrophic. Unless greatly strengthened, they implode with the sheer weight of the wire, leaving aside what it may hold at its nether end.

Another area of concern, closely related to cranes and strong shackles, is to maintain a generous safety factor to handle, for example, the additional weight inside the exostructure, which fills with water when pressure builds up. If the safety factor merely covers the weight of the submersible in air, the addition of up to two tons of water, in such submersibles as *Beaver IV*, could induce catastrophic failure on the recovery system.

Perry Submarines have an exceptional safety record covering many years of operations all over the globe. Their philosophy has been to place lead/acid batteries *outside* the exostructure, so that by releasing them, positive buoyancy can be quickly achieved in an emergency. However, one must bear in mind the old saw 'Ships are alright! it's the men in 'em!' A company marketing submersibles must be concerned with sales and profitability, and they cannot control the good housekeeping of the buyer, in particular with regard to maintenance. To illustrate the importance of training and maintenance, a prestigious European builder of both military submarines and submersible vehicles made a sale to a Far Eastern buyer, providing expert maintenance personnel, to instruct the buyer's crew and engineers. Quoting from the submersible builder's report yields a vivid picture of the wrong way to approach a new venture. 'To the same extent, as nobody would wish to repair or to maintain an aircraft by a bicycle workshop, the sophisticated submarine cannot be maintained or repaired by unqualified personnel on a tiny shipyard without a minimum of mechanical tools and workshop machines. There was practically nobody available to do the simplest work or to assist our specialists and to learn from them for future maintenance work.'

The two submersible accidents which achieved considerable press comment in 1973, tragically marred by the loss of two lives, have had a seismic effect on a host of considerations related to operational safety. The danger is that legislation engendered by Senators and M.P.'s, with little technical knowledge, could result in the laying up of upwards of 100 useful submersibles. A major problem is that rescue systems suitable for military submarines, such as the US Navy Deep Submergence Rescue Vehicle, built by Lockheed, or the McCann Chamber, are not compatible with small commercial or scientific submersibles. There is one major exception to this difficult situation, concerning Life Support, and this is the great work of Lt. Commander M. R. Todd, R.N. (Retd.), in perfecting rescues from 150 m (500 ft) or more. Consultants who advise both submersible operators and the Insurance Industry feel strongly that the Todd Hood Inflation System is an essential addition to the safety of submersibles. This, of course, has no applications for deep submergence systems in distress beyond Continental Shelf depths, but since most undersea vehicle work is currently within these depths, the Todd System is relevant.

Strongly opposing views on various aspects of submersible equipment and safety demonstrate that the hope of concensus is still somewhat remote. Cable strengths, corrosion and safety factors have already been touched upon. Another debate concerns the use of acrylic viewports vis à vis glass viewports. Some hold the view that glass shatters under pressure, whereas acrylics crack. The optical distortion suffered by acrylic lenses at great depths was studied by the US Naval Undersea Center who performed a successful simulated pressure test on a ceramic glass viewport down to 13,700 m (45,000 ft). Since viewports are a vital element of manned undersea systems, optical distortion and the ability to withstand dynamic and static stresses sufficiently well for wide safety margins, are of great importance. One of the very few fatalities involving one-atmosphere undersea vehicles, was caused by a freakish impact on an unprotected 3,175 cm ($1\frac{1}{4}$ in) thick acrylic viewport fitted to the *Nekton Beta* which was in course of salvaging a motor boat. The severing of a lifting cable released the motor boat, which tracked with tragic accuracy onto the viewport. This failure led to electrocution resulting from the failure of a fuse to exercise its function. The pilot had remained at 55–58 m (180–190 ft) in order 'to avoid confusion on the surface,' and this seemingly sensible contradiction to normal undersea safety led to the fatality. It is said that the same man seldom, if ever, makes the same mistake twice, and so it is that each accident is analysed by experts and amateurs alike, in order to prevent repetition, and such matters are discussed in depth by members of the Society for Undersea Technology and the Marine Technology Society, with great emphasis on the forthcoming Volume II of its *Guidelines*. Here an excursion into the mysterious interface between the classification societies' role and Underwriters' consultants seems relevant. The classification societies such as Lloyd's Register of Shipping, the American Bureau of Shipping, Bureau Veritas and others are primarily concerned with engineering, embracing pressure hulls, lifting rings, communications systems and other integral parts of a submersible system. A glance at the rules of any one of these societies will show clearly that pressure tests, defined safe depths, maintenance, correctly placed and spaced penetrations, are their particular concerns. Where consultants arrive and classification societies retire is when environmental problems, such as sea states, salinity or lack of it, and local conditions obtain. For example, corrosion is much more rapid in the Arctic than in temperate waters, and the emergence of a submersible or submarine into a tongue of fresh water has a drastic effect on buoyancy, as the crew of *Aluminaut* discovered during sea trials in Long Island sound, while crossing the offshore mouth of the Connecticut River.

Buoyancy tanks are a feature of submersibles, whether one-atmosphere 'dry' or 'wet' underwater transport systems. Here again a variety of materials have an equal variety of supporters. An important consideration with the smaller dry atmosphere vehicles which utilise fibreglass in the interests of the lightness of this material, is that fibreglass cracks on impact, on such obstructions as wrecks, harbour walls or oil rig risers. If the submersible has two saddle tanks and one of them is ruptured on impact, the submersible pilot will induce an emergency ascent, which gravity demands is sideways, so that on surfacing the hatch is half submerged and cannot be opened.

This brings us to problems of 'attitude' which are inherent in military submarines. Future thinking demands urgent pre-planning, when the reality of nuclear submarine oil tankers is looming more distinctly on the horizons of Alaska than they have before, lit by the new political nature of the struggle for oil. A real problem with large military submarines with relatively small hydroplanes at side and rear, is the steady maintenance of lateral 'attitude', at slow speeds, in particular fore and aft. If such large undersea systems, and much larger ones are being discussed urgently, and operate under ice, 'attitude' is a crucial factor where tolerances of only a few feet are available, above the sail and below the keel. Recent papers have discussed undersea tanker terminals in Arctic waters, but no planning has yet taken place with regard to salvage operations requiring submarine tugs powerful enough to drag many times their own bulk to safe ice-free waters.

The debate over power sources is both vigorous and pungent sometimes with noxious and toxic gases. The most popular system is the lead-acid battery. If these are carried within the pressure hull there is a real danger of gas rendering the crew unconscious, in addition the batteries cannot be jettisoned, whether inside the pressure hull or the exostructure. Nickel Cadmium batteries are more common to business aircraft, but a phenomenon known as 'thermal runaway' has been noted which would be even more dangerous to a submersible than to an aircraft. When nickel cadmium batteries are squeezed into a tight corner or sphere, they lend themselves to over-heating which sets up a chain reaction in an over-tired battery, leading to fire. Other forms of power sources for manned submersibles are sufficiently exotic to be outside the debate on safety. These include a variety of systems such as salt water batteries, nickel silver and other combinations not yet economically viable to underwater use.

A communication system between the surface support vessel and the submersible is of the greatest importance. The system devised and used by Vickers Oceanics during the *Pisces III* recovery operation was demonstrated under stress conditions, to be very efficient. Under quite different and much more moderate night conditions, the *Deepstar* was swept by a strong current miles from her 'mother ship' during an operation well out of voice communication range. A small submersible in a large sea area presents a minuscule radar target. The experienced crew inflated a rubber buoy and lit flares, which were more than skilfully sighted by the surface support crew. Those involved, learning through accidents, now have a 'balloon' rising up to 150 m (500 ft) for direction-finding up to a range of 25 miles.

One of the principle uses of submersible vehicles has been mineral coring and sampling. Here another hazard presents itself. If heavy samples are placed in a basket with the aid of a manipulator arm, the system will become negatively buoyant, and fail to respond to any efforts to ascend, except for costly emergency jettison of valuable equipment, including the motors. What must be pre-planned is the carriage of expendable extra weight which may be jettisoned as the samples are gathered, maintaining a safe balance to achieve instant positive buoyancy when the work has been completed, or when an emergency situation develops.

Advocates of safety measures hold that flooding valves are vital to the safety of pilots and crews. If a submersible is trapped at a depth where pressure is great against a one-atmosphere system, the crew cannot exit by reason of pressure on the hatch. The carrying of scuba equipment or the Todd Escape Hoods is, in these sorts of circumstances, where operations are at Continental Shelf depths of around 180 m (600 ft), absolutely essential in the interests of safety. Whilst such situations are happily rare, negative buoyance is common, and droppable ballast is as essential as it is in the gondola of a balloon. There the analogy ceases abruptly. Submersibles are not respecters of gravity, in the sense that they can lie on the sea bed at any angle, or upside down. If the droppable ballast is from the keel, and the vehicle is upside down, it cannot jettison ballast. Several Diving Saucers have this problem to solve.

Whilst we are still on the seabed upside down, it is relevant to consider the comfort of the crew. In popular imagination, submersible vehicles are miraculously spacious on the lines of Jules Verne's *Nautilus*, in spite of their small size. The truth, however, is quite different, and sufficient padding that can be moved to new sites within the pressure hull in emergency is essential to crew comfort, and in the case of injury, perhaps survival. Emergency concentrated rations of food and sufficient water to equate the life support time are also a compulsory extra. This problem has already been met in reality and not just in rhetoric.

On the subject of jettison and the desire for positive buoyancy at the right time, a word must be said concerning equipment. Books about submersible vehicles frequently extol the smoothness of the exostructure and make no mention of the fearsome array carried on the 'bows', of T/V systems, coring devices, lights and simulations of the human arm. Aside from becoming 'snaggable mechanical appendages', to use the Marine Technology Society's words in the 1968 *Guidelines*, keen attention must be focused on the vulnerability of equipment to pressure, when it has been purchased subsequent to the original submersible vehicle. An ancillary system fitted outside the pressure hull unless it is pressure compensated to the same pressures as its 'host,' can implode, and have the same result as the detonation of a mine. The International Standards Organisation is looking at such important notions as international standardisation of undersea engineering equipment and diving equipment as well. Emergency scuba gear in a stricken submersible or habitat is not very interesting to a crewman who finds that the gas bottle with Middle Eastern markings does not mate with his British or American gear.

The two most recent and also dramatic rescues of vehicles trapped by a lack of fairing have demonstrated also that lifting rings are no respecters of gravity in the upside-down conditions that may have to be met in submersible accidents, hence advocating a wide choice of lifting rings, in order that one would demonstrate a signal lack of appreciation of the designing engineer's tolerance of the high grade steels used in pressure hulls.

This debate inevitably leads to the choice of materials used, or usable, for the construction of deep ocean vehicles as well as those designed for 'Conshelf' use. But before plunging into this debate, of whether glass is superior to plastic and steel to aluminium, a close look at visibility is needed, in depth. Water transparency varies very greatly from area to area from temperate to tropical, and from cold to Arctic waters. View-ports remain in the same positions, lights at the same power and speeds at the same rate, in spite of wide-ranging differences in visibility that have to be met within operating submersible work-boats. It is well to bear in mind the view expressed by Dr. Harold Edgerton of the Massachusetts Institute of Technology, who opined that 'Under the sea the weather is always bad.' The light-scattering and absorption effects of the water, plankton, churning sand and mud, act as fog over land. A 'diving saucer', operating in deep water in the Mediterranean, faced a terrifying mud-avalanche, when it settled on a gently sloping undersea mountain side. The gentle slope turned out to be mud-filled 'steps', and the mud displaced by the propellers started a chain reaction of vast proportions. The submersible was forced further and further down, in spite of jettisoning all available droppable ballast. Finally the depth gauge showed that stability of depth was being maintained and an agonising period of time later, that the submersible was rising.

Lifting inert objects from depths beyond human intervention, inhibited by pressure, poses particular difficulties. 'Mother ships' and support vessels with over-the-stern launch and recovery systems cannot lift heavy objects from great depths. Their normal function, with wave compensating gear, is to launch or recover their submersible protege at an angle over the stern. The arc of the lifting gear must be narrow, or strength must be sacrificed. This means that the lifting and recovery of an object much heavier than the one actually launched becomes impossible because of insufficient over-hang. In the case of *Pisces III*, the cable ship, *John Cabot* was fortuitously available, with sufficient strong cable and sufficient overhang to lift the damaged submersible from the surface. This was not the first occasion when seabed rescue became an acute need. It does seem therefore, that the somewhat haphazard borrowing from neighbourly owners of cable ships, and other manned and unmanned systems, requires rationalisation. The problems have been better solved in outer space, but that medium sped out on an ocean of money, sadly lacking in the much more vital saving of mankind's fast depleting larder. A central agency, such as the unique Intelligence Department at Lloyd's, could perhaps be adapted to being a rapporteur of submersible operations in deep water or potentially hazardous areas, so that a rescue submersible or deep water salvage vessel was always available, Continent by Continent.

Another source of accidents and therefore debate among the 'cognoscenti,' is the length of propeller tunnels. *Deep Quest* was stuck on the seabed to give sufficient time for the world's press to make a suitable but short-lived 'puff', and was 'rescued' by the diminutive *Nekton*, with the addition of a knife taped to a manipulator arm. Publicity of this kind is legitimate in an industry that has cried for recognition for so many years, but the facts were at variance with the supposed danger, since *Deep Quest* could have jettisoned her motors to free herself. However, the accidental injection of a nylon rope in the propeller tunnel, accentuates the battle for speed, since the longer (and safer) the tunnel the slower the speed. Operators are loath to use mesh guards against nylon rope ends, since here again speed suffers. The debate continues, and will be unresolved until a single safe power source is discovered that relates lightness in weight to power and compares very favourably in cost and maintenance to lead/acid batteries.

Fires in submersibles are happily rare, since smoke in the confines of a pressure hull can lead quickly to severe breathing problems. An example of this type of accident occurred in July 1970, 480 m (1,575 ft) below surface in North American Rockwell's *Beaver IV*. Fire broke out in a junction box, but caused no catastrophic damage beyond the necessity of re-wiring the electrical circuits. Salt water condensation breaks down insulation and this causes electrical leakage, which can lead to short circuits and thus fires. Another feature needing careful attention is that both rubber and bakelite absorb moisture, and thus become conductors, when they should be orchestrators of safe circuits. The US Naval Research Laboratory has produced many inventive and dynamic processes and systems, and amongst these is a butyl alcohol coating which drives out moisture already in the insulation, as well as preventing further intrusions of unwanted water.

The brief reference to Jules Verne makes further mention of this extraordinary man essential to grasp the whole panorama of undersea vehicles in the right perspective. The fact that the US Navy's historic nuclear-powered voyage under the Polar ice fields in 1968 bore the same name as Captain Nemo's command in *Twenty Thousand Leagues* is happily not accidental. Verne's extraordinarily accurate forecasts actually created the concept of submersibles which now seems quite common place. The previous mention of undersea tankers leads to a new field of activity under the ice, where hydrocarbons are evidence that polar regions were once fecund tropical jungles. It has become essential for mankind to bridge the ten year gap between the use of oil and other power sources. As the latter-day *Nautilus* found out, operating a large submarine between surface ice and the sea bed is no sinecure when 'attitude' is brought into great prominence. The problems that must be faced are similar to navigating a large airship under a viaduct in a stiff breeze, not necessarily head to wind. Since ice-breaker tankers are virtually self-defeating, in terms of cost, for the removal of hard won hydrocarbons in Arctic regions, and heated pipelines over the perma-frost hold similarly daunting barriers to profitability, and the survival of migrating animals, then Arctic submarines gain viability.

Unhappily, we have no present-day Jules Verne to show the way and the pitfalls, to the safe extraction of oil and minerals from frozen and semi-frozen waters, and experience of such pitfalls is thin on the ground, to say the least. The abruptness with which undersea engineers have had to accept this great challenge bodes ill for safety, if one is allowed the analogy of airship disasters and the demise of a system reliant on fair and gentle winds. It is evident that the challenge must be met, and to forecast the utilisation of undersea oil completions, in relation to submarine tankers does not require great powers of concentration or research. The rescue of a stricken nuclear tanker full of crude oil under an ice canopy, however, poses a hitherto unimagined salvage enigma. Promotion of autonomous underwater drilling systems, with no surface power systems, betray a misunderstanding of the small amount of power that can be generated from present day undersea power sources, where a three-man submersible is lucky to have a cruising speed of 2 to 3 knots in comparatively still water, for a comparatively short time. To subject this amount of energy to the towage of a large inert object is patently out of the question, as is the use of submersible vehicles for deep sea drilling unless powered from the surface. The problem begs urgently for a solution, as oil and minerals become more scarce, and it must be borne in mind that the lower surface of ice floes are as irregular as the upper surface, and that the seabed is seldom featureless except in the abyssal plains.

Reports by the pilots and crews of submersibles, and underwater scientists confirm that the present use of charts originally made through the good offices of Captain Cook and successively updated, are highly inaccurate. Confirmation of this state of affairs is acquired with great ease by scanning the ship casualty columns of Lloyd's List, where an average of three groundings a week are recorded. However, the magnitude of the problem is quite beyond the capabilities of submarine work-boats, although drones may be towed in arrow formation for surveying frequently-used inshore sea passages.

Here a mention should be made of the fierce debate between the operators of manned vehicles and those of unmanned systems, such as the US Navy's C.U.R.V. vehicle which played a prominent part in the Vickers' *Pisces III* rescue operations. It is undeniable that the further men are from the seabed the safer they feel, but against this potent argument must be the use of the human eye, brain and ears, and the dexterity of highly developed manipulator arms, for example. There are also two kinds of safety and two kinds of salvage. The first and most vital is life salvage, but the second can also be vital. Through careless abuse of the designer's instructions, two Rebikoff *Pegasus* wet submersibles lie beside the wreck of the *Andrea Doria*, and through other causes, which might include the bafflement of acoustic control systems by thermoclines and deep scattering layers, two expensive US Navy drones lie on the bed of the Pacific. Thus, in a sense, irretrievable losses of drones predominate, if only numerically, over manned systems, where life salvage has a stronger pull than any lifting gear, and promotes more accuracy than any surface operated seabed TV scanner.

In the heady days of 1968 and 1969, when money poured into any system that moved, or claimed to move, a number of submersible vehicles surfed along the cash flow for a brief exotic mayfly life. At this time, when insurance was considered a brake on profits, the writer circularised all the operators with a view to collating 'safe dive hours.' The response was such that the project had to be abandoned after fourteen submersibles of all types reported over 10,000 safe dive hours in the fair weather months of 1969. This figure excluded surface operating time, and proves once again that the main danger area for submersibles, if there is one, is on the surface.

The cable ship *John Cabot* hauls the beleaguered *Pisces III* from the sea during the rescue operation (*Courtesy The Press Assoc.*)

INDEX

INDEX

N

O

P

T

U

V